SYSTEMATIC METHODS
OF CHEMICAL
PROCESS DESIGN

PRENTICE HALL INTERNATIONAL SERIES
IN THE PHYSICAL AND CHEMICAL ENGINEERING SCIENCES

NEAL R. AMUNDSON, SERIES EDITOR, *University of Houston*

ADVISORY EDITORS

ANDREAS ACRIVOS, *Stanford University*
JOHN DAHLER, *University of Minnesota*
H. SCOTT FOGLER, *University of Michigan*
THOMAS J. HANRATTY, *University of Illinois*
JOHN M. PRAUSNITZ, *University of California*
L. E. SCRIVEN, *University of Minnesota*

BALZHISER, SAMUELS, AND ELLIASSEN   *Chemical Engineering Thermodynamics*
BIEGLER, GROSSMANN, AND WESTERBERG   *Systematic Methods of Chemical Process Design*
CROWL and LOUVAR   *Chemical Process Safety*
DENN   *Process Fluid Mechanics*
FOGLER   *Elements of Chemical Reaction Engineering, 2nd Edition*
HANNA AND SANDALL   *Computational Methods in Chemical Engineering*
HIMMELBLAU   *Basic Principles and Calculations in Chemical Engineering, 6th edition*
HINES AND MADDOX   *Mass Transfer*
KYLE   *Chemical and Process Thermodynamics, 2nd edition*
NEWMAN   *Electrochemical Systems, 2nd edition*
PAPANASTASIOU   *Applied Fluid Mechanics*
PRAUSNITZ, LICHTENTHALER, and DE AZEVEDO   *Molecular Thermodynamics
    of Fluid-Phase Equilibria, 2nd edition*
PRENTICE   *Electrochemical Engineering Principles*
STEPHANOPOULOS   *Chemical Process Control*
TESTER AND MODELL   *Thermodynamics and Its Applications, 3rd edition*

SYSTEMATIC METHODS OF CHEMICAL PROCESS DESIGN

L.T. Biegler, I.E. Grossmann, and A.W. Westerberg
Carnegie Mellon University

To join a Prentice Hall PTR Internet mailing list, point to:
http://www.prenhall.com/register

Prentice Hall PTR
Upper Saddle River, New Jersey 07458
http://www.prenhall.com

Library of Congress Cataloging-in-Publication Data
Biegler, Lorenz T.
    Systematic methods of chemical process design / L. T. Biegler, I. E.
Grossmann, and A. W. Westerberg.
        p.   cm.
    Includes bibliographical references and index.
    ISBN 0-13-492422-3
    1. Chemical processes.   I. Grossmann, Ignacio E.   II. Westerberg,
Arthur W.   III. Title.
TP155.7.B47 1997
660'.28—dc21   96-52100
                                                                                          CIP

Acquisitions editor: Bernard Goodwin
Cover design director: Jerry Votta
Manufacturing manager: Alexis R. Heydt
Marketing manager: Miles Williams
Compositor/Production services: Pine Tree Composition, Inc.

Reprinted with Corrections December, 1999

        ©1997 by Prentice Hall PTR
        Prentice-Hall, Inc.
        Upper Saddle River, New Jersey 07458

The publisher offers discounts on this book when ordered in
bulk quantities. For more information contact:

    Corporate Sales Department
    Prentice Hall PTR
    One Lake Street
    Upper Saddle River, New Jersey 07458

    Phone: 800-382-3419
    Fax: 201-236-7141
    email: corpsales@prenhall.com

All rights reserved. No part of this book may be reproduced,
in any form or by any means, without permission in writing
from the publisher.

Printed in the United States of America
10 9 8 7 6 5 4 3

ISBN: 0-13-492422-3

Prentice-Hall International (UK) Limited, *London*
Prentice-Hall of Australia Pty. Limited, *Sydney*
Prentice-Hall Canada Inc., *Toronto*
Prentice-Hall Hispanoamericana, S.A., *Mexico*
Prentice-Hall of India Private Limited, *New Delhi*
Prentice-Hall of Japan, Inc., *Tokyo*
Prentice-Hall Asia Pte. Ltd., *Singapore*
Editora Prentice-Hall do Brasil, Ltda., *Rio de Janeiro*

To my parents, to Lynne and to Matthew

*In memory of my father, to my mother, to Blanca
and to Claudia, Andrew and Thomas*

*In memory of my parents, to Barbara
and to Ken and Karl*

To all our students

CONTENTS

Preface xiii
Foreword xvii

1 Introduction to Process Design 1

    1.1 The Preliminary Design Step for Chemical Processes 1
    1.2 A Scenario for Chemical Process Design 3
    1.3 The Synthesis Step 6
    1.4 Design in a Team 8
    1.5 Converting Ill-Posed Problems to Well-Posed Ones 10
    1.6 A Case Study Process Design Problem 13
    1.7 A Roadmap for This Book 18
    References 20
    Exercises 21

I PRELIMINARY ANALYSIS AND EVALUATION OF PROCESSES 23

2 Overview of Flowsheet Synthesis 25

    2.1 Introduction 25
    2.2 Basic Steps in Flowsheet Synthesis 26
    2.3 Decomposition Strategies for Process Synthesis 36
    2.4 Synthesis of an Ethyl Alcohol Process: A Case Study 39
    2.5 Summary 50
    References 51
    Exercises 51

3 Mass and Energy Balances 55

    3.1 Introduction 55
    3.2 Developing Unit Models for Linear Mass Balances 57

3.3 Linear Mass Balances 85
3.4 Setting Temperature and Pressure Levels from the Mass Balance 94
3.5 Energy Balances 98
3.6 Summary 104
References 104
Exercises 105

4 Equipment Sizing and Costing — 110

4.1 Introduction 110
4.2 Equipment Sizing Procedures 111
4.3 Cost Estimation 132
4.4 Summary 138
References 139
Exercises 139

5 Economic Evaluation — 142

5.1 Introduction 142
5.2 Simple Measures to Estimate Earnings and Return on Investment 144
5.3 Time Value of Money 147
5.4 Cost Comparison after Taxes 155
5.5 Detailed Discounted Cash Flow Calculations 162
5.6 Inflation 169
5.7 Assessing Investment Risk 170
5.8 Summary and Reference Guide 173
Exercises 174

6 Design and Scheduling of Batch Processes — 180

6.1 Introduction 180
6.2 Single Product Batch Plants 180
6.3 Multiple Product Batch Plants 184
6.4 Transfer Policies 186
6.5 Parallel Units and Intermediate Storage 187
6.6 Sizing of Vessels in Batch Plants 190
6.7 Inventories 193
6.8 Synthesis of Flowshop Plants 195
References 199
Exercises 199

II ANALYSIS WITH RIGOROUS PROCESS MODELS — 205

7 Unit Equation Models — 207

7.1 Introduction 208
7.2 Thermodynamic Options for Process Simulation 210

Contents

    7.3 Flash Calculations   217
    7.4 Distillation Calculations   224
    7.5 Other Unit Operations   232
    7.6 Summary and Future Directions   239
          References and Further Reading   240
          Exercises   242

8 General Concepts of Simulation for Process Design   243

    8.1 Introduction   243
    8.2 Process Simulation Modes   245
    8.3 Methods for Solving Nonlinear Equations   254
    8.4 Recycle Partitioning and Tearing   271
    8.5 Simulation Examples   285
    8.6 Summary and Suggestions for Further Reading   289
          References   291
          Exercises   292

9 Process Flowsheet Optimization   295

    9.1 Description of Problem   295
    9.2 Introduction to Constrained Nonlinear Programming   297
    9.3 Derivation of Successive Quadratic Programming (SQP)   307
    9.4 Process Optimization with Modular Simulators   314
    9.5 Equation-Oriented Process Optimization   321
    9.6 Summary and Conclusions   331
          References   332
          Exercises   334

III BASIC CONCEPTS IN PROCESS SYNTHESIS   339

10 Heat and Power Integration   341

    10.1 The Basic Heat Exchanger Network Synthesis (HENS) Problem   342
    10.2 Refrigeration Cycles   373
          References   382
          Exercises   382

11 Ideal Distillation Systems   387

    11.1 Separating a Mixture of *n*-Pentane, *n*-Hexane, and *n*-Heptane   387
    11.2 Separating a Five-Component Alcohol Mixture   395
          References   401
          Exercises   401

12 Heat Integrated Distillation Processes — 408

12.1 Heat Flows in Distillation   408
References   425
Exercises   425

13 Geometric Techniques for the Synthesis of Reactor Networks — 429

13.1 Introduction   430
13.2 Graphical Techniques for Simple Reacting Systems   432
13.3 Geometric Concepts for Attainable Regions   438
13.4 Reaction Invariants and Reactor Network Synthesis   447
13.5 Chapter Summary and Guide to Further Reading   450
References   452
Exercises   453

14 Separating Azeotropic Mixtures — 455

14.1 Separating a Mixture of n-Butanol and Water   456
14.2 Separating a Mixture of Acetone, Chloroform, and Benzene   464
14.3 Sketching Distillation and the Closely Related Residue Curves   475
14.4 Separating a Mixture of n-Pentane, Water, Acetone, and Methanol   482
14.5 More Advanced Work   488
References   490
Exercises   490

IV OPTIMIZATION APPROACHES TO PROCESS SYNTHESIS AND DESIGN — 495

15 Basic Concepts for Algorithmic Methods — 497

15.1 Introduction   497
15.2 Problem Representation   498
15.3 Solution Strategies for Tree Representations   503
15.4 Models and Solution Strategies for Network Representations   507
15.5 Alternative Mathematical Programming Formulations   509
15.6 Summary of Mathematical Models   513
15.7 Modeling of Logic Constraints and Logic Inference   514
15.8 Modeling of Disjunctions   519
15.9 Notes and Further Reading   521
References   521
Exercises   523

16 Synthesis of Heat Exchange Networks — 527

16.1 Introduction   527
16.2 Sequential Synthesis   528

Contents xi

16.3 Simultaneous MINLP Model 551
16.4 Comparison of Sequential and Simultaneous Synthesis 559
16.5 Notes and Further Reading 561
References 562
Exercises 563

17 Synthesis of Distillation Sequences 567

17.1 Introduction 567
17.2 Linear Models for Sharp Split Columns 567
17.3 Example of MILP Model for Four-Component Mixture 571
17.4 MILP Model for Distillation Sequences 575
17.5 Heat Integration and Pressure Effects 576
17.6 MILP Model with Continuous Temperatures 578
17.7 MILP Model with Discrete Temperatures 581
17.8 Design and Synthesis with Rigorous Models 587
17.9 Notes and Further Reading 590
References 591
Exercises 592

18 Simultaneous Optimization and Heat Integration 595

18.1 Introduction 595
18.2 Sequential versus Simultaneous Optimization and Heat Integration 596
18.3 Linear Models 601
18.4 Nonlinear Models 604
18.5 Notes and Further Reading 613
References 614
Exercises 615

19 Optimization Techniques for Reactor Network Synthesis 618

19.1 Introduction 618
19.2 Reactor Network Synthesis with Targeting Formulations 620
19.3 Reactor Network Synthesis in Process Flowsheets 645
19.4 Summary and Further Reading 656
References 658
Exercises 660

20 Structural Optimization of Process Flowsheets 663

20.1 Introduction 663
20.2 Flowsheet Superstructures 663
20.3 Mixed-Integer Optimization Models 666
20.4 MILP Approximation 667
20.5 MILP Model for the Synthesis of Utility Plants 669

20.6　Modeling/Decomposition Strategy　672
20.7　Notes and Further Reading　686
　　　References　686
　　　Exercises　687

21　Process Flexibility　690

21.1　Motivating Example　691
21.2　Mathematical Formulations for Flexibility Analysis　697
21.3　Flexibility Test Problem　698
21.4　Flexibility Index Problem　699
21.5　Vertex Solution Methods　701
21.6　Example with Nonvertex Critical Point　702
21.7　Active Set Method　704
21.8　Active Set Method for Nonvertex Example　707
21.9　Special Cases for Flexibility Analysis　709
21.10　Optimal Design under Uncertainty　712
21.11　Notes and Further Reading　713
　　　References　714
　　　Exercises　715

22　Optimal Design and Scheduling for Multiproduct Batch Plants　719

22.1　Introduction　719
22.2　Horizon Constraints for Flowshop Plants—Single-Product Campaigns　719
22.3　MINLP Design Model for Flowshop Plants—Single-Product Campaigns　722
22.4　MILP Reformulation for Discrete Sizes　725
22.5　NLP Design Model—Mixed-Product Campaigns (UIS)　728
22.6　Cyclic Scheduling in Flowshop Plants　729
22.7　NLP Design Model—Mixed Product Campaigns　735
22.8　State-Task Network for the Scheduling of Multiproduct Batch Plants　736
22.9　Notes and Further Reading　743
　　　References　743
　　　Exercises　745

Appendix A　Summary of Optimization Theory and Methods　748

Appendix B　Smooth Approximations for max {0, $f(x)$}　771

Appendix C　Computer Tools for Preliminary Process Design　773

Author Index　781

Subject Index　786

PREFACE

Process design is one of the more exciting activities that a chemical engineer can perform. It involves creative problem solving and teamwork in which basic knowledge in chemical engineering and economics are applied, commonly through the use of computer-based tools, to devise new process systems or modifications to existing plants. The teaching of process design, however, continues to present a major challenge in academia. There are several reasons for this. Faculty who are not actively engaged in doing research in process systems engineering are generally uncomfortable teaching a design course, unless they have had some industrial experience. Another complicating factor is that process design is still perceived among many academics as a subject that is too practical in nature with little fundamental content. Also, there are relatively few textbooks on process design, both at the undergraduate and graduate levels. Finally, teaching design is difficult because problems tend to be open-ended, with incomplete information, and requiring decision making.

Fortunately, process design, and more generally, process systems engineering, has undergone a dramatic change over the last 20 years. During this period many new fundamental and significant advances have taken place. The more or less ad hoc analysis of flowsheets has been replaced by systematic numerical solution techniques that are now widely implemented in computer modeling systems and simulation packages for both preliminary and detailed design. The largely arbitrary selection of parameters in process flowsheets has been replaced by the use of modern optimization strategies. The intuitive development of structures of process flowsheets has been largely replaced by systematic synthesis methods, both in the form of conceptual insights and in the form of advanced discrete optimization techniques.

It is from the perspective of the above advances in process design that this textbook has been written: to teach modern and systematic approaches to design. The emphasis is on the application of strategies for preliminary design, on the systematic development of representations for process synthesis, and on the development of mathematical models for simulation and optimization for their use in computer-based solution techniques. The main aim in learning these techniques is to be able to synthesize and design process flowsheets, understanding the decisions involved in the reaction, separation, and heat integration subsystems, as well as their interactions and economic implications. The applications deal mostly with large-scale continuous processes, although some introduction to multiproduct batch processes is given. Also, while economics is used as the main measure for evaluation, a brief exposure to operability and discussion on multiple criteria (safety, environmental impact) is covered.

The book consists of 22 chapters, organized into four major parts: I: *Preliminary Analysis and Evaluation of Processes*, II: *Analysis with Rigorous Process Models*, III: *Basic Concepts in Process Synthesis*, IV: *Optimization Approaches to Process Synthesis and Design.* An introductory chapter is also presented to give a broader view of process design. The textbook is aimed at senior undergraduate and graduate students in chemical engineering. At the undergraduate level it is intended to be a textbook for the senior design course. Chapters 1 to 11 (except 9) could be typically covered in such a course. Chapters 9 and 15 to 17 of Part IV can be used as part of an undergraduate optimization course. At the graduate level, Chapters 9 to 22 and Appendix A can be used as a basis for an advanced process systems engineering course. Chapters 10 to 22 (Parts III and IV) are aimed specifically at a graduate course in process synthesis. Each chapter contains a set of exercises and references to representative publications. Design practitioners who wish to learn about modern design techniques should find this book useful as a reference text.

It is important to note that this book is not meant to be a research monograph. All the material presented here has been developed and taught extensively in courses at Carnegie Mellon University. For instance, a portion of Part I was first developed by Art Westerberg in 1978, and has gradually evolved since then into lecture notes that are currently used in the Senior Undergraduate Design course. Part II was developed first in the early 1980s for a graduate course taught by Art Westerberg on Advanced Process Engineering. Its current form reflects the lecture notes used by Larry Biegler for an advanced undergraduate/graduate level course on computational design methods. Part III corresponds to lecture notes used by Art Westerberg in a current graduate course on Process Systems Engineering. A portion of Part IV was first developed by Ignacio Grossmann in a course on Special Topics on Advanced Process Enginneering course in 1985. In its present form it is being used in the graduate course on Process Systems Engineering. Also note that all the chapters include exercises. Some of these require the use of spreadsheets and modeling systems for optimization (see Appendix A).

The authors would like to acknowledge the many individuals that made this book possible. We express our gratitude to Professor John Anderson for having encour-aged us to undertake the task of writing this textbook. Larry Biegler is grateful to the Department of Chemical Engineering for releasing him of teaching duties for one semester to write this book. Ignacio Grossmann is grateful to the School of Chemical Engineering at Cornell University and to the Centre for Process Systems Engineering at Imperial College for having provided time and financial support for his sabbatical leaves in 1986–1987, and 1993–1994, respectively, in which most of the chapters on Part IV were written. Art Westerberg is grateful to the University of Edinburgh for the time and support he received to prepare portions of this book. The three authors are indebted to the following individuals who have provided us extensive feedback on the book: Dr. Alberto Bandoni, Dr. Mark Daichendt, Professor

Truls Gundersen, Dr. Zdravko Kravanja, Dr. Antonis Kokossis, Dr. Guillermo Rotstein, and Professor Ross Swaney. We are also grateful to all our current graduate students at Carnegie Mellon who helped us in the proofreading of the manuscript. Finally, we are most grateful to Dolores Dlugokecki and Laura Shaheen for their help and patience in typing and correcting many of the versions of our manuscript.

Lorenz T. Biegler
Ignacio E. Grossmann
Arthur W. Westerberg
Department of Chemical Engineering
Carnegie Mellon University
Pittsburgh, PA

FOREWORD

Design is perhaps the quintessential engineering activity. Based on mathematics, basic science, engineering science, and flavored by the humanities and social science, engineering design is the devising of an artifact, system, or process to best meet a stated objective. Engineering design involves development of specifications and criteria, and the synthesis, analysis, construction, testing, and evaluation of alternative solutions to best meet the desired criteria in light of safety, reliability, economic, aesthetic, ethical, and social considerations. Engineering accreditation bodies recognize the fundamental importance of design through requirements that modern design theories, methodologies, and open-ended, creative design experiences be integrated into all engineering programs.

Chemical process design is the subject of this book. Chemical processes are primarily concerned with making materials from which other articles are manufactured. Materials made by chemical processes span the range from metals and ceramics to fibers and fuels, from resins and refrigerants to elastomers and explosives, from paper and polymers to pharmaceuticals and preservatives, from crop protectants and container plastics to computer chips and catalysts, colorants, solvents, intermediates, foods, clean water, and on and on. These materials in turn are made by batch, continuous, and sometimes biological processes on scales from a few grams to billions of kilograms per year.

Chemical processes are also unique among engineered artifacts in that often they are simultaneously capital cost intensive and operating expense intensive, are designed for very long lifetimes, and sometimes are not readily adaptable to the production of materials much different from those for which they were designed. The potential of many years of continuing incurred costs underlines the importance of achieving the very best manufacturing process possible. Furthermore, although optimization is an integral part of each stage in the entire chemical process innovation cycle from chemistry development through plant construction and operation, the process design itself has a disproportional impact on ultimate economic performance. It has been estimated that decisions reached during process design, an activity which accounts for perhaps two or three percent of the project cost, fix approximately eighty percent of the capital and operating expenses of the final plant. This impact is too great to be left to chance and is the impetus for the development of systematic methods for chemical processes design.

This book describes such systematic methods for a number of chemical process design activities including the synthesis, analysis, evaluation, and optimization of chemical process alternatives. It is unique among currently available texts in the field in both its breadth of coverage and its use of optimization as a fundamental design paradigm. The typical introductory process design material on individual equipment sizing and costing is followed with discussion of modern process simulation and optimization techniques which enable a better understanding of the sensitivity of design parameters on initial capital costs, continuing operating costs, and the overall economic attractiveness of any given flowsheet. This is followed by a discussion of a number of basic systematic methods by which various sections of a process flowsheet are generated in the first place. A proficiency in such alternative invention is becoming a critical process engineering skill. Finally, the last part of the book describes a novel approach to process alternative generation based on the application of algorithmic mathematical optimization techniques to the making of structural design decisions. It is an advanced synthesis approach that coupled with ever increasing computational capability may very well revolutionize the practice of chemical process design.

Jeffrey J. Siirola
Research Fellow
Eastman Chemical Company
Kingsport, Tennessee

INTRODUCTION TO PROCESS DESIGN

1

The goal of the engineer is to design and produce artifacts and systems that are beneficial to mankind. In design we get to express our creativity in discovering what, why, and how we should devise new things. Engineers design, construct, and manufacture many different types of complex physical artifacts such as cars, consumer electronics, space shuttles, highway systems, refineries, robots, heart-lung machines, and new heating systems inside an existing high-rise building. We as chemical engineers create processes to manufacture chemicals. How we can attack such a large and complex problem is the subject of this book.

While the book deals with systematic methods for process design, there are a number of broader issues that are largely qualitative in nature and that are important to recognize. In this chapter we discuss some of these general issues in relation to chemical process design. The objective here is to give an overview of the steps involved in the design process, as well as a general idea of the complexity of the design activity. Finally, we stress the importance of synthesis, formation of teams, and generation of alternatives in process design.

1.1 THE PRELIMINARY DESIGN STEP FOR CHEMICAL PROCESSES

Design is a complex and varied activity. A single person might design the shelving in a home office, while it takes thousands of persons to design a new aircraft. The design of the next automobile model is largely a routine activity, well understood by its participants, but designing the first space shuttle was a new experience for the NASA design team. The design of a new personal computer must be done in a few months or else the product will miss its niche in the marketplace; personal computers are totally out of date in two to three years. In contrast, a refinery will have a lifetime of decades, during which it will be repeatedly modified and improved. A consumer product manufacturer will sell thousands of toasters; an architectural company will design only one John Hancock Building in Boston. All of these diverse characteristics for design problems lead to different strategies to carry out design.

In this text we shall emphasize preliminary design for chemical processes. Ideas for these processes can come from almost anywhere. Our sales team can discover a customer need for a material, with properties not covered by any product currently on the market. We may have a new catalyst that can dramatically reduce manufacturing costs for a chemical our competitor produces. Our research team may have a new monomer whose properties look promising for producing a polymer for car bumpers. Management may want us to discover a process where we can use up the surplus feedstock the company is currently producing.

In the preliminary design step we develop and evaluate a conceptual flowsheet for a specific chemical process. This task also requires us to generate and analyze a number of suitable alternative process flowsheets. We describe each flowsheet in terms of the types of equipment (e.g., heat exchangers, pumps, distillation columns, reactors) in it and how we have interconnected that equipment. We use mass and energy balances, supplemented with physical property correlations and rate expressions, to analyze our processes, that is, to estimate the flows, temperatures, and pressures of all the streams in the flowsheet. We also estimate investment and operating costs using simple correlations that approximate the actual costs. We sketch each process and list the flows, temperatures, and pressures of all the streams on process flow diagrams (PFDs), on two blueprint-size sheets of paper. Our report from this step allows management to decide if the project has enough economic potential for them to continue to study it. Moreover, given the competition due to simultaneous consideration of many corporate projects, we should not be surprised at a decision to drop the project. In fact, skilled designers who have watched a project fail for unanticipated reasons adopt the mindset that it is their goal to prove a process will fail. When all such proofs elude them, then the project just may be one that can succeed. In the generation, search and evaluation of alternative designs, we will see in Chapter 2 that this approach, in fact, leads to efficient and powerful design strategies.

Preliminary design is but one step of many in the life cycle of a chemical process. To appreciate the role that process design plays in practice, we also examine a typical sequence of activities that lead to the design and construction of a chemical process, starting at the beginning with the activities of those who run the company. Design activities that lead to plant construction and subsequent operation pass through several stages, which include preliminary design, basic process design, detailed engineering, and, finally, startup and operation. Our activity in the preliminary design step involves a team of two to five people. At the other extreme, several hundred people may be involved during plant construction.

The next section presents a corporate scenario for design and includes the role of preliminary process design. Section 1.3 discusses the synthesis step for preliminary design while section 1.4 discusses the design team. Section 1.5 then provides some directions for addressing the synthesis activity. A process design case study is introduced in section 1.6 to illustrate these concepts. Finally, section 1.7 concludes this chapter with an outline of the text.

1.2 A SCENARIO FOR CHEMICAL PROCESS DESIGN

1.2.1 Board of Directors' Design Problem

The Board of Directors for our XYZ Chemical Company have, albeit at a very high level of abstraction, a design problem to solve. They need to decide where best to direct the company and where to place major investments. One of their goals, simply put, is to maximize the generation of wealth using the resources available to them. In carrying out their goal as a chemical company, they will shy away from starting a completely different manufacturing activity, but might pursue an atypical project where the company has a strategic advantage.

1.2.2 Discovery of Possible New Projects

Narrowing the set of projects to those familiar to the company, they investigate the long-term wealth generation capability of various combinations of projects, subject to the constraint that the company will have the needed financing at the right time to implement projects they select. If the company needs more funds, they examine potential ways to raise them; for example, they may consider issuing more stock, selling bonds, and/or simply borrowing the funds. They must also take into account the risk associated with each alternative.

Let us assume, that based on their financial and risk analysis, one project they choose is to revamp their large Gulf Coast facility, to improve its performance and to improve its operation and safety. This project is not one the executive committee initiated. Rather, the Gulf Coast Plant manager may have developed it with a group of technical people at that plant site, putting forth an assessment of it in several earlier reports that made their way to the executive committee.

1.2.3 Feedback and Customer Reaction

The executive committee appoints a small team of assistants to try out the idea on several of their plant managers. They also interview the management team from the Gulf Coast facility and check with the operating personnel there, all of whom like it very much. They find the local community is very supportive. Armed with this information, the executive committee presents its decision to the Board, which, after examining many alternative projects, approves it as one that fits the company goals and has acceptable risks.

1.2.4 Planning and Organizational Design

The executive committee directs the engineering department manager to carry out the study. She must structure a team to carry out the design, construction, and operating procedure improvements. With some experienced persons from previous projects of a similar nature, she devises the criteria for selecting the team members, size, and tools this project will need. She also determines the budget for this effort. The executive committee reviews and approves her detailed plans. The engineering department manager then appoints a design team leader and asks him to propose the other members of the team. The starting team he proposes has engineers experienced in past design projects. It includes an engineer who has run this process for the past four years and the part-time commitment of a plant operator. This team helps to create an understanding of the problem and to propose alternatives for improving the process.

1.2.5 Preliminary Process Design

At this stage the design team generates and evaluates the conceptual flowsheet as well as several alternative designs. Here they apply and refine the design strategies described in this book in order to put together the process flow diagram. Moreover, to enhance their understanding of the design, the process design team models the process using commercially available simulators, and, with plant data, they improve the accuracy of these models. With this understanding, they then propose many alternative process improvements. With each they develop an estimate as to the needed investment and the expected return.

In addition to economic aspects, they may also examine safety and maintenance issues. For instance, they may determine that the plant reactor configuration can be much improved, and, with improved operator training facilities, it can run with improved safety. The team may also examine in a preliminary fashion how the operators will start up and control this process. If the economic evaluation is favorable, then this design could meet with the approval of the executive committee, and we move on to the next decision stage.

1.2.6 Layout and Three Dimensional Modeling

Engineering now sets up a team from within the company and contracts with the UVW Construction Company to take over this project. Directed by a project manager who works for UVW, this team must identify the equipment they must purchase and install to accomplish the changes shown in the process flow diagram (PFD). Both companies agree to place on this team the leader of the process design team, the plant manager (part-time), a control engineer, and the software engineer who will lead the development effort for the operator training facility. The last two are employees of the construction company. The engineering team converts the PFD into a piping and instrumentation diagram (P&ID), from two blueprint-size sheets to 30 blueprint-size sheets. These P&IDs list all equipment, including spares, showing pipe diameters and materials (e.g., carbon steel, hasteloy steel, glass-lined stainless), vessel nozzles, and so forth. For the retrofit of an existing plant, the team also has to determine how the new equipment will fit in the existing layout, using advanced graphically oriented computer programs that aid in visualizing the plant in three dimensions. In addition, control engineers develop the blueprints for all the control system hardware, often using new computer-based control schemes.

The UVW Construction Company develops its own estimate of the costs and presents these to the management of the XYZ Chemical Company, who, after the lawyers work over the contract details, may approve continuing the project. At this point the costs to XYZ Chemical Company are fixed, unless changes are requested.

1.2.7 Construction

The project manager directs the construction of the modifications. This activity could take from three months to several years, and the existing plant may need to be shut down to carry out the modifications. Here speed and correctness of construction is of utmost importance to minimize lost profits. Several dozen people, many of them contract labor, are active during this phase of the project. For the construction of a large chemical plant, the number of people can be in the hundreds.

1.2.8 Startup and Comissioning

Before the XYZ Chemical Company will accept the plant, the UVW Construction Company must demonstrate that the modified process will operate as expected. The UVW team has designed a startup procedure that anticipates all sorts of mistakes (for example, valves could be installed backwards and pumps could be undersized). The first startup is often a "debugging" process. These procedures must insure as safe and expeditious a debugging process as is possible. The startup team thoroughly verifies the connectivity of the process, looks for leaks, and starts up subsets of the equipment first, leading to a full plant startup. When the plant does startup fairly easily and quickly, the XYZ Company can accept delivery of it after UVW successfully operates it for about two weeks.

1.2.9 Plant Operation

All the time the UVW Construction Company has been working on building the plant, the XYZ Company has had a team designing how to operate it. This team interacts closely with the team developing the startup procedure. It has developed operating manuals whose correctness it must now verify. Using experienced engineers working alongside experienced operators, this team learns to run the plant by debugging the manuals and the process if need be. It also decides how to present this material (for example, it is possible today to do it electronically). All the while this activity occurs, the team designing the operator training facility is watching very carefully. It, too, must verify the correctness of what it has created; in particular, it has to be sure that a response from the training facility to an incident is essentially the same as what the process will do. This team must also design how to carry out the training—for example, how often must operators be retrained? What organization will do the retraining? How will the training simulator be maintained?

1.2.10 Debottlenecking

As one runs a plant, one discovers that it can be improved. Often a team is set to work on a process to find ways to increase throughput and/or safety. Making changes to improving process performance is termed *debottlenecking*. They must propagate these changes to the operator training facility and into the manual and process blueprints.

1.2.11 Decommissioning

Finally, all plants will cease to run someday, although many will run for decades. When they do, the company must design and execute a process to decommission the plant.

1.3 THE SYNTHESIS STEP

As we have just seen in the life cycle design scenario, the design process involves both an abstract description of what is wanted and a more detailed (that is, more refined) description in each of the steps of designing, constructing, and operating a process. For example, the board of directors wishes to improve the future value of the company, which is an abstract description of its desires. It generates and selects among a number of alternative actions the company might take; this represents a more detailed or refined description of what they want. This description becomes the abstract description for those working next on this project. In a preliminary process design example, the abstract goal might be to convert excess ethylene into ethyl alcohol. The more refined description will be a preliminary process design to accomplish just that.

We label the process of converting an abstract description into a more refined description a *synthesis* activity, and several steps of that activity are illustrated in Figure 1.1. As we saw earlier, synthesis is repeated over and over again in the course of creating a complete process design. It is used to create the preliminary process design; to create a piping and instrumentation diagram (P&ID) from this description is another cycle through a synthesis process.

Figure 1.1 breaks the synthesis step into several substeps. The first is *concept generation*. Here we identify the different concepts on which to base the design. For our process we must decide if we limit ourselves to the chemistry found in the literature. Will we stay with well-proven processes, or will we look for unconventional solutions? Will we purchase our process as a package from someone else? Are we going to adopt a particular strategy to attacking the design problem?

During the next step, we consider the *generation of alternatives*. Examples of sources for alternative concepts are the library (patent literature, journal articles, encyclopedias of technology), corporate files, consultants, and, of course, brainstorming when any or all of this information is in hand. These information sources should be scoured thoroughly. One will often find fairly detailed descriptions of existing processes to accomplish the design task at hand especially if one is proposing to produce a commodity chemical. In addition, the brainstorming process leads us to question these alternatives and develop new ones.

Armed with the decisions that define our design space and with the means to generate all of the alternative designs, we then consider the next step, *analysis* of each alternative to establish how it performs. For process design, this typically means carrying out mass and energy balances on the process to find what its flows, temperatures, pressures, and so on will be. This information results directly from our decisions on design alternatives. In the next step we have to *evaluate* the process's performance; we can compute its

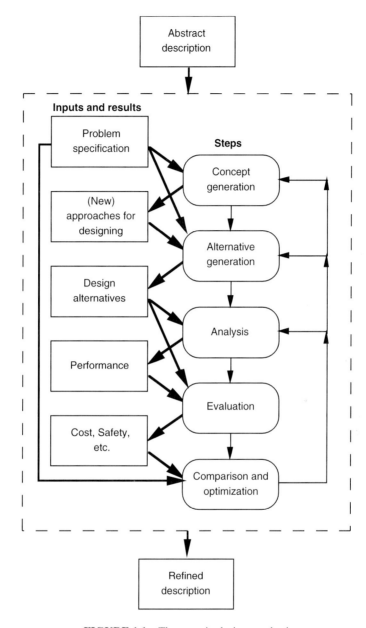

FIGURE 1.1 The steps in design synthesis.

economic worth, its flexibility, its safety, and so on. Finally, *optimization* requires the adjustment and refinement of decisions to improve the design. When we are done, we hope to have the one design that best satisfies all our goals, and we will have transformed an abstract description to a more refined one as a proposed process flowsheet.

Figure 1.1 sets the scope for the design activities and details of these activities are described through this entire text. Before proceeding into more depth on these issues, we first examine and then address some social issues in design. As described in the next section, this background also helps to focus the design tasks in Figure 1.1.

1.4 DESIGN IN A TEAM

Design problems in industry are usually addressed in team situations. As a result, an understanding of group dynamics and activities is essential for the accomplishment of the design task. In particular, these aspects can be critical to the successful development and completion of a design activity. In this section we concentrate on the composition and organization of a design team.

Let us consider as an example the organization of students into teams for a senior design class. This activity is a design problem by itself for which there are many alternatives available. The team size is the first consideration. Depending on the task, the team will likely range from three to five members. Larger teams have the disadvantage that one or two members in the team will often not do their share of the work, but they have the advantage that more can be accomplished if all actively participate. More diversity in the team also enhances the generation of different ideas. On the other hand, a three-person team often suffers from having two of its members form a subgroup and ignore the third member.

A second consideration is compatibility of personality types of team members. When setting up teams, the class members could agree to take a personality test (for example, the Myers-Briggs test). Or, with considerably less effort, each member in the class can attempt to classify him- or herself as one of the four personality types (Amiable, Expressive, Analytical, Driver) described in Table 1.1. Note that each personality type has its own strengths, and there is no intention here to make one seem preferable to another. Amiable and analytical people are more indirect and operate at a slower pace than do the other two types, who will take charge and tell others what to do. A driver will want to start working on the problem immediately while the amiable person will want the team members to get to know each other first.

Amiable and driver members will have problems being in the same team, as will expressive and analytical members. If these types are in the same team, they should be aware of their characteristics and account for them in the team dynamics. Moreover, it has been our experience that, given no guidance, a class of students will have at least one team in which no member is willing to make a decision. As deadlines come, this team will still be exploring alternatives. Consequently, members should be very honest in appraising themselves to be certain it has at least one person in it who is willing to make decisions.

TABLE 1.1 Different Personality Types and Their Behavior within a Team (material from S. Schubert, Leadership Connections Inc., Highland Lake, NJ 07422)

	OPEN (Relationship Oriented) They Emote		
	Amiable	**Expressive**	
	Emphasis: Steadiness; cooperating with others to carry out the tasks	**Emphasis:** Influencing others; forming alliances to accomplish results	
	Pace: Slow and easy; relaxed	**Pace:** Fast	
	Priority: Relationships	**Priority:** Relationships	
	Focus: Getting acquainted and building trust	**Focus:** Interaction; dynamics of relationship	
	Irritation: Pushy, aggressive behavior	**Irritation:** Boring tasks and being alone	
INDIRECT (Slow Pace)	**Specialty:** Support—"We're all in this together so let's work as a team."	**Specialty:** Socializing—"Let me tell what happened to me …"	**DIRECT** (Fast Pace)
They Ask			They Tell
	Analytical	**Driver**	
	Emphasis: Compliance; working with existing circumstances to promote quality in products and services	**Emphasis:** Dominance; shaping the environment by overcoming opposition to accomplish the tasks	
	Pace: Slow; steady; methodical	**Pace:** Fast	
	Priority: The task	**Priority:** The task	
	Focus: The details; the process	**Focus:** Results	
	Irritation: Surprise; unpredictability	**Irritation:** Wasting time; 'touchy-feely' behavior that blocks action	
	Specialty: Processes; systems—"Can you provide documentation for your claims?"	**Specialty:** Being in control—"I want it done right and I want it done now."	
	SELF-CONTAINED (Task Oriented) They Control		

Teams pass through different stages. At first everyone feels good about the team and all seems to be going pretty smoothly. This period often ends abruptly when some team members become angry with each other because not all of them contribute to the same extent. Many teams never get past the angry phase, and the design project obviously suffers. The next stage is tolerance, where team members accept their differences and learn to work together in spite of them. It is not a particularly enjoyable situation, but work gets done. A really successful team passes into a stage where it uses the strengths of its members to its advantage. It allows drivers to drive and invites the amiable members to smooth over its personality problems.

1.5 CONVERTING ILL-POSED PROBLEMS TO WELL-POSED ONES

Having set up a design team, how should they attack a design problem, especially a problem for which they have had no prior experience? Here we consider some ideas that help in carrying out the activities in Figure 1.1.

Starting on a new type of design problem is difficult because the problem is often ill-posed with only a "fuzzy" description of what is desired. Therefore, we first need to focus on a clear problem definition. A design team for a construction company that specializes in turn-key ammonia plants will have little difficulty in making its design problem well-posed. On the other hand, the task of creating an effective design organization to carry out such designs may resist attempts to make it well-posed for years. In this section we consider four steps that help to convert an ill-posed problem into a well-posed one. These steps require us to:

- Establish goals
- Propose tests one can carry out to assess if one is meeting one's goals
- Identify the starting points
- Identify the space of design alternatives

Application of these steps helps to define and capture the nature of our preliminary design problems. In the remainder of this chapter, we will work on these tasks repeatedly.

In the early stages, each of these steps is often best done by involving the design team in a brainstorming approach. Table 1.2 lists some ideas on how to approach brainstorming. Only after the brainstorming step is terminated—which might occur after a preset time of two to three hours, should the team examine each of the items on the list that it constructs and offer comments and criticisms on each. At that time it can attempt to consolidate the items listed, eliminate some, combine others to produce added items, and the like. This activity serves to expand the space of alternatives and then separately, to contract it. Moreover, the brainstorming process may be repeated with a larger team later in the design process to expand the design space based on more information and experience with the design problem at hand.

With the organizational and brainstorming concepts in mind, we now explore the four steps needed to help define the design problem.

ESTABLISHING GOALS

To make this design problem well-posed, the design team first needs to establish a clear definition of its goals. Among the goals that a brainstorming process might generate for this design problem are

Sec. 1.5 Converting Ill-Posed Problems to Well-Posed Ones 11

TABLE 1.2 Brainstorming

Do brainstorming with a team, and populate this team with persons of diverse backgrounds to bring in a variety of views.

Choose a facilitator to keep the process on track. It is very easy for the team dynamics to stray from a brainstorming activity. The facilitator should also capture any key ideas by writing them on a poster board where everyone can see what is written. With adhesive tape, stick the sheets around the room on the wall so all can see each of them.

Never, never allow criticism of any of the ideas raised during brainstorming. Criticizing comes later. The facilitator must identify criticizing and terminate it with a polite: "We will criticize ideas later. Now we want to generate ideas."

Encourage wild ideas. Also encourage participants to take other ideas and add new twists to them. Often a combination of separate off-the-wall ideas leads to a very interesting and novel new idea.

Encourage everyone to participate. A possible mechanism is to stop the team activity for about 15 minutes and have each of the team members list his or her ideas on a separate sheet of poster paper. Immediately after, have each present his or her list to everyone else. Place these sheets on the wall, too.

- Make a profit (otherwise why do this).
- Maximize the profit.
- Minimize operating and investment costs.
- Insure design meets safety standards.
- Create a design we can control easily.
- Maximize the flexibility of the process to feedstock fluctuations.
- Create a design that fits within the space available for a new process plant at the Gulf Coast facility.
- Create a design that does not pollute.

Some of the goals will be constraints; others will be objectives we wish to maximize or minimize. For example, the first (make a profit) is a constraint. We insist that profit is greater than zero. The next two are objectives. Subject to making a profit, we would like then to maximize the profit we make. A team may later narrow the total set to about a half dozen or so, and we see clearly from this that our design will be a compromise on meeting all of the goals. For instance, adding process flexibility will almost certainly reduce our profits, and we might report the maximum profit we can attain for different values of the flexibility, leaving it for our supervisor to decide where she would like to make the trade-off.

PROPOSE TESTS

A test involves the evaluation of any proposed design while enumerating the design alternatives. For our process design, we can propose to evaluate the net present value (covered in Chapter 5) of a proposed project using a precisely defined set of cost estimation methods and correlations (discussed in Chapter 4). Also, we might use a simple profit model to screen among alternatives. Our analysis might be to complete a form that the company provides for project evaluation. Alternatively, we might use a more sophisticated present worth model that we construct using a spreadsheet program such as Lotus 1-2-3 or Excel, and this may involve complex timing of payments and incomes. For instance, the company could partition the investment required for the equipment in annual amounts of 50 percent, 30 percent, and 20 percent during each of the first three years of the project. We might further estimate that product production starts at the end of year three at a 50 percent production rate rising linearly to 100 percent over the next nine months. We could assume that the next year, due to debottlenecking, will provide another 10 percent production for an added 3 percent investment.

It is important to consider these tests from the beginning as they focus the effort required. Also, there is no purpose generating information that no test uses. For example, if one test for safety is to evaluate all the chemicals in the design for toxicity, then the team knows it must identify all species in each design and gather toxicity information for them. If heat exchanger cost estimation is to use a correlation that predicts cost given only the area, the materials of construction, and the pressure, then the team needs to generate this and only this information for exchangers to estimate their costs.

IDENTIFY INITIAL POINT(S)

Identifying where one intends to start the design problem may seem of little importance. However, suppose one has as a goal to climb Mount Everest. Starting one foot below the summit is a very different problem from starting at the base of the mountain. The starting point for our design could be a design carried out two years ago that is still in our files or it may be a patent description. On the other hand, we may choose to start completely from scratch and use several new alternatives as starting points.

IDENTIFY SPACE OF DESIGN ALTERNATIVES

The design team next needs to identify design decisions and their alternative values. Many of the decisions are discrete, such as locating specific unit operations in the flowsheet, while others are continuous, generally made after we settle on the discrete decisions. We often need to work for some time on our design problem to identify the space of design alternatives, as this is a very large, complex problem. Unless we are doing a routine design, several days or weeks could be dedicated to this task.

A typical approach to identifying the space of alternatives is first to develop a base case design. From the decisions made to develop this base case, we identify where we made decisions that led to this particular design. We also list the alternative decisions we could have made, and they could lead to very different designs and follow-up decisions. Here we can also anticipate future decisions that we encounter and explore to complete the synthesis activity. If we do not keep these decisions foremost in our minds, we will fail to appreciate the number of alternatives we really should be investigating for our design. For instance, it is not uncommon for the number of design alternatives for a chemical process (based on the discrete decisions alone) to number 10^{15}—*and it is unlikely that your team picked the best one on its first try.*

To begin the tasks of alternative generation, we have at least four purposes for wanting a base case design.

1. Once we have it, we need to focus our activity of converting our ill-posed design problem into a well-posed one. In particular we develop a description of the design space of all alternatives to carry out the design. Thus, we use the base case to learn about our design space of alternatives. If we do not return to the activity of defining the design space, we may fail to generate the alternatives we need for our problem. We also re-examine the goals and tests we proposed and revise them based on what we have just learned.
2. It may enlighten us with little added effort about important features of this design problem. For example, we might discover that the design of the reactor is crucial, or we might discover that we must preprocess the feed to discover an economic process.
3. The base case provides us a solution for which we can estimate the actual profits. No design with lower profits need be explored if our goal is to find the most profitable design.
4. The base case design gives us a starting point from which to generate improved alternatives.

The more systematic generation of alternatives will be discussed in the next chapter. At this point, however, we illustrate the four-step process of this subsection to generate alternatives with a process design case study. This example process also forms the basis of many of the concepts we introduce in the next three chapters.

1.6. A CASE STUDY PROCESS DESIGN PROBLEM

We illustrate these ideas by considering the following chemical process. The plant manager for our Gulf Coast plant has asked us to determine what we might do to utilize an excess of approximately 75 million kg/yr of ethylene that this facility is producing. In discussions with the head of our process design team, one option is to build a new process to make a product from ethylene that we could sell profitably. Among the possible products, our Sales Department believes it could sell about 150,000 cubic meters of 190 proof ethyl alcohol per year, which would use a significant portion of our available excess ethylene. The head of our team, therefore, requests us to investigate the design of a plant to convert a substantial part of this excess ethylene to 150,000 cubic meters of 190 proof ethanol. He informs us that our ethylene feed is 96 mole percent ethylene, 3% propylene and 1% methane. Note that ethylene put into an ethylene pipeline is typically 99.996% pure so this is a very impure ethylene feed.

A first step for our example problem that we should undertake is to examine the relevant literature about the manufacture of ethanol and, in particular, about its manufacture from ethylene. The reaction is straightforward:

$$\text{CH}_2 = \text{CH}_2 + \text{H}_2\text{O} \rightarrow \text{CH}_3\text{CH}_2\text{OH} \quad (1.1)$$
$$\text{ethylene} + \text{water} \rightarrow \text{ethanol}$$

Two technical encyclopedias for the chemical industry [Kroschwitz and Howe-Grant, 1992; McKetta and Cunningham, 1983] describe a process based on using a high-temperature, high-pressure homogeneous noncatalytic reactor. The reactor temperature typically ranges between 535 K to 575 K, and the pressure is 1000 psia (about 68 atm or 69 bar). These same articles report reactor conversion to be about 5 to 7 mole percent. The ratio of water to ethylene in the feed can be as large as 4 to 1, which is four times that needed by the reaction stoichiometry if all the ethylene were to convert in a single pass through the reactor. However, because of the low conversion per pass, we can choose a smaller water ratio of 0.6 to 1 (Westerberg, 1978), and this reduces the molar flowrates in the process flowsheet.

These articles report a second reaction, the conversion of ethanol to diethylether and water, which is at equilibrium.

$$2\ \text{CH}_3\text{CH}_2\text{OH} \rightarrow \text{C}_2\text{H}_5\text{-O-C}_2\text{H}_5 + \text{H}_2\text{O} \quad (1.2)$$
$$2\ \text{ethanol} \rightarrow \text{diethylether} + \text{water}$$

We are also advised by our chemistry department to keep the mole fraction of methane in the reactor feed to less than 10% to prevent coking at these extreme conditions. Also, they mention that excess water in the reactor serves at least two purposes. One is to push the equilibrium conditions for the first reaction to the product, ethanol, and the second is to push the equilibrium of the second reaction back to the reactant, again ethanol. For a process where methane is present, as here, the water will also serve to dilute the methane and stop it from coking, that is, undergoing decomposition to carbon and hydrogen.

Also, the process produces a trace amount of a four-carbon aldehyde, croton aldehyde, which would be a waste product for us. Moreover, if propylene is present in the feed, it will also react with water to form isopropanol. Its conversion is about 10% of that for ethylene (i.e., to about 0.5 to 0.7% conversion of the propylene in the reactor feed).

Figure 1.2 indicates the species in the feeds and possible products for the reactor in this process. To start the analysis of this process we will need some physical property data. Table 1.3 contains data we might find useful.

The species, arranged in order of increasing boiling point, are shown in the product stream in Figure 1.2. It is worthwhile assessing these data. We note that at one atmosphere methane, ethylene, and propylene boil at very cold temperatures, well below ambient. The critical temperatures for methane and ethylene are also below ambient; thus, we cannot condense methane and ethylene at room temperature. Assume that we can cool mixtures to about 310 K with cooling water, generally the least expensive method we have for cooling. At that temperature, propylene already has a vapor pressure of about 15 atm. That is a fairly high pressure, but not an unthinkable one, to be operating a condenser

Sec. 1.6 A Case Study Process Design Problem

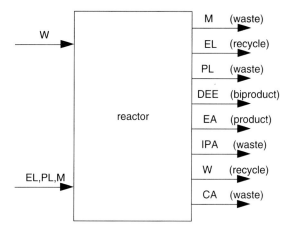

FIGURE 1.2 Components in reactor feeds and products.

for a distillation column. We can imagine having propylene as the top product in a column, but it will be an expensive column. At one atmosphere diethylether boils at 34.6°C, the temperature of a warm summer day, while the others boil well above ambient. Water is a notable outlier when it comes to its critical conditions. Note its critical pressure of 217.6 atm is three to four times that of the other species.

In the last section we considered process design goals and tests proposed to evaluate alternative designs. These can be applied directly to this case study. We now consider some initial starting points and quickly sketch a possible design for the ethylene-to-ethyl alcohol process. This helps us to think about our design as we work towards developing

TABLE 1.3 Physical Property Data for Species

Species	W water	EA ethyl-alcohol	EL ethylene	DEE diethyl-ether	M methane	PL propylene	IPA isopropyl-alcohol	CA croton aldehyde
Formula	H_2O	CH_3CH_2OH	$CH_2=CH_2$	$(C_2H_5)_2O$	CH_4	$CH_3CH=CH_2$	$CH_3CH\text{-}OHCH_3$	$CH_3CH=CH\ CH=O$
MW	18.02	46.07	28.05	74.12	16.04	42.08	60.10	70.09
Sp. Gr.	1.0	0.789	0.56	0.708	—	0.609	0.785	
Melt Pt, °C	0	−114.5	−169.2	116.3(α)	−182.5	−185.3	−89.5	159–160
BP, °C	100	78.4	−103.7	34.6	−161.5	−47.7	82.4	
ΔH_v (kcal/mo)	539.55	204.3	115.4		121.9	104.6	159.4	
VP A[1]	8.10765	8.04494	6.74756	7.4021	6.61184	6.81960	6.66040	
VP B	1750.286	1554.3	585.00	1391.4	389.93	785.00	813.055	
VP C	235.0	222.65	255.00	273.16	266.00	247.00	132.93	
t_c, °C	374.14	243.5	9.6	193.8	−82.1	91.4	235.16	
P_c, atm	217.6	63.1	50.7	35.5	45.8	45.4	47.02	

[1] $VP(mm\ Hg) = 10^{A-B/(C+t(°C))}$ where VP is vapor pressure and t is temperature.

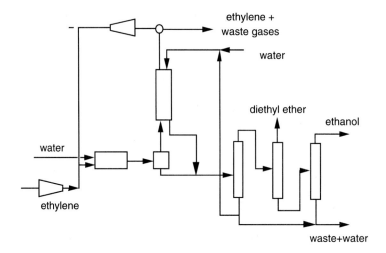

FIGURE 1.3 Typical process for converting ethylene to ethyl alcohol.

our base case. We discover the literature reports a typical design for this process. Figure 1.3 sketches such a process where the ethylene feed is relatively pure.

In this flowsheet, water and ethylene mix with an ethylene recycle stream and enter the reactor. Only 7% of the ethylene reacts, and the literature suggests we should feed 0.6 moles of water per mole of ethylene. The reactor effluent therefore contains large amounts of unreacted reactants, water, and ethylene. It also contains ethyl alcohol and diethylether in significant amounts. In a heat exchanger (not shown) we cool the reactor effluent into the two-phase region, while holding the stream at high pressure. We do not want to lose pressure as we are going to recycle large parts of this stream back to our high pressure reactor. We will have to compress the vapor recycle stream to bring it back to the reactor pressure, and compressors are expensive. In the flash unit following the reactor, we separate the liquid phase from the vapor phase.

The vapor from the flash unit is largely ethylene but contains significant amounts of diethylether and some ethyl alcohol. To recover the ethyl alcohol from this vapor stream, we scrub it by passing it against water in an absorber. We usually choose to run an absorber as cold as is economically possible, so we operate this unit near ambient temperature, which we can reach using cooling water. To remove any light contaminants that we trap when recovering the ethylene, we split off a small part of the recycle as a bleed or purge stream. Depending on the species in it, we may be able to use this stream as fuel. Having passed through several units—the reactor, a heat exchanger to cool it, the flash unit, and the absorber—we find the ethylene recycle is at a lower pressure by a few atmospheres than the reactor (which we know from earlier operates at about 68 atm). We compress the vapor recycle to increase its pressure to that needed so we can return it to the reactor.

The liquid stream from the flash unit is largely water, ethyl alcohol, and diethylether. We send both this stream and the water stream from the scrubber to a series of distillation columns. The first column removes the bulk of the water as a lower product. The second separates out diethylether. The third column recovers 190 proof ethyl alcohol as a top product from the remaining water. Finally, the trace amount of croton aldehyde will exit largely in the first water stream.

Now, let's consider some *process alternatives*. First, how should we alter the above flowsheet to account for our ethylene feed, which contains 3 mole % propylene and 1 mole % methane? We need to remove the propylene and methane from the process. We can either separate out one or both of these species before the ethylene enters the reactor, or we can let either or both of them enter the reactor and remove them and their possible products after the reactor. Figure 1.4 illustrates some of the alternatives possible.

Methane is difficult to separate from ethylene, especially if we chose to use distillation. We would have a top distillation product of methane. We note that the critical temperature of methane is $-81.2°C$. To form reflux we would have to condense methane at extremely cold temperatures even if we operate at high pressure. We probably would not choose to do this.

We might also consider separating the methane and ethylene using membranes; for this we need to bring the ethylene up to the pressure of the reactor, about 68 atm. We would do this using a compressor, a fairly expensive option. Here a typical membrane would work by putting a mixture at high pressure on one side so that the smaller molecule, methane, preferentially passes through the membrane, exiting at much lower pressures. The larger molecule, ethylene, then proceeds at high pressure to the reactor. One worry for membranes is just how sharply we can carry out the separation. Would we lose a lot of the ethylene with the methane, for example, or would we still have significant amounts of methane left with the ethylene? Other methods we might consider include adsorption and absorption.

On the other hand, we are permitted to let methane into the reactor up to 10 mole %. As will be discussed in Chapter 2, we can elect to remove methane by letting it enter with the ethylene and build up in the recycle that recovers the ethylene. We then remove a small part of that recycle stream as a purge stream.

Finally, to separate propylene from ethylene using distillation again requires refrigeration to form a top reflux of ethylene. Membranes are not so appealing because now ethylene passes through on the low pressure side and recompression costs would likely rule this option out. As a result, we also let the propylene enter the reactor where a small part of it converts to isopropyl alcohol. We note that this compound boils only 4°C higher than ethyl alcohol, which could give us separation difficulties when we try to recover our final product.

We have suggested ways to create several options above, but we have not been methodical in our description and exploration of the design space. We will discuss more systematic approaches to the synthesis step extensively in the next chapter. Nevertheless, in this synthesis procedure we plan to use what we learn at each step to return to our quest to define the search space of alternatives.

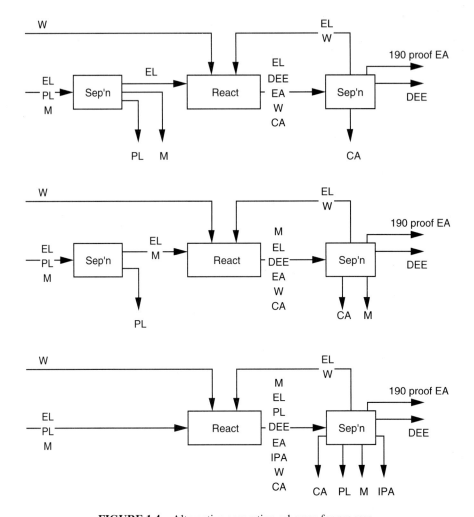

FIGURE 1.4 Alternative separation schemes for process.

1.7 A ROADMAP FOR THIS BOOK

In this chapter we introduced many issues that occur in process design. We illustrated some of them by looking at several, more general design issues. In particular, we concentrated on preliminary process design and showed how this fits into a realistic corporate design scenario. Next, we considered design in a team and discussed the factors relating to team composition and team activities. We sketched an approach to help designers attack problems for which they may have little previous experience. We concluded this chapter with a process design case study, which sets the stage for the process synthesis

Sec. 1.7 A Roadmap for this Book

problem. While Chapter 1 has given us a broad overview of issues in process design, systematic design and synthesis strategies are the main theme of this book.

The hallmark of Part I (Chapters 2 through 6) of the text is to allow quick evaluation among alternatives. You will not get particularly accurate assessments with the techniques advocated. However, you can learn much about the design problem using them, and this learning is a very important first step—as we have argued earlier in this chapter. Chapter 2 of this text discusses several approaches for representing, evaluating, generating, and searching among the many possible flowsheets that can satisfy one's design goals. It introduces strategies for decomposing the design problem into more manageable tasks, decompositions which are experienced based. There are many issues common to all of design no matter the discipline. We try to expose some of them here. On the other hand, the details of the representations and applications of the strategies we discuss here differ significantly from their application in other domains.

Preliminary process design requires us to evaluate alternative flowsheets quickly. Chapter 3 presents a simple hand calculation method to carry out mass and energy balances to set flows, temperatures, and pressures throughout a proposed flowsheet. These methods cannot be particularly accurate, but they allow a quick assessment to learn about a design problem. Chapter 4 tells us how to estimate the equipment and operating costs associated with such a design, information we will need if we wish to assess its economic value.

Chapter 5 discusses the details of assessing the economic value to the company of a design. We discover that it is the flow of cash versus time that we must use for evaluation. Chapter 6 ends this first section by looking at how the design of batch processes adds the wrinkle of scheduling the use of the equipment to decide what equipment to buy.

Chapters 7 through 9 form Part II of the text. Part II tells us how we can analyze our process alternatives to get much more accurate answers. The computations implied here are so extensive they must be done using the computer. In Chapter 7 we look at detailed modeling of many of the unit operations we find in our processes. Chapter 8 then examines the solution strategies of models for complete processes. These models are built by connecting together the unit operation models we described in Chapter 7. Here we discuss the characteristics of commercially available simulation tools called "flowsheeting systems" for carrying out mass and energy balance calculations for arbitrarily configured processes. Finally, in Chapter 9 we look at the tools available to improve the operation of a design by applying optimization to it.

Up to this point we have not provided extensive methodology for creating and searching among the myriad of alternatives that exist when designing a process. The five chapters (10 to 14) forming Part III present many basic concepts useful in inventing the better alternatives. Chapter 10 looks at how we can heat integrate processes, looking first at the synthesis of heat exchanger networks. Below ambient heat integration involves heat pumps, and we develop insights for designing them. Chapter 11 concentrates on designing distillation-based systems to separate reasonably well behaved liquid mixtures. Distillation columns are major consumers of heat. We put heat into their reboilers and remove it from their condensers. In Chapter 12, we consider how the ideas we discussed in Chapter 10 on heat integration apply specifically to managing this heat passing through distillation columns. Chapters 13 and 14 discuss physical and geometric concepts for two nonlinear subsystems of chemical processes: the synthesis of chemical reactor networks and the design of nonideal, azeotropic separation sequences, respectively.

Part IV of the book looks at the use of advanced optimization methods to search among design alternatives. The main emphasis is the mathematical modeling of synthesis problems. A summary of concepts and algorithms is given in Appendix A. Here many of the models we present in this part of the text use binary (yes-no) as well as continuous variables. We often use such variables to indicate whether a flowsheet will have a particular unit in it or not. These chapters show how to formulate suitable models and how to solve them. Problem formulation can make or break our chances to solve many of them. Chapter 15 discusses the general approach for problem formulation in terms of representation of alternatives and discrete/continuous optimization models. Chapters 16 and 17 revisit the synthesis problems for heat exchanger networks and heat integrated distillation sequences. When these are expressed as mathematical programming problems, we can search rigorously over a very large number of alternatives. In several cases we can guarantee finding the best solution for the problem that has been formulated. In addition, these chapters introduce the concepts of sequential and simultaneous optimization for process synthesis.

In Chapter 18 we present a model that allows us to compute the minimum use of utilities required if the process were to be heat integrated as we are optimizing over the operating levels and sizes for the equipment in the process. Essentially, we embed the optimal heat exchanger synthesis problem within the flowsheet optimization problem. Optimization proves also to be a powerful tool for selecting among the many alternative ways to configure reactors. Chapter 19 shows us how to model and solve reactor synthesis problems using these strategies. Chapter 20 then deals with structural optimization of process flowsheets and describes a decomposition strategy for effectively solving nonlinear discrete optimization problems that integrate several process subsystems together.

Processes have to be flexible. While we all have an intuitive feel for what flexibility is, we still need a precisely defined meaning for flexibility if we wish to use optimization to find the most flexible processes. Chapter 21 provides this rigorous mathematical definition and shows we can use it to design flexible processes. Finally, Chapter 22 returns to the design and scheduling of batch processes, this time with an emphasis on plants that can produce many different products. Consistent with the rest of Part IV, this chapter stresses the use of optimization.

REFERENCES

Kroschwitz, J. I., & Howe-Grant, M. (Eds.). (1992*). Kirk Othmer Encyclopedia of Chemical Technology,* 4th ed., Vol. 9 (pp. 820–826). New York: John Wiley & Sons.

McKetta, J. J., & Cunningham, W. A. (Eds.). (1983). *Encyclopedia of Chemical Processing and Design,* Vol. 9 (pp. 452–455). New York: Marcel Dekker.

Westerberg, A. W. (August, 1978). "Notes for a Course on Chemical Process Design," taught at INTEC, Santa Fe, Argentina.

EXERCISES

1. Consider the design problem a senior design class first faces. It has to form into design groups. For this design problem:
   a. List an appropriate set of at least six goals for this design problem.
   b. Devise tests for at least three of the goals you list in part a. Remember you must be able to evaluate a test now and not after the groups are formed and are operational. You are trying to assess how each of the group-forming options meets the goals without yet having the groups formed.
   c. Describe the search space for this problem if the class has 14 students in it with names $n[1]$, $n[2]$, . . . , $n[14]$. Create one instance of a solution to the design problem, where this solution is one member of the search space.

   Given this instance of a solution, is it obvious to you how you would then apply each of the tests in part b? If it is not, you are missing something in your response to this question.

PART I

PRELIMINARY ANALYSIS AND EVALUATION OF PROCESSES

OVERVIEW OF FLOWSHEET SYNTHESIS 2

In this chapter we introduce many of the technical issues involved in discovering better process flowsheets from among the enormous number of alternatives possible. We also use this discussion to motivate the remainder of the book. Specifically, we examine some basic steps involved in the synthesis of process flowsheets including *gathering information, representation of alternatives, assessment of preliminary designs,* and *search among alternatives.* We complete this chapter with a case study where, using a multilevel hierarchical representation, we synthesize a base case flowsheet for the ethyl alcohol manufacturing process we introduced in the last chapter.

2.1 INTRODUCTION

Preliminary process design is a synthesis activity. A design team carries out a preliminary design to discover better process configurations for the stated design goals. It is an extremely important activity. If it is carried out poorly, the company may decide against what could have been a profitable activity, or it may find itself saddled with a marginally profitable process that requires constant revamping to keep up with the competition. Moreover, while the design activity itself is not costly relative to the entire project cost, the decisions from the design team impact the project in major ways and over its entire life, which could be decades.

Many industrial studies have compared the monies spent on process design and construction projects to the fraction of the costs committed as the projects progress. Typical results from such studies indicate that, during the preliminary design step, a company will have spent about 15 to 20% of the total funds it will devote to the project. However, the decisions that the preliminary design team makes fix about 80% of the subsequent costs the project will incur. In other words, no matter how well the company carries out the remaining activities, the best it can do is make improvements in about 20% of the costs for the project. To appreciate the plausibility of these observations, think of the impact of the decision to use a particular raw material and reaction step in a process. This decision is at the heart of the process and everything else follows from it. Once made, it fixes the majority of the costs the company will incur in building and starting up the process.

To illustrate the impact of the design we consider a particular chemical process, the manufacture of methyl acetate by the Eastman Chemical Company. In 1985 Eastman Chemicals received the Kirkpatrick Award in Chemical Engineering [Chemical Engineering Magazine, 1985] for developing a radically new process to manufacture methyl acetate. At that time, conventional processes consisted of a reactor followed by half a dozen separation units to purify the product, recover and recycle unreacted raw materials, and isolate wastes. The new Eastman process, on the other hand, carries out all these steps in a single reactive distillation column, and this decision was made *at the preliminary design stage*. The costs for building and operating this new process are only a fraction of the costs for conventional processes. Consequently, none of the conventional processes could compete with it.

Preliminary design involves generating alternatives and, for each, carrying out analyses to determine how it performs, with a value placed on that performance. As seen in Chapter 1, this activity occurs repeatedly as one progresses through a design. As an example, we described an ethyl alcohol process in this chapter. Here a chemical company establishes the goal to use its excess ethylene to produce ethyl alcohol. At the end of the first synthesis step, the design team reports on the best process configuration it has found. This configuration is the starting point for the next step to produce piping and instrumentation diagrams (P&IDs). Here the designers search for better alternatives related to the actual equipment, the materials of construction and the controllers.

Finally, in preliminary design we consider the creation of an entirely new process (termed *grassroots design*) or improve an existing process (a *retrofit design*). In retrofit design the number of possible alternatives is many times larger than for grassroots design, although many of the ideas for grassroots design carry over to the retrofit problem. In fact, one option in retrofit design is to tear down the existing structure and design the entire process from scratch. Consequently, in this chapter and for much of this book, we shall concentrate on grassroots design.

For the preliminary design problem, we can take advantage of many systematic approaches to this problem. In next section, we present an overview of the basic steps in flowsheet synthesis. Following this section, we focus on more structured, hierarchical decomposition strategies that guide the decisions that lead to an initial base case design. In section 2.4 we then return to the ethyl alcohol case study and illustrate these basic steps to synthesize the flowsheet. Section 2.5 summarizes the chapter with a bridge to the more detailed analyses presented later in the book.

2.2 BASIC STEPS IN FLOWSHEET SYNTHESIS

In this section we present an overview of the basic steps required to carry out the synthesis of a chemical process. From the first chapter we learned that, even for simple problems, the number of alternatives is generally enormous, and our goal will be to discover good alternatives without an exhaustive search. In this chapter and throughout the rest of the book, we consider the technical steps to discover and evaluate better flowsheet alternatives. The first step is to *gather relevant information*. This step helps to uncover existing process alternatives. Next, the process alternatives need to be *represented* in a concise way for decision making. To do this, we need to develop *criteria to assess and evaluate our designs* by deciding which measures to use, such as economic worth and safety. As the design problem offers so many alternative solutions, we will also need to develop systematic methods to *generate and search among these alternatives*. We shall discuss each of these issues briefly in the remainder of this section. Based on this discussion, we then develop structured decomposition strategies to guide the search process.

2.2.1 Gathering Information

It is difficult to overstress the need to search thoroughly for relevant information. Seldom is a design problem entirely new; many parts of it will be well analyzed somewhere in the literature, and it would be shame to overlook such previous work. The obvious places to look are in the technical journals and encyclopedias, handbooks, textbooks, and so forth. Most libraries provide electronic searching over available indices to aid this process, such as Chemical Abstracts. Most computer-based indices list articles back to the mid–1980s, although much useful information may also predate these computer-based indices. The search for information also includes the patent literature. Here a company reveals some of its industrial knowledge in exchange for its exclusive ownership for several years (e.g., seventeen years in the United States). Thus, aside from using the patent literature to find what others have done, it also must be searched thoroughly as a defensive measure to avoid legal problems later.

In addition, companies use consultants who know the real value of the literature. They also join organizations that carry out studies for their member companies. For heat exchanger information, two such organizations are Heat Transfer Research Institute (HTRI) in the United States and Heat Transfer and Fluid Flow Service (HTFS) in Britain, while the Fractionation Research Institute (FRI) provides information on distillation. Other organizations, such as SRI International, carry out detailed design studies for most of the conventional petrochemical and refinery processes.

Finally, the World Wide Web is a resource that can only improve with time. Many companies maintain information about themselves on the Web. This information allows us to begin a general search and to ask more specific questions. Indeed, the Web provides a path to find much of the other information we have discussed above. Most companies have their web address as www.*company-name*.com. For example, to find the DuPont company, try www.dupont.com as the web address.

2.2.2 Representing Alternatives

Representation of alternative decisions for the process is intimately tied to the way we intend to generate and search among these alternatives. For example, an obvious representation of the ethyl alcohol process from Chapter 1 is the complete flowsheet in Figure 2.1,

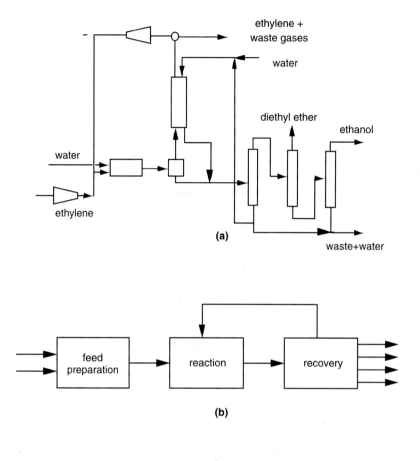

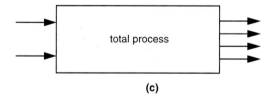

FIGURE 2.1 Flowsheet and different aggregations.

which shows all the equipment and how it is interlinked. To simplify this representation we might aggregate equipment to represent a higher level function such as "feed preparation," "reaction" and "recovery," as shown in Figure 2.1b. We may even aggregate the entire flowsheet into a single object. In creating a representation, the goal is to provide a relevant but concise depiction of the design space that allows an easier recognition and evaluation of available alternatives.

Sec. 2.2 Basic Steps in Flowsheet Synthesis

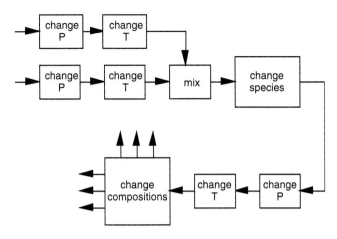

FIGURE 2.2 Representing processes using tasks.

For instance, in addition to thinking of the unit operations in a process, we can base our representation of alternatives on the "tasks" that occur in the process, such as heating, reacting, and separation. Figure 2.2 shows such a representation for the ethyl alcohol process; there will usually be many different alternatives for this association of tasks to equipment. This representation is also very useful for batch processes, where many of these tasks occur in the same piece of equipment but at differing times, as we shall discuss in Chapters 6 and 22.

Finally, for process subsystems, more specialized representations are in common use. For the synthesis of heat exchanger networks, for instance, we represent the flow of heat in a process using a plot of temperature versus the amount of heat transferred as shown in Figure 2.3. In Chapter 10, we will use this type of representation to discover the least amount of utilities we will need to heat and cool a given set of process streams. This

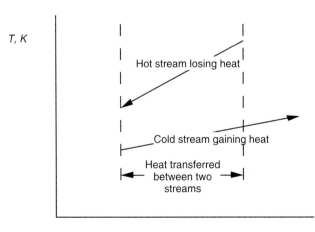

FIGURE 2.3 Representing heat exchange between streams.

representation does not even look like a process flowsheet, but it does describe the alternative ways to exchange heat among numerous process streams.

Another way to represent a process is to show its transitions in the space of chemical compositions. Representing changes in composition space is useful for the synthesis of reactor networks (Chapter 13) and nonideal separation processes (Chapter 14). For instance, Figure 2.4 shows such a representation as a ternary composition diagram. In this space we can describe transitions from raw material compositions to product compositions through reaction, separation, mixing, and heating. At a later stage we can map these transitions into equipment; we may even discover new types of equipment with this representation.

There are many very different representations we can use to think about our design problem and to describe alternatives for it. It can also take years to discover a useful representation and present its implications for design. A useful representation is, therefore, a significant intellectual contribution to design, and, with time, often forms the subject matter of the undergraduate courses taught in a discipline. The McCabe-Thiele diagram is one such example; anyone involved in distillation uses the insights provided by this diagram to see the impact of design decisions one might make for a column.

2.2.3 Criteria for Assessing Preliminary Designs

How much is a design worth to our company? To respond we need to assess the performance of a design alternative and a value for that performance. We use the equations of physics to establish how a process performs, including mass and energy balances to establish stream flows, temperatures, and pressures. We assess the value of a design when we ask if it will be profitable. Here performance evaluation determines how *economic, safe, environmentally benign, safe, flexible, controllable,* and so on a process is. Moreover, different evaluations generally correspond to conflicting goals for a design and increasing the value for one usually requires decreasing the value for another. In principle we would like to convert each criterion into an impact on a single measure—for example, the economics of the process—so we could have a single measure of process worth. But this is not always possible. Some basic criteria evaluated at the preliminary design stage include the following.

Economic evaluation in preliminary design requires us to establish the cost of equipment and the costs associated with purchasing utilities. These methods assume we have completed the mass and energy balances, either approximately from Chapter 3 or

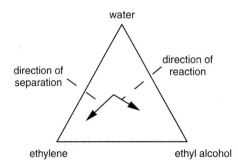

FIGURE 2.4 Representation in composition space showing reaction and vapor–liquid separation directions for a given composition.

more rigorously from Chapter 7. Chapter 5 then discusses how to convert these numbers into cash flows which a company can use to assess the worth of the project when comparing it to its competing projects.

Environmental concerns involve satisfying the very large number of regulations the government imposes on the operation of a process. Where the plant is built determines which government has jurisdiction and, therefore, which regulations the plant will have to meet. One set of regulations may limit pollution a process can pass into the air, a different set limits pollution into the waterways, and a third limits solids into landfill. To understand the existing regulations and follow the many new ones requires the efforts of several persons in a company. Moreover, additional difficulties occur at the design stage in handling small (trace) amounts of hazardous components.

Safety analysis attempts to determine whether any reasonable combination of events leads to unsafe situations: fires, explosions, or releases of toxic chemicals. The U.S. government now requires that each process operating in the United States be the subject of a periodic study to determine and then reduce its potential hazards. These studies are called HAZOP (hazard and operability) studies, and they are very methodical and thus very laborious. A team of process experts looks at every unit, every pipe, every valve, every controller—in other words, at every identifiable part—of the process and asks what would happen if that part were to fail. The team then asks what would happen if two parts were to fail in either order or together. They then repeat for three events at a time, each time considering a larger space of possibilities.

Flexibility in process design requires the manufacture of specified products in spite of variations in the feeds it handles, in the temperature of cooling water from summer to winter, in the heat transfer coefficients as heat exchangers become fouled with use, or other variations. One example of a flexible process is a petroleum refinery, which must tolerate differences in the crude oils it processes. Most refineries receive their feed crudes from pipelines or ocean tankers from oil fields around the world. These processes require flexible operation, but they must still exercise care as they cannot process all crudes that might come their way. The company's profits depend crucially on knowing which crudes they can process at any given time and on deciding a suitable planning strategy. More precise definitions and analysis of flexibility are presented in Chapter 21.

Finally, **controllability** deals with the ability to operate the process satisfactorily while undergoing dynamic changes from one operating condition to another, or while recovering from disturbances. Often, we cannot exactly characterize the disturbances, which makes this analysis even more difficult. Moreover, between the desired states, the process may move momentarily through undesired operating conditions, or become dangerously unstable. While some methods exist for this type of analysis, it often requires the repeated solution of detailed dynamic models and is still an active research topic.

These criteria and many others help to assess the value of a process alternative. In the early stages of design, performance evaluations must be fast as we are likely applying them to a very large number of alternatives. They must also be based on little information at first as we do not yet know much about our alternatives. On the other hand, if an evaluation is very expensive to make, we must leave it for only those alternatives that survive the simpler evaluations. For example, we carry out a full HAZOP study only for the final alternative for a design.

2.2.4 Generating and Searching among Alternatives

To find the better design alternatives we first need to have a method to generate them. Different generation schemes depend heavily on the representations we use, as we see from our earlier discussion. The availability of a concise representation is essential for the generation and description of these alternatives. For simple design problems, we can often see explicitly how to generate all the alternatives and determine their number ahead of time. Nevertheless, a huge number of alternatives is likely and we may not be able to generate and evaluate all of them. Moreover, for more difficult problems, we only know how to generate alternatives implicitly, for example, as a variation of an existing alternative.

To see this combinatorial explosion for even a simple problem, we consider a simple heat exchanger example.

EXAMPLE 2.1 Generating Alternatives for a Heat Exchanger Network

For this example, we choose to exchange heat among three hot streams—$H[1]$, $H[2]$, and $H[3]$—that we wish to cool and three cold streams—$C[1]$, $C[2]$, and $C[3]$—that we wish to heat. A convenient *representation* of alternative heat exchanger networks is a matrix where streams $H[1]$ to $H[3]$ label the rows and $C[1]$ to $C[3]$ the columns. We place a dot in row $H[i]$ and column $C[j]$ to indicate the existence of a heat exchanger between streams $H[i]$ and $C[j]$.

One alternative is to place no dots in the matrix—the null network. There are nine locations in which to place a single dot. The matrix on the left side of Figure 2.5 is one such option.

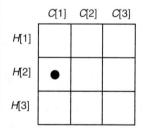
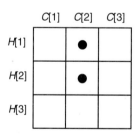

FIGURE 2.5 Enumerating heat exchange alternatives.

When we place two dots, as in the matrix on the right side of Figure 2.5, we can either line them up—meaning that one of the streams exchanges heat with two others—or we can place them so four different streams are involved. For the former, there are three ways we could place two dots involving $H[1]$: exchanging with $C[1]$ and $C[2]$, with $C[1]$ and $C[3]$, or with $C[2]$ and $C[3]$. $H[1]$ could meet the two streams in either order or in parallel. Thus, stream $H[1]$ has nine possible ways that it could exchange heat with two of the cold streams. Each of the six streams could be the common stream, giving us another 54 alternatives. For the case when no streams are in common in the exchanges, we need to select two of the hot and two of the cold streams for the network. There are three ways to pick two streams (as we just saw above). Once we have picked them, there are two ways to pair the hot with the cold. Thus, there are $3 \times 3 \times 2 = 18$ more alternatives for this case. We have already enumerated 82 alternatives.

Sec. 2.2   Basic Steps in Flowsheet Synthesis

> We next place three dots and enumerate where they can be located. When two or three are lined up, we get alternative sequences in which the common stream meets the other streams, including combinations that meet some or all of them in parallel. We continue with four dots, five dots, and finally six dots. Unless we rule them out, there can also be alternatives where a stream such as $H[1]$ meets $C[1]$, then $C[2]$, and then $C[1]$ again. As a result, we could enumerate thousands of alternatives for this apparently simple problem.

Evaluating and searching among alternatives requires the application of systematic approaches. Here we briefly describe the following methodologies, which have been developed and applied in process synthesis.

Total enumeration of an explicit space is the most obvious. Here we generate and evaluate every alternative design. We locate the better alternatives by directly comparing the evaluations. This option is feasible only if the total number is small enough, based on the computer or human resources required to conduct the evaluation.

A more coordinated search involves a **tree search in the space of design decisions** (see, for example, Figures 2.8 and 2.9). At every node point on the tree we record the assessment and decisions prior to branching further. At some point a completed design is created; to examine further alternatives, we can backtrack to any earlier node and make an alternate decision. Moreover, a partial evaluation of a choice along a new branch may prove that choice inferior to one already made. In this way, we can prune the search space and, based on a partial evaluation, decide against exploring further along the branch. This strategy leads to the systematic branch and bound algorithm, presented in detail in Chapter 15.

Evolutionary methods follow from the generation of a good base case design. Designers can then make many small changes, a few at a time, to improve the design incrementally. Also, they can use the insights obtained when evaluating the current design to see where improvements might be possible. They may select the types of small changes they will allow a priori, in which case this approach might be automated.

Another approach to searching large spaces is to postulate a **superstructure of decisions** that contains all the alternatives to be considered for a design. Figure 2.6 shows a superstructure for a heat exchanger network where a hot stream, $H[1]$, exchanges heat with three cold streams, $C[1]$ to $C[3]$. By removing different connections shown in this network, we can have $H[1]$ exchange with none, one, two, or three of the cold streams. It can pass through the exchangers in series and/or in parallel. If we create this superstructure and optimize it based on our evaluation criteria, we would find the best alternative embedded within the superstructure. The use of superstructure optimization appears often in Part IV of this text as a method to determine better alternatives for a design.

Another aid to looking for better designs is to **establish targets for the design**. These have been especially useful in designing heat recovery and reactor networks. In the synthesis of heat exchanger networks (see Example 2.1), Chapter 10 shows that it is possible to compute the minimum amount of utility heating and cooling for this design problem *before* one invents any network that solves this problem. These utility requirements become the targets for our design, and we can reject any design requiring more than these

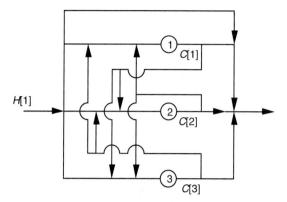

FIGURE 2.6 Superstructure for heat exchangers.

target amounts when designing a heat integration network. Moreover, in Chapter 10 we will discover methods to generate exchanger networks directly that are guaranteed to meet these targets.

Finally, related to the creation of design representations, one of the most powerful ways to reduce the size of the space is through **problem abstraction**. Here the search for better design alternatives begins by formulating a less detailed problem statement and attempting to solve this more abstract problem first. In this more abstract space we make decisions that affect whole families of alternatives. Moreover, a suitable abstraction will group parts of the problem together which behave similarly.

EXAMPLE 2.2 Ethanol Process Alternatives

To illustrate the concepts of abstraction and tree searching, we consider the development of a separation process for the mixture of species that can exit the reactor in the ethylene to ethyl alcohol process. Figure 2.7, repeated from Chapter 1, shows these species.

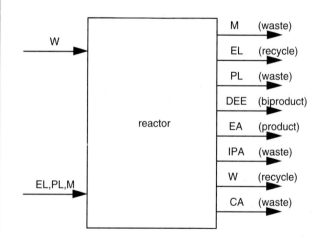

FIGURE 2.7 Species leaving reactor in ethylene to ethyl alcohol process.

Sec. 2.2   Basic Steps in Flowsheet Synthesis   35

We invent a separation process for these species by enumerating over all possible separation technologies and all possible ways to split these species using these technologies. Our list of technologies includes distillation, flash, absorption, extractive distillation and adsorption (we could certainly think of more, but this list will suffice to make the point). We then generate a tree of alternatives, sketched in Figure 2.8, to carry out the required separations. Branching from the top node are all possible separation tasks using all possible separation methods to accomplish them. The leftmost separation task removes methane from the remaining species using distillation. We are left with a mixture without methane to which we again attach all possible separation tasks using our available methods. Also, we see that we can generate an enormous number of alternatives with this decision tree.

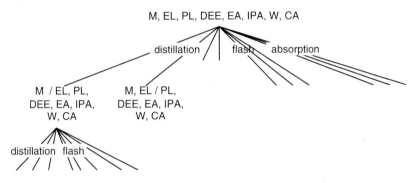

FIGURE 2.8 Developing separation alternatives.

Rather than solve the problem of separating individual species, we now organize them based on their normal boiling points. What we call the noncondensibles—ethane, ethylene and propylene—condense at temperatures well below ambient even if we operate at high pressure. The remaining species, diethyl ether, isopropylalcohol, water, and crotonaldehyde are classified as condensibles. At this higher level of abstraction we first look for a design to separate noncondensibles from condensibles and our separation alternatives reduce to those shown in Figure 2.9. There are fewer alternatives to consider here than in Figure 2.8, and we have partitioned our separation problem into two much smaller subproblems. Note that the flowsheet shown in Figure 2.1 for the separation system uses a flash unit followed by an absorption unit using water to separate the noncondensibles from the condensibles. The classification in this example helps to explain this particular design.

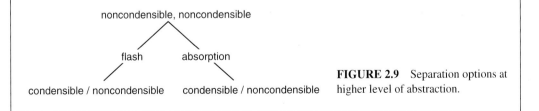

FIGURE 2.9 Separation options at higher level of abstraction.

2.3 DECOMPOSITION STRATEGIES FOR PROCESS SYNTHESIS

Because of the explosion of alternatives in considering the overall process synthesis problem, several studies (for example, Daichendt and Grossmann, 1997; Douglas, 1988; Linnhoff and Ahmad, 1983; Siirola, Powers, and Rudd, 1971) have placed a partial order or decomposition on the decisions we should make when developing the process flowsheet. These studies also lead to a decision hierarchy in generating and exploring alternatives for process synthesis. For example, if we first consider the reactions in the process, we greatly influence all subsequent design decisions we will make, because they limit which of the available raw materials we can use effectively. In addition, the reactor conditions determine the necessity for recycle or raw materials and product recovery.

Next, we consider a set of decisions to connect the various sources of chemical species with the various targets. Our target streams are the products, by-products, waste streams, and the feed to the reactor. Our sources are the raw materials and the effluent from the reactor, and we need to decide to which targets these sources go. These decisions determine our separation tasks for the flowsheet.

The final step is the design of the energy network, and here we consider options such as cooling the reactor effluent to preheat its feed or using the condenser of a distillation column to preheat another column's feed.

2.3.1 Bounding Strategies for Process Synthesis

It is not hard to see that the later decisions in this decision hierarchy also have less impact on the final process economics. Moreover, one we cannot easily make many of these later decisions without having made the earlier ones first and sequencing the decisions in this manner directs how we discover design alternatives.

To assess the impact of these decisions, we apply a search strategy that uses bounds on our evaluation criteria (for example, profit). These bounds eliminate unfavorable process alternatives (Daichendt and Grossmann, 1997) and are especially effective with early decisions that make big differences in our evaluations. This approach suggests we look first at the reactions we use, then the separation processes we use, and so forth. We also use abstraction to partition the separation problem. Figure 2.10 shows how this search might proceed if we are looking for designs that maximize profit.

Our original space might be that of all designs that are subject only to the stoichiometry of the reaction. The value of the products less the cost of the raw materials needed to produce them would give us a bound on the maximum profit possible, denoted here as $100. From the chemistry we discover next that 5% of one of the reactants will form waste because of reactor selectivity. We eliminate (with a hatched line) all designs not subject to this 5% loss, denoted here by region 1. We next complete a computation that shows that this loss of reactant reduces the maximum profit possible to $90, which represents a new, lower estimate for the maximum profit possible for any design.

Sec. 2.3   Decomposition Strategies for Process Synthesis   37

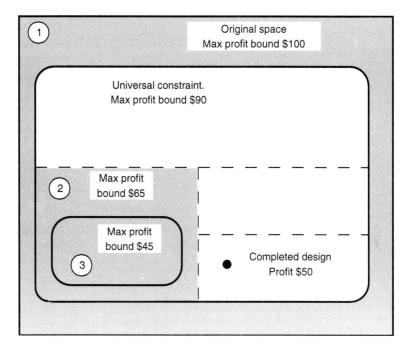

FIGURE 2.10   Searching for the best designs.

As our next option, we decide to use distillation to purify the feed. This decision is not universal so we break our designs into two sets: those that use distillation to purify the feed and those that do not. The evaluation of these two sets requires a search among all of the process alternatives we consider. The horizontal dashed line midway down corresponds to this decision. For the distillation set of designs (those below the dashed line) we make another partitioning decision that breaks the space into two subpartitions to the left and right of the vertical dashed line and discover a maximum profit bound of $65 for region 2 on the left. We explore the right subset of designs further and discover a constraint that further partitions this subset into two subsets. For one of these subsets, we complete a design, finding that this design has an actual projected profit of $50. It is not an optimal design, but it is complete and we know its profit. If we want the most profitable design, we need not accept any design that is less profitable. We return to the previous left subset of designs and further discover a constraint for all designs within it. We eliminate designs not obeying this constraint and obtain region 3. In this region the maximum profit bound is only $45. This bound is below the profit we found for a complete design, and we can eliminate region 3 altogether. Only designs in the right subspace remain for exploration.

Finally, while we will not always use rigorous bound estimates to eliminate regions, we can often estimate the value of a typical design in a region, and, if is too small, we can eliminate the region on the assumption that a rigorous bound would not be much better. Using typical design values must be used with caution, however.

2.3.2 A Hierarchical Decomposition for Process Synthesis

To guide the selection of process alternatives, Douglas (1988) formalized a decision hierarchy as a set of levels, where more detail in the process flowsheet is successively added to the problem. These levels are classified according to the following process decisions:

> Level 1: Batch versus continuous
> Level 2: Input–output structure of the flowsheet
> Level 3: Recycle structure of flowsheet
> Level 4: Separation system synthesis
>   4a: Vapor recovery
>   4b: Liquid recovery
> Level 5: Heat recovery network

In the *first level*, we consider batch processes only if at least one of the following holds. These are characteristic of pharmaceutical, food, and specialty plastics processes.

- We must get the process operational in a few months. The product is one where the first company to market wins an enormous competitive advantage.
- We need only a few days production for a year's supply.
- We have little design information and the process is sensitive to upsets and variations.
- The product will likely have a total lifetime of one to two years before some other product will come out that replaces it.
- The value of the product overwhelms the cost to manufacture it.

In almost all other cases, we should consider using a continuous process. Even for very small processes, continuous processes will prove to be less expensive in terms of equipment and operating costs. Dedicated continuous processes often put batch processes out of business.

In *level 2*, we consider the number of raw material and product streams and their overall relation to the process. We also consider the presence of by-products and inert components in the process and how they participate in the reaction chemistry. An important question is the recovery of these compounds. At this level, a process recycle may be needed for the reactor, and the designer needs to consider the addition of purge streams to avoid the buildup of inert components or by-products.

Level 3 further explores the recycle structure of the flowsheet and focuses more closely on the reactor itself. We consider the number of separate reactor networks in the flowsheet and their interactions through recycle streams. We also consider the effects of reactor conditions on the rest of the flowsheet. These could include the effect of inerts as a diluent in the reactor feed and the effects of equilibrium in choosing pressure, excess components, and adiabatic operation for the reactor. A more detailed discussion of these decisions is also presented in Chapters 13 and 19.

Level 4 is divided into two decision stages: vapor and liquid recovery. Raw materials from this step will be recycled to the reactor while products and by-products are generally processed further and removed. At this level we are concerned both with the selection and placement of separation units. In vapor recovery, the more expensive stage, we also need to consider the effect of purge streams and the removal of components based on their value and their effect on the reactor if they are recycled. Clearly, a purge stream represents a no-cost separation, but, as we will see in the next section, it has a tremendous effect on the process. In the liquid recovery stage, we prefer to use distillation, as this is often the least expensive separation. Design decisions at this stage include sequencing of the separators and determining their operating conditions. Detailed discussion of these decisions is deferred to Chapters 11 and 14.

Finally, *level 5* deals with the heat recovery network once all of the other flowsheeting decisions have been made. A thorough presentation on these synthesis methods begins in Part III of this text.

The Douglas hierarchy is structured in a direct top-down strategy, and a single pass application of this strategy tends to ignore some strong interactions between the levels. Moreover, the interactions between levels can be considered systematically with more powerful search strategies. For instance, the interactions with the heat recovery network and the flowsheet are explored in Chapter 18, where they are treated through optimization strategies. Furthermore, these approaches can be used in a branch and bound strategy, with a search tree that is based on hierarchical decomposition, as discussed in Daichendt and Grossmann (1997).

Despite some of these limitations, the decision hierarchy of Douglas (1988) has the benefit of guiding the decisions that generate candidate flowsheets. These are especially useful to generate base case designs and also uncover many of the likely flowsheet alternatives. In the next section we apply this decision hierarchy as well as bounding and tree search concepts to our ethyl alcohol case study.

2.4 SYNTHESIS OF AN ETHYL ALCOHOL PROCESS: A CASE STUDY

In this section we apply the concepts of our bounding strategy and the Douglas hierarchy to develop a base case process flowsheet for the ethanol process. We begin by determining a bound on the capital and operating costs. If this leads to a favorable economic decision, we next apply the decision hierarchy to generate and assess the flowsheet alternatives for this process.

2.4.1 Maximum Potential Profit

Before we begin the generation and search among alternatives, we first need to develop a simple economic bound for this process. We would not design our process if it is not profitable. Therefore, we first compute the maximum potential profit. This computation is universally true if we have one set of raw materials and follow only one set of reaction chemistry to produce our product, a situation that applies to our ethyl alcohol process. For this process, a bound on the maximum potential profit would be the difference in value between the product ethyl alcohol and the least amount of raw materials we would need to create this product. Reaction stoichiometry, a few physical properties, and prices for ethylene, water, and ethyl alcohol are all we need for this analysis.

The price of the product and raw material can be obtained from a variety of sources, either within the company or on the market. For instance, the price of 190 proof ethyl alcohol is found in the Chemical Marketing Reporter (formerly the Oil, Paint and Drug Reporter), which provides market prices received for commodity chemicals in the recent past. Table 2.1 gives the prices reported in the July 17, 1995, issue.

The prices for ethyl alcohol and ethyl ether apply directly to this process, but ethylene typically is sold as 99.996 mole % pure while our ethylene feedstock is only 96% pure. Consequently, for our example problem, our manufacturing group has given us a price of 0.18/lb, a value that appears to be in line with the above. Using these prices, we estimate an upper bound on gross profits as follows:

1. 150,000 m^3/yr of ethanol product translates into 39.6 million gallons/yr. Using the above prices, the value for this much ethyl alcohol would range from \$101 million to \$111 million per year.

2. We now need to determine the number of moles of ethyl alcohol that are in 150,000 m^3 of 190 proof ethyl alcohol, to compute how much water and ethylene we will consume to make it. Lange's Handbook (11th ed., pp. 10–142) tabulates the density of ethyl alcohol and water solutions versus weight fraction of ethyl alcohol. This same handbook tells us that 190 proof ethyl alcohol is 85.44 mole % ethyl alcohol and 14.56 mole % water. Therefore, the weight of one kmole of 190 proof ethyl alcohol solution is

$$0.8544 \text{ kmole} \times 46.07 \frac{\text{kg EA}}{\text{kmole EA}} + 0.1456 \text{ kmole} \times 18.02 \frac{\text{kg W}}{\text{kmole W}} = 41.99 \text{ kg} \quad (2.1)$$

The weight fraction of ethyl alcohol is then

TABLE 2.1 Prices for Chemicals from Chemical Marketing Reporter, July 17, 1995

	Price Range	Comment
ethyl alcohol	\$2.55–2.80/gal	190 proof, USP tax free, tanks, delivered. E.
ethyl ether	\$0.575/lb	refined tanks fob
ethylene	0.28–0.30/lb	contract, delivered

Sec. 2.4 Synthesis of an Ethyl Alcohol Process: A Case Study

$$\frac{0.8544 \times 46.07}{41.99} = 0.937 \qquad (2.2)$$

for which Lange's reports a density of 0.810 gm/ml or 810 kg/m^3. The amount of ethyl alcohol is therefore

$$\frac{0.937 \dfrac{\text{kg EA}}{\text{kg solution}} \times 150{,}000 \dfrac{m^3 \text{solution}}{\text{yr}} \times 810 \dfrac{\text{kg solution}}{m^3 \text{ solution}}}{46.07 \dfrac{\text{kg EA}}{\text{kmole EA}}}$$
$$= 2{,}471{,}000 \dfrac{\text{kmole EA}}{\text{yr}} \qquad (2.3)$$

Assuming 100 percent conversion of ethylene to ethyl alcohol, we compute the total weight of feed we need as follows.

$$2{,}471{,}000 \frac{\text{kmole EL}}{\text{yr}} \times 28.05 \frac{\text{kg EL}}{\text{kmole EL}} = 69{,}310{,}000 \frac{\text{kg EL}}{\text{yr}} \qquad (2.4)$$

$$\frac{3}{96} \times 2{,}471{,}000 \frac{\text{kmole PL}}{\text{yr}} \times 42.08 \frac{\text{kg PL}}{\text{kmole PL}} = 3{,}249{,}000 \frac{\text{kg PL}}{\text{yr}} \qquad (2.5)$$

and

$$\frac{1}{96} \times 2{,}471{,}000 \frac{\text{kmole M}}{\text{yr}} \times 16.04 \frac{\text{kg M}}{\text{kmole M}} = 412{,}900 \frac{\text{kg M}}{\text{yr}} \qquad (2.6)$$

or a total of 72,980,000 kg/yr. The cost of this feed is

$$72{,}908{,}000 \frac{\text{kg}}{\text{yr}} \times 2.2046 \frac{\text{lbm}}{\text{kg}} \times 0.18 \frac{\$}{\text{lbm}} = 28{,}960{,}000 \frac{\$}{\text{yr}} \qquad (2.7)$$

3. Assuming the cost of the water we feed to the process is negligible, we see a maximum profit of about $72 to $82 million per year. This maximum profit has to cover our annual operating costs and our annualized costs for investing in equipment for the process. Assuming a five-year payout time and an eight-year depreciable life (ignoring time value of money), we can convert dollars in investment into annualized dollars by dividing roughly by 3. Thus, we need a process where

$$\frac{\text{equipment costs}}{3} + \text{annual operating cost} \leq \$72 \text{ to } 82 \text{ million/yr} \qquad (2.8)$$

This equation can also be justified by a more detailed cash flow analysis (see Douglas, 1988).

The maximum potential profit calculation indicates that the process is economically favorable so we continue with our design. Note, if the potential maximum profit had been very small or negative, we would have been able to stop, reporting that no profitable design exists. Note also that our maximum potential profit estimate is very strongly affected by these prices. If we try to establish how much these prices might change, we can establish the range of maximum profits we might see for this process.

For instance, from a marketing study we could have a 25% probability that the minimum ethyl alcohol price decreases by 20%, as well as a further 25% probability that maximum ethyl alcohol price could be 10% higher. Also, our most likely price is at the midpoint of the current price range with a 50% probability. Further, our engineering manager suggests that, if the price of ethyl alcohol reduces by 20% below the minimum, the cost of the ethylene feed will be discounted by 10%; if the price is 10% higher, the cost of the ethylene feedstock will be 15% higher. Repeating the maximum profit calculations leads to the following table (Table 2.2).

From the table, the most probable estimate for maximum potential profits is given by adding the maximum potential profits times their respective probabilities, yielding:

$$0.25 \times 54,700,000 + 0.5 \times 77,000,000 + 0.25 \times 88,800,000 = 77,400,000 \frac{\$}{yr} \qquad (2.9)$$

which is a number not too far from our original estimate.

Another scenario we might consider is that the price of the ethyl alcohol drops 20% below its minimum while the cost of the ethylene feed increases by 15%. In this case the maximum potential profit can drop to

$$[0.8 \times 101,000,000 - 1.15 \times 28,960,000]\frac{\$}{yr} = 47,500,000 \frac{\$}{yr} \qquad (2.10)$$

substantially less than we estimated above. The point we want to make here is that the maximum profit is quite sensitive to the price estimates, and the decision to proceed with the process design hinges on these.

TABLE 2.2 Sensitivity of Maximum Profit Based on Price Changes

Probability	Ethanol and Ethylene Prices	Maximum Profit
25%	0.8 × $2.55 0.9 × $0.18	$54,700,000/yr
50%	$2.675 $0.18	$77,000,000/yr
25%	1.1 × $2.80 1.15 × $0.18	$88,800,000/yr

Sec. 2.4   Synthesis of an Ethyl Alcohol Process: A Case Study   43

2.4.2 Developing a Flowsheet with Hierarchical Decomposition

We now develop a base case for our design by progressing through the decision hierarchy developed in section 2.3. Moreover, we base the levels of decision making by successively refining models of the process. Nevertheless, the model we shall consider is still a very simple one and can be set up and solved using a spreadsheet program such as Lotus 1-2-3 or Excel. We recommend that one set up such a model at this stage in order to record the decisions, make rough estimates of costs, and prepare for more detailed designs that will be analyzed in Chapters 3 and 4. We now proceed through each of the levels in the Douglas hierarchy.

LEVEL 1: BATCH VS. CONTINUOUS

None of reasons for choosing a batch process in section 2.3 holds for our ethyl alcohol process. We may be in a rush to develop this process, but 190 proof ethyl alcohol is already in the market and we are going to be just one more producer. We note that we need to convert a continuously flowing supply of ethylene throughout the year so we are not going to produce the full year's supply in a few days. Ethyl alcohol has been a commodity chemical for decades and will continue to be; we are not dealing with a product having a short life in the market. Finally, the cost to produce ethyl alcohol sets its price, so we have to be a cost-effective producer to sell it to anyone. We decide, therefore, to consider manufacturing ethyl alcohol using a continuous process.

LEVEL 2: INPUT OUTPUT STRUCTURE OF FLOWSHEET

The ethylene feed contains 3 mole % propylene and 1% methane. Also the conversion per pass of ethylene to ethanol is low (7%) so we need to consider the effect of process recycles and the presence of inert components and impurities. As noted in Chapter 1, both propylene and methane eventually need to be removed from the process. These species can either be removed before the ethylene enters the reactor, or we can let either or both of them enter the reactor and remove them (and their possible products) after the reactor. The resulting options from Chapter 1 are shown again in Figure 2.11 to show some of the alternatives possible. However, because of the difficulties and expense of separating both methane and propylene in the feed, we choose to let both components enter the reactor as shown in the third option in Figure 2.11. As we refine the model, we may opt to return to the first two alternatives in Figure 2.11 and evaluate them as part of our bounding strategy.

From the specification of the reactor conditions, we are permitted to let methane build up to 10 mole percent in the reactor feed stream. By letting the methane enter with the ethylene and build up in the recycle, we then need to remove a small part of that recycle stream as a purge stream. As we will see in Level 4, the split fraction of this stream has a significant impact on the recycle loop.

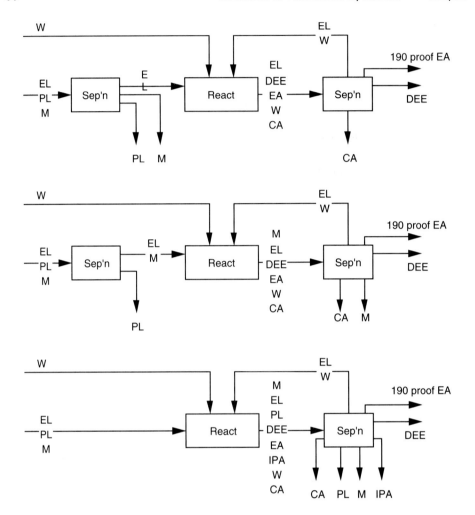

FIGURE 2.11 Alternative separation schemes for process.

LEVEL 3: RECYCLE STRUCTURE OF FLOWSHEET

At this level we focus on the details of reaction chemistry and the reactor network. Converting raw materials fed to the reactor into undesired by-products is one of the most costly losses we can have in a process. Moreover, these by-products may pollute the environment, forcing us to design costly clean up measures to recover or destroy them.

To illustrate the selectivity losses, suppose we consider a process with the following chemistry.

$$\text{reaction 1:} \quad A + B \to C$$
$$\text{reaction 2:} \quad 2A \to D \tag{2.11}$$

Sec. 2.4   Synthesis of an Ethyl Alcohol Process: A Case Study   45

where C is the desired product and D is a waste product. Suppose further that 50% of species A converts in the reactor to C while 10% converts to D. Total conversion for A is 0.5 + 0.1 or 0.6. Selectivity in the conversion of species A is the fraction of A that converts to desired product over the total conversion of A in the reactor, given by:

$$\text{selectivity for } A \text{ to produce } C = \frac{0.5}{0.5 + 0.1} = 0.8333 \qquad (2.12)$$

We can modify our bound for maximum potential profit to account for selectivity losses if we can estimate a lower bound on the losses we will suffer. In making this calculation we assume we will recover and recycle all unreacted A back to the reactor so no unreacted A escapes being converted.

If only one chemical route is considered to manufacture our products, as in our ethyl alcohol process, and we know the selectivity losses in the reactor, then we should account for these universally across all designs. In this process we can convert the ethylene, in principle, entirely to ethyl alcohol. However, the ethyl alcohol undergoes a further reaction where it converts to diethyl ether:

$$2 \text{ CH3CH2OH} \rightarrow \text{C2H5-O-C2H5} + \text{H2O} \qquad (2.13)$$

$$2 \text{ ethyl alcohol} \quad \rightarrow \quad \text{diethylether} + \text{water}$$

Here two ethyl alcohol molecules react to produce one molecule each of diethyl ether and water. The literature says that this reaction is equilibrium limited, which leads to the following equation:

$$\frac{(\hat{a}_{C_2H_5OH})^2}{(\hat{a}_{H_2O})(\hat{a}_{(C_2H_5)_2O})} = K_{eq}(T) = e^{-\Delta G_{rxn}/RT} \qquad (2.14)$$

where the quantities in parentheses are component activities (related to compositions). As a result, if we recycle all this diethyl ether back to the reactor, it will build up in the reactor feed until it suppresses this reaction. At steady state the ether that we recycle is the amount in equilibrium with the water and ethyl alcohol in the reactor effluent, and the reactor will produce no further diethyl ether. On the other hand, if we do not recycle, we can produce diethyl ether as a by-product we can sell, but this will lead to selectivity losses for ethyl alcohol.

Because we choose to recycle the diethyl ether, there need be no loss of reactants to undesired products; thus our estimate for the maximum potential profit still stands.

LEVEL 4: SEPARATION SYSTEM SYNTHESIS

Next, we design a base case separation process. In section 2.3 we looked at reducing the size of a search space by using problem abstraction. Moreover, in Example 2.2, we grouped the species, leaving the ethyl alcohol reaction process into two groups: noncondensible and condensible. These correspond directly to the vapor and liquid recovery steps in this decision hierarchy. Methane, ethylene, and propylene fall into the former class, while diethyl ether, ethyl alcohol, isopropyl alcohol, water, and crotonaldehyde fall

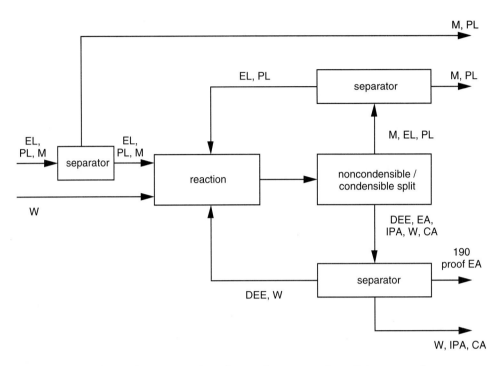

FIGURE 2.12 Abstract view of separation system after splitting noncondensibles from condensibles.

into the latter. To create our design we look for methods to separate noncondensible from condensible species. For vapor recovery we list two separation methods as applicable: using a flash followed by absorption. Whatever method or combination of methods that we use, we decide to separate the noncondensible from the condensibles as the first step in separating these species. Figure 2.12 gives an abstract view of the resulting process flowsheet where we include structure wherever any of the species can exist.

In Chapter 1 and in level 2, we already debated among the options for treating the feed, and we decided to let both the methane and the propylene enter with the ethylene. Methane, propylene, and isopropyl alcohol exit the reactor, and we decide to split the condensibles from the noncondensibles. We now have the problem of removing methane from the ethylene recycle stream.

Vapor Recovery. Using the arguments from Chapter 1, we could use membranes to remove methane from the feed. Alternately, we could try adsorption or distillation (but this would require refrigeration) as further alternatives. Instead, for this base case, we let the methane build up in the recycle and remove a fraction of the recycle using a purge stream. We form a purge stream by splitting the recycle stream into two parts and directly removing one of the parts from the process. For example, we could remove 2% of the purge while recycling the remaining 98%. We also have to consider what we can do

Sec. 2.4 Synthesis of an Ethyl Alcohol Process: A Case Study

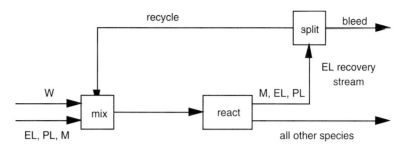

FIGURE 2.13 Abstract flow diagram for process where methane is removed using a purge stream.

with the purge stream if we do produce it. If it is combustible, we might use it as fuel, or we might flare it if it is environmentally safe, as in this process.

In the purge stream we need to minimize the loss of valuable reactant and product molecules. The smaller the purge split, the more the methane will build up in the recycle and the less ethylene we lose. So it appears we should want to split off very little of the recycle. However, there is a cost. The smaller the fraction we remove, the larger the flow of the recycle stream. To gauge the effect of this split, let's design our process using a 7% conversion of the ethylene to ethyl alcohol per reactor pass. We further assume that the conversion per pass for the propylene to isopropyl alcohol is only 0.7%. We assume that after the reactor we separate completely all the methane and unreacted ethylene and propylene from all the other species we produce. Figure 2.13 shows an abstract flow diagram for our flowsheet.

To determine the split fraction for the purge stream, we define μ_{ij}^k as the molar flowrate of species k leaving unit i in the jth output stream of that unit. Let b be the fraction of the ethylene recovery stream we remove as the purge stream. With these definitions, we can write the following recycle loop material balances to compute the size of the recycle stream and amount of ethylene we will lose as a function of fraction of the recycle we purge from the process. Table 2.3 shows the results we then compute using these equations.

As shown later in Chapter 3, the methane balance is given by:

$$\mu_{mix}^M = 0.01 \text{ kmol} + \mu_{split,recycle}^M = 0.01 \text{ kmol} + (1-b) \times \mu_{EL \text{ react, EL recovery}}^M \quad (2.15)$$
$$= 0.1 \text{ kmol} + (1-b) \times \mu_{mix}^M$$

or

$$\mu_{mix}^M = \frac{0.01 \text{ kmol}}{b} \quad (2.16)$$

and this determines the molar flowrate for methane in the reactor feed. Similarly a balance on the ethylene

$$\mu_{mix}^{EL} = 0.96 \text{ kmol} + 0.93 \, \mu_{mix}^{EL}(1-b) \quad (2.17)$$

TABLE 2.3 Flows in kmol for Purge Stream Analysis (Basis: 1 kmol of ethylene feed—computed using a spreadsheet program)

b, purge fraction	M mixer outlet	EL mixer outlet	PL mixer outlet	M feed %	EL purge	PL purge	Total Recycle
0.001	10	13.534	3.753	28.24	0.0125	0.0037	33.842
0.002	5	13.359	3.339	16.83	0.0248	0.0066	28.147
0.003	3.333	13.189	3.006	12.15	0.0368	0.0089	25.875
0.004	2.5	13.022	2.734	9.59	0.0484	0.0109	24.504
0.005	2	12.860	2.507	7.97	0.0599	0.0124	23.517
0.006	1.667	12.702	2.315	6.86	0.0709	0.0138	22.739
0.007	1.428	12.547	2.150	6.04	0.0817	0.0149	22.089
0.008	1.25	12.397	2.007	5.41	0.0922	0.0159	21.526
0.009	1.111	12.250	1.882	4.92	0.1025	0.0168	21.027
0.010	1.0	12.106	1.772	4.51	0.1126	0.0176	20.575
0.011	0.909	11.966	1.674	4.18	0.1224	0.0183	20.162
0.012	0.833	11.828	1.586	3.90	0.1320	0.0189	19.779
0.013	0.1333	11.694	1.507	3.66	0.1414	0.0195	19.421

gives the molar flowrate:

$$\mu_{mix}^{EL} = (0.96 \text{ kmol})/(0.93b + 0.07) \tag{2.18}$$

and finally for propylene we have the molar flowrate:

$$\mu_{mix}^{PL} = (0.03 \text{ kmol})/(0.993b + 0.007) \tag{2.19}$$

These balances also account for the flowrate of water into the reactor (at 0.6 times the flowrate of ethylene). From the ratios of these flowrates, the mole fractions in Table 2.3 are straightforward to determine.

When b, the fraction we purge, is 0.004 or higher, we satisfy the constraint that methane is less than 10% of the reactor feed. As predicted above, ethylene purge loss decreases from about 11% to 1.2% of that in the feed as we decrease the purge fraction, b, from 1% to 0.1%. However, we get this decrease at a cost: The recycle flow increases almost 65%, substantially increasing the size of equipment we need to handle it. This especially applies to the compressor in the recycle stream, which may have large capital and operating costs.

The trade-off for the purge stream is therefore the loss of ethylene, which forces us to purchase more feed to make 150,000 m^3/yr of product versus additional compression costs in the recycle. If we can estimate a lower bound on the compressor investment and operating costs as a function of b as well as the cost for losing ethylene in the purge, we can tabulate these costs versus b and subtract them from the maximum potential profit. This result leads to a reduced and improved estimate of the upper bound on profit. If any

Sec. 2.4 Synthesis of an Ethyl Alcohol Process: A Case Study

of these bounds were to become negative, we could eliminate designs for those values of b from further consideration.

In addition, purge streams are required to regulate trace amounts of contaminants that no one has thought of—in any process. These species range from being very heavy to very light. A process must not trap them but must provide a path (through purge streams) for them to escape.

Finally, we elect to separate the noncondensibles from the condensibles as the first step following the reactor. Further analysis shows, however, that the flash separation is not sharp and some diethyl ether and ethyl alcohol exit in the vapor product stream with the noncondensibles. It appears that the absorber used in the flowsheet we found in the literature prevents the ethyl alcohol and diethyl ether from recycling and then being lost in the purge stream. In this absorber, we pass the vapor from the flash against a water stream which captures the diethyl ether and ethyl alcohol. This water stream is then further purified in the liquid recovery step.

Liquid Recovery. The liquid separation system processes the liquid from the flash and the absorber and isolates 190 proof ethyl alcohol as product. All the species appear suitable for separating using distillation. (At some time we should worry about whether water and diethyl ether may form two liquid phases.) The bulk of the initial stream will be water, and we may want to remove most of the water and crotonaldehyde first. Then, in order of decreasing volatility, we are left with diethyl ether, ethyl alcohol, isopropyl alcohol, and perhaps some residual water. Our product is in the middle. Therefore, we next separate off the diethyl ether, which we intend to recycle. In the last column, we separate the ethyl alcohol very close to its alcohol/water azeotropic composition (190 proof) from the isopropyl alcohol and any remaining water.

We would like to recycle the bulk of the water from the first column either to the reactor or the absorber, but it contains crotonaldehyde that builds up in a recycle. To control this we could purge off some of the water that we would send to a waste treatment plant. Alternatively, we could use adsorption or other separations to remove some of the aldehyde. As this option could be expensive, we pick the purge option and use this as our base case. This is shown in Figure 2.14. The bound on this case can also be used to eliminate any future alternatives we might examine that are not as good.

LEVEL 5 AND BEYOND

In progressing further in our design, we continue to look for constraints to add that partition the design space. The decisions we made for the separation system partition the space, and we continue to look for constraints in order to refine our maximum profit estimates for all designs in that partition. At this point we also consider the design of a heat exchanger network for energy recovery as well as further refinement of the flowsheet. Nevertheless, in Figure 2.14 we have a first design to analyze, and the next two chapters will show us how to determine mass and energy balances and investment and operating cost estimates for this base case.

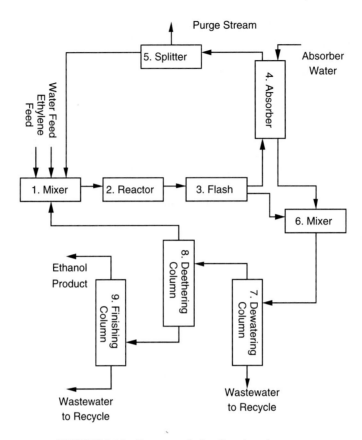

FIGURE 2.14 Base case design for ethanol process.

2.5 SUMMARY

This chapter introduces the technical concepts needed to develop process flowsheets for preliminary design. Here we outline the basic steps in the synthesis process:

- Gathering information
- Representing alternatives
- Developing criteria for assessing preliminary designs
- Generating and searching among alternatives

This discussion also sets the stage for more detailed presentation of these topics in Parts III and IV of this book and illustrates some of the challenges in dealing with huge numbers of process alternatives. These challenges also prompt the discussion of decomposition and bounding strategies for process synthesis, which leads to a decision hierarchy in the generation and search of alternatives. The decision hierarchy helps to keep the synthesis problem manageable and quickly leads to the generation of good base case designs. In this chapter, we examined bounding strategies and the hierarchical decomposition of Douglas. Both of these were illustrated and applied to develop the base case flowsheet in our ethyl alcohol process case study.

Now that we have a base case flowsheet as well as some knowledge of flowsheet alternatives for this process, we proceed in the next three chapters to evaluate this flowsheet and assess its technical and economic feasibility. In the next chapter we develop quick shortcut calculations for mass and energy balances. These will be used to determine flowrates, temperatures, pressures, and heat duties for our process. In Chapter 4 we consider sizing and costing of the units in the flowsheet in order to determine both capital and operating costs. This information is then used in Chapter 5 for the economic evaluation of the preliminary process design.

REFERENCES

1985 Kirkpatrick Chemical Engineering Achievement Award. (1985, December 9). *Chemical Engineering Magazine,* 92(25), 79.

Daichendt, M. M., & Grossmann, I. E. (to appear 1997). Integration of hierarchical decomposition and mathematical programming for the synthesis of process flowsheets. *Comp. Chem. Engng.*

Douglas, J. M. (1988). *Conceptual Design of Processes.* New York: McGraw-Hill.

Linnhoff, B., & Ahmad, S. (1983, November). *Towards Total Process Synthesis,* Paper 26d. Annual Meeting, AIChE, Washington, DC.

Reid, R. C., Prausnitz, J. M., & Poling, B. E. (1987). *The Properties of Gases and Liquids*, 4th ed. New York: McGraw-Hill.

Siirola, J. J., Powers, G. J., & Rudd, D. F. (1971). Synthesis of systems designs: III. Toward a process concept generator. *AIChE J.,* 17(3), 677–682.

Smith, J. M., & Van Ness, H. C. (1987). *Introduction to Chemical Engineering Thermodynamics*, 4th ed. New York: McGraw-Hill.

EXERCISES

1. Prove that the ratio of methane to ethylene in a purge stream must exceed that same ratio in the ethylene feed for the purge stream to work in removing methane from our ethylene-to-ethyl alcohol process.
2. Go to the library and discover at least twenty articles, books, and patents relevant to the manufacture of ethyl alcohol from ethylene. Create a World Wide Web page

using HTML in which you summarize five articles you deem most relevant and list the references for the remaining ones. Allow your instructor and others in the class to have access to this page. (Do not scan in any of the articles as that is illegal. You would be "distributing" a scanned in article, which is against copyright laws.)

3. Consider the manufacture of styrene from ethylbenzene. The reactions that occur are

$$C_6H_5\text{-}C_2H_5 \longrightarrow C_6H_5\text{-}C_2H_3 + H_2 \quad [ST1]$$
ethylbenzene $\longrightarrow$ styrene + hydrogen

$$C_6H_5\text{-}C_2H_5 \longrightarrow C_2H_4 + C_6H_6 \quad [ST2]$$
ethylbenzene $\longrightarrow$ ethylene + benzene

$$C_6H_5\text{-}C_2H_5 + H_2 \longrightarrow CH_4 + C_6H_5\text{-}CH_3 \quad [ST3]$$
ethylbenzene + hydrogen $\longrightarrow$ methane + toluene

$$C_6H_5\text{-}C_2H_5 \longrightarrow \text{tar} \quad [ST4]$$
ethylbenzene $\longrightarrow$ tar

$$CH_4 + 2\,H_2O \longrightarrow CO_2 + 4\,H_2 \quad [ST5]$$
methane + water $\longrightarrow$ carbon dioxide + hydrogen

Assume you are given the selectivities in the styrene process (e.g., 90% of the ethylbenzene converts to styrene, 5% converts to benzene, 3% converts to toluene, and the rest decomposes to CO_2 and hydrogen).

   a. Tabulate several of the physical properties (as in Table 1.3, Chapter 1) for all the species you would expect in this process. Comment on these species. Which boil at very low temperatures, which at very high temperatures? Classify all species as being reactants, products, by-products and waste for this process.
   b. Find prices for those species having commercial value. If all the ethylbenzene could be converted to product, what is the maximum gross profit attainable?
   c. Using the selectivities above, adjust the maximum gross profit attainable. These are assumed selectivities. You would have to find better values in the literature or in the data built up in a corporate file on this process to carry out this analysis accurately.
   d. Let all the prices vary by as much as 10%. What are the ranges for the maximum and minimum gross profit bounds in parts b and c?
   e. Suppose only $x\%$ of the ethyl benzene converts per pass in the reactor. Argue that this process would require a purge stream or something equivalent. Explain your answer clearly. Suggest alternatives to using a purge stream. For $x = 70\%$, compute the recycle rate for the unconverted ethyl benzene as a function of the fraction, b, that one elects to purge.

4. Find information on the manufacture of methanol in the literature. Choose one chemical route and repeat the type of analyses asked for in the previous problem for the ethylbenzene process.

Exercises

5. Using a thermodynamic analysis, we will lead you through steps that will allow you to show that the equilibrium conversion expected at the conditions indicated in the literature is about 8 to 46% of the ethylene, depending on the temperature. You should consider the two reactions:

$$EL(g) + W(g) \rightarrow EA(g) \tag{2.22}$$

$$2 \, EA(g) \rightarrow DEE(g) + W(g) \tag{2.23}$$

Assume the reactor feed is 1 mole of ethylene, 0.6 moles of water, and 0.15 moles of methane. Assume the pressure is 1000 psia and the temperature 550 K at the reactor exit. You should consider using a spreadsheet to carry out these computations.

  a. Using standard Gibbs free energies of formation (see, for example, Table 15–1, Smith and Van Ness, 4th ed., pp. 512–513 [1987], or the tables at the end of Reid et al. [1987]), compute the change in the standard Gibbs free energy for both reactions. You should get numbers at 298 K of about −7782 J/mol (1860 cal/mol) and −14390 J/mol (−3440 cal/mol).
  b. Using your answers in part a, evaluate the equilibrium K values for the two reactions at 1 atm and 298 K. The equation for reaction 1 is:

$$K(1 \text{ atm}, 298 \text{ K}) = \exp\left(\frac{-\Delta G_R(1 \text{ atm}, 298 \text{ K})}{R \times 298 \text{ K}}\right) \tag{2.24}$$

  c. Calculate the value for the two equilibrium constants at the temperature of interest. An approximate equation (obtained by assuming the enthalpy of reaction does not change with temperature) for reaction 2.22 to do this calculation is:

$$K(1 \text{ atm}, T \text{ K}) = K(1 \text{ atm}, 298 \text{ K}) \times \exp\left(\frac{-\Delta H(1 \text{ atm}, 298 \text{ K})}{R}\left(\frac{1}{T} - \frac{1}{298 \text{ K}}\right)\right) \tag{2.25}$$

  d. Write the material balances for each of the species present as being the amount in the feed less the amount formed by each of the reactions, each represented by its extent of conversion, typically written with the symbol ξ_j for reaction j. (The extent of conversion is the number of times the reaction occurs as written. For example, if the first reaction occurs 0.53 times, then 0.53 mols each of ethylene and water convert to form 0.53 mols of ethyl alcohol.) Compute the mole fractions of the products in terms of these two extents.
  e. The definition of the equilibrium constant for the first reaction is:

$$K \equiv \frac{\hat{a}_{EA}}{\hat{a}_{EL}\hat{a}_W} = \frac{\hat{f}_{EA}/f^0_{EA}}{\hat{f}_{EL}/f^0_{EL} \cdot \hat{f}_W/f^0_W} = \frac{\hat{f}_{EA}}{\hat{f}_{EL}\hat{f}_W} \cdot \frac{f^0_{EL} f^0_W}{f^0_{EA}}$$

$$= \frac{y_{EA}\phi_{EA}P}{y_{EL}\phi_{EL}P \times y_W\phi_W P} \cdot \frac{1 \text{ atm} \times 1 \text{ atm}}{1 \text{ atm}} \tag{2.26}$$

where $\hat{a}_i$ is the activity for species i in the mixture, $\hat{f}_i$ is the fugacity of species i in the mixture, and ϕ_i the fugacity coefficient at the temperature and pressure of the mixture. f_i^0 is the standard state fugacity of pure species i, which, by definition, is 1 atm for each of the species at the temperature of the system. Note there is a pressure dependence for this equation when we convert to mole fractions as the reaction changes the number of total moles present by creating one mole of product from two moles of reactants. As written you must state pressure in atm. Assume the fugacity coefficients are unity—which you should note is questionable at 1000 psia.

Set these expressions to the values you computed for the equilibrium constants in part c. Adjust the reaction extents until these two equations are satisfied. Report the fraction conversion for ethylene and water—the numbers should be around 19% and 23% respectively. If these numbers are correct, then the reactor in the process reported in the literature is not near to equilibrium. A new catalyst could change the economics of this process significantly.

(Hint: you are solving two simultaneous nonlinear equations in two unknowns. Your spreadsheet program should have a solver capability to aid you to do this quickly.)

f. If you have done all these calculations using a spreadsheet, then change the temperature for the reactor outlet, ranging it from 500 K to 600 K, to see the impact of temperature.

g. If you are ambitious, include equations to compute the fugacity coefficients for these species and solve again.

MASS AND ENERGY BALANCES 3

The previous chapters introduced a systematic strategy for generating candidate flowsheets. This chapter deals with the development of simple, fast, and useful methods for evaluating the behavior of a candidate flowsheet. Often the rules involved in this process lead to the elimination of several undesirable alternatives. The remaining alternatives, however, require a more detailed evaluation and this task forms the basis of the next three chapters. In particular, this chapter develops simple strategies for obtaining mass and energy balances for a candidate flowsheet. This task is one of the most necessary and the most time-consuming for flowsheet evaluation. Still, with the simplifications introduced in this chapter, the mass and energy balance can be calculated quickly and a great deal of insight is gained in the process. Nevertheless, the simplifications in this chapter do lead to inaccuracies in the final flowsheet that need to be corrected with more detailed models. These will be discussed in Chapters 7 and 8.

3.1 INTRODUCTION

In order to evaluate the conceptual flowsheet presented in the previous chapters, we need to consider the detailed and time-consuming task of heat and mass balances. This precedes the later tasks of plant equipment sizing and economic evaluation. Solution of mass and energy balances has typically been covered in detail as a first course in the chemical engineering curriculum. Therefore, we assume the reader is familiar with the basic concepts. On the other hand, this chapter develops the evaluation of this task from a systematic viewpoint that exploits a number of approximations in order to reduce the problem size and to simplify the calculations in a hierarchical manner. With these approximations, we clearly sacrifice some accuracy in evaluating the flowsheet. However, the goal of this strategy is to develop simple relations among the key flowsheet variables that allow us to gain some insight into the candidate design and calculate a complete mass and energy balance simply and quickly for further evaluations.

For more detailed mass and energy balances, on the other hand, there are many computer programs, or process simulators that perform these tasks in a more rigorous way. These are described in Chapter 7 and listed in Appendix C. Typically a candidate flowsheet model can be defined as a large set of nonlinear equations describing:

1. The connectivity of the units of the flowsheet through process streams
2. The specific equations for each unit; these usually deal with internal mass and energy balances as well as equilibrium relationships
3. Underlying physical property relationships that define enthalpies, equilibrium constants, and other transport and thermodynamic properties.

Taken together, these equations can number in the many thousands. To deal with them directly, two methods for flowsheet simulation, the *modular* and *equation-oriented* modes, have been developed and incorporated into engineering practice. While a complete description of these modes is deferred to Chapter 8, a little background is also useful here.

In the *modular mode*, a clear separation is made between the three equation categories described above. In particular, physical property relations are first separated and accessed as standard procedures. Unit procedures that incorporate the specific unit equations are then constructed with the aid of physical property procedures. These unit procedures or modules remain self-contained by calculating desired unit outputs (e.g., effluent streams and calculated capacities) once all of the unit inputs are specified (e.g., feed streams and performance requirements). Finally, the connectivity equations are considered implicitly by solving each module at a time, then proceeding to the next. Here an iterative procedure is introduced when information recycle or recycle streams are present in the flowsheet.

In the *equation-oriented mode,* on the other hand, we combine all of the process equations (mass and energy balances, equipment performance, thermodynamics and transport, kinetic expressions, and other relationships) into a large, sparse (few variables in each equation) equation set. This set is then solved simultaneously, frequently by using a Newton-type equation solver (see Chapter 8) after first partitioning the equation system to determine independent subsets. The advantage of this approach is that more efficient solution strategies are employed than in the modular mode. On the other hand, specific knowledge about process units is easier to incorporate in the modular mode (e.g., initializing the variables) and a more reliable calculation procedure can result.

Simulation strategies of rigorous models will be covered in more detail in Part II. In this chapter, on the other hand, we simplify the nonlinear equations (categories 1, 2, and 3) through the following approximations. First, we assume ideal solutions in all of our calculations. This greatly simplifies our equilibrium and energy balance calculations. Second, we assume that most streams are available as saturated vapor or liquid. This assump-

Sec. 3.2 Developing Unit Models for Linear Mass Balances

tion is generally valid for equilibrium staged operations and it allows us to set temperature and pressure levels *before* the more tedious energy balance. Finally, we structure the unit calculations so that the flowsheet can be represented as a *linear* system of component equations. This leads to a rapid calculation procedure for the mass balance alone, after which the energy balance can be performed.

The next section outlines these assumptions and applies them to each individual process unit. Following this, the linear mass balance algorithm for the overall flowsheet is described in section 3.3. This is followed by setting temperature and pressure for levels in section 3.4. Finally, the concepts developed in each section will be combined and an energy balance will be calculated in section 3.5, where the concepts will be applied to the ethanol flowsheet introduced in the previous chapter.

3.2 DEVELOPING UNIT MODELS FOR LINEAR MASS BALANCES

Once temperature and pressure are fixed in the feed and output streams, we can develop a linear set of equations for each process unit and thereby solve the entire flowsheet with these equations. Thus, our overall strategy will be:

1. Fix temperature and pressure for all process streams.
2. Approximate each unit with split fractions representing outlet molar flows *linearly* related to inlet molar flows.
3. Combine the linear equations and solve the overall mass balance.
4. Recalculate stream temperatures and pressures from equilibrium relationships.
5. If there are no large changes in temperature and pressure go to step 6, else, go to step 1.
6. Given all temperatures and pressures, perform the energy balance and evaluate heat duties.

In order to follow this decomposition, we assume that all vapor and liquid streams have ideal equilibrium relationships (particularly in step 2) and that, unless stated otherwise, all streams are at saturated conditions. With these assumptions physical properties can be calculated easily from standard handbook data. In this text, we rely on Reid et al. (1987) as our data source. The advantages of this approach are that calculations are very easy to set up and solve with few iterations (usually no more than two) required for convergence of a preliminary design.

Consider the flowsheet shown in Figure 3.1, with the units shown as rectangles connected by input and output streams. In this section we construct linear model approximations for the following units:

- Mixer
- Splitter

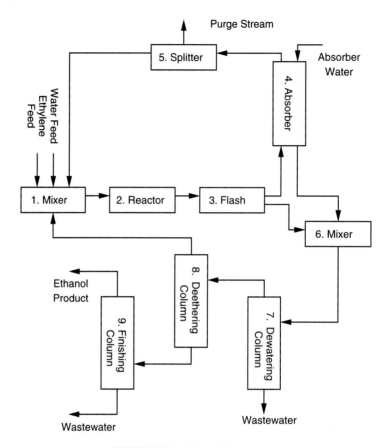

FIGURE 3.1 Ethanol flowsheet.

- Reactor
- Flash
- Distillation column
- Absorber
- Stripper

The above list contains a comprehensive set of mass balance units and in the next section we will show how to put the flowsheet in Figure 3.1 together with them. Additional information on the shortcut separation units can also be found in Douglas (1988) and Perry et al. (1984). To construct the linear unit models, we label the stream vector of molar flows μ_{ij} as the jth output stream of unit i. μ_{ij}^k is the flowrate of component k in this stream. Also, if there is only one outlet stream in unit i, the j subscript is suppressed. Note that with this notation, we express stream composition in terms of molar flows instead of mole fractions, as this preserves linearity of the equations. For example for Unit 2

Sec. 3.2 Developing Unit Models for Linear Mass Balances

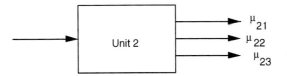

FIGURE 3.2 Components: hydrogen, methane, carbon dioxide.

in Figure 3.2 above, $\mu_{22}CH_4$ refers to the molar flowrate of CH_4 in the second effluent stream.

3.2.1 Linear Mass Balances for Simple Units

Equations for the following units can be written simply as follows.

MIXER UNIT

This unit (Figure 3.3) merely sums all of the inlet streams as a single output stream with the following mass balance equations. Given, upstream units i_1, i_2, ... that feed into the mixer with the j_1th outlet from unit i_1, the j_2th outlet from unit i_2, etc., for component k, μ_M is written as:

$$\mu_M^k = \sum_\ell \mu_{i\ell,j}^k$$

SPLITTER UNIT

The splitter unit (Figure 3.4) divides a given feed stream into specified fractions ξ_j for each output stream j. Note that all output streams have the same compositions as the feed stream. Thus, for NS output streams we have NS $-$ 1 degrees of freedom in choosing ξ_j and write the equations:

$$\mu_{Sj}^k = \xi_j \mu_{IN}^k, j = 1,...NS - 1 \qquad \mu_{S,NS}^k = (1 - \Sigma_{j=1}^{NS-1}\xi_j) \mu_{IN}^k$$

REACTOR (FIXED CONVERSION MODEL)

For linear mass balances, we assume that the reactor model can be simplified by specifying the molar conversion of the NR parallel reactions in advance (Figure 3.5). As a result, the mass balance equations remain linear and relatively easy to solve. For each reaction r, we define a limiting component $l(r)$, and normalized stoichiometric coefficients

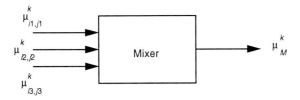

FIGURE 3.3 Mixer unit.

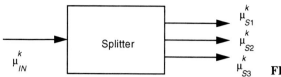

FIGURE 3.4 Splitter unit.

$\gamma_{r,k} = (C_{r,k}/C_{r,l(r)})$, $r = 1$, NR for each component k, where the coefficients $C_{k,r}$ appear in the specified reactions. We also adopt the convention:

$$\gamma_{r,k} = \begin{pmatrix} > 0, \text{ prod } k \\ < 0, \text{ } k \text{ reactant} \\ = 0, \text{ } k \text{ inert} \end{pmatrix}$$

Defining the fraction converted per pass based on limiting reactant as η_r, $r = 1$, NR, gives us:

$$\mu_R^k = \mu_{IN}^k + \sum_{r=1}^{NR} \gamma_{r,k}\, \eta_r \mu_{IN}^{l(r)}$$

The equations for the fixed conversion reactor model are best illustrated by example.

EXAMPLE 3.1

Consider the following reactions where CH_4 is considered the limiting reactant in the first reaction, and C_2H_6 is the limiting reactant in the second, with conversions per pass specified at 60% and 80% for the first and second reactions:

$$CH_4 + 2O_2 \rightarrow CO_2 + 2H_2O \quad \eta_1 = 0.6$$
$$C_2H_6 + 7/2\, O_2 \rightarrow 2CO_2 + 3H_2O \quad \eta_2 = 0.8$$

which leads to the following table of normalized coefficients, $\gamma_{r,k}$

r	k = CH_4	O_2	C_2H_6	CO_2	H_2O
1	−1	−2	0	1	2
2	0	−7/2	−1	2	3

The equations for the limiting reactants can be written as:

$$\mu_R^{CH_4} = \mu_{IN}^{CH_4} - 0.6\, \mu_{IN}^{CH_4} = 0.4\, \mu_{IN}^{CH_4}$$
$$\mu_R^{C_2H_6} = \mu_{IN}^{C_2H_6} - 0.8\, \mu_{IN}^{C_2H_6} = 0.2\, \mu_{IN}^{C_2H_6}$$

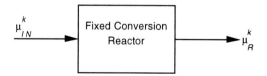

FIGURE 3.5 Reactor unit.

Sec. 3.2 Developing Unit Models for Linear Mass Balances

with the remaining components defined by the following relations:

$$\mu_R^{O_2} = \mu_{IN}^{O_2} - 2(0.6)\,\mu_{IN}^{CH_4} - 7/2(0.8)\,\mu_{IN}^{C_2H_6}$$

$$\mu_R^{H_2O} = \mu_{IN}^{H_2O} + 2(0.6)\,\mu_{IN}^{CH_4} + 3(0.8)\,\mu_{IN}^{C_2H_6}$$

$$\mu_R^{CO_2} = \mu_{IN}^{CO_2} + (0.6)\,\mu_{IN}^{CH_4} + 2(0.8)\,\mu_{IN}^{C_2H_6}$$

For reaction mechanisms that have series as well as parallel components, this approach can be generalized simply by defining additional reactor units and solving these in series.

3.2.2 Calculation of Flash Units—the "Building Block" Unit in Process Flowsheets

This calculation is the most fundamental and important one in a flowsheet. Aside from the physical separation unit itself, it is the building block for deriving linear models for equilibrium-staged separations such as distillation and absorption. These calculation procedures will also be used later for setting pressures and temperatures around the flowsheet. We first consider the simple phase separation unit described in Figure 3.6, as well as a number of calculation procedures for this unit.

To develop the flash model, we first define an overhead split fraction $\xi_k = v_k/f_k$ for each of the *ncomp* components k. We further identify component n as a key component (for which a given recovery can be obtained) and also define $\phi = V/F$ for specified vaporization of the feed. As specifications, the variables, ξ_n, ϕ, P,T, and Q (heat supplied to flash unit), can be specified. If we now write the equations for the flash unit:

$$f_k = l_k + v_k \qquad (k = 1, \ldots ncomp)$$

$$v_k/V = K(x, P, T)\, l_k/L \qquad (k = 1, \ldots ncomp)$$

$$\sum_k l_k = L \qquad \sum_k v_k = V$$

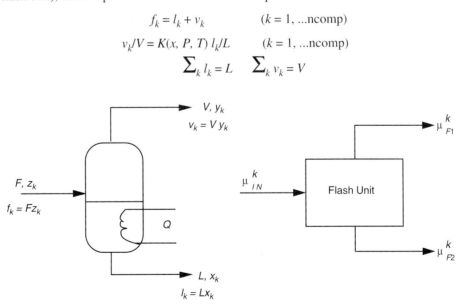

FIGURE 3.6 Liquid-vapor flash unit.

we find that for a specified feed, the (number of variables) − (number of equations) = 2 degrees of freedom. This means that we can completely specify the condition of the flash unit if we select two of the variables. Since we have not yet considered energy balances, we defer specifications on Q and now consider the following cases:

Case 1 ξ_n specified (key comp overhead recovery) and T or P specified
Case 2 T and P specified (isothermal flash)
Case 3 ϕ specified and T or P specified

The first case is very useful for the shortcut methods in this chapter, but is not used for more detailed models. Cases 2 and 3 are needed for analyzing design and operating conditions.

We now consider some approximations for vapor/liquid phase equilibrium. Equating the mixture fugacities in each phase leads to a reasonably general expression at low to moderate pressures:

$$\phi_k y_k P = \gamma_k x_k f^0_k \quad \text{for } k = 1, \text{ncomp}$$

where ϕ_k is a vapor fugacity coefficient, γ_k is the liquid activity coefficient, and f^0_k is the pure component fugacity. For process calculations, it is often convenient to represent the equilibrium relation as: $y_k = K_k x_k$, with the K value, $K_k = (\gamma_k f^0_k / \phi_k P)$. For our shortcut calculations, we assume ideal behavior which leads to the following assumptions:

$$\phi_k = 1, \gamma_k = 1, f^0_k = P^0_k \text{ (vapor pressure)}$$

Antoine equation for vapor pressure: $\ln P^0_k = A_k - B_k / (T + C_k)$

where the Antione equation is a representative correlation with coefficients that can be found, for example, in Reid et al. (1987). These assumptions lead to Raoult's Law:

$$y_k P = x_k P^0_k \text{ or more simply, } y_k / x_k = P^0_k / P = K_k.$$

With respect to key components, we can now define a *relative* volatility:

$$\alpha_{k/n} = K_k / K_n = P^0_k / P^0_n$$

which, for ideal systems, is independent of P and is much less sensitive to T than K_k is. Note that component k can be nonvolatile, in which case $\alpha_{k/n} \to 0$. On the other hand, if component k is noncondensible, $\alpha_{k/n} \to \infty$. We can now rederive and simplify the flash equations. Let:

$$\alpha_{k/n} = \frac{K_k}{K_n} = \frac{y_k / x_k}{y_n / x_n} \frac{V/L}{V/L} = \frac{v_k / l_k}{v_n / l_n}$$

We now reintroduce the split fractions and define:

$$v_k = \xi_k f_k \text{ and } l_k = (1 - \xi_k) f_k$$

Substituting, these definitions into the above equation gives us:

Sec. 3.2 Developing Unit Models for Linear Mass Balances

$$K_k = \xi_k \, L/(V(1-\xi_k)) \text{ as well as } \alpha_{k/n} = \frac{\xi_k/(1-\xi_k)}{\xi_n/(1-\xi_n)}$$

at equilibrium. Rearranging this expression gives:

$$\xi_k = \frac{\alpha_{k/n}\,\xi_n}{1 + (\alpha_{k/n} - 1)\,\xi_n} \text{ for each } k$$

and we have now defined the recovery of each component in terms of the key component recovery. Note also that the limiting cases of nonvolatile ($\alpha_{k/n} \to 0$, $\xi_k \to 0$) and noncondensible ($\alpha_{k/n} \to \infty$, $\xi_k \to 1$) components are also observed.

With specification of key component recovery, an additional specification is still required (two degrees of freedom). Implicit in the above expression is that a correct value of temperature (T) was known in advance in order to calculate the relative volatilities. Given that we have specified T or P, how do we calculate the corresponding value of P or T? Moreover, if we have specified T or P directly, how do we use the above equation to determine the corresponding key component recovery? Here we need to consider a bubble (or dew point) equation that also needs to be satisfied at equilibrium. At the bubble point (for the saturated liquid effluent stream) we have:

$$\sum y_i = \sum K_i \, x_i = 1$$

or in terms of relative volatilities:

$$1/K_n = \sum (K_i/K_n)\, x_i = \sum \alpha_{i/n}\, x_i = \overline{\alpha}$$

where $\overline{\alpha}$ is defined as an average relative volatility. Using this definition allows us to redefine the K-value as:

$$\frac{P_k^0}{P} = K_k = \frac{\alpha_{k/n}}{\overline{\alpha}}$$

which forms a simplified bubble point equation. For T fixed and P unknown, we can calculate a value of P directly from:

$$P = \frac{\overline{\alpha}}{\alpha_{k/n}} \, P_k^0(T)$$

On the other hand, for P fixed and T unknown, the value of T can be calculated approximately from:

$$P_k^0(T) = \alpha_{k/n}\, P/\overline{\alpha}$$

To reduce approximation errors, we choose the index k to be the most abundant component in the liquid phase.

With the above equations we can now develop the following algorithms for the three most commonly specified flash problems.

Case 1: ξ_n and P (or T) Fixed

a. For a specified ξ_n and P (or T), guess T (or P).
b. Calculate K_k, $\alpha_{k/n}$ at specified T.
c. Evaluate $\xi_k = \alpha_{k/n}\,\xi_n/(1 + (\alpha_{k/n} - 1)\xi_n)$ for each component k.
d. Reconstruct a mass balance and calculate mole fractions.

$$v_k = \xi_k f_k \qquad y_k = v_k/\Sigma v_i$$
$$l_k = (1 - \xi_k) f_k \qquad x_k = l_k/\Sigma l_i$$

e. For T fixed, $P = \dfrac{\overline{\alpha}}{\alpha_{k/n}}\, P_k^0(T)$.

For P fixed, solve for T from $P_k^0(T) = \alpha_{k/n} P/\overline{\alpha}$.

Case 2: T and P Fixed

a. For a specified T and P, pick a key component n and guess ξ_n.
Follow steps b, c, and d of algorithm for Case 1.
e. If the bubble point equation is satisfied: $\alpha = P\alpha_{k/n}/P_k^0$, stop. Otherwise, reguess ξ_n, and go to step c. (Simple iterative methods, such as the secant algorithm in Chapter 8, can be used to obtain convergence for ξ_n.)

Case 3: ϕ and P (or T) Fixed

a. For a specified $\phi = V/F$ and P (or T)
b. Guess T (or P), calculate $\alpha_{k/n}$, K_k and define $\theta = K_n \phi/(1 - \phi) = v_n/l_n$
Define $\xi_n = \theta/(1 + \theta)$.
Then follow steps c and d of the previous algorithm.
e. If the bubble point equation is satisfied: $\alpha = P\alpha_{k/n}/P_k^0$, stop. Otherwise, reguess T (or P), and go to step b. (Simple iterative methods, such as the secant algorithm can be used to obtain convergence for ξ_n.)

These algorithms have been stated very concisely. Each of these algorithms will be illustrated by the following examples.

EXAMPLE 3.2   Flash Calculation

Consider the mixture with the components, flowrates, boiling points, and Antoine coefficients given in the following table.

Sec. 3.2 Developing Unit Models for Linear Mass Balances

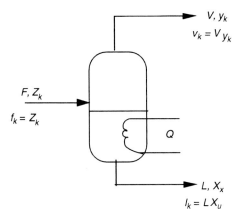

comp., k	f_k		Boiling Point(K)	A_k	B_k	C_k
Benzene	30	kmol/hr	353	15.9008	2788.51	−52.34
Toluene	50	kmol/hr	383	16.0137	3096.52	−53.67
O-xylene	40	kmol/hr	418	16.1156	3395.57	−59.44

Here we choose toluene as the key component ($n = 2$) because of its intermediate volatility.

Case 1: Fixed $\xi_2 = 0.9$ and $P = 1$ bar

If we assume that $\alpha_{k/n}$ remains constant over the temperature range, we can do a direct calculation without iteration.

  a. Specify $\xi_2 = 0.9$, $P = 1$ bar and guess $T = 390$ K.
  b. Calculate relative volatilities $\alpha_{k/n} = P_k^0/P_n^0$ (same as above problem).
  $\alpha_{1/2} = 2.305$
  $\alpha_{3/2} = 0.381$
  c. Calculate recoveries of nonkey components.
  $\xi_1 = 0.954$
  $\xi_3 = 0.774$
  d. Solve for mass balance and evaluate mole fractions.
  $v_1 = 28.62 \quad \ell_1 = 1.38 \quad x_1 = 0.089$
  $v_2 = 45 \quad \ell_2 = 5 \quad x_2 = 0.325$
  $v_3 = 30.96 \quad \ell_3 = 9.04 \quad x_3 = 0.586$
  e. Evaluate bubble point equation.

$$P_3^0 = 344.7 \neq \frac{P\alpha_{k/n}}{\bar{\alpha}} = \frac{(750)(0.381)}{0.752} = 380 \text{ mm Hg}$$

but $T(P^0 = 380) = 393$ K (estimate of T is close enough)

Case 2: Flash Calculation at 1 Bar and 390 K

Following the algorithm above, we note the following steps:

a. $T = 390$ K, $P = 1$ bar. Guess $\xi_2 = 0.9$.

b. From Antione equation, determine vapor pressures at 390 K:
$$\ln P_k^0 = A_k - B_k/(C_k + T)$$
$$\alpha_{k/n} = P_k^0/P_n^0$$
$$\alpha_{1/2} = 2083.8/904.1 = 2.305$$
$$\alpha_{3/2} = 344.7/904.1 = 0.381$$

c. Solve for remaining recoveries:

$$\xi_1 = \frac{(2.305)(0.9)}{1 + (1.305)(0.9)} = 0.954$$

$$\xi_3 = \frac{(0.381)(0.9)}{1 - (0.619)(0.9)} = 0.774$$

d. Solve for mass balance and mole fractions:

$v_1 = 30(0.954) = 28.62$ $\ell_1 = 1.38$ $x_1 = 0.089$
$v_2 = 50(0.9) = 45$ $\ell_2 = 5$ $x_2 = 0.324$
$v_3 = 40(0.774) = 30.96$ $\ell_3 = 9.04$ $x_3 = 0.586$

e. Check the bubble point equation:

$$\frac{P\alpha_{k/n}}{P_k^0} = \frac{(750)(.381)}{(344.7)} = 0.82$$

but $\alpha = \Sigma x_k \alpha_{k/n} = 0.752$

Go to step c with ξ_2 reguessed at 0.80:

$\xi_1 = 0.902$
$\xi_3 = 0.604$

d. $\ell_1 = 2.94$ $x_1 = 0.102$
$\ell_2 = 10$ $x_2 = 0.347$
$\ell_3 = 15.84$ $x_3 = 0.55$

e. $\alpha = 0.792$

$$\frac{P\alpha_{k/n}}{P_k^0} = 0.82$$

(Close enough for rough estimate: $\xi_n = 0.8$ @ $P = 0.96$ bar.)

Case 3: Vapor Fraction = 0.8 and P = 1 Bar

a. $\phi = 0.8$, $P = 1$ bar, guess $T = 390$ K.

b. Evaluate K values, relative volatilities and key component recovery:
$\alpha_{1/2} = 2.305$ $K_1 = 2.778$

Sec. 3.2 Developing Unit Models for Linear Mass Balances 67

$\alpha_{1/2} = 1.0 \quad K_2 = 1.205$
$\alpha_{3/2} = 0.381 \quad K_3 = 0.460$

$$\theta = (1.250)\left(\frac{0.8}{0.2}\right) = 4.82 \quad \xi_2 = 0.828 = \frac{\theta}{1+\theta}$$

c. Evaluate nonkey component recoveries:
$\xi_1 = (2.305)(0.828)/(1 + 1.305(0.828)) = 0.917$
$\xi_3 = (0.381)(0.828)/(1 - 0.619(0.828)) = 0.647$

d. Solve mass balances and evaluate mole fractions:
$v_1 = 27.5 \quad \ell_1 = 2.5 \quad x_1 = 0.099$
$v_2 = 41.4 \quad \ell_2 = 8.6 \quad x_2 = 0.341$
$v_3 = 25.9 \quad \ell_3 = 14.1 \quad x_3 = 0.560$

e. $\bar{\alpha} = 0.782$

$\dfrac{P\alpha_{k/n}}{\bar{\alpha}} = 365.4 \qquad T(\text{for } P_3^0 = 365.4 \text{ mm Hg}) = 391.9 \text{ K} \sim 390 \text{ K}$

(Answer is close enough for rough estimate.)

BUBBLE AND DEW POINT CALCULATIONS

The algorithms presented above allow rapid calculation of flash separators. However, in the limiting cases of the bubble point ($\phi = 0$) or the dewpoint ($\phi = 1$), these algorithms can be further simplified and are given in Figures 3.7 and 3.8.

Here $\xi_k = 0$, $\ell_k = f_k$ and $x_k = z_k$
For P fixed, calculate T directly from $P_n^0(T) = P/\bar{\alpha}_n$
For T fixed, calculate P from $P = \bar{\alpha}_n P_n^0(T)$
In both cases, n is chosen as the most abundant component.

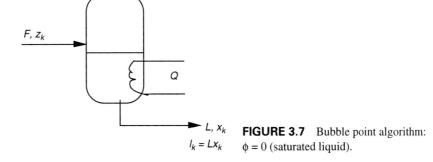

FIGURE 3.7 Bubble point algorithm: $\phi = 0$ (saturated liquid).

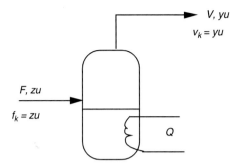

FIGURE 3.8 Dew point algorithm: $\phi = 1$ (saturated vapor).

Here $\xi_k = 1$, $v_k = f_k$ and $y_k = z_k$.

For this case, we derive a dew point equation based on: $y_k = z_k$

Here $\sum x_k = 1 \Rightarrow K_n \sum \dfrac{y_k}{K_k} = \sum \dfrac{y_k}{\alpha_{k/n}} = K_n$

Select as $k = n$ the most abundant vapor component. Then $\sum \dfrac{y_k}{\alpha_{k/n}} = \dfrac{P_n^0}{P}$ and:

For T fixed $\quad P = P_n^0(T) / \left(\sum \dfrac{y_k}{\alpha_{k/n}} \right)$

For P fixed $\quad P_n^0(T) = P \left(\sum \dfrac{y_k}{\alpha_{k/n}} \right)$ and solve directly for T

Again a key assumption for this last equation is that $\alpha_{k/n}$ remains fairly constant with temperature.

UPPER LIMITS OF PRESSURE AND TEMPERATURE IN VAPOR LIQUID EQUILIBRIUM

Of course, the above simplified flash calculations (as with more detailed calculations) cannot be applied at or above the critical region. At the critical point, we have equal densities for the vapor and liquid phases. If we examine the phase diagram for mixtures, illustrated in Figure 3.9, we note some unusual behavior not described by the flash algorithms. For example, in the region of isobaric retrograde condensation, above the critical pressure, increasing temperature will lead to liquefaction. Similarly, for the region of isothermal retrograde condensation, above the critical temperature, an increase in pressure will lead to increased vaporization.

Calculations in the neighborhood of the critical point still remain important challenges for detailed phase equilibrium algorithms. For the purpose of our simplified design calculations, we will simply avoid critical regions by using the following guideline to test the existence of a liquid phase. Here we define a pseudocritical temperature for a mixture as: $T_c^m = \Sigma_k x_k T_c^k$, where T_c^k is the critical temperature of component k. Here we use liquid mole fractions because these give more realistic estimates of critical temperatures for mixtures.

Sec. 3.2   Developing Unit Models for Linear Mass Balances

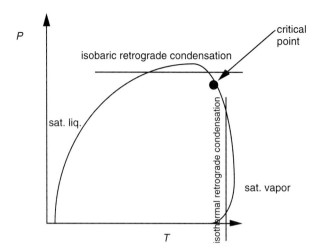

FIGURE 3.9 Phase diagram for retrograde condensation.

EXAMPLE 3.3

Consider the mixture of the previous flash calculation, we would like to determine if the critical point of this mixture is above 392 K and 1 bar, the point at which we would like to flash this mixture. From handbook data we have:

	T_c	x_k
Benzene	56.2K	0.099
Toluene	592	0.341
O-xylene	630	0.560

and the mixture critical point is $T_c^m = 610$ K, well above the flash specification (392 K).

EXAMPLE 3.4

Consider the following H_2/H_2O system with mole fractions, z_k, where we know a liquid phase exists at room temperature (300K) and pressure (1 bar). From handbook values we have the following properties:

	T crit.	K_k (300 K, 1 bar)	z_k
1. H_2	33.2	645.1	0.75
2. H_2O	647.3	0.035	0.25

Here if we set water as the key component, we have $\alpha_{1/2} = 18,400$. Assume that $\xi_2 = 0.01$, we calculate $\xi_1 = 0.994$ and the following mass balance can be obtained as a rough guess:

$\ell_1 = 0.5$ $v_1 = 74.5$ $x_1 = 0.02$
$\ell_2 = 24.78$ $v_2 = 0.25$ $x_2 = 0.98$

with a mixture critical value of $T_c^m = 645$. Note that if we had used the feed composition for the estimated critical temperature we would have $T_c^m = 186.7K$, which is much lower than the desired flash temperature.

3.2.4 Distillation Models

In this subsection we establish split fractions based on simple shortcut methods for distillation. Distillation operations can be described as a cascade of equilibrium trays with each one solved as a flash unit (Figure 3.10). The feed stream enters at an intermediate tray; at the bottom, liquid product is removed, a reboiler vaporizes the liquid stream on the lowest stage, and counter-current liquid and vapor streams are set up in the distillation column. Similarly, vapor leaving the top tray is condensed and overhead product is removed, with the remaining liquid returned or refluxed back to the top tray. Detailed calculation of the tray-by-tray behavior of a distillation column will not be considered at this stage in the design, but will be deferred to Chapter 7. Instead, we will make a number of approximations using limiting column behavior (total reflux) in order to obtain linear mass balance models and relevant equipment parameters.

First, let's identify the degrees of freedom available for mass balance in a distillation column. For determining the *mass balance*, it turns out that if we know the overhead split fractions, ξ_{lk}, ξ_{hk} (where *lk* and *hk* refer to light and heavy key components, respectively) and the overhead column pressure, we have already fully specified the column equations. So why are there only three degrees of freedom in a column mass balance, regardless of the number of distillation trays? Intuitively, we can think of the top of the column, which further refines the light key with ξ_{lk} and P_T specified (as in a flash unit), and the bottom of the column, which further refines the heavy key with ξ_{hk} and P_B specified (as in a flash unit). Thus, we have four specifications. But since $P_T + \Delta P = P_B$ where ΔP is the column pressure drop, there are only three independent degrees of freedom.

CALCULATING LINEAR SPLIT FRACTIONS

To further specify a distillation column and derive the component recoveries in the linear mass balance equations, we classify five types of components:

1. Components lighter than the light key
2. Light key component
3. Components between keys (distributed components)
4. Heavy key component
5. Components heavier than the heavy key

Sec. 3.2 Developing Unit Models for Linear Mass Balances

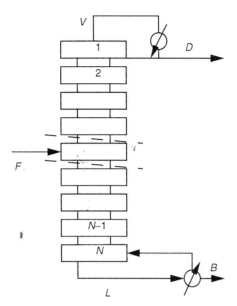

FIGURE 3.10 Tray-by-tray representation of distillation column.

As with the flash unit we will assume that ideal behavior exists in our simplified column. From the flash unit, we *know* $\alpha_{k/n}$ is independent of pressure, and we *assume* it is independent of temperature in ideal situations. Moreover, in order to do the mass balance, we must know the split fractions of the distributed components. After the mass balance, we also need to consider the number of trays and the temperatures in the column. To find this, we use the Fenske (1932) equation for total reflux. This equation is easily derived and gives an approximate product distribution as well as an estimate of the minimum number of trays. Consider the total reflux case shown in Figure 3.11. Here we note that the feed and bottom streams are negligible compared to the total reflux flow and can be ignored.

Starting at the reboiler, we note from the mass balance of vapor and liquid streams above the reboiler that $x_{k,N-1} = y_{k,R}$. Also, from the equilibrium relation:

$$\frac{y_{\ell k,R}}{y_{hk,R}} = \alpha_{\ell k/hk} \frac{x_{\ell k,R}}{x_{hk,R}}$$

At stage $N-1$, we can again write the equilibrium expression:

$$y_{lk,N-1}/y_{hk,N-1} = \alpha_{lk/hk}(x_{lk,N-1}/x_{hk,N-1}) = (\alpha_{lk/hk})^2 \, x_{lk,R}/x_{hk,R}$$

Similarly, at stage $N-2$ we have the relation:

$$y_{lk,N-2}/y_{hk,N-2} = (\alpha_{lk/hk})^3 \, x_{lk,R}/x_{hk,R}$$

Finally, since $x_{k,j-2} = y_{k,j-1}$ for every stage j, we can write:

$$x_{lk,D}/x_{hk,D} = y_{lk,1}/y_{hk,1} = (\alpha_{lk/hk})^{N_m} \, x_{lk,R}/x_{hk,R}$$

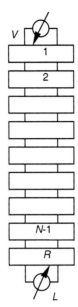

FIGURE 3.11 Tray-by-tray representation at total reflux.

where N_m is the minimum number of equilibrium stages. Writing in terms of distillate and bottoms flowrates and defining split fractions for these yields:

$$(d_{lk}/D)/(d_{hk}/D) = (\alpha_{lk/hk})^{N_m}(b_{lk}/B)/(b_{hk}/B)$$

and with $\xi_k = d_k/f_k$ we rearrange the above expression to yield:

$$\frac{d_{\ell k}}{b_{hk}} = \alpha^{N_m}\frac{d_{hk}}{b_{hk}} \Rightarrow \frac{\xi_{\ell k}}{1-\xi_{\ell k}} = \alpha_{\ell k/hk}^{N_m}\frac{\xi_{hk}}{1-\xi_{hk}}$$

If we have specified the light and heavy key recoveries, then the minimum number of stages is given directly by the Fenske equation:

$$N_m = \ln[\{\xi_{lk}(1-\xi_{hk})\}/\{\xi_{hk}(1-\xi_{lk})\}] / \ln \alpha_{lk/hk}$$

Once we know N_m, all of the other component split fractions can be obtained simply by substituting k for l_k in the above expressions. With minor rearrangement, we have:

$$\xi_k = \frac{\alpha_k^{N_m}\xi_{hk}}{1+(\alpha_k^{N_m}-1)\xi_{hk}}$$

Note that this equation reduces to the split fraction for the flash unit when $N_m = 1$. Moreover, while the above equation applies to all components, we will simplify our analysis and apply this equation to distributed components only. This follows because key component split fractions, ξ_{lk} and ξ_{hk}, will be specified close to one and zero, respectively. Hence, for all but the distributed components, we can assume:

Sec. 3.2 Developing Unit Models for Linear Mass Balances

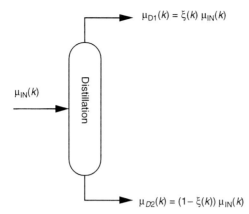

FIGURE 3.12 Mass balance for distillation.

Component type	ξ_k
1. Lighter than light key	1, ($\alpha_{k/hk} > 1$, as $N_m \to \infty$, $\xi_k = 1$)
2. Light key	ξ_{lk} fixed (e.g., 0.99)
3. Distributed component	from equation for ξ_k
4. Heavy key	ξ_{hk} fixed (e.g., 0.01)
5. Heavier than heavy key	0, ($\alpha_{k/hk} < 1$, as $N_m \to \infty$, $\xi_k = 0$)

Once these split fractions are calculated, the linear mass balance for the distillation column is straightforward (Figure 3.12).

SETTING COLUMN PRESSURES AND TEMPERATURES

In addition to specifying recoveries of key components, we also need to set an appropriate pressure (or temperature) for the top of the column. To do this, we first need to explore the contraints on these specifications. These are primarily dictated by the cooling water temperature (T_{cw}) in the condenser and the steam supply temperature (T_{st}) in the reboiler. Consider Figure 3.13 with a total condenser and reboiler and with temperatures marked in different column locations.

Since we know that the column pressure is lower at the top than the bottom, and that the more volatile (low boiling components) are also higher in concentration at the top, we note the following temperature relationships:

$$T_{cw} \leq T_{bub,C} \leq T_{dew,C} \leq T_{bub,R} \leq T_{dew,R} \leq T_{st}$$

Column pressure can be selected so that the following constraints hold:

1. Select condenser pressure so that $T_{bub,C} \geq T_{cw}$ (about 30°C) + ΔT (about 5 K) ~ 310 K.
2. Select condenser pressure so that all bubble point temperatures are below the critical temperature of a mixture, i.e.: $T_{bub} \leq T_{cm} = \Sigma\, T_c^k x_{k,D}$.

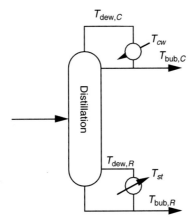

FIGURE 3.13 Setting column pressure and temperature.

3. From the bubble point equation, we note T_{bub} increases with P and we prefer to choose P to be above one atmosphere. Thus, $P \geq \overline{\alpha}_n P_n^o(T_{bub}) \geq 1$ atm. (Below 1 atm, thicker vessel walls and additional safety precautions are required to avoid air leaks and explosion hazards.)

These constraints can be difficult to meet when we have both noncondensible (very low boiling) components or nonvolatile (very high boiling) components in the system. One common way to still satisfy the above pressure restrictions is to consider partial condensers and reboilers for noncondensible and nonvolatile components, respectively. Mass balances with these additional devices can be determined through an additional flash calculation. Consider first the partial condenser shown in Figure 3.14.

Calculating the mass balance and temperatures around the partial condenser can be greatly simplified by noting that the product streams are at saturated liquid and vapor and can be obtained through a simple flash calculation, once the product flows and compositions (d_k) are specified. From this we note that the partial condenser can be represented schematically in Figure 3.15.

From this, a direct way to calculate the mass balance involves the following scheme:

1. Relate D to L through a predetermined reflux ratio ($R = L/D$). This can be determined from shortcut methods (Fenske, Underwood, Gilliland equations) discussed in the next chapter.

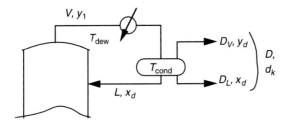

FIGURE 3.14 Partial condenser.

Sec. 3.2 Developing Unit Models for Linear Mass Balances

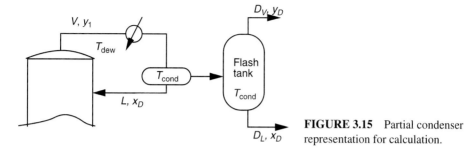

FIGURE 3.15 Partial condenser representation for calculation.

2. To obtain T_{cond}, do a Case 3 flash calculation on the flash tank with P and $\phi = D_v/D$ specified to get T_{cond}, y_D and x_D. Note that the feed to this tank is given by d_k. (The vapor fraction of the product, ϕ, can be specified, for example, by the fraction of noncondensible components in the product.)
3. Calculate L, V, and the dewpoint composition, y_1, in V, from the mass balance equations:
$$V = (1+R)\,D = D + L$$
$$Vy_1 = D_V y_D + (D_L + L)\,x_D$$
4. To find T_{dew}, perform a dew point calculation for V with P and y_1 specified. These temperatures will be useful for sizing the condenser as well as for the energy balance.

Partial reboilers can also be analyzed in a simpler manner as shown in Figure 3.16.

Note that the dew point exiting the reboiler is the highest temperature in the column. To avoid excessively high temperatures a partial reboiler effectively adds an extra equilibrium stage. To calculate the difference in temperatures, the dewpoint temperature in a total reboiler is given by:

$$P_n^0(T_{dew}) = P'\left(\sum_k \frac{y_k}{\alpha_{k/n}}\right)$$

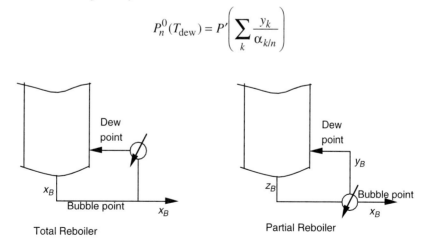

FIGURE 3.16 Reboiler configurations.

where n is the most plentiful component and $P' = P + \Delta P$. Here the composition, y_k, is the same as the bottoms product and there is a large contribution in the summation from high-boiling components. With a partial condenser, on the other hand, the composition, y_k, is not as rich in these components—both P_n^0 and T_{dew} are lower. Similarly, the bubble point temperature for the reboiler product can be calculated from the bubble point equation.

EXAMPLE 3.5   Distillation

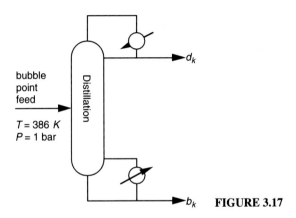

FIGURE 3.17

Consider the separation of a benzene, toluene, ortho-xylene mixture where we would like to recover 99% of the benzene overhead and 99.5% of the o-xylene in the bottoms stream. We therefore choose benzene and o-xylene as light and heavy keys, respectively, and note the following data for the feed.

Component	Flow (kgmol/h)	K(386, 1bar)	$\alpha_{lk/hk}$
Benzene	20	2.52	6.209
Toluene	30	1.079	2.662
O-Xylene	50	0.405	1.0

For $\xi_1 = 0.99$ and $\xi_3 = 0.005$, the minimum number of trays (at total reflux) is given by the Fenske equation:

$$N_m = \ln\left\{\frac{0.99}{1-0.99} \cdot \frac{1-0.005}{0.005}\right\} / \ln(6.209) = 5.41$$

The split fraction for the distributed component (toluene) is given by:

$$\xi_2 = \frac{\alpha_{2/3}^{N_m}\xi_3}{1+\left(\alpha_{2/3}^{N_m}-1\right)\xi_3} = 0.501$$

and the mass balance can be calculated directly from the split fractions:

Sec. 3.2 Developing Unit Models for Linear Mass Balances

$$d_1 = 19.8 \quad b_1 = 0.2$$
$$d_2 = 15.03 \quad b_2 = 14.97$$
$$d_3 = 0.25 \quad b_3 = 49.75$$

Now, to determine the pressure and temperature at the top of the column with a total condenser, we choose benzene as the most plentiful component and perform a bubble point calculation. Here:

$$x_1 = 0.564 \quad \alpha_{1/1} = 1.0$$
$$x_2 = 0.422 \quad \alpha_{2/1} = 0.428$$
$$x_3 = 0.007 \quad \alpha_{3/1} = 0.161$$
$$\overline{\alpha} = \Sigma\, x_i\, \alpha_{i/1} = 0.746$$

and from the bubble point equation, $P_n^0(T) = P/\overline{\alpha}$, we have:

$$P_1^0(T) = 750/0.746 = 1005.4 \text{ mm Hg} \Rightarrow T = 362.6 \text{ K from Antoine equation}$$

So the distillate temperature is 362.6 K, well above cooling water temperature; so far, the pressure specification of 1 bar seems appropriate. The overhead vapor temperature can be obtained from a dew point calculation as follows. Again, choose $n = 1$ as the most plentiful component and evaluate:

$$P_n^0(T) = P\sum (y_k/\alpha_{k/n}) \approx (750 \text{ mm}) \left(\frac{0.564}{1.0} + \frac{0.422}{0.428} + \frac{0.007}{0.161} \right)$$

so that we have:

$$P_1^0(T) = 1195.1 \Rightarrow T = 368.7 \text{ K}$$

(overhead vapor temp. from Antoine equation)

To determine the bottom temperatures with a total reboiler, we now choose o-xylene as the most plentiful component and evaluate the bottom mole fractions:

$$b_1 = 0.2 \quad x_1 = 0.0031 \quad \alpha_{1/3} = 6.209$$
$$b_2 = 14.97 \quad x_2 = 0.231 \quad \alpha_{2/3} = 2.662$$
$$b_3 = 49.75 \quad x_3 = 0.766 \quad \alpha_{3/3} = 1.0$$

The bottoms product temperature is given directly from the bubble point equation:

$$P_3^0(T) = \frac{P}{\overline{\alpha}_3} = \frac{750}{1.400} \text{ mm} = 535.6 \text{ mm}$$
$$\Rightarrow T = 404.8 \text{ K bottoms temp.}$$

and the vapor exiting the total reboiler has a temperature that can be calculated from the dew point equation:

$$P_3^0(T) = P\, (\Sigma\, y_k/\alpha_{k/3}) = 640 \text{ mm}$$
$$\Rightarrow T = 411.2 \text{ K (highest temp. in column)}$$

Note that in order to perform this separation, steam must be supplied to the reboiler above this temperature.

Now how does the condenser temperature change if we had a partial condenser? First, we need to know the reflux ratio and the required vapor fraction of the overhead product. If we have a reflux ratio, $R = 20$, then with the specified distillate flowrate, $D = 35.08$, we have the following liquid and vapor streams: $L = 701.6$ and $V = 736.7$. For this reflux ratio, the highest condenser temperature corresponds to a vapor product. If we vaporize all of D ($\phi = 1$), the product temperature is obtained from the dew point calculation.

$$P_1^0(T) = P\left(\sum (y_{Dk}/\alpha_{k/1})\right)$$
$$\Rightarrow T = 368.7 \text{ K} \quad (\text{temp. of } D)$$

At this temperature, the corresponding bubble point composition of the reflux stream is given by:

$y_1 = 0.564 \quad K_1 = 1.593 \quad x_1 = 0.341 \quad \alpha_{1/1} = 1.0$

$y_2 = 0.422 \quad K_2 = 0.647 \quad x_2 = 0.629 \quad \alpha_{2/1} = 0.406$

$y_3 = 0.007 \quad K_3 = 0.226 \quad x_3 = 0.030 \quad \alpha_{3/1} = 0.142$

(Note that α doesn't change much over this temperature range.) Finally, we calculate the composition of the overhead vapor stream from the following mass balance:

$$Vy_1 = Dy_D + L x_L$$
$$y_1 = [(35.08) y_D + (701.6)x_L]/736.7$$

which yields:

$$y_{1,1} = 0.364$$
$$y_{2,2} = 0.641$$
$$y_{3,3} = 0.0298$$

A dew point calculation for this stream leads to:

$$P_1^0(T) = P$$
$$\sum (y_{1,k}/\alpha_{k/1}) = (750)(2.15)$$
$$= 1614.5 \text{ mm}$$
$$\Rightarrow T_{dew} = 379.8 \text{ K}$$

Note that because of the simplification introduced for partial condensers, this example was done *very* quickly without iteration. Here we assumed that the relative volatilities remained constant and therefore all calculations are noniterative.

Effect of Pressure on Separations. Before concluding this subsection, we note the effect of increasing pressure on the difficulty of the separation. Under an ideal assumption, we see that α is not directly affected by pressure. However, it is indirectly related because bubble point temperatures change significantly with pressure and thus lead to significant differences in relative volatilities. Therefore, as P becomes large, so do the

Sec. 3.2   Developing Unit Models for Linear Mass Balances   79

partial pressures of the overhead product as well as the overhead temperature. Moreover, for ideal systems: $\alpha_{k/n} = P_k^0/P_n^0 \to 1$ and this increases the difficulty of the separation.

EXAMPLE 3.6

To illustrate the effect of increasing column pressure, we consider the separation of a mixture of 50 mol/hr C_3H_8 (1) and 50 mol/hr C_3H_6 (2) at a pressure of 1.1 bar and a bubble point feed temperature of 230 K. Under these conditions, $P_1^0 = 930.5$ mm, $P_2^0 = 724.1$ mm and $\alpha_{1/2} = 1.285$. If we set the recoveries of these two components at $\xi_1 = 0.99$ and $\xi_2 = 0.01$, we find out that the minimum number of trays at total reflux is:

$$N_m = \ell n \left[\frac{0.99}{0.01} \cdot \frac{0.99}{0.01} \right] / \ell n \, \alpha_{1/2} = 36.65$$

Now if we increase the pressure tenfold to $P = 10.94$ bar, we have a bubble point feed temperature of 300 K and $P_1^0 = 8975.6$ mm, $P_2^0 = 7458.5$ mm and $\alpha_{1/2} = 1.203$. As a result, for the same recoveries, the separation becomes more difficult and the minimum number of trays increases to $N_m = 49.72$.

3.2.4 Gas Absorption with Plate Absorbers

As with distillation, gas absorption can be modeled approximately as a cascade of equilibrium trays. The assumption of equilibrium stages is weaker here, and as with distillation, we will seek to correct this in the next chapter through the use of tray efficiencies. In this subsection we will model two similar gas-liquid separations, absorption and stripping. Absorption represents a vapor recovery operation where a desired component is transferred from a gas to the liquid phase through countercurrent mass transfer (modeled here through a series of equilibrium stages). In the stripping operation we have the reverse situation—the desired component is transferred from the liquid to the gas phase. For both operations, we will make ideal equilibrium tray assumptions regarding absorption and stripping in order to yield split fractions and a linear mass balance quickly.

For these systems, we note that four degrees of freedom are available for specifying the mass balance, once the vapor feed stream is given. For absorption this follows, because we can specify pressure (P) on an equilibrium tray (say, the top tray) and the other pressures are related to it. We also specify the number of equilibrium trays (N) for a desired recovery of key component (or vice versa). Finally, we need to specify both the temperature (T_0) and flowrate (L_0) of the absorbing liquid stream. (For the stripping operation, two degrees of freedom must be specifed for the corresponding gas stream.)

Consider the absorption unit with the notation illustrated in Figure 3.18. Given that these four specifications are made, we can now derive the mass balance relationships.

At each equilibrium stage i, we have the arrangement shown in Figure 3.19.

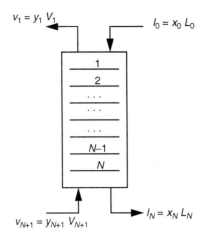

FIGURE 3.18 Absorber model.

If we drop the superscript k and assume only the molar flows for the key component we have, after rearrangement: $l_i = (L_i/K_i V_i) v_i = A_i v_i$ and we define $(L_i/K_i V_i)$ as an absorption factor, A_i, for a given stage.

Next, we form the mass balance between stages, starting from the top of the absorber with the relations:

$$\ell_1 + v_1 = \ell_0 + v_2$$

or

$$(A_1 + 1) v_1 = \ell_0 + v_2$$
$$v_2 = (A_1 + 1) v_1 - \ell_0$$

For each stage i we also have:

$$v_{i+1} = l_i + v_i - l_{i-1}$$
$$v_{i+1} = (A_i + 1) v_i - A_{i-1} v_{i-1}$$

So by induction, we have:

$$v_3 = (A_2 + 1) v_2 - A_1 v_1 \quad \text{(and substituting } v_2\text{)}$$
$$= (A_2 + 1)(A_1 + 1) v_1 - (A_2 + 1) \ell_0 - A_1 v_1$$
$$= (A_2 A_1 + A_2 + 1) v_1 - (A_2 + 1) \ell_0$$
$$v_4 = (A_3 + 1) v_3 - A_2 v_2 \quad \text{(substituting } v_3 \text{ and } v_2\text{)}$$
$$= (A_3 + 1)(A_2 A_1 + A_2 + 1) v_1 - (A_3 + 1)(A_2 + 1) \ell_0$$
$$- A_2(A_1 + 1) v_1 - A_2 \ell_0$$

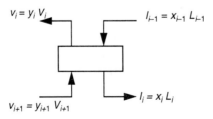

FIGURE 3.19 Absorber equilibrium stage.

Sec. 3.2 Developing Unit Models for Linear Mass Balances

$$= [A_3 A_2 A_1 + A_3 A_2 + A_3 + 1] v_1$$
$$- [A_3 A_2 + A_3 + 1] l_0$$

and we end up with:

$$v_{N+1} = [1 + A_N + A_N A_{N-1} + A_N A_{N-1} A_{N-2} + \ldots$$
$$A_N A_{N-1} A_{N-2} \ldots A_1] v_1$$
$$- [1 + A_N + A_N A_{N-1} + \ldots + A_N A_{N-1} \ldots A_2] l_0$$

To simplify these expressions we make two assumptions:

1. Define an effective constant absorption factor, A_E, that remains constant for all stages. This leaves:

$$v_{N+1} = \sum_{i=0}^{N} (A_E)^i v_1 - \sum_{i=0}^{N-1} (A_E)^i l_0$$

2. Define

$$\beta_N = \sum_{i=0}^{N} (A_E)^i$$

$$(1 - A_E)\beta_N = \sum_{i=0}^{N} (A_E)^i - \sum_{i=1}^{N+1} (A_E)^i$$

$$\beta_N = \frac{1 - A_E^{N+1}}{1 - A_E}$$

which simplifies the previous relationship for v_{N+1} to:

$$v_{N+1} = \beta_N v_1 - \beta_{N-1} l_0$$

and ℓ_N can be obtained by overall mass balance:

$$\ell_N = v_{N+1} + l_0 - v_1$$

The overall A_E can be defined for two stages by the following mass balance:

$$v_3 = (A_E^2 + A_E + 1) v_1 - (A_E + 1) l_0$$
$$= (A_2 A_1 + A_2 + 1) v_1 - (A_2 + 1) l_0$$

From the quadratic formula, if we knew A_2 and A_1, where A_1 could represent the absorption factor at the column top and A_2 is evaluated at bottom of an N stage absorber, we can define an effective factor by the Edmister formula (Edmister, 1943):

$$A_E = (A_2(1 + A_1) + 1/4)^{1/2} - 1/2$$

Finally, we can define a recovery fraction, r, for the key component (n) and from the mass balance equations we can calculate the number of trays. Here, we have:

$$v_1^n = (1 - r) v_{N+1}^n$$

and
$$v_{N+1}^n = \beta_N(1-r)v_{N+1}^n - \beta_{N-1}\ell_0^n$$
which can be rewritten and rearranged as:
$$v_{N+1}^n = \frac{1-A_E^{N+1}}{1-A_E}(1-r)v_{N+1}^n - \frac{1-A_E^N}{1-A_E}\ell_0^n$$

$$(1-A_E)v_{N+1}^n = (1-A_E)^{N+1}(1-r)v_{N+1}^n - (1-A_E)^N \ell_0^n$$

$$\ell_0^n + (r - A_E)v_{N+1}^n = A_E^N[\ell_0^n - A_E(1-r)v_{N+1}^n]$$

$$N = \ell n\left\{\frac{\ell_0^n + (r-A_E)v_{N+1}^n}{\ell_0^n - A_E(1-r)v_{N+1}^n}\right\}/\ell n\{A_E\}$$

This relation is known as the Kremser equation (Kremser, 1930) and gives us a simple design method based on the recovery of a key component. Note that if none of the key component appears in the liquid feed stream, then the above equation simplifies as we have $l_0 = 0$ and:

$$N = \ell n[(r-A_E)/A_E(r-1)]/\ell n\{A_E\}$$

Now to choose the four degrees of freedom that allow the calculation of a mass balance, we specify: (1) r, the recovery of the key component n; (2) overhead column pressure; (3) solvent temperature (For our approximations, we will assume that the absorber operates isothermally at this temperature.); (4) the absorption factor, A_E at 1.4 as a guideline (Douglas, 1988; p. 427) for specifying the "optimum" liquid flowrate. With these specifications, the split fractions for the linear mass balance are calculated from the following algorithm.

Absorption Algorithm

1. Select key component n, fix recovery (typically, $r = 0.99$) fix P and solvent temperature.

2. Calculate L_0 from
$$A_E = \frac{L_0}{V_{N+1} K_n} = 1.4$$
$$L_0 = 1.4 V_{N+1} \frac{P_N^0(T)}{P}$$

Note from this expression that L_0 decreases with increasing pressure and decreasing temperature.

3. **a.** Calculate the number of stages from the Kremser equation:
$$N = \ell n\left(\frac{rv_{N+1}^n + \ell_0^n - A_E v_{N+1}^n}{\ell_0^n - A_E(1-r)v_{N+1}^n}\right)/\ell n\{A_E\}$$

Sec. 3.2 Developing Unit Models for Linear Mass Balances

(Note that if $r = 0.99$ and $\ell_0^n = 0$ then $N = 10$)

b. Prepare the mass balance by calculating absorption factors and aggregate terms for all of the remaining components by:

$$A^k = \frac{L_0}{V_{N+1}} \frac{P}{P_k^0(T)} \quad k \neq n$$

or

$$A^k = \frac{1.4}{\alpha_{k/n}}$$

and for β_N^k, β_{N-1}^k with $\quad \beta_N^k = [1 - (A^k)^{N+1}]/(1 - A^k)$

4. Complete the mass balance for all components:

$$v_1^k = \frac{v_{N+1}^k}{\beta_N^k} + \frac{\beta_{N-1}^k}{\beta_N^k} \ell_0^k$$

$$\ell_N^k = \left(1 - \frac{\beta_{N-1}^k}{\beta_N^k}\right) \ell_0^k + \left(1 - \frac{1}{\beta_N^k}\right) v_{N+1}^k$$

5. If necessary, readjust P or T and return to step 1 under the following conditions

a. If the temperature of ℓ_N is too high (check with the bubble point equation), increase L_0. If the final design has significant temperature changes between the top and bottom of the column, use an effective absorption factor calculated with the Edmister equation.

b. If too much solvent vaporizes in v_1, increase P or decrease T.

c. If too many undesirable components are absorbed, increase T, decrease P, or select a more suitable solvent for absorption.

EXAMPLE 3.6 Absorption

Consider the absorption problem with the specifications given in Figure 3.20:

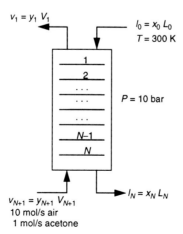

FIGURE 3.20 Absorption example.

With a solvent (water) temperature of 300 K and pressure of 10 bar, we also choose a recovery of acetone at $r = 0.95$. Setting the absorption factor, $A_E = 1.4$, we can calculate the required water flowrate:

$$L_0 = 1.4\, V_{N+1} K_n(T) = 1.4\,(11)\frac{\left(P^0_{Ac}(300)\right)}{10} = 0.51 \text{ mol/sec}$$

Also, the number of equilibrium stages can be calculated from the Kremser equation:

$$N = \ln\left\{\frac{r - A_E}{(r-1)A_E}\right\} / \ln(A_E) = 5.53$$

Now to complete the mass balance, we know the recovery of acetone and because air is noncondensible, $A_{air} \sim 0$ and $\beta^{air}_{N-1} = \beta^{air}_N = 1$, and the flowrates for air are known as well. To estimate the mass balance for the entrained water, we have:

$$\alpha_{W/Ac} = P^0_W(300)/P^0_{Ac}(300) = 0.106 \qquad A_W = 1.4\,/\,\alpha_{W/Ac} = 13.24$$

$$\beta^W_{N-1} = 1.307 \cdot 10^5 \qquad\qquad \beta^W_N = 1.73 \cdot 10^6$$

Substituting these values into the mass balance equations yields the following flowrates and mole fractions for the exiting streams.

	v_1	ℓ_N
	10 mol/sec Air	0.0 mol/s Air
	0.05 mol/s Ac	0.95 mol/s Ac
	0.038 mol/s W	0.472 mol/s W
	$y_{Ac} = 0.005$	$x_{Ac} = 0.668$
	$y_{Air} = 0.991$	$x_{Air} = 0.0$
	$y_W = 0.004$	$x_W = 0.332$

STRIPPER MODEL: A SIMPLE REFORMULATION

We conclude this subsection with a simple derivation of the stripper model. The stripper can be viewed as an "absorber in reverse" as shown in Figure 3.21.

Again, the same equilibrium relations hold on each stage:

$$K^k_i \frac{\ell^k_i}{L_i} = \frac{v^k_i}{V_i}$$

and we can relate the vapor flowrate to the liquid flowrate through a stripping coefficient, $S_i = 1/A_i$.

$$v^k_i = \left(\frac{V_i K^k_i}{L^k_i}\right)\ell^k_i = S_i\,\ell_i$$

Sec. 3.3 Linear Mass Balances

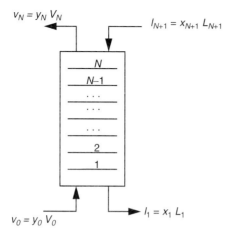

FIGURE 3.21 Stripper model.

In the stripping operation we choose a key component n in the liquid feed with a specified recovery r. If we now reconsider the derivation for the absorber and replace A_i with S_i and v_i with ℓ_i, we can derive the analogous Kremser equation for the stripping unit.

$$N = \ell n\left[\frac{r\ell_{N+1} + v_0 - S_E \ell_{N+1}}{v_0 - S_E(1-r)\ell_{N+1}}\right] / \ell n\, \{S_E\}$$

As with the absorber we specify an effective stripping factor, $S_E = 1.4$. The vapor stream is then given by:

$$V = 1.4\, L_i^k\, \frac{P}{P_n^0(T)}$$

and we calculate the mass balance using the same algorithm as for the absorber. Again, for $r = 0.99$ and $S_E = 1.4$, we have a stripper with ten theoretical trays. Also, from the above relation we see that running the stripper at lower pressure or higher temperature will also minimize the molar vapor flow for a specified recovery.

3.3 LINEAR MASS BALANCES

In the previous section, we developed split fraction models for a wide variety of "mass balance" units (i.e., separators, mixers, and reactors). In this section we further develop and combine this information in order to analyze the ethanol process shown below. Therefore, in this section we also follow the algorithm presented below.

Linear Mass Balance Algorithm

1. Guess P and T levels in the flowsheet. Specify recoveries, split fractions, and so on (use degrees of freedom for each unit).
2. Determine coefficients for linear models in each unit ($\alpha_{k/n}$, β, N_m, ξ).

86 Mass and Energy Balances Chap. 3

3. Set up linear equations and solve for flowrates of each component.
4. Check guessed values from step 1.
   a. Calculate P and T from flowrates. If different from step 1, go to step 2.
   b. If flowsheet does not meet specs, change T, P, or modify flowsheet.

3.3.1 Using the Linear Mass Balance Algorithm

This information now allow us to establish heat balances, cooling and heating duties, and opportunities for heat integration. First, we consider the ethanol flowsheet from Chapter 2 and create the following block diagram (Figure 3.22) for the mass balance. Note that units such as pumps, compressors (i.e., pressure "changers"), and heat exchangers (temperature "changers") have been removed because they do not affect the mass balance. Now let's march around the flowsheet and consider each unit in the flowsheet separately. Here, we will establish the split fractions following the methods presented in the previous section.

As a basis, we choose 100 mol/sec for μ_{02} (ethylene feed). The components for the flowsheet (methane, ethylene, propylene, diethyl ether, ethanol, isopropanol, and water) are represented with the index set: $k = M, EL, PL, DEE, EA, IPA, W$. Also, since only a small amount of crotonaldehyde is produced and it is removed in μ_{92} as the heaviest component, we will neglect this component in the mass balance. We start with linear equations for the units shown in Figure 3.22.

1. **Mixer**
$$\mu_{01} + \mu_{02} + \mu_{51} + \mu_{81} = \mu_1$$

2. **Reactor** Here we have the following reactions:

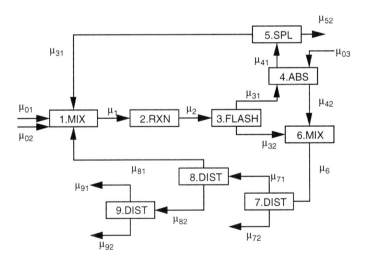

FIGURE 3.22 Flowsheet representation for linear mass balance.

Sec. 3.3 Linear Mass Balances

$$EL + W \rightarrow EA$$
$$PL + W \rightarrow IPA$$
$$2\ EA \leftrightarrow DEE + W$$

For the equilibrium reaction at the specified inlet temperature (590 K) and pressure (69 bars), we can maintain an equilibrium level of diethyl ether in the recycle loop according to the following expression:

$$(DEE)(W)/(EA)^2 = 0.2$$

The remaining reactions consist of the following fixed conversions with limiting reactants, *EL* and *PL*, respectively:

$$7\% \text{ conversion/pass } EL \text{ to } EA\ (\eta_1)$$
$$0.7\% \text{ conversion/pass } PL \text{ to } IPA\ (\eta_2)$$

The mass balance for the reactor can be written as:

$\mu_2^M = \mu_1^M$ (inert component)
$\mu_2^{EL} = (1 - \eta_1)\mu_1^{EL}$ (limiting component, first reaction)
$\mu_2^{PL} = (1 - \eta_2)\mu_1^{EL}$ (limiting component, second reaction)
$\mu_2^{DEE} = 0.2\,(\mu_2^{EA2}/\mu_2^W)$ (equilibrium condition)

and solved for the remaining components:

$$\mu_2^{EA} = \eta_1 \mu_1^{EL} + \mu_1^{EA}$$
$$\mu_2^{IPA} = \eta_2 \mu_1^{PL} + \mu_1^{IPA}$$
$$\mu_2^W = \mu_1^W - \eta_1 \mu_1^{EL} - \eta_2 \mu_1^{PL}$$

From Chapter 2 we have $\mu_1^W = 0.6\,\mu_1^{EL}$. (Note that the limiting component in the first reaction is actually the water. However, since the conversion of *EL* is very low and because *W* participates in multiple reactions, we choose *EL* as the key component to make the calculations easier.)

3. **Flash Unit** Here we want to take the reactor effluent to cooling water temperature and separate the liquid product from reactant gases. We assume a pressure drop of 0.5 bar from the reactor and operate the flash unit at 68.5 bar. Given the component list, we choose *DEE* as the intermediate key component and examine the relative volatilities of the component list at cooling water temperature.

comp, k	M	EL	PL	DEE	EA	IPA	W
$P^0(T=310)$	211000 mm Hg	55500	11360	824	114.5	75.1	47.1
$\alpha_{k/DEE}$	256.1	67.3	13.8	1.0	0.138	0.091	0.057
ξ_k	0.996	0.985	0.932	0.5	0.121	0.083	0.054

At this point, however, we don't know the feed component flows, so we need to assume that $\xi_n = 0.5$ for DEE and calculate the other split fractions from:

$$\xi_k = \frac{\alpha_{k/n}\xi_n}{1 + (\alpha_{k/n} - 1)\xi_n}$$

These split fractions are also given in the above table and we are now able to write the following linear mass balances:

$$\mu_{31}^M = \xi_M \mu_2^M = 0.996 \mu_2^M; \quad \mu_{32}^M = 0.004 \mu_2^M$$
$$\mu_{31}^{EL} = 0.985 \mu_2^{EL}; \quad \mu_{32}^{EL} = 0.015 \mu_2^{EL}$$
$$\mu_{31}^{PL} = 0.932 \mu_2^{PL}; \quad \mu_{32}^{PL} = 0.068 \mu_2^{PL}$$
$$\mu_{31}^{DEE} = 0.5 \mu_2^{PL}; \quad \mu_{32}^{DEE} = 0.5 \mu_2^M$$
$$\mu_{31}^{EA} = 0.121 \mu_2^{EA}; \quad \mu_{32}^{EA} = 0.879 \mu_2^{EA}$$
$$\mu_{31}^{IPA} = 0.083 \mu_2^{IPA}; \quad \mu_{32}^{IPA} = 0.917 \mu_2^{IPA}$$
$$\mu_{31}^W = 0.054 \mu_2^W; \quad \mu_{32}^W = 0.946 \mu_2^W$$

Note that because we *assume* a key component recovery we don't need to know feed rates. At a later point, however, when the flowrates are established, we need to check if this assumption corresponds to our desired temperature and pressure specification. Also note that for noncondensible gases (e.g., hydrogen, methane) the solubility in liquid is overestimated with ideal thermodynamics.

4. **Absorber** The mass balance model for the absorber has four degrees of freedom: P, T, key component recovery, and liquid rate. Here, we choose the liquid rate by using the *heuristic* that $A = L_0/V_{N+1}K_{EA} = 1.4$ and we also want to run the absorber at *low* temperature and *high* pressure. (Why?) So we choose $P = 68$ bar (again assume 0.5 bar pressure drop from the flash unit) and $T = 310$ K (cooling water). Our valuable component is the ethanol product so with a 99% recovery into the liquid phase, we have: $\xi_n = 0.99$, $n = EA$. Using our heuristic, the water flowrate is:

$$K_{EA} = P_n^0(310)/P = 2.25 \cdot 10^{-3}$$
$$L_0 = (V_{N+1} K_{EA})(1.4) = 3.15 \cdot 10^{-3} \mu_{31}$$

Because this is a very small liquid stream, we need to see how much water we lose in the overhead vapor and if this evaporation is acceptable. For $\xi_n = 0.99$, $A_{EA} = 1.4$, the number of equilibrium stages for the absorber is:

$$N = \ell n \left\{ \frac{r \, v_{N+1}^{EA} + l_0^{EA} - A v_{N+1}^{EA}}{l_0^{EA} - A_{EA}(1-r) v_{N+1}^{EA}} \right\} / \ell n \, A_{EA} = 10$$

Using this to determine the split fractions for the other components leads to:

$$A_k = \frac{1.4}{\alpha_{k/EA}} = \frac{L}{VK_k}; \quad \beta_N^k = \frac{1-(A_k)^{N+1}}{1-A_k} \quad \text{and} \quad \beta_{N-1}^k = \frac{1-(A_k)^N}{1-A_k}$$

Now we have:

$$v_1^k = \frac{v_{N+1}^k}{\beta_N^k} + \frac{\beta_{N-1}^k}{\beta_N^k} \ell_0^k$$

$$\ell_N^k = \left(1 - \frac{\beta_{N-1}^k}{\beta_N^k}\right) \ell_0^k + \left(1 - \frac{1}{\beta_N^k}\right) v_{N+1}^k$$

Sec. 3.3 Linear Mass Balances

To complete the mass balance, split fractions need to be calculated and we also need to consider the vaporization of the solvent. At $T = 310$:

$$\alpha_{W/EA} = 47.1/114.5 = 0.41 \qquad A^W = 1.4/\alpha_{W/EA} = 3.415$$
$$\beta_N^W = 3.05 \cdot 10^5 \qquad \beta_{N-1}^W = 8.93 \cdot 10^4$$

From the mass balance equations we see that $\beta_{N-1}^W/\beta_N^W = 0.293$, which is the fraction of solvent lost in the overhead vapor. Because this large fraction is likely to violate our assumption of isothermal operation, we need to reconsider our operating parameters. To improve operation we can further increase P or decrease T, but these are already at their respective limits without incurring additional capital cost (compression or refrigeration). Instead, we can operate close to isothermal conditions by increasing the solvent rate. Here we increase the effective absorption factor to 10, say, and obtain at $P = 68$ bar and $T = 310$:

$$A_{EA} = 10 = \frac{L}{VK_{EA}} \quad \text{and} \quad L_0 = 0.0225 \, \mu_{31}$$

and

$$N = \ell n\left\{\frac{r - A_{EA}}{-A_{EA}(1-r)}\right\} / \ell n \, A_{EA} = 1.95 \text{ stages}$$

Solving for the solvent split fractions yields:

$$A_W = \frac{10}{\alpha_{W/EA}} = 24.39 \qquad \begin{array}{l} \beta_N^W = 528.7 \\ \beta_{N-1}^W = 21.68 \end{array}$$

and the loss of water in the overhead vapor is $\beta_{N-1}^W/\beta_N^W = 0.041$, which is now acceptable for isothermal operation. (Note that by increasing the solvent flowrate in this ideal calculation, we do not change the *amount* of water vaporized in the overhead stream. Only the *fraction* vaporized is changed so that the absorber operates close to the inlet water temperature.)

We are now ready to calculate the remaining ξ_k in the vapor and liquid streams.

Comp	$\alpha_{k/n}$	A_k	β_N	β_{N-1}	ξ_{41}	ξ_{42}
M	1854	$5.4 \, 10^{-3}$	1	1	1.0	0
EL	486.3	0.021	1.021	1.021	0.979	0.021
PL	99.5	0.101	1.11	1.10	0.901	0.099
DEE	7.24	1.38	4.17	2.30	0.24	0.76
EA	1.0	10	98.92	9.79	0.01	0.99
IPA	0.79	12.66	153.2	12.02	$6.5 \cdot 10^{-3}$	0.993
W	0.41	24.4	529.1	21.6	$1.9 \cdot 10^{-3}$	0.998

For water we have

$$\mu^W_{41} = \xi_{41}\mu^W_{31} + \beta^W_{N-1}/\beta^W_N \mu^W_{03} = \xi_{41}\mu^W_{31} + \beta^W_{N-1}/\beta^W_N$$

$$(A_{EA}K_{EA})\mu_{31}$$

$$= 0.0019\,\mu^W_{31} + 0.041\,\mu^W_{03} = 0.0019\,\mu^W_{31} + 0.00092\,\mu_{31}$$

$$\mu^W_{42} = \xi_{42}\mu^W_{31} + (1 - \beta^W_{N-1}/\beta^W_N)(A_{EA}K_{EA})\mu_{31} = 0.998\,\mu^W_{31} + 0.999\,\mu_{31}$$

and for the remaining components, we have:

$$\mu^k_{41} = \xi^k_{41}\mu^k_{31} \text{ and } \mu^k_{42} = \xi^k_{42}\mu^k_{31}$$

5. **Splitter** For this unit, we need to specify the purge rate ξ, for the recycle stream. The function of the purge stream is to avoid an accumulation of inert components and impurities. For this process, we determine the purge rate by enforcing a constraint that the mole fraction of methane in the recycle be less than 10%. From the mass balance we have:

$$\mu_{52} = \xi\mu_{41}$$
$$\mu_{51} = (1 - \xi)\mu_{41}$$

To find ξ we need to enforce the methane constraint and perform a rough estimate of a mass balance around the recycle loop from the following approach.

Assume *EA, IPA,* and *DEE* are negligible in the recycle, as the first two are products to be separated and the last is in a small amount at equilibrium. Now to calculate the mole fraction of methane with the remaining components, $\mu^M_1/(\mu^M_1 + \mu^{PL}_1 + \mu^{EL}_1 + \mu^W_1)$, we need to estimate the flowrates of ethylene, propylene, methane, and water, we write the following equations:

$$EL: \quad \mu^{EL}_1 = \mu^{EL}_2(1 - \xi) + 96$$
$$= 0.93\,\mu_1(EL)(1 - \xi) + 96$$
$$= 96/(0.07 + 0.93\xi)$$

$$PL: \quad \mu^{PL}_1 = \mu^{PL}_2(1 - \xi) + 3$$
$$= 0.993\,\mu^{PL}_1(1 - \xi) + 3$$
$$= 3/(0.007 + 0.993\xi)$$

$$M: \quad \mu^M_1 = (1 - \xi)\mu^M_1 + 1 = 1/\xi$$

$$W: \quad \mu^W_1 = 0.6\,\mu^W_1 = 57.6/(0.07 + 0.93\xi) \text{ (approximate estimate)}$$

Substituting the flowrates into the methane constraint:

$$\mu^M_1/(\mu^M_1 + \mu^{PL}_1 + \mu^{EL}_1 + \mu^W_1) = 0.1$$

yields the equation:

$$[153.6/(0.07 + 0.93\xi) + 3/(0.007 + 0.993\xi) + 1/\xi] = 10/\xi$$

which can be solved by trial and error to get $\xi = 0.0038$. Since the methane mole fraction should be less than 10%, choose a larger purge fraction, $\xi = 0.005$ and:

$$\mu_{52} = 0.005\,\mu_{41}$$
$$\mu_{51} = 0.995\,\mu_{41}$$

Sec. 3.3 Linear Mass Balances

6. **Mixer** Split fractions are easily determined for this unit from:

$$\mu_{42}^k + \mu_{32}^k = \mu_6^k$$

7. **Dewatering Distillation** The purpose of this unit is to remove 90% of the water from downstream separations. We operate this column at low pressure since the lightest component in large amounts is *DEE*. Here we would like to recover 99.5% of the *EA* overhead; thus, we have split fractions for the key components, *EA* and *W*, as $\xi_{EA} = 0.995$ and $\xi_W = 0.1$. Components *M, EL, PL*, and *DEE* are lighter than the light key, and the remaining component, *IPA*, is distributed between *EA* and *W*. Also, we would like to run this column with cooling water (at $T = 310$ K), so a partial condenser may be needed for trace lowboiling components *M, EL*, and *PL*. To perform this separation, we have $\alpha_{EA/W} = 2.44$, and from the Fenske equation:

$$N_m = \ell n\ [(0.995)(0.9)/(0.005)(0.10)]/\ell n(2.44) = 8.4 \text{ trays}$$

The distributed component *IPA* has its split fraction is calculated from $\alpha_{IPA/W} = 1.93$ and from the rearrangement of the Fenske equation:

$$\xi_{IPA} = \frac{\alpha_{IPA/W}^{N_m} \xi_W}{1 + \left(\alpha_{IPA/W}^{N_m} - 1\right)\xi_W} = 0.96$$

This leads to the following component split fractions for the column mass balance equations:

Components	M	EL	PL	DEE	EA	IPA	W
ξ_k	1.0	1.0	1.0	1.0	0.995	0.96	0.1

$$\mu_{71}^k = \xi_k \mu_6^k \text{ and } \mu_{72}^k = (1-\xi_k)\mu_6^k$$

8. **De-ethering Column** In this column, diethyl ether from the ethanol-rich stream is removed overhead and returned to the recycle loop. Here we simply specify a tight specification for recoveries (99.5%) between adjacent components, *EA* and *DEE*, and the resulting split fractions and mass balance equations become:

Components	M	EL	PL	DEE	EA	IPA	W
ξ_k	1.0	1.0	1.0	0.995	0.005	0.0	0.0

$$\mu_{81}^k = \xi_k \mu_{71}^k \text{ and } \mu_{82}^k = (1-\xi_k)\mu_{71}^k$$

9. **Final Azeotropic Separation** This last column is used to obtain ethanol product at the azeotrope composition (85.4% *EA*, 14.5% *W*). We treat this azeotrope and specify a recovery of $\xi_{az} = 0.995$. In addition, there is a further constraint that the product contain no more than 0.1 mol% *IPA* (the adjacent heavy key). However, in order to specify a recovery for IPA, we need to know the incoming flowrates first.

3.3.2 Solving Linear Mass Balance Equations

Now that we have split fractions for each component and each unit we are in a position to write the overall mass balance. If we consider the recycle part of the flowsheet in Figure 3.23, we have two recycles (10 streams, 7 components; with T and P this leads to 90 equations). To solve, however, we know:

1. All units *except the reactor* have independent split fractions for each component (they relate inlet and outlet flows of each component *separately*). Here there is no interaction among components.
2. The reactor mass balance relates component flows to limiting components in reaction.

Therefore, for the recycle mass balance, we consider the limiting components first. We could write all equations for *EL* (with superscript suppressed) and then solve:

$$\mu_1 = \mu_{81} + \mu_{01} + \mu_{51}$$
$$\mu_2 = 0.93\,\mu_1$$
$$\mu_{31} = 0.985\,\mu_2$$
$$\mu_{32} = 0.015\,\mu_2$$
$$\mu_{41} = 0.979\,\mu_{31},\ \mu_{42} = 0.021\,\mu_{31}$$
$$\mu_{51} = 0.995\,\mu_{41}$$
$$\mu_{52} = 0.005\,\mu_{41}$$
$$\mu_6 = \mu_{32} + \mu_{42}$$
$$\mu_{81} = \mu_{71}$$

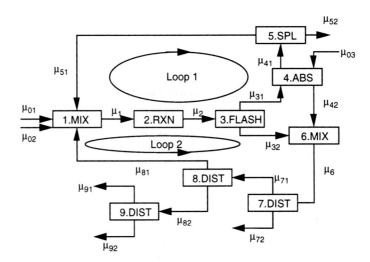

FIGURE 3.23 Recycle loops for mass balance.

Sec. 3.3 Linear Mass Balances

But because of the above two properties, following the *tearing algorithm* given below gives a much easier method.

1. Choose tear streams that break all recycle loops in flowsheet (typically the reactor inlet).
2. Trace path backwards from reactor inlet until all loops are covered (end up at reactor inlet again).
3. Fill all streams by using split fractions and moving forward from the reactor feed.

To illustrate this, we start with the reactor inlet as the tear stream and write the loop equations for the two limiting components:

Trace path for *EL* along both recycle loops

$$\mu_1^{EL} = \mu_{01}^{EL} + \mu_{51}^{EL} + \mu_{81}^{EL}$$
$$\mu_1^{EL} = \mu_{01}^{EL} + (0.995)(.979)(0.985)(0.93)\,\mu_1^{EL}$$
$$+(1)(1)(0.021(0.985) + 0.015)(0.93)\,\mu_1^{EL}$$
$$\mu_1^{EL} = 96 + 0.9255\,\mu_1^{EL} \rightarrow \mu_1^{EL} = 1289 \text{ gmol/s}$$

Trace path for *PL* along both recycle loops

$$\mu_1^{PL} = \mu_{01}^{PL} + \mu_{51}^{PL} + \mu_{81}^{PL}$$
$$= 3 + (0.995)(0.901)(0.932)(0.993)\,\mu_1^{PL}$$
$$+ (1)(1)(0.099(0.932) + 0.068)(0.993)\,\mu_1^{PL}$$
$$\mu_1^{PL} = 268.6 \text{ gmol/sec.}$$

Once we have the reactor inlet flowrates, we can recover the other component flows at the reactor inlet as well. For *EA*, for example, we trace a path along both recycle loops:

$$\mu_1 = \mu_{01} + \mu_{51} + \mu_{81}$$
$$= (0.995)(0.01)(0.121)(\mu_2)$$
$$+ (0.005)(0.995)(0.879 + 0.121(0.99))\,\mu_2$$

and

$$\mu_2^{EA} = \mu_1^{EA} + \eta_1\,\mu_1^{EL}$$
$$= \mu_1^{EA} + 90.2$$
$$\mu_1^{EA} = 0.556/0.994 = 0.56 \text{ gmol } EA/\text{s}$$

The remaining recycle streams can be calculated simply by moving forward from the reactor and applying the known split fractions. For example, the ethanol flowrates are:

$$\mu_2 = 90.8$$
$$\mu_{31} = 10.99$$
$$\mu_{32} = 79.81$$
$$\mu_{41} = 0.11$$
$$\mu_{42} = 10.88$$
$$\mu_{51} = 0.1093$$
$$\mu_{52} = 0.0005$$

$\mu_6 = 90.68$
$\mu_{71} = 90.23$
$\mu_{72} = 0.45$
$\mu_{81} = 0.45$
$\mu_{82} = 89.77$
$\mu_{91} = 89.33$
$\mu_{92} = 0.45$

The last two streams were not part of the recycle loops and were calculated separately, once the azeotropic column feed was known. The remaining components are calculated in a similar way and the final mass balance is given in Table 3.1.

3.4 SETTING TEMPERATURE AND PRESSURE LEVELS FROM THE MASS BALANCE

Now that the mass balance has been calculated, we set the remaining temperature and pressure levels so that unit outlet streams remain at saturated liquid or vapor. Here we need to be concerned with the following questions:

- Check if the saturated stream is below the critical point.
- Is the specified recovery achieved in the flash units?
- Do distillation columns require partial or total condensers in order to allow cooling water?
- Are steam temperatures adequate to drive the reboilers in the distillation columns?

With these questions, let's now check a selection of the units in the flowsheet (Figure 3.23) to verify the mass balance specifications.

3. Flash Unit From the mass balance, we first examine the validity of the recovery for diethyl ether, $\xi_{DEE} = 0.5$. The mole fractions for the feed and effluent streams are:

	z_k	y_k	x_k	$T_c^k(K)$
M	0.08	0.1187	0.001	190.6
EL	0.491	0.7038	0.0235	282.4
PL	0.109	0.1481	0.0237	365.0
DEE	0.001	0.0007	0.0016	466.7
EA	0.037	0.0065	0.1045	516.2
IPA	0.0008	$9.3 \cdot 10^{-5}$	0.0022	508.3
W	0.279	0.0219	0.843	647.3

and from the liquid mole fractions, we have: $T_c^m = \Sigma\, x_k\, T_c^k = 616.9$ K. To determine the flash temperature, we note that at $T = 310$ K, we have $\alpha_{DEE} = 1.949$ and

TABLE 3.1 Mass and Energy Balance for Ethanol Process Flowsheet

	μ_{01}	μ_{02}	μ_1	μ_2	μ_{31}	μ_{32}	μ_{41}	μ_{42}	μ_{03}
Methane (gmol/s)	1	0	200	200	199.2	0.8	199.2	0	0
Ethylene	96	0	1289	1198.77	1180.78	17.98	1155.99	24.796	0
Propylene	3	0	268.6	266.71	248.58	18.136	223.97	24.609	0
Diethyl Ether	0	0	0	2.421	1.210	1.2108	0.2906	0.9202	0
Ethanol	0	0	0.56	90.79	10.98	79.80	0.1098	10.87	0
Isopropanol	0	0	0	1.8802	0.156	1.724	0.001018	0.1550	0
Water	0	771.797	773.4	680.72	36.75	643.97	1.610	72.896	37.747
Total	100	771.797	2531.56	2441.31	1677.68	763.62	1581.177	134.25	37.747
Temperature, K	300	300	590	590	393	393	381.57	338.7	310
Pressure, bar	1	1	69	69	68.5	68.5	68	68	68
Vap. Frac	1	0	1	1	1	0	1	0	0
Enthalpy, kcal/s	1198.85	−52097.04	−21683.63	−22689.24	115515.18	−47920.28	13439.75	−5324.42	−2544.97

	μ_{51}	μ_{52}	μ_6	μ_{71}	μ_{72}	μ_{81}	μ_{82}	μ_{91}	μ_{92}
Methane (gmol/s)	198.204	0.996	0.8	0.8	0	0.8	0	0	0
Ethylene	1150.21	5.780	42.778	42.778	0	42.7781	0	0	0
Propylene	222.85	1.1198	42.746	42.746	0	42.7466	0	0	0
Diethyl Ether	0.2891	0.00145	2.131	2.131	0	2.1205	0.01065	0.01065	0
Ethanol	0.1093	0.000549	90.680	90.226	0.4534	0.451	89.775	89.3267	0.4489
Isopropanol	0.001013	5.09323E-06	1.879	1.804	0.075	0	1.804	0.1046	1.6994
Water	1.6024	0.00805	716.867	71.68	645.18	0	71.686	15.1490	56.537
Total	1573.27	7.9058	897.882	252.173	645.70	88.896	163.277	104.591	58.686
Temperature, K	381.57	381.57	372	310	480	310	418	350	383
Pressure, bar	67.5	67.5	68	17.56	18.06	10.7	11.2	1	1.5
Vap. Frac	1	1	0	0	0	1	0	0	0
Enthalpy, kcal/s	13372.55	67.197	−53244.70	−10436.14	−42629.37	590.10	−10576.78	−6787.79	−3930.30

$\alpha_{W/DEE} = 0.057$. Basing the flash calculation on the most abundant component (W) leads to: $P_W^0(T) = P\,\alpha_{W/DEE}/\bar{\alpha} = 1502$ mm, which corresponds to a temperature of 393 K. This is acceptable, because the temperature lies between the critical estimate (616.9 K) and cooling water temperature (310 K).

4. **Absorber** Again, we check that the operation is below the critical temperature from the liquid stream composition. This leads to an estimate of $T_c^m = 591.1$ K. Since water is the most plentiful component, we determine the bubble point for the liquid stream from the bubble point equation: $P_k^0(T) = P\,\alpha_{k/n}/\bar{\alpha}_n$ with $k = W$ (the most plentiful component) and $n = DEE$. Using the relative volatilities evaluated at $T = 310$ we have:

$$\bar{\alpha}_n = 0.223,\ \alpha_{k/n} = 0.000841\ \text{and}\ P_k^0(T) = 192\ \text{mm Hg}$$

which corresponds to a temperature of $T_{42} = 338.7$ K (below critical).

For stream 41, we evaluate the dewpoint for the vapor mixture in the table. Using the same relative volatilities at 310 K with $n = EL$ (the most plentiful component) we evaluate the dew point equation: $P_n^0(T) = P\left(\sum y_k/\alpha_{k/n}\right)$ with $P = 68$ bar. This gives us $P_{EL}^0(T) = 13736$ mm Hg, which corresponds to $T_{41} = 382$ K.

7. **De-watering Column (Pre-rectifier)** This column contains a considerable number of light components. While its main function is to remove the water from μ_6, we can consider two options, a total condenser and a partial condenser. If we assume that the condenser operates with cooling water, we choose $T_{con} = 310$ K. (Why?) For the two options we have:

   a. *Total condenser* From stream 71 and basing the calculation on $n = EA$ (the most plentiful component), we have:

   $$P = P_n^0(310)\,\bar{\alpha} = 17.56\ \text{bar}.$$

   b. *Partial condenser* To separate the light components, we assume $\xi_{DEE} = 0.05$ in the vapor. We now perform a flash calculation of μ_{71} with $T = 310$ K. This leads to the following flows in the vapor and liquid product:

Comp.	M	EL	PL	DEE	EA	IPA	W
μ_{71}	0.8	42.78	42.74	2.131	90.22	1.894	71.67
liquid	0.021	9.433	24.80	2.025	89.57	1.793	71.47
vapor	0.778	33.34	17.94	0.106	0.651	0.101	0.202

   Basing the relative volatilities at 310 K with $n = EA$ (most plentiful), we determine the bubble point of the liquid phase:

   $$P = P_n^0(310)\,\bar{\alpha} = (113.9\ \text{mm Hg})(28.93) = 4.39\ \text{bar}.$$

   Since the overhead stream must be refined further in unit 8, we choose the total condenser option since it operates at higher pressure (and consequently allows unit 8 to operate at a high pressure without additional equipment).

   c. *Reboiler* We choose a pressure drop of 0.5 bar in the column and set the reboiler pressure to 18.06 bar. From Table 3.1 we note that μ_{72} is over 99.9%

Sec. 3.4   Setting Temperature and Pressure Levels from the Mass Balance   97

water, so we know that the temperature of μ_{72} is the boiling point of water at the specified pressure, $T_{72} = 480$ K.

8. **De-ethering Column**   For this unit we separate light components from the ethanol product, and because the overhead stream returns to the (vapor) recycle loop, we choose a partial condenser with saturated vapor product. If we assume that the condenser operates with cooling water and choose $T_{con} = 310$ K, we can calculate the pressure from the dew point equation:

$$P = P_n^0(T)/(\sum y_k/\alpha_{k/n}) = (55347 \text{ mm Hg})/(6.9) = 10.7 \text{ bar}$$

where $n = EL$, the most plentiful component. Note that this pressure is below the one for unit 7.

Reboiler   Again, we choose a pressure drop of 0.5 bar in the column and set the reboiler pressure to 11.2 bar. From Table 3.1 we note that ethanol is the most plentiful component in μ_{82} and we perform a bubble point calculation at the specified pressure. Choosing $n = EA$, we have from:

$$P_n^0(T) = P/\overline{\alpha} = (11.2 \text{ bar})/(1.638) = 5128 \text{ mm Hg}$$

which corresponds to a temperature of $T_{82} = 418$ K.

9. **Finishing Column**   The last column corresponds to a simple split at 1 bar, and from Table 3.1, we see that the overhead composition is 99.9% azeotropic composition of *EA/W*. The boiling point of this mixture at 1 bar is about $T_{91} = 350$ K. Similarly, the bottom stream composition is mostly water (96%). If we perform a bubble point calculation for the bottom stream at 1.5 bar, with $n = W$, we have:

$$P_n^0(T) = P/\overline{\alpha} = (1.5 \text{ bar})/(1.037) = 1084 \text{ mm Hg}$$

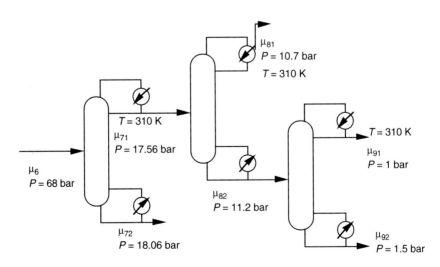

FIGURE 3.24   Column temperatures and pressures.

which corresponds to a temperature of $T_{92} = 383$ K. For all of these streams, it is easy to verify that these temperatures are below the critical temperature estimates for these mixtures. Finally, note that by selecting cooling water temperatures and appropriate choices for the condensers, we have a decreasing cascade of pressures for the distillation columns, as shown in Figure 3.24.

To summarize this section, consider the temperature and pressure values for Table 3.1. Note that stream μ_6 does not have a temperature assignment yet because it deals with the adiabatic mixing of two liquid streams. Otherwise, the assumptions of saturated liquid and vapor have been used to complete the table.

3.5 ENERGY BALANCES

Our final task for this chapter is to complete the energy balance. For most of the streams we have already specified temperatures and pressures by assuming saturated streams. We now need to evaluate the heat contents of all of the streams in order to determine heating and cooling duties for all of the heat exchangers in the flowsheet. Moreover, once these heat duties are known, we are able to consider heat integration among the process streams. This will be explored further in Chapter 10. Finally, to deliver these heat duties we must also consider the temperatures of the heat transfer media in order to size the heat exchangers and avoid crossovers. As we will see in the next chapter, heat exchangers will be sized with a 10 K temperature difference for heat exchange above ambient conditions and a 5 K temperature difference for heat exchange below ambient conditions.

As with the assumptions for the mass balance, we also assume ideal properties for evaluating the energy balance of the process streams. Moreover, we neglect kinetic and potential energies for these streams and consider only enthalpy changes. As our standard reference state for enthalpy, where $\Delta H = 0$, we consider $P_0 = 1$ atm, $T_0 = 298$ K, and elemental species. Moreover, for these preliminary calculations, we assume no ΔH of mixing or pressure effect on ΔH. We are now ready to consider the enthalpy changes for several cases.

3.5.1 Enthalpies for Vapor Mixtures

To calculate enthalpies of vapor phase mixtures we consider the evolution of enthalpy changes given in Figure 3.25.

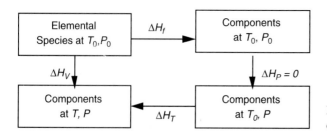

FIGURE 3.25 Evolution of enthalpy changes.

Sec. 3.5 Energy Balances

Here we define ΔH_V as the desired enthalpy change from our standard state. This can be represented by the heat of formation of the components (ΔH_f) and the enthalpy associated with temperature change (ΔH_T). As seen in Figure 3.25, pressure changes do not lead to enthalpy changes under the ideal assumption. Here the general formula for gas mixture specific enthalpy is:

$$\Delta H_v(T,y) = \Delta H_f + \Delta H_T = \Sigma_k y_k H_{f,k}(T_1) + \sum_k y_k \int_{T_1}^{T_2} C_{p,k}^0(T)dT$$

where $H_{f,k}(T_1)$ is the heat of formation for component k at T_1 and temperature dependent heat capacities for component k are represented by $C_{p,k}^0(T)$. Two representative cases for the enthalpy balance are given below.

HEAT EXCHANGER—TEMPERATURE CHANGE, NO COMPOSITION CHANGE (FIGURE 3.26)

Using the expression for vapor enthalpy, the energy balance can be made by ignoring heats of formation, as these cancel. The heat duty for the heat exchanger can be calculated from:

$$(\mu \Delta H)_{in} + Q = (\mu \Delta H)_{out} \text{ and } Q = \mu \left(\sum_k y_k \int_{T_1}^{T_2} C_{p,k}^0(T)dT \right)$$

GAS PHASE CHANGE DUE TO REACTION (FIGURE 3.27)

Here we define $Q_R = \mu_2 \Delta H_v(T,y_2) - \mu_1 \Delta H_v(T,y_1)$ and adopt the convention that if heat is added, $Q_R > 0$ and the reaction is endothermic. Otherwise, if heat is removed, $Q_R < 0$ and the reaction is exothermic. Note that the heat of reaction is automatically included because:

$$\Delta H_v^k = H_{f,k}^0 + \int_{T_0}^{T} C_{p,k}^0(T)dT$$

This approach only requires μ_1 and μ_2 and not the specific reactions in the unit.

3.5.2 Enthalpies for Liquid Mixtures

Enthalpies for liquid mixtures are evaluated directly from the ideal vapor enthalpy and subtracting the heat of vaporization at the saturation conditions. Figure 3.28 describes the

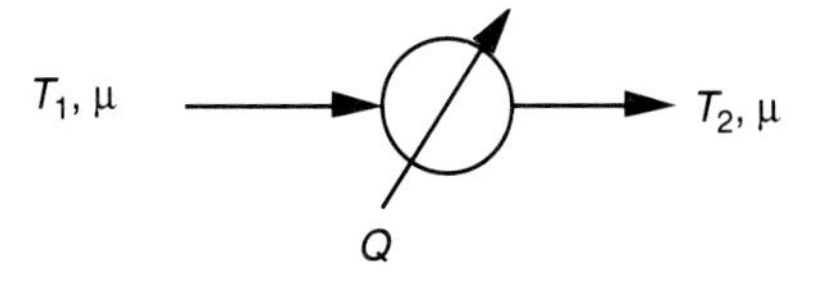

FIGURE 3.26 Heat exchanger.

FIGURE 3.27 Heat of reaction.

calculation of ΔH_L starting from standard conditions. This quantity can be defined for each component k by:

$$\Delta H_L^k(T) = \Delta H_{f,k}^0 + \int_{T_0}^{T} C_{p,k}^0(T)dT - \Delta H_{vap}^k$$

Note here that we do not need liquid heat capacities, but we do need $\Delta H_{vap}^k(T)$. The dependence on temperature can be found through the Watson correlation (Figure 3.28):

$$\Delta H_{vap}^k(T) = \Delta H_{vap}^k(T_b) [(T_c^k - T)/(T_c^k - T_b)]^\eta$$

where T_c^k is the critical temperature, T_b^k is the atmospheric boiling point for component k, and $\Delta H_{vap}^k(T_b^k)$ is the known heat of vaporization at this temperature. In the absence of other information the exponent η can be estimated at 0.38. With this correlation, we have a monotonic decrease of $\Delta H_{vap}^k(T)$ with increasing temperature, and $\Delta H_{vap}^k(T_c^k) = 0$ at the critical point.

Therefore, for liquid mixtures the specific stream enthalpy is estimated by:

$$\Delta H_L(T,x) = \sum_k x_k \left(H_{f,k}^0 + \int_{T_0}^{T} C_{p,k}^0(\tau) d\tau - \Delta H_{vap}^k(T) \right)$$

and for a two-phase mixture with vapor fraction, ϕ, we have the specific enthalpy:

$$\Delta H(T,z) = \phi \, \Delta H_V(T,y) + (1 - \phi) \, \Delta H_L(T,x)$$

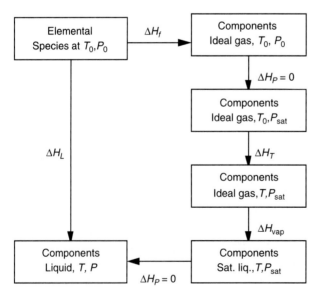

FIGURE 3.28 Enthalpy of liquids.

Sec. 3.5 Energy Balances

To illustrate these concepts we return to the mass balance of the previous sections and consider the enthalpies around some key units. The next examples show how to complete the mass balance for the ethanol process.

EXAMPLE 3.7 Evaluate the total enthalpy of the liquid stream μ_{42} exiting the absorber.

From Table 3.1 stream 42 has a temperature of 381.7, $P = 68$ bar, and the molar flowrates are:

Comp.	M	EL	PL	DEE	EA	IPA	W
μ_{42}	0.	24.80	24.61	0.920	10.87	0.155	72.90

From the liquid enthalpy equation, the total enthalpy is given by:

$$\Delta H_L(T,x) = \mu_{42} \sum_k x_k \left(\Delta H_{f,k}^0 + \int_{T_0}^T C_{pk}^0(\tau) \, d\tau - \Delta H_{vap}^k(T) \right)$$

or equivalently:

$$\Delta H_L(T,x) = \sum_k \mu_{42}^k \left(\Delta H_{f,k}^0 + \int_{T_0}^T C_{pk}^0(\tau) \, d\tau - \Delta H_{vap}^k(T) \right)$$

where

$$\int_{T_0}^T C_{pk}^0(\tau) \, d\tau = A_k(T - T_0) + B_k(T^2 - T_0^2)/2 + C_k(T^3 - T_0^3)/3 + D_k(T^4 - T_0^4)/4$$

$$\Delta H_{vap}^k(T) = \Delta H_{vap}^k(T_b) \left[(T_c^k - T)/(T_c^k - T_b) \right]^{0.38}$$

and the heat capacity coefficients A_k, B_k, C_k, and D_k, as well as $\Delta H_{f,k}^0$, $\Delta H_{vap}^k(T_b)$, T_c^k, and T_b^k can be obtained from handbook values (e.g., Reid et al., 1987), as shown in Table 3.2. Choosing a reference temperature of $T_0 = 298$ K and evaluating the above formulas leaves an enthalpy for stream 42 of -5324 kcal/s.

Using the information in Table 3.2, we can complete all of the enthalpy entries in Table 3.1. From these we can test several assumptions about our approximations.

TABLE 3.2 Enthalpy Constants

Comp. k	A_k (cal/K gmol)	B_k	C_k	D_k	$\Delta H_{f,k}^0$ (kcal/gmol)	T_b^k (K)	T_c^k (K)	$\Delta H_{vap}^k(T_b)$ (cal/gmol)
M	4.598	1.25E−02	2.86E−06	−2.70E−09	−17.89	111.7	190.6	1955
EL	0.909	3.74E−02	−1.99E−05	4.19E−09	12.5	184.5	282.4	3237
PL	0.886	5.60E−02	−2.77E−05	5.27E−09	4.88	225.4	365	4400
DEE	5.117	8.02E−02	−2.47E−05	−2.24E−09	−60.28	307.7	466.7	6380
EA	2.153	5.11E−02	−2.00E−05	3.28E−10	−56.12	351.5	516.2	9260
IPA	7.745	4.50E−02	1.53E−05	−2.21E−08	−65.11	355.4	508.3	9520
W	7.701	4.60E−04	2.52E−06	−8.59E−10	−57.8	373.2	647.3	9717

1. *Heat duty for the reactor* Comparing the enthalpies for streams 1 and 2, we have:

$$Q_R = \mu_2 \Delta H_v(T, y_2) - \mu_1 \Delta H_v(T, y_1) = (-22689.24) - (-21683.64) = -1005.6 \text{ kcal/s}$$

First, we confirm that the reaction is exothermic and that over 1000 kcal/s of heat are available for energy integration in the rest of the process.

2. *Energy balance for columns* Note that by calculating the enthalpies of streams 03, 32, and 41, we can assess the accuracy of our assumptions of saturated streams for our "adiabatic" absorber. Here we see a slight violation of the energy balance. Denoting Q as the amount of energy that needs to be removed in order to balance the reboiler gives us the following equations:

$$\Delta H_{L,03} + \Delta H_{V,31} = \Delta H_{L,42} + \Delta H_{V,41} + Q$$

$$-2545 + 11515.2 = 13440 - 5324 + Q$$

and solving for $Q = -854.2$ kcal/s. This difference can be explained by the inconsistencies of the ideal approximations for both the energy balance and phase equilibrium, approximations in our bubble and dew calculations and, most importantly, from the isothermal assumptions in the Kremser equation. Similar violations occur in our shortcut distillation columns.

3.5.3 Adiabatic Flash Calculations

We conclude this section with a description of an important set of process calculations that are a special class of the flash calculations considered in section 3.2. In operations where the system is defined by a known enthalpy and pressure (or temperature), the remaining quantities need to be calculated by an iterative process. Here we need to determine the state of the system (liquid, vapor, or mixed) as well as the temperature (or pressure, if the system is non-ideal). For these calculations we first determine the bubble and dew points of the mixture and the enthalpies for both. Then, if the specified enthalpy lies between bubble and dew point enthalpies, a flash calculation is required and a vapor fraction or component recovery of the resulting two-phase mixture needs to be found that satisfies the specification. Flash calculations can be performed systematically from the following procedure:

Adiabatic Flash Algorithm

1. For a given enthalpy specification (ΔH_{spec}) and pressure, P, calculate the bubble and dew point temperatures and the enthalpies associated with them.
   - If $\Delta H_{spec} > \Delta H_{dew}$, then the mixture is all vapor, and we solve for T from $\Delta H_V(T) = \Delta H_{spec}$.
   - If $\Delta H_{spec} < \Delta H_{bub}$, then the mixture is all liquid, and we solve for T from $\Delta H_L(T) = \Delta H_{spec}$.
2. Otherwise, if $\Delta H_{dew} \geq \Delta H_{spec} \geq \Delta H_{bub}$, guess ξ_n (or ϕ).
3. Perform a flash calculation with ξ_n (or ϕ) and P specified to obtain y_k, x_k, and T. Calculate $\Delta H(T) = \phi \Delta H_V(T) + (1 - \phi) \Delta H_L(T)$.

Sec. 3.5   Energy Balances   103

4. If $f = \Delta H_{spec} - \Delta H(T) = 0$, stop. Otherwise, if $f > 0$, reguess a higher ξ_n (or ϕ), else guess a lower ξ_n (or ϕ). Go back to step 2. This iteration can be accelerated by secant or Newton methods for f and ξ_n (or ϕ).

These examples tend to be very tedious and it helps to program them on the computer. To illustrate this procedure, we consider two small examples. The second example is particularly useful, as it completes the energy balance for the ethanol flowsheet.

EXAMPLE 3.8

Consider a 50/50 liquid mixture of benzene and toluene flowing at 100 gmol/s at 300 K and 1 bar. If heat is added to this stream at a rate of 3600 kJ/s, what is the temperature of the benzene/toluene mixture?

From the relations for liquid enthalpy, we have $\Delta H_L(300) = -847557.9$ cal/s for the benzene/ toluene stream. If we add $Q = 3600$ kJ/s $= 860.42$ kcal/s to this stream, we want to match an enthalpy of 12862.7 cal/s for the outlet stream.

If we make a rough guess of $T = 370$, then:

$$P_B^0(370) = 1238.9 \text{ mm Hg} \quad P_T^0(370) = 505.13 \text{ mm Hg} \quad \alpha_{B/T} = 2.453$$

If we now assume that $\alpha_{B/T}$ remains fairly constant with T, we can guess the key component recovery, ξ_T and calculate ξ_B, ϕ and T from:

$$\xi_B = \alpha_{B/T} \xi_T / (1 + (\alpha_{B/T} - 1) \xi_T)$$

$$\phi = 50(\xi_B + \xi_T)/100$$

$$P_T^0(T) = P / \overline{\alpha}$$

With this information, we can calculate $\Delta H(T) = \phi \Delta H_V(T) + (1 - \phi) \Delta H_L(T)$ and compare with the specified enthalpy. Starting with the ξ_T, we have the following iterations:

ξ_T	ξ_B	T	ϕ	$\Delta H(T)$	
0.7	0.851	370.1	0.776	68439	
0.6	0.786	369.5	0.693	20876	
0.57	0.765	369.4	0.667	4895	
0.585	0.776	369.5	0.680	12993	~12862

Thus at the solution, the stream is 68% vaporized with a temperature of 369.5 K. Note that in determining this enthalpy balance, heats of formation are not required. Why?

EXAMPLE 3.9

To complete the energy balance for the ethanol process, we note that μ_6 results from the adiabatic mixing of two liquid streams, μ_{32} and μ_{42}. From the energy balance from these streams we need to find the temperature of μ_6 that matches the following specification:

$$\Delta H_6 = \Delta H_{32} + \Delta H_{42} = -47920 - 5324 = -53245 \text{ kcal/s}$$

Because the inlet streams are high pressure liquids, we first guess a rough average temperature (say, 370 K) and evaluate the liquid phase enthalpy using the handbook values given above and the expression for liquid phase enthalpy ($\Delta H_L(370) = -53259$ kcal/s). From this value, we see that we are already fairly close. Further temperature guesses show that the enthalpy balance is satisfied with a liquid stream at $T_6 = 372$ K.

3.6 SUMMARY

This chapter presents systematic shortcut strategies for calculating quickly a mass and energy balance for a proposed flowsheet. This approach makes several ideal assumptions, including the use of Raoult's Law for vapor liquid equilibrium. Additional assumptions include the use of relative volatilities that are assumed to be pressure and (relatively) temperature insensitive. As a result, the process calculations for mass balances, temperature and pressure specifications, and energy balances can be solved in a sequential, decoupled manner with few iterations on the desired specifications. As a result, the calculations can easily be performed by hand or through the use of simple spreadsheet programs. In fact, all of the calculations in this chapter were aided by small Excel spreadsheets.

The main result of this chapter is the methodology to generate the mass energy balance table for the ethanol process introduced in the previous chapter. While the values in this table are only approximate (due to our ideal assumptions), they give a qualitative description of the relevant flowrates, temperatures, pressures, and heat contents. These form the necessary ingredients for further economic evaluation of the flowsheet that will be covered in the next two chapters. It should also be noted that because of the simple nature of the mass balance expressions, it is relatively easy to explore trends with respect to recoveries, purities, and other design variables. Again, these parametric studies can be accelerated through the use of simple spreadsheets.

Finally, in several examples in this chapter, we questioned and tested the accuracy of the ideal assumptions. It should be clear to the reader that relaxing these assumptions can greatly complicate the mass and energy balance calculations, so that they elude the hand calculation approach covered here. More rigorous approaches toward nonideal processes will be pursued in Part II of this text through the use of computer algorithms.

REFERENCES

Douglas, J. M. (1988). *Conceptual Design of Chemical Processes*. New York: McGraw-Hill.

Edmister, W. (1943). Design for hydrocarbon absorption and stripping. *Ind. Eng. Chem.*, **35**, 837.

Fenske, R. (1932). *Ind. Eng., Chem.*, **24**, 482.

Kremser, A. (1930). *Natl. Petrol. News*, **22** (21), 42.

Perry, R. H., Green, D. W., Maloney, J. O. (Eds.). (1984). *Perry's Chemical Engineers' Handbook,* 6th ed. New York: McGraw-Hill.

Reid, R. C., Prausnitz, J. M., & Poling, B. E. (1987). *The Properties of Gases and Liquids.* New York: McGraw-Hill.

EXERCISES

1. A simplified flowsheet for the Union Carbide oxo process is given below.
   a. Determine the overall conversion of propylene to n-butyraldehyde for purge rates of 1% and 0.1%.

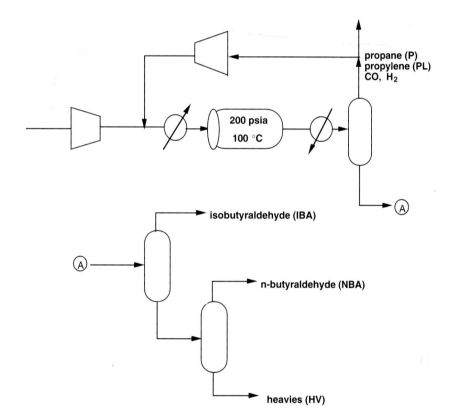

Reaction Mechanism

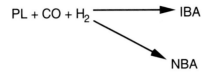

80% PL converted
IBA/NBA ratio = 0.1

IBA + NBA ⟶ HV

1% conversion

Assume all of the separation steps (distillation towers) give perfect splits for the components shown.

   b. How does the propane flowrate change at the reactor inlet when the purge rate goes from 0.1% to 1%?

2. Assume the feed into a flash tank consists of 25 moles pentane, 40 moles cis-2-butene and 35 moles n-butane.

   a. Find the recovery of n-butane when the pressure is 200 kPa and temperature is 300K.

   b. Calculate dew and bubble point temperatures at 200 kPa. What are the bubble and dew compositions?

   c. If the flash tank operates at 100 kPa, at what temperature could you recover 60% cis-2-butene in the vapor?

3. It is desired to separate propylene from trans-2-butene in a distillation column. The feed stream is available as saturated liquid at 15 bar, and has the following composition

propylene	45 gmol/s	
trans-2-butene	10	"
cis-2-butene	15	"
1-butene	6	"
ethylene	5	"
propane	4	"

It is desired to recover 99.5% of the propylene in the distillate and 99% of trans-2-butene in the bottom stream.

   a. Determine the temperature of the feed stream and the minimum number of plates that are required for the column.

   b. Determine the mole fraction composition of the distillate and the bottoms.

   c. If cooling water at 90°F is to be used in the condenser with ΔT min = 10°C, what would be the lowest pressures at which the column should operate if the distillate is obtained as either saturated liquid or saturated vapor? Also, for these two cases, what would be the maximum temperatures in a total reboiler?

Exercises

4. It is proposed to use an absorption column to recover 99.2% of acetone from a gas stream at 2 bar, 300°K, that has the following composition: 94.3 gmol/s air, 5.0 gmol/s acetone, 0.7 gmol/s formaldehyde.
   a. If water is to be used as the solvent, estimate its required flowrate for the following conditions:

P column	T water
2 bar	300°K
2 bar	330°K
10 bar	300°K
10 bar	330°K

   b. Estimate the number of theoretical trays required for this column.
   c. Assuming the absorber will operate at 2 bar, and with the temperature of the water at 300°K, calculate the mass balance for the column, and estimate the temperature of the outlet liquid stream.

5. Given a saturated liquid stream of 30 gmol/sec propane and 70 gmol/sec 1-butene at 10 bar,
   a. Find the vapor fraction of this stream if it is throttled down to 2 bar.
   b. Find the heat load to vaporize 60% of the stream at 10 bar.

6. Consider a distillation column with a feed of 10 gmol/s benzene, 20 gmol/s o-xylene and 15 gmol/s toluene at 1 atm and 230°F.
   a. For a benzene recovery of 98%, what is the minimum number of trays if the ratio of benzene to o-xylene in the overhead is 100? Find the bottoms and tops compositions.
   b. Periodically a small amount of H_2S appears in the feed. Since it is undesirable to have this component in the product, explain qualitatively how you would design and operate this column to separate the H_2S.
   c. An overhead product of 55% benzene, 40% toluene, and 5% o-xylene is recovered as saturated liquid. Can cooling water be used if the column operates at 2 bar?

7. Separate the following feed stream:

   50 gmol/s hexane $T = 350$ K
   30 gmol/s pentane $p = 150$ kPa
   20 gmol/s octane

   so that pentane and octane have overhead recoveries of 0.99 and 0.02, respectively. The overhead pressure is 100 kPa.
   a. Find the top and bottoms compositions.
   b. *Estimate* the condenser temperature if the distillate is all vapor.
   c. If the reflux ratio is 0.2, find the reboiler duty if the feed is 20% vaporized.

8. Consider the ammonia process given below. A feed of 20% N_2, 78% H_2, and 2% CH_4 is mixed with two recycle streams and enters a reactor. Here, conversion per pass of N_2 to NH_3 is 45% according to the reaction:

$$N_2 + 3H_2 \rightarrow 2NH_3$$

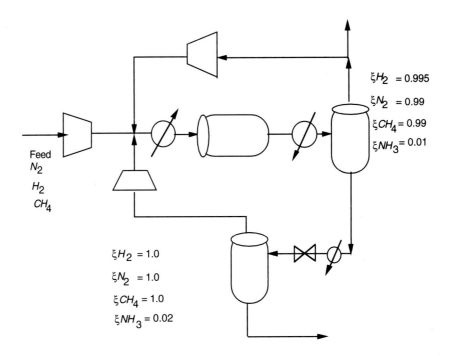

The ammonia product is recovered by flashing the reactor effluent in two stages and recycling the overhead vapor. If the purge fraction is 5% for the high pressure recycle, what is the methane concentration in the reactor feed?

9. Given the following feed stream at 1 atm:

Component	Flowrate (gmol/sec)	Vapor Pressure (mm Hg @ 310K)
n-butane	100	2588
diethyl ether	5	824
n-butanol	2	14.3
water	1	46.5

  a. Design an absorber to recover 90% of the ether in the liquid stream. Find the theoretical number of trays as well as the flowrates of the other components.
  b. How would you increase the water composition in the vapor phase?

10. Toluene (C_7H_8) is to be converted thermally to benzene (C_6H_6) in a hydrodealkylation reactor. While the main reaction is:

$$C_7H_8 + H_2 \rightarrow C_6H_6 + CH_4$$

an unavoidable side reaction produces biphenyl:

$$2\,C_6H_6 \rightarrow C_{12}H_{10} + H_2$$

Conversion to benzene is 75% and 2% of the benzene present reacts to form biphenyl. The flowsheet consists of the reactor followed by a single flash tank. There are no recycles. Given the data and flowsheet below, find the flowrates of the reactor effluent, the vapor product, and the liquid product. (To avoid trial and error calculations, assume a 50% recovery of benzene in the flash).

Components	Feed	Relative Volatility (100°F)
hydrogen	2045.9	infinite
methane	3020.8	infinite
benzene	46.2	1.0
toluene	362.0	0.32
biphenyl	1.0	0.068

11. Given the following feed stream:

	lbmol/hr	
methane	20	
methanol	70	10 atm, 350°K
water	60	

   a. Design an absorber to recover 95% of the methanol using water as the solvent. Specify all of the stream flowrates around the absorber.
   b. Explain qualitatively how the column design will change if the heavy oil solvent ($K_{oil}(350\ K) = 0.01$) is used for the same recovery specification of methanol.

12. Design a distillation column to separate 40 lbmol/hr of propane and 60 lbmol/hr of propylene.
   a. For 99% recovery of propylene and 95% of propane, how many stages are required? What are the top and bottom compositions?
   b. Estimate the pressure ranges for which cooling water may be used in the condenser of this column.

EQUIPMENT SIZING AND COSTING   4

In the previous chapter we developed the tools for a preliminary mass and energy balance of our candidate flowsheet. This task provided us with important data for economic evaluation of the process. In this chapter we will build on these concepts and pursue the next step of determining equipment sizes, capacities, and costs. As in the previous chapter, we will use approximations in order to perform the calculations quickly and establish qualitative trends for screening process alternatives. In particular, direct, noniterative correlations will be applied for equipment sizing and a well established method developed by Guthrie (1969) will then be used for costing this equipment. With this information, we are then able to complete an economic analysis, which will be discussed in the next chapter.

4.1 INTRODUCTION

Economic analysis of a candidate flowsheet requires knowledge of capital and operating costs. The former, in turn, are based on equipment sizes and capacities and their associated costs. Pikulik and Diaz (1977) noted that capital cost estimates can be classified into the following categories, based on the accuracy of the estimate:

Estimate	Error
Order-of-magnitude estimate	< 40% (error)
Study estimate	< 25%
Preliminary estimate	< 12%
Definitive estimate	< 6%
Detailed estimate	< 3%

Sec. 4.2 Equipment Sizing Procedures   111

Moreover, the difficulty and expense of obtaining more accurate estimates easily increases by orders of magnitude and frequently can be justified only within the final design stages. Douglas (1988) observes that for candidate flowsheet screening and preliminary design, an order-of-magnitude estimate is sufficient. Therefore, we will concentrate on simplified sizing and costing correlations in order to allow rapid determination of cost estimates at the 25 to 40% level of accuracy. Once we have obtained the process flows and heat duties through a mass and energy balance, we are ready to begin with investment and operating costs. Here we proceed in two steps:

1. *Physical sizing of equipment units.* This includes the calculation of all physical attributes (capacity, height, cross sectional area, pressure rating, materials of construction, etc.) that allow a unique costing of this unit.
2. *Cost estimation of the unit.* Here the sized equipment will be costed using power law correlations developed in Guthrie (1969). In addition to unit capital costs we will also consider operating costs such as utility charges. This information, together with the feedstock costs and product sales, will be used in the subsequent economic evaluation.

In the remainder of the chapter we will consider sizing and cost models for all of the process units analyzed so far. The next section will develop shortcut correlations for the sizing of these units. Section 4.3 will then describe Guthrie's cost estimation as applied to these units. Both sizing and costing will be illustrated by numerous examples. Finally, the last section will summarize the chapter and set the stage for the economic analysis in Chapter 5.

4.2 EQUIPMENT SIZING PROCEDURES

This section presents an overview of quick calculations for equipment sizing. Basic procedures will cover the following units:

- Vessels
- Heat transfer equipment
- Columns, distillation and absorption
- Compressors, pumps, refrigeration

All of these calculations require flowrates, temperatures, pressures, and heat duties from the flowsheet mass and energy balance, and these sizing calculations will determine the capacities needed for the cost correlations developed in the next section. In addition, we will develop the concept of material and pressure factors (MPF) used to evaluate particular instances of equipment beyond a basic configuration. This concept is an empirical factor developed by Guthrie as part of the costing process. As shown in section 4.3, the MPF multiplies the base cost in the evaluation of the final equipment cost.

4.2.1 Vessel Sizing

Vessels include flash drums, storage tanks, decanters, and some reactors. Unless specified otherwise by particular unit requirements, these will be sized by the following criteria.

1. Select vessel volume (V) based on a five-minute liquid holdup time with an equal volume added for vapor flows. Thus, the formula is given by:

$$V = 2 [F_L \tau / \rho_L] \qquad (4.1)$$

   where F_L is the liquid flowrate leaving the vessel (as in a flash drum), ρ_L is the liquid density, and τ is a residence time, typically set to five minutes. Specification of this residence time is dictated by maintaining a liquid buffer for on/off switching times for pumps.

2. In addition, we make a few assumptions:
   - For general costing purposes, the aspect ratio, L/D, will be assumed to be four. (This is the optimal ratio if the bottom and top caps are four times as expensive as sides.)
   - If diameter is greater than four feet (1.2 m), size the unit as a horizontal vessel. (This requires more space but less cost for structural support.)
   - As a safety factor choose the vessel (gauge) pressure to be 50% higher than the actual process pressure from the mass and energy balance. From this we also observe the appropriate pressure factors in Guthrie's method when costing the vessel.
   - For the desired temperature range, consider the required materials of construction as shown in Table 4.1. Observe the appropriate material factors in Guthrie's method when costing the vessel.

TABLE 4.1   Materials of Construction

High Temperature Service		Low Temperature Service	
T_{max}(°F)	Steel	T_{min}(°F)	Steel
950	Carbon steel (CS)	−50	**Carbon steel (CS)**
1150	*502 stainless steels*	−75	**Nickel steel (A203)**
1300	*410 stainless steels*	−320	**Nickel steel (A353)**
	330 stainless steel	−425	**Stainless steels (SS)**
1500	*430, 446 stainless steels*		**(302, 304, 310, 347)**
	Stainless steels (SS)		
	(304, 321, 347, 316)		
	Hastelloy C, X		
	Inconel		
2000	*446 stainless steels*		
	Cast stainless, HC		

(Recommended Steels, for *corrosion resistance* and **strength**; Perry's Handbook, 1984)

Sec. 4.2   Equipment Sizing Procedures

TABLE 4.2   Guthrie Material and Pressure Factors for Pressure Vessels

Shell Material	MPF = $F_m F_p$	
	Clad, F_m	Solid, F_m
Carbon Steel (CS)	1.00	1.00
Stainless 316 (SS)	2.25	3.67
Monel	3.89	6.34
Titanium	4.23	7.89

	Vessel Pressure (psig)									
Up to	*50*	*100*	*200*	*300*	*400*	*500*	*600*	*700*	*800*	*900*	*1000*
F_p	1.00	1.05	1.15	1.20	1.35	1.45	1.60	1.80	1.90	2.30	2.50

A partial list of recommended steels for materials of construction, compatible with Guthrie's factors, is given in Table 4.1. These apply not just to pressure vessels but also to the remaining equipment items. For more information, consult Perry's Handbook (Chapter 23, 1984).

In Guthrie, the basic configuration for pressure vessels is given by a carbon steel vessel with a 50 psig design pressure, and average nozzles and manways. For vertical construction, this includes the shell and two heads, the skirt, base ring and lugs, and possible tray supports. For horizontal construction, this includes the shell and two heads and two saddles. The material and pressure factor for various types of vessels is given in Table 4.2. In addition, various types of vessel linings are costed in Guthrie (1969, Figure 5).

4.2.2   Heat Transfer Equipment

Consider the countercurrent, shell and tube heat exchanger shown in Figure 4.1. Sizing equations for these heat exchangers can be found from the following equation:

$$Q = UA \, \Delta T_{lm} \quad (4.2)$$

where Q is the heat duty, known from the energy balance, A is the required area, the log mean temperature (ΔT_{lm}) is given by:

$$\Delta T_{lm} = [(T_1 - t_2) - (T_2 - t_1)] / \ln\{(T_1 - t_2)/(T_2 - t_1)\} \quad (4.3)$$

and the overall heat transfer coefficients can be estimated from Table 4.3. Again for sizing and costing, we need to observe the design criteria for temperature and pressure ($P_{rated} = 1.5 \, P_{actual}$) and observe the appropriate pressure and material factors in costing the exchanger.

Note that phase changes in heat exchangers lead to changes in U and need to be considered more carefully. In this case, we split the exchanger into serial units and, as

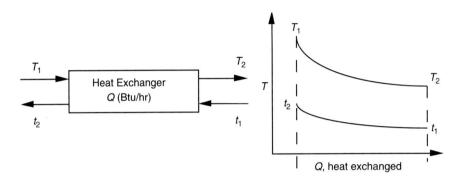

FIGURE 4.1 Heat exchanger temperatures.

TABLE 4.3 Typical Overall Heat Transfer Coefficients

Shell side	Tube side	Design U
Liquid-liquid media		
Cutback asphalt	Water	10–20
Demineralized water	Water	300–500
Fuel oil	Water	15–25
Fuel oil	Oil	10–15
Gasoline	Water	60–100
Heavy oils	Heavy oils	10–40
Heavy oils	Water	15–50
Hydrogen-rich reformer stream	Hydrogen-rich reformer stream	90–120
Kerosene or gas oil	Water	25–50
Kerosene or gas oil	Oil	20–35
Kerosene or jet fuels	Trichlorethylene	40–50
Jacket water	Water	230–300
Lube oil (low viscosity)	Water	25–50
Lube oil (high viscosity)	Water	40–80
Lube oil	Oil	11–20
Naphtha	Water	50–70
Naphtha	Oil	25–35
Organic solvents	Water	50–150
Organic solvents	Brine	35–90
Organic solvents	Organic solvents	20–60
Tall oil derivatives, vegetable oil, etc.	Water	20–50
Water	Caustic soda solutions (10–30%)	100–250
Water	Water	200–250
Wax distillate	Water	15–25
Wax distillate	Oil	13–23

Sec. 4.2   Equipment Sizing Procedures

TABLE 4.3 *(Continued)*

Shell side	Tube side	Design U
Condensing vapor-liquid media		
Alcohol vapor	Water	100–200
Asphalt (450°F)	Dowtherm vapor	40–60
Dowtherm vapor	Tall oil and derivatives	60–80
Dowtherm vapor	Dowtherm liquid	80–120
Gas-plant tar	Steam	40–50
High-boiling hydrocarbons V	Water	20–50
Low-boiling hydrocarbons A	Water	80–200
Hydrocarbon vapors (partial condenser)	Oil	25–40
Organic solvents A	Water	100–200
Organic solvents high NC, A	Water or brine	20–60
Organic solvents low NC, V	Water or brine	50–120
Kerosene	Water	30–65
Kerosene	Oil	20–30
Naphtha	Water	50–75
Naphtha	Oil	20–30
Stabilizer reflux vapors	Water	80–120
Steam	Feed water	400–1000
Steam	No. 6 fuel oil	15–25
Steam	No. 2 fuel oil	60–90
Sulfur dioxide	Water	150–200
Tall-oil derivatives, vegetable oils (vapor)	Water	20–50
Gas-liquid media		
Air N$_2$, etc. (compressed)	Water or brine	40–80
Air, N$_2$, etc., A	Water or brine	10–50
Water or brine	Air, N$_2$ (compressed)	20–40
Water or brine	Air, N$_2$, etc., A	5–20
Water	Hydrogen containing natural-gas mixtures	80–125
Vaporizers		
Anhydrous ammonia	Steam condensing	150–300
Chlorine	Steam condensing	150–300
Chlorine	Light heat-transfer oil	40–60
Propane, butane, etc.	Steam condensing	200–300
Water	Steam condensing	250–400

(U = Btu/ft^2-hr-°F; data from Perry's Handbook, 1984)   × 5.7 = W/m^2 K
NC = noncondensable gas present, V = vacuum, A = atmospheric pressure

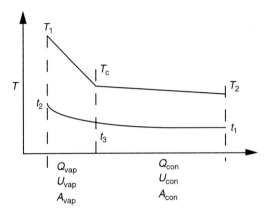

FIGURE 4.2 Sizing heat exchangers with intermediate phase changes.

shown in Figure 4.2, calculate U and A for vapor media and for condensing media separately. Thus, the total area is given by:

$$A_{vap} = Q_{vap} / \{U_{vap} [(T_1 - t_2) - (T_c - t_3)]/ \ln\{(T_1 - t_2)/(T_c - t_3)\}\}$$
$$A_{con} = Q_{con} / \{U_{con} [(T_c - t_3) - (T_2 - t_1)]/ \ln\{(T_c - t_3)/(T_2 - t_1)\}\} \quad (4.4)$$

$$\text{and } A_{total} = A_{vap} + A_{con}$$

Finally, we choose 10,000 ft² (or ~1000 m^2) as the maximum exchanger area. If more heat exchange area is required, we simply use multiple heat exchangers in parallel.

While this simplified method is adequate for preliminary designs, we note that detailed sizing of heat exchangers is much more complicated. See Welty, Wicks, and Wilson (1984) and Peters and Timmerhaus (1980) for a more detailed treatment on the sizing of heat exchangers. In Guthrie, the basic configuration for heat exchangers is given by a carbon steel floating head exchanger with a 150 psig design pressure, and this includes complete fabrication. The material and pressure factors for various types of heat exchangers are given in Table 4.4.

4.2.3 Furnaces and Direct Fired Heaters

Capacities and sizes for furnaces and direct fired heaters will not be obtained directly for a preliminary design. Instead we follow Guthrie and base the cost of these units on the heat duty. Observe that pressure and material factors, as well as design types, still need to be considered for costing. Here the basic configuration for furnaces is given by a process heater with a box or A-frame construction, carbon steel tubes, and a 500 psig design pressure. This includes complete field erection. The material and pressure factors for various types of furnaces are given in Table 4.5.

Similarly, in Guthrie the basic configuration for direct fired heaters is given by a process heater with cylindrical construction, carbon steel tubes, and a 500 psig design

TABLE 4.4 Guthrie Material and Pressure Factors for Heat Exchangers

$$\text{MPF} = F_m (F_p + F_d)$$

Design Type	F_d
Kettle Reboiler	1.35
Floating Head	1.00
U tube	0.85
Fixed tube sheet	0.80

Vessel Pressure (psig)

	Up to	150	300	400	800	1000
F_p		0.00	0.10	0.25	0.52	0.55

Shell/Tube Materials, F_m

Surface Area (ft²)	CS/CS	CS/Brass	CS/SS	SS/SS	CS/Monel	Monel/Monel	CS/Ti	Ti/Ti
Up to 100	1.00	1.05	1.54	2.50	2.00	3.20	4.10	10.28
100 to 500	1.00	1.10	1.78	3.10	2.30	3.50	5.20	10.60
500 to 1000	1.00	1.15	2.25	3.26	2.50	3.65	6.15	10.75
1000 to 5000	1.00	1.30	2.81	3.75	3.10	4.25	8.95	13.05

TABLE 4.5 Guthrie Material and Pressure Factors for Furnaces

$$\text{MPF} = F_m + F_p + F_d$$

Design Type	F_d
Process Heater	1.00
Pyrolysis	1.10
Reformer (without catalyst)	1.35

Vessel Pressure (psig)

	Up to	500	1000	1500	2000	2500	3000
F_p		0.00	0.10	0.15	0.25	0.40	0.60

Radiant Tube Material F_m

Carbon Steel	0.00
Chrome/Moly	0.35
Stainless Steel	0.75

TABLE 4.6 Guthrie Material and Pressure Factors for Direct Fired Heaters

$$\text{MPF} = F_m + F_p + F_d$$

Design Type	F_d
Cylindrical	1.00
Dowtherm	1.33

	Vessel Pressure (psig)		
Up to	500	1000	1500
F_p	0.00	0.15	0.20

Radiant Tube Material F_m	
Carbon Steel	0.00
Chrome/Moly	0.45
Stainless Steel	0.50

pressure. This also includes complete field erection. The material and pressure factors for various types of direct fired heaters are given in Table 4.6.

4.2.4 Reactors

For reactor sizing we assume a given space velocity (s in hr^{-1}) based on a liquid or gas molar flowrate, μ. Then we have:

$$s = 1/\tau = \mu / (\rho \, V_{cat}) \tag{4.5}$$

where ρ is the molar density at standard temperature and pressure (1 atm, 273 K) and V_{cat} is the volume of catalyst. The total volume, V, is then calculated based on the void fraction, ε, of the catalyst (assume 50%). In this case, we have:

$$V = V_{cat}/(1 - \varepsilon) = 2 \, V_{cat} \tag{4.6}$$

Depending on reactor conditions, we can then cost the reactor as a pressure vessel, heat exchanger, or furnace. Also, for these units use the appropriate material and pressure factors in Guthrie's method.

4.2.5 Distillation Columns

To apply costing for distillation columns, we need to calculate the height, diameter, and number of trays in the tray stack. In particular, tray stacks are defined in Guthrie with the basic configuration given by a 24" tray with carbon steel of either plate, sieve, or grid type. This includes all fittings and supports. The material and pressure factors for various types of tray stacks are given in Table 4.7.

Sec. 4.2   Equipment Sizing Procedures

TABLE 4.7 Guthrie Material and Pressure Factors for Tray Stacks

MPF = $F_m + F_s + F_t$	
Tray Type	F_t
Grid (no downcomer)	0.0
Plate	0.0
Sieve	0.0
Valve or trough	0.4
Bubble Cap	1.8
Koch Kascade	3.9

Tray Spacing, F_s			
(inch)	24"	18"	12"
F_s	1.0	1.4	2.2

Tray Material F_m	
Carbon Steel	0.0
Stainless Steel	1.7
Monel	8.9

However, in order to cost the vessel, tray stack, and heat exchangers, we first need to calculate the number of theoretical trays and the reflux ratio. As discussed in Chapter 3, shortcut calculations for these can be performed through the Fenske equation (for minimum number of theoretical trays), the Underwood equation (for minimum reflux ratio), and the Gilliland correlation that allows us to obtain the actual reflux ratio and tray number. Using the Underwood equation, however, involves an iterative procedure and will be deferred to Part III. Instead, we apply a simple, direct correlation that has been developed for (nearly) ideal systems (Westerberg, 1978). This allows us to determine qualitative trends rapidly for preliminary design.

1. *Determining Tray Number and Reflux Ratio*
   The following direct procedure can be applied to determine the desired quantities.
   a. From the mass and energy balance calculations we have the relative volatilities and the top and bottom recoveries: $\alpha_{lk/hk}$, $\beta_{lk} = \xi_{lk}$, $\beta_{hk} = 1 - \xi_{hk}$.
   b. Calculate tray number and reflux ratio from the following correlation:

   $$N_i = 12.3/\{(\alpha_{lk/hk} - 1)^{2/3}(1-\beta_i)^{1/6}\} \text{ and } R_i = 1.38/\{(\alpha_{lk/hk}-1)^{0.9}(1-\beta_i)^{0.1}\}$$

   for both $i = lk, hk$. Then the number of theoretical plates is:

   $$N_T = \gamma_N \max_i (N_i) + (1 - \gamma_N) \min_i (N_i)$$

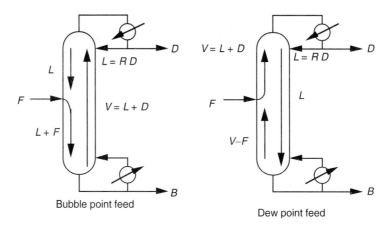

$$F_{lv} = L'/V' \ (\rho_g/\rho_l)^{0.5}$$

FIGURE 4.3 Internal liquid and vapor column flowrates.

and the reflux ratio is:

$$R = \gamma_R \max_i (R_i) + (1 - \gamma_R) \min_i (R_i)$$

where γ_R and γ_N are arbitrary weights (set to 0.8).

2. *Calculate Column Diameter*

    From the reflux ratio and the state of the feed, we can now calculate flow rates in the distillation column. Based on these flowrates and relationships for flooding velocity we can then calculate the diameters.

    a. Figure 4.3 illustrates the flowrates for two feed conditions. For two-phase feed the flowrates can be calculated in an analogous manner.
    b. To determine the diameter we design the column to run at 80% of the flooding velocity. At the flooding velocity, the vapor flowrate is so high that no net liquid flow occurs and entrainment begins. Flooding relations are represented in Figure 4.4.

    Here we define U_{nf} as the linear flooding velocity (in ft/sec) and from Figure 4.4, U_{nf} is given by:

$$U_{nf} = C_{sbf}((\rho_l - \rho_g)/\rho_g)^{0.5}(20/\sigma)^{0.2} \qquad (4.7)$$

ρ_g and ρ_l are the gas and liquid mass densities, respectively, and σ is the liquid surface tension in dynes/cm. (For many hydrocarbons we use $\sigma = 20$ dynes/cm). Douglas (1988) provides a simplification of the diameter calculation by noting that the C_{sb} remains fairly constant for F_{lv} values of 0.01 to 0.2, a fairly wide range. Also by noting that $\sigma = 20$ dynes/cm for hydrocarbons and $(\rho_l - \rho_g) \sim \rho_l$, the flooding velocity is given by:

Sec. 4.2 Equipment Sizing Procedures   **121**

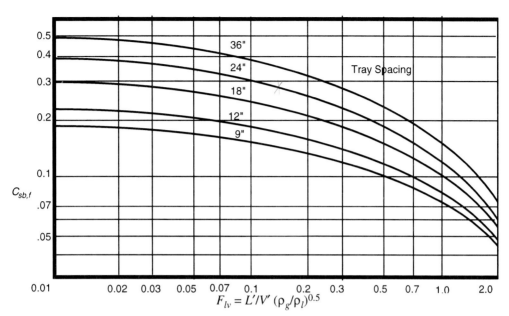

FIGURE 4.4 Flooding limits for bubble cap and perforated trays. L'/V' is the liquid/gas mass ratio at the point of consideration. (Data taken from Fair, 1961.)

$$U_{nf} = C_{sb} (\rho_l/\rho_g)^{0.5} \qquad (4.8)$$

For a typical 24" tray spacing, C_{sb} is about 0.33 ft/s in this range. The column diameter is then given by the cross sectional area:

$$A = \pi D^2/4 = V'/(0.8\, U_{nf}\, \varepsilon\, \rho_g) \qquad (4.9)$$

where ε is the fraction of the area available for vapor flow (about 0.6 for bubble cap trays, 0.75 for sieve trays). In this text, we allow the maximum column diameter to be 20 ft (6 m). Any larger calculated diameters require the column to be split into two columns run in parallel.

3. *Determine the Tray Stack Height*

    The number of actual trays is given by N_T/η where the efficiency (η) is assumed to be 80%. Also, assuming a two-foot (0.6 m) tray spacing, the tray stack height is easily calculated. Finally, we choose a maximum height of 200 feet (60 m). A larger calculated height will require that the column be split into two with liquid and vapor flows running between them.

4. *Calculate Heat Duties for Reboiler and Condenser*

    For a total condenser, we know from the energy balance that $Q_{cond} = H_V - H_L$, where H_V and H_L are the total stream enthalpies for the vapor and liquid streams around the condenser. The reboiler duty can then be calculated either directly from the vapor flow or from a total energy balance around the column.

5. *Costing of the Distillation Column*

In order to apply Guthrie's method and obtain the costing information, we now need only to group the capacities of the following components: the empty vessel, the tray stack, and the heat transfer equipment (condensers and reboilers). The overall sizing procedure is best illustrated with a small example.

EXAMPLE 4.1

Consider the separation of acetone and water at 100 kPa as shown in Figure 4.5.

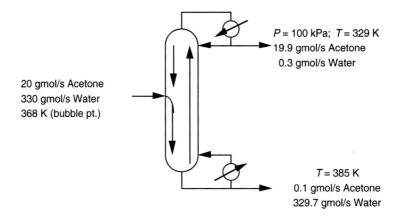

FIGURE 4.5 Distillation example.

At $T = 368$ K we have a relative volatility, $\alpha_{A/W} = 3.896$, and key component overhead recoveries of $\xi_A = 0.995$ and $\xi_W = 0.001$ ($\beta_A = 0.995$ and $\beta_W = 0.999$). From the tray number correlations we have:

$$N_A = 12.3/\{(3.896 - 1)^{2/3} (1 - 0.995)^{1/6}\} = 14.64 \qquad (4.10)$$

$$N_W = 12.3/\{(3.896 - 1)^{2/3} (1 - 0.999)^{1/6}\} = 19.14$$

$$\text{and } N_T = 0.8(N_W) + 0.2 N_A = 18.24$$

with the actual tray number $N = N_T = 18.24/0.8 = 23$.

The reflux ratio is calculated by:

$$R_A = 1.38/\{(3.896 - 1)^{0.9} (1 - 0.995)^{0.1}\} = 0.9 \qquad (4.11)$$

$$R_W = 1.38/\{(3.896 - 1)^{0.9} (1 - 0.999)^{0.1}\} = 1.06$$

$$\text{with } R = 0.8 R_W + 0.2 R_A = 1.025$$

Now the column height is calculated by considering the following components:

Sec. 4.2 Equipment Sizing Procedures

Tray stack = $(N-1)$ (0.6m) (24 in. spacing)	13.2 m
Extra feed space	1.5 m
Disengagement space (top & bottom)	3.0 m
Skirt height	1.5 m
Total height	19.2 m (~ 63 ft.)

A rough condenser and reboiler sizing can be done by noting the flowrates and, from handbook data, the enthalpies.

$$D = 20.2 \text{ gmol/s}$$

$$L = R\,D = 1.025\,(20.2) = 20.7 \text{ gmol/s}$$

$$V = L + D = 40.9 \text{ gmol/s}$$

$$\Delta H_{vapA} = 30.2 \text{ kJ/gmol}$$

$$\Delta H_{vapW} = 40.7 \text{ kJ/gmol}$$

As a result, the condenser duty is:

$$Q_{cond} = (40.9/20.2)[19.9(30.2) + 0.3(40.7)] = 1241 \text{ kW} \tag{4.12}$$

and assuming pure water in the bottom stream, the reboiler duty is:

$$Q_{reb} = 40.9\,(40.7) = 1665 \text{ kW} \tag{4.13}$$

Now the column diameter can be calculated separately for the top and bottom of the column. For the bottom section below the feed we first compute liquid and vapor densities: $\rho_l = 10^6$ g/m³, and assuming $\Delta P = 50$ kPa, then from the ideal gas law, we have

$$\rho_g = PM/RT = (150 \text{ kPa})(18 \text{ g/gmol})/(8.314 \text{ J/gmol K})(385 \text{ K}) = 870 \text{ g/m}^3 \tag{4.14}$$

Next, the mass flow rates can be calculated by:

$$L' = L\,M = (370.7)\,(18) = 6672 \text{ g/s}$$

$$V' = V\,M = (40.9)\,(18) = 736 \text{ g/s} \tag{4.15}$$

$$\text{and } \sigma = 0.07 \text{ N/m} = 70 \text{ dynes/cm}$$

This does not satisfy the assumptions of the simplified correlation, so from the flooding curve we calculate the abscissa, the dimensionless flow parameter:

$$F_{lv} = (L'/V')(\rho_g/\rho_L)^{0.5} = 0.267 \tag{4.16}$$

and for 24" tray spacing we obtain the ordinate, the capacity parameter C_{sb}, is about 0.25. Here:

$$C_{sb}\{= U_{nf}\,(20/\sigma)^{0.2}\,[\rho_g/\rho_l - \rho_g)]^{0.5}\} = 0.25 \text{ ft/sec} \tag{4.17}$$

Solving for the flooding velocity yields:

$$U_{nf} = C_{sb}\,(\sigma/20)^{0.2}\,[\rho_l/\rho_g - 1]^{0.5} = 10.9 \text{ ft/s} = 3.3 \text{ m/s} \tag{4.18}$$

and at 80% flooding we have: $U = 2.65$ m/s as the gas velocity through the net area.

The diameter is then calculated from

$$V' = \rho_g\,U\,\varepsilon\,\pi\,D^2/4 \tag{4.19}$$

and for bubble cap trays $\varepsilon = 0.6$, so the diameter is

$$D = 0.82 \text{ m } (\sim 2.7 \text{ ft}) \text{ for the bottom column section}$$

Repeating this calculation for the top of the column leads to

$$D = 0.66 \text{ m} = 2.2 \text{ ft}$$

and since the difference between these diameters is not large, we choose the larger diameter for the entire column. With this information we are ready to determine costs for the vessel and the tray stack.

4.2.6 Absorbers

Column sizes for absorbers are calculated as in the previous example for distillation columns. However, here N_T is derived from the Kremser equation (see Chapter 3) and we use a very low efficiency (20%) as equilibrium on a tray is usually a very poor assumption. Height and diameter for the vessel and the tray stack are costed in the same way as for distillation columns. Again, we assume a 24" tray spacing. Packed columns can also be costed using the data in Guthrie. Information on the costs of various packings is given in Figure 5 of Guthrie (1969).

4.2.7 Pumps

For pumping (pressure increases) in liquids we define the theoretical work as $V \Delta P$, since the specific volume remains (nearly) constant. The brake horsepower can be written as:

$$W_b = \mu \, (P_2 - P_1)/(\rho \, \eta_p \, \eta_m) \tag{4.20}$$

where η_p is the pump efficiency (assume to be 0.5) and η_m is the motor efficiency (assume to be 0.9).

In particular, centrifugal pumps are defined in Guthrie with the basic configuration cast iron unit operating below 250°F and a suction pressure of 150 psig. This includes the driver and coupling as well as the base plate. The material and pressure factor for centrifugal pumps is given in Table 4.8. Cost correlations for reciprocating pumps are also given in Guthrie (1969, Figure 7).

4.2.8 Compressors and Turbines

In sizing compressors and turbines (essentially, compressors running in reverse) we make several ideal assumptions on gas compression (Figure 4.6). We divide our discussion into two categories: centrifugal compressors, with relatively high capacities and low compression ratios, and reciprocating compressors, with low capacities and high compression ratios.

Sec. 4.2　Equipment Sizing Procedures

TABLE 4.8　Guthrie MPFs for Centrifugal Pumps and Drivers

MPF = $F_m F_o$			
Material Type	F_m		
Cast iron	1.00		
Bronze	1.28		
Stainless	1.93		
Hastelloy C	2.89		
Monel	3.23		
Nickel	3.48		
Titanium	8.98		
Operating Limits, F_o			
Max Suction Press.	150	500	1000 (psig)
Max Temperature	250	550	850 (°F)
F_o	1.0	1.5	2.9

For an adiabatic compressor, the ideal compression work can be calculated from the change in enthalpy:

$$W = \mu \, [H_V(P_2, T_2) - H_V(P_1, T_1)] \quad (4.21)$$

where μ is the molar flowrate (e.g., gmol/s) and the gas enthalpy is given by H_V.

Assuming an ideal system, this equation can be written as:

$$W = \mu \, C_p \, (T_2 - T_1) = \mu \, (\gamma/(\gamma-1)) R \, (T_2 - T_1) \text{ (for ideal gas)} \quad (4.22)$$

where C_p is the constant pressure heat capacity, $\gamma = C_p/C_v$ is 1.4 for an ideal system, and the gas constant, $R = 8.314$ J/gmol K. Assuming an ideal, isentropic, adiabatic expansion, we can calculate T_2 from the pressure ratio (P_2/P_1) using

$$T_2 = T_1 \, (P_2/P_1)^{(\gamma-1)/\gamma} \quad (4.23)$$

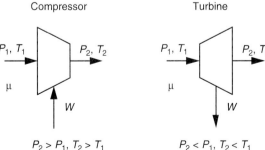

FIGURE 4.6　Compressor and turbine configurations.

and substituting back into the above expression gives the theoretical power for an ideal system:

$$W = \mu \, (\gamma/(\gamma - 1)) R \, T_1 \, [(P_2/P_1)^{(\gamma-1)/\gamma} - 1] \qquad (4.24)$$

For efficiencies of compressors or turbines, we choose $\eta_c = 0.8$ for compression and expansion work efficiency. If a shaft-driven electric motor is the compressor driver, we assume the motor efficiency is $\eta_m = 0.9$. If a turbine is the driver the efficiency is $\eta_m = 0.8$. Thus the actual (or brake) horsepower for a compressor is:

$$W_b = W/(\eta_m \, \eta_c) = 1.39 \, (W) \text{ (if motor driven) or } 1.562 \, (W) \text{ (if turbine driven)}$$

In addition, we limit compressor sizes to a maximum horsepower = 10,000 hp (about 7.5 MW).

Various types of compressor configurations are specified in Guthrie. The basic configuration is a centrifugal compressor with a carbon steel circuit and a maximum pressure of 1000 psig. This includes the motor driver and coupling as well as the base plate. The material and pressure factor for various types of compressors is given in Table 4.9.

4.2.9 Staged Compressors

Staged compressors are useful to perform a given service (a desired increase in gas pressure) with less work. This is accomplished by allowing intercooling after each compression stage. As a result, for ideal systems the compression work required per mole is illustrated in Figure 4.7 from $W = \int_{P_1}^{P_2} V \, dp$.

Note that isothermal compression requires constant heat removal to keep the system at T_0. While it is physically unrealistic, it is easy to see that isothermal compression is a limiting case of staged compression (as the number of stages goes to infinity). Intercooling in staged compression requires the configuration in Figure 4.8.

Moreover, for a fixed number of compressors, N, it can be shown that the minimum work occurs when all compression ratios are equal, i.e:

TABLE 4.9 Guthrie Material and Pressure Factors for Compressors

Design Type	MPF = F_d
	F_d
Centrifugal/motor	1.00
Reciprocating/steam	1.07
Centrifugal/turbine	1.15
Reciprocating/motor	1.29
Reciprocating/gas engine	1.82

Sec. 4.2   Equipment Sizing Procedures

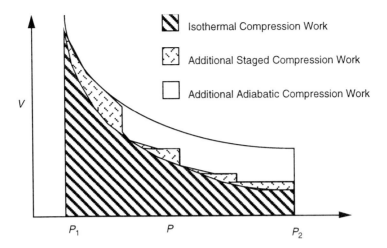

FIGURE 4.7   Work required ($\int V\, dp$) for different compression sequences.

$$P_1/P_0 = P_2/P_1 = P_3/P_2 = P_4/P_3 = \ldots = P_N/P_{N-1} = (P_N/P_0)^{1/N} \qquad (4.25)$$

and the work required is given by:

$$W = \mu\, N\, (\gamma/(\gamma - 1))R\, T_0\, [(P_N/P_0)^{(\gamma-1)/\gamma N} - 1] \qquad (4.26)$$

and as $N \to \infty$, we obtain the expression for isothermal work:

$$W = \mu\, R\, T_0\, \ln (P_N/P_0) \qquad (4.27)$$

We now see that there is a trade-off that must be dealt with in staged compression. Minimum work is obtained as the number of stages becomes large, but this leads to unrealistic capital costs. On the other hand, maximum work occurs with a single (adiabatic) compression stage. Rather than finding the optimum number of stages, which is case specific, for preliminary designs we invoke a guideline that the compression ratio will be $(P_i/P_{i-1}) = 2.5$ (A practical limit for centrifugal compressors is a compression ratio of five.) Having established these relations, let's see how they work in a small example.

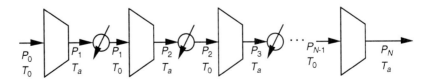

FIGURE 4.8   Compression sequence with intercooling.

> **EXAMPLE 4.2 Compression Sizing**
>
> Find the work to compress 10 gmol/s of an ideal gas at 298 K from $P_0 = 100$ kPa to $P_N = 1500$ kPa using adiabatic compression, isothermal compression, and staged compression.
> For adiabatic compression, we have
>
> $$W = \mu \, (\gamma/(\gamma - 1)) R \, T_0 \, [(P_N/P_0)^{(\gamma-1)/\gamma} - 1] = 101.26 \text{ kW} \qquad (4.28)$$
>
> For isothermal compression we have:
>
> $$W = \mu \, R \, T_0 \ln(P_N/P_0) = 66.86 \text{ kW} \qquad (4.29)$$
>
> And for staged compression we choose a compression ratio at approximately 2.5. Thus, the number of stages is derived from:
>
> $$(P_N/P_0)^{1/N} = 2.5 \qquad (4.30)$$
>
> So,
>
> $$N \sim 3 \text{ and } (P_3/P_2) = (P_2/P_1) = (P_1/P_0) = 2.47.$$
>
> From the staged relation we have:
>
> $$W = \mu \, N \, (\gamma/(\gamma - 1)) R \, T_0 \, [(P_N/P_0)^{(\gamma-1)/\gamma N} - 1] = 76.53 \text{ kW} \qquad (4.31)$$
>
> Finally, for staged compression the outlet temperature from each compressor is:
>
> $$T_a = (P_1/P_0)^{(\gamma-1)/\gamma} T_0 = 386 \text{ K} \qquad (4.32)$$
>
> and the heat duty required for each exchanger is:
>
> $$Q_{\text{int}} = \mu \, C_p \, (T_1 - T_0) = 25.6 \text{ kW}. \qquad (4.33)$$

4.2.10 Reciprocating Compressors

Reciprocating compressors perform work and effect a pressure change through a mechanical change of volume through a piston and cylinder. Unlike centrifugal compressors they are best selected for low capacities and high changes in pressure.

As can be seen from the compression cycle in Figure 4.9, theoretical power can be calculated from:

$$W = \mu \, (\gamma/(\gamma - 1)) R \, T_0 \, [(P_N/P_0)^{(\gamma-1)/\gamma} - 1] \, / \, [1 - (c(P_2/P_1)^{1/\gamma} - 1)] \qquad (4.34)$$

where $c = V_4/(V_2 - V_4)$ is a clearance factor between 0.05 and 0.10 and we assume a compression efficiency of $\eta_c = 0.9$. Selecting among centrifugal and reciprocating compressors depends on the gas flowrate and the desired pressure increase. A discussion on appropriate selection regimes is given in Perry's Handbook (1984).

4.2.11 Refrigeration

If a process stream needs to operate below about 300 K, some sort of refrigeration is required and a refrigeration cycle needs to be considered. Often, refrigeration can be "purchased" from an off-site facility, if available. Otherwise, a separate refrigeration facility

Sec. 4.2   Equipment Sizing Procedures

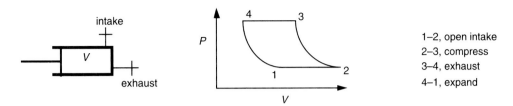

FIGURE 4.9 Compression cycle for reciprocating compressor.

needs to be constructed. In either case, because both compression and cooling water are required in a refrigeration cycle, refrigerating a stream is far more expensive (on an energy basis) than lowering the temperature with cooling water, or even raising the temperature with steam. Consequently, refrigeration is generally not a desirable option and other process alternatives should be considered first.

Given that a refrigeration system needs to be designed, we first consider the refrigeration cycle and the pressure-enthalpy diagram pictured in Figure 4.10. Here Q is the heat absorbed from the process stream at a subambient temperature and Q' is the heat per unit mass of refrigerant. As with staged compression, we see from the diagram that there is a trade-off between capital and operating costs in choosing the *number* of refrigeration cycles. Here a single cycle requires the maximum work and cooling water, Q_c, while a large number of cycles require minimum work and Q_c. To relate the W and heat rejected for refrigeration (Q), we define a *coefficient of performance*, $CP = Q'/W'$. As with staged compression, we apply a general guideline and select $CP \sim 4$ for design purposes. Thus, in a typical cycle, we have:

$$W = Q/4, \text{ and } Q_c = W + Q \sim 5/4\, Q \qquad (4.35)$$

and for the compressor driven with an electric motor,

$$W_b = W / \eta_m \eta_c = W / 0.72 \qquad (4.36)$$

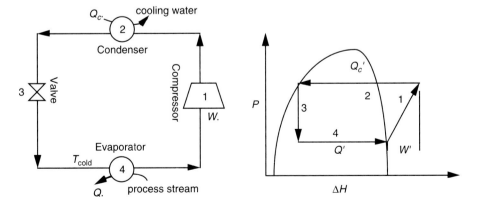

FIGURE 4.10 Refrigeration cycle and phase diagram.

In order to analyze multiple cycles and choose refrigerants for each cycle, we first consider the following temperature constraints:

1. Refrigerant (R) must remain below its critical point in the condenser, say:

$$T_{cond,max} = 0.9 T_c^R. \tag{4.37}$$

If cooling water is used here, then we also know that:

$$T_{cond,max} > T_{CW} (\sim 300 \text{ K}) + \Delta T_{min} \tag{4.38}$$

2. In the evaporator, refrigerant and pressure should be chosen so that $T_{evap} > T_{boil,R}$. Also, the evaporator pressure should be chosen greater than 1 atm. to avoid air leaks into the evaporator.

3. Finally, we choose $\Delta T_{min} \sim 5\text{K}$, for both the evaporator and condenser heat exchangers.

EXAMPLE 4.3

Suppose we want to cool air as a process stream to 180K. Consider the refrigerants:

R	T_{boil}(K)	$0.9 T_c$(K)
Ethylene	169	254
Propane	231	332

We know that ethylene will go down to 180 K but not up to 300 K. The opposite holds for propane. Therefore, we need at least two stages: one propane, one ethylene.

Stage 1

Here R = propane and we obtain the following cooling curves:

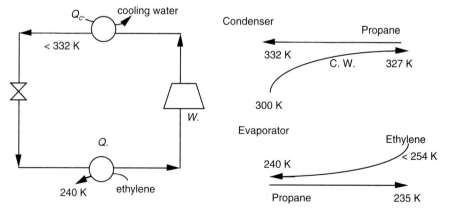

FIGURE 4.11

Sec. 4.2   Equipment Sizing Procedures   131

Stage 2

Here R = ethylene and we obtain the following cooling curves:

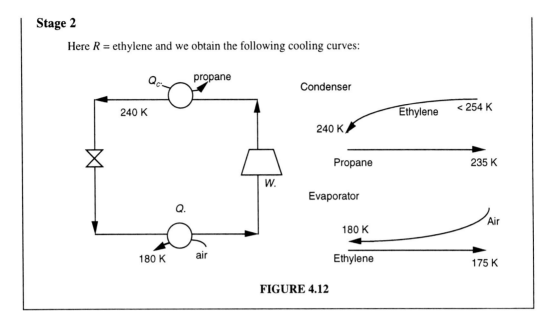

FIGURE 4.12

SHORTCUT MODEL

In Example 4.3, we did not evaluate the work requirements for each cycle. However, we noticed a large temperature change (> 60 K) for each cycle. We now analyze each cycle and develop simplified sizing relationships. If CP is the same for all N cycles, we know that $Q_{i-1} = Q_{c,i} = (1 + 1/CP)Q_i$ and we can write:

$$W = \sum_i (Q_i/CP) = \sum_i (Q/CP)(1 + 1/CP)^{i-1} \qquad (4.39)$$

By expanding this series and telescoping we get:

$$W = Q[(1+1/CP)^N - 1] \qquad (4.40)$$

and, using our guideline, for $CP = 4$, we have:

$$W = Q[(5/4)^N - 1] \qquad (4.41)$$

$$Q_c = (5/4)^N Q$$

$$W_b = W/(\eta_m \eta_c)$$

With this guideline, we assume $\Delta T = 30$K/cycle for our shortcut system. Thus, if $T_{cold} = 180$, we need about $(300 - 180)/30 = 4$ cycles. This method gives us a quick way of estimating utilities Q_c (cooling water duty) and W_b (electricity). Similarly, capital costs for the refrigeration cycle (heat exchangers and compressor) could be determined using Guthrie's method.

Alternatively, mechanical refrigeration configurations are specified directly in Guthrie. The basic configuration includes centrifugal compression, evaporators, con-

TABLE 4.10 Guthrie Equipment Factor for Mechanical Refrigeration

MPF = F_t	
Evaporator Temperature	F_t
40°F/ 278 K	1.0
20°F/ 266 K	1.95
0°F/ 255 K	2.25
−20°F/ 244 K	3.95
−40°F/ 233 K	4.54

densers, field erection, and subcontractor indirect costs. The basis of the refrigeration unit is for an evaporator temperature of 40°F (278 K). The equipment factor for other refrigeration cycles is given in Table 4.10.

EXAMPLE 4.4

Design a refrigeration system to condense 500 gmol/s of ethylene to 240 K.

From the handbook we have ΔH_{vap} = 9.336 kJ/gmol for ethylene, so the total duty is Q = 4668 kW. The number of stages is:

$$N = [300 - 240]/30 = 2 \text{ stages}$$

$$W = [(5/4)^2 - 1] \, Q = 2626 \text{ kW}$$

$$W_b = [W/0.72] = 3647 \text{ kW}$$

$$Q_c = [(5/4)^2 \, Q] = 7294 \text{ kW}$$

Assuming the following data and prices:

Electricity:	1.2¢/kWh,	8640 hrs/yr
Cooling water:	2¢/1000 gal,	ΔT_{rise} = 22K

we have cooling water and electricity expenses of $13,000/yr and $378,120/yr, respectively.

4.3 COST ESTIMATION

For preliminary design calculations, we note that equipment costs (C) increase nonlinearly with equipment size (S) or capacity. This behavior can often be captured with a power law expression: $C = C_0(S/S_0)^\alpha$, where the exponent is less than one, often about 0.6 to 0.7, and S_0 and C_0 are the base capacities and costs, respectively. This nonlinear cost behavior is reflected in an *economy of scale*, where the incremental costs decrease with larger capacities.

Sec. 4.3   Cost Estimation   133

Why is this the case? For pressure vessels, for instance, the service capacity depends on volume (V), while the cost depends on the weight (W) of the metal (proportional to surface area). For example, for a spherical vessel, we have:

$$V = \pi/6\, D^3 \text{ and } W = \rho_M t\, (\pi D^2) \tag{4.42}$$

where t is the vessel thickness and ρ_M is the metal density. In terms of volume, we have:

$$D = (6V/\pi)^{1/3} \text{ and } W = \rho_M t\, (\pi^{1/3} (6V)^{2/3}) \tag{4.43}$$

with the vessel cost as: $C \propto W = k\, V^{2/3}$.

For cylindrical pressure vessels, we adopt a more general form used by Guthrie: $C = C_0\, (L/L_0)^\alpha (D/D_0)^\beta$. Correlations for pressure vessels are given in Table 4.11. Guthrie also considers separate correlations for storage vessels of various geometries.

In Guthrie (1969), costs are plotted on charts with log–log scales so that $C = C_0(S/S_0)^\alpha$ is represented as $\log C = \log [C_0/S_0^\alpha] + \alpha \log S$. Note that the slope is given by the exponent α. Deviations on costs on some units vary by about 20% in Guthrie. However, for preliminary design, we will only use the median data. Data for the correlations taken from Guthrie are given in Table 4.12.

The cost data in Tables 4.11 and 4.12 are given in terms of mid-1968 prices. In order to update these costs, we apply an update factor to account for inflation. The update factor is defined by:

$$\text{UF} = \frac{\text{present cost index}}{\text{base cost index}}$$

For cost updating we will use the Chemical Engineering (CE) Plant Index reported in *Chemical Engineering* magazine. Representative indices are given below:

Year	CI
1957–59	100
1968 1/2	115 (Guthrie's article)
1970 1/2	126 (Guthrie's book)
1983	316
1993	359
1995	381

4.3.1 Guthrie's Modular Method

To account for numerous direct and indirect costs associated with the cost of equipment, Guthrie proposed a simple factoring method for add-on costs. A typical cost module (with representative numbers) is given below.

1. Free on board equipment (FOB)—   100
    (Base cost, BC, or equipment cost, E, from graph)

TABLE 4.11 Base Costs for Pressure Vessels

Equipment Type	C_0($)	L_0(ft)	D_0(ft)	α	β	MF2/MF4/MF6/MF8/MF10
Vertical fabrication $1 \leq D \leq 10$ ft, $4 \leq L \leq 100$ ft	1000	4.0	3.0	0.81	1.05	4.23/4.12/4.07/4.06/4.02
Horizontal fabrication $1 \leq D \leq 10$ ft, $4 \leq L \leq 100$ ft	690	4.0	3.0	0.78	0.98	3.18/3.06/3.01/2.99/2.96
Tray stacks $2 \leq D \leq 10$ ft, $1 \leq L \leq 500$ ft	180	10.0	2.0	0.97	1.45	1.0/1.0/1.0/1.0/1.0

(Data from Guthrie, 1969)

    2. Installation
        a. Piping instruments, etc.   62.2
        b. Labor (L)   58.0
    3. Shipping, taxes, supervision   74.9
        Total cost   295.1

TABLE 4.12 Base Costs for Process Equipment

Equipment Type	C_0(10^3)	S_0	Range(S)	α	MF2/MF4/MF6/MF8/MF10
Process furnaces S = Absorbed duty (10^6Btu/hr)	100	30	10–300	0.83	2.27/2.19/2.16/2.15/2.13
Direct fired heaters S = Absorbed duty (10^6Btu/hr)	20	5	1–40	0.77	2.23/2.15/2.13/2.12/2.10
Heat exchanger Shell and tube, S = Area (ft^2)	5	400	100–10^4	0.65	3.29/3.18/3.14/3.12/3.09
Heat exchanger Shell and tube, S = Area (ft^2)	0.3	5.5	2–100	0.024	1.83/1.83/1.83/1.83/1.83
Air coolers S = [Calculated area (ft^2)/15.5]	3	200	100–10^4	0.82	2.31/2.21/2.18/2.16/2.15
Centrifugal pumps	0.39	10	10–2·10^3	0.17	3.38/3.28/3.24/3.23/3.20
	0.65	2·10^3	2·10^3–2·10^4	0.36	3.38/3.28/3.24/3.23/3.20
	1.5	2·10^4	2·10^4–2·10^5	0.64	3.38/3.28/3.24/3.23/3.20
S = C/H factor (gpm × psi)					
Compressors S = brake horsepower	23	100	30–10^4	0.77	3.11/3.01/2.97/2.96/2.93
Refrigeration S = ton refrigeration (12,000 Btu/hr removed)	60	200	50–3000	0.70	1.42

(Data from Guthrie, 1969)

Sec. 4.3   Cost Estimation   135

As a result, we define the Bare Module Cost = BC × MF. Here the module factor (MF) is 2.95 (a typical value); that is, the equipment cost is almost three times the base cost. This module factor is also affected by the base cost. Consequently, in Tables 4.11 and 4.12 we give module factors for the following base costs (BC in 1968 prices):

MF2	Up to $200,000
MF4	$200,000 to $400,000
MF6	$400,000 to $600,000
MF8	$600,000 to $800,000
MF10	$800,000 to $1,000,000

Moreover, for special materials and high pressures, we have already defined materials and pressure correction factors (MPF) for various types of equipment. Here the bare module cost is modified by the following factors:

Uninstalled cost   = (BC) (MPF)
Installation = (BC) (MF) − BC = BC (MF − 1)
(this is usually calculated on a carbon steel basis)
Total installed cost   = BC (MPF + MF − 1)
Updated bare module cost = UF (BC) (MPF + MF − 1)

Finally, we do not treat contingency costs and indirect capital costs as Guthrie does. Instead, as discussed in the next chapter, for preliminary designs we apply overall indirect cost factors and a flat 25% contingency rate after all the equipment is costed. From this description, let's consider the above examples again.

COSTING FOR SIZING EXAMPLES

First Example 4.2 is reconsidered in order to determine the 1993 costs of the compressors and heat exchangers. For the three stage compression we assume that the compressed air is desired at the inlet temperature, 298 K and therefore need to find the costs of three identical compressors and heat exchangers:

Compressor costs are calculated from the individual capacities (W = 76.53/3 kW = 25.51 kW):

$$W_b = W/0.72 = 35.43 \text{ kW} = 47.4 \text{ hp} \tag{4.44}$$

From Table 4.12, the base cost is estimated at $23,000(47.4/100)^{0.77}$ ~$12,940 for a centrifugal compressor with electric motor. As a result, both F_d and MPF = 1 and the module factor (MF) is 3.11. The bare module costs for the three compressors are:

$$\text{BMC} = 3(\text{UF})(\text{MPF} + \text{MF} - 1)(\text{BC}) = 3(3.12)(3.11)(12,940) = \$376,600. \tag{4.45}$$

Assuming a service factor of 0.904(365) = 330 days the electricity cost at 10¢/kWh is about $60,600 /yr.

The heat exchangers, on the other hand, each have a heat duty of Q_{int} = 2.56kW and from Table 4.2 with a water (shell) / air(tube) system, the overall heat transfer coefficient is U = 20–40 Btu/hr-ft^2-°F; we choose the lower value (why?) as U = 20 Btu/hr-ft^2-°F = 114 W/m^2 K. Assuming cooling water available at 295 K and an allowable discharge at 317 K, we calculate the log mean temperature difference to get:

$$\Delta T_{lm} \frac{(386 - 317) - (298 - 295)}{\ln\left[\frac{(386 - 317)}{(298 - 295)}\right]} = 21.05 \text{ K} \qquad (4.46)$$

$$A_{int} = Q_{int} / [U \Delta T_{lm}]$$
$$= (25{,}600 \text{ W})/[(114 \text{ W/m}^2 \text{ K}) (21.05 \text{ K})]$$
$$= 10.67 \text{ m}^2 = 115 \text{ ft}^2$$

From Table 4.12 we obtain a base cost (BC) of $300(115/5.5)^{0.024}$ = $323. For a carbon steel, floating head exchanger with a pressure factor of 0.25 (why?) we have a materials and pressure factor (MPF) of 1.25 and a module factor (MF) of 1.85. Also, the update factor (UF) is (395/115) = 3.12. The bare module cost for the two exchangers is:

$$\text{BMC} = 2 \text{ (UF)(MPF + MF − 1)(BC)} \qquad (4.47)$$
$$= 2(3.12)(1.25 + 1.85 − 1)(323) = \$4230$$

Assuming a cooling water cost of 5.2¢/1000 gal. = $1.398 · 10^{-8}/g and a temperature rise of (317–295) = 22 K, with a service factor of 0.904, the utility cost of both exchangers is:

Cooling Water Cost = $1.398 · 10^{-8}/g x (Flow = Q/Cp ΔT)
$$= \$1.398 \cdot 10^{-8}/\text{g } [2(25{,}600 \text{ W})/ (4.184 \text{ J }/(\text{g–K}) \text{ 22 K})]$$
$$= \$ 7.77 \cdot 10^{-6} /\text{s}$$

Cost/yr = 0.904 (86400 s/day)(365 days/yr) ($7.77 · 10^{-6}/s = $222 /yr

For Example 4.1 (see Figure 4.13) we can calculate the costs updated to 1993 prices. From this example we have the following data:

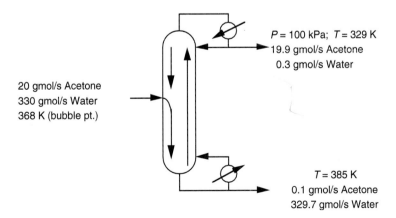

FIGURE 4.13 Distillation column example.

Sec. 4.3 Cost Estimation

$$\text{Column diameter} = 0.82 \text{ m } (2.7 \text{ ft.})$$
$$\text{Column height} = 19.2 \text{ m } (63 \text{ ft.})$$
$$\text{Tray Stack Height} = 13.2 \text{ m } (24 \text{ in. spacing})$$

First, we find the cost of the column vessel itself. From Table 4.11 we have an FOB cost (BC) of about $8350. Assuming carbon steel construction, we have F_m and F_p as well as the MPF equal to 1.0 (why?). The resulting module factor (MF) is 4.23 and the update factor is UF = 359/115 = 3.12. The bare module cost (BMC) is then obtained from:

$$\text{BMC(vessel)} = \text{UF } (\text{MF} + \text{MPF} - 1) (\text{BC}) = \$110,000. \quad (4.48)$$

The tray stack is also calculated from Table 4.11 with L = 43.3 ft. (13.2 m) and D = 2.7 ft. (0.82 m) we have BC = $1150. Assuming bubble cap trays with 24″ spacing, we have MPF $(F_s + F_m + F_t)$ = 2.8. Note there is no module factor for tray stacks. As a result we have the following cost:

$$\text{BMC(vessel)} = \text{UF } (\text{MPF}) (\text{BC}) = \$ 10,000 \quad (4.49)$$

Now the column condenser requires both utility costs and capital costs. The utilities can be calculated first. From the above example,

$$Q_{\text{cond}} = (40.9/20.2) [19.9(30.2) + 0.3(40.7)] = 1242 \text{ kW} \quad (4.50)$$

and we assume the following for cooling water:

$$C_{pw} = 75.3 \text{ J/gmol K}$$
$$T_{\text{out}} = 319 \text{ K}$$
$$T_{\text{in}} = 300 \text{ K}$$
$$\mu_w = Q_c/C_{pw} (T_{\text{out}} - T_{\text{in}}) = 863 \text{ gmol/s}$$
$$\text{Price} = 5.2¢/1000 \text{ gal.} = \$2.47 \cdot 10^{-7}/\text{gmol}$$
$$\text{Service factor} = 0.904$$
$$\text{Days of operation} = 0.904 (365) = 330 \text{ days yr.}$$

As a result, the annual cooling water utility cost is given by:

$$(\$2.47 \cdot 10^{-7}) (863) (3600) (24) (330) = \$6080/\text{yr}. \quad (4.51)$$

The condenser can be sized and costed as follows. The overall heat transfer coefficients can be estimated from Table 4.2, as well as Perry's Handbook. For an acetone-water (shell) / water (tube) system, we have U = 100 – 200 Btu/hr. ft^2°F and we select U = (100) (5.678) = 567.8 W/m^2 K. Also, from the example we have:

$$\Delta T_{\text{lm}} = [(329 - 300) - (329 - 319)]/ \ln(29/10) = 17.8 \text{ K}$$
$$A = Q_c/(U \Delta T_{\text{lm}}) = 122 \text{ m}^2 \sim 1300 \text{ ft}^2 < 10,000 \text{ ft}^2 \text{ (max.)} \quad (4.52)$$

From Table 4.12, the base cost (BC) = $10,800 and for a floating head, carbon steel heat exchanger MPF = 1.0 and the module factor (MF) = 3.29. Hence, the bare module cost is:

$$BMC = 3.12 \, (10800) \, (3.29) = \$110700 \sim \$111,000 \qquad (4.53)$$

Finally, the reboiler can be costed in a similar manner. First, the utility costs are computed from the heat duty in the above sizing example:

$$Q_{reb} = 40.9 \, (40.7) \, (\text{water}) = 1665 \text{ kW}. \qquad (4.54)$$

and we assume steam is available at 150 psig with the following characteristics:

$$T_{steam} = 459 \text{ K and } \Delta H_{vap} = 3587 \text{ J/gmol} \qquad (4.55)$$

so that μ_s = 463.8 gmol/s. If we are given a steam price of $4/1000 lbs and a condensate credit of $1.2/1000 lbs, we apply a net price = $2.8/1000 lbs or $1.11·10^{-4}/gmol. The annual utility cost with a service factor of 0.904 is then:

$$\text{steam cost} = (\$1.11 \cdot 10^{-4}) \, (463.8) \, (3600) \, (24) \, (330) = \$1.468 \cdot 10^6/\text{yr}. \qquad (4.56)$$

Sizing the reboiler first requires an overall heat transfer coefficient. For a water (shell) / steam (tube) system we have from Table 4.2, U = 250 – 400 Btu/hr ft^2°F and we select U = 250 = 1420 W/m^2 K. Also, ΔT_{lm} = (459 – 385) = 74 K and so the area is:

$$A_{reb} = Q_{reb}/U \, \Delta T_{lm} = 15.8 \text{ m}^2 = 170 \text{ ft}^2 \qquad (4.57)$$

From Table 4.12, we have BC = $2900, MPF = 1.45 (for a slightly higher pressure and carbon steel kettle reboiler) and MF = 3.29. The resulting bare module cost becomes:

$$BMC = (3.12) \, [(3.29 + 1.45 - 1) \, (2900)] = \$33,840 \sim 34,000 \qquad (4.58)$$

In summary, the column has the following capital and utility costs.

Capital Costs

Vessel (19.2m x 0.78m)	$ 110,000
Tray stack (13.2m x 0.78m)	10,000
Condenser	111,000
Reboiler	34,000
Total	$ 265,000

Utility Costs

Cooling water	$ 6,000/yr
Steam @ 150 psig	$ 1,468,000/yr.
Total	$ 1,474,000/yr

4.4 SUMMARY

This chapter was devoted to sizing and costing calculations for preliminary process design. Our goal in developing these calculations was to obtain rough estimates quickly and to observe qualitative trends for candidate designs. As a result, our estimated capital costs will be accurate within 25 to 40%. This is considered sufficient for preliminary design.

In the next chapter the bare module costs and the operating costs will be combined into an overall assessment of the plant economics. Here a key consideration will be the application of appropriate economic metrics in order to evaluate alternative designs.

REFERENCES

Douglas, J. M. (1988). *Conceptual Design of Chemical Processes.* New York: McGraw-Hill.

Fair, J. R. (1961, September). *Petro. Chem. Eng.*, **33** (10), 45.

Guthrie, K. M. (1969, March 24). Capital cost estimating. *Chemical Engineering,* 114.

Reid, R. C., Prausnitz, J. M., & Poling, B. E. (1987). *The Properties of Gases and Liquids.* New York: McGraw-Hill.

Perry, R. H., Green, D. W., & Maloney, J. O. (Eds.). (1984). *Perry's Chemical Engineers' Handbook*, 6th ed. New York: McGraw-Hill.

Peters, M., & Timmerhaus, K. (1980). *Plant Design and Economics for Chemical Engineers.* New York: McGraw-Hill.

Pikulik, A., & Diaz, H. E. (1977). "Cost Estimating Major Process Equipment." *Chemical Engineering,* **84** (21), 106.

Welty, J., Wicks, C., & Wilson, R. (1984). *Fundamentals of Momentum, Heat and Mass Transfer.* New York: Wiley.

Westerberg, A. W. (1978, August). *Notes for a Course on Chemical Process Design.* Taught at INTEC, Santa Fe, Argentina.

EXERCISES

1. A gas stream of 1 kgmol/s consisting of 50% mol H_2, 50% mol CH_4 is available at 100 kPa, 300°K. If the stream is compressed up to a pressure of 3000 kPa and delivered at 350°K, determine the required investment and annual operating cost for the two following cases:
   a. Two compression stages with intercooling and a final cooler
   b. Three compression stages with intercooling and a final cooler

(Use only simple enthalpy balances.)

Data Investment cost: bare module cost updated to January 1994 prices (Guthrie's method)

Service Factor:	0.904
Driver:	electric motor
Cost electricity:	3¢/kwh
Cooling water:	inlet 303°K ; outlet 325°K
Cost cooling water:	5.5¢/1000 gal.

Minimum temperature approach: 10°K

2. A mixture of 50 gmols/s n-butyraldehyde (NBA), 30 gmol/s iso-butyraldehyde (IBH), and 20 gmol/s isobutanol (IBA) are to be separated at 1 bar. Assume that the feed is 50% vaporized as it enters the column.

   Assume overhead recoveries for IBH and IBA of 99% and 1%, respectively.
   a. Find the overhead and bottoms compositions, the theoretical and actual number of trays, and the reflux ratio for this column.
   b. Find the column height and diameter. Determine the column cost from Guthrie's article for July 1968. Use carbon steel for all parts.
   c. Using a *very simple* enthalpy balance, size the condenser in this column.
      Assume cooling water entering at 310 K with a 15 K temperature rise. Find the cost of the condenser from Guthrie's article.

3. 300 gmol/sec of a 70/30 mole % mixture of benzene and xylene are separated at 2 atm. The light and heavy recoveries are 0.99 and 0.01, respectively.
   Potentially useful information:

	ΔH_{vap} (kJ/gmol)	P_{vap} (350 K) kPa	T_{boil}(K)
Benzene	29.32	91.5	353.3
O-xylene	34.94	11.17	417.6

   a. Find the theoretical number of trays, the reflux ratio, and the column height if 24″ trays are used.
   b. For a reflux ratio of 0.5, find the reboiler and condenser duties if the column in part a) has bubble point feed.

4. Starting from the relationship for work in a single centrifugal compressor derive the equation for N compressors with intercooled stages. State all assumptions in the derivation.
   a. Show that equal compression ratios are optimal with intercooling.
   b. Derive an analogous equation for a multicompressor system *without* intercooling.
   c. What is the actual compressor horsepower to compress 40 gmol/sec. of propane from 300 K and 1 atm to 10 atm with intercooling? What is the final temperature? State all assumptions.

Exercises

5. 160 gmol/s of propane requires a cooling load 200 kW to cool the stream to 260 K.
   a. How many stages of refrigeration are required and which refrigerant should be used in each cycle? (Choose refrigerants from problem 6.)
   b. If $\Delta T_{min} = 5$ K, choose operating pressures for the refrigeration cycles.
   c. What is the total compressor work and cooling water duty if the coefficient of performance is 4?

6. A stream of n-butane needs to be cooled from 300 K to 250 K. The change in heat content for this stream is 300 kW.

Possible Refrigerants	Boiling point (K)	Critical Temperature
Ethane	184.5	304
Propane	231.1	370
Isobutane	261.3	408

   a. How many stages of refrigeration are required? Which refrigerants among the above should be used in each stage to maintain the lowest cycle pressures above 1 atm?
   b. If $\Delta T_{min} = 5$ K, choose the operating pressures in the refrigeration cycles using the coolants in part a).
   c. For a coefficient of performance of 5, find the compressor work and the cooling water duty for this refrigeration system.

7. A 50 gmol/s stream of nitrogen needs to be compress from 1 atm and 310 K to 35 atm. The stream must also be delivered at 650 K. Assume a constant C_p of 7 cal/gmol for nitrogen.
   a. For a staged compression system with intercooling (back to 310 K), calculate the work and the amount of heating and cooling required. Assume an average compression ratio of 2.5.
   b. Your boss believes that you can deliver the stream at 650 K more cheaply by avoiding intercooling in the compression. Is she right or wrong? How can you make this argument quantitatively?

ECONOMIC EVALUATION   5

In Chapters 2, 3, and 4, respectively, we selected a flowsheet, performed a mass and energy balance, and calculated the equipment capacities and operating costs. We are now in a position to evaluate the profitability of the process. In this chapter, we classify costs and revenues for the process and organize these into capital and operating expenses. Next, we consider simple measures of profitability so that the advantage of a design alternative can be assessed quickly. We also consider more detailed forms of comparison that require the time value of money. With this concept we are in a position to consider the effects of taxes and depreciation, along with operating and capital expenses, and to introduce the net present value, a widely accepted measure of profitability. Finally, we use the tools from this analysis to consider the implications of more detailed cash flows when performing an economic analysis.

5.1   INTRODUCTION

How much does it cost to produce a chemical? How do we measure process profitability? To consider these questions we need to assess the costs of building and operating the chemical process. Now that the costs of the capital equipment are known and the utility and raw material requirements have been determined, we need a systematic accounting strategy to evaluate the overall profitability of the process. In Chapter 2, the simple measure of maximum potential profit was introduced. This chapter extends many of the concepts associated with these calculations and justifies some of the assumptions behind the simple gross profit calculation. In this section we introduce a few working definitions for capital and operating costs, which were calculated based on the methods in Chapter 4. In the next section we derive simple measures for assessing profitability. Section 5.3 then introduces the concept of time value of money and develops more accurate evaluation mea-

Sec. 5.1   Introduction   143

sures based on this concept, while section 5.4 extends these concepts to include taxes and depreciation. This analysis is based on income and payment streams that represent cash flows. Detailed cash flow analysis is then covered in section 5.5. We also consider inflation and investment risk in sections 5.6 and 5.7. Finally, section 5.8 summarizes the chapter with a guide to further reading.

To begin an economic evaluation of a process, we first define some terms and classify a number of items. Costs associated with the process can be divided into:

- **Fixed Costs**—Direct investment as well as overhead and management associated with this investment. In particular, we are interested here in *capital investment* costs, which are incurred initially at the start of the project.
- **Variable Costs**—Raw material, labor, utilities, and other costs that are dependent on operations. Here we are primarily concerned with *manufacturing costs,* which are continuous expenses, given on an annual basis.

A typical distribution of these costs is given in sections 5.1.1 and 5.1.2.

5.1.1 Capital Investment

This item represents all of expenses made at the beginning of the plant life. Included in this initial expense are the costs to build and start up the process. The total capital investment is given by fixed and working capital. Further classification of these categories is given below.

Fixed capital represents the cost of building the physical process itself and can be further classified into:

- **Manufacturing capital**—Bare Module Cost (BMC, see Chapter 3) of equipment as well as a 25% contingency on this figure.
- **Nonmanufacturing capital**—Buildings, service, land (typically 40% of BMC).

Working capital represents funds required to operate the plant due to delays in payment and maintenance of inventories. As these funds are replaced by additional revenues, the working capital represents the money available to fill the tanks and meet the initial payroll and expenses. This varies from reference to reference and is usually 10 to 20% of the total (fixed and working) investment cost. We will standardize on the following:

- Raw material and product inventories (typically 7 days)
- Goods in process (e.g., catalyst)
- Accounts receivable (30-day lag in payment) = 1 month manufacturing production cost = 10 to 20% total investment with depreciation.

Douglas (1988) also suggests a simpler form:

$$\text{Working Capital} = 0.15 \text{ (Total Investment)} = 0.194 \text{ (Fixed Investment)}$$

5.1.2 Manufacturing Costs

These costs include all expenses that are made on a continuous basis over the life of the plant. They involve expenses that directly relate to the day-to-day operation of the plant as well as indirect expenses such as taxes, insurance, and depreciation. A typical classification of manufacturing costs is given by:

- **Raw Materials**—represent feedstocks for the process that are consumed on a continuous basis.
- **Credits**—include usable purge gases (fuel) as well as utilities (steam, electricity) and by-products that are generated on a continuous basis.
- **Direct Expenses**—include labor, supervision, payroll (typically, 20% of labor and supervision), utilities (electricity, steam, cooling water), maintenance (repair), supplies (2% of fixed investment), and royalties (typically on a licensed operation or on a catalyst).
- **Indirect Expenses**—include depreciation (8%/year), local taxes, and insurance (3%/year).

The percentages given above represent typical values that can vary from project to project.

One item that needs further mention is depreciation. This can be considered to be a cost prorated throughout equipment life. For instance, a $20,000 car depreciated at 10% per year has a book value of $14,000 after 3 years. However, this expense is never really incurred and is actually a fictitious cost since nobody pays or receives it. It is used, however, in some simple economic measures for comparison evaluations. Moreover, the real purpose of depreciation costs lies in the calculation of taxes and deduction for depreciation write-offs.

In the next section we discuss simple measures for the economic evaluation of projects. These measures are used to assess quickly the profitability of a project. However, they do not consider the timing of payments and incomes and do not always yield an accurate profitability measure.

5.2 SIMPLE MEASURES TO ESTIMATE EARNINGS AND RETURN ON INVESTMENT

In this section we discuss some simple and quick ways to assess process profitability. While they are easy to use and are common in process engineering, they all have serious shortcomings and need to be considered cautiously. We will see, for instance, that unfavorable processes have favorable simple economic measures and vice versa. What follows below is a brief listing and illustration of these measures.

We define:

Sec. 5.2 Simple Measures to Estimate Earnings and Return on Investment

- Gross profit = Gross sales − manufacturing cost
- Net profit before taxes = Gross profit − SARE (Sales, Administration, Research, & Engineering) expenses (10% sales)
- Net annual earnings = Net profit before taxes − taxes on net profit

With these items and those defined above we have the following simple economic measures:

- **Return on investment (ROI)**   is the (net annual earnings)/ (fixed and working capital). A typical minimum desired ROI is about 15% (or 30% before taxes). However, ROI does not take time value of money (i.e., the timing of expenses and incomes) into account. It is only useful for a mature plant project when startup cost is not significant.
- **Payout time**   is the (total capital investment)/ (net annual profit before taxes + annual depreciation). Note that the depreciation that was part of the manufacturing cost is added back and cancelled. Therefore, this measure represents the total time to recover investment based on the net income *without* depreciation. Like the ROI, the payout time does not take time value of money (i.e., the timing of expenses and incomes) into account.
- **Proceeds for dollar outlay (PDO)**   is the (total net income over life)/(total investment). This measure is calculated *without* including depreciation. Aside from neglecting the timing of payments, PDO does not consider the length of the project.
- **Annual proceeds per dollar outlay (APDO)**   is PDO divided by the project life. It has the same shortcomings as the above measures and favors short quick-return projects over long steady projects.
- **Average income on initial cost (AIIC)**   is the net profit before taxes (including depreciation) divided by (fixed and working) capital. This measure has the same characteristics as payout time, but also includes investment recovery as a cost.

To illustrate how these measures are used, consider the following economic process evaluation.

EXAMPLE 5.1 Simple Economic Measures for Process Evaluation

Consider the following process with a capacity of $120 \cdot 10^6$ lb/yr and a product price of 20¢/lb. The economic information for this plant is given by:

Fixed capital	$\$15 \cdot 10^6$
Working capital	$\$3 \cdot 10^6$
Fixed and Working Capital	**$\$18 \cdot 10^6$**
Raw material (@ 8¢/lb prod)	$\$9.6 \cdot 10^6$/yr
Utilities (@ 1.2¢/lb prod)	$\$1.44 \cdot 10^6$/yr
Labor (@ 1.5¢/lb prod)	$\$1.8 \cdot 10^6$/yr

Maintenance (6% yr f.c.) $900,000/yr
Supplies (2% yr f.c.) $300,000/yr
Depreciation (8%/yr)
(or straightline over ~12 yrs) $1.2 · 10^6/yr
Taxes, insurance (3%/yr) $450,000/yr
Total Manufacturing Cost   **$15.69 · 10^6/yr**
(13.1¢/lb.)

Gross Sales (120 · 10^6) (0.2) = $24 · 10^6/yr
Manufacturing Cost   − $15.69 · 10^6/yr
Gross Profit $8.31 · 10^6/yr
SARE Expenses (at 10% sales)   − $2.4 · 10^6/yr
Net Profit before Taxes = $5.91 · 10^6/yr
Taxes (50% net profit)   − $2.96 · 10^6/yr
Net Profit after Taxes $2.95 · 10^6/yr

Using the economic measures defined above, the following evaluation can be made of the plant.

ROI	= 2.95 · 10^6 / 18 · 10^6 =	16.4 %
Payout Time	= 18 · 10^6 / (5.91 · 10^6 + 1.2 · 10^6) =	2.53 years
PDO	= (5.91 · 10^6 + 1.2 · 10^6) 12 / 18 · 10^6 =	4.74
APDO	= 4.74/12 =	0.395
AIIC	= 5.91 · 10^6 / 18 · 10^6 =	0.328

While the above measures are easy to calculate, they lead to inconsistent results when trying to compare alternative projects. One area of disagreement occurs when a person decides to invest a lot of money with a modest return or a little money with large return.

EXAMPLE 5.2   Comparison of Project Alternatives

Consider the following two 5-year projects with the following economic data:

	1	2
Fixed capital	2.5 · 10^6	250,000
Working capital	500,000	50,000
Net income before taxes	10^6	200,000
Depreciation	500,000	50,000
ROI (Pretax)	10^6/3 · 10^6 = 0.33	200,000/300,000 = 0.66
Payout period	3 · 10^6/1.5 · 10^6 = 2 yrs.	300,000/250,000 = 1.2 yrs.
PDO	(1.5 · 10^6) (5)/3 · 10^6 = 2.5	(250,000)(5)/300,000 = 4.17
APDO	0.5	0.83
AIIC	10^6/3 · 10^6 = 0.33	200,000/300,000 = 0.66

For all indicators, alternative 2 is better. However, we know that by paying $2.7 · 10^6 more we make $450,000 more per year. Clearly, we need a better basis for comparison.

5.3 TIME VALUE OF MONEY

The simple economic indicators given above are often not a good basis of comparison. Hence, we have to deal with a more rigorous analysis. To consider the schedule of payments and income, we know that value of money changes due to:

1. Interest, which reflects rent paid on the use of money.
2. Returns received from competing investments. Consequently, the investment must compensate the loss of opportunity to invest elsewhere.
3. Inflation, which can be compensated in the interest rate and will be considered in section 5.6.

What is the correct interest rate for a company to choose? We can argue that this is the rate that the company receives for its money when the money is sitting in reserve. Based on its history, a company may know it can virtually always invest in projects somewhere, with a guaranteed rate, say 10%. If it has the mechanisms to move money into and out of such projects with enough fluidity, then it can justifiably use 10% as its "bank interest rate." It often terms this interest rate as the least acceptable return for any project.

For this economic analysis first need to consider the effect of a compounded interest rate. We define P as the present value of a sum and S as its future worth. Compounding the interest after a one-year period gives:

$$S = (1 + i) P$$

So a $1000 investment ($P$) with a 10% interest rate has a future worth (S) after one year of $1100. For multiple compounding periods (n) we derive:

Year	S	Interest end year
0	P	iP
1	$P + iP$	$i(1 + i)P$
2	$(P + iP)(1 + i)$	$i(1 + i)^2 P$
—	—	—

$$\Rightarrow S = P(1 + i)^n \quad (5.1)$$

Similarly, the present value of a future value is given by

$$P = S/(1 + i)^n \quad (5.2)$$

and $1/(1 + i)^n$ is known as a discount factor. For example, if the future worth is $S = 10^6$ in 100 years with a compound interest, what is present value of the principal (P) now? Here, $P = 10^6/(1 + i)^{100}$, and for $i = 0.05$, $P = \$7604$. On the other hand, if $i = 0.2$, P is only $0.012.

5.3.1 Nominal and Effective Interest Rates

The above relations for P and S hold only if the compounding period coincides with the nominal rate per period (e.g., annually). Interest rates for multiple compounding per year can actually yield a slightly higher effective rate than the nominal one. This is modeled by the following relations:

i, nominal interest rate
n, number of periods (years) for nominal rate
m, number of compounding intervals/nominal period

$$S = P(1 + i/m)^{mn} \tag{5.3}$$

For example, if we have $i = 6\%$ compounded quarterly, then for $m = 4$ compounding intervals per period and $n = 1$ year, we calculate

$$(1 + i/m)^{mn} = (1 + .06/4)^4 = 1.0614$$

giving an effective rate is 6.14%.

5.3.2 Continuous Interest

If we now take the number of compounding intervals to infinity, then we approach a limit for the effective interest rate. This concept is useful for process economics as we are continuously making payments or receiving revenues. Since the process is continuously producing income or incurring expenses, what would an effective rate be?

$$\lim_{m \to \infty} S = P \lim_{m \to \infty} (1 + i/m)^{mn} = P \lim_{m \to \infty} (1 + i/m)^{(m/i)in}$$

Since:

$$\lim_{x \to \infty} (1 + 1/x)^x \to e^x, \text{ we have } \lim_{m \to \infty} S = P e^{in} \tag{5.4}$$

and for continuous compounding, the effective rate = $e^{in} - 1$ (e.g., 6% nominal = 6.18%).

5.3.3 Annuities

In order to find the present value (P) of distributing an equal payment on a regular basis (R), we consider the following timeline and derive the relations shown in Figure 5.1.

We assume that this payment is at end of period (e.g., mortgage, loan, or life insurance premium). Applying the discount factor to each payment yields:

$$P = R/(1+i) + R/(1+i)^2 + \cdots = R \sum_{k=1}^{n} 1/(1+i)^k$$

By telescoping the series we have to discount on all annuities:

Sec. 5.3 Time Value of Money

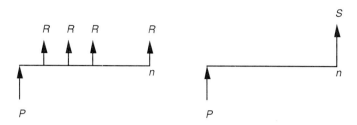

FIGURE 5.1 Timelines for payments.

$$P[1-(1+i)] = R\sum_{k=1}^{n} 1/(1+i)^k - R\sum_{k=0}^{n-1} 1/(1+i)^k = R[1/(1+i)^n - 1]$$

or after simplification, we have:

$$R = P\, i/[1 - (1+i)^{-n}] \tag{5.5}$$

where the term multiplying P represents the capital recovery factor. In terms of future yield, we apply the relation between P and S and obtain:

$$R = P\, i/[1 - (1+i)^{-n}] = (S/(1+i)^n)\, i/[1 - (1+i)^{-n}]$$

$$R = Si\, /[(1+i)^n - 1] \tag{5.6}$$

Also, an annuity of n payments timed at the beginning of the period shown in Figure 5.2 leads to the relations:

$$R = Si\, / [(1+i)^{n+1} - (1+i)] \tag{5.7}$$

$$R = Pi\, /[(1+i) - (1+i)^{1-n}] \tag{5.8}$$

EXAMPLE 5.3 Annuity Payments

Consider a $10,000 loan borrowed at present to be repaid in 60 monthly installments at the end of each month. If the nominal (annual) rate is 12%, what is the monthly payment?

Here $i = 0.01$, $n = 60$ and $P = \$10,000$. Applying the capital recovery factor from Eq. (5.5) we have:

$$R = P\, i/[1 - (1+i)^{-n}] = \$222.4\text{ /month}$$

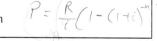

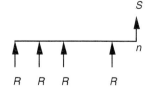

FIGURE 5.2 Present and future value of annuities.

EXAMPLE 5.4 Future Value of Regular Payments

Consider a life insurance policy with a lump sum payment starting at 65. Monthly payments start at 21 by paying a premium of $10 at the beginning of each month. If we assume a nominal rate of 3%, what is the value of the lump sum?
Assume $i = 0.03/12 = 2.5 \cdot 10^{-3}$, $n = 44 \cdot 12 = 528$. By paying at the beginning of the month, from (5.7), we have:

$$S = R[(1 + i)^{n+1} - (1 + i)]/i = \$10976.$$

5.3.4 Continuous Payment over a Fixed Period

This payment schedule simulates the expenditures for continuous production. Here we increase the number of pay intervals (m) to infinity.

$$P = R[1 - (1 + i)^{-n}]/i = (\overline{R}/m)[1 - (1 + i/m)^{-mn}]/(i/m)$$

where $\overline{R}$ is the average yearly payment $\int R dt = \overline{R}$. Taking the limit as m goes to infinity,

$$\lim_{m \to \infty} P = \overline{R}[1 - e^{-in}]/i \qquad (5.9)$$

Moreover, since $S = P e^{in}$, we have:

$$S = \overline{R}[e^{in} - 1]/i \qquad (5.10)$$

EXAMPLE 5.5 Continuous Payments

The (continuous) energy bill for a boiler is prorated at $1000 per month. Assume $i = 0.10$ per year, what is the present value of energy cost for a two-year operation? Here $n = 24$, $\overline{R} = 1000$ and $i = 0.00833$ from Eq. (5.9):

$$P = \overline{R}[1 - e^{-in}]/i = \$21752$$

5.3.5 Perpetuities

Consider the present value of an expenditure that needs to be made for an infinite time period. To fund such a payment schedule, the interest that accrues in each interval needs to support each payment. Therefore, if we need to supply continuous utility indefinitely then the annuity becomes:

$$P = \overline{R}[1 - e^{-in}]/i = \overline{R}/i \text{ as } n \to \infty \qquad (5.11)$$

Now if the payment interval is made after a multiple (z) of compounding periods, then over z years, interest earned on P should pay for C, as shown in Figure 5.3.

Sec. 5.3 Time Value of Money

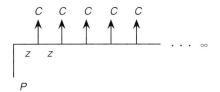

FIGURE 5.3 Replacement cost into perpetuity.

$$C = P(1+i)^z - P \quad \text{or} \quad P = C/[(1+i)^z - 1] \tag{5.12}$$

This case occurs in the periodic replacement of process equipment. Here C is the replacement cost for equipment (cost − salvage value) and C_0 is the original price. The capitalized cost for the equipment, K is given by:

$$K = C_0 + C/[(1+i)^z - 1] = C_o + P \tag{5.13}$$

The use of perpetuity calculations is also useful for comparing equipment with different lives.

EXAMPLE 5.6 Comparison of Two Reactors

Consider a stainless steel and a carbon steel reactor with the following data:

	Reactor A (SS)	Reactor B (CS)
Original cost (C_0)	10,000	5,000
Life	8	3
Replacement (C) (C_0 − salvage)	8,000	5,000
K (@ i = 10%)	$16,995	$20,105

Based on the capitalized cost into perpetuity, reactor A is actually cheaper.

5.3.6 Using Time Value of Money for Cost and Project Comparisons

To evaluate the profitability of a process we will use discounted cash flow calculations based on the concepts outlined above. Even here there are different criteria for comparing projects. We will consider three approaches:

1. Net present value (NPV) of project with a given rate of return (i)
2. Annualized payments with a given rate of return (i)
3. Calculated rate of return (i^*) with NPV = 0.

The first criterion (NPV) gives the present value of all payments and provides a basis of comparison for projects with different payment schedules but similar lifetimes.

The project with the highest NPV profit or lowest NPV cost is superior. The method of annualized payments has the same benefits as the NPV but also allows comparison of projects with different lifetimes.

The rate of return calculations can be interpreted as the interest rate that can be compared with a competing investment. Here a typical investment (e.g., bond or savings account) has an NPV = 0 for its rate of return. The higher rate of return is clearly favored, but this criterion does not consider the magnitudes in the investment. Sometimes it is useful to calculate rate of return to compare projects, as the discounted cash flow (DCF) rate of return does not need to be specified in advance (see section 5.5). However, the rate of return calculation is only useful for projects with both costs and income.

EXAMPLE 5.7 Project Comparison

Consider two investments both with a 5-year lifetime to be evaluated before taxes.

	A	B
Capital, fixed & working	$3 \cdot 10^6$	$3 \cdot 10^5$
Income before taxes ($/yr.)	10^6	200,000

For project A)

$$\text{NPV} = -3 \cdot 10^6 + 10^6 [1 - (1+i)^{-5}]/i$$

and for project B)

$$\text{NPV} = -300{,}000 + 200{,}000 [1 - (1+i)^{-5}]/i.$$

Using a variety of interest rates yields a set of NPVs. If we set NPV = 0 and iterate for the value of i, we obtain i^*.

	A	B
$i = 10\%$	$790,800	$458,200
$i = 20\%$	$ −9,387	$298,120
i^* (NPV = 0)	19%	60%

Superiority of the project obviously depends on the rate of return that is selected. Thus, a reasonable i based on competing investments is needed for an NPV calculation. Moreover, a high rate of return favors projects with income payments at beginning.

Instead of comparing present values, we can also compare income on annualized basis. This leads to:

A. $R = 10^6 - 3 \cdot 10^6 \, i/[1 - (1+i)^{-5}]$
B. $R = 200{,}000 - 300{,}000 \, i /[1 - (1+i)^{-5}]$

For projects with same lives, the conclusions are same as the NPV calculation with fixed i.

	A	B
$i = 10\%$	$208,600	$120,870
$i = 20\%$	$ −3,139	$ 99,685

Sec. 5.3   Time Value of Money

EXAMPLE 5.8   Cost Comparison for Equal Lifetimes

Consider the problem of buying an old car with a higher operating cost or a new car with a lower operating cost, given the data below.

	Old	New
Price	$2,000	$12,000 (includes trade-in)
Operating cost	$1,000/yr.	$ 300/yr.

Using a project life of 5 years with $i = 6\%$ (this is the investment rate for money you didn't spend) we express the NPV of each project as follows:

Old) NPV = $2,000 + 1,000[\{1 - (1 + i)^{-5}\}/i]$ = $6,212 ($1,475/yr.)
New) NPV = $13,000 + 300[\{1 - (1 + i)^{-5}\}/i]$ = $14,263 ($3,386/yr.)

Despite the higher operating cost, the old car has a lower NPV.

5.3.7   Cost Comparison for Different Project Lives

To deal with project comparisons that have different lives, we have three alternative approaches:

1. Project each project life into perpetuity, then do an NPV calculation.
2. Put both project lives on the same time basis (use least common multiple, LCM) then do NPV calculations.
3. Convert all income and costs to an annualized basis.

The results of these alternatives will be similar, although the selections may differ slightly depending on the timing of the payments.

EXAMPLE 5.9   Cost Comparison with Different Lives

Consider two pumps, of carbon steel and stainless steel, respectively, with different operating lives. Based on the data below and a 10% rate of return, which pump is more economical? We now consider three different ways to assess this.

	CS	SS
Purchased price (C_0)	$5,000	$8,000
Salvage value ($C_0 - C$)	$ 0	$2,000
Operating cost/year (R)	$ 200	$ 150
Operating life	4 years	8 years

1. Compare projects into perpetuity.

   Consider the payment schedule in Figure 5.4, with each project life (z) being repeated endlessly.

   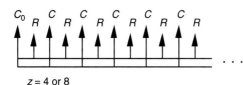

   FIGURE 5.4 Payments into perpetuity.

   Using the discount factors derived above, the net present value of each project becomes:

   $$NPV = C_0 + R/i + C/[(1+i)^z - 1]$$

   and the table below summarizes the calculations:

	CS	SS
C_0	$ 5,000	$ 8,000
R	$ 200	$ 150
C	$ 5,000	$ 6,000
z	4	8
NPV	$17,773	$14,747

2. Common life for both projects

   The least common multiple of both projects, LCM(4,8) = 8 and the cost schedule for an 8-year period is given in Figure 5.5 for each of these projects.

   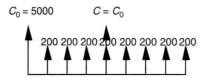

   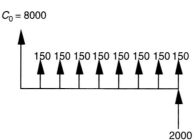

   FIGURE 5.5 Least common multiple payments.

   Using the discount factors, the NPV calculations are given by:

   CS) NPV = $5{,}000 + 200\,[1 - (1+i)^{-8}]/i + 5{,}000/(1+i)^4 = \$9{,}482$

   SS) NPV = $8{,}000 + 150\,[1 - (1+i)^{-8}]/i - 2{,}000/(1+i)^8 = \$7{,}867$

3. Annualized costs for each project
   A simpler strategy is to calculate the NPV for each project over its lifetime and to convert this amount to an annualized basis. In this way we have:

   $$NPV = C_0 + R[1-(1+i)^{-n}]/i - (C_0 - C)/(1+i)^z$$
   $$X = NPV\, i/[1-(1+i)^{-z}]$$

   and with the data for these two projects we have:

	CS	SS
life, z	4	8
NPV	$5,634	$7,867
X	$1,777	$1,475

Note that the NPV by itself provides a misleading comparison if the project life is different. However, since we account for changing project lives in all three of these methods, any of these methods should give the correct decision. Among the three methods the first and third incorporate essentially the same results while the second may differ slightly due to the timing of payments at the end of the project life. The importance of these end payments should be considered carefully in the project comparison.

5.4 COST COMPARISON AFTER TAXES

In the previous sections, we considered several before-tax profitability measures. These were used with or without depreciation. In after-tax calculations, depreciation plays an important and a complicating role. Moreover, it is only in the after-tax calculations that depreciation has an unambiguous meaning. Depreciation can be treated as a yearly expense that accounts for obsolescence or wear and tear of equipment. Here, a company faces the dilemma of showing a high net worth to its stockholders (with no depreciation) and, on the other hand, writing off a large depreciation for taxes. We will consider the latter case and will see that profitability is maximized when depreciation can be done quickly.

5.4.1 Depreciation as a Tax Incentive

What is depreciation? Many people suggest it is the amount of money we must put aside yearly to replace a major piece of equipment when it comes to the end of its normal operating life. However, a tangible result of depreciation is the effect on the payment of taxes. Note that a company is allowed to deduct operating expenses in the year they occur. Thus, it can deduct wages paid, utilities bought, and so on, directly from income to arrive at a net income on which it has to pay income taxes to the government. However, the government will not allow a company to deduct all of the money paid for major capital goods in the year in which it buys these goods. Rather, the government requires a company to deduct this investment cost over a period of years. Why would they make this distinction? Both are outflows of money needed to run the business.

One way to respond is as follows. If a company invests in an asset that will not lose value, such as a Rembrandt painting or a piece of undeveloped land, it can later sell the painting or land to recover its money. Such an investment is a trade of dollars for something else of value that, in principle, can be turned back into dollars. The company has neither gained nor lost value by making the investment. Such an expenditure has nothing to do with "expenses" for the company, therefore. In contrast, wages, once paid, are irretrievable. They are true expenses.

A major piece of equipment is somewhat like an investment, except it slowly loses value as it wears out. For some time after purchasing it, the company could, in principle, sell it and recover a portion of its value. The government does not deem the amount it could recover to be an expense until the amount is irretrievable.

We can compute the impact of depreciation on taxes with the following example. Consider a company with a $\$10^7$/yr profit before taxes. With a 50% tax rate and without depreciation, taxes are $\$5 \cdot 10^6$ per year. With a capital depreciation of, say, $\$10^6$ the taxes now become $0.5 \, (10^7 - 10^6) = \$4.5 \cdot 10^6$, and the after tax profit $= \$10^7 - \$4.5 \cdot 10^6 = \$5.5 \cdot 10^6$. Depreciation is thus a source of tax savings.

In general, for a profit π, depreciation D, and a tax rate t, we have

$$\text{taxes} = \pi \, t \text{ (no depreciation)}$$

$$\text{taxes} = (\pi - D)t \text{ (with depreciation)}$$

and

$$\text{tax credit} = Dt.$$

Commonly used depreciation methods are influenced by accuracy, simplicity, and profitability—and, of course, depend on what is legally allowed by the taxing authority. Here we concentrate on two popular depreciation methods:

1. *Straight line*—simple, equal write-offs during project life
2. *Declining balance*—early write-offs in project life

with a brief statement of the 1986 US tax laws, which represent a combination of the two. We first define the following notation and then develop the equations for each method.

C_I, initial unit cost
C_S, salvage value
C_D, depreciable value (replacement cost) $(C_I - C_S)$
n_t, tax life
n, total useful life $(n_t \leq n)$
f_j, depreciation factor, depends on method
D_j, depreciation in year j
B_j, book value in year j, $B_j = C_I - \sum_{k=1}^{j-1} D_k$

Sec. 5.4   Cost Comparison After Taxes

Straight line depreciation discounts an equal amount each year. Here the depreciation factor is constant and is given by: $f_j = 1/n_t$, $j = 1, n_t$ with $D_j = C_D f_j = C_D / n_t$.

For example, if $C_I = \$6 \cdot 10^6$, $C_S = \$1 \cdot 10^6$, and $n_t = 6$ yr, then we have $C_D = \$5 \cdot 10^6$ and $D_j = \$833{,}333$ for each year j. The present value (PV) of this tax credit with $i = 0.10$ (after tax rate of return) and $t = 0.5$ (tax rate) is given by:

$$PV = \sum_{j=1}^{n_t} D_j\, t/(1+i)^j = C_D\, t/n_t \sum_{j=1}^{n_t} 1/(1+i)^j = C_D\, t\, [1 - (1+i)^{-n_t}]/(i\, n_t)$$

$$= \$1.815 \cdot 10^6 \text{ tax credit}$$

On the other hand, the *declining balance method* depreciates only on the book value of the capital item. Here the annual depreciation is given by:

$$D_j = B_j f_j = \left(C_I - \sum_{k=1}^{j-1} D_k\right) f_j$$

and the salvage value is not considered. If we set $f_j = 2/n_t$ (twice the straight-line factor) then we have the *double declining balance method*. Developing these expressions we have:

$$D_1 = C_I f$$
$$D_2 = C_I (1-f) f$$
$$D_j = C_I (1-f)^{j-1} f$$

and similarly:

$$B_j = (1-f)^{j-1} C_I$$

Using the double declining method, the present value of the depreciation tax savings is given by:

$$PV = \sum_{j=1}^{n_t} D_j t/(1+i)^j = C_I t f/(1-f) \sum_{j=1}^{n_t} [(1-f)/(1+i)]^j$$

Expanding and telescoping the series and simplifying the equations for $f = 2/n_t$ gives:

$$PV = C_I t f [1 - \{(1-f)/(1+i)\}^{n_t}] /(i+f)$$
$$= (2 C_I t /n_t) [(1 - \{(1 - 2/n_t)/(1+i)\}^{n_t}] /(i + 2/n_t)$$

EXAMPLE 5.10   Depreciation with Double Declining Balance

Find the present value of the depreciation on an initial investment of $C_I = \$6 \cdot 10^6$ over a 6-year tax life and a rate of return of 10%. The tax rate is 50%. From

$$D_j = C_I (1-f)^{j-1} f$$
$$B_j = (1-f)^{j-1} C_I$$
$$PV = (2 C_I t / n_t) [1 - \{(1 - 2/n_t)/(1+i)\}^{n_t}]/(i + 2/n_t)$$

we have the following figures:

j	B_j	D_j
1	$6 \cdot 10^6$	$2 \cdot 10^6$
2	$4 \cdot 10^6$	$1.33 \cdot 10^6$
3	$2.67 \cdot 10^6$	$0.89 \cdot 10^6$
4	$1.78 \cdot 10^6$	$0.59 \cdot 10^6$
5	$1.19 \cdot 10^6$	$0.40 \cdot 10^6$
6	$0.79 \cdot 10^6$	$0.26 \cdot 10^6$

$$PV = 2 (\$ 6 \cdot 10^6)(0.5)/6 [1 - (0.667/1.1)^6] / 0.433$$
$$= \$ 2.193 \cdot 10^6 \text{ tax savings}$$

Note that since we depreciate faster, more is written off at beginning and the tax credit is higher.

5.4.2 Tax Reform Act of 1986

With the enactment of tax laws, the government defines the expected depreciable life for capital equipment, not the company. The 1986 tax law has categorized depreciable assets into 3-, 5-, 7-, 10-, 15-, and 20-year life classes. It has prepared extensive lists of what types of assets fall into each class. For example, process equipment, computers, copiers, cars, and light duty trucks fall into the 5-year class. Office furniture, cellular phones, and fax machines are in the 7-year class. Being in a class does not mean the item will last that long or that it will not last longer. It is simply the defined life by the government.

The 1986 tax reform act also lowers the *federal* tax rate from a previous 48% (we assume about 50% when we add in state taxes) to 34% and also introduces a few changes into depreciation calculations. In particular, half-year conventions are considered at beginning and end of project life and a double declining balance method is used in the first half, switching to a straight line in the remaining lifetime. This depreciation schedule is known as the Modified Accelerated Cost Recovery System (MACRS). For a 5-year life, depreciation is therefore calculated from the following table:

Year, j	f_j
1	0.20
2	0.32
3	0.192
4	0.1152
5	0.1152
6	0.0576

Sec. 5.4 Cost Comparison After Taxes

The present value of the tax savings is calculated directly from:

$$D_j = C_I f_j$$

$$PV = \sum_{j=1}^{n_t} D_j t / (1+i)^j$$

EXAMPLE 5.11 MACRS Depreciation

Find the present value of the depreciation on an initial investment of $C_I = \$6 \cdot 10^6$ over a 5-year tax life (with half years at beginning and at end) and a rate of return of 10%. The tax rate is 34% and from the MACRS depreciation method

j	f_j	D_j	B_j
1	0.20	$1.20 \cdot 10^6$	$6.0 \cdot 10^6$
2	0.32	$1.92 \cdot 10^6$	$4.80 \cdot 10^6$
3	0.192	$1.152 \cdot 10^6$	$2.88 \cdot 10^6$
4	0.1152	$0.69 \cdot 10^6$	$1.728 \cdot 10^6$
5	0.1152	$0.69 \cdot 10^6$	$1.038 \cdot 10^6$
6	0.0576	$0.348 \cdot 10^6$	$0.348 \cdot 10^6$

we have a tax savings of $PV = \$ 1.577 \cdot 10^6$.

Note that while this method combines both of the previous methods, the distribution over the tax life as well as the different tax rate does not allow a direct comparison of methods. For simplicity we will use straightline depreciation for our economic evaluations.

5.4.3 Net Present Value after Taxes

To complete this section, we consider all of the sources of income and expenditures in the discounted cash flow calculations. We consider the following items in our combined calculation:

C_I, fixed capital investment
C_S, salvage value
C_w, working capital
R, receipts (sales/year)
X, expenses (manufacturing cost w/o depreciation)
D, depreciation/year
t, tax rate

Economic Evaluation   Chap. 5

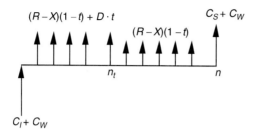

FIGURE 5.6 Combined cash flow for process.

i, after tax rate of return
n, useful plant life
n_t, depreciation life (tax purposes)

and make the following definitions,

profit/year = $R - X$
taxes = $(R - X - D)t$
after tax profit = $(R - X)(1 - t) + Dt$

with the cash flow schedule given in Figure 5.6.

The after tax, net present value (or venture worth) is given by:

$$NPV = -(C_I + C_w) + \sum_{j=1}^{n}(R-X)_j(1-t)/(1+i)^j \\ + \sum_{j=1}^{n_t} D_j t/(1+i)^j + (C_S + C_w)/(1+i)^n \qquad (5.14)$$

Now to compare alternative projects we can either:

1. Find highest *NPV* for given i
2. Find highest i for *NPV* = 0
3. Find greatest annualized value.

On the other hand, if we compare only costs then the expression for *NPV* becomes:

$$NPV(\text{cost}) = C_I + \sum_{j=1}^{n} X_j(1-t)/(1+i)^j - \sum_{j=1}^{n_t} D_j t/(1+i)^j - C_S/(1+i)^n \qquad (5.15)$$

and we choose the project with the lowest cost.

Sec. 5.4   Cost Comparison After Taxes

EXAMPLE 5.12   Project Evaluation

Consider the process in Example 5.1 where we have a capacity of $120 \cdot 10^6$ lb/yr with a product price of 20¢/lb. As shown previously the process has a fixed investment, $C_I = \$15 \cdot 10^6$; a salvage value, $C_S = 0$; and working capital, $C_w = \$3 \cdot 10^6$. Assuming a project life, $n = 15$ years; a tax life, $n_t = 12$ years; a rate of return, $i = 0.1$; and a tax rate of 50% (to cover additional state and local income taxes), what is the *NPV* of this project?

Manufacturing cost was calculated previously at $\$14.49 \cdot 10^6$/year (without depreciation) and if we assume straight line depreciation we have: $D = C_I/n_t = \$1.25 \cdot 10^6$/year. In addition, gross sales = $0.2 (120 \cdot 10^6) = \$24 \cdot 10^6$/yr and SARE expenses (@ 10% of sales) = $\$2.4 \cdot 10^6$/yr. Consequently, we have:

Revenues: $R = \$24 \cdot 10^6$/yr
Total Expenses: $X = (14.49 + 2.4) \cdot 10^6$/yr = $\$16.89 \cdot 10^6$/yr

Because $(R - X)_j$ and D_j are constant, equation (5.14) simplifies to:

$$NPV = -(C_I + C_w) + (R - X)(1 - t)[1 - (1 + i)^{-n}]/i + Dt[1 - (1 + i)^{-n_t}]/i + (C_S + C_w)/(1 + i)^n$$

$$= \$14.016 \cdot 10^6 \qquad (5.16)$$

If we convert the *NPV* to an annualized basis over a 15-year lifetime, we obtain:

$$NPV \, i/[1 - (1 + i)^{-n}] = \$1.84 \cdot 10^6.$$

Finally, to find the rate of return when the *NPV* is zero, we use the above expression for *NPV* and find i.

$$NPV = 0 \text{ when } i = 22.1\%.$$

To conclude, we reconsider this example with an *NPV* calculation in terms of continuous interest. Here the future worth and annuities are calculated by

$$S = Pe^{in}$$

and

$$A = P\,i/[1 - e^{-in}]$$

respectively. Now if R, X, and D are functions of time (τ), we can write the NPV as:

$$NPV = -(C_I + C_w) + \int_0^n (R(\tau) - X(\tau))(1-t)e^{-i\tau}d\tau + \int_0^{n_t} D(\tau)t\, e^{-i\tau}d\tau + (C_S + C_w)e^{-in} \quad (5.17)$$

If R, X, and D remain constant, the integral can be simplified to yield:

$$NPV = -(C_I + C_w) + (R - X)(1-t)[1 - e^{-in}]/i + Dt[1 - e^{in_t}]/i + (C_S + C_w)\,e^{-in} \quad (5.18)$$

Using the data from the above example gives us $NPV = \$14.65 \cdot 10^6$, which is close to the $\$14.016 \cdot 10^6$ obtained with the conventional method.

As a result of this small difference, we will standardize our calculations by using the conventional method with operating costs and revenues based on full year periods with payments timed at the end of these periods.

5.5 DETAILED DISCOUNTED CASH FLOW CALCULATIONS

Until now, we have developed and used closed form relations for our economic analysis. However, in realistic situations the timing of payments and incomes is often irregular and complicated. As a result, more detailed and complex cash flow calculations are required. In this section we illustrate these calculations in a spreadsheet format and discuss their implications. This also allows us to consider additional complications in cash flows, such as inflation and economic risk.

5.5.1 Selecting Major Projects

How a company selects among major projects is the same as how you might select among investments to make. Because the timing of payments and incomes may be complicated, we consider this analysis through an example.

EXAMPLE 5.13 Choosing among Three Investments

On January 1, you have $10,000 saved in an account that pays 5.5% interest per year, compounded monthly. You have three investment opportunities available to you over the next year. First, you could invest $8000 at the end of January and will be paid back your investment plus 12% annual interest (compounded monthly) at the end of six months. The second is an investment of $9000 at the end of April for six months with an interest rate of 18%, while the third is an investment of $12,000 at the end of August for three months, paying 14% interest. Table 5.1 summarizes the investments available to you.

TABLE 5.1 Bank Account and Investment Opportunities for Example 5.13

Investment Alternative	Amount of Investment	Start Time, at End of	Duration, Months	Annual Interest (Compounded Monthly)
bank account	$10,000			5.5%
1	$ 8,000	January	6	12%
2	$ 9,000	April	6	18%
3	$12,000	August	3	14%

Because the interest rate is 5.5% annually from Eq. (5.1) the interest you will be paid is:

$$\$10,000 \times 0.055 \frac{\$}{\$yr} \times \frac{1}{12} yr = \$45.83$$

The interest adds to your savings account, making your account worth $10,045.83. If the principal and this interest were to remain in the bank for another month, you would be paid another

$$\$10,045.83 \times 0.055 \frac{\$}{\$yr} \times \frac{1}{12} yr = \$46.04$$

Sec. 5.5   Detailed Discounted Cash Flow Calculations

in interest. Table 5.2 lists the amount of money you will have in the bank versus the month if you were to leave the $10,000 and accumulated interest in the account. By the end of twelve months you would have $10,564.08.

TABLE 5.2   Analysis of Investment Alternatives for Example 5.13

Month	Bank Account		Investment 1		Investment 2		Investment 3	
	Cash Flow	Accum Flow	Cash Flow	Accum Flow	Cash Flow	Accum Flow	Cash Flow	Accum Flow
	$10,000.00	$10,000.00	$ —		$ —		$ —	
Jan		$10,045.83	$(8,000.00)	$(8,000.00)	$ —		$ —	
Feb		$10,091.88		$(8,036.67)	$ —		$ —	
Mar		$10,138.13		$(8,073.50)	$ —		$ —	
Apr		$10,184.60		$(8,110.50)	$(9,000.00)	$(9,000.00)	$ —	
May		$10,231.28		$(8,147.68)		$(9,041.25)	$ —	
Jun		$10,278.17		$(8,185.02)		$(9,082.69)	$ —	
Jul		$10,325.28	$8,492.16	$ 269.62		$(9,124.32)	$ —	
Aug		$10,372.60		$ 270.86		$(9,166.14)	$(12,000.00)	$(12,000.00)
Sep		$10,420.14		$ 272.10		$(9,208.15)		$(12,055.00)
Oct		$10,467.90		$ 273.35	$9,840.99	$ 590.64		$(12,110.25)
Nov		$10,515.88		$ 274.60		$ 593.34	$12,424.92	$ 259.16
Dec		$10,564.08		$ 275.86		$ 596.06		$ 260.35

Now the first investment represents an outflow of money from your account of $8000 at the end of January. We list this outflow at the end of January in column 3 of Table 5.2 as a negative amount of money (we show negative numbers by enclosing them in parentheses—accountants often use this practice when presenting financial statements). As the money is no longer in your account, you will lose the bank interest on it. At the end of one month (the end of February) you will lose

$$\$8000 \times 0.055 \frac{\$}{\$yr} \times \frac{1}{12} \text{yr} = \$36.67$$

We can account for this loss by accumulating this amount with the $8000 withdrawal from your account, listing this total in column 4 at the end of February as the "value" (shown as a negative number) of this investment to your bank account at that time. The fourth column is thus the adjustment you have to make to your bank account if you were to make this investment.

We get precisely this amount by adding the entries for February in columns 2 and 4 of Table 5.2. The first column of Table 5.3, labeled "Invest 1", is the sum, entry by entry, of these two columns in Table 5.2. It shows the amount of money in your bank account at the end of each month if you were to make investment 1.

Note that you are paid back on investment 1 six months after making it, i.e., at the end of July. We compute with Eq. (5.3) the amount you are paid back as the original principal plus 12% annual interest compounded monthly. The amount we are to be paid for our $8000 investment after six months would be

$$\$8000 \times (1 + \frac{0.12}{12})^6 = \$8492.16$$

We show this amount being paid back to you at the end of July in column 3 of Table 5.2. We continue to assess the value of this investment to your bank account in column 4. The last

TABLE 5.3 Analysis of Investment Combinations to Find Better Ones

	Invest 1	Invest 2	Invest 3	Invest 1&2	Invest 1&3	Invest 2&3	Invest 1,2,&3
	$10,000.00	$10,000.00	$ 10,000.00	$10,000.00	$10,000.00	$ 10,000.00	$ 10,000.00
Jan	$ 2,045.83	$10,045.83	$ 10,045.83	$ 2,045.83	$ 2,045.83	$ 10,045.83	$ 2,045.83
Feb	$ 2,055.21	$10,091.88	$ 10,091.88	$ 2,055.21	$ 2,055.21	$ 10,091.88	$ 2,055.21
Mar	$ 2,064.63	$10,138.13	$ 10,138.13	$ 2,064.63	$ 2,064.63	$ 10,138.13	$ 2,064.63
Apr	$ 2,074.09	$ 1,184.60	$ 10,184.60	$ (6,925.91)	$ 2,074.09	$ 1,184.60	$ (6,925.91)
May	$ 2,083.60	$ 1,190.03	$ 10,231.28	$ (6,957.65)	$ 2,083.60	$ 1,190.03	$ (6,957.65)
Jun	$ 2,093.15	$ 1,195.48	$ 10,278.17	$ (6,989.54)	$ 2,093.15	$ 1,195.48	$ (6,989.54)
Jul	$10,594.90	$ 1,200.96	$ 10,325.28	$ 1,470.59	$10,594.90	$ 1,200.96	$ 1,470.59
Aug	$10,643.46	$ 1,206.46	$ (1,627.40)	$ 1,477.33	$ (1,356.54)	$(10,793.54)	$(10,522.67)
Sep	$10,692.25	$ 1,211.99	$ (1,634.86)	$ 1,484.10	$ (1,362.75)	$(10,843.01)	$(10,570.90)
Oct	$10,741.25	$11,058.54	$ (1,642.35)	$11,331.89	$ (1,369.00)	$ (1,051.71)	$ (778.36)
Nov	$10,790.48	$11,109.22	$10,775.04	$11,383.83	$11,049.64	$ 11,368.39	$ 11,642.99
Dec	$10,839.94	$11,160.14	$10,824.43	$11,436.00	$11,100.29	$ 11,420.49	$ 11,696.35
Without loans	$10,839.94	$11,160.14	reject	reject	reject	reject	reject
With loans	$10,839.94	$11,160.14	$10,775.89	$11,262.99	$11,063.89	$ 11,153.54	$ 11,256.39

entry in column 4 of Table 5.2 is the extra amount of money you will have in your bank account by making this investment: $275.86. Column 1 of Table 5.3 shows your bank account at the end of the year if you make this investment.: $10,839.94, while column 2 in Table 5.2 shows what you would have if you do not: $10,564.08, the difference being $275.86. We can carry out similar analyses for the other two investments and show the adjustment required to your bank account for each of these investments in Table 5.2 while column 2 in Table 5.3 shows what would be in your bank account if you made investment 2 and column 3 if you made investment 3.

We see a problem with investment 3 in Table 5.3. It makes your bank account negative for the months of August through October. No bank will allow you to overdraw an account. Instead, the bank will insist that you take out a loan to cover this overdrawn amount. To prevent your account from being overdrawn, let's say that you take out a $2000 loan for this three month period. A loan is just another cash flow except you get a deposit into your account first and then a somewhat larger withdrawal later. Table 5.4 analyzes the impact this loan would have on your bank account during the year. It is computed exactly as we have computed the adjustments for the investments. You receive $2000 at the end of August and have to pay back $2094.75 at the end of November. The impact of the loan at the end of the year to your bank account is the last number in this column: a negative $48.54. As expected, loans cost money.

Is investment 3 worthwhile? It will gain us $260.45 but cannot be done unless we take out a loan that will cost us $48.54. The net gain is the difference: $211.91. Investment 3 with a loan does give us a benefit. The three investments, if made by themselves, would give us a benefit of $275.86, $596.06, and $211.91, respectively, compared to keeping the money in the bank. Investment 2 would be the best to make if we were to make only one. However, we can attempt to make investments that are combinations of these three: investments 1 and 2; 1 and 3; 2 and 3; or 1, 2, and 3. Columns 4 through 7 in Table 5.3 indicate the effects on our bank account for each of these. All cause us to overdraw our account.

We can propose loans necessary to prevent our bank account from being overdrawn. Each of these is analyzed in Table 5.4. We can simply use the final costs shown for December to adjust the final bank account amounts in Table 5.3 for each of these alternatives. Among them, investments 1 and 2 together with the appropriate loan will give us the maximum amount in our account by the end of December. Based on this analysis, this combination of investments and

Sec. 5.5 Detailed Discounted Cash Flow Calculations

TABLE 5.4 Analysis of Minimum Loans Required to Overcome Negative Bank Balances for Investment Alternatives

	Loan 3	Accumulated Cash Flow	Loan 1&2	Accumulated Cash Flow	Loan 1&3	Accumulated Cash Flow	Loan 2&4	Accumulated Cash Flow
Jan		$ —		$ —		$ —		$ —
Feb		$ —		$ —		$ —		$ —
Mar		$ —		$ —		$ —		$ —
Apr		$ —	$ 7,000.00	$ 7,000.00		$ —		$ —
May		$ —		$ 7,032.08		$ —		$ —
Jun		$ —		$ 7,064.31		$ —		$ —
Jul		$ —	$(7,265.79)	$ (169.10)		$ —		$ —
Aug	$ 2,000.00	$2,000.00		$ (169.88)	$ 1,500.00	$1,500.00	$ 11,000.00	$ 11,000.00
Sep		$2,009.17		$ (170.66)		$1,506.88		$ 11,050.42
Oct		$2,018.38		$ (171.44)		$1,513.78		$ 11,101.06
Nov	$(2,075.95)	$ (48.32)		$ (172.22)	$(1,556.95)	$ (36.24)	$(11,417.68)	$ (265.73)
Dec		$ (48.54)		$ (173.01)		$ (36.40)		$ (266.95)

loans is our best choice. Moreover, we can make the search problem for the best investment combination more complicated by allowing us to pick the start times for the investments, shifting them up to two months earlier or later. This search problem is now much harder, but the idea that we are looking for the best combination of investments is still valid.

This cash flow analysis can be generalized to accommodate all of the payment and income streams that we discussed in sections 5.3 and 5.4. We also see that a company makes decisions for its investments in a similar manner and this requires the same complex cash flow analyses, but often with many more payment and income streams. Nevertheless, the following conclusions can be drawn from this example:

1. A company cannot spend more money than it has. It can, however, raise money through loans and stock and bond offerings to increase this amount of available money, each of which will have an associated cost. A company should raise money in this manner only if it can make more than this money will cost. We saw above that we could make an investment by borrowing substantially less than the investment, which is often why a loan is worthwhile even if it comes with a high interest rate.
2. The company really needs to understand the flow of cash versus time into and out. Only then can it correctly choose how to combine them so as not to overdraw its account.
3. A loan for a company is simply a cash flow versus time for a company and can be analyzed like any other cash flow versus time.

Finally, let us look at the assessment of investment 1 again, but this time we shall do it graphically. In Figure 5.7 we plot the value of the bank account versus time if we were to leave all the money in the bank; this is the upper, slowly rising dashed line. We also plot the value to the bank account of making investment 1, the lower dashed line, and finally, we plot their sum as the solid line.

Investment 1 has a negative value for some time before returning to a slightly net positive value. One geometric interpretation we can make is that the *area* under the plot for the bank account represents our ability to invest elsewhere. If we make investment 1, we subtract the *area* for investment 1 from this ability. An area has two dimensions: its height, which represents the

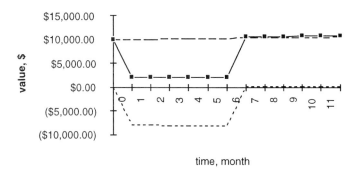

FIGURE 5.7 Graphical representation useful in assessing investment 1.

amount of money involved, and its width, which represents how long we need that amount of money.

Assessing the different combinations of investing above has been the equivalent of attempting to pack the areas of the investments into the area made available by the original bank account. The loans we proposed also increase this area. Also, the larger the negative area for an investment, the more it takes away from our ability to invest elsewhere. Therefore, both dimensions have an impact, and we can rate an investment by this area, representing the product of its amount and its duration.

5.5.2 Assessing the Value of a Project

What if we had about 100 major investments to examine rather than just the three we did here? We need some way to screen out quickly the ones that are not very good. There are 2^N different combinations of investments we can make given N investments, and 2^{100} is a very large number. We first compute the present value for these 100 investments. For each one with a negative present value, we will earn more by leaving the money in the bank. Thus we screen out any with a negative or zero present value.

We would also like to screen out investments that make money but not enough. We need a measure that allows us to assess the "quality" of an investment. Its present value is not enough for us to do that by itself. What if we had two investments having the same duration and the same present value, but the second required twice the investment? Our intuition tells us this latter project is not as good as the first. It will take more cash which we could use to invest in something else. Another project might require the same investment and result in the same present value, but it might take twice as long to give us that present value. It, too, seems less desirable as a project than the first. So, somewhere in our analysis we must relate the increase in the present value, the time it takes to produce this present value, and the investment we need to make. We prefer larger present values, shorter times, and smaller investments.

One type of measure we can propose is to divide the present value ($) of the project by a time (yr) characterizing the length of the project and by a measure of the investment

($), getting a rate (1/yr) that we would like to maximize. Another would be to form the reciprocal, getting a time (yr) that we would like to minimize.

For a measure of the first type, divide the present value accounting for all investments and income by the present value of the investment without income and by the number of years it took to get that final present value. As an example, look at investment 1 in Table 5.2. At the time it is made, the investment alone has a present value of −$8000. By the time it is paid back half a year later, it increases our present value by (discounting it back to the start of the investment)

$$\$269.62/(1 + 0.055/12)^6 = \$206.05$$

This measure becomes

$$\$206.05/(\$8000 \cdot 0.5 \text{ yr}) = 0.0515/\text{yr}.$$

We are increasing the present value of the company on average at a rate of about 5.15 % per year of the present value of the investments made. If we doubled the investment and had the same present value for the project, we would halve this rate, making it fairly obvious that it is not as good. If we double the time, we also get half the value for this measure. A company could rate all its projects with such a computation and rank order them, eliminating all those that fall below a cutoff value. Only those that remain would then be passed to upper management as candidate projects. When selecting among several incompatible projects that provide the same final "service," we might use this measure as our objective function.

As we noted before, *payout time* is a reciprocal form for a measure that ignores time value of money. It asks how long it will take to recover one's investment. *Breakeven time* (BET) is a measure that accounts for the time value of money. It is the time at which the present value of the project just becomes positive and stays positive thereafter. We mark the breakeven time in Figure 5.8, which is a plot of accumulated cash flow for a typical project.

We can note that the denominator of our first form of rating function has the units of area on a plot of accumulated present value against time. That observation suggests

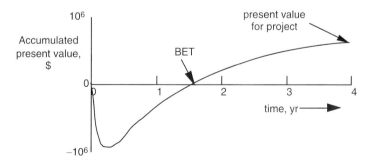

FIGURE 5.8 Accumulated present value for the cash flows for a typical project.

that we might further consider our earlier discussion about using the area of the accumulated present value versus time plot of an investment to aid in assessing its value. We might divide the final present value of the investment by this area.

We can see two areas in Figure 5.8 that relate to our project cash flow analysis. There is a negative area before the project starts to produce a positive present value. If the project will make money, there is also a positive area thereafter. Negative areas reduce investment ability for the company; positive areas enhance it. We could perhaps have two measures; the first divides the present value by the absolute value of the negative area, measuring the reduction in investment ability caused by the project until it makes money. We would want this number to be small. The second divides the present value by the positive area, measuring the increase in investment ability caused by the project over its life. We would like this number to be large. Both would have the units of rate (1/time).

5.5.3 Discussion

When all is said and done about rating a project, the true measure is the present value of the project as it evolves versus time, as we have shown in Figure 5.8. The whole plot is the measure as well as the final present value. Management has to combine different projects (by adding their respective curves) to maximize the present value of the company while not overspending the amount of funds available for investing. Some projects, while not best in any of the above measures, may just fit together so that in combination they are best for the company.

The final present value measure can be used by itself for comparing small incompatible projects where we are not concerned about their impact on our ability to invest. The choice between two pumps in Example 5.9 is such a case. Suppose the service we wish from the pump is 6 years. The carbon steel pump has a service life of 4 years; the stainless has a service life of 8 years. We develop a scenario that depicts the entire flow of cash to provide service for 6 years and evaluate the present value of each. We would need to buy a second carbon steel pump in 4 years. We would use the second pump only 2 years. We would have to discover if we could then sell it for a salvage value and, if so, enter that cash inflow into its scenario. We will also have to check if the salvage value increases for the stainless pump after only 6 years of service and account for that cash flow.

If there are many large projects and we want to reduce the number that we have to consider in combination, we need to find a useful measure that characterizes how much the project will reduce the company's ability to invest elsewhere. This measure will allow us to eliminate the poor ones directly.

Most proposed projects will not have positive economics when they are analyzed. A measure that rates a project in isolation becomes very useful because it allows us to stop working on a project when we discover it will not give an adequate value for this measure. A company can move the passing value, called a hurdle, up and down to reflect the company's economic situation. If times are tough, it might reject looking at projects that have breakeven times exceeding 1 to 2 years. When times are good economically, it may be willing to look at projects with longer breakeven times.

One company has proposed reevaluating breakeven time for a project every month as the project proceeds. A significant change flags management that something could be going wrong with the project. Suppose that BET for a project under way in 1995 has been January 2000 for several months and then it becomes March 2006. Management will want to understand why. Is there a new competitor so the sales price for the product cannot be as high as previously thought? Is there a technical snag that will delay the project startup time? Or what?

In the next two sections we look at how to account for two additional factors: inflation and investment risk. Again, these can be handled with the above cash flow analysis.

5.6 INFLATION

Prices for goods and services tend to increase from year to year. What cost a dollar in 1960 now costs over four dollars. This yearly rate at which prices increase is the inflation rate. For example, if the inflation rate is 4%/yr, then it is expected that next year the same item we pay $1 for today will cost $1.04. How does one handle inflation in the context of an economic analysis?

There is a straightforward way to account for inflation. We simply use the inflation rate to adjust the prices received or paid for goods and services over the life of a project and then compute the cash flows using these adjusted prices. The present value analysis changes only in that the cash flow amounts are different because they recognize the existence of inflation. This is seen directly through the following example.

EXAMPLE 5.14   Inflation Calculation

Assume inflation is running at 3% per year (use annual compounding formula). How much money do you need to put into the bank today at an interest rate of 6% compounded annually to buy furniture in 2 years that costs $1000 today? The furniture will cost $1000*(1.03)² = $1060.90 in 2 years. The present value of this amount of money is discovered by solving

$$\$1060.90 = P(1 + 0.06)^2 \Rightarrow P = \$944.20$$

which is the amount of money we need in the bank today to have $1060.90 in two years. We note that we need less than $1000 because bank interest is more than the rate of inflation. We can combine all this into an equation, getting

$$P = \$1000 \frac{(1 + i_{\text{inf}})^n}{(1 + i)^{nm}} = \$1000 \frac{(1 + 0.03)^2}{(1 + 0.06)^2} = \$944.20$$

We inflate with the numerator and discount with the denominator. This formula should put to rest a common belief that one simply subtracts the inflation rate from the interest rate to account for inflation. Using the difference in interest and inflation rates is only an approximate way to handle inflation. To show that it is approximately right, let us consider the case of no inflation and discount $1000 in 2 years using the approximate bank interest less inflation rate of $(6 - 3)\% = 3\%$. We would get

$$P \frac{\$1000}{(1 + 0.03)^2} = \$942.60$$

which is close to the previous answer.

5.7 ASSESSING INVESTMENT RISK

Let us reconsider Example 5.13 again where we had $10,000 in the bank and the opportunity to make any of three different investments over the next year. Our analysis assumed that the investments were without risk. What if there is a 2% chance that the first investment would pay only 50% of what was due and a 0.5% chance the investment would pay nothing back to us? We might further suppose that the other two investments are safe and have negligible probabilities that they will not be paid back in full. How do we account for these possible failures of investment 1 in our decision making?

First, we have to decide if reduced payment or nonpayment affects our other decisions. We would not know in time to alter any decisions to make investment 2 if investment 1 failed. However, we could reconsider our decision to make investment 3, because we should receive our repayment on investment 1 at the end of July while we would not make investment 3 until the end of August. If we had made both investments 1 and 2, we will have to borrow more money to survive the risk in investment 1. We will avoid borrowing large amounts of money by not allowing investment 3 if investment 1 pays only half or none of it back to us when it is due.

For each alternative where investment 1 is part of our strategy, we must generate and evaluate added alternatives that correspond to full or partial failure of that investment. We appear to have eight new alternatives to analyze, namely:

- Alternatives 1 and 2: Make only investment 1. There are two new possibilities: (a) 50% repayment and (b) no repayment.
- Alternatives 3 and 4: Make investments 1 and 2. Again we have two possibilities leading to different amounts of money we will have to borrow.
- Alternatives 5 and 6: Make investments 1 and 3. If investment 1 fails in any way, we will not make investment 3. Not making investment 3 means these two alternatives become the same as alternatives 1 and 2.
- Alternatives 7 and 8: Make all three investments. If investment 1 fails in any way, we again will not make investment 3. Not making investment 3 means these two alternatives become the same as alternatives 3 and 4 above where we make only investments 1 and 2.

The analysis needed for these four alternatives is similar to the alternatives we did earlier for our risk-free investment alternatives. To illustrate, we show in Table 5.5 the

Sec. 5.7   Assessing Investment Risk

TABLE 5.5 Evaluation of Investing in 1 and 2, with Investment Failing to Make Any Repayment

	Invest 1 No repayment	Invest 1 (0%) and 2	Loan		Invest 1(0%), 2 and Loan
	$—	$10,000.00	$—		$10,000.00
Jan	($8,000)	$(8,000.00)	$ 2,045.83	$—	$ 2,045.83
Feb		$(8,036.67)	$ 2,055.21	$—	$ 2,055.21
Mar		$(8.073.50)	$ 2,064.63	$—	$ 2,064.63
Apr		$(8,110.50)	$ (6,925.91)	$ 7,000.00	$7,000.00	$ 74.09
May		$(8,147.68)	$ (6,957.65)		$7,032.08	$ 74.43
Jun		$(8,185.02)	$ (6,989.54)		$7,064.31	$ 74.77
Jul		$(8,222.54)	$ (7,021.58)		$7,096.69	$ 75.12
Aug		$(8,260.22)	$ (7,053.76)		$7,129.22	$ 75.46
Sep		$(8,298.08)	$ (7.086.09)		$7,161.89	$ 75.81
Oct		$(8,336.12)	$ 2,722.42	$(7,541.68)	$ (346.96)	$ 2,375.46
Nov		$(8,374.32)	$ 2,734.90		$ (348.55)	$ 2,386.35
Dec		$(8,412.70)	$ 2,747.44		$ (350.15)	$ 2,397.29

analysis needed for the alternative of investing in 1 and 2, with investment 1 failing to make any repayment. The amount in our bank account will be only $2397 at the end of December, almost $9000 less than the value of $11262.99 it had when investment 1 did not fail. Table 5.6 summarizes the results for all previous and these new alternatives. The last column is the "expected value" of the bank account at the end of December for each of the eight decisions alternatives we might make. To illustrate, we compute it for investment alternative 2 as follows:

$$\$10,839.94 \times 0.975 + \$6495.66 \times 0.02 + \$2151.37 \times 0.005 = \$10,709.61$$

We see that alternative 3 has the highest expected value for the bank account at the end of December. Based on this criterion we would choose it. It has the nice feature that it avoids investment 1 altogether.

However, a person willing to take risks might chose alternative 5 because, if investment 1 does pay back, it produces the highest value: $11,262.99. That is $102.85 more than investment 3. Admittedly that is not much of an incentive over the safety of investment 3, but, with a 97.5% probability of success, many people would be willing to take such a chance. A very conservative person, on the other hand, would never invest in investment 1, even if it was part of an alternative where the expected value for the bank account was much higher.

There are many risks that a company can face. The future prices for the goods we manufacture may be much lower than we predict. The cost for the manufacturing plant may be much higher than we compute because we are unaware of a by-product that we will produce in the reactor, requiring us to do some very expensive retrofitting. We may discover that the separation process we designed, in fact, does not work. Not all risks are negative. For instance, we may find a competitor decides not to enter the market.

TABLE 5.6 Summary of All Investment Alternatives (with Appropriate Loans Taken to Prevent Ever Having a Negative Bank Balance)

Alternative	Description (Number in Parentheses Shows Percent Repayment on Investment 1)	Value of Bank Account in December	Probability Alternative Occurs	Expected Value of Bank Account
1	No investments	$10,564.08		
2	Investment 1 (100%) Investment 1 (50%) Investment 1 (0%)	$10,839.94 $6495.66 $2151.37	97.5% 2% 0.5%	$10,709.61
3	Investment 2	$11,160.14		$11,160.14
4	Investment 3	$10,775.89		$10,775.89
5	Investment 1 (100%) and 2 Investment 1 (50%) and 2 Investment 1 (0%) and 2	$11,262.99 $6845.57 $2397.29	97.5% 2% 0.5%	$11,130.31
6	Investment 1 (100%)and 3, Investment 1 (50%), cancel 3 Investment 1 (0%), cancel 3	$11,063.89 $6495.66 $2151.37	97.5% 2% 0.5%	$10,927.96
7	Investment 2 and 3	$11,153.54		$11,153.54
8	Investment 1 (100%), 2 and 3 Investment 1 (50%), 2, cancel 3 Investment 1 (0%), 2, cancel 3	$11,256.39 $6845.57 $2397.29	97.5% 2% 0.5%	$11,123.88

Finally, if we are unable to enumerate all the likely outcomes for an investment, how can one account for risk? One possible approach is to adjust the hurdle rate needed for each project to account for its perceived risk, with more conservative rates set for what the company views to be riskier projects. If the hurdle is in terms of breakeven time, the company may ask for an estimated breakeven time of 6 months for a risky project while accepting a breakeven time of 2 to 3 years for a project having negligible risk.

From this discussion, we can draw the following conclusions about risk analysis:

1. If there is risk associated with any of our decisions, we should develop the alternative outcomes possible—if we can—and evaluate each in a manner similar to the way we evaluated alternatives when not accounting for risk.
2. Bad outcomes from earlier decisions will almost certainly alter later decisions; we will likely have to establish policies for how we will handle such situations.
3. For each alternate decision, there will be a whole range of outcomes with associated probabilities each will occur. We may assess the expected value of the outcome for the various decision alternatives. All of these results are then input to our decision making.

4. By itself, the cash flow analysis does not tell us what to do to account for risk. We must also add in our feelings about taking risks to make our decisions. If we feel conservative at decision time, we will likely try to pick the best of the worst outcomes that have a non-negligible, say 10%, probability of occurring. We may attempt to stay neutral and pick the decisions that lead to the highest expected value for the outcome.
5. Our willingness to take a risk will change depending on the economic situation we are facing. If we are in a period of high optimism about the economy and our future, we will take more risks. If we see only downsizing for the next few years, we will be very conservative.

In summary, to deal with risk, we should enumerate all the possible outcomes and evaluate the consequences of each of them, if we are able to do so. Then we must choose our actions according to how conservative we feel at the moment. Both elements are part of dealing with risk.

5.8 SUMMARY AND REFERENCE GUIDE

Economic evaluation represents the key performance measure for making project decisions. Moreover, the synthesis and analysis steps described in the previous chapters were geared toward making this evaluation. This chapter first presents concepts related to overall manufacturing and capital costs, along with the indirect costs that are incurred in the project. To evaluate the success of this project we then derived simple measures that could be evaluated quickly. These measures help to assess the economic feasibility of a project and to compare competing projects. For more information on detailed calculation of these expenses and economic measures, refer to:

Baasel, W. D. (1976). *Preliminary Chemical Engineering Plant Design.* New York: Elsevier.

Douglas, J. M. (1988). *Conceptual Design of Chemical Processes.* New York: McGraw-Hill.

Peters, M., & Timmerhaus, K. (1980). *Plant Design and Economics for Chemical Engineers.* New York: McGraw-Hill.

On the other hand, more detailed evaluations are needed for an accurate representation of the project economics over its lifetime. This evaluation requires the concept of time value of money and cash flows. Here these concepts were translated to closed-form expressions that allow the evaluation of net present values and rates of return. Moreover, these expressions allow us to compare project costs, evaluate project profitability, and even influence market selling prices. Also included are the effects of both taxes and depreciation. Finally, we consider the extension of this analysis to more complex income and payment streams. These lead to more complicated cash flow analyses that are best performed with the aid of spreadsheets. This analysis is especially important in order to assess the factors of risk and inflation on the project.

Rather than provide an extensive treatment on economic evaluation, this chapter has focused on its application to chemical processes at the preliminary design stage. For a more sophisiticated treatment of this topic, there is a very broad literature on engineering economics and many excellent textbooks cover the topics of this chapter in great detail. A selection of these is given below.

Au, T. (1983). *Engineering Economics for Capital Investment Analysis.* Boston: Allyn & Bacon.

Grant, E., & Ireson, W. G. (1982). *Principles of Engineering Economy.* New York: Wiley.

Jelen, F. C., & Black, J. H. (1983). *Cost and Optimization Engineering.* New York: McGraw-Hill.

Kurtz, M. (1984). *Handbook of Engineering Economics.* New York: McGraw-Hill.

Park, C. S. (1993). *Contemporary Engineering Economics.* Reading, MA: Addison-Wesley.

EXERCISES

1. Consider the economic evaluation of the melamine process described below.
   a. Estimate the working capital and determine the annual proceeds per dollar outlay (APDO) and payout time for a melamine plant given below. Melamine sells for 20¢/lb.

Manufacturing Cost Worksheet for Melamine

Cost Category	Item	Unit Consumption	Unit Price	Unit Cost
Raw materials	Urea	3.3 tons/ton	$50/ton	$165/ton
	Ammonia, 99%	0.1 tons/ton	$60/ton	6
By-product credit	Ammonia	1.1 tons/ton	$30/ton	−33
Utilities	Steam, 400 psig	14.5 tons/ton	$ 1/ton	14.5
	Electricity	1,900 kwh/ton	0.5¢/kwh	9.5
	Cooling water	94,000 gal/ton	2¢/1,000 gal	2
Labor	Operating & supervision	4 people/shift	$4.00/hr/man + 150%	25
Fixed charges	Maintenance	4% of capital/yr.	$240/ton	9.5
	Depreciation	11% of capital/yr	$240/ton	26.5
	Insurance & taxes	3% of capital/yr	$240/ton	7
Total estimated manufacturing cost		$232/ton 11.6¢/lb		
Basis		25,000,000 lb/yr (38 tons/day or 1.6 tons/hr) Battery-limits plant erected on Gulf Coast, requiring an investment of $3,000,000.		

Exercises

    b. What is the payout time for the above plant if it runs at 70% capacity? Assume that fixed charges, labor, and total capital are the same as for full capacity.

2. Determine the present value of the following items assuming annual interest rates of 10% and 20%:
   a. $8,000 earned 6 years from now
   b. A payment of $15,000 at the end of each year for a period of 10 years

3. You are going to borrow $15,000 for 3 years from the bank to pay what you still owe on your car. The bank charges you 11% interest. What will your monthly payment be?

4. Consider the following investment opportunities:

	Project A	Project B
Fixed investment	$250,000	$450,000
Salvage value	0	50,000
Working capital	40,000	80,000
Annual product sales ($/yr)	200,000	250,000
Operating expense ($/yr)	10,000	110,000
Economic life (yrs)	4	6
Lifetime for tax purposes (yrs)	3	3

Assume straight line depreciation, an after-tax interest rate of 12%, and a 52% federal-state income tax rate.
   a. Which, if either, of the projects do you recommend?
   b. What is the rate of return on project B?

5. A 5-year-old machine costs $15,000 when new and is being depreciated on a straight line basis to a zero salvage value in 5 more years (10 years total life). The operating expenses for this machine are $2,500 as of the end of each year. At the end of its life, it will be replaced by a new machine that costs $22,000, will last 10 years, and have operating costs of $1,500/year. Should we replace it now instead of waiting for 5 years? The interest rate is 10%/year and the tax rate 50%. What is the current book value of the old machine?

6. A manufacturing process has the following financial information:

    Fixed capital $15,000,000
    Working capital $ 4,500,000
    Salvage value $ 2,000,000
    Manufacturing cost $13,000,000/yr
    Revenues $20,000,000/yr
    SARE expenses $ 2,000,000/yr

Assume a tax life of 7 years, straight line depreciation and a total life of 10 years, with a DCF rate of return at 15% and a tax rate of 52 %.
   a. What is the net present value of the process before taxes?
   b. What is the net present value of the process after taxes?

7. If inflation is 3% per year, what would be the ratio of the cost now for an item to its cost two decades ago? Use continuous compounding. Does this ratio surprise you?
8. You have just won three million dollars in the lottery in June. The state tells you it will send you a check at the end of the next 240 months (20 years) for $12,500. The first payment to you will be June 30. Note that 240 times $12,500 is $3,000,000. Assume bank interest is 6%.
   a. Let time zero be June 30. What is the present value at time zero of your winnings?
   b. You assume you will stop your job and live only on this income. Based on that assumption, you estimate that you will have to pay about 45% of the winnings you receive for each year in federal and state income taxes. You are required to pay estimated taxes on this income in four equal payments in April, June, September, and January (of the next year)—yes, these are not evenly spaced payments. Assume these payments occur at the end of the month (they actually occur on the 15th). What is the present value of your three million dollars in winnings after taxes?
   c. Do you find this answer disheartening? Should the state be taken to court for false advertising?
9. A person with a bachelor's degree in chemical engineering might make a starting salary of $40,000 per year in 1996. Estimate what the starting salary might be in 2001? in 2006? in 2016? State your assumptions.
10. Develop the MACRS tables for the following options.
    a. 7-year life using 200% acceleration schedule. (If you do it right, year 5 will have a factor of 8.92%.)
    b. 7-year life using a 150% acceleration schedule (year 5 is 9.30%).
    c. 10-year life using a 150% acceleration schedule (year 8 is 8.74%).
11. Consider the following investment opportunities.

	Project A	Project B
Fixed investment	$250,000	$450,000
Salvage value	0	50,000
Working capital	40,000	80,000
Annual product sales ($/yr)	200,000	250,000
Operating expense ($/yr)	10,000	110,000
Economic life (yrs)	4	6
Depreciation life (yrs)	3	3

Assume straight line depreciation (assume you can only depreciate a half-year's worth for the first and last year), a "bank interest" rate of 12%/yr compounded monthly, and a 50% federal-state income tax rate.
   a. Which, if either, of the projects do you recommend?

b. Determine the "bank interest" rate for each project that would make its present value exactly zero.

12. After starting your first job, you are investigating some housing options. You plan to move after 5 years anyway and your investments (i.e., savings) currently yield 5%.
    a. To buy a $100,000 house with a $10,000 down payment, you are able to secure a mortgage loan for $90,000 at 10% over 30 years. What is the monthly payment on this mortgage loan?
    b. Assume that your $10,000 down payment will lead to an equity of $15,000 in 5 years and that the combined mortgage and tax payments come to $850/month. Is this better than renting an apartment for $750/month?

13. Look at the cash flows in the following table. Note that each case corresponds to an outflow of cash of $1,000,000 and an inflow of $1,200,000 over the course of the year.
    a. Without analyzing them, which cash flow(s) would you prefer and why. (Please make your best guess for this part of the question before you go to part b to see how well your intuition corresponds to the results in part b.)
    b. Once you have completed part a, calculate the present value for each of them. The cash flows occur at the end of the month indicated. "Bank interest" is 11%/yr and compounding is monthly. Now which cash flow would you prefer? Did your intuition give you the same preference ordering as your calculations now do?

Month	Case A	Case B	Case C	Case D
0	($1,000,000)	($1,000,000)	($500,000)	($500,000)
1				
2				
3	$300,000		$300,000	
4				
5				
6	$300,000		$300,000 + ($500,000) = ($200,000)	($500,000)
7				
8				
9	$300,000		$300,000	
10				
11				
12	$300,000	$1,200,000	$300,000	$1,200,000

14. You have just completed a preliminary design for a chemical process. The total investment required is $250,000,000. You can depreciate this investment over 10 years. You have estimated annual operating costs to be 8% of this amount per year.

What should be your gross income at full production for you to have a zero present value in 5 years? Carefully explain all your assumptions. The bank interest for the company is 15% per year.

15. You are part of a small company employing 50 people. Which of the investments in Table 5.7 should your company make if they are all risk-free? Bank interest for your company is 5% per year compounded monthly. All cash flows are at the end of the month indicated. Your company has $1,200,000 in reserves. To explain the first project, you have to make an investment of $200,000 at the end of month 12. There are also monthly expenses of $50,000 paid at the end of months 13, 14, 15, 16, and 17. You receive a cash inflow of $90,000 per month for months 18, 19, 20, 21, 22, 23, and 24.

TABLE 5.7 Competing projects

Description	Amount of Money	First Month	Last Month
Project 1			
Investment	($200,000)	12	
Expenses	($50,000)/month	13	17
Net profit	$90,000/month	18	24
Project 2			
Expenses	($40,000)/month	0	10
Investment	($500,000)	5	
Working capital	($200,000)	9	
Income	$160,000/month	10	24
Working capital	$200,000	24	
Project 3			
Expenses	($20,000)/month	0	5
Investment	($500,000)	3	
Income	$120,000/month	6	12

16. Your company can borrow money in increments of $500,000 for six months at a 12%/yr interest rate, compounded monthly. Now which investments should you choose in the previous problem?

17. If you can move each of the investments in the Table 5.7 forward or backward by as much as 5 months, which should you then make and when? (Time zero is a year into the future so a project starting at time zero can be started earlier if desired) (Hint: Try plotting the impact on the money in the bank of the project as we did in Table 5.2 versus the month—see Figure 5.7. Then cut the plots out as areas. Subject to how much you can move them around, try to pack them under the available cash curve in the best way.)

18. The second project in Table 5.7 has a 10% chance of having an income of $100,000 per month rather than $160,000 per month for months 10 to 24. It has a 5% chance of that income being $200,000 per month.

a. What are the best, worst, and most probable present values of this second project?
b. If you are very conservative, which projects would you pick?
c. If you are extremely optimistic, which would you pick?

19. Consider Eq. (5.14), the formula we developed earlier to compute the present value of a prototypical project. Modify the formula to allow for c compounding periods per year. Use your result to recompute the present value for the example when c is equal to 4 periods per year.

Are your answers close to that which we computed for Example 5.12 when we compounded annually?

DESIGN AND SCHEDULING OF BATCH PROCESSES

6

6.1 INTRODUCTION

While many chemicals are manufactured in large scale continuous processes, it is also the case that chemicals are often manufactured in batch processes, especially if the production volumes are rather small. With the recent trend of building small flexible plants that are close to the markets of consumption, there has been renewed interest in batch processes.

Batch processes are used in the manufacture of specialty chemicals, pharmaceutical products, food, and certain types of polymers (Reeve, 1992). Since commonly the production volumes are low, batch plants are often multiproduct facilities in which the various products share the same pieces of equipment. This requires that the production in these plants be scheduled. Specifically, one has to decide the order in which products will be produced and the time allocation for each of them. This in turn also implies that at the design stage one has to anticipate how the production will be scheduled and this can have a large economic impact as we will see in this chapter (see Reklaitis, 1990; Rippin, 1993).

The major objective in this chapter will be to introduce basic scheduling and design concepts for batch processes. We will first describe a simple batch plant to introduce the concepts of recipes and Gantt charts. We will then describe the major types of scheduling policies and the computation of their cycle times. Next, we will present a preliminary design procedure for sizing and discuss the major effects for inventories. Finally, alternatives for the synthesis of these types of plants will be described.

6.2 SINGLE PRODUCT BATCH PLANTS

Batch processes are commonly used to manufacture specialty chemicals with relatively short life cycles. For this reason a common solution is that the manufacturing will follow a recipe specified by a set of processing tasks with fixed operating conditions and fixed pro-

Sec. 6.2   Single Product Batch Plants   181

cessing times. Recipes are also common in the production of pharmaceuticals and food products because of regulatory requirements. There are cases, however, when operating conditions and processing lengths can be modified, such as in the case of solvents. In this chapter, for simplicity, we will restrict ourselves to the case of batch processes that are specified through recipes. As we will see, even under this simplification, the design is not entirely trivial due to the need of anticipating operational issues, mostly related to scheduling.

Figure 6.1 presents a simple example of a batch process for manufacturing a single product. Note that it consists of four major pieces of equipment that are operated in batch mode: reactor, mixing tank, centrifuge, tray dryer. The pumps and the cooler are equipment that operate in semi-continuous mode. Initially we will assume that a single product is produced. This is accomplished by performing the following tasks that correspond to the recipe described below:

Processing Recipe

1. Mix raw materials A and B. Heat to 80°C and react during 4 hours to form product C.
2. Mix with solvent D for 1 hour at ambient conditions.
3. Centrifuge to separate solid product C for 2 hours.
4. Dry in a tray for 1 hour at 60°C.

Note that each of the above tasks is performed in each of the four batch equipment of Figure 6.1. We can represent in a chart, denoted as a Gantt chart, the time activities involved at each stage of the processing as seen in Figure 6.2a. In this chart we have shown with thick lines the times for emptying and filling. Since these are commonly much

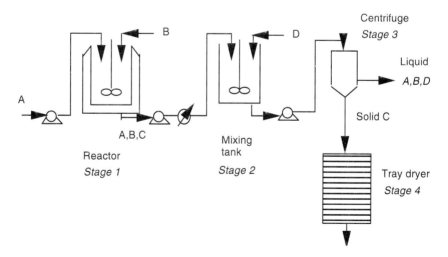

FIGURE 6.1 Simple example of batch process.

182 Design and Scheduling of Batch Processes Chap. 6

shorter than the processing times, we will neglect them, which then gives rise to the simpler Gantt chart of Figure 6.2b.

Since we will manufacture many batches or lots, one of the first decisions we need to make is whether we will use a non-overlapping or an overlapping operation as shown in Figure 6.3. In the non-overlapping operation, each batch is processed until the preceding one is completed. In this way no two batches are manufactured simultaneously. In the overlapping operation, on the other hand, we eliminate the idle times as much as possible, which then leads to the simultaneous production of batches. For instance, after 7 hours,

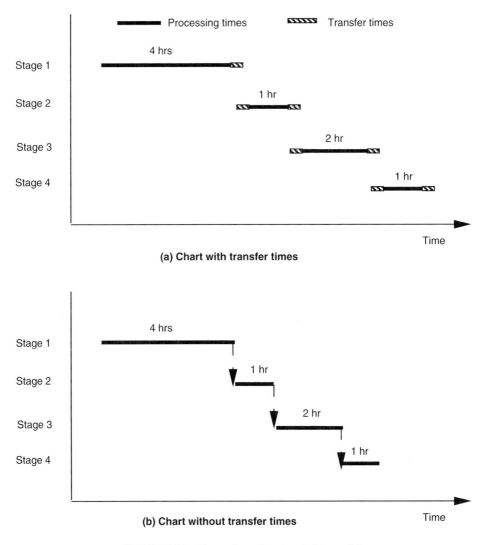

FIGURE 6.2 Gantt charts for plant in Figure 6.1.

Sec. 6.2   Single Product Batch Plants

the first batch has been completed in the third stage, while the second batch has been processed 75% of the time in stage 1.

From Figure 6.3 it is clear that the overlapping mode of operation is more efficient because the idle times are greatly reduced. In fact, stage 1 has no idle time, it operates without interruption. Also, what Figure 6.3b suggests is that stage 1 represents the bottleneck for manufacturing successive batches.

The above observation can be quantified with the following definition of cycle time, CT,

$$CT = t_f - t_s$$

where t_s and t_f are the initial and final times of each operating cycle. So, for instance, in Figure 6.3a we have for each stage:

$$CT_1 = (8 + t_{s1}) - t_{s1} = 8 \text{ hours}$$
$$CT_2 = (8 + t_{s2}) - t_{s2} = 8 \text{ hours}$$
$$CT_3 = (8 + t_{s3}) - t_{s3} = 8 \text{ hours}$$
$$CT_4 = (8 + t_{s4}) - t_{s4} = 8 \text{ hours}$$

where $t_{s1}, t_{s2}, t_{s3},$ and t_{s4} are the initial times at each stage. It is clear that all stages operate with identical cycle times of 8 hours.

For the case of Figure 6.3b, the cycle times for each stage are as follows:

$$CT_1 = (4 + t_{s1}) - t_{s1} = 4 \text{ hours}$$
$$CT_2 = (4 + t_{s2}) - t_{s2} = 4 \text{ hours}$$
$$CT_3 = (4 + t_{s3}) - t_{s3} = 4 \text{ hours}$$
$$CT_4 = (4 + t_{s4}) - t_{s4} = 4 \text{ hours}$$

Thus, the cycle time is 4 hours for all stages. In this way for Figure 6.3a $CT = 8$ hours implies every 8 hours a batch is manufactured, while for Figure 6.3b with $CT = 4$ hours, a batch is completed every 4 hours.

From the above example, it clearly follows that the cycle times for a single product plant are given in general as follows:

- Cycle time non-overlapping operation

$$CT = \sum_{j=1}^{M} \tau_j \tag{6.1}$$

- Cycle time overlapping operation

$$CT = \max_{j=1,M} \{\tau_j\} \tag{6.2}$$

where τ_j is the processing time in stage j. The above equations can easily be verified with our examples. It should also be mentioned that the scheduling term *makespan* corresponds to the total time required to produce a given number of batches. From Figure 6.3a it can be seen that the makespan for producing two batches is 16 hours; for Figure 6.3b it is 12 hours.

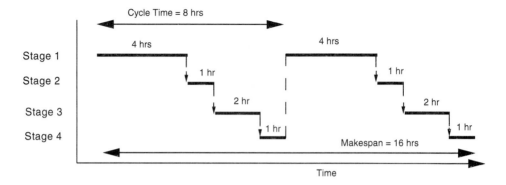

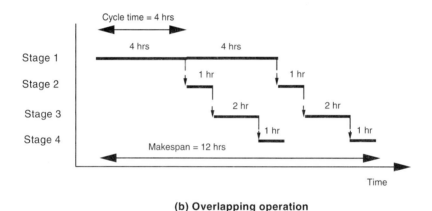

FIGURE 6.3 Non-overlapping and overlapping modes of operation.

6.3 MULTIPLE PRODUCT BATCH PLANTS

When a batch process is used to manufacture two or more products, two major limiting types of plants can arise: flowshop plants in which all products require all stages following the same sequence of operations, and jobshop plants where not all products require all stages and/or follow the same sequence (see Figure 6.4). Note that in Figure 6.4a all three products follow the same processing sequence, while in Figure 6.4b the three products follow different paths. The greater the similarity in the products being produced, the closer a real plant will approach a flowshop, and vice versa—the more dissimilar, the more it will approach a jobshop. It should also be noted that flowshop plants are often denoted as "multiproduct plants", while jobshop plants are denoted as "multipurpose plants."

Sec. 6.3 Multiple Product Batch Plants

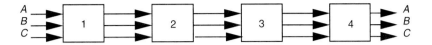

(a) Flowshop plant

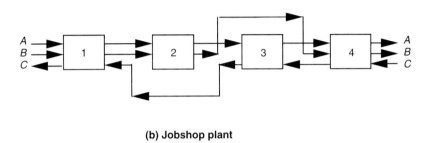

(b) Jobshop plant

FIGURE 6.4 Flowshop and jobshop plants.

Another important issue in flowshop plants is the type of production campaign that is used for manufacturing a prespecified number of batches for the various products. To illustrate this point consider the manufacturing of three batches each of products A and B in a plant consisting of two stages. The processing times are given in Table 6.1.

It should be noted that the case of batch plants with multiple products, it is not generally possible to obtain closed form expressions for the cycle times.

As seen in Figure 6.5a, one option is to use single-product campaigns (SPC) in which all batches of a given product are manufactured before switching to another product. The other option, shown in Figure 6.5b, is to use mixed-product campaigns (MPC) in which the various batches are produced according to some selected sequence (e.g., ABABAB). Note that the makespan for the campaign in Figure 6.5a is 29 hours, while for Figure 6.5b it is 25 hours. The cycle time for the sequence AAABBB in Figure 6.5a is 25 hours; for ABABAB in Figure 6.5b it is 21 hours. This might suggest that mixed product campaigns are more efficient. This might not necessarily be the case if the cleanup times or changeovers that might be needed are significant when switching from one product to

TABLE 6.1 Processing Times for Two-Product Plant (Processing Times, hrs)

	Stage 1	Stage 2
A	5	2
B	2	4

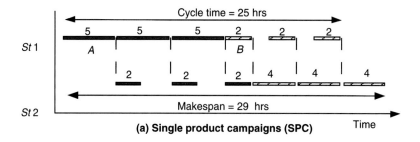
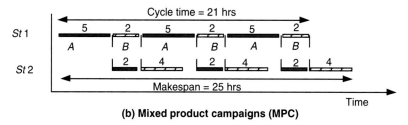

FIGURE 6.5 Schedules for single and mixed-product campaigns.

another. For instance, if in our example the cleanup times are all 1 hour, then it can be seen in Figure 6.6 that the makespan is increased from 25 hours to 30 hours and the cycle time from 21 hours to 27 hours.

6.4 TRANSFER POLICIES

In the previous section we have assumed that the batch at any stage would be transferred immediately to the next stage. Thus, it is known as zero-wait (ZW) transfer and is commonly used when no intermediate storage vessel is available or when it cannot be held further inside the current vessel (e.g., due to chemical reaction). The zero-wait transfer, as it turns out, is the most restrictive policy. The option at the other extreme is unlimited intermediate storage (UIS) in which it is assumed that the batch can be stored without any capacity limit in the storage vessel. Finally, an intermediate transfer option is known as no-intermediate storage (NIS), which allows the possibility of holding the material inside the vessel.

To illustrate the effect of the various transfer policies, consider a flowshop plant consisting of three stages for producing products A and B. Let us assume we would like to manufacture the same number of batches of each product using a sequence ABAB . . . and that the processing times are as given in Table 6.2.

From Figure 6.7 it is easy to verify that the cycle times for each pair AB are as follows:

Sec. 6.5 Parallel Units and Intermediate Storage

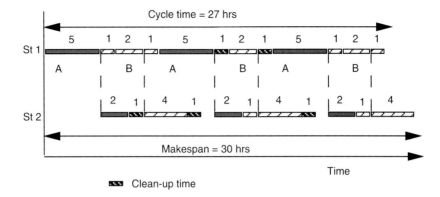

FIGURE 6.6 Effect of cleanup time on cycle time.

ZW: 11 hours
NIS: 10 hours
UIS: 9 hours

Thus, as we anticipated, the ZW transfer required the longest cycle time and UIS the shortest. In practice, plants will normally have a mixture of the three transfer policies.

Finally, it is worth mentioning that the cycle time for UIS can be determined from the following equation (see exercise 4):

$$CT_{UIS} = \max_{j=1..M} \left\{ \sum_{i=1}^{N} n_i \tau_{ij} \right\} \quad (6.3)$$

where τ_{ij} is the processing time of product i for stage j, n_i is the number of batches for product i, and M and N are the number of stages and products, respectively.

6.5 PARALLEL UNITS AND INTERMEDIATE STORAGE

In the previous section the examples have dealt with simple sequential flowshop plants that involve one unit per stage. As we will see in this section, adding intermediate storage tanks between stages or adding parallel units operating out of cycle can increase the efficiency of equipment utilization.

TABLE 6.2 Processing Times for Example on Transfer Policies (hrs)

	Stage 1	Stage 2	Stage 3
A	6	4	3
B	3	2	2

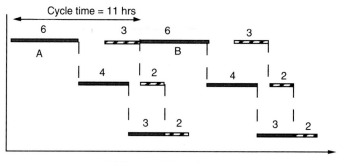

(a) Zero-wait transfer

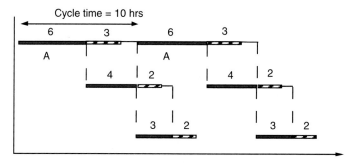

(b) No intermediate storage

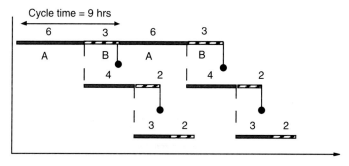

(c) Unlimited intermediate storage

FIGURE 6.7 Cycle times for various transfer policies.

Sec. 6.5   Parallel Units and Intermediate Storage

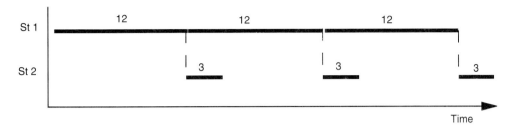

FIGURE 6.8 Gantt chart for fermentation plant.

As an example, consider the fermentation plant in Figure 6.8 in which stage 1 (fermenter) takes 12 hours compared to only 3 hours for stage 2 (separation). For simplicity, we assume zero-wait transfer and that the size of the batch in each stage is the same (1000 kg).

It is clear that the cycle time for each batch in Figure 6.8 is 12 hours applying Eq. (6.2). Since stage 1 is the bottleneck, we might consider adding a unit in parallel in that stage. With this additional unit the plant can be operated as shown in Figure 6.9 in which the cycle time has been reduced to 6 hours. The equation for cycle time with ZW transfer and parallel units, $NP_j, j = 1 \ldots M$, is the following,

$$CT = \max_{j=1..M} \{\tau_{ij} / NP_j\} \qquad (6.4)$$

Applied to our example in Figure 6.9, this leads to $CT = \max \{12/2, 3\} = 6$ hours. Note that if a large number of batches are to be produced, then to produce the same amount we can reduce the batch size to 500 kg since the cycle time has been halved.

The other alternative in Figure 6.8 is to introduce intermediate storage between stages. This has the effect of decoupling the two stages so that each stage can operate with different cycle times and batch sizes. As seen in Figure 6.10, stage 1 has a cycle time of

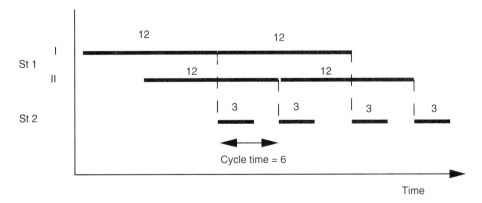

FIGURE 6.9 Plant with parallel units in fermenter.

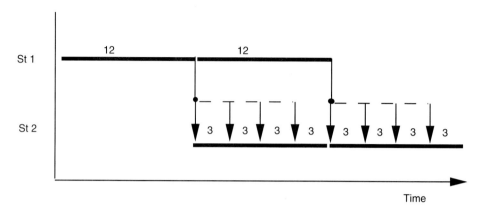

FIGURE 6.10 Fermentation plant with intermediate storage.

12 hours and handles batches of 1000 kg; stage 2 has a cycle time of 3 hours and handles batches of 250 kg. Thus, for every batch in stage j, four batches can be processed in stage 2. In this case it is also easy to verify that the intermediate storage must hold up to three batches (i.e., 750 kg) and that all the idle times have been eliminated.

6.6 SIZING OF VESSELS IN BATCH PLANTS

We will consider first the equipment sizing for the case of single product plants, and we will illustrate the ideas through an example problem.

Assume we have a two-stage plant and we want to produce 500,000 lb/yr. of product C. The plant is assumed to operate 6000 hours per year. The recipe for producing product C is as follows:

1. Mix 1 lb A, 1 lb B, and react for 4 hours to form C. The yield is 40% in weight and the density of the mixture, ρ_m, is 60 lb/ft^3.
2. Add 1 lb solvent and separate by centrifuge during 1 hour to recover 95% of product C. The density of the mixture, ρ_m, is 65 lb/ft^3.

Figure 6.11 shows all the relevant elements for the mass balance according to the above recipe. To perform the equipment sizing it is convenient to define size factors, S_j, for each stage j:

S_j = volume vessel j required to produce 1 lb of final product.

Sec. 6.6   Sizing of Vessels in Batch Plants

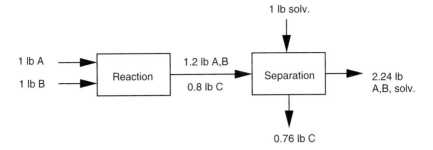

FIGURE 6.11   Mass balance information for batch plant.

For our example, the specific volume for stage 1 is $v = 1/\rho_m = 0.0166$ ft^3/lb mix. In this way we have

$$S_1 = 0.0166 \frac{\text{ft}^3}{\text{lb mix}} \frac{2 \text{ lb mix}}{0.76 \text{ lb prod}} = 0.0438 \frac{\text{ft}^3}{\text{lb prod}} \quad (6.5)$$

Similarly, for stage 2 the specific volume is $v = 0.0153$ ft.^{3}/lb.mix, thus the size factor is

$$S_2 = 0.0153 \frac{\text{ft}^3}{\text{lb mix}} \frac{3 \text{ lb mix}}{0.76 \text{ lb prod}} = 0.0604 \frac{\text{ft}^3}{\text{lb prod}} \quad (6.6)$$

If we use one unit per stage and operate with zero-wait transfer, the cycle time from Eq. (6.2) is:

$$CT = \max \{4,1\} = 4 \text{ hours} \quad (6.7)$$

This, then, implies that the number of batches to be processed in 6000 hours is

$$\text{no. batches} = \frac{6000 \text{ hrs.}}{4 \text{ hrs./batch}} = 1500 \text{ batches} \quad (6.8)$$

Since the product demand is 500,000 lb, the batch size of the final product is

$$B = \frac{500,000 \text{ lb}}{1500} = 333 \text{ lb} \quad (6.9)$$

We can then easily compute the volumes of the two vessels:

$$V_1 = S_1 B = 0.0438 \frac{\text{ft}^3}{\text{lb}} 333 \text{ lb} = 14.6 \text{ ft}^3$$
$$V_2 = S_2 B = 0.0604 \frac{\text{ft}^3}{\text{lb}} 333 \text{ lb} = 20.1 \text{ ft}^3 \quad (6.10)$$

Since the bottleneck is in stage 1, we might consider placing two units operating in parallel out-of-phase. The cycle time from Eq. (6.4) is then:

$$CT = \max \{4/2, 1\} = 2 \text{ hours} \tag{6.11}$$

This implies we can produce twice as many batches—3000 each of 166 lb, or half the original batch size. In this way the sizes are as follows:

$$V_1 = 7.3 \text{ ft}^3, \ V_2 = 10 \text{ ft}^3 \tag{6.12}$$

Although the total volume (24.6 ft^3) is smaller than in the case of 1 unit per stage (34.7 ft^3), we require a total of 3 vessels, 2 in stage 1 and 1 in stage 2. Depending on the cost correlation we may or may not achieve a reduction in the investment cost.

We will consider next the equipment sizing for the case of plants for multiple products, and again use a simple example to illustrate the main ideas (see Flatz, 1980, for an alternative treatment).

Let us consider a plant consisting of two stages that manufactures two products, A and B. The demands are 500,000 lb/yr. for A and 300,000 lb/yr. for B, and the production time considered is 6000 hours. Data on processing times, size factors, and cleanup times are given in Table 6.3. In order to perform the sizing, we need to specify the production schedule. There are many alternatives, some of which you will analyze in exercise 5. Here we will consider the simplest case, namely single product campaigns. Even here, however, we need to specify the length of the production cycle. We will select arbitrarily a production cycle of 1000 hours (42 days), which implies that over one year the cycle will be repeated six times. The choice of length of cycle has implications for inventories as we will see in section 6.7.

From Figure 6.12 it is clear that the effective time for production in each cycle is 992 hours. The main question is how to allocate the production of A and B (i.e., selecting t_A, t_B in Figure 6.12) during this time horizon. A simple solution is to use as a heuristic the same batch size for all products. The batch size B_i of product i is given by:

$$B_i = \frac{\text{production } i}{\text{no. batches } i} = \frac{\text{production } i}{t_i / CT_i} \tag{6.13}$$

where t_i and CT_i are the total production time and cycle time for each product, respectively. The production of A and B in each campaign is 500,000/6 = 83,333 lb and 300,000/6 = 50,000 lb, respectively. Applying the heuristic of equating the batch sizes and constraining the production times to 992 hours yields the two equations,

TABLE 6.3 Data for Sizing Two-Product Plant

	Processing Times (hr.)		Size Factors (ft^3/lb prod)	
	Stage 1	*Stage 2*	*Stage 1*	*Stage 2*
A	8	3	0.08	0.05
B	6	3	0.09	0.04

Cleanup times: 4 hours A to B, B to A

Sec. 6.7 Inventories   193

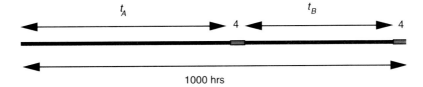

FIGURE 6.12 Time allocation for production of A and B.

$$\frac{83,333}{t_A/8} = \frac{50,000}{t_B/6}$$ (6.14)

$$t_A + t_B = 992$$

whose solution is $t_A = 684$ hours, $t_B = 308$ hours, and hence $B_A = B_B = 974$ lb. It is easy to show that for N products the generalization to the above equations will lead to a system of N linear equations (see exercise 6).

Given the batch size we can then compute the required volumes for each product in the two stages ($V_{ij} = S_{ij} B$):

	Volumes V_{ij} (ft³)	
	Stage 1	Stage 2
A	77.9	48.7
B	87.7	39.0

Finally, the largest volumes to be selected in each stage are given by:

$$V_j = \max_{i=1,N} \{V_{ij}\}$$ (6.15)

with which $V_1 = 87.7$ ft³, $V_2 = 48.7$ ft³.

6.7 INVENTORIES

An important issue in batch design and operation is the selection of the production cycle. The main trade-off involved is the fraction of transition or cleanup times versus inventories. The shorter the production cycle, the less inventory we need to carry since products are available more frequently, but the fraction of the transitions becomes greater; conversely, the longer the production cycle, the smaller the fraction of transitions. However, in this case inventories will increase because products are produced less frequently.

In the example of the previous section we can determine the inventory profiles as shown in Figure 6.13. The details are as follows.

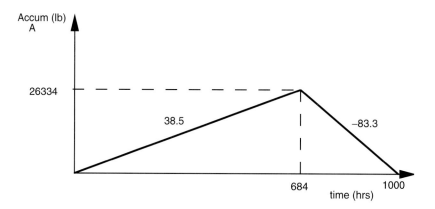

FIGURE 6.13 Inventory profile for product A.

The demand rates of the two products are the following:

$$d_A = 83{,}333/1000 = 83.3 \text{ lb/hr.}$$
$$d_B = 50{,}000/1000 = 50 \text{ lb/hr.} \tag{6.16}$$

while the production rates are:

$$p_A = \frac{83{,}333}{684} = 121.8 \text{ lb/hr.}$$
$$p_A = \frac{50{,}000}{308} = 162.3 \text{ lb/hr.} \tag{6.17}$$

The inventory profile of A can then be obtained as follows:

1. 0–684 hrs. Accumulation rate = $p_A - d_A$ = 121.8 – 83.3 = 38.5 lb/yr.
2. 684–1,000 hrs. Depletion rate = $-d_A$ = – 83.3 lb/hr.

Figure 6.13 shows this profile. For product B the procedure is similar (accumulation: 688 – 996 hrs; depletion: 996 – 688 hrs.) and the corresponding profile is shown in Figure 6.14.

The annual inventory cost can be calculated by determining the average inventory and knowing the corresponding unit cost. The average inventory is given by calculating the areas under the curve in Figures 6.13 and 6.14 and dividing them by the length of the production cycle, 1000 hours. The average inventory of product A is:

$$I_A = \frac{1000 \ (26334)}{2 \ (1000)} = 13{,}167 \text{ lb} \tag{6.18}$$

while the average inventory of product B is:

Sec. 6.8 Synthesis of Flowshop Plants   195

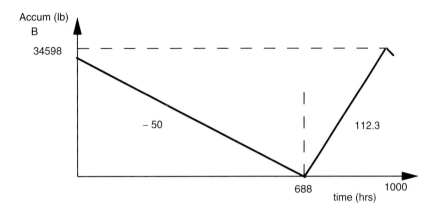

FIGURE 6.14 Inventory profile product B.

$$I_B = \frac{1000\,(34600)}{2\,(1000)} = 17,300 \text{ lb} \quad (6.19)$$

If the inventory cost is $1.25/lb yr, the total inventory cost is:

$$\begin{aligned} C_{inv} &= 1.25\,(13{,}167 + 17{,}300) \\ &= \$38{,}084/\text{yr} \end{aligned} \quad (6.20)$$

The main variable affecting this cost is often the length of the production cycle (see exercise 5).

6.8 SYNTHESIS OF FLOWSHOP PLANTS

Having introduced the main concepts involved in the scheduling and sizing of batch processes, we will outline in this section some of the major alternatives that must be generated and evaluated at the synthesis stage of the design. For most problems the number of alternatives is very large. Since the economic trade-offs for most of the alternatives are generally complex, there is a need to resort to systematic optimization approaches such as those given in Chapter 22. Here we will limit ourselves to discussing the alternatives for flowshop plants. For a more comprehensive treatment of this topic see Yeh and Reklaitis (1987).

For the economic evaluation of the alternatives and their comparison the net present value NPV is used (see Chapter 5) and given as follows:

$$\begin{aligned} NPV = &-CI + (R - CO - C_{inv})(1 - tx)[(1 - (1 + i)^n)/i] \\ &+ (CI/n)tx[(1 - (1 + i)^n)/i] + sCI/(1 + i)^n \end{aligned} \quad (6.21)$$

where R is the annual revenue of the products, CI the investment cost, CO is the operating cost, C_{inv} the inventory cost, i the interest rate, n the length of the project life, tx the tax rate and s the fraction of investment for salvage value. Note that since the amounts to be produced are specified and the production is performed by a recipe, the revenue R and the operating cost CO are constant. Therefore, if the only objective is to compare alternatives there is no need to evaluate these terms.

The three major decision levels and their corresponding items are the following:

1. Structural level
   a. Assignments of tasks to equipment
   b. Number of parallel units or intermediate storage
2. Sizing level
   Equipment sizing
3. Scheduling level
   a. Nature of production campaigns, transfer policies
   b. Length of production cycles
   c. Sequencing of products

At the structural level the assignment of tasks to equipment is one of the decisions that can have the greatest impact in the scheduling and economics. To illustrate this point,

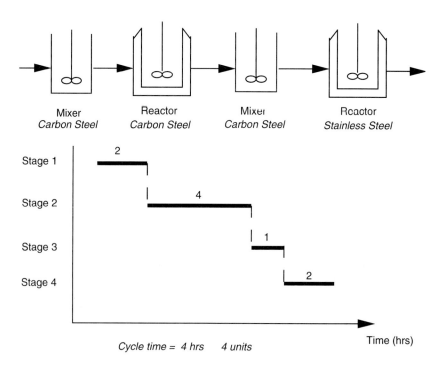

FIGURE 6.15 Design alternative with assignment of one equipment to each task.

Sec. 6.8 Synthesis of Flowshop Plants

consider as an example the case of a single product batch process that involves the following four processing tasks:

Task 1:   Mixing, 2 hours
Task 2:   Reaction, 4 hours
Task 3:   Mixing, 1 hour
Task 4:   Reaction, 2 hours

The simplest alternative is to assign each task to one processing equipment as shown in Figure 6.15. Note that the two mixing tasks take place in simple vessels with an agitator, while the reactions take place in jacketed vessels. Also, except for the second reactor, which must be made of stainless steel, the three remaining units are made of carbon steel. As seen in Figure 6.15, the cycle time is 4 hours assuming zero wait transfer.

A second alternative is to assign tasks 3 and 4 to one single piece of equipment, namely to the stainless steel reactor as shown in Figure 6.16. Note that in this alternative the cycle time remains unchanged in 4 hours despite the fact that we have eliminated one piece of equipment. This alternative is clearly superior to the one in Figure 6.15. Thus, a simple design guideline that we can postulate is: "Merge adjacent tasks whose sum of processing times does not exceed the cycle time."

Finally, a third alternative that we can consider is shown in Figure 6.17. All tasks have been merged in one pierce of equipment—the jacketed stainless steel vessel that can

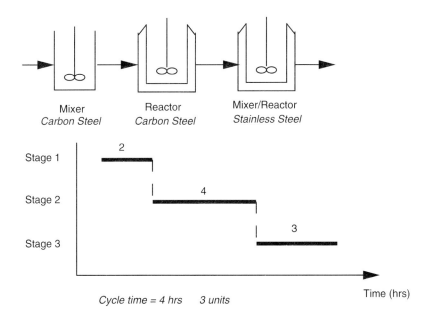

FIGURE 6.16 Design alternative with merging of tasks 3 and 4.

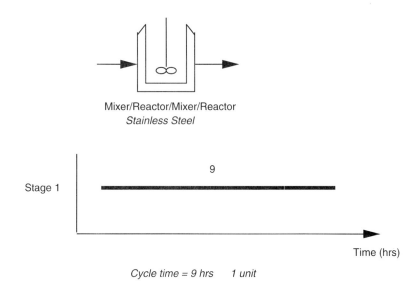

FIGURE 6.17 Design alternative with complete merging of all tasks.

perform the four tasks. The trade-off here is that while we only require one unit, the cycle time increases to 12 hours, and thus a much larger stainless steel vessel is required.

It should be noted that in some cases merging of tasks requires new equipment to meet the materials of construction requirement and to perform all the required functions. For instance, if one task requires a jacketed carbon steel and the other task a simple stainless steel vessel, the merged tasks require a jacketed stainless steel vessel.

The other major structural decision is the assignment of intermediate storage between stages and the selection of number of units in parallel. As was shown in section 6.5, these decisions also commonly have a great impact in the scheduling. The choice of intermediate storage is usually dictated by feasibility of keeping intermediate material in storage. This alternative tends to be favored whenever there is a stage with a much larger processing time (see Figure 6.8). The alternative for placing parallel units operating out of phase is favored when there is a requirement for maintaining batch integrity. Generally, the trade-off here is a smaller number of bigger pieces of equipment versus a larger number of smaller pieces.

The sizing outlined in section 6.6 is used as a heuristic to select the same batch size for all products. It should be noted that the more the size factors differ between products, the worse this heuristic sizing becomes.

Finally, the scheduling level involves deciding the type of campaign (single products versus mixed product), the transfer policy (ZW, NIS, or UIS), the length of the production cycle, and the sequencing of the products. If cleanup or transition times are large, single product campaigns are favored; otherwise, the reverse is true. Also, a very useful aid here are the Gantt charts, since they clearly indicate the extent to which idle times are

REFERENCES

Flatz, W. (1980). Equipment sizing multiproduct plant. *Chemical Engineering*, **87**(6.4), 71.

Reeve, A. (1992). Batch control, the recipe for success? *Process Engineering*, **73**(6.1), 33.

Reklaitis, G. V. (1990). Progress and issues in computer-aided batch process design. *FOCAPD Proceedings*, Elsevier, New York, 275.

Rippin, D. W. T. (1993). Batch process systems engineering: A retrospective and prospective review. *Computers Chem. Engng.*, **17**, Suppl., S1-S13.

Yeh, N. C., & Reklaitis, G. V. (1987). Synthesis and sizing of batch/semicontinuous processes. *Computers Chem. Engng.*, **11**, 639.

EXERCISES

1. A given batch plant produces one single product for which stage 1 requires 8 hours/batch; stage 2, 4 hours per batch; and stage 3, 7 hours per batch. If zero-wait transfer is used, what is the cycle time? How many parallel units should be placed in each stage to reduce the cycle time to 4 hours?

2. Given the processing times for three products A, B, C, below, determine with a Gantt chart the makespan and cycle time for manufacturing two batches of A, 1 of B, and 1 of C for the following cases:
   a. Zero-wait policy with sequence *AABC* and sequence *BAAC*.
   b. Same as (a) but with no intermediate storage policy (NIS).
   c. Same as (a) but with unlimited intermediate storage policy (UIS).

	Processing Times (hr)		
	Stage 1	*Stage 2*	*Stage 3*
A	5	4	3
B	3	1	3
C	4	3	2
	Zero cleanup times.		

3. Given is a product A that is to be manufactured in four processing stages. Determine with a Gantt chart the makespan and cycle time for the manufacturing of three batches of A for the following cases:
   a. Zero-wait policy with one unit per stage.

b. Zero-wait policy with two parallel units in stage 3 and one unit in stages 1, 2, 4.
c. Zero-wait policy with one unit per stage but with merging of tasks in stages 1 and 2.

	Processing times (hr)			
	Stage 1	Stage 2	Stage 3	Stage 4
A	4	3	6	2

4. Derive Eq. (6.3) for the cycle time for a jobshop plant consisting of one unit per stage and with unlimited intermediate storage (UIS) transfer.

5. For the example given in section 6.6 and Table 6.3, compute the size of the two vessels and the average inventories for the following lengths of production cycles: (a) 50 hrs, (b) 500 hrs, (c) 2000 hrs.

6. Show that the time allocation t_i, for N products, $I = 1, 2, \ldots, N$, in single product campaigns can be determined through a system of N linear equations in N unknowns t_i, assuming the same batch size is used for all products (see Eqs. (6.13) and (6.14)), and that the production requirements and cycle times are given for each product i.

7. Determine the size of the vessels of a multiproduct batch plant that consists of three stages for manufacturing products A and B. Only one vessel is to be used in each stage. Consider the two following cases:
   a. Production cycles of 500 hrs consisting of two campaigns: one for A and one for B.
   b. Cyclic sequence of production $AABAABAAB\ldots$

   Data

   Demands: A: 600,000 kg/yr

   B: 300,000 kg/yr

   Horizon time = 6000 hrs

	Processing Times (hr)		
	Stage 1	Stage 2	Stage 3
A	4	2	3
B	3	2	5

	Size factors (kg)		
	Stage 1	Stage 2	Stage 3
A	2	5	3
B	1.5	6	2

Note: Assume that both products have the same batch size.

Exercises

8. Consider a flowshop plant that is to be designed for manufacturing four different products. Data on demands, processing times, and other parameters are given below.

   a. Determine the design and its net present value for the case that each task is assigned to a separate unit, and the plant is operated with 8 cycles during the year using single-product campaigns with zero-wait transfer.

   b. Propose a design that can improve the net present value of the alternative in (a).

Data

	Product Demands (kg/yr)	Net profit ($/kg)*
A	400,000	0.60
B	200,000	0.65
C	200,000	0.70
D	600,000	0.55

*Accounts for raw material cost, processing cost and indirect costs. Does not account for inventory.

Operating time per year = 8000 hrs

Product cannot be held in inventory for more than 90 days.

Inventory cost = $2.40/kg per yr

Interest Rate: 10%
Tax rate: 45%
Service Life: 10 years
Depreciation: Straight line with no salvage value

Production Recipe

The four products require the following processing steps:

Step 1. Reaction.
   Mix solutions $F1$ and $F2$, and heat at 40°C. Solution $F3$ is formed with x weight percentage of product.

   Equipment: Stainless steel jacketed vessel with agitator
   Storage not allowed

Step 2. Recovery of product with solvent.
   Mix $F3$ and solvent $F4$ in equal volume for 30 minutes to recover product from $F3$. Mixture is allowed to settle for 2 hours to form $F3$ and $F4$ phases. $F3$ phase is drained ($F5$) and sent to wastewater treatment. 95% of product is recovered in phase $F4$ (stream $F6$).

Equipment: Stainless steel vessel with agitator.
Storage allowed

Step 3. Purification of solvent with water.
Mix $F6$ with 2.5 volume of water ($F7$) for 20 minutes. Mixture is allowed to settle for 90 minutes to form $F6$ and water phases. Water phase is drained ($F8$) and sent to wastewater treatment. 98% of product is recovered in phase $F6$ (stream $F9$).

Equipment: Cast iron vessel with agitator.
Storage allowed

Step 4. Crystallization.
$F9$ is cooled to 15°C. The mixture is aged for a specified length of time giving a slurry of product crystals with 95% recovery.

Equipment: Cast iron jacketed vessel with agitator.
Storage not allowed

Step 5. Centrifuge.
The slurry $F10$ is centrifuged for 50 minutes to give a solution with $y\%$ weight of product. The liquid $F11$ is sent to a solvent recovery unit.

Equipment: Automatic basket centrifuge.

Specific data for each product

Addition of solution $F2$ (kg) for 1 kg of $F1$.

A	B	C	D
0.4	0.6	0.7	0.5

Weight % (x) of product formed in step 1.

A	B	C	D
8	9	6.5	7

Weight % (y) of product in final solution.

A	B	C	D
45	38	55	42

The following densities can be assumed to be the same for the manufacturing of the four products.

Specific gravity (kg/L)
F1   0.8
F2   1.0
F4   0.7
F7   1.0

Exercises

Processing times (hrs)

Step 1 Reaction

A	B	C	D
4.5	5.5	3.75	7.25

Step 4. Crystallization

A	B	C	D
3.75	1.5	5.75	8.5

Cleanup Times

It is assumed that they are the same for each piece of equipment. However, cleanup times depend on the sequence of products according to the following (time in hrs):

	A	B	C	D
A	0	0.2	0.5	2
B	0.2	0	0.5	2
C	0.5	0.5	0	0.5
D	2	2	0.5	0

Equipment Cost: Cost = Fixed charge + a*(Volume)**b

Equipment	Fixed charge($)	a($)	b

Min size = 2000 liters, Max size = 20,000 liters, increments 2000 liters

Equipment	Fixed charge($)	a($)	b
Stainless steel jacketed/agitator	105,000	650	0.6
Stainless steel agitator	82,000	550	0.6
Cast iron jacketed/agitator	65,000	350	0.6
Cast iron agitator	48,000	280	0.6

Min size = 3000 liters, Max size = 15,000 liters, increments 3000 liters

Equipment	Fixed charge($)	a($)	b
Centrifuge	150,000	350	0.8

Min size = 1000 liters, Max size = 10,000 liters, increments 1000 liters

Equipment	Fixed charge($)	a($)	b
Cast iron storage vessel	22,000	120	0.6
Stainless steel storage vessel	35,000	120	0.6

PART II

ANALYSIS WITH RIGOROUS PROCESS MODELS

UNIT EQUATION MODELS 7

This chapter provides a summary of detailed unit operations models that are appropriate for modern computer-aided design and analysis tools. In Part I, emphasis was placed on preliminary analysis and process evaluation. As a result, shortcut models were used to develop a qualitative understanding of a process flowsheet and the impact of design decisions. Moreover, the quantitative, economic metrics for characterizing and evaluating design decisions were developed for both continuous and batch processes. These concepts extend into Part II, but here we will consider more detailed design models and evaluation strategies. This part covers Chapters 7, 8, and 9 and deals with a description of detailed process models, methods for solving these models, and flowsheet optimization strategies for determining optimal levels of continuous variables.

In Chapter 7 we increase the level of detail for the unit operations models considered in Chapter 3 in order to provide more accurate models of design units. Just as in Chapter 3, the purpose of these models is to provide a mass and energy balance for evaluation of the process flowsheet. Consequently, many of the assumptions used for the shortcut models will be removed and more detailed concepts on nonideal behavior and the development of larger, nonlinear models will be presented. In particular, this chapter introduces models for nonideal physical properties and shows how these are embedded within more rigorous process models. In addition, we consider detailed phase equilibrium and separation models, which are considerably larger and more difficult than previous shortcut models.

As a result, these models are no longer appropriate for hand calculations and the numerical methods described in Chapter 8 must be applied to these models. In Chapter 8 we describe two popular simulation strategies, the modular and equation based modes, and discuss numerical algorithms that relate to both. Both modes require the solution of nonlinear equations and basic derivations are provided for popular methods with these simulation modes. In addition, some discussion is provided on flowsheet partitioning and tearing that is required for the modular mode, and, analogously, sparse matrix decomposition that applies to the equation based mode.

With strategies available for process modeling and simulation, we next consider the systematic determination of the best equipment parameters and operating levels in a candidate flowsheet. These require the application of continuous variable *optimization* strategies, or *nonlinear programming*. As in Chapter 8, Chapter 9 develops flowsheet optimization algorithms for both modular and equation based process simulation modes and also provides some background on nonlinear programming theory. Moreover, these concepts also help to set the stage for the advanced optimization approaches presented in Part IV of this text.

Practical examples and case studies are used to highlight all of the concepts presented in the next three chapters, and these are often drawn from industrial applications. From these we hope to illustrate both the complexity of the applications and the effectiveness of the modeling and solution strategies. As a result, these three chapters set the stage for an understanding of modern computer-aided simulation and optimization tools in process engineering.

7.1 INTRODUCTION

The development of mass and energy balance models is a basic component upon which process evaluation and design decisions need to be made. As in Chapter 3, we consider the candidate flowsheet model as a large set of nonlinear equations that describe

1. Connectivity of the units in the flowsheet through process streams.
2. The specific equations of each unit, which are described by conservation laws as well as constitutive equations for that unit.
3. Underlying data and relationships that relate to physical properties and serve as building blocks for each unit operation model.

In this chapter we focus on topics 2 and 3 and present a more detailed representation of the unit operations models. To do this, we reconsider the approach taken in Chapter 3. In that chapter we decoupled the relations between the mass balance, temperature and pressure specifications, and the energy balance. This allowed us to execute the mass balance first, specify the temperature and pressure levels by assuming saturated output streams, and then calculate the energy balance and energy duties once temperatures and pressures were fixed. These calculations were made possible by assuming:

- Ideal behavior in phase equilibrium
- Relative volatilities "nearly" independent of temperature
- Ideal behavior for energy balances

Sec. 7.1 Introduction

- Noninteracting components in unit operations (except for reactors)
- Fixed conversion reactor models
- Simplifications in applying shortcut calculations

The main goal of this chapter is to relax all of these assumptions and present reasonably accurate unit behavior for developing mass and energy balances. Specifically, we consider the influence of nonideal equilibrium behavior and the derivation of more detailed models. Nevertheless, the treatment of detailed unit models is necessarily brief and is motivated by design decisions. Indeed, the primary perspective of this chapter is to gain a better understanding of the level of modeling detail used in computer-aided simulation tools. More complete descriptions of these models are referenced in the last section.

In particular, we will model separation units entirely through phase equilibrium relations and mass and energy balances. These equilibrium staged models rely only on thermodynamic concepts. Moreover, reliance on thermodynamic concepts is a key point in the development of most detailed *design* models for process flowsheets. This assumption greatly simplifies the calculations, as only thermodynamic properties need to be incorporated into the physical property database, and transport properties need not be considered in detail. Moreover, this assumption allows the mass and energy balances to be obtained without knowledge of the capacity and geometry of the units. Consequently, the sizing and costing calculations of Chapter 3 can be performed after these detailed models are executed; this provides a further simplification of the design evaluation. However, this assumption is not without drawbacks, and we caution that these *design* models are usually appropriate for modeling *new units* (or "grassroots" units) where geometries and capacities can be specified reasonably freely and easily. On the other hand, accurate simulation of *existing units* is frequently governed by geometric considerations and requires the development of more detailed *performance* models, that are beyond the scope of this text. Moreover, we caution that thermodynamic-based design models may be inadequate for many complex separations that are currently considered with more accurate mass transfer models. However, these separation models are also beyond the scope of this text.

In the next section, we provide a brief summary of nonideal thermodynamic relations that are commonly used for process simulation tools. These are classified into phase equilibrium relations, relations for specific enthalpy and entropy, and relations for specific volume and other less commonly used physical properties. Here commonly used thermodynamic models are summarized and guidelines are given for their use.

The third section deals with nonideal flash calculations. These "building block" calculations refer not only to single-stage phase separations but also apply to analysis of any process stream where the phase condition needs to be determined. By incorporating nonideal equilibrium relations, a more complex flash model results than was addressed in Chapter 3. Here we derive this model and discuss direct and nested solution strategies for these flash models. The fourth section extends the flash model to equilibrium-staged separations. In particular, we discuss the derivation of distillation models along with methods for their solution. It should be noted that this model easily extends beyond conventional distillation columns to cover complex column configurations. Again, direct and nested modes for the solution of these models will be discussed. Extensions to other equilibrium stage separation operations such as absorption and extraction will also be outlined.

The fifth section deals with unit models that are less detailed than the ones described above and include transfer and exchange operations carried out by pumps, compressors, and heat exchangers of various types. We retain the motivation of *design* calculations and assume that sizing and costing can be done once the mass and energy balance is fixed. Consequently, the mass and energy balance models themselves will be largely unaffected by geometric considerations. In this section we also consider reactor models briefly, with the same set of assumptions. The last section summarizes the chapter and presents some future directions for flowsheet modeling. These address some of the shortcomings exhibited by the models in this chapter but at the expense of more computationally intensive models.

7.2 THERMODYNAMIC OPTIONS FOR PROCESS SIMULATION

This section provides a brief summary of thermodynamic relationships that are required for the formulation of nonideal, equilibrium-based process models. Clearly, treatment of this broad area will be incomplete and somewhat superficial, as a large (and burgeoning) literature is devoted to this topic. Instead, we consider a qualitative description of physical property models that are available in current process simulators. Supporting these models, one finds a tremendous amount of effort devoted to the construction and verification of physical property data banks, based on careful experimentation. The models themselves are based on concepts of solution thermodynamics as discussed, for example, in Smith and VanNess (1987) and VanNess and Abbott (1982). A summary of thermodynamic options is presented in Reid et al. (1987) and exhaustive details of the physical property options can be found in the user manuals of most process simulators. Built on top of this are robust numerical procedures for the calculation of thermodynamic and transport properties. Nevertheless, within a process simulator, this is often presented to the user simply as a set of options, often with few guidelines (or knowledge of the consequences) for their selection.

In this section there is no attempt at providing a complete survey of these options, just a basic understanding of these relationships. We start by concentrating on thermodynamic calculations that support nonideal phase equilibrium, through chemical potentials and fugacities, and then continue with applications to the calculation of other thermodynamic quantities, especially partial molar enthalpies and volumes. Once covered, these thermodynamic and physical property calculations provide the basic building blocks for the detailed unit operations models which follow.

7.2.1 Phase Equilibrium

Phase equilibrium is determined when the Gibbs free energy for the overall system is at a minimum. Here, underlying relationships for phase equilibrium are derived from a minimization of the Gibbs free energy of the system. Given a mixture of n moles with NC

Sec. 7.2 Thermodynamic Options for Process Simulation

components, if we have equilibrium between NP phases and n_{ip} moles for each component i in phase p, this can be expressed by the following problem:

$$\text{Min } n G = \sum\sum n_{ip} \mu_{ip} \qquad (7.1)$$

$$\text{s.t. } \sum_p n_{ip} = n_i, \ i = 1, \ldots NC$$

$$n_{ip} \geq 0$$

where n_i is the total number of moles for component i, G is the Gibbs energy *per mole* of the system, and the chemical potential of component i is defined by

$$\mu_i = [\partial(n G)/\partial n_i] \text{ with } T, P, \text{ and } n_j \ (j \neq i) \text{ constant} \qquad (7.2)$$

For nonempty phases, the solution of this optimization problem is given by equality of the chemical potentials across phases, that is:

$$\mu_{i1} = \mu_{i2} \ldots \mu_{i,NP} \qquad i = 1, \ldots NC \qquad (7.3)$$

To describe the chemical potential, we define a mixture fugacity for each of these phases and components according to:

$$d\mu_{ip} = RT \, d \ln f_{ip} \qquad (7.4)$$

and integrating from the same initial condition (say, μ') for all phases gives:

$$\mu_{ip} - \mu' = RT \ln (f_{ip}/f') \qquad (7.5)$$

Simplifying this expression shows that the mixture fugacities must also be the same in all phases:

$$f_{i1} = f_{i2} \ldots f_{i,NP} \qquad i = 1, \ldots NC \qquad (7.6)$$

Confining ourselves to vapor-liquid equilibrium (VLE), we now specialize the fugacities to particular cases. For the vapor phase, we introduce a fugacity coefficient defined by:

$$\phi_i = f_{iv} / (y_i P) \qquad (7.7)$$

where y_i is the mole fraction of component i in the vapor mixture and P is the total pressure. For the liquid phase, we define an activity a_i as well as the activity coefficient γ_i according to:

$$\gamma_i = a_i / x_i = f_{il} / (x_i f_{il}^0) \qquad (7.8)$$

where f_{il}^0 is the pure component fugacity. This pure component fugacity is further defined by:

$$f_{il}^0 = f_{il}(T, P, x_i = 1) = P_i^0(T) \, \phi_i \, (x_i = 1, P_i^0, T) \exp[\int_{P_i^0}^{P} V_{il}(T, P)/RT \, dp] \qquad (7.9)$$

where the exponential of the volume integral in this expression is known as the *Poynting correction factor*. Equating the mixture fugacities in each phase now leads to a reasonably general expression:

$$\phi_i y_i P = \gamma_i x_i f_{il}^0 \quad \text{for } i = 1, NC \tag{7.10}$$

and we can define K values, $K_i = \gamma_i f_{il}^0 / (\phi_i P)$, that will be used for the flash calculations in the next section.

As was assumed in Chapter 3, there are a number of simplifications that can be made to the above expressions for the ideal case:

- For an ideal solution in the liquid phase, the activity coefficient $\gamma_i = 1$.
- For an ideal solution in the vapor phase, the mixture fugacity $f_{iv} = f_{iv}^0$.
- For a mixture of ideal gases in the vapor phase, $\phi_i = 1$.
- For negligible liquid molar volumes or for low pressures, the volume integral is negligible and the Poynting factor is unity.

The nonideal cases can be characterized by violations of the above simplifications. Violation of the first assumption is the most common and we frequently expect nonideality in the liquid phase. The second assumption is valid for most chemical systems up to moderate pressure levels and we will not consider any modifications of this assumption in this text. The third and fourth assumptions are valid for low to medium pressures. In considering nonideality in phase equilibrium, we first consider nonideality in the liquid phase when the third and fourth assumptions are valid. Then we consider higher pressure systems where nonidealities need to be considered for the vapor phase as well.

7.2.2 Liquid Activity Coefficient Models

Departures from ideality can be represented by defining departure functions or *excess thermodynamic quantities*. For molar Gibbs free energy we define:

$$G = G^{id} + G^E \tag{7.11}$$

or

$$\begin{aligned} G^E/RT &= G/RT - G^{id}/RT = \sum x_i \ln(f_{il}/f_{il}^0) - \sum x_i \ln x_i \\ G^E/RT &= \sum x_i \ln(f_{il}/(x_i f_{il}^0)) = \sum x_i \ln \gamma_i \end{aligned} \tag{7.12}$$

where G^{id} is the molar Gibbs free energy for the ideal system and G^E is the excess molar Gibbs free energy. The activity coefficient can also be treated as a partial molar quantity of G^E/RT:

$$\ln \gamma_i = [\partial(n\, G^E/RT)/\partial n_i] \text{ with } T, P, \text{ and } n_j\, (j \neq i) \text{ constant} \tag{7.13}$$

and after some manipulation, we can obtain, for component i, a direct relationship between G^E/RT and $\ln \gamma_i$ from the following equation:

$$\ln \gamma_i = G^E/RT + \partial(G^E/RT)/\partial x_i - \sum_k x_k \partial(G^E/RT)/\partial x_k \tag{7.14}$$

EXAMPLE 7.1

For a binary system, consider the simplest excess function, the two-suffix Margules model, $G^E/RT = A\, x_1 x_2$. What are the activity coefficients for this model?

Applying the expression:

$$\ln \gamma_i = G^E/RT + \partial(G^E/RT)/\partial x_i - \sum_k x_k\, \partial(G^E/RT)/\partial x_k \qquad (7.15)$$

leads to

$$\ln \gamma_1 = A\, x_1 x_2 + A\, x_2 - 2(A\, x_1 x_2) = A\, x_2(1 - x_1) = A\, x_2^2 \qquad (7.16)$$

$$\ln \gamma_2 = A\, x_1 x_2 + A\, x_1 - 2(A\, x_1 x_2) = A\, x_1(1 - x_2) = A\, x_1^2 \qquad (7.17)$$

The Margules model in Example 7.1, however, applies only to nearly ideal systems with molecules of similar sizes. Similarly, other models derived before computer simulation tools were developed (e.g., regular solution theory and the van Laar equations) have relatively simple forms and are largely restricted to nonpolar, hydrocarbon mixtures; these are less widely used than current methods.

For process simulation, the popular liquid activity coefficient models estimate multicomponent activity using only binary interactions among molecules. This assumption is valid for nonelectrolyte mixtures where there are only short-range (two-body) interactions in the mixture. A great advantage to this approach is that relatively little data are needed to model complex mixtures accurately. Current liquid activity coefficient models include the Wilson equation:

$$G^E/RT = -\sum_i x_i \ln(\sum_j x_j \Lambda_{ij}) \qquad (7.18)$$

with binary parameters Λ_{ij}, and the *NRTL* (non random two-liquid) equation:

$$G^E/RT = \sum_i x_i\, [(\sum_j \tau_{ji}\, G_{ji}\, x_j)/(\sum_k G_{ki}\, x_k)] \qquad (7.19)$$

with related binary parameters τ_{ji} and G_{ji} that can be derived from simpler forms. Both models have parameters that often need to be estimated from experimental data, although Reid et al. (1987) discuss approximations to these parameters that yield reasonable results. Of these two models, the Wilson equation is more accurate for homogeneous mixtures and it is computationally the least expensive of all of the methods in this section. However, it is functionally inadequate to deal with equilibrium between two liquid phases (LLE) or with two liquids and a vapor phase (VLLE). The NRTL equation must be used in this case.

The UNIQUAC (Universal Quasi Chemical) model also handles vapor liquid and liquid–liquid phase equilibrium. It is mathematically more complicated than NRTL but requires fewer adjustable parameters, which are also less dependent on temperature. In addition, this model is applicable to a wider range of components. The UNIQUAC model is given by:

$$G^E/RT = \sum_i x_i \ln(\Phi_i/x_i) + (\zeta/2) \sum_i q_i x_i \ln(\theta_i/\Phi_i) - \sum_i q_i x_i \ln(\sum_j \theta_j \tau_{ji}) \qquad (7.20)$$

where ζ is typically set to 10 and all of the parameters except τ_{ji} are calculated from pure component properties. The first two terms in this model represent *combinatorial* contributions due to differences in size and shape of the molecule mixtures and are based only on pure component information. The last term is a *residual* contribution to the excess molar Gibbs energy, is based on energy interactions between molecules, and requires binary interaction parameters τ_{ji}. As a result the activity coefficient can be represent by both parts as:

$$ln\ \gamma_i = ln\ \gamma_i^C + ln\ \gamma_i^R \qquad (7.21)$$

A further extension of these models is given by group contribution methods. Here the models contain parameters that characterize interactions between pairs of structural groups in the molecule (e.g., methyl, -OH, ketone, olefin). This information can then be used to predict activity coefficients in molecules with similar structural groups, for which data may not be available. This essentially describes the UNIFAC (UNIQUAC Functional-Group Activity Coefficient) model, which starts with the UNIQUAC equations and retains the combinatorial (or pure component) parts. Here the residual activity coefficient is substituted with a linear combination of group residual activity coefficients:

$$ln\ \gamma_i^R = \sum_k v_k^i (ln\ \Gamma_k - ln\ \Gamma_k^l) \qquad (7.22)$$

where v_k^i are the numbers of individual groups, Γ_k is the group activity coefficient for group k in the molecule and Γ_k^l is the residual activity in a reference solution l. Both Γ_k and Γ_k^l are given by

$$\Gamma_k = Q_k [\ 1 - ln\ (\sum_m \theta_m \Psi_{mk}) - \sum_m \{\theta_m \Psi_{km} / (\sum_n \theta_n \Psi_{nm})\}] \qquad (7.23)$$

where Q_k is a surface area parameter for each structural group m. Here θ_m represents the area fraction of group m and Ψ_{nm} is the group interaction parameter. Both sets of parameters are governed by further equations related to the mole fractions of the structural groups and their interaction energies, respectively.

This approach is accurate for nonelectrolyte systems for VLE, LLE, and VLLE applications. It is especially useful when binary data are missing and need to be estimated. Recent studies have also extended this approach to polymer and electrolyte systems and the methods enjoy wide use in process simulation applications. More information on the theoretical background of the UNIFAC method and its application can be found in Reid et al. (1987) and Fredenslund et al. (1977).

7.2.3 Equation of State (EOS) Models

The above activity coefficient models represent phase behavior for liquids. We now consider a generalized set of equation of state (EOS) models that can model the behavior in *both* the liquid and vapor phases. In addition, these models are especially important at "higher" pressures where we observe a departure from ideal gas behavior in the vapor phase. These equations need to be applied both for the calculation of vapor phase fugacity and for the Poynting correction factor for the pressure effect on the liquid phase. Common

Sec. 7.2 Thermodynamic Options for Process Simulation

models for nonideal gases are the cubic equations of state; two popular instances of these are the Soave-Redlich-Kwong (SRK) equation:

$$P = RT/(V - b) - a/(V^2 + bV) \tag{7.24}$$

and the Peng Robinson (PR) equation:

$$P = RT/(V - b) - a/(V^2 + 2bV - b^2) \tag{7.25}$$

The parameters a and b are related to reduced temperatures and pressures as well as an acentric factor and these can be derived for each component. For mixtures with components (i,j) with compositions z_i, quadratic mixing rules are often used:

$$a_M = \Sigma_i \Sigma_j z_i z_j (1 - C_{ij}) (a_i a_j)^{1/2} \tag{7.26}$$
$$b_M = 1/2 \, \Sigma_i \Sigma_j z_i z_j (1 + D_{ij}) (b_i + b_j)$$

to substitute for a and b in the equations of state, with adjustable binary parameters C_{ij} and D_{ij}. These equations are useful for pure component and vapor fugacities and they can also be used to estimate the *liquid* activities at equilibrium. Since the cubic equations permit multiple solutions for molar volume, one defines the largest root for the vapor phase (V^V) and the smallest for the liquid phase (V^L). Here we also define the fugacity coefficients for both phases:

$$\phi_{iv} = f_{iv} / (y_i P) \qquad \phi_{il} = f_{il} / (x_i P) \tag{7.27}$$

along with K values, $K_i = y_i / x_i = \phi_{il} / \phi_{iv}$. By defining compressibility factors ($Z = PV/RT$) for both phases we have:

$$Z^L = P \, V^L/RT \qquad Z^V = P \, V^V/RT \tag{7.28}$$

and this gives a direct assessment of the departure from ideal gas behavior, where $Z = 1$. We can estimate the fugacity coefficients for both phases from

$$RT \ln \phi_{il} = \int_{V^L}^{\infty} [(\partial P(T, V, x_i)/\partial n_i) - RT/V] \, dV - RT \ln Z^L \tag{7.29}$$

$$RT \ln \phi_{iv} = \int_{V^V}^{\infty} [(\partial P(T, V, y_i)/\partial n_i) - RT/V] \, dV - RT \ln Z^V \tag{7.30}$$

This approach is used widely for hydrocarbon mixtures, including natural gas and petroleum applications, but it is not useful for strongly polar or hydrogen bonded mixtures where the assumption of simple mixing is poor. Nevertheless, numerous modifications have been made to the mixing rules and equations of state to extend them to a wider range of mixtures, including polar solutions and dimeric liquids.

7.2.4 Enthalpy and Density Calculations

While the above fugacity models were applied to phase equilibrium, the thermodynamic concepts for deviations from ideality can also be applied in a straightforward way to other nonideal properties for process modeling. In particular, for the unit operations models based on thermodynamic data, we are interested in estimating the enthalpy (ΔH), volume (ΔV) (or density) and entropy (ΔS). All of these can be represented by excess molar quantities, as with Gibbs free energy, and can be written as:

$$\Delta H = \Delta H^{id} + \Delta H^E$$
$$\Delta S = \Delta S^{id} + \Delta S^E \qquad (7.31)$$
$$V = V^{id} + V^E$$

Here the *id* superscript deals with the pure component quantities using ideal mixing rules. From the properties of thermodynamic partial derivatives, the excess Gibbs energies presented above can be used directly for the following excess properties:

$$V^E = (\partial G^E/\partial P)_T$$
$$\Delta S^E = -(\partial G^E/\partial T)_P \qquad (7.32)$$
$$\Delta H^E = \Delta G^E + T \Delta S^E$$

or

$$\Delta H^E/RT = -T(\partial [G^E/RT]/\partial T)_P$$

EXAMPLE 7.2

Find the excess quantities for the UNIQUAC model, assuming all of the parameters are temperature and pressure independent.

The UNIQUAC model is given by:

$$G^E/RT = \sum_i x_i \ln(\Phi_i/x_i) + (\zeta/2) \sum_i q_i x_i \ln(\theta_i/\Phi_i) - \sum_i q_i x_i \ln(\sum_j \theta_j \tau_{ji}) \qquad (7.33)$$

Using the above relations the excess quantities are:

$$V^E = (\partial G^E/\partial P)_T = 0$$
$$\Delta H^E = \Delta G^E + T \Delta S^E = 0 \quad \text{or}$$
$$\Delta H^E/RT = -T(\partial [G^E/RT]/\partial T)_P = 0 \qquad (7.34)$$
$$\Delta S^E = -(\partial G^E/\partial T)_P$$
$$= -R[\sum_i x_i \ln(\Phi_i/x_i) + (\zeta/2) \sum_i q_i x_i \ln(\theta_i/\Phi_i)$$
$$- \sum_i q_i x_i \ln(\sum_j \theta_j \tau_{ji})]$$

Therefore, all of the thermodynamic options that were developed for phase equilibrium can be extended directly to calculation of enthalpies, densities, and entropies. In the next sections, we will describe where these quantities are needed.

7.2.5 Implementation in Process Simulators

This section describes only a small fraction of physical property options that are available to the user within current process simulation tools. The above survey avoids giving a long list

Sec. 7.3   Flash Calculations   217

of options but should give the reader an appreciation of the breadth of models available for physical property estimation. Currently used models are not mathematically simple nor are they inexpensive to calculate, although these have been automated so that they can be accessed easily. Nevertheless, their selection and use should not be done carelessly, nor should this aspect of process simulation be taken for granted. It is therefore hoped that this section provides some background and guidelines for proper selection of these options.

The primary application for these nonideal models is in phase equilibrium calculations (also referred to as flash calculations) as these are the basic building blocks for thermodynamics-based unit operations models. These models also apply directly to energy balances and other process calculations. Moreover, in terms of numbers of equations and fraction of computational effort, calculating these properties represents a significant part (up to 80%) of the simulation and modeling task. In the remaining sections of this chapter we will develop more detailed models based on thermodynamic concepts and we will see how they interact with the physical property calculations described in this section.

7.3 FLASH CALCULATIONS

In process simulation programs, flash calculations represent the most frequently invoked and most basic sets of calculations. A flash calculation is required to determine the state of any process stream following a physical or chemical transformation. This occurs after the addition or removal of heat, a change in pressure or a change in composition due to reaction. In this section we consider the derivation of the nonideal flash problem and two common approaches for its solution. Unlike Chapter 3, we make no simplifications in the model to allow for a simplified solution procedure. Consequently, the solution of this model requires the numerical algorithms developed in Chapter 8.

7.3.1 Derivation of Flash Model

Consider the phase separation operation represented in Figure 7.1 with the same notation as in Chapter 3.

In Chapter 3, we developed a linear split fraction model for this unit based on the molar flows for NC components i in the feed, vapor, and liquid streams, f_i, v_i and l_i, respectively. Here we assume that the state of the feed stream is completely defined so that we know the inlet flowrate, mole fractions (z_i), and enthalpy. By defining the mole fractions as $x_i = l_i /(\Sigma_i l_i)$ and $y_i = v_i /(\Sigma_i v_i)$ we obtain a minimal set of mass balances:

$$f_i = v_i + l_i, \quad i = 1, \ldots NC \tag{7.35}$$

equilibrium equations:

$$\gamma_i(l, T) f_i^0(T, P) x_i = \phi_i(v, T) P y_i, \quad i = 1, \ldots NC \tag{7.36}$$

and an enthalpy balance:

$$F H_f(f, T, P) + Q = V H_v(v, T, P) + L H_l(l, T, P) \tag{7.37}$$

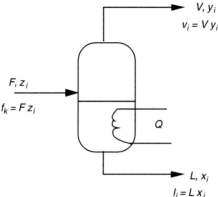

FIGURE 7.1 Flash unit.

which gives us ($2NC + 1$) equations for the ($2\ NC + 3$) variables, v_i, l_i, T, P, and Q. As in Chapter 2, we therefore have two degrees of freedom to specify the flash problem.

However, when a phase disappears, a model derived from mass balances in molar flows leads to undefined compositions for dewpoint and bubble point conditions. Moreover, since nonlinear phase equilibrium relations are composition dependent, we now develop a slightly different flash model in terms of total flows and mole fractions. Following the minimal description above, the mass balance over the unit is given by:

$$z_i F = V y_i + L x_i, \quad i = 1, \ldots NC \quad (7.38)$$

and an enthalpy balance yields

$$F H_f + Q = V H_v(y, T, P) + L H_l(x, T, P) \quad (7.39)$$

Equilibrium expressions are given by:

$$y_i = K_i x_i, \quad i = 1, \ldots NC \quad (7.40)$$

with physical property definitions from section 7.2 used to define the K values:

$$K_i = \gamma_i(x, T) f^0{}_i(T, P) / (\phi_i(y, T) P), \quad i = 1, \ldots NC \quad (7.41)$$

We therefore have $3\ NC + 5$ variables (y_i, K_i, x_i, L, P, Q, T, V) and only $3\ NC + 1$ equations so far. Note that we have not specified any conditions yet on the conditions of x_i and y_i (e.g., that mole fractions sum to one). Interestingly, this choice needs to be made carefully since spurious roots are introduced even with some obvious choices.

Because neither liquid nor vapor mole fraction is specified we can include an overall mass balance:

$$F = L + V \quad (7.42)$$

Now consider the simpler case where T and P are specified. This decouples the enthalpy balance and allows Q to be calculated once the mass balance is solved. Combining

Sec. 7.3 Flash Calculations

the overall mass balance equation with the component mass balance and equilibrium expressions above leads to the following relations for the mole fractions:

$$x_i = \frac{z_i}{1+(K_i-1)\frac{V}{F}} \quad y_i = \frac{K_i z_i}{1+(K_i-1)\frac{V}{F}} \quad (7.43)$$

Now we need an additional specification on either set of mole fractions to obtain a model with the required two degrees of freedom.

Consider the two obvious choices: $\sum x_i = 1$ or $\sum y_i = 1$. Now for the first choice we have:

$$\sum_i x_i = \sum_i [F z_i /(F + (K_i - 1)V)] = 1 \quad (7.44)$$

and the flash model is trivially satisfied *for every flash problem* if we set $x_i = z_i$ and $V = 0$. Similarly, if we use $\sum y_i = 1$ we find that:

$$\sum y_i = \sum [F K_i z_i /(F + (K_i - 1)V)] = 1 \quad (7.45)$$

and the flash model is trivially satisfied *for every flash problem* if we set $y_i = z_i$ and $V = F$.

Clearly, either equation leads to spurious solutions (at the feed composition) that are completely unrelated to the true solution of the flash problem. To eliminate the trivial solutions we consider an alternate specification from Rachford and Rice (1952). By taking the difference of $\sum x_i = 1$ and $\sum y_i = 1$ we have:

$$\sum y_i - \sum x_i = 0. \quad (7.46)$$

Note that this new specification, along with the overall mass balance, still leads to the correct specifications on the mole fractions. Applying this condition to the relations for the mole fractions leads to:

$$\sum y_i - \sum x_i = \sum \left[F(K_i - 1)z_i /(F + (K_i - 1)V) \right] = 0 \quad (7.47)$$

and we see that the above spurious roots cannot solve this equation. In fact, $x_i = z_i$ or $y_i = z_i$ are allowable solutions only under the (quite appropriate) condition that $K_i = 1$ and we have an azeotropic mixture.

7.3.2 Strategies for Flash Calculations

The flash model can be given concisely by:

$$z_i F = V y_i + L x_i, \quad i = 1, \ldots NC$$
$$y_i = K_i x_i, \quad i = 1, \ldots NC$$
$$K_i = \gamma_i(x, T) f_i^0(T, P) / (\phi_i(y, T) P), \quad i = 1, \ldots NC \quad (7.48)$$
$$F = V + L$$
$$\sum y_i - \sum x_i = 0$$
$$F H_f + Q = V H_v(y, T, P) + L H_l(x, T, P).$$

and now leaves two degrees of freedom to be specified. While many alternatives are possible for design calculations, flash calculations are often solved for degrees of freedom chosen among the variables (*V/F*, *Q*, *P*, and *T*).

The simplest case is given by the (*P*, *T*) flash since this requires no iteration for the enthalpy balance. For this case, the flash problem can be solved by the *TP* flash calculation sequence.

TP Flash Calculation Sequence

1. For fixed z_i (make sure $\Sigma z_i = 1$) and *F*, specify *T*, *P*. Proceed if between bubble and dew points. (For composition-dependent nonidealities, provide an initial guess for x_i and y_i.)
2. Guess *V/F*.
3. Calculate $K_i = \gamma_i(x, T) f_i^0(T, P) / (\phi_i(y, T) P)$.
4. Calculate $x_i = z_i/(1 + (K_i - 1) V/F)$ and $y_i = K_i x_i$.
5. Evaluate the implicit relation $\psi(V/F) = \Sigma x_i - \Sigma y_i$. If $\psi(V/F)$ is zero (or within a small tolerance), STOP. Else, go to 6.
6. Update the guess for *V/F* and go to 3.

EXAMPLE 7.3 TP Flash

Consider a mixture of 40 mol % methanol, 20 mol % propanol and 40 mol % acetone. Perform a TP flash calculation at 1 atm and 343 K (70 C).

For ease of demonstration we model this mixture with the two-suffix Margules equation, estimated from Holmes and van Winkle (1970) and Reid et al. (1987). For a ternary mixture, we have:

$$G^E/RT = -0.0753 \, x_1 x_2 + 0.6495 \, x_1 x_3 + 0.557 \, x_2 x_3 \quad (7.49)$$

with activity coefficients given by:

$$\ln \gamma_1 = -0.0753 \, x_2^2 + 0.6495 \, x_3^2 + 0.0172 \, x_2 x_3$$
$$\ln \gamma_2 = -0.0753 \, x_1^2 + 0.557 x_3^2 - 0.1678 \, x_1 x_3 \quad (7.50)$$
$$\ln \gamma_3 = 0.6495 \, x_1^2 + 0.557 \, x_2^2 + 1.2818 \, x_2 x_1$$

Using the Antoine constants, the vapor pressures are given by $\ln P_i^0 = A_i - B_i/(C_i + T)$ with P_i^0 in mm *Hg*, *T* in *K*, and the following data:

	Methanol	Propanol	Acetone
A_i	18.5874	17.5439	16.6513
B_i	3626.55	3166.38	2940.46
C_i	-34.29	-80.15	-35.93

Since the phase equilibrium occurs at low pressure, the activity coefficients represent the only source of nonideality and the *K* values are given by $K_i = \gamma_i P_i^0/P$. Applying the flash calculation se-

Sec. 7.3 Flash Calculations

quence given above with a secant method for V/F, and starting from $V/F = 0.5$, we obtain convergence to a tolerance of 10^{-6} for $\psi(V/F)$ in 14 iterations. For this mixture at this temperature and pressure, $V/F = 0.8639$ and the compositions, K values, and activity coefficients are given by:

	y_i	x_i	K_i	γ_i
Methanol	0.4107	0.3319	1.2374	1.00642
Propanol	0.1555	0.4824	0.3224	1.00376
Acetone	0.4337	0.1857	0.3362	1.5014

TP specifications are most common for narrow boiling mixtures, where all of the components have boiling points in a narrow range, such as benzene and toluene. Here V/F can vary between zero and one, with a small range in temperature. This case is common for mixtures separated by distillation. On the other hand, for wide-boiling mixtures (such as air and water) the TP specification in the *flash calculation sequence* works poorly because the equilibrium temperature varies widely for small changes in V/F. These mixtures are commonly separated by absorption and the specifications $(V/F, T)$ and $(V/F, P)$ are used. Otherwise, the algorithm is similar to the *flash calculation sequence* presented above. Also, for these cases, note that the enthalpy balance is not needed in the iteration loop.

Finally, when a specification on the heat input, Q, is made (as in an adiabatic flash), then an enthalpy balance is imposed and needs to be incorporated into the flash algorithm. Often the enthalpy balance is treated by guessing the temperature, say, and solving the TP flash in an inner loop. The enthalpy is then calculated, matched to the heat input, and the temperature is reguessed in the outer loop. This calculation sequence makes the flash calculation much more time consuming. Alternatively, all of the equations flash model can be solved simultaneously using the Newton or Broyden method developed in Chapter 8. With this simultaneous approach both wide and narrow boiling mixtures can be handled in a straightforward way. However, for all of these methods, nonideal thermodynamic routines need to be called frequently and this increases the computational expense.

EXAMPLE 7.4 PQ Flash

Consider a liquid feed mixture of 40 mol % methanol, 20 mol % propanol and 40 mol % acetone at 373 K and 10 atm. Perform an adiabatic flash calculation at 1 atm.

Using the physical property information from the previous example and heat capacities in Reid et al. (1987), we note that $\Delta H^E = 0$ and that ideal enthalpy relations can be chosen for both liquid and vapor phases. The vapor and liquid enthalpies can therefore be calculated from the relations developed in Chapter 2:

$$\Delta H_v(T, y) = \sum_i y_i \left(H_{f,i}^0 + \int_{T_0}^T C_{pi}^0(\tau)\, d\tau \right)$$

$$\Delta H_L(T, x) = \sum_i x_i \left(H_{f,i}^0 + \int_{T_0}^T C_{pi}^0(\tau)\, d\tau - \Delta H_{\text{vap}}^i(T) \right) \tag{7.51}$$

$$\Delta H_{\text{vap}}^i (T) = \Delta H_{\text{vap}}^i (T_b)[(T_c^i - T)/(T_c^i - T_b^i)]^{0.38}$$

$$C_{pi}(T) = a_i + b_i T + c_i T^2 + d_i T^3$$

based on a reference temperature of 298 K and with the following data for heat capacities in cal/gmol-K.

	Methanol	Propanol	Acetone
a_i	5.052	0.59	1.505
b_i	0.01694	0.07942	0.06224
c_i	$6.179 \cdot 10^{-6}$	$-4.431 \cdot 10^{-5}$	$-2.992 \cdot 10^{-5}$
d_i	$-6.811 \cdot 10^{-9}$	$1.026 \cdot 10^{-8}$	$4.867 \cdot 10^{-9}$
ΔH^i_{vap}	8426.	9980.	6960.
T^i_b	337.8	370.4	329.4
T^i_c	512.6	536.7	508.1

The initial liquid feed enthalpy is -6.331 kcal/gmol and starting from a guess of 343 K, we execute the TP flash algorithm as the inner loop. In an outer loop we match the specific enthalpy for the liquid and vapor streams to the feed enthalpy and reguess the temperature. The adiabatic flash calculation converges in about five outer iterations to a temperature of 334.58 K with $V/F = 0.1782$. The results of the adiabatic flash are given below:

	y_i	x_i	K_i	γ_i
Methanol	0.3874	0.4027	1.9621	1.0877
Propanol	0.0526	0.2320	0.2266	1.0510
Acetone	0.5600	0.3653	0.5330	1.2905

INSIDE-OUT METHOD FOR FLASH CALCULATIONS

The *flash calculation sequences* developed above suffer from two drawbacks:

- They are designed either for wide boiling or narrow boiling mixtures and perform poorly for the opposite cases.
- They require frequent calls to evaluate nonideal thermodynamic functions, especially when the enthalpy balance needs to be incorporated in the flash calculation.

To address these concerns, Boston and Britt (1978) developed an "inside-out" algorithm that greatly accelerates the solution of flash problems. In an outer loop, this approach matches the nonideal physical property equations to simplified expressions for K values and enthalpies (similar to those used in Chapter 2) and then uses these expressions to solve the flash equations in an inner loop. The solution of these equations is then used to update the simplified expressions and the procedure terminates once the simplified expressions match the actual nonideal ones in the outer loop.

Sec. 7.3 Flash Calculations

To illustrate the advantages of the inside-out algorithm, we consider the PQ flash with the flash equations given above. Boston (1980) further suggests the following simplifications for the inner loop:

$$K_i = \alpha_i K_b$$
$$\ln(K_b) = A + B(1/T - 1/T^*)$$
$$H'_v = C + D(T - T^*) \quad (7.52)$$
$$H'_l = E + F(T - T^*)$$

where the parameters A, B, C, D, E, F, and α_i are available for matching with the nonideal expressions for K values and enthalpies (H'_v and H'_l computed on a *mass basis*). K_b is an average K value that is based on a geometric weighting of component K values. Similar to Chapter 3, α_i represents the relative volatilities, and H'_v and H'_l are the ideal gas enthalpies (on a mass basis) with reference temperature T^*. To handle both wide and narrow boiling mixtures in the inner loop, Boston and Britt define an artificial iteration variable, $R \equiv K_b / (K_b + L/V)$. This variable captures the dominance of temperature or V/F for wide and narrow boiling mixtures, respectively, and eliminates the need for separate algorithms for these systems. This is because R can now vary widely both for large changes in T (wide boiling) and L/V (narrow boiling). Now once the parameters (A, B, C, D, E, F, α_i) are fixed from the outer loop, we can derive the following relations through the substitution of the flash equations and the simplified expressions:

$$z_i F = f_i = V y_i + L x_i. \quad (7.53)$$

Using $y_i = K_i x_i$ and by defining $K_i = \alpha_i K_b$, we have:

$$f_i = (VK_i + L) x_i = (\alpha_i V K_b + L) x_i \quad (7.54)$$

Dividing by $(VK_b + L)$ and substituting for R yields:

$$f_i / (VK_b + L) = (\alpha_i V K_b + L) x_i / (VK_b + L)$$
$$f_i / (VK_b + L) = (\alpha_i R + 1 - R) x_i \quad (7.55)$$

We now define a new set of variables:

$$p_i \equiv x_i (VK_b + L) = x_i L / (1 - R) \quad (7.56)$$
$$= f_i / (\alpha_i R + 1 - R).$$

Note that the p_i are determined only from R and quantities specified in the outer loop. From the summation and equilibrium equations we can recover:

$$L = (1 - R) \sum p_i$$
$$V = F - L$$
$$K_b = (\sum p_i / \sum \alpha_i p_i) \quad (7.57)$$
$$x_i = p_i / (VK_b + L)$$

$$y_i = \alpha_i K_b x_i$$

$$T = ((\ln K_b - A)/B + 1/T^*)^{-1}$$

Using R as the iteration variable, the flash calculation is completed by checking the simplified enthalpy balance. The Boston-Britt algorithm can be summarized by the following calculation sequence.

Inside-Out Calculation Sequence

1. Initialize $A, B, C, D, E, F, \alpha_i$.
2. Guess R.
3. Solve for p_i, K_b, T, L, V, x_i, and y_i using the above equations.
4. Convert flow rates to a *mass basis* and evaluate simplified mass enthalpies for the balance equation:
$$\psi(R) = H'_f + Q/F' + (L'/F')(H'_l(x, T, P) - H'_v(y, T, P)) - H'_v(y, T, P).$$
5. If $\psi(R)$ is within a zero tolerance, go to 6. Else, update the guess for R and go to step 3.
6. At first pass, obtain new values of A, B, C, D, E, F, and α_i by comparing with nonideal expressions. Thereafter, update only A, C, E, and α_i by using Broyden's method to match these parameters with the nonideal expressions.

Boston and Britt prefer a mass basis for the enthalpy balance to avoid insensitivity to R (through L/F) when $(H_v - H_l)$ is close to zero on molar terms. This algorithm converges much more quickly than the algorithms developed above and has been incorporated as the standard flash algorithm in commercial process simulators. While this derivation deals only with the PQ flash formulation, several other cases can be derived (see Exercise 4).

To demonstrate this algorithm, Boston and Britt solved a wide variety of nonideal systems including narrow and wide boiling systems, and with Wilson, UNIQUAC, NRTL, and equation of state options. Typical experience on these examples was less than six outer interations (where physical property evaluations are required). Finally, numerical experiments have shown that the above algorithm often can deal with composition-dependent K values even though the simplified expressions are not a function of x_i. For highly nonideal cases, however, Boston (1980) suggests a modification that makes the simplified K values composition-dependent and makes the algorithm more robust.

7.4 DISTILLATION CALCULATIONS

Distillation is perhaps the most detailed and well modeled unit within a process simulator, since it can often be represented accurately by an equilibrium stage model. The distillation column can be modeled as a coupled cascade of flash units and we now consider the detailed phase equilibrium behavior on each tray as well as mass and energy balances

Sec. 7.4   Distillation Calculations   225

among trays. Also, the thermodynamic models and flash algorithms considered in the previous sections therefore have an important influence on the calculation of this unit. In this section we construct a detailed equilibrium stage model for a conventional column and briefly discuss methods to solve these models. The section concludes with a small example to illustrate these concepts.

In contrast to the shortcut models in Chapter 2, we now consider a more detailed tray-by-tray model that extends from the flash calculations in the previous section. Shortcut models are not suitable for detailed modeling because of the assumption of constant relative volatility on all trays and equimolar overflow. Clearly this assumption can be violated for nonideal systems, especially with azeotropes. Moreover, even for nearly ideal systems, shortcut models are based on the concept of key component specifications. However, if we choose different key (and distributed, "between key") components, we can obtain significantly different results for the mass balance. As a result, the shortcut approach for distillation is only approximate at best.

Consider the conventional distillation column shown in Figure 7.2. The model of this distillation column consists of indices, j, for each of the NT trays and NC components, i. As seen from the figure, there is a cascade of trays starting with a reboiler for vapor boilup at the bottom and a vapor condenser at the top. Each tray has a liquid holdup (M_j) and a much smaller vapor holdup with liquid and vapor mole fractions are given by x_{ij} and y_{ij} respectively. Each tray has vapor and liquid flowing from it (L_j and V_j) and is connected to streams above and below. Possibilities at every tray also exist for a vapor or liquid feed (F_j) as well as liquid or vapor products (PL_j or PV_j). Enthalpies are calculated for

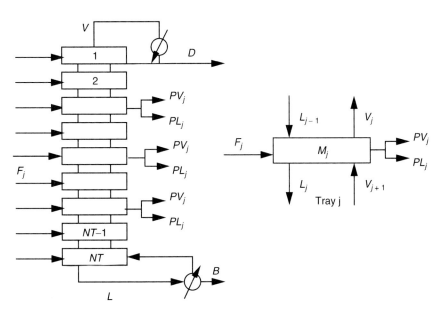

FIGURE 7.2   Schematic of distillation column model.

each of these streams (H_v or H_l, based on tray temperature, T_j); equilibrium expressions relate y_{ij} to x_{ij} on each tray. The column pressure is usually specified (P_j) for each tray although a more complex model can be incorporated that considers tray hydraulics and pressure drops across each tray. Similarly heat sources and sinks (Q_j) can be included for each tray. The distillation model for Figure 7.2 is given by:

Mass balance

$$F_j z_{ij} + L_{j-1} x_{i,j-1} + V_{j+1} y_{i,j+1} - (PL_j + L_j) x_{ij} - (V_j + PV_j) y_{ij} = 0$$
$$i = 1, \ldots NC, j = 1, \ldots NT \tag{7.58}$$

Equilibrium expressions

$$y_{ij} = K_{ij} x_{ij}$$
$$K_{ij} = K(T_j, P_j, x_{ij}) \tag{7.59}$$

Summation equations

$$\sum_i x_{ij} = 1 \quad \sum_i y_{ij} = 1 \quad j = 1, \ldots NT \tag{7.60}$$

Heat balance

$$F_j HF_j + L_{j-1} H_{l,j-1} + V_{j+1} H_{v,j+1} - (PL_j + L_j) H_{lj} - (V_j + PV_j) H_{vj} + Q_j = 0, \ j = 1, \ldots NT$$
$$H_{lj} = h(T_j, P_j, x_j), \ H_{vj} = H(T_j, P_j, y_j), \ HF_j = HF(T_f, P_f, z_j) \tag{7.61}$$

These Mass, Equilibrium, Summation, and Heat (MESH) equations form the standard model for a tray-by-tray distillation model. Note that the thermodynamic properties (K values and specific enthalpies) are expressed as implicit functions that require the physical property models in section 7.2. For the condenser, the balance equations are further simplified to:

Mass balance

$$V_1 y_{i,1} - (DL + L_0) x_{iD} - DV y_{iD} = 0 \quad i = 1, \ldots NC \tag{7.62}$$

Summation equations

$$\sum_i x_{iD} = 1 \quad \sum_i y_{iD} = 1 \tag{7.63}$$

Equilibrium expressions

$$y_{iD} = K_{iD} x_{iD}$$
$$K_{iD} = K(T_D, P_D, x_D) \tag{7.64}$$

Heat balance

$$V_1 H_{v,1} - (DL + L_0) H_{lD} - DV H_{vD} - Q_{con} = 0, \ j = 1, \ldots NT \tag{7.65}$$
$$H_{lD} = h(T_D, P_D, x_D), \ H_{vD} = H(T_D, P_D, y_D)$$

and similarly the reboiler equations are given by:

Mass balance

$$L_{N-1} x_{i,\,N-1} - BL\, x_{iB} - (V_N + BV)\, y_{iB} = 0 \quad (7.66)$$
$$i = 1, \ldots NC,\, j = 1, \ldots NT$$

Equilibrium expressions

$$y_{iB} = K_{iB}\, x_{iB}$$
$$K_{iB} = K(T_B, P_B, x_B) \quad (7.67)$$

Summation equations

$$\sum_i x_{iB} = 1 \quad \sum_i y_{iB} = 1 \quad (7.68)$$

Heat balance

$$L_{N-1} H_{l,\,N-1} - BL\, H_{lB} - (V_N + BV)\, H_{vB} + Q_{\text{reb}} = 0,\ j = 1, \ldots NT \quad (7.69)$$
$$H_{lB} = h(T_B, P_B, x_B),\ H_{vB} = H(T_B, P_B, y_B)$$

For the reboiler and condenser, the Summation and Equilibrium equations are dropped if the overhead and bottom products, D and B, are single phase. The combined systems consists of $(NT + 2)(2NC + 3) + 2$ equations and $(NT + 2)(3NC + 5) + 3$ variables. After specifying the number of trays, feed tray location, and the feed flowrate, composition, and enthalpy ($NT (NC + 1)$ variables), only $NT + 1$ degrees of freedom remain. A common specification for the MESH system is to fix the pressures on the trays and the reflux ratio, $R = L_0/D$.

Many algorithms have been invented to solve the MESH system of equations. In fact, Taylor and Lucia (1995) observe that since the late 1950s at least one new distillation algorithm has been published almost every year. Early methods were devoted to developing decompositions of the MESH equations by fixing a subset of variables and solving for the remaining ones in an inner loop.

For instance, if the temperatures and flowrates are fixed, one can solve for the compositions componentwise using the linearized Mass and Equilibrium equations in an inner loop. In the outer loop, the temperatures and flowrates are adjusted using the Summation and Heat equations. In this scheme, pairing the temperatures with the energy balance leads to the "sumrates" method, applicable for wide boiling mixtures suitable for absorption. On the other hand, pairing the flowrates with the Heat balance leads to the "bubble point" method, more suited to narrow boiling mixtures. A simplification of the bubble point approach occurs in the case of equimolar overflow where the flowrates are fixed (by specifying the reflux ratio) and the tray temperatures are determined by the Summation equations. Here the equimolar overflow assumption is based on heats of vaporization that are assumed the same for all components. In this case the Heat balance is redundant and is deleted.

Solving the Summation and Heat equations simultaneously for the temperatures and flowrates in the outer loop was proposed in the early 1970s, leading to algorithms appropriate for both wide boiling and narrow boiling mixtures. However, a nonlinear equation solver (see Chapter 8) is required for this case. Decomposition strategies for the MESH equations often lead to fast algorithms for conventional distillation columns. For nonideal systems with composition dependent K values, however, the Equilibrium equations become nonlinear in x, which leads to additional computational difficulty and expense. Moreover, additional design specifications such as product purity must be imposed as an outer loop for these algorithms.

A more direct way to deal with these difficulties is to apply Newton-Raphson methods to the total set of MESH equations. This approach was first suggested in the mid–1960s and is now perhaps the most popular method for distillation. Moreover, the Newton approach leads to coordinated strategy for solving a general class of nonideal separation problems. This approach can be summarized for distillation by combining the MESH equations and the vector of variables into a large set of nonlinear equations and variables, $f(w) = 0$. Linearizing these equations about a current point w_k at which the variable vector is specified, we have:

$$f(w_k) + (\partial f/\partial w)(w_{k+1} - w_k) = 0 \qquad (7.70)$$

with w chosen as the next estimation for iteration $k+1$. This value is determined from the solution of the linear equations:

$$(\partial f/\partial w)(w_{k+1} - w_k) = (\partial f/\partial w)\Delta w = -f(w_k) \qquad (7.71)$$

Solving the linear equations requires evaluation of the Jacobian matrix, $(\partial f/\partial w)$, using the partial derivatives from the MESH equations. By grouping the MESH equations according to each stage, the Jacobian matrix becomes block tridiagonal and can be factorized with computational effort that is directly proportional to the number of trays. Moreover, the simultaneous Newton approach easily allows the addition of design specifications without imposing an outer loop for the column calculation. Also, the approach is extended in a straightforward manner to deal with complex column configurations including heat loops and pumparounds, bypass streams and multiply coupled columns.

Nevertheless, there are a few drawbacks to this simultaneous approach. One difficulty comes from obtaining derivatives from the physical property equations for the K values and the equilibrium expressions, especially if $\partial K/\partial x \neq 0$. For highly nonideal systems, accurate derivatives are a necessity for good performance. Fortunately, most process simulators now incorporate analytic partial derivatives for these calculations. The Newton method also requires good initialization procedures—these are often problem dependent and require some skill on the user's part. Automatic initialization strategies generally are based on obtaining good starting points for the Newton method by using simple shortcut calculations or initial application of the decoupling strategies used by earlier distillation algorithms. Nevertheless, even with these intuitively helpful strategies, current distillation algorithms can encounter difficulties, especially for highly nonideal systems.

Finally, inside-out concepts have also led to popular and fast distillation algorithms. Similar to the inside-out flash algorithm, this approach removes the composition dependence for the K values and enthalpies and solves these simplified MESH equations in an inner loop. As discussed above, this calculation is much easier than direct solution of the MESH equations. Again, these simplified quantities are compared with the detailed ther-

Sec. 7.4   Distillation Calculations   229

modynamics in an outer loop and convergence occurs when the simplified properties match with the rigorous ones. As with the flash algorithm, Boston and coworkers demonstrated this approach on a wide variety of equilibrium staged systems including absorbers and distillation columns. This approach can be significantly faster than the decoupled algorithms or the direct Newton solvers. Moreover, for systems that are only mildly nonideal, the inside-out strategy is less sensitive to a good problem initialization.

EXAMPLE 7.5

To illustrate the formulation and solution of the MESH equations we consider the separation of benzene, toluene and o-xylene. Here the problem formulation is modeled in GAMS; the component mixture is nearly ideal and for illustration purposes, we define $\gamma_i = 1$ so that the K values are given by $P_i^0(T)/P$. Similarly, the vapor and liquid enthalpies were calculated using the ideal enthalpy relations given above and developed in Chapter 3. For this separation we have a bubble point feed at 1.2 atm and a flowrate of 50 kg-mols/h. The feed composition is $x_B = 0.55$, $x_T = 0.25$, $x_O = 0.20$ and the feed temperature is therefore 390.4 K. We specify the number of trays at 40 (including the condenser and reboiler) and the column pressure at 1 atm (for simplicity we assume no pressure drops through the column). Also, we specify the feed tray location to be the tenth tray below the condenser. Setting up the MESH equations for this column and accounting for these equations, we need an additional specification for this column and for this we specify the reflux ratio ($R = L_1/D$). For this example we perform a parametric study of the reflux ratio to study its effect on the column performance.

Figures 7.3, 7.4, and 7.5 show the composition and temperature profiles for this column for reflux ratios specified at $R = 0.5$, 1.0, and 2.0. In all cases note that the profiles are nondiffer-

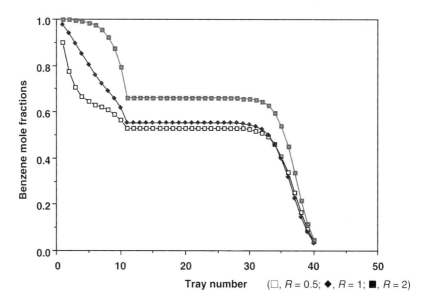

FIGURE 7.3 Benzene composition profiles for different reflux ratios.

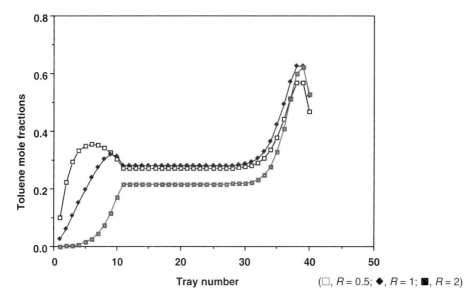

FIGURE 7.4 Toluene composition profiles for different reflux ratios.

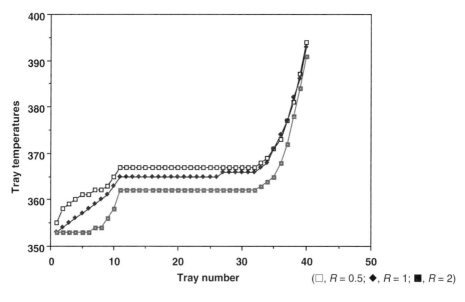

FIGURE 7.5 Temperature profiles for different reflux ratios.

entiable at the feed point and otherwise they remain fairly constant for tray 10 (immediately below the feed) through tray 30. In fact, these trays can be removed without severely affecting the column performance. For the benzene profiles the purity increases substantially as the reflux ratio increases. For the lowest reflux ratio, $x_B = 0.899$. For a reflux ratio of one it becomes $x_B = 0.975$ and for the highest reflux ratio the distillate is almost pure benzene ($x_B = 0.999$).

For the middle component, toluene, mole fractions above the feed decrease with increasing reflux ratio. Below the feed the mole fractions rise steadily and then suddenly dip down due to the mass balance in the reboiler. The bottoms mole fractions increase with increasing reflux ratio. The benzene mole fraction in the bottom stream remains fairly constant at about 0.04. The o-xylene profile, not shown here, is obtained by difference of the benzene and toluene profiles. Finally, the temperature profiles decrease with increasing reflux ratio and approach the boiling point of benzene in the condenser (354 K).

Note that from this example the product purities were not specified directly. If this problem were extended to an optimization framework (see Chapter 9), inequality constraints could be specified for these purities. However, while equations that define the top and bottom purities can be added easily to the MESH equations, adjusting the remaining column specifications is not always easy. For instance, in the above example the number of trays and the feed tray location were fixed and these are discrete variables. To satisfy the purities, only the reflux ratio (and possibly overhead pressure) could be varied but this may not give enough freedom to satisfy the specifications. Instead, we have solved this example in a *simulation* rather than *design* mode with reflux directly specified. By avoiding direct purity specifications we end up with a more time-consuming design procedure, but also avoid convergence failures that occur from unreachable specifications.

Also, as can be seen from this small example, even simple distillation systems can lead to large, nonlinear systems of the MESH equations. Solving these with the above algorithms needs to be done with care and a good understanding of the design or simulation problem. For large columns it is not unusual to encounter columns with several thousand nonlinear equations. To reduce the size of these systems, the composition and temperature profiles in the column (e.g., in Figures 7.3, 7.4, and 7.5) can be approximated with lower order polynomials, rather than an evaluation at each tray. By choosing interpolation points for these lower-order approximations, one can write interpolating equations similar to the MESH equations, but at far fewer points than the number of trays. For units with large numbers of trays (such as superfractionators with over 100 trays) this approach can significantly reduce the problem size and computational burden. This approach is known as *collocation* and a related approach for solving differential equations will be presented in Chapter 19.

This brief summary only gives a sketch of available methods for distillation units. More detailed discussion of nonideal distillation behavior is covered in Chapter 12 and systematic methods for the synthesis of separation sequences are described in Chapter 14. The methods that were outlined above can also be extended readily to more complex systems such as three-phase distillation and reactive distillation. For three-phase distillation, the MESH equations need to be extended to cover an additional liquid phase and a sufficiently general thermodynamic model (e.g., NRTL or UNIQUAC) needs to be selected. On the other hand, the three-phase problem is fraught with additional numerical difficulties. For these problems the Gibbs energy minimization (implicitly solved on each tray) contains local solutions and consequently, nonunique solutions and singular points abound for systems described by the MESH equations. Moreover, trivial solutions (with x_i converging to the feed composition) can occur for poor initialization of the compositions. Thus, while three-phase variations have been developed for the above algorithms, some work still remains in the development of reliable and robust methods. Reactive distillation operations can also be formulated by augmenting the Mass and Heat equations in the MESH system with the appropriate reaction terms. As with multiphase distillation, reactive distillation frequently exhibits more complex nonlinear behavior along with solution multiplicities. Doherty and coworkers have investigated these nonideal systems and have analyzed their behavior with geometric approaches. This approach will be discussed in greater detail in Chapter 14.

Finally, an equilibrium stage model for an absorption or distillation operation is only an approximation of the actual behavior of these systems. The above models ignore mass and heat transfer effects and also do not consider important features such as the column geometry, influences of flows, and transport characteristics on trays. In the past these have been handled by overall column and tray efficiencies and can easily be incorporated into the MESH equations. However, for systems far removed from equilibrium behavior, these efficiencies represent a crude approximation at best. More recently, mass transfer models have been developed to describe these systems more accurately. Taylor and coworkers discuss mass transfer or rate-based models made up of the MERQ (Mass, Equilibrium, Rate, and Energy) equations. These models, however, require additional transport properties for both mass and heat transfer characteristics, as well as phase equilibrium models. Also, uncertain parameters such as interfacial area between phases must be estimated. Nevertheless, rate-based models are already being introduced in commercial applications and their success will spur further development of better mass transfer models and more complete physical property data banks.

7.5 OTHER UNIT OPERATIONS

In Chapter 2, several additional unit operations models were described for evaluating a candidate flowsheet. These include *simple units* such as mixers and splitters; *transfer units* such as valves, pumps, and compressors; *energy exchangers* that include a variety of heat exchanger models; and *process reactors*. For design purposes, the conceptual models for these units remain largely unchanged from the descriptions in Chapter 3, except possibly for process reactors. However, the solution of these unit models is often complicated by the substitution of more detailed thermodynamic relations, developed in section 7.2. This section provides a brief summary of these extensions.

7.5.1 Mixers

The conceptual mixer model (Figure 7.6) remains the same for this unit as it is completely defined by a mass and energy balance. For streams i and components k, we have:

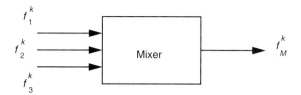

FIGURE 7.6 Mixer model.

$$f_M^k = \Sigma_i f_i^k \quad f_M^k \Delta H_M = \Sigma_i f_i^k \Delta H_i \quad (7.72)$$

Moreover, the downstream pressure is usually given by: $P_M = \text{Min}_l\{P_l\}$. However, an added complication is determination of the downstream temperature T_M. This requires an adiabatic flash calculation with detailed thermodynamic models.

7.5.2 Splitters

Again, the splitter unit (Figure 7.7) divides a given feed stream into specified fractions ξ_i for each output stream i. Because the output streams have the same compositions and intensive properties as the input stream, no additional calculations are required. This is the simplest unit in a flowsheet simulator and we only need to write the equations:

$$f_i^k = \xi_i f_{IN}^k, \; i = 1, \ldots N-1 \quad f_{NS}^k = (1 - \Sigma_i^{NS-1} \xi_i) f_{IN}^k \quad (7.73)$$

7.5.3 Pumps

For preliminary design calculations (Figure 7.8) the inlet and outlet pressures (or ΔP) are normally specified and therefore the compositions and pressure of the outlet stream are directly specified. To complete the definition of the outlet stream, we again define the theoretical work as $V \Delta P$, since the specific volume of the liquid remains (nearly) constant. The brake horsepower can be written as:

$$W_b = f(P_2 - P_1)/(\rho \, \eta_p \, \eta_m) \quad (7.74)$$

where η_p and η_m are the pump and motor efficiencies described in Chapter 3. This $V \Delta P$ work is added to the stream energy and thus specifies the molar enthalpy of the outlet stream. From this relation, the temperature is calculated using the nonideal enthalpy models outlined in section 7.2. Additional detailed sizing and costing can then be applied once the stream conditions and the work requirements of these units have been calculated. These additional calculations are beyond the scope of this chapter.

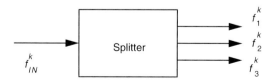

FIGURE 7.7 Splitter model.

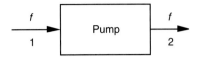

FIGURE 7.8 Pump model.

7.5.4 Compressors and Turbines

Mass and energy balances for compressors and turbines (Figure 7.9) are usually made for preliminary design by a direct specification of the outlet pressure or pressure change. For an isentropic compressor, the ideal compression work can be calculated from the change in enthalpy of the stream, calculated by holding the entropy constant. The temperatures and enthalpies are calculated after the iterative calculation:

$$\Delta S(T_1, P_1, f) = \Delta S(T_2, P_2, f) \tag{7.75}$$

where f is the molar flowrate. The theoretical work is then given by:

$$W_T = [H_V(P_2, T_2, f) - H_V(P_1, T_1, f)] \tag{7.76}$$

where the entropies and enthalpies are calculated using nonideal thermodynamic models. The actual work is then calculated using adjustments from isentropic behavior through an isentropic efficiency and a motor efficiency, both specified by the user so that: $W_b = W_T / \eta_m \eta_s$ for the compressor and $W_b = \eta_m \eta_s W_T$ for the turbine. Additional detailed sizing and costing can then be applied once the stream conditions and the work requirements of these units have been calculated. These are beyond the scope of this chapter.

7.5.5 Heat Transfer Equipment

For heat exchangers (Figure 7.10), the simplest units are those with three process temperatures specified and the fourth is calculated by closing the energy balance.

For a countercurrent, shell and tube heat exchanger, for instance, with T_1, T_2, and T_3 specified, this is given by:

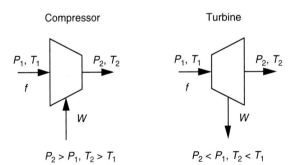

FIGURE 7.9 Compressor and turbine model.

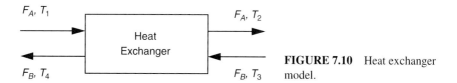

FIGURE 7.10 Heat exchanger model.

$$Q = H(F_A, T_2, P_2) - H(F_A, T_1, P_1) = H(F_B, T_3, P_3) - H(F_B, T_4, P_4) \qquad (7.77)$$

and T_4 is solved for iteratively from the nonideal enthalpy balance. Sizing equations for these heat exchangers can be found from the following equation:

$$Q = UA\, \Delta T_{lm} \qquad (7.78)$$

where Q is the heat duty, known from the energy balance, A is the required area, the log mean temperature (ΔT_{lf}) is given by:

$$\Delta T_{lm} = [(T_1 - T_4) - (T_2 - T_3)] / \ln\{(T_1 - T_4)/(T_2 - T_3)\} \qquad (7.79)$$

and the overall heat transfer coefficient, U, is often specified by the user. Should phase changes occur between the inlet and outlet, a more accurate sizing of the exchanger requires a partitioning into multiple exchangers for the subcooled, two-phase, and superheated portions. The boundaries of these partitions are determined by finding the bubble and dew points of the multiphase stream. More detailed heat exchanger calculations can be performed through the following models.

If no phase changes occur in the heat exchanger:

- The overall heat transfer coefficient can be calculated by estimating tube and shell side resistances with heat transfer correlations and combining them.
- Geometry of the heat exchanger can be dealt with simply by calculating an appropriate geometric factor to give: $Q = F\, U\, A\, \Delta T_{lm}$. This approach covers a wide range of exchanger calculations and is often used for multiple pass shell and tube heat exchangers with shell side baffles.

If phase changes occur in the heat exchanger:

- More detailed (and time-consuming) calculations need to be made by computing the internal temperature profiles directly. Here a multipoint boundary value problem is formulated and the shell and tubeside differential equations are integrated along the length of the heat exchanger. Nonideal enthalpies also need to be calculated within the solution of the differential equations.

Modern process simulators allow for all of these options and therefore permit the calculation of quite detailed heat exchanger designs. See Welty et al. (1984) for a survey of models. Fortunately for many preliminary designs (especially in processes where raw

FIGURE 7.11 Reactor model.

material conversion to product dominates the design objective), simple heat exchanger models are adequate to determine an accurate mass and energy balance and also to give a good approximation for the area requirements for heat exchange.

7.5.6 Reactor Models

In process simulators, reactor models (Figure 7.11) are often greatly simplified. A major reason for this lies in the fact that physical properties are almost entirely based on thermodynamic concepts and there is no general database for reaction kinetics. Moreover, for many new and even existing processes, the reaction kinetics are simply not known or are too difficult to obtain at the design stage.

As a result, simplified models are often used within flowsheeting tools. However, far more can be done to exploit the character of the process when the reactor performance is modeled accurately. Process synthesis approaches that deal with reactor networks are discussed later in Chapters 13 and 19. Nevertheless, for process simulation, reactor models can be classified into the following three types:

- Stoichiometric reactors
- Equilibrium-based reactors
- Specific kinetic models

The simplest types, *stoichiometric reactors*, are similar to the linear reactor models that were described in Chapter 3. Here we specify the molar conversion of the NR parallel reactions in advance. This requires that for each reaction r, we define a limiting component $l(r)$, and normalized stoichiometric coefficients $\gamma_{r,k} = (C_{r,k}/C_{r,l(r)})$, $r = 1$, NR for each component k, where the coefficients $C_{r,k}$ appear in the specified reactions. Defining the fraction converted per pass based on limiting reactant as η_r, $r = 1$, NR, gives us:

$$f_R^k = f_{IN}^k + \sum_{r=1}^{NR} \gamma_{r,k} \, \eta_r \, f_{IN}^{l(r)} \qquad (7.80)$$

With an outlet pressure specification and nonideal thermodynamic models, an energy balance can also be completed for stoichiometric reactors according to:

$$Q_R = \Delta H(T_R, f_R) - \Delta H(T_{IN}, f_{IN}) \qquad (7.81)$$

and this allows us either to specify the outlet temperature and calculate the appropriate reactor heat duty, or specify this heat duty (say, adiabatic) and calculate the outlet temperature.

Sec. 7.5   Other Unit Operations

Equilibrium reactor models provide a better description for many industrial reactors and still allow thermodynamic calculations that are compatible with process simulation databases. For a single reaction,

$$aA + bB \rightarrow cC + dD \tag{7.82}$$

the equilibrium conversions can be given directly from:

$$(f_C)^c (f_D)^d / [(f_A)^a (f_B)^b] = K = \exp(-\Delta G_{rxn}(T)/RT) \tag{7.83}$$

where f_i is the fugacity of component i and where, for the above reaction, ΔG_{rxn} is given by:

$$\Delta G_{rxn} = (c\Delta G_{f,C} + d\Delta G_{f,D}) - (a\Delta G_{f,A} + b\Delta G_{f,B}) \tag{7.84}$$

and $\Delta G_{f,i}$ are the free energies of formation that can be evaluated as a function of temperature. For gas phase reactions at "low" pressure, the fugacity can be replaced by the partial pressures and the expression becomes:

$$(P_C)^c (P_D)^d / [(P_A)^a (P_B)^b] = K \tag{7.85}$$

or in terms of mole fractions:

$$(y_C)^c (y_D)^d / [(y_A)^a (y_B)^b] = K\, P^{(a+b-c-d)} \tag{7.86}$$

where P is the total pressure of the system.

EXAMPLE 7.6

Consider the water gas reaction:

$$CO + H_2O \leftrightarrow CO_2 + H_2 \tag{7.87}$$

at a pressure of 5 atm and a temperature of 600 K. What is the equilibrium concentration?
The Gibbs energy of reaction can be determined by:

$$\Delta G_{f,CO_2} = -94.26 \text{ kcal/gmol} \qquad \Delta G_{f,CO} = -32.81 \text{ kcal/gmol} \tag{7.88}$$

$$\Delta G_{f,H_2} = 0. \text{ kcal/gmol} \qquad \Delta G_{f,H_2O} = -54.64 \text{ kcal/gmol}$$

at 298 K, and therefore $\Delta G_{rxn} = -6.81$ kcal/gmol at 298 K. Assume that the temperature correction of ΔG_{rxn} to 600 K is negligible (see Exercise 10) for this reaction and therefore the equilibrium constant is given by:

$$K = \exp(-\Delta G_{rxn}/RT) = 306.9 \tag{7.89}$$

Starting with equal amount of CO and H_2O at 5 atm, the equilibrium expression is given by:

$$(P_{CO_2})(P_{H_2}) / [(P_{CO})(P_{H_2O})] = K \tag{7.90}$$

and since the total number of moles is conserved we have, for a reaction extent ξ,

$$\xi^2 / [1-\xi]^2 = K \tag{7.91}$$

we have:

$$\xi = K^{1/2} / (1 + K^{1/2}) = 0.946 \quad (7.92)$$

This leaves:

$$\begin{aligned}
P_{CO_2} &= P\xi = 2.365 \text{ atm} & y_{CO_2} &= \xi = 0.473 \\
P_{H_2} &= P\xi = 2.365 \text{ atm} & y_{H_2} &= \xi = 0.473 \\
P_{CO} &= P(1-\xi) = 0.135 \text{ atm} & y_{CO} &= (1-\xi) = 0.027 \\
P_{H_2O} &= P(1-\xi) = 0.135 \text{ atm} & y_{H_2O} &= (1-\xi) = 0.027
\end{aligned} \quad (7.93)$$

For multiple reactions, calculating the equilibrium conversion becomes more complex. Here, the the Gibbs energy of the system must be minimized directly subject to constraints on the mass (or element) balance. Again, this equilibrium conversion calculation can be carried out using only thermodynamic data. The resulting optimization problem is therefore:

$$\begin{aligned}
&\text{Min } \sum_i n_i [\Delta G_{f,i} + RT \ln (f_i/f_i^0)] \\
&\text{s.t. } \sum_i n_i a_{ik} = A_k, \; k = 1, \ldots NE \\
&n_i \geq 0
\end{aligned} \quad (7.94)$$

where f_i^0 is the standard state fugacity ($f_i^0 = 1$), n_i are the moles of species i in the system, a_{ik} is the number of atoms of element k in species i and A_k is the number of moles of the NE elements, k, in the system. For gas phase reactions, we can simplify the above problem by noting that the fugacity can be written as: $f_i = y_i \phi_i P$, which leads to:

$$\begin{aligned}
&\text{Min } \sum_i n_i [\Delta G_{f,i}(T) + RT (\ln n_i + \ln \phi_i + \ln P - \ln (\sum n_j))] \\
&\text{s.t. } \sum_i n_i a_{ik} = A_k, \; k = 1, \ldots NE \\
&n_i \geq 0
\end{aligned} \quad (7.95)$$

By accessing the appropriate nonideal thermodynamic models for $\Delta G_{f,i}$ and ϕ_i, this minimization problem can be solved with the nonlinear programming algorithms discussed in Chapter 9. Moreover, more complex cases of these equilibrium reactors, with multiple phases as well as reactions can also be addressed with current process simulators.

Finally, *specific kinetic models* are sometimes incorporated within process simulations. The most common models are the ideal reactor models such as plug flow reactors (PFRs) and continuous stirred tank reactors (CSTRs). For a reactor stream with an inlet concentration c_0 and flowrate F_0, the PFR equation is given by:

$$d(Fc)/dV = r(c, T), \quad c(0) = c_0 \quad (7.96)$$

where c is the vector of molar concentrations, V is the reactor volume, and $r(c)$ is the vector of reaction rates. For continuous stirred tank reactors (CSTR), the outlet concentration is given by:

$$F c - F_0 c_0 = V r(c, T) \quad (7.97)$$

Note that for both reactors, the vector of reaction rates (reaction rate for each species) needs to be specified. This task is frequently left up to the user, if kinetic expressions are available for the reacting system. Moreover, these equations also require thermodynamic models for the calculation of enthalpies for the energy balance around the reactor. As with the stoichiometric models, this is necessary to determine the temperatures for a given heat load specification, or vice versa.

Of course, many more detailed reactor models could be developed. However, these are considerably more expensive computationally and are usually used for "off-line" studies, rather than integrating them directly into the flowsheet. More detail on these reactor models and their role in reactor network synthesis is presented in Chapters 13 and 19.

7.6 SUMMARY AND FUTURE DIRECTIONS

This chapter provides a concise summary of detailed unit operations models frequently used in computer-aided process design tools. These process simulation tools are essential for the analysis and evaluation of candidate flowsheets. In the next chapter we continue the discussion of process simulation by describing the overall calculation strategy for the simulation of a process flowsheet. In particular, we will present and describe the algorithms needed to solve the process models given in this chapter. Moreover, we will discuss the integration of these models to simulate the entire flowsheet.

At the present time, most detailed unit operations for preliminary process design are based on thermodynamic models. Consequently, section 7.2 was devoted to a concise overview of these models for nonideal process behavior. The motivating problem for this discussion was phase equilibrium, which allowed us to include nonidealities both in the liquid and the gas phases. Popular thermodynamic models include equation of state (EOS) models for hydrocarbon mixtures and liquid activity coefficient models for nonideal, nonelectrolyte solutions. For the liquid activity coefficient models, model parameters frequently need to be determined from VLE or VLLE data; in the absence of these data, group contribution methods using the UNIFAC model have been very successful. The nonidealities that are described by these models can also be used directly in calculations of specific volumes, enthalpies, and entropies. However, we note that the nonideal models in this section need to be chosen with care because:

- They are far more complicated than ideal models and incur a much greater computational cost for process calculations.
- They are defined for specific mixture classes and often yield highly inaccurate results if not selected appropriately.

To develop the unit operations models, we note that the states of process streams are determined entirely by their thermodynamic properties. These properties and nonideal models for them are also considered sufficient for many of the unit operations in preliminary design. Separations are usually assumed to consist of equilibrium stage models, with efficiencies used to determine the actual column capacities. Simple mixing and splitting operations are similar to those developed in Chapter 3, except that now nonideal models are used to complete the energy balance. Similarly, transfer operations, including heat exchangers, pumps, and compressors, are altered slightly to accommodate nonideal thermodynamic models. These modifications are adequate to determine a reasonably accurate mass and energy balance for a candidate process flowsheet. Nevertheless, detailed sizing and costing for these units have not been covered in this chapter. Instead, for preliminary design we will rely on the simplified strategies developed in Chapter 4. To develop more detailed designs, there is a wealth of literature devoted to each unit operation and its coverage is clearly beyond the scope of this text. The reader is instead encouraged to consult the unit operations texts listed at the end of this chapter.

Finally, a number of research advances are related to unit operations modeling that are starting to be incorporated in process simulation tools. Certainly, more detailed reactor models have been incorporated into process flowsheets whenever the need arises and a good kinetic model is available for a specific process. In addition, mass transfer models are becoming well developed for absorption and nonideal distillation processes. These models are essential when tray efficiencies cannot adequately describe deviations from an equilibrium model. In fact, several process simulators have already incorporated these rate based models as standard models.

As a longer-term horizon, there are numerous advances in molecular dynamics and statistical mechanics that are leading to important breakthroughs in physical property modeling when no experimental data are available. While these methods are still too computationally expensive to incorporate directly within a process simulator, they are becoming useful in filling in the gaps present for many nonideal model parameters. As a result these approaches will also play a bigger role in the development of future process simulation strategies.

REFERENCES AND FURTHER READING

This chapter provides only a brief description of modeling concepts and elements used in process design. As a result, it is necessarily incomplete for all of these elements. For each section, a broad literature exists and this needs to be consulted for relevant details of the process models and their application to a particular design problem. An incomplete list of survey references is given below.

Further information on thermodynamic models, flash calculations, and their use in process simulation can be found in:

Fredenslund, A., Rasmussen, P., & Gmehling, J. (1977). *Vapor-Liquid Equilibria Using UNIFAC : A Group Contribution Method.* New York: Elsevier Scientific.

Gmehling, J., & Onken, U. (1988). *The Dortmund Data Bank: A Computerized System for*

Retrieval, Correlation, and Prediction of Thermodynamic Properties of Mixture. DECHEMA.

Hirata, M., & Ohe, S. (1975). *Computer Aided Data Book of Vapor Liquid Equilibria* New York: American Elsevier.

Holmes, M. J., & van Winkle, M. (1970). "Prediction of Ternary Vapor Liquid Equilibria from Binary Data," *Ind. Eng. Chem.*, 62(1), 21.

Rachford, H. H. & Rice, J. D. (1952). *J. Petrol. Technol.*, 4(10), 20.

Reid, R. C., Prausnitz, J. M., & Poling, B. E. (1987). *The Properties of Gases and Liquids.* New York: McGraw-Hill.

Smith, J. M., & van Ness, H. C. (1987). *Introduction to Chemical Engineering Thermodynamics.* New York: McGraw-Hill.

van Ness, H. C., & Abbott, M. M. (1982). *Classical Thermodynamics of Nonelectrolyte Solutions: With Applications to Phase Equilibria.* New York: McGraw-Hill.

Further details on the inside-out method can be found in:

Boston, J. F. (1980). Inside-out algorithms for multicomponent separation process calculations. In *Computer Applications to Chemical Engineering,* Squires and Reklaitis (eds.), ACS Symposium Series 124, 35.

Boston, J. F., & Britt, H. I. (1978). A radically different formulation for solving phase equilibrium problems." *Comp. and Chem. Engr.,* 2, 109.

Reviews and detailed descriptions for distillation modeling for process simulation can be found in:

Taylor, R., & Lucia, A. (1995). Modeling and analysis of multicomponent separation processes. In *FOCAPD IV,* Biegler and Doherty (eds.), AIChE Symp. Ser #304, 19.

Wang, J. C., & Wang, Y. L. (1980). A review on the modeling and simulation of multistaged separation processes. In *Proc. FOCAPD,* Mah and Seider (eds.), Engineering Foundation, Vol. II, 121.

Finally, there are several standard texts for unit operations models, including:

Coulson, J. M., & Richardson, J. F. (1968). *Chemical Engineering: Vol. 2—Unit Operations.* Oxford: Pergamon Press.

Geankoplis, C. J. (1978). *Transport Processes and Unit Operations.* Boston: Allyn and Bacon.

Henley, E. J., & Seader, J. D. (1981). *Equilibrium Stage Separation Operations in Chemical Engineering.* New York: Wiley.

McCabe, W. L., Smith, J. C., & Harriott, P. (1992). *Unit Operations of Chemical Engineering,* New York: McGraw-Hill.

Green, D. W. (ed.). (1984). *Perry's Chemical Engineers' Handbook.* New York: McGraw-Hill.

Welty, J. R., Wicks, C. E., & Wilson, R. E. (1984). *Fundamentals of Momentum, Heat and Mass Transfer*. New York: Wiley.

Further description of the unit models and physical property options can also be found in the documentation for the process simulators themselves. Three useful references are:

ASPEN Plus User's Guide
HYSIM User's Guide
Pro/II User's Manual

EXERCISES

1. For the multicomponent two suffix Margules model, derive the expressions for activity coefficients used for the methanol, propanol, acetone example.
2. Derive the equation for the fugacity coefficients used in the equation of state models.
3. Simplify the flash equations and the TP flash algorithm to develop bubble and dewpoint algorithms. Find the bubble and dewpoints for the 40 mol % methanol, 20 mol % propanol, and 40 mol % acetone system at one atm.
4. Fill in the steps in the derivation of the inside-out algorithm. Show that for a TP flash, the Boston-Britt model is related to the PQ flash algorithm presented in this chapter.
5. Using the GAMS case study model as a guide, solve the benzene, toluene, o-xylene column for 30 trays with stage 15 as the feedtray location. Vary this location and comment on the change in the distillate composition for a reflux ratio = 5.
6. Resolve the benzene, toluene, o-xylene column example and plot the liquid and vapor flowrates for a reflux ratio = 5. Is equimolar overflow a good assumption for this system?
7. Modify the MESH equations to deal with an equimolar overflow assumption. How are the equations simplified?
8. Apply the shortcut models developed in Chapters 3 and 4 to the benzene, toluene, o-xylene column example. Which specifications would you make to compare this model to the tray-to-tray model?
9. Resolve the example with the equilibrium reaction:

$$CO + H_2O \leftrightarrow CO_2 + H_2$$

and show that the Gibbs free energy minimization yields the same result as for the relation:

$$(f_C)^c (f_D)^d / [(f_A)^a (f_B)^b] = K = \exp(-\Delta G_{rxn}(T)/RT)$$

10. For the water gas shift example, show that the temperature correction for ΔG_{rxn} is not negligible for a temperature change from 298 K to 600 K. Resolve Example 7.6 with this correction.

GENERAL CONCEPTS OF SIMULATION FOR PROCESS DESIGN

8

In Part I, assumptions and model simplifications were made to analyze candidate flowsheets easily. These include ideal thermodynamic behavior, simplified split fraction models for noninteracting components, and saturated streams for most exit streams. With these assumptions the analysis tasks could be decomposed and smaller problems leading to a mass balance, temperature and pressure specification, and the energy balance, could be performed sequentially. In many cases, these calculations could be done by hand or with the help of a spreadsheet. In Chapter 7, we considered more detailed design models and noted that by removing the assumptions of Part I, flowsheet analysis or *simulation* becomes much more complicated. In that chapter, relatively little discussion was devoted to the daunting tasks of solving these detailed models. Because the mass and energy balances are tightly coupled we need to consider large-scale numerical methods. This chapter provides a concise description of the simulation problem along with solution strategies and methods needed to tackle it.

8.1 INTRODUCTION

In Chapter 2, we performed mass and energy balances by:

- "Tearing" the flowsheet, usually at reactor feed
- Choosing split fractions for all units
- Solving linear mass balance equations
- Setting temperatures and pressures based on bubble and dewpoints
- Calculating heating and cooling duties

While this approach gives an easy *decomposition* of tasks and gives a qualitative understanding of the process, the results are not accurate for more detailed designs. Because of the need for more detailed models, such as the ones described in Chapter 7, we need to consider the solution of the mass and energy balance (along with temperature and pressure specifications) in a simultaneous manner. Using the nonideal thermodynamic and detailed unit operations models in the previous chapter, a typical flowsheet consists of 10,000 to 100,000 equations, and often more than this. Clearly, much more advanced computer tools are required. Therefore, to perform the flowsheet analysis and evaluation, we rely on process simulation software, or *process simulators* (a list of commercial simulators is given in Appendix C). These computer tools embody and extend the models in Chapter 7. Moreover, these simulators have additional subsystems devoted to them, including a graphical user interface, extensive interactive diagnostic options, and a variety of reporting features, in addition to the core simulator. In fact, the core simulator itself consists of several hundred thousand lines of code and is carefully maintained and extended on a continuous basis, often by a software vendor devoted to this purpose.

Current process simulators can be classified as *modular* or *equation-oriented*. In the *equation-oriented mode*, the process equations (unit, stream connectivity, and sometimes thermodynamics models) are assembled and solved simultaneously. In the *modular mode*, unit and thermodynamic models remain self-contained as subprograms or procedures. These are then called at a higher level in order to converge the stream connectivity equations represented in the flowsheet topology. The modular mode has a longer development history and is the more popular mode for design work. While it is easier to construct and debug, these simulators are relatively inflexible for a wide variety of user specifications. On the other hand, with the application of more sophisticated numerical methods and software engineering concepts, the equation-oriented mode has seen considerable development over the last ten years. Primary applications for this mode are for on-line modeling and optimization.

Process simulation tools have an interesting development that spans almost forty years. In the 1950s, as stand-alone models were developed for individual units, it made sense to string these units along as subprograms. Executed in sequence these units or modules form a flowsheet, with iterations on unknown recycle or "tear" streams. This approach was known as *sequential modular*. Perhaps the first of these efforts was the Flexible Flowsheet developed in 1958 at M. W. Kellogg Corp. In the 1960s a great amount of effort was expended toward the development of sequential modular flowsheeting packages, with most petrochemical companies developing their own in-house packages and devoting large groups to their development and maintenance. On the other hand, several academic researchers developed more fundamental methods for equation-based or *equation-oriented* simulators. This approach allowed for more flexibility in deriving flowsheet decomposition and solution strategies. In this decade, many of the architectural concepts of current process simulators were fixed and many of the poorer strategies were tried and discarded.

In the 1970s, more advanced methods were developed for the decomposition and solution of modular flowsheets, leading to concepts of simultaneous modular flowsheeting. Here, while the unit models remained intact, the solution of the flowsheet streams was performed in a global or simultaneous manner instead of in sequence. Also, better

Sec. 8.2   Process Simulation Modes   245

unit algorithms and more general models (such as solids handling) were incorporated along with more sophisticated numerical methods. This development was also motivated by the ASPEN project at the Massachusetts Institute of Technology.

In the 1980s and 1990s, the equation-oriented simulation mode saw considerable industrial development, especially for on-line modeling and optimization. In addition, modern software engineering concepts led to the development of user friendly interfaces and even more powerful algorithms. Finally, the rapid advance of computer hardware led to a variety of personal computer based products and a much wider user community for simulation tools. The rapid growth and development of these sophisticated tools caused most chemical manufacturers to standardize on vendor-supported software, and therefore to support the development of only a few process simulation packages. Currently, the most popular *modular* simulators include ASPEN/PLUS from Aspen Technology, Inc., HYSIM and HYSYS from Hyprotech, Ltd., and PRO/II from Simulations Sciences, Inc. Equation-oriented simulators include SPEEDUP from Aspen Technologies as well as a number of packages that deal with real-time modeling and optimization (DMO and RTOPT from Aspen Technology and NOVA from DOT Products, Inc.). These programs are listed in Appendix C. Concise reviews of these simulation packages, and many others, are given in the *Chemical Process Software Guide,* published annually by AIChE. A summary of some of the characteristics of these codes is also given in Biegler (1989).

In this chapter we will describe the main concepts of both modular and equation-oriented process simulation. The descriptions here will focus on basic ideas, which actually become much more detailed in particular implementations of current simulators. For more information on these implementations and application, the user is strongly urged to consult the software manuals for a specific process simulator. The next section provides more detail into the structure of both equation-oriented and modular simulators and illustrates these with a small process example. Section 8.3 then provides a concise review of methods for solving nonlinear equations, which are essential for both simulation modes. Section 8.4 then provides some information on flowsheet decomposition or "tearing". This background is most useful for the modular mode. The concepts of both sections will be highlighted with illustrative examples. Section 8.5 presents a brief application of these concepts to our small flowsheeting example, and section 8.6 summarizes the chapter.

8.2 PROCESS SIMULATION MODES

In order to provide a clearer description of process simulation strategies, we first present a simple process flowsheet.

8.2.1 Flowsheeting Example

Consider the process flowsheet shown in Figure 8.1. For illustration purposes we consider the modification of a small process, initially proposed by Williams and Otto (1960) as a typical chemical process simulation. This process has also been used in numerous process optimization studies.

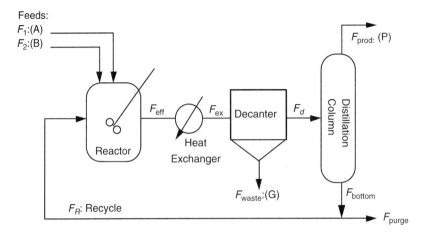

FIGURE 8.1 Williams and Otto flowsheet.

Feed streams with pure species A and B are mixed with a recycle stream and enter a continuous stirred tank reactor, where the following reactions take place.

$$A + B \rightarrow C$$
$$C + B \rightarrow P + E \quad (8.1)$$
$$P + C \rightarrow G$$

Here C is an intermediate, P is the main product, E is a by-product, and G is an oily waste product. Both C and E can be sold for their fuel values, while G must be disposed of at a cost. The plant consists of the reactor, a heat exchanger to cool the reactor effluent, a decanter to separate the waste product G from reactants and other products, and a distillation column to separate product P. Due to the formation of an azeotrope, some of the product (equivalent to 10 wt% of the mass flowrate of component E) is retained in the column bottoms. Most of this bottom product is recycled to the reactor, and the rest is used as fuel. The plant model can be defined without an energy balance and we further simplify this problem to consider only isothermal reactions for the manufacture of compound P. The rest of these units are also simplified greatly in order to keep the example small and illustrate simulation concepts with fewer complications. The topological information for this flowsheet is given by:

Unit	Type	Input	Output
1	Reactor	F_1, F_2, F_R	F_{eff}
2	Heat Exch.	F_{eff}	F_{ex}
3	Decanter	F_{ex}	F_d, F_{waste}
4	Column	F_d	F_{prod}, F_{bottom}
5	Splitter	F_{bottom}	F_{purge}, F_R

Sec. 8.2 Process Simulation Modes 247

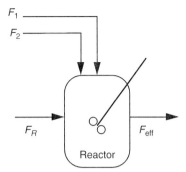

FIGURE 8.2 Williams-Otto reactor.

We now consider the unit models in the order executed in the flowsheet. All of the streams (F) are given in mass flowrates instead of the molar flowrates (μ) used in Part I.

REACTOR MODEL (FIGURE 8.2)

The rate vector for components A, B, C, P, E, and G is given by elementary kinetics based on mass fractions. For simplicity we assume an isothermal reactor (with temperature pre-specified at 674°R). The equations for this reactor are given by:

$$F^A_{\text{eff}} = (F^A_1 + F^A_R) - (k_1 X_A X_B) V \rho$$

$$F^B_{\text{eff}} = (F^B_2 + F^B_R) - (k_1 X_A + k_2 X_C) X_B V \rho$$

$$F^C_{\text{eff}} = F^C_R + (2k_1 X_A X_B - 2k_2 X_B X_C - k_3 X_P X_C) V \rho$$

$$F^E_{\text{eff}} = F^E_R + (2k_2 X_B X_C) V \rho \qquad (8.2)$$

$$F^P_{\text{eff}} = F^P_R + (k_2 X_B X_C - 0.5 k_3 X_P X_C) V \rho$$

$$F^G_{\text{eff}} = F^G_R + (1.5 k_3 X_P X_C) V \rho$$

$$X_j = F^j_{\text{eff}} / (F^A_{\text{eff}} + F^B_{\text{eff}} + F^C_{\text{eff}} + F^E_{\text{eff}} + F^P_{\text{eff}} + F^G_{\text{eff}}), j = A, B, C, E, G, P$$

where the rate constants are given by:

$$k_1 = 5.9755 \cdot 10^9 \exp(-12000/T) \ h^{-1} \ (\text{wt fraction})^{-1}$$

$$k_2 = 2.5962 \cdot 10^{12} \exp(-15000/T) \ h^{-1} \ (\text{wt fraction})^{-1} \qquad (8.3)$$

$$k_3 = 9.6283 \cdot 10^{15} \exp(-20000/T) \ h^{-1} \ (\text{wt fraction})^{-1}$$

and X_j is the weight fraction of component j, V is the volume of the reactor vessel, T is the reactor temperature, and ρ is the density of the mixture.

HEAT EXCHANGER MODEL (FIGURE 8.3)

Since there is no energy balance, the equations for this unit are direct input and output relations:

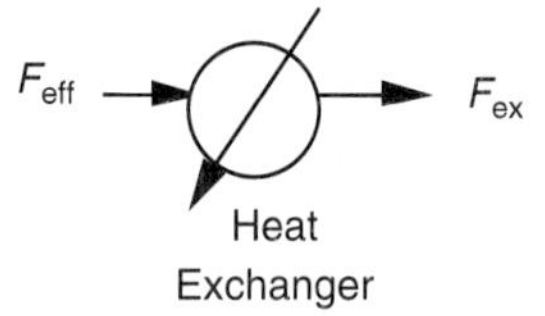

FIGURE 8.3 Williams-Otto exchanger.

$$F_{ex}^j = F_{eff}^j, \quad j = A, B, C, E, G, P \tag{8.4}$$

DECANTER (FIGURE 8.4)

This unit assumes a perfect separation between component G and the rest of the components, so the equations can be written as:

$$\begin{aligned} F_d^j &= F_{ex}^j, \quad j = A, B, C, E, P \\ F_d^G &= 0 \\ F_{waste}^G &= F_{ex}^G \\ F_{waste}^j &= 0, \quad j = A, B, C, E, P \end{aligned} \tag{8.5}$$

DISTILLATION COLUMN (FIGURE 8.5)

This unit assumes a pure separation of product P overhead but also assumes that some of the product is retained below due to the formation of an azeotrope, leading to the following equations.

$$\begin{aligned} F_{bottom}^j &= F_d^j \quad j = A, B, C, E \\ F_{prod}^j &= 0, \quad j = A, B, C, E \\ F_{bottom}^P &= 0.1 \, F_d^E \\ F_{prod}^P &= F_d^P - 0.1 \, F_d^E \end{aligned} \tag{8.6}$$

FLOW SPLITTER (FIGURE 8.6)

Equations for this unit are given by:

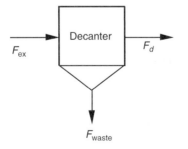

FIGURE 8.4 Williams-Otto decanter.

Sec. 8.2 Process Simulation Modes

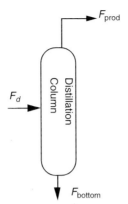

FIGURE 8.5 Williams-Otto column.

$$F^j_{\text{purge}} = \eta \, F^j_{\text{bottom}}, j = A, B, C, E, P$$
$$F^j_R = (1 - \eta) \, F^j_{\text{bottom}}, j = A, B, C, E, P \tag{8.7}$$

Despite the simplifications in this process model, we obtain a system of 58 variables and 54 equations. Also, note that this system cannot be solved sequentially because of the recycle stream, and the reactor equations themselves need to be solved simultaneously.

In particular, the system has four degrees of freedom and the specification of these variables leads to different kinds of simulation problems. For instance, if we specify the feed flowrates of A and B (F_1 and F_2), the reactor volume (V), and the split fraction (η), we have a *performance* or *rating problem* that deals with an existing design or process. These are considered "normal" inputs to the process as the calculation sequence follows the material flow in the process. On the other hand, if we specify four outlet flowrate specifications (say F^P_{prod}, F^A_{purge}, F^B_{purge}, and F^E_{purge}) we term this a *design problem* and we need to calculate the "normal" inputs from these specifications. Intuitively, one can see that solving this rating problem is easier than the design problem. In fact, for some values of the design specifications, there may not be a solution to the flowsheet. Nevertheless, both types of problems need to be considered for design calculations by process simulation tools. With this description of the Williams-Otto process, we now consider the solution strategies for this process by the modular and the equation-oriented modes.

8.2.2 The Modular Mode

For detailed flowsheet simulation for design and analysis, the modular mode is currently the most popular among commercial process simulators. Here the unit models are encapsulated as procedures where the output streams (and other calculated information) are evaluated from input streams and desired design parameters. These procedures are then solved in a se-

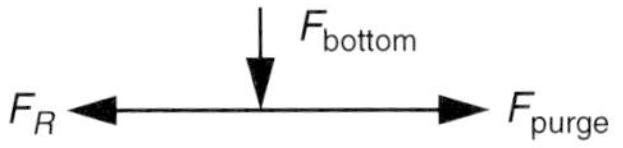

FIGURE 8.6 Williams-Otto splitter.

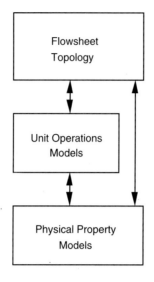

FIGURE 8.7 Structure of modular simulators.

quence that roughly parallels the flow of material on the actual process. Process simulators are generally constructed in a hierarchy with three levels, as shown in Figure 8.7.

The top level deals with the flowsheet topology, where the main task is to sequence the unit modules, initialize the flowsheet, identify the recycle loops and the tear streams, and ensure the convergence of these streams for the overall mass and energy balance of the flowsheet. The middle level deals with the unit operations procedures and represents a library of unit models, each solved with a specialized calculation procedure. Inputs from the top level include the input streams and parameters to each unit, and outputs from the unit (streams and parameters) are fed back to the top level once the unit is calculated. The library of units includes separators, reactors, and transfer units, as described in Chapter 7. Finally, the lowest level deals with the physical property models. These include the thermodynamic models presented in Chapter 7 for phase equilibrium, enthalpy, entropy, density, and so on. This level is accessed frequently by the unit operations procedures and can also be accessed by the top level for flowsheet initialization and stream calculations. Each level is largely self-contained with little communication with the other levels. This allows the simulator to concentrate on one task at a time.

At each level, a key task is the solution of sets of nonlinear equations, f, with unknowns, x, given generally as: $f(x) = 0$. From Chapter 7, these equations can represent phase equilibrium calculations that involve nonideal thermodynamic models. Also, the unit operations themselves consist of nonlinear mass and energy balances that are coupled with these thermodynamic relations. Solving these systems requires an iterative solution procedure that is beyond the range of hand calculations and simple spreadsheet tools. Frequently, the solution algorithms are tailored to the particular structure of the unit operation. This is especially the case for distillation and absorption calculations (see Chapter 7). Consequently, section 8.3 will present an overview of methods to solve nonlinear equations.

In addition, the flowsheet topology or recycle level deals with the structural decomposition of the flowsheet and the sequencing of the units. Here we need to identify recycle loops and identify tear streams. Once these tear streams are specified, the units can be executed directly in sequence. As shown in Chapter 3, a good tear stream choice is frequently the reactor feed. Once identified, the stream values must be determined through an iterative process. To solve this, methods for solving nonlinear equations can be applied directly. Moreover, for the particular problem of recycle convergence, we can consider a more specific *fixed point relation*: $x = g(x)$, where x is the guessed tear stream flowrate vector and $g(x)$ is the corresponding calculated stream flowrate vector. Fixed point methods will also be covered in section 8.3.

Related to recycle convergence is the often difficult problem of identifying the tear streams. Here a set of streams needs to be found that breaks all of the recycle loops. One option, which has been explored in the process engineering literature, is to choose *all* streams as tear streams. This approach has some interesting characteristics but requires sophisticated convergence algorithms. On the other hand, there are computational advantages to keeping the number of tear streams small and choosing them so that they do not interact adversely during the convergence process. Therefore, we need to have a systematic strategy to determine which streams to tear. Methods for tear stream selection are briefly described in section 8.4.

We now reconsider the small example described above and discuss how this example can be solved with a modular strategy.

SOLVING THE WILLIAMS-OTTO FLOWSHEET IN MODULAR MODE

In the modular mode we group the process equations within each unit and execute these units in sequence. The first task in solving this flowsheet is to identify the streams that break all of the recycle loops. Since the flowsheet has only one recycle loop, any of the streams within that loop can be used as a tear stream. In keeping with the convention in Chapter 2, we choose the reactor inlet stream and consider the process, as shown in Figure 8.8.

Here we solve the units according to the flowsheet topology table (reactor, heat exchanger, decanter, distillation column, splitter) where the output streams of each unit are calculated from the inputs. For each module, we need to make sure that all of the inputs are specified. We specify the feed flowrates F_1 and F_2 to the process, the volume of the reactor, and the purge fraction to the splitter. Here we initialize the problem by guessing the flowrates for F_R and evaluating a calculated value for this stream. Executing the sequence of units, starting from the reactor, we also obtain the input stream to each unit.

The flowsheet convergence problem is then given by the fixed point equation:

$$F_R = g(F_R) \tag{8.8}$$

where $g(F_R)$ is found implicitly after executing the sequence of units (or a flowsheet pass), and the vector values for F_R are determined iteratively by making several flowsheet passes. These flowsheet passes are the dominant expense of the simulation and the recycle convergence algorithm determines the efficiency of the flowsheet simulation. Solution of this flowsheet at the top (or recycle) level is straightforward if we assume that all of the

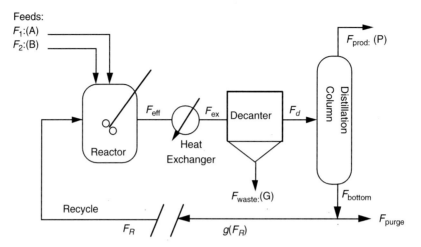

FIGURE 8.8 Flowsheet solved in modular mode.

output streams could be determined readily within each unit. For this process, we require a robust iterative solution scheme for the reactor equations. These algorithms are covered in section 8.3. Moreover, the structure of the flowsheet is a simple single loop and much more complicated topologies are frequently encountered. For such flowsheets we will see in section 8.4 that the determination of "good" tear streams is often far from trivial.

8.2.3 The Equation-Oriented Mode

In the modular mode, equations for each unit were kept distinct; each module was characterized by a specialized procedure to solve the unit equations and a restricted set of inputs to that module (e.g., input stream and procedure specific input parameters). Neither of these characteristics is part of the equation-oriented simulation mode. Instead, we combine the flowsheet topology equations (e.g., stream connectivity) with the unit equations (and, if possible, the physical property equations) into one large equation set. This problem structure allows us much more flexibility in specifying independent variables as parameters and to solve for the remaining ones. Moreover, the solution of this equation set is performed by a general purpose nonlinear equation solver. In virtually all cases, a very efficient Newton-Raphson solver is used to converge the nonlinear equations, as will be discussed in section 8.3. Figure 8.9 illustrates the problem structure for equation-oriented simulation. Note that, because of their number and nonlinearity, physical property models are frequently left as distinct procedures and are kept separate from the unit operations and connectivity equations.

With the modular mode, we were concerned with exploiting the flowsheet topology through stream tearing and specialized unit procedures for equation solving. In contrast, with the equation-oriented simulation we apply large-scale, simultaneous solution strate-

Sec. 8.2 Process Simulation Modes

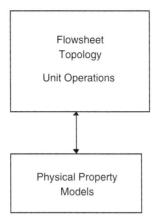

FIGURE 8.9 Structure of equation-oriented mode.

gies directly to the equations for the entire flowsheet. Such large systems of equations have a *sparse* structure, in that a small fraction of the total number of variables participate in any single equation. Exploiting this concept is a key feature of equation-oriented simulators.

Because of their structure, equation-oriented simulators tend to converge process flowsheets much faster than their modular counterparts. However, modular simulators are easy to initialize because they execute the process units in sequence according to the structure of the flowsheet. This leads to a reasonably good and "safe" starting point for solving the flowsheet. Equation-oriented simulators have no analogous initialization schemes that arise naturally from the flowsheet structure and considerable effort can be required to initialize these problems (essentially, the equations need to be grouped into a modular structure in order to get a good problem initialization). In addition, because a general purpose equation solver is used, it is harder to incorporate the unit structure of the equations into the solution procedure. Similarly, it takes more effort to construct and to debug an equation-oriented simulation.

Nevertheless, both modular and equation-oriented modes have clear advantages on different types of flowsheeting problems. Both modes are constructed from specialized concepts for decomposition and nonlinear equation solving, which will be explored in later sections of this chapter. To conclude this section, we illustrate how the equation-oriented mode is applied to the Williams-Otto process.

SOLVING THE WILLIAMS-OTTO FLOWSHEET IN EQUATION-ORIENTED MODE

In the equation mode we combine all of the process equations and solve them simultaneously. As derived above, the equations for this flowsheet are given by:

$$F^A_{\text{eff}} = (F^A_1 + F^A_R) - (k_1 X_A X_B) V \rho$$

$$F^B_{\text{eff}} = (F^B_2 + F^B_R) - (k_1 X_A + k_2 X_C) X_B V \rho$$

$$F^C_{eff} = F^C_R + (2k_1 X_A X_B - 2k_2 X_B X_C - k_3 X_P X_C)\, V\rho$$

$$F^E_{eff} = F^E_R + (2k_2\, X_B X_C)\, V\rho$$

$$F^P_{eff} = F^P_R + (k_2 X_B X_C - 0.5 k_3 X_P X_C) V\rho \tag{8.9}$$

$$F^G_{eff} = F^G_R + (1.5 k_3\, X_P X_C)\, V\rho$$

$$X_j = F^j_{eff}/(F^A_{eff} + F^B_{eff} + F^C_{eff} + F^E_{eff} + F^P_{eff} + F^G_{eff}),\ j = A, B, C, E, G, P$$

$$F^j_{ex} = F^j_{eff},\quad j = A, B, C, E, G, P \tag{8.10}$$

$$F^j_d = F^j_{ex},\quad j = A, B, C, E, P$$
$$F^G_d = 0$$
$$F^G_{waste} = F^G_{ex} \tag{8.11}$$
$$F^j_{waste} = 0,\quad j = A, B, C, E, P$$

$$F^j_{bottom} = F^j_d,\quad j = A, B, C, E$$
$$F^j_{prod} = 0,\quad j = A, B, C, E$$
$$F^P_{bottom} = 0.1\, F^E_d \tag{8.12}$$
$$F^P_{prod} = F^P_d - 0.1\, F^E_d$$

$$F^j_{purge} = \eta\, F^j_{bottom}\quad j = A, B, C, E, P$$
$$F^j_R = (1 - \eta)\, F^j_{bottom}\quad j = A, B, C, E, P \tag{8.13}$$

The structure of these equations has a strong impact on the efficiency of the solution process. In particular, note that very few variables appear in a given equation (usually two or three) and this *sparsity* property needs to be exploited. Since the structure of the problem is exploited at the *equation level*, there is also scope for specifying the four degrees of freedom. Also, the equations for this problem can be simplified considerably. For instance, as seen from the model, the equations and variables corresponding to the heat exchanger and decanter can be eliminated trivially.

In this section, we have considered characteristics of both modular and equation-oriented modes. Both of these require the solution of nonlinear equations as well as decomposition principles. These will be considered in the next two sections.

8.3 METHODS FOR SOLVING NONLINEAR EQUATIONS

Solving algebraic nonlinear equations is the primary task in steady state process simulation. In both modular and equation-oriented modes, the unit operations, physical property, and flowsheet topology equations are constructed and need to be solved reliably. These problems can be stated in *standard form: solve $f(x) = 0$*, or in *fixed point form: $x = g(x)$*.

Sec. 8.3   Methods for Solving Nonlinear Equations

Both forms are equivalent and the methods developed in this section can be applied to either form. For instance, to convert to standard form, we can write

$$f(x) = x - g(x) = 0 \tag{8.14}$$

and to convert to fixed point form, we can write, for example:

$$x = x + h(f(x)) = g(x) \tag{8.15}$$

where we can choose $h(.)$ as any function, where $h(y) = 0$ if and only if $y = 0$. Moreover, we will see that the fixed point form is easier to work with for recycle convergence in the modular mode.

This section is divided into two main parts. In the first, we deal with Newton-type methods expressed in the standard form. The Newton-Raphson method is the most widely used for solving nonlinear equations. For process simulation, it is the core algorithm for the equation-oriented mode and is also used often in solving unit operation equations, particularly for detailed separation models. In addition, we will also introduce quasi-Newton or Broyden methods. The second section deals with first-order fixed-point methods. Unlike the Newton method, these methods do not require derivative information from the equations, but are also slower to converge. These methods are used to converge recycle streams and can also be used in calculation procedures where derivatives are difficult to obtain.

8.3.1 Newton-Type Methods

Consider the problem in standard form, $f(x) = 0$, where x is a vector of n real variables and $f()$ is a vector of n real functions. If we have a guess for the variables at a given point, say x', then we can take a Taylor series expansion about x' in order to extrapolate to the solution point, x^*. We can write each element of the vector function f as:

$$f_i(x^*) \equiv 0 = f_i(x') + \partial f_i(x')/\partial x^T (x^* - x') \\ + 1/2 \, (x^* - x')^T \, \partial^2 f_i(x')/\partial x^2 \, (x^* - x') + \ldots \; i = 1,\ldots n \tag{8.16}$$

or

$$f_i(x^*) \equiv 0 = f_i(x') + \nabla f_i(x')^T (x^* - x') \\ + 1/2 \, (x^* - x')^T \, \nabla^2 f_i(x') \, (x^* - x') + \ldots \; i = 1,\ldots n \tag{8.17}$$

Here $\nabla f_i(x)$ and $\nabla^2 f_i(x)$ are the gradient vector and Hessian matrix of the function $f_i(x)$, respectively. If we truncate this series to only the first two terms, we have:

$$f(x' + p) \equiv 0 = f(x') + J(x') p \tag{8.18}$$

where we have defined the vector $p = (x^* - x')$ as a search direction and the matrix J with elements

$$\{J\}_{ij} = \frac{\partial f_i}{\partial x_j} \tag{8.19}$$

for row i and column j of matrix J. We call this matrix the *Jacobian*. If the Jacobian matrix J is nonsingular, we can solve for p directly and this is a *linear approximation* to the solution of the nonlinear equations.

$$p = - (J(x'))^{-1} f(x') \qquad (8.20)$$

This relation allows us to develop a recursive strategy for finding the solution vector x^*. Here we start with an initial guess x^0 and using k as an iteration counter, we find the solution by:

$$p^k = - (J^k)^{-1} f(x^k) \qquad x^{k+1} = x^k + p^k \qquad (8.21)$$

where $J^k \equiv J(x^k)$. These recursion formulas can be formalized in the following basic algorithm for Newton's method.

Algorithm

0. Guess x^0, $k = 0$.
1. Calculate $f(x^k)$, J^k.
2. Calculate $p^k = -(J^k)^{-1} f(x^k)$.
3. Set $x^{k+1} = x^k + p^k$.
4. Check convergence: If $f(x^k)^T f(x^k) \leq \varepsilon_1$ and $p^{kT} p^k \leq \varepsilon_2$, stop. Here ε_1 and ε_2 are tolerances set close to zero.
5. Otherwise, set $k = k + 1$, go to 1.

Newton's method has some very desirable convergence properties. In particular, it has a fast rate of convergence close to the solution. More precisely, Newton's method converges at a quadratic rate, given by the relation:

$$\frac{\left\| x^k - x^* \right\|}{\left\| x^{k-1} - x^* \right\|^2} \leq K \qquad (8.22)$$

where, e.g., $\|x\| \equiv (x^T x)^{1/2}$ is the Euclidean norm and defines the length of a given vector x. One way to interpret this relation is to think of the case where $K = 1$ and we have one digit of accuracy for x^{k-1}, that is, $\|x^{k-1} - x^*\| = 0.1$. Then, at the next iteration, we have two digits of accuracy, then four, then eight, and so on.

On the other hand, this fast rate of convergence occurs only if the method performs reliably. And this method can fail on difficult simulation problems. Sufficient conditions for convergence of the above Newton algorithm are given qualitatively as:

- The functions, $f(x)$ and $J(x)$ exist and are bounded for all values of x.
- The initial guess, x^0, must be close to the solution.
- The matrix $J(x)$ must be nonsingular for all values of x.

8.3.2 Bounded Functions and Derivatives

By inspection, we can rewrite the equations to avoid division by zero and undefined functions. In addition, new variables can be specified through additional equations as well. To illustrate, we consider two small examples:

1. To solve $f(t) = 10 - e^{3/t} = 0$, we notice that for t close to zero, the exponential term becomes very large and so does its derivative, $-3\, e^{3/t}$. Instead, we define a new variable $x = 3/t$ and add the equation: $x\,t - 3 = 0$. This now leads to a larger set of equations but with bounded functions; we therefore solve:

$$f_1(x) = 10 - e^x = 0 \tag{8.23}$$
$$f_2(x) = x\,t - 3 = 0$$

where the Jacobian matrix is:

$$J(x) = \begin{bmatrix} -e^x & 0 \\ t & x \end{bmatrix} \tag{8.24}$$

Note that both the functions and J remain bounded and defined for finite values of x. Nevertheless, J may still be singular for certain values of x and t.

2. For the problem, $f(x) = \ln x - 5 = 0$, the logarithm is undefined for nonpositive values of x. This problem can be rewritten by introducing a new variable and equation. Here we let $x_2 = \ln x_1$ or $f_1 = x_1 - \exp(x_2) = 0$. The equation system becomes:

$$\begin{aligned} f_1 &= x_1 - \exp(x_2) = 0 \\ f_2 &= x_2 - 5 = 0 \end{aligned} \tag{8.25}$$

with the Jacobian matrix given by:

$$J(x) = \begin{bmatrix} 1 & -e^{x_2} \\ 0 & 1 \end{bmatrix} \tag{8.26}$$

Again, these functions are defined and bounded for finite values of the variables.

8.3.3 Closeness to Solution

In general, ensuring a starting point "close" to the solution is not practical. Consequently, if we start from a poor guess, we need to control the length of the Newton step to ensure that progress is made toward the solution. We therefore modify the Newton step so that we have a new point that is only a fraction of the step predicted by the Newton iteration. This is given by:

$$x^{k+1} = x^k + \alpha\, p^k$$

where α is a fraction between zero and one and p^k is the direction predicted by the Newton iteration. Of course, if $\alpha = 1$, we recover the full Newton step. We now consider a strategy for choosing the stepsize, α, automatically. Moreover, an approach like this is needed in order to provide reliable convergence for Newton's method.

Let's define an objective function $\phi(x) = 1/2\, f(x)^T f(x)$ and seek to mininimize $\phi(x)$. Using the Newton direction with α, the step size, we have $x^{k+1} = x^k + \alpha\, p^k$, and from the Taylor series expansion of $\phi(x)$:

$$\phi(x^{k+1}) = \phi(x^k) + \alpha\, d\phi/d\alpha + \alpha^2/2\, d^2\phi/d\alpha^2 + \ldots \tag{8.27}$$

or

$$\phi(x^{k+1}) = \phi(x^k) + \nabla\phi(x^k)^T(\alpha\, p^k) + \alpha^2/2\, p^{k\,T}\nabla^2\phi(x^k)\, p^k + \ldots \tag{8.28}$$

Here we again define $J^k \equiv J(x^k)$, $\{J(x^k)\}_{ij} \equiv \partial f_i/\partial x_j$ and from this we have the derivative of $\phi(x)$:

$$\nabla\phi(x^k)^T = f(x^k)^T J^k \tag{8.29}$$

and the Newton step

$$p^k = -(J^k)^{-1} f(x^k). \tag{8.30}$$

Postmultiplying the derivative of $\phi(x)$ by p^k and substituting for the Newton step gives the relation:

$$\nabla\phi(x^k)^T p^k = -(f(x^k)^T J^k (J^k)^{-1} f(x^k)) = -f(x^k)^T f(x^k) = -2\,\phi(x^k) < 0 \tag{8.31}$$

Now if we take $\alpha \to 0$ in the Taylor expansion, we have:

$$\phi(x^{k+1}) - \phi(x^k) \approx -2\,\alpha\,\phi(x^k) < 0 \tag{8.32}$$

so for a step size α sufficiently small, we know that the Newton step will reduce $\phi(x)$. This important *descent* property will be used to derive our algorithm and find an improved point for $\phi(x)$.

We could now minimize $\phi(x^k + \alpha p^k)$ and find an optimal value of α along p^k. But this can become expensive in terms of function evaluations, especially since the particular direction will change at later iterations. Instead, we will choose a stepsize α^k that gives us only a sufficient reduction for $\phi(x)$. This approach is known as the *Armijo line search*. To develop this, we consider Figure 8.10 below.

Starting at the origin, we note the negative slope at $\alpha = 0$ and also note that there is a value for the step size for which $\phi(x^k + \alpha\, p^k)$ is minimized. Instead of a direct minimization for α, we define a sufficient condition for reduction when $\phi(x^k + \alpha\, p^k)$ is below the Armijo chord, that is.:

$$\phi(x^k + \alpha\, p^k) - \phi(x^k) \leq -2\,\delta\,\alpha\,\phi(x^k), \tag{8.33}$$

where δ is the fraction of the slope (typically specified between zero and 1/2) that defines this chord. In this way, we insure that a satisfactory reduction for $\phi(x)$ is obtained that is at

Sec. 8.3   Methods for Solving Nonlinear Equations

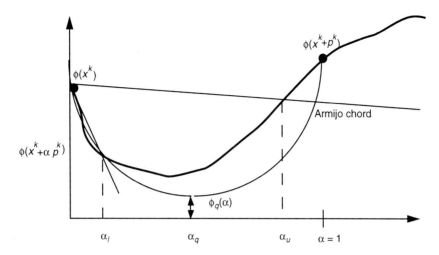

FIGURE 8.10   Schematic for Armijo line search.

least a fraction, δ, of the rate of reduction at the current point, x^k. If this relation is satisfied for a sufficiently large value of α, then we take this step. On the other hand, consider the case in Figure 8.10, where $\alpha = 1$. Here the value of $\phi(x^k + p^k)$ is above the Armijo chord, there is no reduction of $\phi(x)$, and we need to choose a smaller step. We also need to make sure that this step is not too short (in the range between, say, α_l and α_u) so that a large enough move is taken in x (otherwise, the moves in x would shrink to zero before the equations are converged). To do this, we perform a quadratic interpolation for α by defining an interpolating function $\phi_q(\alpha)$ based on three parameters, the values of $\phi(x)$ at a base point, x^k; at the new trial point, $x^k + \alpha p^k$; and the slope at the base point $d\phi_q(0)/d\alpha = -2\phi(x^k)$. The minimization of $\phi_q(\alpha)$ can be done analytically (see exercise 7) and leads to a new value for α (shown as α_q) which, with appropriate safeguards, lies in the desired range. Based on these properties, we now state the Armijo line search algorithm. This requires the following substitution for step 3 in the Newton algorithm given above.

Armijo Line Search Method

a. Set $\alpha = 1$.
b. Evaluate $\phi(x^k + \alpha p^k)$.
c. If $\phi(x^k + \alpha p^k) - \phi(x^k) \leq -2\delta\alpha\phi(x^k)$, the step size is found. Set $x^{k+1} = x^k + \alpha p^k$ and go to step 4 in the Newton algorithm. Otherwise, continue with step d.
d. Let $\lambda = \max\{\eta, \alpha_q\}$, where $\alpha_q = \alpha\phi(x^k)/((2\alpha - 1)\phi(x^k) + \phi(x^k + \alpha p^k))$ set $\alpha = \lambda\alpha$ and go to b.

Typically, both δ and η are set to 0.1. This procedure adds robustness and reliability to Newton's method, particularly if the starting point is poor. However, if a step size is not found after, say, five passes through this algorithm, the Newton direction, p^k, may be very poor due to ill-conditioning of the problem (i.e., $J(x^k)$ close to singular). This leads to a line search failure, and examination of the equations is usually indicated. In the extreme case, if $J(x^k)$ is singular, then the Newton step does not exist and failure occurs for the Newton algorithm. We will consider remedies for this condition next.

8.3.4 Treating Singularity of the Jacobian—Modifying the Newton Step

If the Jacobian is singular or nearly singular (and thus ill-conditioned), the Newton step is (nearly) orthogonal to the direction of steepest descent for $\phi(x)$. The direction of steepest descent is defined by $-\nabla\phi(x)$ and for a small step, this gives the greatest reduction in the function $\phi(x)$. As a result, we could consider a steepest descent instead of Newton direction when the Newton direction is poor. The steepest descent step is given by:

$$p^{sd} = -\nabla\phi(x^k) = -J(x^k)^T f(x^k) \qquad (8.34)$$

This step has a descent property but has only a linear rate of convergence, defined by:

$$\|x^k - x^*\| < \|x^{k-1} - x^*\| \qquad (8.35)$$

An advantage of the steepest descent methods is that as long as $p^{sd} = -J(x^k)^T f(x^k) \neq 0$, an improved point will be found, even if J is singular. However, the performance of this method can be very slow.

As a compromise we consider methods where we combine the steepest descent and Newton directions. Two of these strategies are the *Levenberg-Marquardt method* and the *Powell dogleg method*. In the former method, we combine both steps and solve the following linear system to get the search direction:

$$(J(x^k)^T J(x^k) + \lambda I) p^k = -J(x^k)^T f(x^k) \qquad (8.36)$$

where λ is a scalar nonnegative parameter that adjusts direction and length of step. For $\lambda = 0$, we obtain Newton's method directly:

$$p^k = -(J(x^k)^T J(x^k))^{-1} J(x^k)^T f(x^k) = -J(x^k)^{-1} f(x^k). \qquad (8.37)$$

On the other hand if λ becomes large and dominates $J(x^k)^T J(x^k)$, the system of equations approaches:

$$p^k = -(\lambda I)^{-1} J(x^k)^T f(x^k) = -(J^k)^T f(x^k)/\lambda, \qquad (8.38)$$

which is the steepest descent step with a very small step size. With an intermediate value of λ, we obtain a search direction that lies on the arc between the steepest descent and the Newton steps, as shown in the left side of Figure 8.11.

A disadvantage of the Levenberg-Marquardt method is that a different linear system must be solved every time that λ is changed. This can be expensive, as the algorithm may require several guesses for λ before choosing an appropriate step. Instead, we consider an algorithm that uses a combination of the Newton and steepest descent steps and chooses a search direction between them automatically. This *dogleg method* is illustrated on the

Sec. 8.3 Methods for Solving Nonlinear Equations

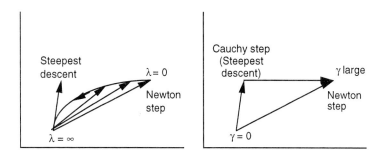

FIGURE 8.11 Comparison of Levenberg-Marquardt and dogleg steps.

right in Figure 8.11 and was developed by Powell. Here the largest step is the Newton step and smaller steps follow a linear combination of the steepest descent and Newton steps. For steps that are smaller than the given steepest descent step, the steepest descent direction is still retained.

To develop this method we first need to find the proper length (given by the scalar, β) along the steepest descent direction,

$$p^{sd} = -J^T f(x^k). \tag{8.39}$$

For this we consider the minimization of a quadratic model function formed from the linearized equations along the steepest descent step:

$$\text{Min}_\beta \; 1/2 \, (f(x^k) + \beta \, J \, p^{sd})^T (f(x^k) + \beta \, J \, p^{sd}) \tag{8.40}$$

Substituting the definition for the steepest descent direction, we have:

$$\beta = [f(x^k)^T J \, J^T f(x^k)] / [f(x^k)^T J \, (J^T J) \, J^T f(x^k)] = \|p^{sd}\|^2 / \|J \, p^{sd}\|^2 \tag{8.41}$$

The step $\beta \, p^{sd}$ is known as the *Cauchy step*, and it can be shown that the length of this step is never greater than the length of the Newton step: $p^N = -J^{-1} f(x^k)$. For a desired steplength γ, of the overall step, we can calculate the search direction for the *Powell dogleg method* as follows. Here if we wish to adjust the steplength γ automatically, the search direction, p, can be determined according to:

- for $\gamma \leq \beta \, \|p^{sd}\|$, $p = \gamma \, p^{sd} / \|p^{sd}\|$
- for $\gamma \geq \|p^N\|$, $p = p^N$
- for $\|p^N\| > \gamma > \beta \, \|p^{sd}\|$, $p = \eta \, p^N + (1-\eta) \beta \, p^{sd}$ where
  $\eta = (\gamma - \beta \, \|p^{sd}\|) / (\|p^N\| - \beta \, \|p^{sd}\|)$

Note that if the allowable steplength, γ, is small, we choose the steepest descent direction; if it is large, we choose a Newton step. For values of γ between the Newton and Cauchy steps, however, we choose a linear combination of these steps as seen in Figure 8.11. Since this approach requires only two predetermined directions and simple stepsize determinations, it is much less expensive than the Levenberg-Marquardt method. Moreover, in cases where the Jacobian is ill-conditioned, the Newton step becomes very large and this method simply defaults to taking Cauchy steps with steplengths of γ.

Finally, it should be noted that both Levenberg-Marquardt and dogleg approaches fall into a general class of algorithms known as *trust region methods*. For these problems, the steplength γ corresponds to the size of the region around x^k for which we *trust* the quadratic model in p (based on a linearization of $f(x)$, i.e., $1/2(f(x^k)+J\,p)^T(f(x^k)+J\,p))$ to be an accurate representation of $\phi(x)$. An approximate minimization of this quadratic model requires an adjustment of either λ or η at each iteration by the Levenberg-Marquardt or the dogleg methods, respectively. While trust region methods can be more expensive than the Armijo line search strategy, they have much stronger convergence characteristics, particularly for problems that are ill-conditioned.

8.3.5 Treating Singularity of the Jacobian—Continuation Methods

For singular or severely ill-conditioned Jacobians, we can also consider the class of continuation methods. Unlike trust region methods, we do not attempt to solve the equations by driving $f(x)$ to zero. Instead, we evaluate the functions at some initial guess, $f(x_0)$ and then solve a simpler problem, say: $f(x) - 0.9\,f(x_0) = 0$. We hope that this will not require x to change very much and our equation solver (say Newton's method) will not have difficulty solving this problem. If we succeed in solving this modified problem with 0.9, we reduce this *continuation parameter* to 0.8 and repeat, finally reducing it to 0, at which point we have solved our original equation. One can see two issues here for this approach:

- How fast can one reduce the continuation parameter?
- How much more expensive is this method than those the approaches developed above?

Our use of a fixed parameter is a form of the algebraic continuation method. There are several modifications to this method that include switching the continuation parameter with a variable upon encountering a singular Jacobian. Replacement with this parameter can lead to a nonsingular Jacobian and this increases the likelihood of success on more difficult problems, but not without an increase in computational cost.

8.3.6 Methods That Do Not Require Derivatives

The methods we have considered thus far require the calculation of a Jacobian matrix at each iteration. This is frequently the most time-consuming activity for some problems, especially if nested nonlinear procedures are used. A simple alternative to an exact calculation of the derivatives is to use a finite difference approximation, given by:

$$\left(\frac{\partial f}{\partial x_j}\right)_{x^k} = \frac{f(x^k + h e_j) - f(x^k)}{h} \tag{8.42}$$

Sec. 8.3 Methods for Solving Nonlinear Equations

where each element i of the vector e_j is given by: $(e_j)_i = 0$ if $i \neq j$ or $= 1$ if $i = j$, and h is a scalar normally chosen from 10^{-6} to 10^{-3}. This approach requires an additional n function evaluations/iteration.

On the other hand, we can also consider the class of *Quasi-Newton methods* where the Jacobian is approximated based on differences in x and $f(x)$, obtained from previous iterations. Here the motivation is to avoid evaluation (and decomposition) of the Jacobian matrix. The basis for this derivation can be seen by considering a single equation with a single variable, as shown in Figure 8.12.

If we apply Newton's method to the system starting from x^a, we obtain the new point x_c from the tangent to the curve at x_a, given by the thick line in Figure 8.12 and the relation:

$$\text{Newton step: } x_c = x_a - f(x^a)/f'(x^a) \tag{8.43}$$

where $f'(x)$ is the slope. If this derivative, $f'(x)$, is not readily available, we can approximate this term by a difference between two points, say x_a and x_b. From the thin line in Figure 8.12, the next point is given by x_d and this results from a secant that is drawn between x_a and x_b. The secant formula to obtain x_d is given by:

$$\text{Secant step: } x_d = x_a - f(x^a) \left[\frac{x_b - x_a}{f(x_b) - f(x_a)} \right] \tag{8.44}$$

Moreover, we can define a *secant relation* so that for some scalar, **B**, we have:

$$\boldsymbol{B}(x_b - x_a) = f(x_b) - f(x_a) \qquad x_d = x_a - \boldsymbol{B}^{-1} f(x^a) \tag{8.45}$$

For the multivariable case, we need to consider additional conditions to obtain a secant step. Here we define a matrix B that substitutes for the Jacobian matrix and again satisfies the secant relation, so that

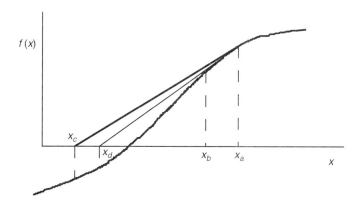

FIGURE 8.12 Comparison of Newton and secant methods for single equation.

$$B^{k+1}(x^{k+1} - x^k) = f(x^{k+1}) - f(x^k) \qquad (8.46)$$

and assuming that $f(x^{k+1}) \approx 0$, B^k can be substituted to calculate the change in x:

$$x^{k+1} = x^k - (B^k)^{-1} f(x^k) \qquad (8.47)$$

However, for the multivariable case, the secant relation alone is not enough to define B. Therefore, given a matrix B^k, we calculate the *least change* for B^{k+1} from B^k that satisfies the secant formula. This is a constrained minimization problem posed by Dennis and Schnabel (1983) and it can be written as:

$$\begin{aligned} \text{Min} \quad & \left\| B^{k+1} - B^k \right\|_F \\ \text{s.t.} \quad & B^{k+1} s = y \end{aligned} \qquad (8.48)$$

where $y = f(x^{k+1}) - f(x^k)$, $s = x^{k+1} - x^k$ and $\|B\|_F$ is the Frobenius norm given by $[\sum_i \sum_j B_{ij}]^{1/2}$. This problem can be stated and solved more easily with scalar variables. Let $b_{ij} = (B^k)_{ij}$, $\bar{b}_{ij} = (B^{k+1})_{ij}$ and y_i and s_j be the elements of vectors y and s, respectively. Then we have

$$\begin{aligned} \text{Min} \quad & \sum_i \sum_j \left(\bar{b}_{ij} - b_{ij}\right)^2 \\ \text{s.t.} \quad & \sum_j \bar{b}_{ij} s_j = y_i \quad i = 1, \ldots n \end{aligned} \qquad (8.49)$$

and we would like to find the best values of $\bar{b}_{ij}$ that make up the elements of the updated matrix B^{k+1}. From the definition in Appendix A and as discussed later in Chapter 9, this problem can be shown to be strictly convex and has a unique minimum. Applying the concepts in Appendix A, we form the corresponding Lagrange function:

$$L = \sum_i \sum_j \left(\bar{b}_{ij} - b_{ij}\right)^2 + \sum_i \lambda_i \left(\sum_j \bar{b}_{ij} s_j - y_i\right) \qquad (8.50)$$

and the stationary conditions of this function are:

$$\partial L/\partial \bar{b}_{ij} = 2(\bar{b}_{ij} - b_{ij}) + \lambda_i s_j = 0 \implies \bar{b}_{ij} = b_{ij} - \lambda_i s_j/2 \qquad (8.51)$$

To find λ_i, we apply secant relation again:

$$\begin{aligned} y_i &= \sum_j \bar{b}_{ij} s_j = \sum_j b_{ij} s_j - \frac{\lambda_i}{2} \sum s_j^2 \\ \frac{\lambda_i}{2} &= \frac{\sum b_{ij} s_j - y_i}{\sum s_j^2} \end{aligned} \qquad (8.52)$$

Now, substituting for $\lambda_i/2$ into the stationary condition for $\bar{b}_{ij}$, and writing in matrix form leads to *Broyden's formula:*

Sec. 8.3   Methods for Solving Nonlinear Equations

$$B^{k+1} = B^k + \frac{(y - B^k s)s^T}{s^T s} \tag{8.53}$$

With this relation we can calculate the new search direction by solving the linear system:

$$B^{k+1} p^{k+1} = -f(x^{k+1})$$

directly. However, we can also calculate p^{k+1} explicitly by updating the *inverse* of B^{k+1} through a modification of Broyden's formula. Here we apply the Sherman Morrison Woodbury formula for a square matrix A with an update using vectors x and v:

$$(A + xv^T)^{-1} = A^{-1} - \frac{A^{-1} xv^T A^{-1}}{1 + v^T A^{-1} x} \tag{8.54}$$

Since the matrix xv^T has only one nonzero eigenvalue, it has a rank of one, and we term the relation $(A + xv^T)$ a *rank one update* to A. Now, by noting that

$$A = B^k \qquad A + xv^T = B^{k+1} \tag{8.55}$$

$$x = (y - B^k s)/s^T s \qquad v = s$$

after simplifying, we have for $H^k = (B^k)^{-1}$

$$H^{k+1} = H^k + \frac{(s - H^k y)s^T H^k}{s^T H^k y} \tag{8.56}$$

The Broyden algorithm can now be stated as follows:

1. Guess x^0 and B^0 (e.g., $= J^0$ or I) and calculate H^0 (e.g., $(J^0)^{-1}$).
2. If $k = 0$, go to 3, otherwise calculate $f(x^k)$, $y = f(x^k) - f(x^{k-1})$, $s = x^k - x^{k-1}$ and H^k or B^k from either (8.56) or (8.53)
3. Calculate the search direction by $p^k = -H^k f(x^k)$ or by solving $B^k p^k = -f(x^k)$.
4. If $\|p^k\| \le \varepsilon_1$, and $\|f(x^k)\| \le \varepsilon_2$ stop. Else, find a stepsize α and update the variables so that: $x^{k+1} = x^k + \alpha p^k$.
5. Set $k = k+1$, go to 2.

The Broyden method has been used widely in process simulation, especially when the number of equations is fairly small. For instance, this approach is used for inside-out flash calculations and for recycle convergence in flowsheets. The rank one update formulas for Broyden's method that approximate the Jacobian ensure fast convergence. In fact, this method converges superlinearly, as defined by:

$$\lim_{k \to \infty} \frac{\|x^{k+1} - x^*\|}{\|x^k - x^*\|} \to 0 \tag{8.57}$$

which is slower than Newton's method but significantly faster than steepest descent.

On the other hand, both H^k and B^k are generally dense matrices, although recent studies have considered specialized update formulas that take advantage of sparse structures. In addition, both matrices can become ill-conditioned (independently) through the rank one updates. To remedy this, a more stable procedure would be to update the factors that are formed from a matrix decomposition of B^k. In particular, Broyden update formulas have been developed for the LU factors or the QR factors of B^k (Dennis and Schnabel, 1983). Finally, there is no guarantee that the Broyden method generates a descent direction. As a result, the Armijo inequality may not hold even though line searches can be applied. In addition, variations of the trust region methods and the dogleg method have also been reported. However, many implementations in process engineering simply use full Broyden steps unless the residuals increase by a large amount.

To conclude the discussion of these methods, we present a small example on solving nonlinear equations. In addition, this will help to illustrate some of the steps used in constructing our algorithms.

EXAMPLE 8.1

Using Newton's method with an Armijo line search, solve the following system of equations:

$$f_1 = 2x_1^2 + x_2^2 - 6 = 0 \tag{8.58}$$

$$f_2 = x_1 + 2x_2 - 3.5 = 0 \tag{8.59}$$

1. We first consider the formulation of Newton's method from a starting point close to the solution. Here we expect very good performance and little difficulty with convergence. The Newton iteration is given by:

$$x^{k+1} = x^k - (J^k)^{-1} f(x^k) \tag{8.60}$$

and the Jacobian matrix and its inverse are given as:

$$J^k = \begin{bmatrix} 4x_1 & 2x_2 \\ 1 & 2 \end{bmatrix} \quad J^{-1} = (8x_1 - 2x_2)^{-1} \begin{bmatrix} 2 & -2x_2 \\ -1 & 4x_1 \end{bmatrix} \tag{8.61}$$

Multiplying these matrices in the Newton iteration leads to the following recurrence relations,

$$\begin{aligned} x_1^{k+1} &= x_1^k + \alpha^k p_1 \\ x_2^{k+1} &= x_2^k + \alpha^k p_2 \\ p_1 &= -\{(2f_1(x^k) - 2x_2^k f_2(x^k))/(8x_1^k - 2x_2^k)\} \\ p_2 &= -\{(-f_1(x^k) + 4x_1^k f_2(x^k))/(8x_1^k - 2x_2^k)\} \end{aligned} \tag{8.62}$$

Here the stepsize, α^k, at each iteration, k, is determined by the Armijo line search. Starting from $x^0 = [2., 1.]^T$, we obtain the following values for x^k and we see that the constraint violations quickly reduce with full steps. The problem is essentially converged after three iterations with a solution of $x_1^* = 1.59586$ and $x_2^* = 0.95206$. Note that because we start reasonably close to the solution, $\alpha^k = 1$ for all of the steps.

Sec. 8.3   Methods for Solving Nonlinear Equations

k	x_1^k	x_2^k	ϕ^k	α^k
0	2.00000	1.00000	4.6250	1.0000
1	1.64285	0.92857	$3.3853 \cdot 10^{-2}$	1.0000
2	1.59674	0.95162	$1.1444 \cdot 10^{-5}$	1.0000
3	1.59586	0.95206	$1.5194 \cdot 10^{-12}$	1.0000

2. On the other hand, if we start from $x_1^0 = x_2^0 = 0$, the Jacobian matrix, J^k, is singular and the Newton step is not defined. Instead we generate a steepest descent or Cauchy step based on the description above. At this starting point we have:

$$[f_1(x^0) \, f_2(x^0)]^T = [-6, -3.5]^T \tag{8.63}$$

and the steepest descent step is given by:

$$p^{sd} = -j_k^T f(x^k) = [3.5, 7]^T \tag{8.64}$$

Also, the stepsize that is based on minimization of a quadratic model is given by:

$$\beta = \|p^{sd}\|^2 / \|J \, p^{sd}\|^2 = 0.1789 \tag{8.65}$$

and we therefore obtain the next point:

$$x^1 = x^0 + \beta p^{sd} = [0.6263, 1.2527]^T \tag{8.66}$$

From x^1 we can apply Newton's method with an Armijo line search and we obtain the following values for x^k and the step sizes for convergence. The problem is essentially converged after four iterations.

k	x_1^k	x_2^k	ϕ^k	α^k
0	0.62630	1.25270	6.71535	0.10000
1	0.88058	1.14397	4.98623	0.54801
2	1.51683	0.91667	0.16698	1.00000
3	1.59853	0.95073	$1.05270 \cdot 10^{-4}$	1.00000
4	1.59586	0.95206	$1.27799 \cdot 10^{-10}$	1.00000
5	1.59586	0.95206	$1.90022 \cdot 10^{-22}$	1.00000

There are a number of excellent library codes (e.g., IMSL library, NaG library, Harwell library) that incorporate these strategies and are very reliable and efficient for nonlinear equation solving. For instance, the MINPACK codes from Netlib combine the above concepts within a family of excellent trust region methods. These codes are highly recommended for solving moderate-sized nonlinear systems of equations.

8.3.7 First-Order Methods

We conclude this section with a brief presentation of first-order methods. These methods do not evaluate or approximate the Jacobian matrix and are much simpler in structure. On the other hand, convergence is only at a linear rate, and this can be very slow. We develop these methods in a fixed point form: $x = g(x)$, where x and $g(x)$ are vectors of n stream variables. These methods are most commonly used to converge recycle streams, and here x represents a guessed tear stream and $g(x)$ is the calculated value after executing the units around the flowsheet.

8.3.8 Direct Substitution Methods

The simplest fixed point method is *direct substitution*. Here we define $x^{k+1} = g(x^k)$ with an initial guess x^0. The convergence properties for the n dimensional case can be derived from the contraction mapping theorem (see Dennis and Schnabel, 1983; p. 93). For the fixed point function, consider the Taylor series expansion:

$$g(x^k) = g(x^{k-1}) + \left(\frac{\partial g}{\partial x}\right)^T_{x^{k-1}} (x^k - x^{k-1}) + \ldots \tag{8.67}$$

and if we assume that $\partial g/\partial x$ doesn't vanish, it is the dominant term near the solution, x^*. We also assume it is fairly constant near x^*, then:

$$x^{k+1} - x^k = g(x^k) - g(x^{k-1}) = \left(\frac{\partial g}{\partial x}\right)^T_{x^{k-1}} (x^k - x^{k-1}) \tag{8.68}$$

and for

$$x^{k+1} - x^k = \Delta x^{k+1} = \Gamma \Delta x^k \quad \text{with} \quad \Gamma = \left(\frac{\partial g}{\partial x}\right)^T \tag{8.69}$$

we can write the normed expressions:

$$\| \Delta x^{k+1} \| \leq \| \Gamma \| \, \| \Delta x^k \|. \tag{8.70}$$

From this expression we can show a linear convergence rate, but the speed of these iterations is related to $\|\Gamma\|$. If we use the Euclidean norm, then $\|\Gamma\| = |\lambda|^{max}$, which is the largest eigenvalue of Γ in magnitude. Now by recurring the iterations for k we can develop the following relation:

$$\| \Delta x^k \| \leq (|\lambda|^{max})^k \| \Delta x^0 \|. \tag{8.71}$$

and a necessary and sufficient condition for convergence is that $|\lambda|^{max} < 1$. This relation is known as a *contraction mapping* if $|\lambda|^{max} < 1$. Moreover, the speed of convergence depends on how close $|\lambda|^{max}$ is to zero. Here we can estimate the number of iterations (n_{iter}) to reach $\| \Delta x^n \| \leq \delta$ (some zero tolerance), from the relation:

$$n_{iter} \geq ln[\delta/\| \Delta x^0 \|]/ln \, |\lambda|^{max} \tag{8.72}$$

For example, if $\delta = 10^{-4}$ and $\|\Delta x^0\| = 1$, we have the following iteration counts, for:

Sec. 8.3   Methods for Solving Nonlinear Equations   **269**

$$|\lambda|^{max} = 0.1, n = 4$$
$$|\lambda|^{max} = 0.5, n = 14 \qquad (8.73)$$
$$|\lambda|^{max} = 0.99, n = 916$$

8.3.9 Relaxation (Acceleration) Methods

For problems where $|\lambda|^{max}$ is close to one, direct substitution is limited and converges slowly. Instead, we can alter the fixed point function $g(x)$ so that it reduces $|\lambda|^{max}$. The general idea is to modify the fixed point function to:

$$x^{k+1} = h(x^k) \equiv \omega\, g(x^k) + (1 - \omega)\, x^k \qquad (8.74)$$

where ω is chosen adaptively depending on the changes in x and $g(x)$. The two more common fixed point methods for recycle convergence are the *dominant eigenvalue* (Orbach and Crowe, 1971) method and the *Wegstein* (1958) *iteration* .

In the *dominant eigenvalue method (DEM)* we obtain an estimate of $|\lambda|^{max}$ by monitoring the ratio:

$$|\lambda|^{max} \approx \frac{\|\Delta x^k\|}{\|\Delta x^{k-1}\|} \qquad (8.75)$$

after, say, 5 iterations. Now from the transformation of the fixed point equation, we have:

$$\Delta x^{k+1} = x^{k+1} - x^k = h(x^k) - h(x^{k-1}) \approx \frac{\partial h}{\partial x}(x^k - x^{k-1}) = \Phi(x^k - x^{k-1}) \qquad (8.76)$$

where $\Phi = \partial h/\partial x = \omega \Gamma + (1 - \omega)\, I$. We now choose the relaxation factor ω to minimize $|\lambda|^{max}$ for Φ. Note that if ω is one, we have direct substitution, for $0 < \omega < 1$ we have an interpolation or damping and for $\omega > 1$, we have an extrapolation. To choose an optimum value for ω, we consider the largest eigenvalue for Φ, given by:

$$\det(\Phi - \theta I) = 0 \qquad (8.77)$$

Substituting for Φ gives the relation:

$$\det[\omega(\Gamma - (\omega - 1 + \theta)/\omega\, I)] = 0 \qquad (8.78)$$

From this expression, we note that $(\omega - 1 + \theta)/\omega$ corresponds to the eigenvalue of Γ and so we have: $\theta = 1 + \omega(\lambda - 1)$. To find $|\theta|^{max}$, we note that this value is determined by the largest and smallest eigenvalues for Γ as well as the relaxation factor. In fact, if we plot $|\theta|^{max}$ by ω, one can show that the optimum $\omega*$ occurs when

$$(1 + \omega(\lambda^{min} - 1))^2 = (1 + \omega(\lambda^{max} - 1))^2 \rightarrow \omega* = 2/(2 - \lambda^{max} - \lambda^{min}). \qquad (8.79)$$

While λ^{max} can be estimated from changes in x, λ^{min} is not easy to estimate, and for DEM we make an important assumption. If we assume that $\lambda^{max}, \lambda^{min} > 0$ and that $\lambda^{min} \approx \lambda^{max}$. We have:

$$\omega* = 1/(1 - \lambda^{max}) \qquad (8.80)$$

Note that if this assumption is violated and the minimum and maximum eigenvalues of Φ are far apart, DEM may not converge. This approach has also been extended to the *generalized dominant eigenvalue method* (GDEM) (Crowe and Nishio, 1975) where several eigenvalues are estimated and are used to determine the next step. While GDEM is a more complex algorithm, it overcomes the assumption that $\lambda^{\min} \approx \lambda^{\max}$.

On the other hand, the *Wegstein method* obtains the relaxation factor by applying a secant method independently to each component of x. From above we have for component x_i:

$$x_i^{k+1} = x_i^k - f_i(x^k)[x_i^k - x_i^{k-1}]/[f_i(x^k) - f_i(x^{k-1})] \qquad (8.81)$$

Now, by defining $f_i(x^k) = x_i^k - g_i(x^k)$ and $s_i = [g_i(x^k) - g_i(x^{k-1})]/[x_i^k - x_i^{k-1}]$, we have:

$$\begin{aligned}
x_i^{k+1} &= x_i^k - f_i(x^k)[x_i^k - x_i^{k-1}]/[f_i(x^k) - f_i(x^{k-1})] \\
&= x_i^k - \{x_i^k - g_i(x^k)\}[x_i^k - x_i^{k-1}]/[x_i^k - g_i(x^k) - x_i^{k-1} + g_i(x^{k-1})] \\
&= x_i^k - \{x_i^k - g_i(x^k)\}[x_i^k - x_i^{k-1}]/[x_i^k - x_i^{k-1} + g_i(x^{k-1}) - g_i(x^k)] \qquad (8.82) \\
&= x_i^k - \{x_i^k - g_i(x^k)\}/[1 - s_i] \\
&= \omega_i\, g(x^k)_i + (1 - \omega_i)\, x_i^k
\end{aligned}$$

where $\omega_i = 1/[1 - s_i]$. This approach works well on flowsheets where the components do not interact strongly (e.g., single recycle without reactors). On the other hand, interacting recycle loops and components can cause difficulties for this method.

To ensure stable performance, the relaxation factors for both DEM and the Wegstein method are normally bounded and safeguarded so that large extrapolations are avoided. The algorithm for fixed point methods can be summarized by:

1. Start with x^0 and $g(x^0)$.
2. Execute a fixed number of direct substitution iterations (usually 2 to 5) and check convergence at each iteration.
3. *Dominant eigenvalue method:* Apply the acceleration (8.80) with a bounded value of ω to find the next point and go to 2.

   Wegstein: Apply the acceleration (8.82) with a bounded value of ω_i to find the next point. Iterate until convergence.

To conclude this section, we illustrate the application of first order fixed point methods with a small example. In particular, we are interested in the method's performance and in the estimation of a convergence rate.

EXAMPLE 8.2

Solve the fixed point problem given by:

$$x_1 = 1 - 0.5 \exp(0.7(1 - x_2) - 1)$$

$$x_2 = 2 - 0.3 \exp(0.5(x_1 + x_2))$$

Sec. 8.4   Recycle Partitioning and Tearing

using a direct substitution method, starting from $x_1 = 0.8$, and $x_2 = 0.8$. Estimate the maximum eigenvalue based on the sequence of iterates.

Using direct substitution, $x^{k+1} = g(x^k)$, we obtain the following iterates:

k	x_1^k	x_2^k
0	0.8	0.8
1	0.7884	0.3323
2	0.8542	1.3376
3	1.8325	1.1894
4	1.8389	1.1755
5	1.8373	1.1786
6	1.8376	1.1780

and this method converges to $x_1 = 0.8376$ and $x_2 = 1.1781$ in 6 iterations with $\|\Delta x_k\| < 10^{-3}$. From these iterates, we can estimate the maximum eigenvalue from:

$$|\lambda|^{max} = \|x^5 - x^4\| / \|x^4 - x^3\| = 0.226 \tag{8.83}$$

Also, from $\|\Delta x^5\| = 0.00346$ and $\delta = 10^{-3}$, we can estimate the number of iterations required for direct substitution as:

$$n_{iter} = ln(\delta / \|\Delta x^5\|) / ln|\lambda|^{max} = 1 \tag{8.84}$$

The fixed point methods developed for recycle convergence are strongly influenced by the structure of the flowsheet and the choice of the tear streams. In the next section, we will analyze their selection and briefly highlight some popular criteria for flowsheet tearing.

8.4   RECYCLE PARTITIONING AND TEARING

We will investigate three issues in this section: *partitioning, precedence ordering,* and *tearing*. We shall define these concepts by applying them to an example flowsheet found in the literature (Leesley, 1982, p. 624) as shown in Figure 8.13. We would like to solve this as efficiently as possible. Note that the units A, B, C, D, and E are in a recycle loop and will certainly have to be computed together. With a still closer look, we see we must add units F and G to this group. It appears that this group of seven units can be solved first as we see no streams recycling from later units back to any of these units. The units that we have to solve as a group are called partitions and finding these groups is called *partitioning*, while the order we must solve them is *precedence ordering*. The grouping is unique although the ordering may not be unique and depends on the particular flowsheet.

This example is simple enough that we would have little trouble seeing the partitions and the ordering for them. However, some flowsheets have hundreds of units in

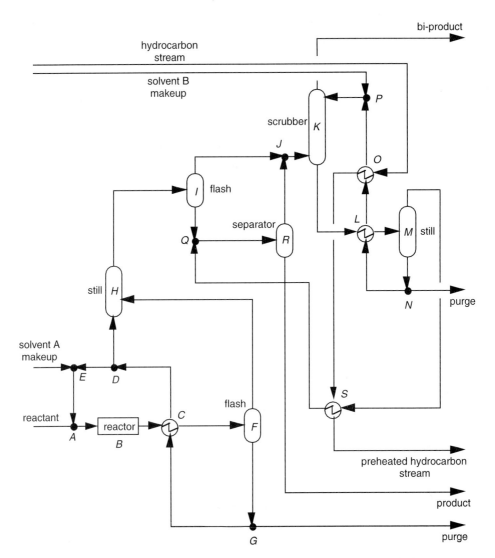

FIGURE 8.13 Example flowsheet for partitioning and precedence ordering.

them, and it is difficult to find the partitions in them and the ordering for those partitions. Fortunately a simple algorithm exists to find the partitions and precedence ordering (Sargent and Westerberg, 1964). Rather than presenting the algorithm, we apply it here and generalize from this example.

We may start with any unit, e.g., unit I, and put it on a list called *list 1*.

List 1: I

Sec. 8.4   Recycle Partitioning and Tearing   273

We extend list 1 by tracing output streams starting with the last object and continue until we find a unit repeating or until there is no other output to trace. This leads to the following trace for list 1.

$$\text{List 1: } IJKLMNL$$

We discovered this sequence by noting that I has an output to J, which has an output to K, which has an output to L, and so on. However, unit L repeats in the sequence and there is a loop that traces from L to M to N to L. Therefore, these units must be in a group and we merge them and treat them as a single entry on List 1: $IJK\{LMN\}$.

We continue tracing the output paths and obtain:

$$\text{List 1: } IJK\{LMN\}OPK$$

Again, we observe a repeating unit in unit K and there is a loop from K to group $\{LMN\}$ through O and P to K. Grouping the units in this loop, leads to the following list:

$$\text{List 1: } IJ\{KLMNOP\}$$

We continue tracing these outputs to obtain:

$$\text{List 1: } IJ\{KLMNOP\}SQRJ$$

Here unit J repeats, giving

$$\text{List 1: } I\{JKLMNOPSQR\}$$

When we try to continue tracing outputs, we discover that the units in the last group have no streams leaving from them to other units in the flowsheet. We remove this group from list 1 and place it on list 2.

$$\text{List 2: } \{JKLMNOPSQR\}$$

We cross off all these units from the flowsheet. We are done analyzing them. Returning to list 1

$$\text{List 1: } I$$

we look for more outputs from unit I. None exist that do not go to units removed from the flowsheet already. We remove unit I from list 1, place it at the *head* of list 2, and cross it off the flowsheet.

$$\text{List 2: } I\{JKLMNOPSQR\}$$
$$\text{List 1: }$$

List 1 is empty. Pick any remaining unit in the flowsheet and place it onto list 1, say unit F.

$$\text{List 1: } F$$

Tracing the outputs we get

$$\text{List 1: } FH$$

and we stop, as *H* has no outputs except to units we already crossed out (and put onto list 2). We therefore remove *H* from list 1 and place it at the head of list 2.

$$\text{List 2: } HI\{JKLMNOPSQR\}$$
$$\text{List 1: } F$$

Start tracing from *F* again we get:

$$\text{List 1: } FGCDEABC$$

and unit *C* repeats. Grouping it with the units between its two occurrences leads to:

$$\text{List 1: } FG\{CDEAB\}$$

We continue to trace and obtain:

$$\text{List 1: } FG\{CDEAB\}F$$

and by grouping *F* and *G* with the other units

$$\text{List 1: } \{FGCDEAB\}$$

we find there are no other outputs to trace. We now remove this last group from List 1, place it at the head of list 2, and remove these units from the flowsheet.

$$\text{List 2: } \{FGCDEAB\}HI\{JKLMNOPSQR\}$$

$$\text{List 1:}$$

There are no more units to place in list 1 so we are done. List 2 is our list of partitions in a precedence ordering. We can first solve the partition $\{FGCDEAB\}$, then unit *H*, then unit *I*, and finally the remaining partition $\{JKLMNOPSQR\}$.

This algorithm works no matter which unit we start with on list 1. It gives a unique set of partitions—that is, the units grouped together. However, the precedence order among the partitions may not always be unique, although it is in this case.

8.4.1 Tearing

The next issue is how we might solve each of the partitions containing more than a single unit. We had two such partitions in the problem in Figure 8.13. The first partition is relatively simple, and we leave it as an exercise. Instead, we illustrate an approach to tearing by examining the second, larger partition and repeat this flowsheet partition in Figure 8.14.

We see a number of units in this part of the flowsheet for which a single stream enters and a single stream leaves. We remove these units in Figure 8.14 as they add nothing to the topology of the underlying network. Finally, we straighten out the lines and redraw it as Figure 8.15, and we label the streams in Figure 8.16.

Comparing Figures 8.15 and 8.16, we see that if we were to choose to tear stream 8 (the connection between units *S* and *K*) we could tear any one of the actual streams along the path between those two units. For small problems like these, a good tear set can be

Sec. 8.4 Recycle Partitioning and Tearing

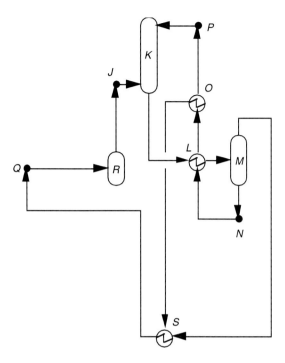

FIGURE 8.14 The second partition for the flowsheet in Figure 8.13. All streams and units outside this partition are removed.

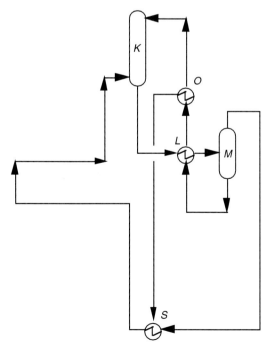

FIGURE 8.15 A reduction of the second partition shown in Figure 8.14. This reduction is formed by removing all units that have a single input/single output stream.

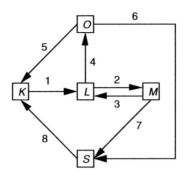

FIGURE 8.16 The underlying topology for the partition in Figure 8.15. The streams are now labeled to aid our analysis of this partition for tearing.

seen by inspection. As the partitioned subsets get bigger, a systematic procedure needs to be applied. The choice of a good tear set is also important because the performance of fixed point algorithms is greatly affected by the choice of tear stream.

Now, we consider a single general tearing approach and use this to place other popular tearing approaches into perspective. The approach treats tear set selection as an optimization problem with binary (0–1) variables or as an integer program. This particular integer programming formulation, devised by Pho and Lapidus (1973), is known as a *set covering problem,* and it allows for considerable flexibility in selecting desirable tear sets. Moreover, the integer programming formulation allows us to interpret a wide range of methods based on graph theory in a more compact way. We therefore treat the selection of tear streams as a minimization problem (e.g., minimize the number of tear streams or tear variables) subject to the constraint that all recycle loops must be broken at least once. Before formulating this problem we first need to identify all of the process loops in order to formulate the contraints. Again, this will be presented through an example, rather than through the formal statement of an algorithm.

EXAMPLE 8.3 Loop Finding

Consider the flowsheet partition in Figure 8.16. We now start with any unit in the partition, for example, unit K.

$$K - (1) \longrightarrow L - (2) \longrightarrow M - (3) \longrightarrow L \qquad (8.85)$$

We note that unit L repeats and the two streams, 2 and 3, which connect the two appearances of unit L are placed on a list of loops, List 3.

$$\text{List 3: } \{2,3\}$$

We then start with the unit just before the repeated one and trace any alternate paths from it.

$$K - (1) \longrightarrow L - (2) \longrightarrow M - (3) \longrightarrow L$$
$$- (7) \longrightarrow S - (8) \longrightarrow K \qquad (8.86)$$

Now K repeats and we place streams $\{1,2,7,8\}$ on the list of loops.

$$\text{List 3: } \{2,3\}, \{1,2,7,8\}$$

Sec. 8.4   Recycle Partitioning and Tearing

If we back up to S and look for an alternate path leaving from it, we find there is none. If we back up to unit M, we again there is no additional unexplored paths. On the other hand, if we back up to L we find another path and this is given by:.

$$K - (1) \longrightarrow L - (2) \longrightarrow M - (3) \longrightarrow L$$
$$| \qquad\qquad |$$
$$| \qquad\qquad -(7) \longrightarrow S - (8) \longrightarrow K \qquad (8.87)$$
$$|$$
$$- (4) \longrightarrow O - (5) \longrightarrow K$$

Here K repeats and we place $\{1,4,5\}$ on the list of loops.

$$\text{List 3: } \{2,3\}, \{1,2,7,8\}, \{1,4,5\}$$

Now if we back up to unit O on the last branch we can identify alternate paths which include:

$$K - (1) \longrightarrow L - (2) \longrightarrow M - (3) \longrightarrow L$$
$$| \qquad\qquad |$$
$$| \qquad\qquad -(7) \longrightarrow S - (8) \longrightarrow K \qquad (8.88)$$
$$|$$
$$- (4) \longrightarrow O - (5) \longrightarrow K$$
$$|$$
$$-(6) \longrightarrow S - (8) \longrightarrow K$$

Again K repeats and we place $\{1,4,6,8\}$ on the list of loops.

$$\text{List 3: } \{2,3\}, \{1,2,7,8\}, \{1,4,5\}, \{1,4,6,8\}$$

Returning to S, to O, to L, and finally to K, we find that none of these units have any alternate paths emanating from them. Since we have returned to the first unit on the list, we are done and there are four loops for this partition. These are listed in a loop incidence array as shown in Table 8.1.

TABLE 8.1 Loop Incidence Array for Partition

	Stream							
Loop	1	2	3	4	5	6	7	8
1		x	x					
2	x	x					x	x
3	x			x	x			
4	x			x		x		x

The loop incidence array (e.g., Table 8.1) is used to initialize a loop matrix, A, with elements:

$$a_{ij} = 1 \text{ if stream } j \text{ is in loop } i$$
$$= 0 \text{ otherwise}$$

The structure of this matrix is identical to the loop incidence array. We define the selection of tear streams through an integer variable, y_j, for each stream j: optimal values of these variables determine:

$y_j = 1$ if stream j is a tear stream

$\quad\; = 0$ otherwise

To ensure that each recycle loop is broken at least once by the tear streams, we write the following constraints for each loop i.

$$\sum_{j=1}^{n} a_{ij} y_j \geq 1 \quad i = 1, L \tag{8.89}$$

where L is the number of loops and n is the number of streams. Once we have the loop equations, we formulate a cost function for tear set selection:

$$\Sigma_j w_j y_j \tag{8.90}$$

and we assign a weight w_j to the cost of tearing stream j. This cost is frequently dictated by the type of recycle convergence problem. Three popular choices for weights are:

- Choose $w_j = 1$ and weight all streams equally so that we minimize the number of tear streams. This approach leads to many tear set candidates. This choice is the most common case and is the objective posed by Barkeley and Motard (1972).
- Choose $w_j = n_j$ where n_j is the number of variables in the jth tear stream. This is the objective chosen by Christensen and Rudd (1969).
- Choose $w_j = \Sigma_i a_{ij}$. If we sum over the loop constraints, we obtain coefficients that indicate the number of loops that are broken by the tear stream j. Breaking a loop more than once causes a "delay" in the tear variable iteration for the fixed point algorithms and much poorer performance. By minimizing the number of multiply broken loops we seek a *nonredundant set* of tear equations for better performance. This is the objective chosen by Upadhye and Grens (1975) and Westerberg and Motard (1981).

The set covering problem is given by:

$$\begin{aligned} \underset{y_j}{\text{Min}} & \sum_{j=1}^{n} w_j y_j \\ \text{s.t.} & \sum_{j=1}^{n} a_{ij} y_j \geq 1 \quad i = 1, L \\ & y_j = \{0, 1\} \end{aligned} \tag{8.91}$$

Solution to this integer problem is combinatorial and an upper bound on the number of alternatives is 2^n cases. However, simple reduction rules can make this problem and the resulting solution effort much smaller. We apply these rules (Garfinkel and Nemhauser, 1972) to the set covering problem and then search among the remaining integer variables

Sec. 8.4 Recycle Partitioning and Tearing

that are left. To facilitate the solution, the most common approach is a branch and bound search (see Chapter 15) although more efficient algorithms have been specialized to this problem.

We define r_i as the row vector i of matrix A and c_j as the column vector j of A. The following properties can be used to reduce the problem size.

- If r_i has only a single nonzero element, $(r_i)_k$, set $y_k = 1$ and choose k as tear stream. Delete this row and column, as it is a self-loop.
- If row k dominates row ℓ (all of the instances in row ℓ are also in row k), then delete r_k (a tear stream for r_ℓ automatically satisfies r_k.) This is a covered loop.
- If c_k dominates c_j and $w_k \leq w_j$ or for some set of columns, S, $\Sigma_{k \in S} c_k$ dominates c_j and $\Sigma_{k \in S} w_k \leq w_j$, then delete column j, as y_k will always contain the optimal solution.

These rules are applied systematically to reduce the loop matrix. If these rules offer no further improvement, then we need to initiate a combinatorial search on the remaining tear streams. It should also be noted that the optimal solutions generated by this reduction and search procedure are not unique if the inequalities for w_j are not strict. Consequently, this approach will find an optimal tear set but other solutions may work equally well.

EXAMPLE 8.4   Stream Tearing

We now consider the flowsheet partition from Example 8.3. From Table 8.1, we obtain the loop matrix directly as shown in Table 8.2. For this problem we consider two cases:

1. Minimize the number of tear streams
2. Minimize the number of times the loops are torn

and use the above reduction properties for this matrix.

TABLE 8.2   Loop Matrix, A, for Flowsheet Partition in Figure 8.16

Loop	Stream							
	1	2	3	4	5	6	7	8
1		1	1					
2	1	1					1	1
3	1			1	1			
4	1			1		1		1

1. *Minimize the number of tear streams.* Here we specify all of the stream weights in the objective function as, $w_j = 1$. From Table 8.2, we see that no rows dominate and none can be removed yet. On the other hand,

- Column 2 dominates column 3
- Column 4 dominates columns 5 and 6
- Column 1 dominates columns 4, 7, and 8

Deleting columns 3, 4, 5, 6, 7, and 8 leads to the following reduced table:

	Stream	
Loop	1	2
1		1
2	1	1
3	1	
4	1	

Now since rows 2 and 4 dominate row 3, we delete rows 2 and 4 and obtain a minimal representation for this system.

	Stream	
Loop	1	2
1		1
3	1	

Both rows have only single elements and streams 1 and 2 need to be selected as tear streams to break these loops. This is the minimum number of tear streams. However, note that loop 2 with streams {1, 2, 7, 8} is torn twice. From the information in Figure 8.16, we would first guess stream 2 and then compute unit M. That gives us stream 3 and a guess for stream 1 allows us to compute unit L. Continuing around the flowsheet, we find the order for computing the units is MLOSK, as shown in Figure 8.17.

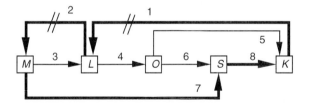

FIGURE 8.17 Double tearing a loop.

The impact of double tearing loop 2, which is highlighted by the thick lines in Figure 8.17, is now evident. To solve, we guess streams 1 and 2 and compute all the units once through in the order shown. The new value that unit L computes for stream 2 impacts the next computation for unit M, but this computation is based on the old value for stream 1. The new value for stream 1 will impact the next computation for L but not for unit M. In fact, it impacts unit L and downstream units when we compute through the units the second time. Its new value will not impact unit M until we compute that unit a third time. This delays the transfer of information around this loop by one pass through the unit computations and this slows the rate of convergence for successive substitution.

2. *Minimize the number of times the loops are torn.* Here the weights of the objective function are given by $w_j = \Sigma_i\, a_{ij}$, which is the column sum for each stream in the loop matrix. Because the cost coefficients are different, the row and column reductions for the previous prob-

Sec. 8.4 Recycle Partitioning and Tearing

lem do not apply. Instead, we make the following observations about the problem, presented in Table 8.3.

For this problem we note that again there are no dominating rows, but that combinations of columns dominate others. Here we note that

- Columns 3 and 7 dominate column 2
- Columns 6 and 7 dominate column 8
- Columns 5, 6, and 7 dominate column 1
- Columns 5 and 6 dominate column 4

TABLE 8.3 Loop Matrix, A, for Minimizing Number of Loop Tearings

Loop	Stream							
	1 $w_1 = 3$	2 $w_2 = 2$	3 $w_3 = 1$	4 $w_4 = 2$	5 $w_5 = 1$	6 $w_6 = 1$	7 $w_7 = 1$	8 $w_8 = 2$
1		1	1					
2	1	1					1	1
3	1			1	1			
4	1			1		1		1

And this leads to the reduced matrix:

Loop	Stream			
	3 $w_3 = 1$	5 $w_5 = 1$	6 $w_6 = 1$	7 $w_7 = 1$
1	1			
2				1
3		1		
4			1	

Since there is only a single element in each row, we have an optimal solution ($\Sigma_j w_j y_j = 4$) with streams 3, 5, 6, and 7 that tear each of the loops only once. Note however, that there are several optimal solutions to this problem. For instance, we can tear streams 1 and 3, which we see by inspection is also optimal. In fact, we can identify a family of optimal solutions given by {1,3}, {3,5,8}, {3,5,6,7}, {2,5,6}, {2,4}, {3,4,7}, which are all nonredundant.

Choosing tear set {1,3}, we see that we can compute unit L, which provides us with streams 2 and 4. We can compute units O and M in either order next, giving us streams 5, 6, and 7. That allows us to compute S and finally unit K. Figure 8.18 shows the partial ordering that characterizes this precedence ordering for these units. We can see why the order is not necessarily unique. It can be either the ordering LOMSK or the ordering LMOSK.

Suppose we choose to solve our flowsheet using successive substitution. We choose a tear set and guess values for each stream in it. Suppose we choose tear set {1,3}. We can then compute the units in the order LOMSK. We now have newly computed values for streams 1 and 3. We simply use them and start through the units again, repeating LMOSK until convergence.

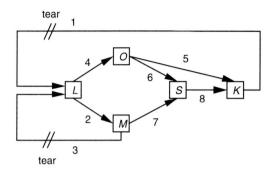

FIGURE 8.18 Precedence order for partition in Figure 8.1 when tearing streams 1 and 3.

It is interesting to note that these families of tear sets (for example, iterating with {1,3} or {3, 5, 6, 7} yields the same stream variable values if a direct substitution algorithm is applied. Both Upadhye and Grens (1975) and Westerberg and Motard (1981) developed graph theoretical algorithms to identify the family of nonredundant tear sets and therefore to generate tear streams that lead to faster convergence.

EFFECT OF TEARING STRATEGIES ON NEWTON-TYPE METHODS

Lastly, we consider the case where a Newton or quasi-Newton algorithm is applied to converge a modular flowsheet. In this case we form (or approximate) the Jacobian matrix for the tear stream equations. We rewrite these equations as:

$$x = g(x) \quad \text{or} \quad f(x) = x - g(x) = 0 \tag{8.92}$$

where x refers to the values of the tear streams and $g(x)$ refers to the calculated value after the loop units are calculated. These equations are then solved using Newton-Raphson or Broyden iterations applied to $f(x) = 0$.

An extreme approach to solving the recycle equations is to tear *all of the streams* in the recycle loops. Applied to the flowsheet partition in Figure 8.16, we form the equations for each. For example, for unit K, we have:

$$S1 = G(S5, S8) \quad \text{or} \quad F(S1, S5, S8) = S1 - G(S5, S8) = 0 \tag{8.93}$$

Here we define the vector SJ as the values for stream J and $G(*,*)$ represents the implicit functions that relate the output of a unit to its inputs. Writing similar equations around all of these units in Figure 8.16 leads to a system of stream equations. Linearizing this system leads to the equations that define the Newton step, given in Figure 8.19. As can be seen from the unit equations, the diagonal entries are the identity matrix while the off diagonal blocks refer to the Jacobians, $\partial G/\partial SI$, with respect to the input streams, SI.

To appreciate the effect of the tear set selection we note from Figure 8.16 that an unconverged tear stream J corresponds to $FJ \neq 0$. On the other hand, if a stream is a directly calculated output from a unit, then the corresponding right hand side is zero. Therefore, if *all* of the streams in Figure 8.16 were torn, all of the entries on the right hand side in Figure 8.19 would be nonzero vectors. On the other hand, if only streams 1 and 3 were torn, then only $F1$ and $F3$ would be nonzero, as shown in Figure 8.20.

Sec. 8.4 Recycle Partitioning and Tearing

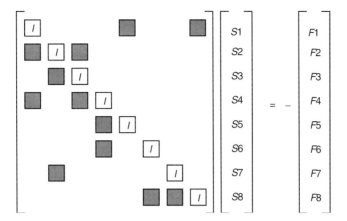

FIGURE 8.19 Linearized equations for flowsheet partition.

We now confine ourselves to the case where all units are *linear*. In this case, the first-order fixed point strategies are still affected by tear set selection and by the conditioning (and eigenvalues) of the unit matrices. On the other hand, for a linear system, Newton's method converges the flowsheet recycles in just one iteration, regardless of the location of nonzero elements on the right hand side. From this we can generalize an important observation:

> *As long as all of the recycle loops are torn, the choice of tear streams has little effect on the convergence rate of either the Broyden or Newton methods.*

In this case a reasonable criterion for tear set selection is motivated by rearranging the rows and columns in Figure 8.19 to reveal the structure of a recycle convergence strat-

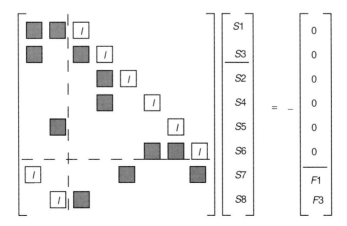

FIGURE 8.20 Linearized equations with $S1$ and $S3$ as tear streams.

egy. In the application of a Newton or Broyden method, the individual unit Jacobians may not be available directly. Instead, approximations to these are obtained from finite difference perturbation or from the quasi-Newton formula. Therefore, there is little need to retain the larger linear system of Figure 8.19.

Instead, we separate the tear streams and tear equations and permute the remaining stream variables and equations to block lower triangular form. For instance, if we choose $S1$ and $S3$ as tear streams and hold these fixed, it is easy to see that the diagonal streams can be calculated directly from streams that are determined from $S1$ and $S3$. Consequently, streams $S2$, $S4$, $S5$, $S6$, $S7$, and $S8$ are implicit functions of $S1$ and $S3$ and can be removed symbolically from this equation system, and this leads to a much smaller system of equations to solve with only $S1$ and $S3$ as stream variables. Since the Jacobian matrices are constructed by finite difference, an approach with fewer variables is easier to implement.

Therefore, we see that for Newton or Broyden methods, it is desirable to choose the minimum number of *stream variables* that breaks all recycle loops.

8.4.2 Decomposition for Equation-Oriented Simulation

Since equation-oriented simulation considers the entire set of flowsheet equations and adopts a simultaneous strategy for their solution, there would appear to be less of a need for analysis of the structure of the flowsheet. In fact, decomposition strategies are very much a part of this simulation mode, but these are introduced later during the equation solving stage. Here we recall that Newton's method was the most efficient and widely used method for equation solving. Moreover, several modifications could be introduced to ensure convergence over a wide range of nonlinear problems.

Now, as equation-oriented simulation problems become large, the dominant cost is the computation of the Newton step through solution of a set of linear equations:

$$J(x^k)\, p^k = -f(x^k)$$

For large-scale flowsheeting problems, we see from section 8.2 that the equations and the matrix J have a sparse structure. For problems with more than a few hundred variables, it is important to exploit this structure both for efficient decomposition of J and solution of the linear equations, and for storage of the decomposed matrix. Note that if the sparse structure is not exploited for a system of n equations, the number of matrix elements to be stored is n^2. Also, the computational effort to decompose these matrices is proportional to n^3. Consequently, even for relatively small systems of 1000 variables and equations, the computational resources can be very expensive. Instead, if we realize that most of these elements are zero (and the decomposition can be organized so that they remain zero during the solution process), then in many cases, both the storage and computational effort for calculating the Newton step can be made to increase only *linearly* with the problem size, at best.

There is a large literature devoted to sparse matrix methods, and their presentation and comparison is beyond the scope of this text (although references to further reading are given in the last section). Moreover, several excellent algorithms and software packages are widely available and easy to apply to process simulation problems. In general, these methods can be classified into *specialized* and *general* structures. In the former case, we

Sec. 8.5   Simulation Examples   285

refer to matrices that have a regular structure that does not change with problem size; examples include block banded matrices with nonzero elements clustered about the diagonal, almost block diagonal matrices, and matrices with a block bordered structure. Here restricted pivoting criteria can be applied and the creation and storage of matrix *fill-in* (new nonzero elements that are created as a result of pivoting and row elimination operations) are easier to analyze and manage. Decomposition of *general* structures requires an analysis of the structure and determination of a pivot sequence that reduces fill-in, conserves storage, and yields an efficient matrix decomposition. For these general methods, a number of heuristic pivoting strategies have been proposed and these are embodied in several general purpose sparse matrix routines.

As a result, the (Newton-based) algorithm for solution of nonlinear equations and the decomposition methods for sparse matrix decomposition of the linear system are the key features in an equation-oriented simulator. In the next section we will consider the application of both simulation modes to the Williams-Otto process described in section 8.2.

8.5 SIMULATION EXAMPLES

We now return to the Williams-Otto process described in section 8.2 and consider the solution of this example using the two simulation modes. We simulate this flowsheet for the following specifications:

$$F_1 = 6582 \text{ lb/h (all A)}$$
$$F_2 = 14{,}995.6 \text{ lb/h (all B)} \quad (8.94)$$
$$V = 1000. \text{ ft}^3$$
$$\eta = 0.1$$

Also the constants $\rho = 50$ lb/ft^3 and $T = 674°$R are given. Using the flowsheet reproduced below, we first apply the modular mode to the solution and then follow with a treatment with the equation-oriented mode.

8.5.1 Solution with Modular Mode

As mentioned above, we choose the reactor feed as the tear stream and solve the units according to the flowsheet topology table (reactor, heat exchanger, decanter, distillation column, splitter) where the output streams of each unit are calculated from the inputs. Figure 8.21 illustrates the process. Here we specify the feed flowrates F_1 and F_2 to the process, the volume of the reactor and the purge fraction to the splitter and we initialize the problem by guessing the flowrates for F_R and converge the flowsheet with a direct substitution approach:

$$F_R = g(F_R) \quad (8.95)$$

From the description of the flowsheet we see that the most difficult equations to solve are the ones that calculate the reactor output from its input streams. This set of equations is solved with a Newton-Raphson method modified to keep the variables within specified

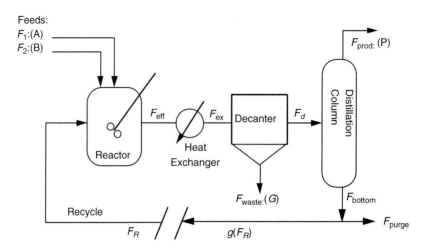

FIGURE 8.21 Flowsheet solved in modular mode.

bounds (e.g., nonnegative). The output streams of all of the other units can be calculated from direct assignment functions of the input streams. This flowsheet was set up in the GAMS (Brooke, Kendrick, and Meeraus, 1992) modeling environment and the reactor module was solved with the MINOS (Murtagh and Saunders, 1982) algorithm, with the recycle stream converged with direct substitution.

Starting with a guessed recycle stream of $F_{R,i} = 2000$ for each component i ($i = A, B, C, E, P$) and with guesses of weight fractions in the reactor $X_i = 1$, the flowsheet was converged after 96 flowsheet passes to a relative recycle tolerance of 10^{-4}. Within the GAMS environment, this required about 8.0 CPU secs on an IBM RS/6000 and the average number of Newton iterations for the reactor unit was about five. The iteration history is shown in Figure 8.22 and from the slope of this graph we see that the maximum eigenvalue for

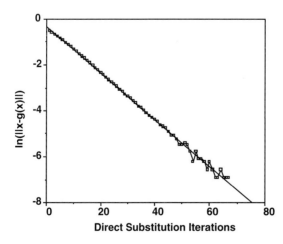

FIGURE 8.22 Convergence history for modular simulation of Williams-Otto flowsheet.

Sec. 8.5 Simulation Examples

converging this flowsheet is 0.903. In this case, the relaxation schemes discussed above will be very useful for accelerating the convergence of this flowsheet.

At the solution, an abbreviated mass balance for this flowsheet is given by:

	F_R	F_{eff}	X	F_{prod}
A	6.0	9.0	$5.7738 \cdot 10^{-4}$	0.
B	3228.0	3586.0	0.244	0.
C	1.0	1.1	$8.2008 \cdot 10^{-5}$	0.
E	8646.0	9607	0.654	0.
P	865.0	1201	0.082	240.3
G	0.0	282.0	0.019	0

8.5.2 Solving the Williams-Otto Flowsheet in Equation-Oriented Mode

In the equation-oriented mode we combine all of the process equations and solve them simultaneously. From the equations given for this flowsheet in section 8.2, we can derive the incidence matrix shown in Figure 8.23. In this figure, each "x" indicates the occurrence of a variable in the corresponding equation, while a period indicates no occurrence of that variable. As can be seen, there are few incidences per equation (usually two or three) and this property is exploited through sparse matrix decomposition in the MINOS solver. This flowsheet model was set up in the GAMS modeling environment with direct use of this Newton-based solver.

If we start with initializing the full set of equations with the recycle stream $F_{R,i}$ at 2000 for each component i ($i = A, B, C, E, P$) and with guesses of weight fractions in the reactor $X_i = 1$ (and zero for the other variables), then the solver has difficulties with these equations and reports a convergence failure. This is not surprising since we are starting from a very poor starting point and a linearization from this point leads to large extrapolations and the evaluation of ill-conditioned and possibly singular matrices.

To remedy this problem we need a problem-based initialization scheme. A natural way to begin is to initialize the flowsheet unit by unit using a modular calculation sequence. This type of initialization scheme is frequently required with equation-oriented simulators and often coupled with careful user intervention at the initialization stage. In this case, if we execute two direct substitution passes with the modular calculation sequence, we end up with a starting point represented by the following partial mass balance:

	F_R	F_{eff}	X	F_{prod}
A	5.8	65.0	0.008	0.
B	440.0	489.0	0.059	0.
C	2.0	2.0	$2.499 \cdot 10^{-4}$	0.
E	2551.0	2834.0	0.345	0.
P	255.0	1240.0	0.151	984.9
G	0.0	3587.0	0.437	0

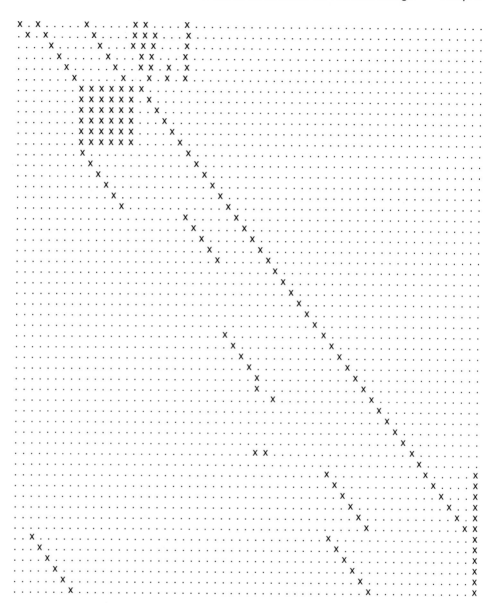

FIGURE 8.23 Incidence matrix for Williams-Otto process.

This initialization requires about 0.45 CPU secs (IBM RS/6000). Note that this starting point is still far from the converged solution, but the nonlinear reactor equations are satisfied at this point. From this starting point, the Newton-based solver (MINOS) requires only 15 iterations and 0.57 CPU secs to converge to the same solution as with the modular mode. Therefore, we see that the equation-oriented simulation mode is about eight times faster than the modular mode for this problem.

For more complex problems, it is hard to generalize from these results. However, qualitatively it is easy to see that:

- simultaneous convergence leads to a much faster solution strategy.
- careful initialization that is often problem specific is required to make the simultaneous strategy work.

With this small example problem, we were able to illustrate the construction of flowsheet models within two popular simulation modes and provide a brief comparison of these modes. For more detailed models, interaction with the physical property routines also plays an important role, as both simulation modes require repeated calls to these calculations. Again, because the equation-oriented mode requires fewer iterations and no inner loop convergence of specific units (e.g., the reactor module) there is an added advantage to this mode.

8.6 SUMMARY AND SUGGESTIONS FOR FURTHER READING

This chapter provides a concise overview of process simulation methods for flowsheet analysis and evaluation. Here we have provided a description and sketched the development of two popular simulation modes: the *modular* approach and the *equation-oriented* approach. A small flowsheeting example based on the Williams-Otto (Williams and Otto, 1960; Ray and Szekely, 1973) process was presented and solved with both modes. From this description we see that the more popular modular mode is more robust for nonlinear process calculations because it requires nested convergence of several calculation loops. These include the solution of physical property equations at the lowest level, convergence of the unit operations at the middle level, and solution of the recycle streams in the flowsheet. This reliability, however, requires considerable computational effort. Moreover, the input-output structure of the modular mode often makes it inflexible to flowsheet design specifications. Satisfying these specifications often requires an additional calculation loop. As a result of these characteristics, the modular mode is used most often for flowsheet design and analysis with detailed process models.

The equation-oriented mode, on the other hand, solves the flowsheet equations simultaneously (at least for the unit operations and recycle convergence). Consequently, solutions are much more efficient and require far less expense. In addition, the simultaneous mode allows arbitrary design specifications to be imposed without additional calculation loops. However, the equation-oriented mode requires a large-scale nonlinear equation solver for the entire flowsheet and careful initialization of the problem is required for successful solution. This initialization is frequently problem specific and often requires careful intervention by the user to get the solution process started. The equation-oriented mode has only recently become popular—this is due mainly to powerful software concepts, tools, and implementations. Most of the applications of this mode have been in real-time optimization where:

- Rapid on-line solution is required for large flowsheets.
- Process models are simpler than detailed design models and are frequently updated with process data.
- Good starting points are available from previous solutions.

A summary of many of these concepts as well as a survey of available packages can be found in Biegler (1989).

Both simulation modes require the solution of nonlinear process equations. Unit operations and physical property models were reviewed in Chapter 7; here we need to combine and solve these for an entire system. The solution strategies considered in this chapter were classified as Newton-based and fixed-point methods. The former type of methods (Newton and quasi-Newton or Broyden) are the most widely used because they have excellent convergence characteristics. Several modifications were also presented to remedy difficulties with poor starting points and singular Jacobians. Moreover, these methods are widely available in a number of software libraries. Excellent implementations of these equation solvers are also available from NETLIB in the MINPACK library. A more complete description and analysis of Newton-type methods is given in Dennis and Schnabel (1983) and Kelley (1995).

The fixed-point methods considered in this chapter do not have the strong convergence properties of Newton-type methods but are suitable when derivatives are difficult to calculate. As a result, they are used most frequently for recycle convergence in the modular mode. Even here, however, the Broyden method is often a better alternative for problems with complex recycle loops. Further descriptions of these methods can be found in Westerberg, Hutchison, Motard, and Winter (1979).

In addition to nonlinear equation solvers, process simulators require decomposition strategies for large flowsheeting problems. These strategies appear at different levels for the modular and the equation-oriented modes. For the modular mode, flowsheet decomposition is performed at the recycle convergence level, where the selection of tear streams and the sequencing of units is the key to an efficient flowsheet simulation. A wide variety of tearing problems can be formulated as set covering problems and solved as integer programs. Above we also illustrated how these problems could be simplified and reduced. A review of recycle tearing strategies is given by Gundersen and Hertzberg (1983), and further description of graph theoretic methods is given in Westerberg et al. (1979).

For the equation-oriented mode, decomposition strategies are usually applied at the linear algebra level, during the solution of Newton steps for the nonlinear equation solver. Here powerful sparse matrix methods have been developed that lead to efficient matrix decomposition and conserve storage of nonzero matrix elements. While a detailed discussion of these methods is beyond the scope of this text, there is a wealth of literature in this area. A classic text in this area is due to Duff, Erisman, and Reid (1986), which discusses the widely used sparse matrix code, MA48. In addition, Stadtherr and coworkers (Coon and Stadtherr, 1995; Zitney and Stadtherr, 1993) have recently developed very efficient sparse matrix codes for large-scale process flowsheeting.

Finally, the simulation concepts presented in this chapter illustrate the necessary tools for the evaluation of a candidate flowsheet for process design. However, even for a fixed flowsheet there are still many degrees of freedom that lead to considerable improvement in the candidate process. The next chapter therefore builds on these simulation concepts, for both the modular and equation-oriented modes, and develops the concepts and methods needed for flowsheet optimization.

REFERENCES

Barkeley, R. W., & Motard, R. (1972). *Chem. Engr. J.*, 3, 265.

Biegler, L. T. (1989). *Chemical Engineering Progress*, **85** (10), 50.

Brooke, A., Kendrick, D., & Meeraus, A. (1992). *GAMS: A User's Guide.* San Franciso, CA: Scientific Press.

Christensen, J. H., & Rudd, D. (1969). *AIChE J.*, 16, 177.

Coon, A. B., & Stadtherr, M. A. (1995). *Comput. Chem. Eng.*, 19, 787.

Crowe, C., & Nishio (1975). *AIChE J.,* 21, 528.

Dennis, J., & Schnabel, R. (1983). *Numerical Methods for Unconstrained Optimization and Nonlinear Equations.* Englewood Cliffs, NJ: Prentice-Hall.

Duff, I., Erisman, A., & Reid, J. (1986). *Direct Methods for Sparse Matrices.* Oxford: Oxford Science Publications.

Garfinkel, R., & Nemhauser, G. L. (1972). *Integer Programming.* New York: Wiley.

Gundersen, T., & Hertzberg, T. (1983). *Comp and Chem. Engr.,* 7, 189.

Kelley, C. T. (1995). *Iterative Methods for Linear and Nonlinear Equations*, Philadelphia: SIAM.

Leesley, M. E. (Ed.). (1982). *Computer-aided Process Plant Design.* Houston: Gulf Pub. Co.

Murtagh, B. A., & Saunders, M. (1982). *Math. Programming Study*, 16, 84.

Orbach, O., & Crowe, C. (1972). *Can. J. Chem Engr.*, 49, 509.

Pho, T. K., & Lapidus, L. (1973). *AIChE J.*, 19, 1170.

Ray, W. H., & Szekely, J. (1973). *Process Optimization*, New York: Wiley.

Sargent, R. W. H., & Westerberg, A. W. (1964). *Trans I Chem E.,* 42, 190.

Upadhye, R. S., & Grens, E. A. (1975). *AIChE J.*, 21, 136.

Wegstein, J. H. (1958). *Comm. ACM*, 1, 9.

Westerberg, A. W., Hutchison, W., Motard, R., & Winter, P. (1979). *Process Flowsheeting.* Cambridge: Cambridge University Press.

Westerberg, A. W., & Motard, R. L. (1981). *AIChE J.*, 27, 725.

Williams, T., & Otto, R. (1960). *AIEE Trans.*, 79, 458.

Zitney, S. E., & Stadtherr, M. A. (1993). *Comp. Chem. Engr.*, 17, 319.

EXERCISES

1. Consider the incidence matrix for the Williams-Otto process. Identify each equation in this matrix and find a pivot sequence for this matrix to use in decomposing the Jacobian for solving the equations with Newton's method.

2. Resolve the Williams and Otto process in the equation-oriented mode with a reactor temperature of 700°R and a purge fraction of 5%.

3. Given the system of equations considered in Example 8.1,

$$f_1 = 2x_1^2 + x_2^2 - 6 = 0$$
$$f_2 = x_1 + 2x_2 - 3.5 = 0$$

   a. Solve this system with Broyden's method (unit step size) using as a starting point $x_1 = 2.0$, $x_2 = 1.0$
   b. Using as a starting point $x_1 = x_2 = 0$, solve the system with the HYBRD code from the MINPACK library in NETLIB. How does this code handle the singular Jacobian?

4. Given is the system of linear equations in n variables x

$$f(x) = b + Ax = 0$$

   where A is a nonsingular matrix. Show the convergence properties of Newton's method and Broyden's method on such a system.

5. Reformulate the following equations so they do not have poles. Why is this necessary?

$$f_1 = \exp(x/(y_2 - 6))/z + 6 = 0$$
$$f_2 = 6 \ln(1/z^2)/t + 6 = 0$$

6. Derive the quadratic rule for stepsize adjustment (α_q) that is used in step d. of the Armijo linesearch.

7. For Broyden's method
   a. Assume that B^0 is symmetric. Derive a symmetric Broyden updating formula of the form: $B^{k+1} = B^k + u\,u^T$ that satisfies the secant relation.
   b. Derive the analogous symmetric inverse update formula without using B^k in the final formula.
   c. Verify that $B^{k+1} = B^k + (y - B^k s)c^T/c^T s$ satisfies the secant relation for an arbitrary vector c.

8. For the flowsheet shown in Figure 8.13, find the two partitions for the flowsheet. Apply the loop tearing algorithm with $w_j = 1$ to the first partition, not considered in this chapter.

9. Show that with a single equation the condition for Newton's method:

$$\left| \frac{f(x)f''(x)}{f(x)^2} \right| < 1$$

comes from the contraction mapping theorem and the relation

$$g(x) = g(y) + g'(\xi)(x - y)$$

where ξ is between x and y (mean value theorem).

10. Show that if $x^{i+1} = g(x^i)$ and g and x^0 satisfy the conditions of the contraction mapping theorem, then:

$$\left|x^{i+1} - x^i\right| \leq L\left|x^i - x^{i-1}\right| \quad i = 1, 2 \ldots$$

and also

$$\left|x^{i+1} - x^i\right| \leq L^i\left|x^1 - x^0\right| \quad i = 0, 1 \ldots$$

where $L < 1$ and represents a bound on $\partial g/\partial x$.

11. **a.** Solve the following system of equations:

$$x_1 = 1 - 0.5 \exp(0.7(1 - x_2))$$
$$x_2 = 2 - 0.3 \exp(0.5(x_1 + x_2))$$

Use as starting point $x_1 = -1$ and $x_2 = -1$, and as criterion of convergence $\|\Delta x^k\|_2 < 0.001$.

b. Estimate $|\lambda|^{max}$ when you complete the fifth iteration and predict the number of iterations required to converge to the tolerance 0.001 with direct substitution.

c. With the estimate of the variables at the fifth iteration predict the next point by using
  i) Dominant eigenvalue method
  ii) Wegstein's method
Which one gives you the better prediction?

12. Partition and precedence order the flowsheet in Figure 8.24 using the algorithm by Sargent and Westerberg. Also, for each group of units determine minimum number of tears and derive the sequence of calculation.

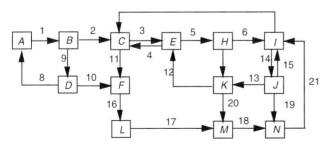

FIGURE 8.24

13. For the two flowsheets shown in Figures 8.25 and 8.26, determine a minimum tear set.

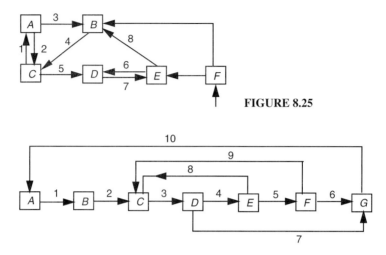

FIGURE 8.25

FIGURE 8.26

14. For the two flowsheets in Figures 8.27 and 8.28, find the members of the nonredundant family of tears.

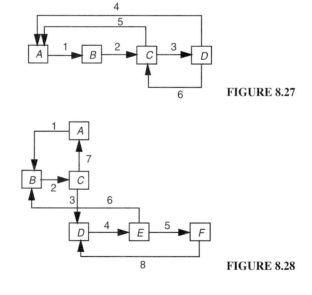

FIGURE 8.27

FIGURE 8.28

PROCESS FLOWSHEET OPTIMIZATION 9

With an understanding of flowsheet simulation and the structure of process models for design, we now begin to consider a key aspect of process design. The purpose of many simulation tasks in engineering is to develop a predictive model that can be used to improve the process. In this chapter we consider systematic improvement or *optimization* strategies for chemical processes with continuous variables. In particular, this chapter develops the Successive Quadratic Programming (SQP) algorithm, which has become a standard method for process flowsheet optimization. This approach builds on previous material required for process simulation, as we derive this method from a Newton-type perspective. In addition, we will develop this strategy for both modular and equation-based process simulation environments and discuss various advantages and disadvantages of each. Moreover, we will consider several small and large scale examples that demonstrate the effectiveness of this approach.

9.1 DESCRIPTION OF PROBLEM

At a practical level, we define the term *optimization* as follows:

Given a system or process, find the best solution to this process within constraints.

To quantify the "best solution" we first need an *objective function* that serves as a quantitative indicator of "goodness" for a particular solution. Typical objectives for process design include capital and operating cost, product yield, overall profit, and so on.

The values of the objective function are determined by manipulation of the *problem variables*. These variables can physically represent equipment sizes and operating conditions (e.g., pressures, temperatures and feed flowrates). Finally, the limits of process operation, product purity, validity of the model, and relationships among the problem variables need to be considered as *constraints* in the process. Similarly, the variable values must be adjusted to satisfy these constraints. Often the problem variables are further classified into *decision variables* that represent *degrees of freedom* in the optimization and *dependent variables* that can be solved from the constraints. In developing the optimization problem, this distinction is important from a conceptual point of view as well as for process problems modeled with modular simulators.

In many cases, the task of finding an improved flowsheet through manipulation of the decision variables is carried out by trial and error (through case study). Instead, with optimization methods we are interested in a *systematic* approach to finding the best flowsheet—and this approach must be as efficient as possible. Related areas that describe the theory and concepts of optimization are referred to as *mathematical programming* and *operations research,* and a large body of research is associated with these areas. Mathematical programming principally deals with characterization of theoretical properties of optimization problems and algorithms, including existence of solutions, convergence to these solutions, and local convergence rates. On the other hand, operations research is concerned with the application and implementation of optimization methods for efficient and reliable use. Finally, in process engineering we are concerned with the application of optimization methods to real-world problems. Here we need to be comfortable with the workings of the optimization algorithm, including the limitations of the methods (i.e., when they can fail). In addition, we need to formulate optimization problems that capture the essence of the actual process, and are tractable and solvable by current optimization methods.

This chapter concentrates on the optimization of systems where the problem variables are allowed to vary continuously in a region. A typical example of this problem lies in adjusting the pressure, temperature, and feed flowrate settings for a process flowsheet, as well as determining the equipment sizes for process units. Optimization problems that have nonlinear objective and/or constraint functions of the problem variables are referred to as *nonlinear programs*, and analysis and solution of this optimization problem is referred to as *nonlinear programming* (NLP). In addition, the optimization problem becomes considerably more difficult if variables are included that take on only integer or binary (0-1) values. These problems are referred to as *mixed integer nonlinear programs* (MINLPs) and they are covered in Chapter 15; process synthesis and optimization applications of these are covered in detail in Chapters 16 to 22.

The next section introduces the nonlinear programming problem and defines the optimality conditions for a solution to this problem. Section 9.3 then explores the Successive Quadratic Programming (SQP) method for solving nonlinear programs. We concentrate on this algorithm because it is frequently used in a wide variety of nonlinear programming applications, both in process engineering and elsewhere. Following this, we discuss in section 9.4 the application of nonlinear programming strategies for the modular simulation mode. In particular, we show that the SQP method leads to very efficient methods for modular simulators. Similar concepts are then explored in section 9.5 for the equation-based simulation mode. A distinguishing feature for this mode is that a large scale opti-

Sec. 9.2   Introduction to Constrained Nonlinear Programming

mization algorithm is required and, in particular, the SQP algorithm must be adapted for this case. Finally, section 9.6 concludes the chapter and provides guides for further reading. Several process examples are also used to illustrate the concepts in this chapter.

9.2   INTRODUCTION TO CONSTRAINED NONLINEAR PROGRAMMING

We consider the nonlinear programming problem, given in general form as:

$$\begin{aligned}\operatorname*{Min}_{x}\quad & f(x)\\ \text{s.t.}\quad & g(x) \leq 0\\ & h(x) = 0\end{aligned} \qquad (9.1)$$

where x is an n vector of continuous variables, $f(x)$ is a scalar objective function, $g(x)$ is an m vector of inequality constraint functions, and $h(x)$ is an meq vector of equality constraint functions. These constraints create a region for the variables x, termed the *feasible region*, and we require $n \geq meq$ in order to have any degrees of freedom for optimization. While Eq. (9.1) will be our standard form for nonlinear programs, the NLP problem can be expressed in a number of different ways. For instance, the signs of the objective function and constraint functions could be changed so that we have:

$$\begin{aligned}\operatorname*{Max}_{x}\quad & q(x)\\ \text{s.t.}\quad & w(x) \geq 0\\ & h(x) = 0\end{aligned} \qquad (9.2)$$

for functions defined by $q(x) = -f(x)$ and $w(x) = -g(x)$. Properties of this nonlinear program (NLP) are summarized in Appendix A. In particular, we will develop methods that will find a *local minimum* point x^* for $f(x)$ for a feasible region defined by the constraint functions; that is, $f(x^*) \leq f(x)$ for all x satisfying the constraints in some neighborhood around x^*. Provided that the feasible region is not empty and the objective function is bounded below on this feasible region, we know that such local solutions exist.

On the other hand, finding and verifying *global solutions* to this NLP will not be dealt with in this chapter. In Appendix A, we see that a local solution to the NLP is also a global solution under the following *sufficient* conditions based on convexity. From Appendix A, we define a convex function $\phi(x)$ for x in some domain X, if and only if it satisfies the relation:

$$\phi(\alpha \xi + (1 - \alpha) \eta) \leq \alpha \phi(\xi) + (1 - \alpha) \phi(\eta) \qquad (9.3)$$

for any α, $0 \leq \alpha \leq 1$, at all points in ξ and η in X. As derived in Appendix A, sufficient conditions for a global solution for the NLP (9.1) are that:

- the solution is a local minimum for the NLP
- $f(x)$ is *convex*
- $g(x)$ are all *convex*
- $h(x)$ are all *linear*

The last two conditions imply that the feasible region is convex, i.e. for all points ξ and η in the feasible region and for all α, $0 \leq \alpha \leq 1$, the point $[\alpha \xi + (1 - \alpha) \eta]$ is also in the region. For process optimization, these properties state that any problem with nonlinear equality constraints is nonconvex and in the absence of additional information, there is *no guarantee that a local optimum is global* if these convexity conditions are not met.

To illustrate these concepts, we consider two nonconvex examples that lead to different kinds of solutions.

EXAMPLE 9.1 Optimal Vessel Dimensions

Consider the optimization of a cylindrical vessel with a specified volume. What is the optimal L/D ratio for this vessel that leads to a minimum cost?

The constrained problem can be formulated as one where we minimize a cost based on the amount of material used to make up the top and bottom of the vessel and the sides of the vessel. For a small wall thickness, the amount of material is proportional to the surface area. The cost per area for the materials is given by C_T and C_S for the top and sides, respectively. The specification for volume is written as a constraint and the NLP is given by:

$$\text{Min} \left\{ C_T \frac{\pi D^2}{2} + C_S \pi D L = \text{cost} \right\}$$

$$\text{s.t.} \quad V - \frac{\pi D^2 L}{4} = 0 \tag{9.4}$$

$$D, L \geq 0$$

Note that for this problem, the feasible region in the variables D and L is nonconvex, because of the nonlinear constraint. We can easily eliminate L from this equation and substitute $L = 4V/\pi D^2$ in the objective function and describe it using the single variable D. Since the constraints has already been incorporated into the objective function we need not consider it further and the problem becomes:

$$\text{Min} \left\{ C_T \frac{\pi D^2}{2} + C_S \frac{4V}{D} = \text{cost} \right\} \text{ with } D \geq 0. \tag{9.5}$$

If the optimum value of D is positive, we can find the minimum by differentiating the cost with respect to D and setting this to zero.

$$\frac{d(\text{cost})}{dD} = C_T \pi D - \frac{4VC_S}{D^2} = 0 \tag{9.6}$$

Solving for variable D leads to the expression below with L obtained from the volume specification:

$$D = \left(\frac{4V}{\pi} \frac{C_S}{C_T}\right)^{1/3} \qquad L = \left(\frac{4V}{\pi}\right)^{1/3} \left(\frac{C_T}{C_S}\right)^{2/3} \qquad (9.7)$$

Moreover, the aspect ratio for the cylinder can be expressed in a compact form: $L/D = C_T/C_S$. If we further examine the cost function, we see that:

$$d^2(\text{cost})/dD^2 = C_T\pi + 8\,V\,C_S/D^3 > 0, \text{ for } D > 0, \qquad (9.8)$$

and by the definitions in Appendix A, this function is convex over the (open) feasible region for D. As a result, the solution to this NLP is a global one and no other (local) solutions exist.

In the next example, however, we have multiple solutions due to nonconvexity.

EXAMPLE 9.2 Minimize Packing Dimensions

Consider three cylindrical objects of equal height but with three different radii, as shown in Figure 9.1 below. What is the box with the smallest perimeter that will contain these three cylinders? Formulate and analyze this nonlinear programming problem.

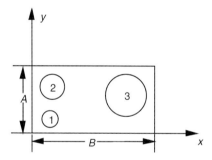

FIGURE 9.1 Illustration of Example 9.2

As decision variables we choose the dimensions of the box, A, B, and the coordinates for the centers of the three cylinders, (x_1, y_1), (x_2, y_2), (x_3, y_3). As specified parameters we have the radii, R_1, R_2, R_3. For this problem we minimize the perimeter $2(A + B)$ and include as constraints the fact that the cylinders remain in the box and can't overlap. As a result we formulate the following nonlinear program:

$$\text{Min } (A + B) \qquad (9.9)$$

$$\text{in box } \begin{cases} x_1, y_1 \geq R_1 & x_1 \leq B - R_1,\ y_1 \leq A - R_1 \\ x_2, y_2 \geq R_2 & x_2 \leq B - R_2,\ y_2 \leq A - R_2 \\ x_3, y_3 \geq R_3 & x_3 \leq B - R_3,\ y_3 \leq A - R_3 \end{cases}$$

$$\text{no overlaps } \begin{cases} (x_1 - x_2)^2 + (y_1 - y_2)^2 \geq (R_1 + R_2)^2 \\ (x_1 - x_3)^2 + (y_1 - y_3)^2 \geq (R_1 + R_3)^2 \\ (x_2 - x_3)^2 + (y_2 - y_3)^2 \geq (R_2 + R_3)^2 \end{cases}$$

$$x_1, x_2, x_3, y_1, y_2, y_3, A, B \geq 0$$

Note that the objective function and the "in box" constraints are linear, and hence, convex. Similarly, the variable bounds are convex as well. The nonconvexities are observed in the nonlinear inequality constraints and this can be verified using the properties in Appendix A (see Exercise 9.1). Because convexity conditions are not satisfied, there is no guarantee of a unique global solution. Indeed, we can imagine intuitively the existence of multiple solutions to this NLP, as follows:

- Find a solution and observe an equivalent solution by turning the box by 90°.
- Use a random arrangement for the cylinders and manually shrink the walls of the box. The solution depends on the initial positions of the cylinders.

Consequently, we see that this problem has many local solutions. This is due to a nonconvex feasible region.

These two examples raise some interesting questions that will be explored next. First, what are the conditions that characterize even a local solution to a nonlinear program? In the first example, once L was eliminated, the constraints became unimportant. On the other hand, in the second example, the NLP solution was completely defined by the constraints. At the solution, these inequality constraints were satisfied as equations and were therefore considered to be *active*. In the remainder of this section we will present the *Kuhn Tucker optimality conditions* to define locally optimal solutions.

Second, the search for NLP solutions is guided by determining the correct active set of constraints and the solution of equations that represent the optimality conditions. In the first example this task was easy as no active constraints were considered, and because the optimal solution could be found analytically from Eq. (9.6). In the second example, we have yet to consider these tasks. These search strategies will be considered when we develop an NLP algorithm in the next section.

9.2.1 Optimality Conditions for Nonlinear Programming

In the remainder of this section we briefly present and discuss the optimality conditions for solution of the nonlinear programming problem (1). These are derived in Appendix A and are presented in detail below. Before presenting these properties, we first consider an intuitive explanation of the optimality conditions.

Consider the contour plot of $f(x)$ in two dimensions as shown in Figure 9.2. By inspection we see that the minimum point is given by x^*. If we consider this plot as a (smooth) valley, then a "ball" rolling in this valley will stop at x^*, the lowest point. At this stationary point we have a zero gradient, $\nabla f(x^*) = 0$, and the second derivatives reveal positive curvature of $f(x)$. In other words, if we move the ball away from x^* in any direction, it will roll back.

Now if we introduce two inequality constraints, $g_1(x) \leq 0$ and $g_2(x) \leq 0$, into the minimization problem, we can visualize this as imposing two "fences" in the valley, as

Sec. 9.2 Introduction to Constrained Nonlinear Programming

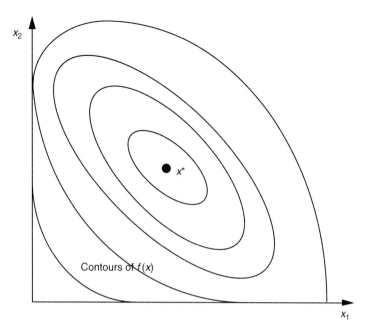

FIGURE 9.2 Contour plot for unconstrained minimum.

shown in Figure 9.3. Again, a ball rolling in the valley within the fences will roll to the lowest allowable point. However, if x^* is at the boundary of a constraint (e.g., $g_1(x^*) = 0$), then this inequality constraint is *active,* the ball is pinned at the fence and we no longer have $\nabla f(x^*) = 0$. Instead, we see that the ball remains stationary because of a balance of "forces": the force of "gravity" $(-\nabla f(x^*))$ and the "normal force" exerted on the ball by the fence $(-\nabla g_1(x^*))$. Also, in Figure 9.3 note that the constraint $g_2(x) \leq 0$ is inactive at x^* and does not participate in this "force balance." In addition to the balance of forces, we expect positive curvature *along the active constraint*; that is, if we move the ball from x^* in any direction along the fence, it will roll back.

Finally, we introduce an equality constraint, $h(x) = 0$, into the problem and we can visualize this as introducing a "rail" into the valley, as shown in Figure 9.4. Now a ball rolling on the rail and within the fence will also stop at the lowest point, x^*. This point will also be characterized by a balance of "forces": the force of "gravity" $(-\nabla f(x^*))$, the "normal force" exerted on the ball by the fence $(-\nabla g_1(x^*))$, and the "normal force" exerted on the ball by the rail $(-\nabla h(x^*))$. In addition to this balance of forces, we expect positive curvature *along the active constraints*. However, in Figure 9.4, we no longer have allowable directions that remain on the active constraints. Instead, the ball remains stationary at the intersection of the rail and the fence—and this condition is sufficient for optimality.

We now generalize these concepts and develop the optimality conditions for constrained minimization. These optimality conditions are referred to as the Kuhn Tucker

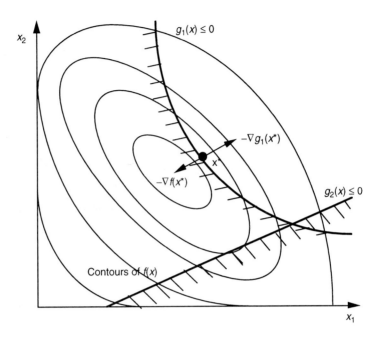

FIGURE 9.3   Constrained minimization with inequalities.

(KT) conditions or Karush Kuhn Tucker (KKT) conditions and were developed independently by Karush (1939) and Kuhn and Tucker (1951). For convenience of notation we define a Lagrange function as:

$$L(x, \mu, \lambda) = f(x) + g(x)^T \mu + h(x)^T \lambda \qquad (9.10)$$

Here the vectors μ and λ act as "weights" for balancing the "forces" shown in Figure 9.4; μ and λ are referred to as *dual variables* or *Kuhn Tucker multipliers*. They are also called *shadow prices* in operations research literature.

The solution of the NLP (9.1) satisfies the following first-order Kuhn Tucker conditions. These conditions are *necessary* for optimality.

1. Linear dependence of gradients ("balance of forces" in Figure 9.4)

$$\nabla L(x^*, \mu^*, \lambda^*) = \nabla f(x^*) + \nabla g(x^*) \mu^* + \nabla h(x^*) \lambda^* = 0 \qquad (9.11)$$

2. Feasibility of NLP solution (within the fences and on the rail in Figure 9.4)

$$g(x^*) \leq 0, \; h(x^*) = 0 \qquad (9.12)$$

3. Complementarity condition; either $\mu_i^* = 0$ or $g_i(x^*) = 0$ (either at the fence boundary or not in Figure 9.4)

$$\mu^{*T} g(x^*) = 0 \qquad (9.13)$$

Sec. 9.2 Introduction to Constrained Nonlinear Programming

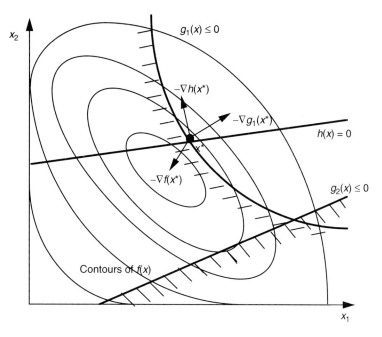

FIGURE 9.4 Constrained minimization with inequalities and equalities.

4. Nonnegativity of inequality constraint multipliers (normal force from "fence" can only act in one direction)

$$\mu^* \geq 0 \qquad (9.14)$$

5. Constraint qualification:
   Active constraint gradients, i.e.:

   $$[\nabla g_A(x^*) \mid \nabla h(x^*)] \text{ for } i \in A, A = \{i \mid g_i(x^*) = 0\}$$

   must be linearly independent.

The first Kuhn Tucker condition Eq. (9.11) describes linear dependence of the gradients of the objective and constraint functions and is derived in Appendix A. The second condition Eq. (9.12) requires that the solution of the NLP, x^*, satisfy all the constraints. The third and fourth conditions Eqs. (9.13, 9.14) relate to complementarity. Here either inequality constraint i is inactive ($g_i(x^*) < 0$) and the corresponding multiplier is zero (i.e., the constraint is ignored in the KT conditions), or, if the constraint is active ($g_i(x^*) = 0$), μ_i can be positive. Finally, in order for a local NLP solution to satisfy the KT conditions, an additional constraint qualification is required. Constraint qualifications take several forms (see Fletcher, 1987), and the one most frequently invoked is that the gradients of the active constraints be linearly independent.

These conditions are only necessary, however, and additional conditions are needed to ensure that x^* is a local solution. So far, the first order conditions define x^* only as a stationary point that satisfies the constraints. For instance, in Example 9.1, the KT conditions Eqs. (9.11–9.14) correspond to setting the gradient of the objective function to zero. To confirm a local optimum for this example, second derivatives have to be evaluated and checked to be positive (or at least nonnegative).

For a multivariable problem, the second derivatives are evaluated in terms of a *Hessian matrix* of a given function. For instance, the Hessian matrix of the objective function, $\nabla_{xx} f(x)$, is made up of elements: $\{\nabla_{xx} f(x)\}_{ij} = \partial^2 f / \partial x_i \partial x_j$. Also, since $\partial^2 f / \partial x_j \partial x_i = \partial^2 f / \partial x_i \partial x_j$, we have $\{\nabla_{xx} f(x)\}_{ij} = \{\nabla_{xx} f(x)\}_{ji}$ and the Hessian matrix is symmetric. Moreover, positive curvature for the contour surface can be evaluated based on the Hessian matrix. For instance, the objective function in Figure 9.2 has positive curvature at x^* if its Hessian matrix is positive definite, i.e.:

$$p^T \nabla_{xx} f(x^*) p > 0 \quad \text{for all vectors } p \neq 0$$

or positive semidefinite:

$$p^T \nabla_{xx} f(x^*) p \geq 0 \quad \text{for all vectors } p \neq 0.$$

For the constrained NLP problem (1), second order conditions are defined using the Hessian matrix of the Lagrange function and by defining nonzero *allowable directions* for the optimization variables based on the active constraints. Starting from the solution x^*, the allowable directions, p, satisfy the active constraints as equalities and therefore remain in the feasible region. Because, the change in x along this direction can be arbitrarily small, these directions must also satisfy linearizations of these constraints and are therefore defined by:

$$\nabla h(x^*)^T p = 0 \tag{9.15}$$

$$\nabla g_i (x^*)^T p = 0 \text{ for } i \in A, A = \{i | g_i (x^*) = 0\}$$

The sufficient (necessary) *second order conditions* require positive (nonnegative) curvature of the Lagrange function in these allowable or "constrained" directions, p. Using the second derivative matrix to define this curvature we express these conditions as:

$$\begin{aligned} p^T \nabla_{xx} L (x^*, \mu^*, \lambda^*) p &> 0 \text{ (sufficient condition)} \\ p^T \nabla_{xx} L (x^*, \mu^*, \lambda^*) p &\geq 0 \text{ (necessary condition)} \end{aligned} \tag{9.16}$$

for all of the allowable directions, p. These second order conditions are also presented in more detail in Appendix A.

EXAMPLE 9.3   Application of Kuhn Tucker Conditions

To illustrate these Kuhn Tucker conditions, we consider two simple examples represented in Figure 9.5.

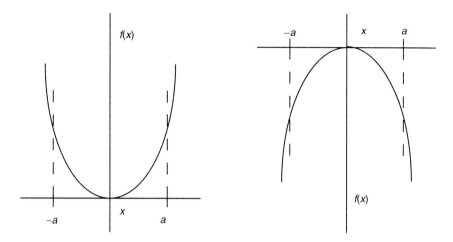

FIGURE 9.5 Illustration of Kuhn Tucker conditions for Example 9.3

First, we consider the single variable problem:

$$\text{Min } x^2 \quad s.t. \quad -a \leq x \leq a, \text{ where } a > 0 \tag{9.17}$$

where $x^* = 0$ is seen by inspection. The Lagrange function for this problem can be written as:

$$L(x, \mu) = x^2 + \mu_1(x - a) + \mu_2(-a - x) \tag{9.18}$$

with the first order Kuhn Tucker conditions Eqs. (9.11–9.14) given by:

$$\nabla L(x, \mu) = 2x + \mu_1 - \mu_2 = 0$$
$$\mu_1(x - a) = 0 \quad \mu_2(-a - x) = 0 \tag{9.19}$$
$$-a \leq x \leq a \quad \mu_1, \mu_2 \geq 0$$

To satisfy the first order conditions Eq. (9.19) we consider three cases: $\mu_1 = \mu_2 = 0$; $\mu_1 > 0$, $\mu_2 = 0$; or $\mu_1 = 0$, $\mu_2 > 0$. Note that the case $\mu_1 > 0$, $\mu_2 > 0$ cannot exist for $a > 0$ (Why?). Satisfying these conditions requires the evaluation of three candidate solutions:

- Upper bound is active, $x = a$, $\mu_1 = -2a$, $\mu_2 = 0$
- Lower bound is active, $x = -a$, $\mu_2 = -2a$, $\mu_1 = 0$
- Neither bound is active, $\mu_2 = 0$, $\mu_1 = 0$, $x = 0$

Clearly only the last case satisfies these conditions because the first two lead to negative values for μ_1 or μ_2. If we evaluate the second order conditions Eq. (9.16) we have allowable directions $p = \Delta x$ with $\Delta x > 0$ and $\Delta x < 0$. Also, we have

$$\nabla_{xx} L(x^*, \mu^*, \lambda^*) = 2 > 0 \quad \text{and}$$
$$p^T \nabla_{xx} L(x^*, \mu^*, \lambda^*) p = 2 \Delta x^2 > 0 \tag{9.20}$$

for all allowable directions. Therefore, the solution $x^* = 0$ satisfies both the sufficient first and second order Kuhn Tucker conditions for a local minimum.

We now consider an interesting variation on this example. As seen in Figure 9.5, suppose we change the sign on the objective function and solve:

$$\text{Min } -x^2 \quad \text{s.t. } -a \le x \le a, \text{ where } a > 0. \tag{9.21}$$

Here the solution, $x^* = a$ or $-a$, is seen by inspection. The Lagrange function for this problem is now written as:

$$L(x, \mu) = -x^2 + \mu_1(x - a) + \mu_2(-a - x) \tag{9.22}$$

with the first order Kuhn Tucker conditions given by:

$$\nabla L(x, \mu) = -2x + \mu_1 - \mu_2 = 0$$

$$\mu_1(x - a) = 0 \quad \mu_2(-a - x) = 0 \tag{9.23}$$

$$-a \le x \le a \quad \mu_1, \mu_2 \ge 0$$

Again, satisfying conditions (9.23) requires the evaluation of three candidate solutions, depending on $\mu_1 = \mu_2 = 0$; $\mu_1 > 0, \mu_2 = 0$; or $\mu_1 = 0, \mu_2 > 0$:

- Upper bound is active, $x = a$, $\mu_1 = 2a$, $\mu_2 = 0$
- Lower bound is active, $x = -a$, $\mu_2 = 2a$, $\mu_1 = 0$
- Neither bound is active, $\mu_2 = 0$, $\mu_1 = 0$, $x = 0$

and all three cases satisfy the first order conditions. We now need to check the second order conditions to discriminate among these points. If we evaluate the second order conditions (16) at $x = 0$, we realize allowable directions $p = \Delta x > 0$ and $-\Delta x$ and we have:

$$p^T \nabla_{xx} L(x, \mu, \lambda) p = -2\Delta x^2 < 0. \tag{9.24}$$

This point does not satisfy the second order conditions. In the other two cases, we invoke a subtle concept. *For $x = a$ or $x = -a$, we require the allowable direction to satisfy the active constraints exactly. Here, any point along the allowable direction, x^* must remain at its bound.* For this problem, however, there are no nonzero allowable directions that satisfy this condition. Consequently, the solution x^* is defined entirely by the active constraint. The condition:

$$p^T \nabla_{xx} L(x^*, \mu^*, \lambda^*) p > 0 \tag{9.25}$$

for all allowable directions, is *vacuously* satisfied—because there are *no* allowable directions.

The first and second order Kuhn Tucker conditions provide a useful tool for identifying local solutions to nonlinear programs. (It should be noted, though, that because second derivatives are often not calculated in process optimization problems, second order conditions are rarely checked.) However, we still need efficient search strategies that locate points that satisfy these conditions. In the next section, we develop a nonlinear programming algorithm called Successive Quadratic Programming (SQP). For process optimization, this algorithm has some desirable features and it has been used widely in many process applications. Moreover, it has proved to be adaptable to several kinds of nonlinear programming problems.

9.3 DERIVATION OF SUCCESSIVE QUADRATIC PROGRAMMING (SQP)

In nonlinear programming applications for process engineering, two approaches are used in virtually all problems: reduced gradient approaches and Successive Quadratic Programming. Both of these are summarized briefly in Appendix A. In particular, Successive Quadratic Programming has emerged as a very popular algorithm for process optimization. A characteristic feature of SQP is that it requires far fewer function evaluations than reduced gradient methods and other competing algorithms. For certain classes of nonlinear programs, such as process flowsheet optimization, this gives SQP a key advantage.

The SQP method can be derived from a direct perspective. Here we consider a modified set of the Kuhn Tucker conditions Eqs. (9.11–9.14) as a set of nonlinear equations in x, μ, and λ. These equations can then be solved with Newton's method (in similar manner as in Chapter 8). As a result, an efficient and reliable method can be developed based on our knowledge of nonlinear equation solvers. This is the essence of SQP and is largely responsible for its desirable performance. In the derivation presented next, we also need to consider some refinements to this algorithm so that it can be applied to the first order Kuhn Tucker conditions directly.

We begin by considering a modification of the Kuhn Tucker conditions. Here, if we know the active set for the inequalities in advance, then we can define $A = \{l | g_l(x^*) = 0\}$ and let $g_A(x)$ be made up of the constraints, $g_l(x)$, $l \in A$. The Kuhn Tucker conditions Eqs. (9.11–9.14) can be simplified by writing:

$$\nabla_x L(x^*, \mu^*, \lambda^*) = \nabla f(x^*) + \nabla g_A(x^*) \mu^* + \nabla h(x^*) \lambda^* = 0$$

$$g_A(x^*) = 0 \tag{9.26}$$

$$h(x^*) = 0$$

and the solution can be obtained by solving these equations for x, μ, and λ. (Note that since the Lagrange function, L, has multiple arguments, its gradient with respect to x is denoted, for clarity, by $\nabla_x L$.) Applying Newton's method to solve the equations (9.26) at iteration i leads to the following set of linear equations that define the Newton step:

$$\begin{bmatrix} \nabla_{xx} L & \nabla g_A & \nabla h \\ \nabla g_A^T & 0 & 0 \\ \nabla h^T & 0 & 0 \end{bmatrix} \begin{bmatrix} \Delta x \\ \Delta \mu \\ \Delta \lambda \end{bmatrix} = - \begin{bmatrix} \nabla_x L(x^i, \mu^i, \lambda^i) \\ g_A(x^i) \\ h(x^i) \end{bmatrix} \tag{9.27}$$

Inspection of the linear system Eq. (9.27) (see Exercise 7) shows that these are simply the Kuhn Tucker conditions of the following optimization problem:

$$\text{Min } \nabla f(x^i)^T d + 1/2\, d^T \nabla_{xx} L(x^i, \mu^i, \lambda^i)\, d$$

$$\text{s.t. } g_A(x^i) + \nabla g_A(x^i)^T d = 0 \tag{9.28}$$

$$h(x^i) + \nabla h(x^i)^T d = 0$$

The NLP (9.28), with a quadratic objective function (in the variable vector d) and linear constraints is called a quadratic program (QP) and if $\nabla_{xx} L(x^i, \mu^i, \lambda^i)$ is positive definite (i.e., $y^T \nabla_{xx} L(x^i, \mu^i, \lambda^i) y > 0$, for all nonzero vectors y), efficient finite step algorithms are available for solving these problems. Solving Eq. (9.28) yields a solution vector d with multipliers μ and λ for g_A and h, respectively. By setting $d = \Delta x$, $\Delta \mu = \mu - \mu^i$ and $\Delta \lambda = \lambda - \lambda^i$, this solution is equivalent to the Newton step in Eq. (9.27).

To relax the problem (9.26) to include the inequalities, $g(x^*) \leq 0$, we generalize the QP (9.28). In this way the QP is easily modified to automatically determine the active set of inequalities, g_A, and here the following QP is solved instead of Eq. (9.28):

$$\text{Min } \nabla f(x^i)^T d + 1/2 \, d^T \nabla_{xx} L(x^i, \mu^i, \lambda^i) \, d$$

$$\text{s.t. } g(x^i) + \nabla g(x^i)^T d \leq 0 \quad (9.29)$$

$$h(x^i) + \nabla h(x^i)^T d = 0$$

This QP generates a search direction in x and also yields reasonable estimates for the Kuhn Tucker multipliers. However, to implement this method we need to evaluate second derivatives of the objective and constraint functions and obtain good initial estimates of μ and λ in order to calculate the Hessian of the Lagrange function ($\nabla_{xx} L$). These two tasks can be serious drawbacks to application with process models.

This approach was originally proposed by Wilson (1963) and applied by Beale (1967). However, in early studies this approach did not work well and was failure prone. A key reason for poor performance is that $\nabla_{xx} L$ may not be positive definite and this leads to a nonconvex QP (9.29) that is difficult to solve with most current QP solvers. To remedy these problems, Han (1977) and Powell (1977) took advantage of advances in the development of quasi Newton methods (9.35) and exact penalty functions (9.36) for solving nonlinear programs. In particular, the Hessian of the Lagrange function can be approximated by a symmetric, positive definite matrix, B^i. This approximation is based on a secant relation and is closely related to Broyden's method for solving nonlinear equations, described in Chapter 8. Here calculation of B^i is based on the difference in the gradient of the Lagrange function from one point to the next.

9.3.1 The BFGS Approximation for $\nabla_{xx}L$

Consider an approximation B^i to $\nabla_{xx} L$ at x^i, where we can update this approximation based on information at a new point x^{i+1} and a *secant relation* given by:

$$B^{i+1} (x^{i+1} - x^i) = \nabla_x L(x^{i+1}, \mu^{i+1}, \lambda^{i+1}) - \nabla_x L(x^i, \mu^{i+1}, \lambda^{i+1})$$

Here we define

$$s = x^{i+1} - x^i,$$

$$y = \nabla_x L(x^{i+1}, \mu^{i+1}, \lambda^{i+1}) - \nabla_x L(x^i, \mu^{i+1}, \lambda^{i+1})$$

and this leads to:

$$B^{i+1} s = y. \quad (9.30)$$

Sec. 9.3 Derivation of Successive Quadratic Programming (SQP)

Note that $\nabla_{xx}L$ is a symmetric matrix and we also want the approximation B^i to be symmetric and positive definite as well. Because of symmetry and positive definiteness, we can define the current approximation as $B^i = JJ^T$, where J is a square, nonsingular matrix. To preserve symmetry, the update to B^i can be given as $B^{i+1} = J_+ J_+^T$ where J_+ is also square and nonsingular. By working with the matrices J and J_+, we will be able to parallel the update of B^i with Broyden's method in Chapter 8, and it will be easier to monitor the symmetry and positive definiteness properties of B^i.

Using the matrix J_+, the secant relation Eq. (9.30) can be split into two parts. From:

$$B^{i+1} s = J_+ J_+^T s = y,$$

we introduce an unknown variable vector v and obtain:

$$J_+ v = y \quad \text{and} \quad J_+^T s = v. \tag{9.31}$$

Now we can obtain an update formula by invoking the same least change strategy used to derive Broyden's method in Chapter 8, and we solve the following nonlinear program for J_+. The least change problem is given by:

$$\text{Min } \| J_+ - J \|_F$$
$$\text{s.t. } J_+ v = y \tag{9.32}$$

where $\| J \|_F$ is the Frobenius norm of matrix J. Solving Eq. (9.32) leads to the Broyden update formula derived in Chapter 8. With our current notation, this is:

$$J_+ = J + (y - J v) v^T / v^T v \tag{9.33}$$

From Eq. (9.33) we can recover an update formula in terms of s, y, and B^i, by using the following identities about v. From (9.31), $J_+ v = y$ and $J_+^T s = v$, we have:

$$v^T v = [y^T (J_+)^{-T}] J_+^T s = s^T y.$$

Also by multiplying J_+^T by s, we have from Eqs. (9.31) and (9.33):

$$v = J_+^T s = J^T s + v (y - J v)^T s / v^T v$$
$$v = J^T s + v [(y^T s - v^T J^T s) / v^T v] \tag{9.34}$$
$$v [1 - (y^T s - v^T J^T s) / y^T s] = J^T s$$
$$v = (s^T y / v^T J^T s) J^T s = \beta J^T s$$

where β and the terms in brackets are scalars.

Finally, from the definitions of B^i and B^{i+1}, Eqs. (9.33) and (9.34), we have:

$$B^{i+1} = (J + (y - J v) v^T / v^T v)(J + (y - J v) v^T / v^T v)^T$$
$$= J J^T + (y y^T - J v v^T J^T) / v^T v \tag{9.35}$$
$$= B^i + y y^T / s^T y - J v v^T J^T / v^T v$$
$$= B^i + y y^T / s^T y - B^i s s^T B^i / s^T B^i s$$

Note that the scalar β cancels in derivation of the update (9.35). From this derivation, we have defined B^i to be a symmetric matrix and this can be verified from Eq. (9.35). Moreover, it can be shown from Eqs. (9.31) and (9.33) that if B^i is positive definite and $s^T y > 0$, then the update, B^{i+1}, is also positive definite. In fact, the condition, $s^T y > 0$, must be checked and satisfied before the update (9.35) can be taken. This update formula is known as the *Broyden-Fletcher-Goldfarb-Shanno (BFGS) update* and the derivation above is due to Dennis and Schnabel (1983). As a result of this updating formula, we have a reasonable approximation to the Hessian matrix that is also positive definite. This leads to a convex QP problem and desirable convergence properties.

9.3.2 Characteristics of SQP Method

As with Newton's method for solving nonlinear equations, the SQP method for nonlinear programming can be characterized by some desirable properties. First, the method converges quickly and requires few function and gradient evaluations. Close to the solution, this can be stated more precisely by the following local convergence rates. Here, if:

- $B^i = \nabla_{xx} L(x^i, \mu^i, \lambda^i)$, then the convergence rate is quadratic, i.e., for a positive constant K, we have:

$$\lim_{i \to \infty} \|x^{i+1} - x^*\| / \|x^i - x^*\|^2 \leq K$$

- B^i is evaluated from a BFGS update and $\nabla_{xx} L(x^*, \mu^*, \lambda^*)$ is positive definite, then the convergence rate is superlinear, that is,

$$\lim_{i \to \infty} \|x^{i+1} - x^*\| / \|x^i - x^*\| = 0$$

- B^i is due to a BFGS update, then the convergence rate is two step superlinear, that is,

$$\lim_{i \to \infty} \|x^{i+1} - x^*\| / \|x^{i-1} - x^*\| = 0$$

As with nonlinear equation solvers, the SQP method can also be modified so that it can converge from starting points far from the solution. In this case, we can introduce a line search algorithm that uses the search direction generated by SQP but modifies the steplength so that: $x^{i+1} = x^i + \alpha\, d$, where α is a scalar, $0 < \alpha \leq 1$. Here α is chosen so that it ensures a decrease of a *merit function* that represents the objective function plus a weighted sum of the constraint infeasibilities. In particular, the *exact penalty function* is a popular choice in most SQP algorithms:

$$P(x, \gamma, \eta) = f(x) + \Sigma_j \gamma_j \max(0, g_j) + \Sigma_j |\eta_j h_j| \qquad (9.36)$$

where the weights are chosen suitably large so that $\gamma_j > \mu_j$, $\eta_j > |\lambda_j|$, and μ_j and λ_j are the current multiplier estimates determined from (QP1) in Table 9.1. Using this merit function, the SQP method, with BFGS updating, is guaranteed to converge to a local solution as long as the objective is bounded below and the QP subproblems are solvable. In addition, several alternative merit functions have been proposed along with additional modifi-

Sec. 9.3   Derivation of Successive Quadratic Programming (SQP)

TABLE 9.1   Basic SQP Algorithm

0. Guess x^0, set $B^0 = I$ (the identity matrix is a default choice). Evaluate $f(x^0)$, $g(x^0)$, and $h(x^0)$.
1. At x^i, evaluate $\nabla f(x^i)$, $\nabla g(x^i)$, $\nabla h(x^i)$. If $i > 0$, calculate s and y.
2. If $i > 0$ and $s^T y > 0$, update B^i using the BFGS formula (9.35).
3. Solve: Min $\nabla f(x^i)^T d + 1/2\, d^T B^i d$ (QP1)
   $\quad\quad\quad d$
   $\quad\quad$ s.t. $g(x^i) + \nabla g(x^i)^T d \leq 0$
   $\quad\quad\quad\quad\, h(x^i) + \nabla h(x^i)^T d = 0$
4. If $\|d\|$ is less than a small tolerance or the Kuhn Tucker conditions (9.26) are within a small tolerance, stop.
5. Find a stepsize α so that $0 < \alpha \leq 1$ and $P(x^i + \alpha\, d) < P(x^i)$. Each trial stepsize requires additional evaluation of $f(x)$, $g(x)$, and $h(x)$.
6. Set $x^{i+1} = x^i + \alpha\, d$, $i = i + 1$ and go to 1.

cations of the SQP algorithm. A concise statement of the SQP algorithm is given in Table 9.1.

EXAMPLE 9.4   Performance of SQP

To illustrate the performance of SQP, we consider the solution of the following small nonlinear program:

$$\text{Min } x_2$$
$$\text{s.t. } -x_2 + 2(x_1)^2 - (x_1)^3 \leq 0 \quad\quad (9.37)$$
$$-x_2 + 2(1-x_1)^2 - (1-x_1)^3 \leq 0$$

The feasible region for Eq. (9.37) is shown in Figure 9.6a along with the countours of the objective function. From inspection we see that $x^* = [0.5, 0.375]$.

Starting from the origin ($x^0 = [0, 0]^T$) and with $B^0 = I$, we linearize the constraints and solve the following quadratic program:

$$\text{Min } d_2 + 1/2\,(d_1^2 + d_2^2)$$
$$\text{s.t. } d_2 \geq 0 \quad\quad (9.38)$$
$$d_1 + d_2 \geq 1$$

From the solution of Eq. (9.38) a search direction is obtained with $d = [1, 0]^T$ with multipliers $\mu_1 = 0$ and $\mu_2 = 1$. The contours of this quadratic function along with the linearized constraints in Eq. (9.38) are shown in Figure 9.6b for the first SQP iteration. A line search along d determines a stepsize of $\alpha = 0.5$ and the new point is $x^1 = [0.5, 0]^T$. Note that this point lies outside of the feasible region. Also, at this new point we see that from:

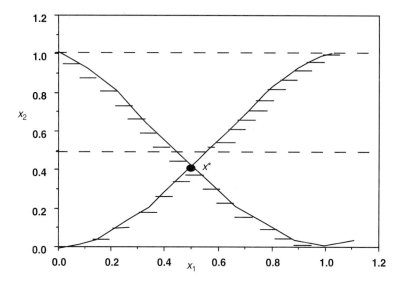

FIGURE 9.6a Contour plots and feasible region for Example 9.4.

$$\nabla_x L(x, \mu) = \begin{bmatrix} 0 \\ 1 \end{bmatrix} + \mu_1 \begin{bmatrix} 4x_1 - 3(x_1)^2 \\ -1 \end{bmatrix} + \mu_2 \begin{bmatrix} -4(1-x_1) + 3(1-x_1)^2 \\ -1 \end{bmatrix}$$

we have:

$$s = x^1 - x^0 = [0.5, 0]^T$$
$$y = \nabla_x L(x^1, \mu^1) - \nabla_x L(x^0, \mu^1)$$
$$= [-1.25, 0]^T - [-1, 0]^T = [-0.25, 0]^T$$

Since $s^T y = -0.125 < 0$, an update of the BFGS approximation cannot be made and we have $B^1 = I$.

We now move to the second iteration and at this point the following QP is solved:

$$\text{Min } d_2 + 1/2 \, (d_1^2 + d_2^2)$$
$$\text{s.t. } -1.25 \, d_1 - d_2 + 0.375 \leq 0 \quad (9.39)$$
$$1.25 \, d_1 - d_2 + 0.375 \leq 0$$

The contours of this quadratic function along with the linearized constraints in Eq. (9.39) are shown in Figure 9.6c for the second SQP iteration. Solution of this QP yields the search direction, $d = [0, 0.375]^T$ and the linesearch allows a full step to be taken so that $x^2 = [0.5, 0.375]^T$. From Eq. (9.39) we also have $\mu_1 = 0.5$ and $\mu_2 = 0.5$, so that at x^2:

$$\nabla_x L(x^2, \mu^2) = \begin{bmatrix} 0 \\ 1 \end{bmatrix} + \mu_1 \begin{bmatrix} 4x_1 - 3(x_1)^2 \\ -1 \end{bmatrix} + \mu_2 \begin{bmatrix} -4(1-x_1) + 3(1-x_1)^2 \\ -1 \end{bmatrix} = \begin{bmatrix} 0 \\ 0 \end{bmatrix}$$

$$g_1(x^2) = -x_2 + 2\,(x_1)^2 - (x_1)^3 = 0$$
$$g_1(x^2) = -x_2 + 2\,(1-x_1)^2 - (1-x_1)^3 = 0,$$

Sec. 9.3 Derivation of Successive Quadratic Programming (SQP)

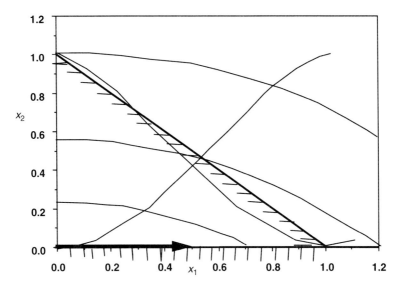

FIGURE 9.6b First SQP iteration for Example 9.4.

that is, the first Kuhn Tucker conditions are satisfied and the algorithm stops with $x^* = x^2$. Note also that since $g_1(x^*) = 0$ and $g_2(x^*) = 0$ there are no allowable directions to test positive curvature (see Eqs. 9.15, 9.16, and Example 9.3) and therefore the second order Kuhn Tucker conditions are satisfied also. A sketch of the constraints, their linearizations, and the search directions for this problem is shown below in Figure 9.6c.

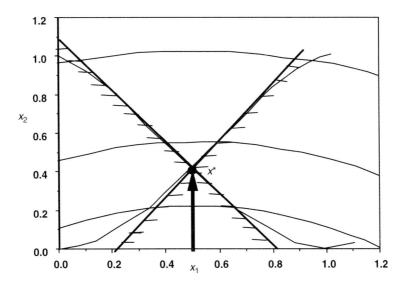

FIGURE 9.6c Second SQP iteration for Example 9.4—convergence to optimal point.

9.3.3 SQP Summary

Since 1977, the SQP algorithm has been analyzed and tested widely both in the numerical analysis and in the process engineering communities. As described above, this algorithm generally requires the fewest function evaluations of current nonlinear programming algorithms. Moreover, as seen in Example 9.4, it does not require feasible points at intermediate iterations and converges to optimal solutions from an infeasible path. Both of these properties make it desirable for flowsheet optimization problems where function evaluations are expensive. Applications of this approach will be seen in the next section.

On the other hand, performance of the SQP algorithm (although not the final solution) is dependent on scaling of the functions and variables. As a result, some care is required to prevent ill-conditioned QP problems. In addition, linearizations of constraints far from the solution lead to QP subproblems that may not have a feasible region. Under these conditions, relaxation strategies for the linearized constraints are usually applied, but they are not always successful (see Exercise 2).

Finally, the SQP algorithm described above is not efficient for large problems (say, over 100 variables) as the BFGS update (9.35) and QP subproblem (in step 3) are factorized and solved with dense linear algebra, which now becomes expensive. For these problems reduced space methods, such as MINOS (Murtagh and Saunders, 1982) described in Appendix A, or large-scale adaptations of SQP, need to be considered.

9.4 PROCESS OPTIMIZATION WITH MODULAR SIMULATORS

In Chapter 8 we defined the modular simulation mode and discussed decomposition and equation-solving strategies for modeling the process flowsheet. In this section we deal with the extension of this approach to flowsheet optimization. In addition to the flowsheet specifications and the equations that determine the mass and energy balance, we can also identify a subset of variables, $\tilde{x}$, that act as degrees of freedom for optimization. These are selected from feed streams, process stream conditions, and input specifications for individual units. For modular simulators we are especially interested in using efficient optimization strategies, such as the SQP strategy in the previous section.

Process optimization problems modeled within the modular simulation mode have a structure represented by Figure 9.7. Here the modules relating to feed processing (FP), reaction (RX), recycle separation (RS), recycle processing (RP), and product recovery (PR) contain the modeling equations and procedures. In this case, we formulate the objective and constraint functions in terms of unit and stream variables in the flowsheet and these are assumed to be implicit functions of the decision variables, $\tilde{x}$ which is a subset of x. Here the *objective function*, $f(x)$, represents processing cost, product yield, or overall profit; *product purities and operating limits* are often represented by inequalities, $g(x)$; and implicit *design specifications* are represented by additional equality constraints, $c(x)$. Since we intend to use a gradient-based algorithm, care must be taken so that the objective and constraints functions are continuous and differentiable. Moreover, for the modular approach, derivatives for the implicit module relationships (with respect to $\tilde{x}$) are not directly available.

Sec. 9.4   Process Optimization with Modular Simulators   315

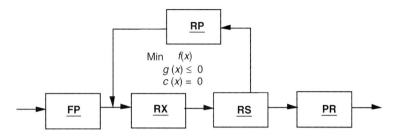

FIGURE 9.7 Structure of modular flowsheet optimization problem.

Often these need to be obtained by finite differences (and additional flowsheet evaluations) or by enhancing the unit models to provide exact derivatives directly.

Flowsheet optimization problems deal with large, arbitrarily complex models but relatively few degrees of freedom. Here, while the number of flowsheet variables could be many thousands, these are "hidden" within the simulator and the degrees of freedom are rarely more than 50 to 100 variables. As discussed in Chapter 8, the modular mode offers several advantages for flowsheet optimization. First, the flowsheeting problem is relatively easy to construct and to initialize, since numerical procedures that are tailored to each unit are applied. Moreover, the flowsheeting model is relatively easy to debug using process concepts intuitive to the process engineer. On the other hand, a drawback to using the modular mode for optimization is that unit models need to be solved repeatedly, and often careful problem definition is required to prevent intermediate failure of these process units.

Early attempts at applying optimization strategies within the modular mode were based on black-box implementations, and these were discouraging. In this simple approach, an optimization algorithm was tied around the process simulator as shown in Figure 9.8. In this black-box mode, the entire flowsheet needs to be solved repeatedly and failure in flowsheet convergence is detrimental to the optimization. Moreover, as gradients are determined by finite difference, they are often corrupted with roundoff errors from flowsheet convergence. This has adverse effects on the optimization strategy. Typically, a flowsheet optimization with ten degrees of freedom requires the equivalent time of several hundred simulations with the black box implementation.

Since the mid 1980s, however, flowsheet optimization for the modular mode has become a widely used industrial tool. This has been made possible by three advances in implementation:

1. The SQP strategy requires few function evaluations and performs very efficiently for process optimization problems with few function evaluations.
2. Intermediate convergence loops, such as recycle streams and implicit unit specifications, can be incorporated as equality constraints in the optimization problem. This is particularly important for loops that were converged with slow fixed point methods in the flowsheet. SQP, on the other hand, converges the equality and inequality constraints simultaneously with the optimization problem.

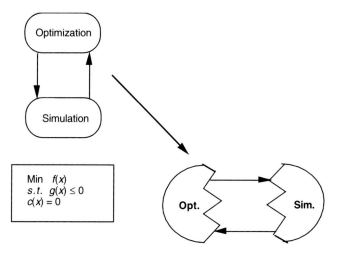

FIGURE 9.8 Evolving from the black-box (left) to the infeasible path approach using SQP.

3. Since SQP is a Newton-type method, it can be incorporated within the modular simulation environment via an "equation solver" block that is frequently used for recycle convergence. As a result, the structure of the simulation environment and the unit operations blocks does not need to be modified.

Consequently, this approach could be incorporated easily within existing modular simulators and could be applied directly to flowsheets modeled within these environments. As shown on the right in Figure 9.8, this approach "breaks open" the simulation problem and incorporates part of it into the nonlinear program. This leads to a strategy that is over an order of magnitude faster than the black-box approach and is far more reliable. A typical application of the SQP optimization strategy on a process flowsheet is shown in Figure 9.9. Here we identify optimization variables, $\tilde{x}$, as well as the tear stream and tear variables, y. As described in Chapter 8, the simulation problem can be described by: $h(y) = y - w(y)$, where $w(y)$ is the calculated tear stream from a full flowsheet pass.

The optimization problem is then formulated as:

$$\text{Min } f(\tilde{x}, y)$$
$$\text{s.t. } h(\tilde{x}, y) = y - w(\tilde{x}, y) = 0$$
$$c(\tilde{x}, y) = 0 \qquad (9.40)$$
$$g(\tilde{x}, y) \leq 0$$

and satisfaction of the tear equations (h) and the design specifications is carried out as part of the optimization problem. This problem can be solved with either the SQP algorithm or

Sec. 9.4 Process Optimization with Modular Simulators

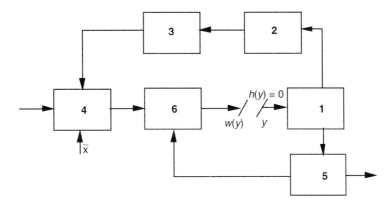

FIGURE 9.9 Typical flowsheet for process optimization.

the reduced gradient algorithm described in Appendix A. Each evaluation of the constraint and objective function requires a full flowsheet pass and additional flowsheet passes are required for the gradient calculations with respect to $\tilde{x}$ and y. Once these are obtained, the SQP method sets up and solves the following QP subproblem:

$$\text{Min}_{d} \; \nabla f(\tilde{x}^i, y^i)^T d + 1/2 \, d^T B^i d \tag{9.41}$$

$$\text{s.t.} \; h(\tilde{x}^i, y^i) + \nabla h(\tilde{x}^i, y^i)^T d = 0$$

$$c(\tilde{x}^i, y^i) + \nabla c(\tilde{x}^i, y^i)^T d = 0$$

$$g(\tilde{x}^i, y^i) + \nabla g(\tilde{x}^i, y^i)^T d \le 0$$

and the search direction, d, is used to update values for $\tilde{x}$ and y through:

$$[\tilde{x}^{i+1\,T}, y^{i+1\,T}] = [\tilde{x}^{iT}, y^{iT}] + \alpha \, d.$$

To illustrate how this approach is applied, we briefly consider the Williams-Otto process described in Chapter 8.

EXAMPLE 9.5 Williams-Otto Flowsheet Optimization

The process simulated in Chapter 8 can be extended to optimization by noting five degrees of freedom: feed flowrates (F_1 and F_2), reactor volume (V), fraction purged (v), and reactor temperature (T). These variables are all bounded and, in addition, an upper bound on the production rate is imposed. These variables are shown in the flowsheet in Figure 9.10.

The objective function is defined as the return on investment (ROI) and is given in terms of the net sales minus fixed charge, raw material, utility, and waste disposal costs. Moreover, be-

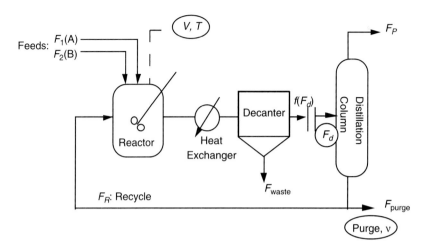

FIGURE 9.10 Williams-Otto flowsheet for optimization.

cause we have modeled the process in the modular mode, the unit equations are the same as those given in Chapter 8. Additional elements of this nonlinear programming problem are the tear equations and variables. In this case, the feed stream to the distillation column was chosen and the tear variables, F_d, represent the flowrates for components A, B, C, E, and P in this stream. The problem (9.40) consists of ten variables and five equality constraints and is given as:

$$\text{Min ROI} = [2207 F_P + 50 F_V - 168. F_A - 252 F_B \quad (9.42)$$

$$- 2.22 F_R - 84 F_{\text{waste}} + 600 V\rho - 1041.6]/ 6 V\rho$$

$$\text{s.t. } F_d - f(F_d) = 0$$

$$0 \leq F_P \leq 4763$$

$$580 \leq T \leq 680$$

$$30 \leq V \leq 100$$

$$0 \leq \nu \leq 0.99$$

$$F_1, F_2, F_R \geq 0$$

Starting and final values for these variables are shown in Table 9.2 and the NLP (42) is difficult to converge from this starting point. Moreover, there are several local solutions and singular points related to this problem. The SQP algorithm described in Table 9.1 found the optimal solution in 196 iterations of Eq. (9.41). This performance is not efficient and it illustrates the need for good problem initialization and scaling. Additional information on this problem is given in Vasantharajan and Biegler (1988).

Sec. 9.4   Process Optimization with Modular Simulators   319

TABLE 9.2   Variable Values for Williams-Otto Optimization

Variable Index	Starting Point	Optimal Values
F_R^A	8,820	46,261.4
F_R^B	39,910	143,281
F_R^C	2,360	7,585.83
F_R^P	7,890	18,826.5
F_R^E	31,660	141,847
F^A	11,540	13,164.3
F_B	31,230	29,991.2
V	60	30
T	610	674.36
v	0.5806	0.8998
ROI	124,624	131,423

EXAMPLE 9.6   Ammonia Synthesis Flowsheet Optimization

A larger scale demonstration of the infeasible path algorithm Eq. (9.40) for flowsheet optimization is given next. Here we consider the ammonia process flowsheet shown in Figure 9.11. Hydrogen and nitrogen feeds are mixed and compressed and then combined with a recycle stream and heated to reactor temperature. Reaction occurs over a multibed reactor (modeled here as an equilibrium reactor) to partially convert the stream to ammonia product. The reactor effluent is then cooled and product is separated using two flash tanks with intercooling. The liquid from the second stage is then flashed at low pressure to yield high purity liquid product. The vapor from the two stage flash forms the recycle and is compressed before mixing with the process feed.

This flowsheet was simulated using the FLOWTRAN simulator, using default economic data provided by the simulator. The objective function maximizes the net present value of the profit at a 15% rate of return and a five-year life. Optimization variables for this process are shown in Figure 9.11 and in Table 9.3; these include the tear variables (tear stream flowrates, pressure and temperature). Constraints on this process include the tear (recycle) equations, upper and lower bounds on the ratio of hydrogen to nitrogen in the reactor feed, reactor temperature limits and purity constraints. The composition of the feed streams is given by:

	Hydrogen Feed	Nitrogen Feed
N_2	5.2%	99.8%
H_2	94.0%	—
CH_4	0.79 %	0.02%
Ar	0.01%	—

However, in this problem we specify the production rate of the ammonia process rather than the feed to the process. As a result, these feed streams are left as decision variables and a production constraint is placed around the entire process. The nonlinear program is given by:

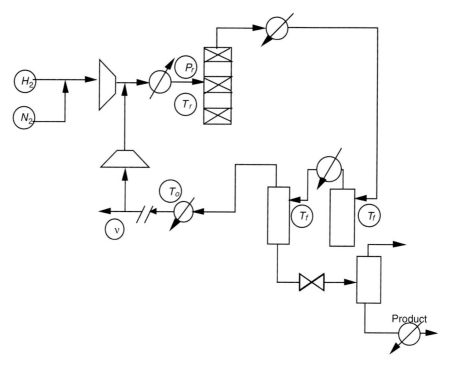

FIGURE 9.11 Ammonia process flowsheet.

Max   {Total Profit @ 15% over five years} (9.43)

s.t. • 10^5 tons NH_3/yr

• Pressure balance

• No liquid in compressors

• $1.8 \leq H_2/N_2 \leq 3.5$

• $T_{react} \leq 1000°$ F

• NH_3 purged ≤ 4.5 lb mol/hr

• NH_3 product purity ≥ 99.9 %

• Tear equations

Using the infeasible path implementation for the SQP algorithm, the ammonia process optimization converges in only five SQP iterations. Moreover, from the starting point for the NLP (given in Table 9.3) it is difficult to converge the flowsheet. As a result, a black-box optimization strategy would have severe difficulties with this problem. On the other hand, the infeasible path optimization strategy requires the equivalent time of only 2.2 base point simulations. Using SQP, the objective function improves from 20.66×10^6 to 24.93×10^6. Optimal values of the decision variables are given in Table 9.3. Additional information on this example is in Lang and Biegler (1987).

TABLE 9.3 Results of Ammonia Synthesis Problem

	Optimum	Starting Point	Lower Bound	Upper Bound
Objective Function(10^6)	24.9286	20.659		
Design Variables				
1. Inlet temp. of reactor (°F)	400	400	400	600
2. Inlet temp. of 1st flash (°F)	65	65	65	100
3. Inlet temp. of 2nd flash (°F)	35	35	35	60
4. Inlet temp. of recycle compressor (°F)	80.52	107	60	400
5. Purge fraction (%)	0.0085	0.01	0.005	0.1
6. Inlet pressure of reactor (psia)	2163.5	2000	1500	4000
7. Flowrate of feed 1 (lb mol/hr)	2629.7	2632.0	2461.4	3000
8. Flowrate of feed 2 (lb mol/hr)	691.78	691.4	643	1000
Tear Variables				
1. Flowrate (lb mol/h)				
N_2	1494.9	1648		
H_2	3618.4	3676		
NH_3	524.2	424.9		
Ar	175.3	143.7		
CH_4	1981.1	1657		
Temperature (°F)	80.52	60		
Pressure (psia)	2080.4	1930		

Based on its effectiveness, the infeasible path strategy has become a widely used tool for modular process simulators. Because it is easy to implement and also straightforward to apply to existing process models, it is used routinely for process design and operation. On the other hand, this strategy still requires unit operations procedures that are robust to input streams and design variables. Moreover, repeated convergence of the unit models at intermediate points can still be expensive. To deal with this issue, we next consider optimization strategies that can be applied to equation oriented simulators. This simulation mode leads to much faster convergence and allows very flexible specifications for the simulation problem. In the next section, we describe these advantages for the optimization problem as well.

9.5 EQUATION-ORIENTED PROCESS OPTIMIZATION

Equation-based process simulation has become popular for complex flowsheets with nested recycle streams and implicit design specifications. As described in Chapter 8, convergence of the unit operations and recycle structure occur simultaneously through a

Newton-Raphson solver. Moreover, in the equation-based mode, exact derivatives are usually available directly and performance of equation solvers and optimization algorithms does not deteriorate due to roundoff errors in gradients. On the other hand, this mode often requires careful formulation and initialization by the user, and this is often carried out by problem specific strategies.

Another trend that we observed in the last two sections is that the more process equations are incorporated into the nonlinear program Eq. (9.1), the larger the NLP becomes that must be tackled by SQP. For process flowsheets the degrees of freedom remain the same but the number of additional variables "seen" by the optimization algorithm increases in size. For instance, with the *black-box* mode, the optimization variables represent the only degrees of freedom, $\tilde{x}$, in the process. With the *infeasible path* approach with modular simulators, tear and additional design variables ($\tilde{x}$, y) are included. Finally, for *equation-based* optimization, all of the stream and unit operations variables (x) that are solved simultaneously need to be incorporated into the optimization problem. Consequently, the nonlinear programming algorithm we apply must be implemented efficiently for large-scale problems.

Unlike optimization for the modular mode, a significant computational cost for equation-based optimization lies not in function evaluations of the process flowsheet, but in the effort expended by the NLP algorithm itself. Here, we are faced with models that are large systems of equations with relatively few degrees of freedom. Computational costs incurred with handling large systems of (linearized) equations tend to dominate, but, as described in Chapter 7, function evaluations from physical property routines also carry a significant computational cost. Hence both the efficiency of the NLP algorithm and the number of function evaluations required are important considerations.

Moreover, the SQP algorithm presented in section 9.3 is not well suited for large problems. While it requires few iterations and function evaluations for convergence, the basic SQP algorithm does not exploit sparsity of the constraint gradients and, in particular, the solution of the QP subproblem is performed with a dense matrix implementation. As a result, the effort to solve this subproblem increases cubically with the problem size. On the other hand, the reduced gradient method (MINOS) described in Appendix A is well suited for many large-scale problems in the equation-oriented mode. While SQP solves quadratic programming subproblems, MINOS (Murtagh and Saunders, 1982) solves linearly constrained NLP subproblems. As a result, it requires many more function evaluations. Nevertheless, it exploits sparsity in the constraint gradients and is implemented with very efficient matrix decomposition procedures. Finally, as a result of this decomposition it solves a nonlinear optimization problem in the reduced space defined by the degrees of freedom for optimization. Since these remain small for flowsheet optimization, MINOS can be very efficient for equation-based flowsheets. On the other hand, SQP has stronger global convergence properties than MINOS and in practice MINOS often has difficulties handling nonlinear constraints that often arise in flowsheet models.

Based on the characteristics of equation-oriented optimization and the relative advantages of SQP and MINOS, we consider a large-scale SQP strategy to incorporate many of the large-scale features in MINOS and also preserve the strong convergence properties of SQP. The resulting SQP algorithm combines these two aspects, works in the

9.5.1 Development of a Large-Scale SQP Strategy

Consider the large scale nonlinear program given by:

$$\text{Min } f(z)$$
$$\text{s.t. } h(z) = 0 \qquad (9.44)$$
$$z_L \le z \le z_U$$

For convenience, we convert the inequality constraints to equalities through the addition of slack variables, $s \ge 0$. The NLP problem is redefined with $z^T = [x^T\ s^T]$ and with n variables and m equality constraints. At iteration i the quadratic programming problem in SQP can be written as:

$$\text{Min } \nabla f(z^i)^T d + 1/2\, d^T B^i d$$
$$\text{s.t. } h(z^i) + \nabla h(z^i)^T d = 0 \qquad (9.45)$$
$$z_L \le z^i + d \le z_U$$

where d is the n dimensional search direction and B^i is the $n \times n$ Hessian of the Lagrangian function or its approximation. For large problems, a BFGS approximation is impractical because it creates a large, dense matrix. Instead, large-scale applications for SQP can be classified into two general approaches: full space and reduced space algorithms.

In the full space approach, the sparse structure of the QP is exploited directly. An advantage of this approach is that the matrix structures of *both* B^i and ∇h are exploited and an efficient factorization can be made. One way to maintain the sparsity of the matrix B^i is by using exact second derivatives for the Lagrange function. This approach is especially well suited to problems with many degrees of freedom, such as in trajectory or shape optimization. However, in addition to the task of providing the second derivatives from the flowsheet, solving the QP can become more difficult if the Hessian matrix is not positive definite. Consequently, a more complex algorithm needs to be derived.

In the reduced space method (rSQP), on the other hand, only the structure of ∇h is exploited and a *projection* of the Hessian matrix is constructed. The order of the projected matrix is equal to the degrees of freedom $(n - m)$ and the matrix can be calculated directly from the exact second derivatives or through a BFGS approximation. This method can be derived from the optimality conditions of the QP (9.45). Ignoring the bound constraints for the moment, this leaves the following linear system:

$$\begin{bmatrix} B & \nabla h \\ \nabla h^T & 0 \end{bmatrix} \begin{bmatrix} d \\ \lambda \end{bmatrix} = -\begin{bmatrix} \nabla f \\ h \end{bmatrix} \qquad (9.46)$$

We now define an $n \times m$ matrix Y and an $n \times (n - m)$ matrix Z that have the properties:

$$\nabla h(z^i)^T Z = 0 \text{ and } \quad [Y \mid Z] \text{ is a nonsingular square matrix.} \quad (9.47)$$

Because of this nonsingular matrix, the search direction can be partitioned into two vector components, d_Y and d_Z, respectively:

$$d = Y d_Y + Z d_Z \quad (9.48)$$

Here the matrix Y is a representation of the range space of $\nabla h(z^i)$ and the vector d_Y contains the variables that are used to satisfy the constraints. On the other hand, the matrix Z is a representation of the null space of $\nabla h(z^i)^T$ and the vector d_Z contains the variables that are used to improve the objective function. By applying the partition of d in Eq. (9.48) and premultiplying the first row of Eq. (9.46) by the transpose of $[Y \mid Z]$, we rewrite optimality conditions as:

$$\begin{bmatrix} Y^T B Y & Y^T B Z & R \\ Z^T B Y & Z^T B Z & 0 \\ R^T & 0 & 0 \end{bmatrix} \begin{bmatrix} d_Y \\ d_Z \\ \lambda \end{bmatrix} = - \begin{bmatrix} Y^T \nabla f \\ Z^T \nabla f \\ h \end{bmatrix} \quad (9.49)$$

where $R = Y^T \nabla h(z^i)$, a square, nonsingular matrix of order m. Note that this linear system is equivalent to the original one, Eq. (9.46), but it leads to an easier decomposition. From the last row of Eq. (9.49) we solve a sparse system of m equations:

$$R^T d_Y = - h(z^i) \quad (9.50)$$

to obtain d_Y. With this solution we solve a set of $(n - m)$ equations for d_Z from the second row of Eq. (9.49):

$$Z^T B Z d_Z = - (Z^T \nabla f(z^i) + Z^T B Y d_Y) \quad (9.51)$$

and this completely defines the search direction. Since both the range and null space steps, d_Y and d_Z, vanish upon convergence, an easier way to calculate the Lagrange multipliers is to neglect the Hessian terms in the first row of Eq. (9.49) and calculate:

$$R \lambda = - Y^T \nabla f(z^i). \quad (9.52)$$

To extend this decomposition to cover variable bounds in Eq. (9.45), the null space step is not determined from Eq. (9.51). Instead, after solving for d_Y from Eq. (9.50), we obtain d_Z by solving the following quadratic program:

$$\text{Min } (Z^T \nabla f(z^i) + Z^T B Y d_Y)^T d_Z + 1/2 \, d_Z^T Z^T B Z d_Z \quad (9.53)$$

$$\text{s.t. } z_L \leq z^i + Y d_Y + Z d_Z \leq z_U$$

with the equality constraints eliminated. Note that only the projected Hessian ($Z^T B Z$) needs to be calculated or approximated with a BFGS update. The "cross term" $Z^T B Y d_Y$ in Eq. (9.53) can be evaluated with exact second derivatives or approximated by finite difference. Often, however, this term is simply set to zero and in cases where d_Y is smaller in magnitude than d_Z, convergence of the SQP method is not affected by neglecting this term.

The reduced Hessian SQP approach has a number of advantages over the basic SQP method. In particular, the basis matrices Y and Z can be chosen so that efficient sparse

Sec. 9.5 Equation-Oriented Process Optimization

matrix factorizations can be used. To determine Y and Z we partition the variables z into $n - m$ independent and m dependent variables, u and v, respectively. This partition is chosen so that v can be determined from the equality constraints once u is fixed. For process optimization, u therefore represents the decision variables for optimization while v represents dependent variables calculated in the flowsheet. Now we partition the sparse system of constraint gradients into:

$$\nabla_z h(z^i)^T = [\nabla_u h(z^i)^T \mid \nabla_v h(z^i)^T] = [N \mid C] \tag{9.54}$$

where C is assumed to be a square, nonsingular matrix of order m. Z is therefore given by:

$$Z = \begin{bmatrix} I \\ -C^{-1} N \end{bmatrix} \tag{9.55}$$

which satisfies $\nabla h(z^i)^T Z = 0$. Y is chosen so that the $n \times n$ matrix $[Y \mid Z]$ is nonsingular and two popular choices for this are the coordinate basis and the orthogonal basis:

$$Y = \begin{bmatrix} 0 \\ I \end{bmatrix} \text{ and } Y = \begin{bmatrix} N^T C^{-T} \\ I \end{bmatrix} \tag{9.56}$$

respectively. With the orthogonal basis, $Y^T Z = 0$, and the range space step, d_Y, is determined by a least squares projection and is of minimum length. This generally leads to fewer SQP iterations and more stable performance. On the other hand, calculation of d_Y is proportional to $(n - m)^3$.

Calculating the range space step with the coordinate basis is much cheaper as it involves only a factorization of the C matrix, that is, $C\,d_Y = -h(z^i)$. In fact, this step is identical to calculating a Newton step for solving the process flowsheet. This property makes the coordinate basis very desirable when implementing SQP strategies to large process models. However, d_Y determined by the coordinate basis may lead to large search directions and safeguards are often required to avoid poor performance of the NLP solver.

The large scale SQP strategy is nonetheless very similar to the one derived in section 9.3. A summary of the rSQP algorithm is presented in Table 9.4. Note that aside from the decomposition and elimination of the equality constraints, many of the components of the basic SQP strategy remain, including the line search method and the BFGS formula applied to the smaller ($Z^T B Z$) matrix. Nevertheless, the key difference between the algorithm in Table 9.4 and the basic SQP algorithm in Table 9.1 is the decomposition step required to find the QP search direction. To illustrate how the range and null space decomposition procedure works, we consider a quadratic program at iteration i in Example 9.7:

EXAMPLE 9.7 An Iteration of the rSQP Algorithm

At iteration i, consider the following quadratic program with $n = 3$ and $m = 2$:

$$\text{Min } (5\,d_1 + d_2 + 4\,d_3) + 1/2\,(d_1^2 + 4\,d_2^2 + 3\,d_3^2)$$

$$\text{s.t. } d_1 + 2\,d_2 = 7$$

TABLE 9.4 Reduced Hessian SQP Algorithm

1. Choose starting point, z^0.
2. At iteration i, evaluate functions and gradients, $\nabla f(z^i)$ and $\nabla h(z^i)$
3. Calculate basis matrices **Y** and **Z**.
4. Solve for step d_Y in Range space using sparse matrix factorizations Eq. (9.50):

$$(\nabla h(z^i)^T Y)\, d_Y = -h(z^i)$$

   and, if needed, calculate the cross term, $Z^T B Y\, d_Y$.
5. Solve small QP Eq. (9.53) for step d_Z in Null space.

$$\text{Min } (Z^T \nabla f(z^i) + Z^T B Y\, d_Y)^T d_Z + 1/2\, d_Z^T\, Z^T B Z\, d_Z$$

$$s.t.\ z_L \le z^i + Y\, d_Y + Z\, d_Z \le z_U$$

6. If the search direction or the Kuhn Tucker error is less than a zero tolerance, stop.
7. Else, calculate the total step $d = Y\, d_Y + Z\, d_Z$.
8. Find a stepsize α so that $0 < \alpha \le 1$ and $P(z^i + \alpha\, d) < P(z^i)$. Each trial stepsize requires additional evaluation of $f(z)$ and $h(z)$.
9. Update projected (small) Hessian ($Z^T B Z$) using the BFGS formula.
10. Set $z^{i+1} = z^i + \alpha\, d$, $i = i + 1$ and go to step 2.

$$2 d_1 + 3 d_3 = 5$$

$$-1 \le d_1 \le 5$$

$$-2 \le d_2 \le 6$$

$$0 \le d_3 \le 4$$

Clearly, the terms in the QP (9.45) can be identified as:

$$\nabla f(z^i)^T = [5,\ 1,\ 4] \quad h(z^i)^T = [-7,\ -5] \text{ and } B^i = \begin{bmatrix} 1 & 0 & 0 \\ 0 & 4 & 0 \\ 0 & 0 & 3 \end{bmatrix}.$$

Bounds for the QP are given by:

$$z_L - z^i = [-1, -2, 0]^T \quad z_U - z^i = [5, 6, 4]^T$$

and the constraint gradients can be partitioned into:

$$\nabla h(z^i)^T = [N\ \ C] = \begin{bmatrix} 1 & 2 & 0 \\ 2 & 0 & 3 \end{bmatrix} \text{ with } C = \begin{bmatrix} 2 & 0 \\ 0 & 3 \end{bmatrix} \text{ and } N = \begin{bmatrix} 1 \\ 2 \end{bmatrix}.$$

From the definition Eq. (9.55) we can evaluate the $n \times (n-m)$ Z matrix as:

$$Z = \begin{bmatrix} I \\ -C^{-1}N \end{bmatrix} = \begin{bmatrix} 1 \\ -1/2 \\ -2/3 \end{bmatrix}$$

and choosing the coordinate basis for Eq. (9.56) yields the following $n \times m$ matrix Y:

$$Y = \begin{bmatrix} 0 \\ I \end{bmatrix} = \begin{bmatrix} 0 & 0 \\ 1 & 0 \\ 0 & 1 \end{bmatrix}$$

Sec. 9.5 Equation-Oriented Process Optimization

and it can readily be verified that $\nabla h(z^i)^T Z = 0$ and that $[Y \mid Z]$ is a nonsingular square matrix. Now to calculate the search direction, $d = Y\, d_Y + Z\, d_Z$, we consider the range and null space component vectors. The m dimensional vector d_Y can be evaluated from Eq. (9.50) and the following relation:

$$\nabla h(z^i)^T Y = R^T = C$$

This leads to:

$$R^T d_Y = C d_Y = -h(z^i) \text{ or } \begin{bmatrix} 2 & 0 \\ 0 & 3 \end{bmatrix} d_Y = \begin{bmatrix} 7 \\ 5 \end{bmatrix}$$

and the range space step is given by: $d_Y = [7/2, 5/3]^T$. For the $(n - m)$ dimensional vector d_Z we need to solve the QP Eq. (9.53). The components that make up this QP can be evaluated as follows:

$$Y d_Y = \begin{bmatrix} 0 & 0 \\ 1 & 0 \\ 0 & 1 \end{bmatrix} \begin{bmatrix} 7/2 \\ 5/3 \end{bmatrix} = \begin{bmatrix} 0 \\ 7/2 \\ 5/3 \end{bmatrix}$$

$$Z^T B Z = [1 \ -1/2 \ -2/3] \begin{bmatrix} 1 & 0 & 0 \\ 0 & 4 & 0 \\ 0 & 0 & 3 \end{bmatrix} \begin{bmatrix} 1 \\ -1/2 \\ -2/3 \end{bmatrix} = 10/3$$

$$Z^T \nabla f(z^i) = [1 \ -1/2 \ -2/3] \begin{bmatrix} 5 \\ 1 \\ 4 \end{bmatrix} = 11/6$$

$$Z^T B Y\, d_Y = [1 \ -1/2 \ -2/3] \begin{bmatrix} 1 & 0 & 0 \\ 0 & 4 & 0 \\ 0 & 0 & 3 \end{bmatrix} \begin{bmatrix} 0 & 0 \\ 1 & 0 \\ 0 & 1 \end{bmatrix} \begin{bmatrix} 7/2 \\ 5/3 \end{bmatrix} = -31/3$$

Combining these terms into the QP Eq. (9.53):

$$\text{Min } (Z^T \nabla f(z^i) + Z^T B Y\, d_Y)^T d_Z + 1/2\, d_Z^T (Z^T B Z)\, d_Z$$
$$\text{s.t. } z_L - z^i - Y d_Y \leq Z d_Z \leq z_U - z^i - Y d_Y$$

yields the following QP for d_Z

$$\text{Min } (-17/2)\, d_Z + (10/6)\, (d_Z)^2$$

$$\text{s.t. } \begin{bmatrix} -1 \\ -11/2 \\ -5/3 \end{bmatrix} \leq \begin{bmatrix} 1 \\ -1/2 \\ -2/3 \end{bmatrix} d_z \leq \begin{bmatrix} 5 \\ 5/2 \\ 7/3 \end{bmatrix}$$

whose solution is $d_Z = 5/2$. Combining the range and null space steps leads to the overall solution vector:

$$d = Y\, d_Y + Z\, d_Z = \begin{bmatrix} 0 \\ 7/2 \\ 5/3 \end{bmatrix} + \begin{bmatrix} 1 \\ -1/2 \\ -2/3 \end{bmatrix} (5/2) = \begin{bmatrix} 5/2 \\ 9/4 \\ 0 \end{bmatrix}.$$

9.5.2 Characteristics of Reduced Hessian SQP

In Example 9.7 we see that the range and null space decomposition of Eqs. (9.50) and (9.53) are equivalent to solving the original QP subproblem Eq. (9.45). Consequently, the reduced Hessian SQP (rSQP) strategy has much in common with the basic SQP strategy. On the other hand, in the rSQP algorithm of Table 9.4, the $Z^T B^i Y\, d_Y$ term and reduced Hessian $Z^T B^i Z$ are calculated directly and not derived from the full Hessian, B^i. Consequently, the full Hessian does not need to be evaluated or approximated.

The local convergence properties are also similar for both SQP and rSQP. Moreover, if the cross term, $Z^T B^i Y\, d_Y$, is included in Eq. (9.53) (say, with a finite difference approximation) then the convergence rate of rSQP actually improves from 2-step to 1-step superlinear, slightly better than the basic SQP method. Another advantage of the reduced strategy is that the actual projected Hessian is expected to be positive definite at a local solution (from the second order optimality conditions), while the full Hessian is not. As a result, using a BFGS approximation for ($Z^T B^i Z$) in Eq. (9.53) leads to much better conditioning and performance than a direct application of B^i in Eq. (9.45).

For instance, for the small Williams-Otto problem in Example 9.5 (and Table 9.2), choosing the tear equations as dependent variables and solving with rSQP (in Table 9.4) requires only *38 iterations with the orthogonal basis for Y (40 iterations with the coordinate basis)* vs. *196 iterations* with the basic SQP method. As a result, rSQP has advantages even for smaller problems.

For large problems, the computational differences of the two methods are dominated by differences in linear algebra calculations. These costs can be summarized by the following relations:

$$\text{Cost for basic SQP} = k_1\, m^3 + k_2\, (n-m)^\beta$$

$$\text{Cost for rSQP} = k_3\, m^\alpha + k_4\, (n-m)^\beta$$

where the constants k_i are of the same order of magnitude. The exponent α deals with the cost of sparse matrix decomposition and is usually between one and two. The exponent β refers to the cost of solving the quadratic program, and depending on the particular QP algorithm selected, this exponent is between two and three. Consequently, for problems where $(n-m)$ is small, the key advantage to rSQP lies in the difference in the first terms, which is due to the sparse elimination of the equality constraints. This leads to performance differences on small process optimization problems (say, 1000 variables and less than 10 degrees of freedom) of over an order of magnitude. Consequently, for problems of this size and larger, the rSQP strategy in Table 9.4 is clearly superior.

EXAMPLE 9.8 Real-Time Optimization with rSQP

In this last example we determine the optimal operating conditions for the Sunoco Hydrocracker Fractionation Plant. This problem represents an existing process and the optimization problem is solved on-line at regular intervals to update the best current operating conditions. These task is termed *real-time optimization*. The fractionation plant separates the effluent stream from a hydrocracking unit and the relevant portion of the plant is shaded in Figure 9.12. The process has 17 hydrocarbon components, six process/utility heat exchangers, two process/process heat exchangers, and the following column models: absorber/stripper (30 trays), debutanizer (20 trays), C3/C4 splitter (20 trays), and a deisobutanizer (33 trays). Further details on the individual units may be found in Bailey et al. (1993).

Sec. 9.5 Equation-Oriented Process Optimization

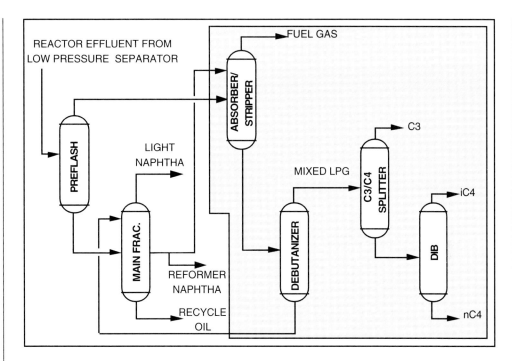

FIGURE 9.12 Sunoco Hydrocracker flowsheet for real-time optimization.

To solve real-time optimization problems a two-step procedure is normally considered. First, one solves a single square *parameter case* in order to fit the model to an operating point. The optimization is performed next, starting from this point. In an on-line system, the solution to the parameter case constitutes the current operating conditions. The process model consists of equality constraints used to represent the individual units and a number of simple bounds that represent actual physical limits on the variables (e.g., nonnegativity constraints on flows and temperatures), as well as bounds on key variables to prevent large changes from the current point. The model consists of 2836 equality constraints and only ten independent variables. It is also reasonably sparse and contains 24123 nonzero Jacobian elements.

The objective function for the on-line optimization includes the energy costs as well as a measure of the value added to the raw materials through processing. The form of the objective function is given below and details on each of the four terms may be found in Bailey et al.

$$P = \sum_{i \in E} z_i C_i^G + \sum_{i \in E} z_i C_i^E + \sum_{m=1}^{Np} \sum_{i \in P_m} z_i C^{P_m} - U$$

where P = profit,
C^G = value of the feed and product streams valued as gasoline,
z_i = stream flowrates
C^E = value of the feed and product streams valued as fuel,

C^{Pm} = value of pure component feed and products, and
U = utility costs.

In addition to the base optimization, four problems were considered in this case study. In Cases 2 and 3 the effect of fouling is simulated by reducing the heat exchange coefficients for the debutanizer and splitter feed/bottoms exchangers. Changing market conditions are reflected by an increase in the price for propane (Case 4) or an increase in the base price for gasoline together with an increase in the octane credit (Case 5). The numerical values for the above parameters are included in Table 9.5.

The rSQP algorithm of Table 9.4 with a coordinate basis (9.56) was applied to this problem and more details of this implementation can be found in Schmid and Biegler (1994). These cases were solved on a DEC 5000/200 using a convergence tolerance of 10^{-8} and results are reported in Table 9.5. Here "infeasible initialization" indicates initialization at a poor starting point while the "parameter initialization" results were obtained using the solution to the parameter case (at current operating conditions) as the initial point. We also compare the results obtained by Bailey et al. using MINOS.

From Table 9.5 we see that rSQP is 8 times faster than MINOS for the parameter case. Moreover, in all cases MINOS requires as many as two orders of magnitude more function evaluations than rSQP does. Since the solution to the parameter case is the starting point in an online system, it is appropriate to compare first the "parameter initialization" results. In Table 9.5 we see an order of magnitude improvement in CPU times when comparing rSQP to MINOS. Finally, Bailey et al. (1993) report only one result for an optimization case which was initialized at the original "infeasible initialization". When this MINOS result is compared to the rSQP result, there is a time difference of almost two orders of magnitude. As a result, it appears that rSQP is less sensitive to poor initial points than MINOS.

TABLE 9.5 Numerical Results for the Sunoco Hydrocracker Fractionation Plant Problem.

	Case 0 Base Parameter	Case 1 Base Optimization	Case 2 Fouling 1	Case 3 Fouling 2	Case 4 Changing Market 1	Case 5 Changing Market 2
Heat Exchange Coefficient (TJ/d$\times$C)						
Debutanizer feed/bottoms	6.565×10^{-4}	6.565×10^{-4}	5.000×10^{-4}	2.000×10^{-4}	6.565×10^{-4}	6.565×10^{-4}
Splitter feed/bottoms	1.030×10^{-3}	1.030×10^{-3}	5.000×10^{-4}	2.000×10^{-4}	1.030×10^{-3}	1.030×10^{-3}
Pricing						
Propane ($/m^3)	180	180	180	180	300	180
Gasoline base price ($/m^3)	300	300	300	300	300	350
Octane credit ($/(RON m^3))	2.5	2.5	2.5	2.5	2.5	10
Profit	230968.96	239277.37	239267.57	236706.82	258913.28	370053.98
Change from base case ($/d, %)	—	8308.41 (3.6%)	8298.61 (3.6%)	5737.86 (2.5%)	27944.32 (12.1%)	139085.02 (60.2%)
Infeasible Initialization						
MINOS						
Iterations (major/minor)	5/275	9/788	—	—	—	—
CPU time (s)	182	5768	—	—	—	—
rSQP						
Iterations	5	20	12	24	17	12
CPU time (s)	23.3	80.1	54.0	93.9	69.8	54.2

TABLE 9.5 Continued

	Case 0 Base Parameter	Case 1 Base Optimization	Case 2 Fouling 1	Case 3 Fouling 2	Case 4 Changing Market 1	Case 5 Changing Market 2
Parameter Initialization						
MINOS						
Iterations (major/minor)	n/a	12 / 132	14 / 120	16 / 156	11 / 166	11 / 76
CPU Time (s)	n/a	462	408	1022	916	309
rSQP						
Iterations	n/a	13	8	18	11	10
CPU time (s)	n/a	58.8	43.8	74.4	52.5	49.7
Time rSQP/Time MNINOS (%)	12.8%	12.7%	10.7%	7.3%	5.7%	16.1%

9.6 SUMMARY AND CONCLUSIONS

This chapter provides a brief introduction to nonlinear programming for process optimization. In particular, process flowsheeting applications that were developed in the previous chapter were considered and optimization strategies for both modular and equation based simulation modes were presented. In addition to providing some basic nonlinear programming concepts as well as reference to reduced gradient algorithms, we highlighted the development of the Successive Quadratic Programming algorithm and its extension to large scale problems. For flowsheet optimization, both for process design and for on-line optimization, SQP has emerged as the most popular algorithm.

A key advantage to SQP is that it requires few iterations (and function and gradient evaluations to converge)—this is due to its Newton-like properties. In fact, from the presentation in this chapter, it is easy to see that SQP is a direct extension of the Newton-Raphson method, generalized from nonlinear equation solving to nonlinear programming. (In the absence of degrees of freedom, SQP actually devolves to a Newton method). As a result, inequality and equality constraints converge simultaneously with the optimization problem and intermediate convergence of the process equations is not required for the NLP.

In the modular simulation mode, SQP can be applied directly to flowsheet optimization problems with the tear equations and design specifications incorporated as design constraints. With the decision and tear variables included within the optimization problem, the NLP rarely exceeds 100 variables. This approach can be implemented very easily within existing process simulators and, as a result, this "infeasible path" strategy is widely used in industry. Two flowsheeting examples were used to demonstrate this approach.

For the equation-based simulation mode, on the other hand, a large scale NLP solver is needed. As with the modular mode, the degrees of freedom remain small but the total number of variables can range from 10,000 to 100,000 and beyond. Consequently, an algorithm that exploits the size and structure of the process model is needed. In this chapter we developed the reduced Hessian SQP (rSQP) strategy, which can be orders of magnitude faster than the basic SQP method and shares many of the large scale features of the MINOS algorithm. To demonstrate the performance of this method, a case study for a real-time process optimization problem was presented.

In later chapters dealing with process synthesis, optimization problems will be extended to include integer variables as well (to form MINLPs). In solving these, we will still solve NLP problems in an inner loop using reduced gradient and SQP strategies. Moreover, there are many other process applications for which SQP has been very successful, including control and dynamics applications, parameter estimation for steady state and dynamic systems, and multiperiod problems. The flexibility and adaptability of this method builds on the decomposition characteristics of SQP that were sketched in this chapter.

9.6.2 Notes for Further Reading

The basic SQP method was developed by Han (1977) and Powell (1977). A comprehensive treatment of the derivation and properties of SQP can be found in the texts by Gill, Murray, and Wright (1981) and Fletcher (1987). The rSQP method has evolved from a number of studies, starting from Murray and Wright (1978). An analysis of the rSQP method is presented in Nocedal and Overton (1985) and an updated analysis of the rSQP method is given in Biegler et al. (1995). Comprehensive numerical studies and comparisons for the SQP method are described in Schittkowski (1987). Studies in process engineering include Berna et al. (1980), Locke et al. (1983), Vasantharajan and Biegler (1988) and Vasatharajan et al. (1990). In particular, a comparison of SQP and rSQP strategies (with different basis representations) is given in Vasantharajan and Biegler (1988). A state of the art implementation of rSQP is discussed in Schmid and Biegler (1994). Finally, sparse full space SQP strategies for large NLPs are discussed in Betts and Huffman (1992), Lucia et al. (1990) and Sargent (1995).

The development of the SQP strategy for modular flowsheets can be found in Biegler and Hughes (1982) and Chen and Stadtherr (1985). Extensions for flowsheet optimization were also proposed in Lang and Biegler (1987) and Kisala et al. (1987). Current implementations of the SQP method in process simulators can be found in the ASPEN, PRO/II, HYSYS, and SPEEDUP simulators. More information on their application can be found in their commercial documentation. It is interesting to note that the SQP method is useful not only for flowsheet optimization, but also as a convergence block to deal with difficult flowsheets.

Finally, the Sunoco Hydrocracker problem was developed by Bailey et al. (1993) and the application of rSQP is given in Schmid and Biegler (1994). In addition, real-time optimization packages such as DMO, NOVA, and RTOPT make use of the large-scale SQP concepts discussed in this chapter.

REFERENCES

Bailey, J. K., Hrymak, A. N., Treiber, S. S., & Hawkins, R. B. (1993). Nonlinear optimization of a Hydrocracker Fractionation Plant. *Comput. chem. Engng.*, **17,** 123.

References

Beale, E. M. L. (1967). Numerical methods. In J. Abadic (Ed.), *Nonlinear Programming* (p. 189). Amsterdam: North Holland.

Berna, T., Locke, M. H., & Westerberg, A. W. (1980). A new approach to optimization of chemical processes. *AIChE J.,* 26, 37.

Betts, J. T., & Huffman, W. P. (1992). Application of sparse nonlinear programming to trajectory optimization. *J. Guid. Control Dyn.*, **15** (1), 198.

Biegler, L. T., & Hughes, R. R. (1982). Infeasible path optimization of sequential modular simulators. *AIChE J.*, **26,** 37.

Biegler, L. T., Nocedal, J., & Schmid, C. (1995). Reduced Hessian strategies for large-scale nonlinear programming. *SIAM Journal of Optimization*, **5** (2), 314.

Bracken, J., & McCormick, G. (1968). *Selected Applications in Nonlinear Programming.* New York: Wiley.

Chen, H-S, & Stadtherr, M. A. (1985). A simultaneous modular approach to process flowsheeting and optimization. *AIChE J.*, **31,** 1843.

Dennis, J. E., & Schnabel, R. B. (1983). *Numerical Methods for Unconstrained Optimization and Nonlinear Equations.* Englewood Cliffs, NJ: Prentice-Hall.

Fletcher, R. (1987). *Practical Methods of Optimization.* New York: Wiley.

Gill, P. E., Murray, W., Saunders, M. A., & Wright, M. H. (1981). *Practical Optimization.* New York: Academic Press.

Han, S-P. (1977). A globally convergent method for nonlinear programming. *JOTA,* **22,** 297.

Karush, N. (1939). MS Thesis, Department of Mathematics, University of Chicago.

Kisala, T. P., Trevino-Lozano, R. A., Boston, J. F., Britt, H. I., & Evans, L. B. (1987). Sequential modular and simultaneous modular strategies for process flowsheet optimization. *Comput. Chem. Eng.*, **11,** 567–579.

Kuhn, H. W., & Tucker, A. W. (1951). Nonlinear programming. In J. Neyman (Ed.), *Proceedings of the Second Berkeley Symposium on Mathematical Statistics and Probability* (p. 481), Berkeley, CA: University of California Press.

Lang, Y-D, & Biegler, L. T. (1987). A unified algorithm for flowsheet optimization. *Comput. chem. Engng.*, **11,** 143.

Liebman, J., Lasdon, L., Shrage, L., & Waren, A. (1984). *Modeling and Optimization with GINO.* Palo Alto: Scientific Press.

Locke, M. H., Edahl, R., & Westerberg, A. W. (1983). An improved Successive Quadratic Programming optimization algorithm for engineering design problems. *AIChE J.*, 29, 5.

Lucia, A., Xu J., & D'Couto, G. C. (1990). Sparse quadratic programming in chemical process optimization. *Ann. Oper. Res.,* **42,** 55.

Murray, W., & Wright, M. (1978). Projected Lagrangian methods based on trajectories of barrier and penalty methods. *SOL Report* 78-23, Stanford University.

Murtagh, B. A., & Saunders, M. A. (1982). A projected Lagrangian algorithm and its implementation for sparse nonlinear constraints. *Math Prog Study*, 16, 84–117.

Nocedal, J., & Overton, M. (1985). Projected Hessian updating algorithms for nonlinearly constrained optimization. *SIAM J. Num. Anal.*, **22**, 5.

Powell, M. J. D. (1977). A fast algorithm for nonlinear constrained optimization calculations. *1977 Dundee Conference on Numerical Analysis*.

Sargent, R. W. H. (1995). A new SQP algorithm for large-scale nonlinear programming, Report C95-36. London: Centre for Process Systems Engineering, Imperial College.

Schmid, C., & Biegler, L. T. (1994). Quadratic programming algorithms for reduced Hessian SQP. *Computers and Chemical Engineering*, **18**, 817.

Schittkowski, K. *More test examples for nonlinear programming codes*. Lecture notes in economics and mathematical systems # 282. Berlin; New York: Springer-Verlag.

Vasantharajan, S., & Biegler, L. T. (1988). Large-scale decomposition for Successive Quadratic Programming. *Computers and Chemical Engineering*, 12, 1089.

Vasantharajan, S., Viswanathan, J., & Biegler, L. T. (1990). Reduced Successive Quadratic Programming implementation for large-scale optimization problems with smaller degrees of freedom. *Comput. chem. Engng.*, 14, 907.

Wilson, R. B. (1963). A simplicial algorithm for concave programming. PhD Thesis, Harvard University.

EXERCISES

1. Show that the NLP represented in Figure 9.4 is convex and has a unique minimum solution.

2. Consider the nonconvex, constrained NLP in Example 9.2. Write the Kuhn Tucker conditions for this problem.
   a. Show that this problem is nonconvex.
   b. What can you say about the optimal active set of inequalities for this problem?
   c. How does the system of Kuhn Tucker conditions lead to multiple NLP solutions?

3. Consider the NLP:

$$\text{Min } x_2$$
$$\text{s.t. } x_1 - x_2^2 + 1 \leq 0$$
$$-x_1 - x_2^2 + 1 \leq 0$$

   a. Sketch the feasible region for this problem
   b. What happens if $x_1 = x_2 = 0$ is chosen as a starting point and SQP or reduced gradient methods are applied?

4. Show the exact penalty function:

$$P(x, \eta) = f(x) + \eta \left(\sum_j \max(0, g_j(x)) + \sum_k |h_k(x)| \right)$$

has a descent direction in d (QP solution) if

$$\eta > \max_{jk} [\mu_j, |\lambda_k|]$$

5. **a.** Show that the solution of the QP

$$\text{Min } a^T x + 1/2\, x^T(B + \rho A^T A)x$$
$$\text{s.t. } Ax = b$$

is independent of ρ.

b. Consider the augmented Lagrange function:

$$L^*(x^k,\lambda) = L(x^k,\lambda) + \rho\, x_k^T A^T A x_k,$$

where $L(x^k,\lambda) = f(x^k) + h(x^k)^T \lambda$, $A = \nabla h(x^k)$ and $Z(x_k)^T A = 0$. Show that if $Z(x_k)^T \nabla_{xx} L(x^k,\lambda) Z(x_k)$ is always positive definite then this function has a positive definite Hessian for ρ sufficiently large and A linearly independent.

c. What are the implications of part b) for using an augmented Lagrangian function as a merit function?

6. While searching for the minimum of

$$f(x) = [x_1^2 + (x_2+1)^2][x_1^2 + (x_2-1)^2]$$

we terminate at the following points
 a. $x^{(1)} = [0,0]^T$
 b. $x^{(2)} = [0,1]^T$
 c. $x^{(3)} = [0,-1]^T$
 d. $x^{(4)} = [1,1]^T$

Classify each point.

7. Show that the Kuhn Tucker conditions of the QP (9.28) correspond to the linear system (9.27), which corresponds to a Newton step for the nonlinear equations (9.26).

8. The following flowsheet is given by:

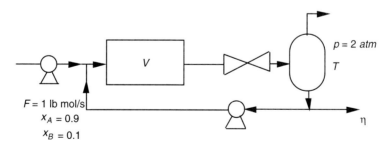

FIGURE 9.13

Reaction $A \Leftrightarrow B$ (plug flow)

$dC_A/dt = k_1 C_A + k_2 C_B$ (C in lb moles/ft^3)

Liquid density = 50 lb/ft³ $k_1 = 0.08/s$
$MW_A = MW_B = 100$ $k_2 = 0.03/s$ } @ 5 atm 900°R

Vapor pressure: $\log_{10} VP_B = 4.665 - 3438/T$

$\log_{10} VP_A = 4.421 - 2816/T$

(VP in atm, T in °R)

Also assume

$0.01 \leq \eta \leq 0.99$

$700°R \leq T \leq 770°R$

$10 \leq V \leq 60 \text{ ft}^3$

and the profit is given by

$$C_B B_{top} - C_u [F_R (900 - T)] - C_R V$$

FR – total reactor effluent (lbmol/hr)

B_{top} – moles B in overhead vapor (lbmol/hr)

$C_B = 0.5$, $C_u = 0.1$, $C_R = 0.01$

a. Formulate the above problem as an equation-based optimization problem. Solve the complete problem using GAMS.

b. Calculate the reduced Hessian at optimum and comment on second order conditions.

9. Show that if $H^i = (B^i)^{-1}$ and $WW^T = H^i$ and $W_+ W_+^T = H^{i+1}$, then the DFP (complementary BFGS) formula can be derived from:

$$\text{Min } \|W_+ - W\|_F$$

$$\text{s.t. } W_+ \gamma = s$$

$$W_+^T y = \gamma$$

10. Consider the alkylation process shown below from Bracken and McCormick (1968):

X_1 = Olefin feed (barrels per day) X_6 = Acid strength (weight percent)

X_2 = Isobutane recycle (barrels per day) X_7 = Motor octane number of alkylate

X_3 = Acid addition rate (1000s pounds/day) X_8 = External isobutane-to-olefin ratio

X_4 = Alkylate yield (barrels/day) X_9 = Acid dilution factor

X_5 = Isobutane input (barrels per day) X_{10} = F-4 performance no. of alkylate

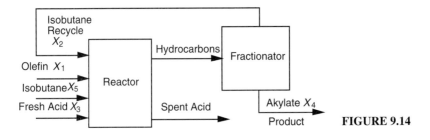

FIGURE 9.14

The alkylation is derived from simple mass balance relationships and regression equations determined from operating data. The first four relationships represent characteristics of the alkylation reactor and are given empirically.

The alkylate field yield, $X4$, is a function of both the olefin feed, $X1$, and the external isobutane to olefin ratio, $X8$. The following relation is developed from a nonlinear regression for temperature between 80 and 90 degrees F and acid strength between 85 and 93 weight percent:

$$X4 = X1*(1.12 + .12167*X8 - 0.0067*X8**2)$$

The motor octane number of the alkylate, $X7$, is a function of $X8$ and the acid strength, $X6$. The nonlinear regression under the same conditions as for $X4$ yields:

$$X7 = 86.35 + 1.098*X8 - 0.038*X8**2 + 0.325*(X6-89.)$$

The acid dilution factor, $X9$, can be expressed as a linear function of the F-4 performance number, $X10$ and is given by:

$$X9 = 35.82 - 0.222*X10$$

Also, $X10$ is expressed as a linear function of the motor octane number, $X7$.

$$X10 = 3*X7 - 133$$

The remaining three constraints represent exact definitions for the remaining variables. The external isobutane to olefin ratio is given by:

$$X8 = (X2 + X5)/X1$$

To prevent potential zero divides it is rewritten as:

$$X8*X1 = X2 + X5$$

The isobutane feed, $X5$, is determined by a volume balance on the system. Here olefins are related to alkylated product and there is a constant 22% volume shrinkage, thus giving $X4 = X1 + X5 - 0.22*X4$ or:

$$X5 = 1.22*X4 - X1$$

Finally, the acid dilution strength ($X6$) is related to the acid addition rate ($X3$), the acid dilution factor ($X9$), and the alkylate yield ($X4$) by the equation, $1000*X3 = X4*X6*X9/(98 - X6)$. Again, we reformulate this equation to eliminate the division and obtain:

$$X6*(X4*X9+1000*X3) = 98000*X3$$

The objective function is a straightforward profit calculation based on the following data:
- Alkylate product value = $0.063/octane-barrel
- Olefin feed cost = $5.04/barrel
- Isobutane feed cost = $3.36/barrel
- Isobutane recycle cost = $0.035/barrel
- Acid addition cost = $10.00/barrel

This yields the objective function to be maximized is therefore the profit ($/day)

$$OBJ = 0.063*X4*X7 - 5.04*X1 - 0.035*X2 - 10*X3 - 3.36*X5$$

The following exercises are based on the description in Liebman et al. (1984).
  a. Set up this NLP problem and solve.
  b. The regression equations presented in section 9.2 are based on operating data and are only approximations; it is assumed that equally accurate expressions actually lie in a band around these expressions. Therefore, in order to consider the effect of this band, Liebman et al. (1984) suggested a relaxation of the regression variables. Replace the variables X4, X7, X9, and X10 with RX4, RX7, RX9, and RX10 in the regression equations (only) and impose the constraints:

$$0.99*X4 \leq RX4 \leq 1.01*X4$$
$$0.99*X7 \leq RX7 \leq 1.01*X7$$
$$0.99*X9 \leq RX9 \leq 1.01*X9$$
$$0.9*X10 \leq RX10 \leq 1.11*X10$$

to allow for the relaxation. Resolve with this fomulation. How would you interpret these results?
  c. Resolve the original formulation as well as the one in part a with the following prices:
   - Alkylate product value = $0.06/octane/barrel
   - Olefin feed cost = $5.00/barrel
   - Isobutane feed cost = $3.50 barrel
   - Isobutane recycle cost = $0.04/barrel
   - Acid addition cost = $9.00/barrel

PART III

BASIC CONCEPTS IN PROCESS SYNTHESIS

HEAT AND POWER INTEGRATION

10

In this chapter we are going to look at the use of heat exchanger equipment to transfer heat from one stream to another to reduce the use of utilities to run a process. Consider the flowsheet for the ethylene to ethyl alcohol plant we proposed in Chapters 1 through 4 (see, for example, Figure 1.3 in Chapter 1). We find we must heat and cool streams for best process operation. For example, as noted in those chapters, the literature suggests we should run the reactor at a very high temperature, about 590 K (ambient is about 300 K). The ethylene feed enters at ambient temperature. It joins the recycle, which comes from an absorber we run as cold as possible—that is, just above room temperature. This merged stream then flows through a multistage compressor with intercooling to bring it to 69 bar. Thus, only the heating of the last compressor stage will preheat the feed, giving it a temperature much closer to ambient than to 590 K. In Chapter 3, we estimated that the flash immediately following the reactor will run at 393 K. Thus, we have the major task of preheating the feed from near 300 K up to almost 600 K and cooling the reactor product back to about 400 K. It would make sense to consider using the heat from the reactor outlet stream to provide much of the heat to preheat the feed.

We see even more opportunities in this flowsheet to exchange heat between process streams. There are several distillation columns, each of which has a condenser and reboiler. We put heat into a reboiler for a column, and, as we shall further discuss shortly, we remove about the same amount of heat from its condenser. Unfortunately the condenser runs at a colder temperature than the reboiler for a column, so, without putting work into our process (in the form of a heat pump), we cannot use the condenser heat to run the reboiler for a column. However, the condenser of one column could well supply heat to the reboiler of another. The condenser might also supply some of the heat we need to preheat the feed to the reactor. There are many alternate ways we could interchange heat in this flowsheet. Our goal in this chapter is to develop insights into how we can find the better ones. For a review of the literature see Gunderson (1988), Linnhoff, et al. (1982), and Linnhoff (1993).

We shall also look at heat integration for processes that operate below ambient temperature. In these processes we must "pump" up the heat to ambient temperatures to reject it from the process. We shall develop some added insights to allow us to discover how best to place these heat pumps in these systems.

We shall start this chapter by examining a carefully posed problem for heat integration. We can call it the *basic HENS* (heat exchanger network synthesis) problem. We choose to study this well-defined problem because it will provide us with several insights into the design of heat exchanger networks. These insights will aid us even when the problem at hand does not conform to the assumptions of the basic HENS problem. An analogy is to study linear programming as an optimization technique. Insights gained in this problem help to understand many of the algorithms we use to solve nonlinear programming problems. We shall then use some of these insights for designing where to place heat pumps for processes operating below ambient temperature. Finally, in Chapter 12 we shall look at the flow of heat in distillation processes.

10.1 THE BASIC HEAT EXCHANGER NETWORK SYNTHESIS (HENS) PROBLEM

Our basic HENS problem is the following. Given

- A set of hot process streams to be cooled and a set of cold process streams to be heated
- The flowrates and the inlet and outlet temperatures for all these process streams
- The heat capacities for each of the streams versus their temperatures as they pass through the heat exchange process
- The available utilities, their temperatures, and their costs per unit of heat provided or removed

determine the heat exchanger network for energy recovery that will minimize the annualized cost of the equipment plus the annual cost of utilities.

The streams we are talking about here are all the streams requiring heating or cooling in a process. The stream from the feed compressor to the reactor in the ethylene to ethanol process is such a stream. The vapor leaving the top of a distillation column that we have to cool to produce reflux and liquid product is also such a stream. We also include the stream between two stages of compression if we intend it to be cooled to enhance compressor performance.

If we know the flowrates and inlet and outlet temperatures, we are assuming we have developed a heat and material balance for a flowsheet. To develop a complete set of balances requires us to set the temperature and pressure levels for all the units. Thus, it is necessary to have carried out our process analysis to this point.

Sec. 10.1   The Basic Heat Exchanger Network Synthesis (HENS) Problem   343

The third required piece of information above requires us to know the pressure for the stream as it passes through the exchanger network. We need only approximate the pressure levels for streams not changing phase, but we have to fix pressures very closely for streams changing phase while they are heating or cooling, as the pressure will set the temperature at which they will give up or require their latent heats. Our basic HENS problem is a restricted problem formulation, but it is a useful one. In later chapters we will want to understand how changes in the stream flows, pressure level, and inlet and outlet temperatures can affect the heat exchanger network we would synthesize.

We shall discover that we can predict several properties for a basic HENS problem before inventing the structure of a network that solves it. For example, we shall discover that we can predict the least amount of utilities we will require. We can also estimate the fewest number of heat exchanges between stream pairs that we will need. Finally we can even estimate the cost of the network. To re-emphasize, we can do all these predictions without inventing the network. Thus, we can use these predictions to aid us to invent a good network. For example, we might predict that we need only heating to run our process. We will first look only for networks that do not have any cooling in them. We might estimate that we need to exchange heat only between ten stream pairs. If our network has exchanges between 20 stream pairs, we will rule it out as a good solution.

We introduce how we can do the first two of these predictions by presenting a small but interesting example HENS problem. We shall look quickly at this example and then return to develop the ideas more fully. Therefore, do not become too concerned if you do not follow everything in the example. We are using it only to introduce the ideas.

EXAMPLE 10.1   A Small but Interesting Problem

Consider the example problem shown in Figure 10.1. It consists of a reactor into which we are feeding two reactant streams. Each is available at 100°F and has to be heated to 580°F. The reaction is slightly exothermic. Thus, the reactor produces an outlet stream at 600°F, which we want to cool to 200°F. We label each stream with FCp (BTU/s), the product of its flowrate F (lb/s), and its heat capacity Cp (BTU/lb °F). For the basic HENS problem, we shall assume that the heat capacities for the streams are not functions of temperature. Thus, we show this product as a fixed number over the entire temperature range for a stream. The following simple heat balance on a stream with a constant FCp computes the amount of heat needed to alter its temperature from T_1 to T_2:

$$Q = FCp(T_2 - T_1) \qquad (10.1)$$

where Q is the amount of heat in BTU/s. Thus, the value of FCp is the BTU/s it takes to change the temperature of the associated stream by one degree. For example, it takes a heat input rate of 1 BTU/s to increase the temperature of the first inlet stream by 1°F. It takes 2 BTU/s to do the same for second inlet stream and 3 BTU/s for the reactor output stream.

We can restate our problem in tabular form as shown in Table 10.1. In this table we label the two *cold* streams to be heated C1 and C2. The reactor output stream is a *hot* stream to be cooled, and we label it H1. We show the total heat available from the stream in the column labeled "Heat out." A negative heat says we need to add the heat to the stream. We provide the

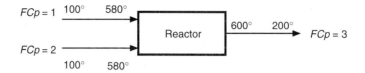

FIGURE 10.1 Example 10.1 of a heat exchanger network synthesis problem.

heating and cooling available from process streams at no charge, in contrast to what it will cost if we use utilities to provide heating and cooling.

TABLE 10.1 Heat Exchanger Synthesis Problem for Example 10.1 in Tabular Form

Stream	T_{in}, °F	T_{out}, °F	FCp, BTU/°F	Heat out, BTU/s	Cost per lb
C1	100	580	1	−480	$0
C2	100	580	2	−960	$0
H1	600	200	3	+1200	$0
				Net = −240	
Utilities					
Steam, S	650	650			High
Hot water, HW	250	>130			Low
Cooling water, CW	80	<125			Moderate

Table 10.1 also lists the utilities available for our problem. The hot utility is steam that condenses, supplying its heat at 650°F. We return the steam condensate at the same temperature to the hot utility system. We also have hot water available for heating at 250°F, which we must return no colder than 130°F. Finally, we have cooling water available at 80°F, which we must not return hotter than 125°F.

Relatively speaking, steam is the most expensive per BTU it provides. Hot water, on the other hand, is often available for free in a process. Many processes produce low temperature heat in excess, and it is often wasted. We have to treat cooling water so it is moderately expensive per pound we use. We will not be developing actual costs for this example so we simply show these costs as ranging from high to low.

Just to show that there are alternative networks to solve this problem, we show three in Figure 10.2. In the top most solution, we use the reactor effluent to heat the lower feed stream, C2, from its inlet, 100°F, to its target temperature, 580°F, requiring 2(580 − 100) = 960 BTU/s to do it. We remove these 960 BTU/s from the hot product stream. It will decrease in temperature by

$$\Delta T = \frac{Q}{FCp} = \frac{960 \text{ BTU/s}}{3 \text{ BTU/s°F}} = 320°F \tag{10.2}$$

to 280°F. We use H1, after it is cooled, to supply heat to the upper feed stream. H1 is now at 280°F. We can heat C1 no hotter than 280°F using H1. We choose to heat C1 only to 260°F, so

Sec. 10.1   The Basic Heat Exchanger Network Synthesis (HENS) Problem   345

there will be an adequate temperature driving force in this exchange. This exchange involves 1(260 − 100) = 160 BTU/s. Removing this amount of heat from H1 cools it to 226.7°F. We use 3(26.7) = 80 BTU/s of cooling water duty to cool H1 to 200°F. Finally we use 1(580 − 260) = 320 BTU/s of steam duty to heat C1 from 260°F to its target of 580°F. We note that the net heat removed from the network is, therefore, 80 − 320 = −240 BTU/s, as predicted in Table 10.1.

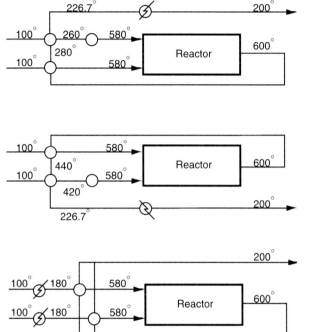

FIGURE 10. 2 Three alternative networks that solve Example 10.1.

The second solution reverses the order we heat the feed streams using H1. We first heat C1 then C2. Coincidentally, H1 reaches 226.7°F after these exchanges, and we need the same amount of utility heating and cooling.

We split H1 in the third solution. One part has an FCp of 1, and we use it to heat C1 while the other part has an FCp of 2, which we use to heat C2. The FCp values exactly match in both these exchanges. Thus, we cool each part of H1 by 400°F (from 600°F to its target 200°F) while we raise the temperature of the corresponding cold stream by exactly 400°F. We can, in fact, heat them to their target temperatures by heating them from 180°F to 580°F. In a pure countercurrent heat exchange, the exchanger will have a constant driving force of 20°F everywhere. When we do this, we find we can use hot water duty equivalent to (1 + 2)(180 − 100) = 240 BTU/s for heating only, which heats both C1 and C2 from 100°F to 180°F. This network does not use cooling water.

The third solution has some interesting advantages over the first two. First of all, it requires no cooling utility. Second, it uses only hot water for heating, a much less expensive source of heat than steam. We need to inject only the net heat required of 240 BTU/s. In the first configuration, we put 320 BTU/s of heat into the network using steam and removed 3(26.7) = 80 BTU/s from the network. The difference is again the net of 240 BTU/s. We have put in an extra

346 Heat and Power Integration Chap. 10

80 BTU/s using steam and removed the same amount using cold utility. We have paid twice for these extra BTU/s—we put in and then took out.

There is a cost for saving on utilities. If we were to size the exchangers, we would find them to be larger for the third alternative as it has smaller temperature driving forces in its exchangers. An economic analysis would aid us in selecting which alternative we prefer.

PREDICTING THE UTILITIES REQUIRED FOR OUR PROCESS

Let us partition our problem into temperature intervals as shown in Table 10.2. To carry out such a partitioning we must fix the minimum temperature difference that we are willing to have in any of the heat exchangers that will be in the final network. For this example, let us choose 10°F. We show vertical lines representing the two cold streams, C1 and C2, on the far left of this table. We then have two columns of temperatures, followed by a column for H1. The right most side of the table is for computing heat balances.

TABLE 10.2 Partitioning the HENS Problem into Temperature Intervals

		Cold Temp	Hot Temp		Heat Leaving Network	
		(590)	600	—		
				|	(−3) (600 − 590)	= −30
—	—	580	(590)	|		
				|	(1 + 2 − 3) (580 − 190)	= 0
|	|	(190)	200	—		
|	|				(1 + 2) (190 − 100)	= 270
|	|	100	(110)			
—	—					
C1	C2			H1	Stream	
1	2			3	FCp for stream	

The two columns, labeled "Cold Temp" and "Hot Temp" respectively, indicate the temperature partitioning we use to decompose our problem. We look first for the hottest temperature among all those listed for the process streams, finding 600°F, which is the inlet temperature for H1. We just selected a minimum temperature driving force for our problem to be 10°F. Thus, we cannot heat any cold stream hotter than 590°F using H1. We shall, therefore, consider 600°F for a hot stream to be equivalent to 590°F for a cold stream. We show this equivalence explicitly by listing 590°F for cold streams adjacent to 600°F for hot streams. The next hottest temperature is the target temperature of 580°F for the two cold streams. Its equivalent hot temperature is 10°F hotter, 590°F. We list these side by side as the next entries in our two temperature columns. Continuing, we find the next hottest temperature to be a "hot" temperature of 200°F (equivalent to a cold temperature of 190°F). Finally, we find a cold temperature of 100°F. We show temperatures that we computed to be equivalent to temperatures found in the problem within parentheses.

These temperatures represent the inlet and outlet temperatures for our streams. We draw vertical lines indicating the range of temperatures for our process streams, C1, C2, and H1. We see the vertical lines for C1 and C2 cover the range from 100°F to 580°F while the vertical line for H1 covers the range from 600°F to 200°F.

Sec. 10.1 The Basic Heat Exchanger Network Synthesis (HENS) Problem   347

The temperatures partition our problem into intevals. The topmost interval is over a range of 10°F, from (590°F) to 580°F on the cold side or from 600°F to (590°F) on the hot side. The next interval is from 580°F to (190°F) on the cold side. Finally, the third interval is from (190°F) to 100°F on the cold side. Each interval has a different set of streams crossing it. The topmost interval has only stream $H1$ crossing it. The second interval has all three streams, while the bottom one has just C1 and C2. We defined the intervals so that each has a different set of streams crossing it.

We next write a heat balance for each interval to determine if it has an excess or a deficiency of heat. The top interval produces 30 BTU/s, as the heat balance to the right of that interval shows in Table 10.2. The next interval is in perfect heat balance. The hot stream has exactly the amount of heat required by the two cold streams over that interval. The bottom interval has a deficiency of 240 BTU/s of heat. Figure 10.3 illustrates this heat flow. We see the top interval rejecting 30 BTU/s of heat, which are certainly hot enough to supply part of the 270 BTU/s of heat needed by the bottom interval, leaving a net of 240 units of heat needed by the bottom interval. It is at the cold end of the problem, and it is cold enough in that interval that we can supply this 240 BTU/s using hot water.

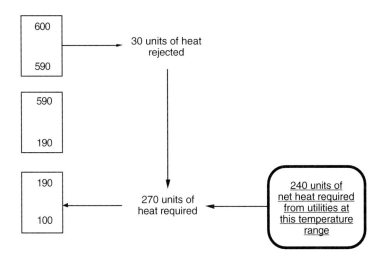

FIGURE 10.3   Flow of heat into and out of intervals for Example 10.1.

These observations are very important. We discovered that we need only heating for this problem. We also discovered that we can supply the heat at temperatures that allow it to be provided by hot water which we can have almost for free.

Figure 10.3 is based on net heats for intervals. To complete our discussion here, we should prove that the net heat needed or produced by an interval is sufficient to characterize it in this analysis. To prove this point we need to prove that we can always transfer within the interval the lesser of (1) the heat the cold streams need and (2) the heat the hot streams have available. Then we only need to consider the net excess or deficiency outside the interval. We will use the following example to develop this proof.

Consider the interval in Figure 10.4 which is based on a minimum driving force of 10°F. In this figure, we have a merged set of hot streams cooling from 200°F to 100°F while we have a merged set of cold streams heating from 90°F to 190°F. Thus, the interval spans 100°F. The hot streams have 500 BTU/s available while the merged set of cold streams requires 400 BTU/s. The lesser of these two amounts is 400 BTU/s. We want to prove that this amount of heat can always be transferred from the hot to the cold streams within the interval.

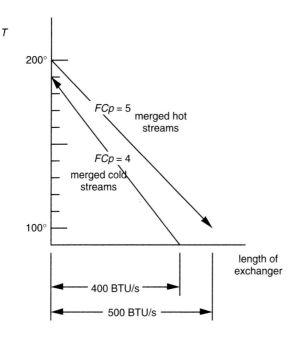

FIGURE 10.4 Proving the lesser of the heat needed and that available can always be transferred within an interval.

We start our transfer at the hot (left) end of both streams. At this end there is a 10°F driving force. The hot streams have an FCp that is larger than the cold. Removing 400 BTU/s from the cold stream cools them to 90°F, while doing the same for the hot streams cools them to 120°F. The driving force increases as we proceed to the right in the exchange. Thus the exchange always has a satisfactory driving force. The heat not transferred is at the cold end of the hot streams. We can pass that heat to a colder interval or to a cold utility. A similar argument follows for the case where the cold streams need more heat than the hot streams have available. The hot streams can always be cooled completely while the cold streams will need added heat, either from a hotter interval or from a hot utility.

In both cases the net heat is all we need to worry about outside the interval, which is what we did when constructing Figure 10.3.

ESTIMATING THE FEWEST MATCHES NEEDED

We can use very simple arguments based on networks to establish an estimate for the fewest matches needed for a heat exchanger network synthesis problem. Figure 10.5 illustrates for Example 10.1. We carry out this analysis after we have decided the amount of heat we will need to transfer to utilitites. For Example 10.1, we need to transfer 240 BTU/s of heat from hot water to design our network. Figure 10.5 has a set of nodes across the top, one for each different heat source. It has a similar set of nodes, one for each heat sink in the problem across the bottom.

Sec. 10.1 The Basic Heat Exchanger Network Synthesis (HENS) Problem

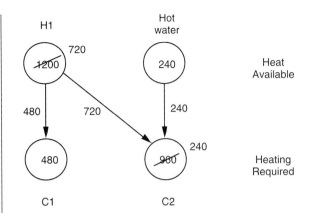

FIGURE 10.5 Estimating the fewest matches required for Example 10.1

Within each node is the amount of heat the source has available or the amount the sink needs. We note that the total of the heat available matches that needed (that is, 1200 + 240 = 480 + 960).

We associate heat sources with sinks by drawing links from a node at the top to one at the bottom. We will not concern ourselves here with whether the temperatures of the streams will actually allow for heat to transfer. Start at the far left. We show heat going from H1 to C1 by linking these two with an arrow from H1 to C1. We label this arrow with the lesser of the amount of heat available and that needed. Here C1 needs less heat that H1 has available so we label the line with 480 BTU/s. We reduce the heat available from H1 by this amount. Thus, H1 now has 720 BTU/s of heat available. We link H1 with the next cold node, C2, and label the line with 720, the lesser of the 720 BTU/s that heat H1 has and the 960 BTU/s that C2 needs. We reduce the heat required by C2 by this 720 BTU/s, bringing it to 240 BTU/s. We match C2 with the next available hot stream, the hot water utility. Because the total of the heat available matches that needed, this last match zeros out the needs for both nodes. The number of matches we just drew is $N - 1$, which is 3 here, where N is the number of nodes in Figure 10.5. Each match eliminated one node until the last, where we eliminated two. In general, we will not eliminate nodes faster (unless a match earlier than the last is exact and removes two nodes, a fortuitous situation). Thus, we should expect that we cannot, in general, complete our network with fewer than $N - 1$ matches.

$N - 1$ is only an estimate. We can sometimes do better and sometimes have to do worse. However, most times we can hit it exactly. Thus, any network we find that has many more matches in it than this should cause us to look for solutions with fewer matches.

Let us examine the solutions for the problem that we posed in Figure 10.2. The first two require both hot and cold utilities, whereas we now know we need only supply heat. The third supplies heat using only hot water. It seems a good candidate. However, it has four exchanges in it, whereas we just estimated we need only three. We need now to wonder if we can find a solution with only three exchanges required.

INVENTING A FIRST SOLUTION

We discovered that we need only hot water to solve our problem. We can use this result to direct us to a first solution. We can supply 240 BTU/s of heat to our cold streams only if we heat the cold end of them. It is evident that we can supply 80 BTU/s to C1 and 160 BTU/s to C2, raising their temperatures to 180°F. We cannot supply the rest of the heat from H1 unless we split H1. Thus, we find ourselves forced into solutions that look like the third one in Figure 10.2. We redraw it here as a network in Figure 10.6.

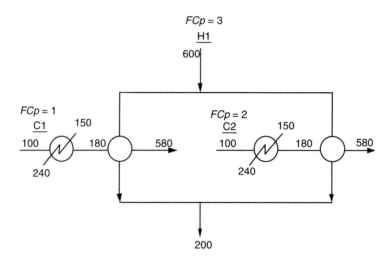

FIGURE 10.6 A first guess at a network for Example 10.1.

Often we can quickly construct a network that will use the minimum utilities we predict. As we have already discovered, however, it has four exchanges in it. Can we reduce the number to three? The next section explores one approach we might use.

DISCOVERING AND BREAKING CYCLES

To reduce the number of exchanges we are going to look for heat flowing in cycles in our solution. We create the matrix in Table 10.3 where we have one column for each of the hot streams and hot utilities and one row for each of the cold streams and cold utilities. In the matrix we place the amount of heat exchanged between heat sources and sinks. For example, we see that 80 BTU/s is exchanged between the hot water and C1, while 160 BTU/s exchanges between hot water and C2. H1 provides 400 and 800 BTU/s each to C1 and C2.

TABLE 10.3 Looking for Cycles in a Network

	HW	H1	Heat into
C1	80–	400+	480
C2	160+	800–	960
Heat from	240	1200	

Around the edges we total the heats listed in each column and row. For example, we total the heats in row 1: 80 and 400 BTU/s. This total is the 480 BTU/s that C1 needs to reach its tar-

Sec. 10.1 The Basic Heat Exchanger Network Synthesis (HENS) Problem

get temperature. In the column for H1, we find 1200 BTU/s, which is the total heat it must give up to be cooled to its target.

We now look for cycles in this matrix. We start with any nonzero entry, say the 80 in the row for C1 and the column for HW. We mark this row as having been explored. We move horizontally, looking for a nonzero entry in an unmarked column. Here we find 400 in the column for H1. We mark this column. We look vertically in this column for an entry we have already included on the path. We find none, so we look next for a nonzero entry in a row that is not marked. We find the 800 in the row for C2. We mark this row. We look horizontally for an entry we already have included on our path. Failing, we then look for a nonzero entry in an unmarked column and find the 160 in column HW. We mark this column. We look vertically in it for a previously visited entry, finding the 80 where we started. We have a cycle. Now we need to discover if we can break it.

As we find the nonzero entries on this cycle, we can mark them with alternating symbols, such as with pluses and minuses. We show such a marking in Table 10.3. Note, the markings alternate everywhere around the cycle. By construction we note there is exactly one plus and one minus marking in each row and in each column belonging to the cycle. For example, the first row has one minus (the 80) and one plus (the 400) in it.

We look among the cells marked with a minus for the smallest heat flow, here 80 BTU/s is less than 800 BTU/s. We select the 80. We subtract this amount of heat flow from every cell marked with a minus and add this same amount to every cell marked with a plus. As shown in Table 10.4, each row and each column involved in the cycle has 80 subtracted once and 80 added once. Thus, the total for each column and for each row shown around the edges is unchanged; that is, the amount of heat from each source and into each sink remains unchanged.

TABLE 10.4 One Set of Results from Breaking a Cycle

	HW	H1	Heat into
C1	0–	480+	480
C2	240+	720–	960
Heat from	240	1200	

The loop now has one heat flow which is zero. It is broken. We must now check if the solution with this loop broken in this way remains feasible. We return to the network in Figure 10.6 and alter the heat flows in each of the exchanges to match those given in Table 10.4. Figure 10.7 results. We analyze this network. We put all 240 BTU/s of heat from the hot water into C2. This heats C2 from 100°F to 220°F. Fortunately, our hot water is available at 250°F so this match is possible. The right branch of H1 supplies the remaining 720 BTU/s that C2 needs to reach its target of 580°F. Removing 720 BTU/s from this branch, with an FC_p of 2 BTU/s/°F, reduces its temperature only to 240°F. We seem to be in trouble. The exchange has a driving force of 20°F throughout so it is feasible; our problem is we did not cool this branch to its target of 200°F.

We continue anyway to see what happens to the rest of the network. We must put the entire 480 BTU/s needed by C1 into C1 by exchanging with the left branch of H1. We can do this exchange, but again the branch does not hit its target of 200°F. This time the branch becomes too cold, reaching 120°F. The exchange is feasible, however, as it has a driving force of 20°F throughout.

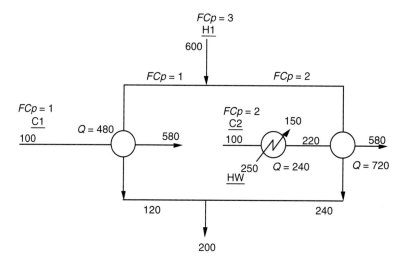

FIGURE 10.7 Breaking a heat loop by removing the heating of C1 using hot water.

What if we mix the two branches of H1, one undercooled and one overcooled? A few seconds of thinking tells us the mixture must be at 200°F, which it is. All exchanges are feasible from a temperature point of view. Only if we should not cool H1 down to 120°F should we rule this network out. An example could be if H1 contains CO_2 and water and starts to condense between 120°F and 200°F. In the liquid phase such a mixture is corrosive.

Note that this solution requires only three heat exchanges and uses a minimum amount of heating supplied entirely by the cheapest heat source. It certainly looks like a candidate one should consider for this problem. It is not one that we would likely invent without being aided by some systematic procedure.

There is second way to break the loop in Table 10.3. We could look for the least amount of heat in cells marked with a plus. Here, we compare 160 and 400, choosing 160. We can then reduce all the cells marked with a plus by 160 and increase all the cells marked with a minus by 160. The heat exchanged between hot water and C2 becomes zero, again breaking the loop. We find in this case that the 240 BTU/s from the hot water must all be used to heat C1. Adding 240 BTU/s to C1 will increase its temperature from 100°F to 340°F. That is too hot to get the heat from hot water that is available at 250°F. Thus, this solution is not feasible.

We note that every loop we find in such a matrix as shown in Table 10.3 has two possible ways to be broken—one corresponding to the entries marked with a minus and one to the entries marked with a plus. None, one, or both may lead to feasible networks. In general, a problem will have many loops in it. The matrix in Table 10.5 has many more loops in it. One is shown by marking it with plus and minus signs. The two ways to break this loop would be to subtract 50 from the minuses and add it to the pluses, eliminating the C2/H4 exchange, or to subtract 60 from the pluses and add it to the minuses, removing the C3/H1 exchange. We leave it to the reader to find the many other loops in this example.

Sec. 10.1 The Basic Heat Exchanger Network Synthesis (HENS) Problem   353

TABLE 10.5 A More Complex Example for Finding and Removing Loops

	H1	H2	H3	H4	H5	HU1
C1	100–		300+		200	
C2		60		50–		125+
C3	60+	200–				
C4			150–	400+		
C5	300				100	
CU1		175+		100		200–

10.1.1 Hohmann/Lockhart Composite Curves

Hohmann (1971), in his PhD working under the guidance of Lockhart, was the first to note that one could compute the minimum utility requirements for a basic heat exchanger network synthesis problem directly from the stream information. To understand his thinking, look at Figure 10.8. We show a countercurrent heat exchanger where the top hot stream is supplying heat to a bottom cold stream. We can plot the temperatures for the two streams in this exchanger against either the position along the exchanger or against the amount of heat transferred, Q. If the heat capacity (and thus the product FCp) for a stream is constant versus temperature, the following equation shows that a plot of T versus Q will be a straight line (the lines will not be straight when plotted against length, however).

$$dT = \frac{1}{FCp} dQ$$

Suppose we have two streams we wish to cool. The first has an FCp of 100 kJ/s and we wish to cool it from 450 K to 375 K. The second has an FCp of 200 kJ/s and we wish

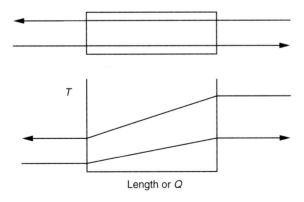

FIGURE 10.8 Countercurrent heat exchange between two streams.

to cool it from 400 K to 350 K. The two streams share a temperature range over which we wish to cool them both: from 400 K to 375 K. Let us suppose that we will use these two streams together to do any heating while they are in this common temperature range. Their combined heat balance will obey

$$Q(T) = (F_1 Cp_1 + F_2 Cp_2)(T - T_{in}) = (100 + 200)\frac{kJ}{s\,K}(T - 400\text{ K})$$
$$= 300\,\frac{kJ}{s\,K}(T - 400\text{ K})$$

as they pass through the exchange. They act like one stream with a combined FCp. When they are in nonoverlapping ranges, we shall use them separately.

Figure 10.9 shows a plot of temperature versus heat flow for both streams over their entire temperature ranges. We start with stream 1. Having an FCp of 100 kJ/s K, it cools from 450 K to 375 K when we remove 7500 kJ/s from it. The right-most arrow shows this cooling. Having an FCp of 200 kJ/s K, stream 2 cools from 400 K to 350 K when we remove 10,000 kJ/s from it, shown by the left-most arrow. We plot stream 1 immediately to the right of stream 2, with the overlapping temperature regions plotted next to each other. The 7500 kJ/s required to cool both streams from 400 K to 375 K is, therefore, the horizontal distance from where stream 1 is at 400 K to where stream 2 is 375 K. Since the plot

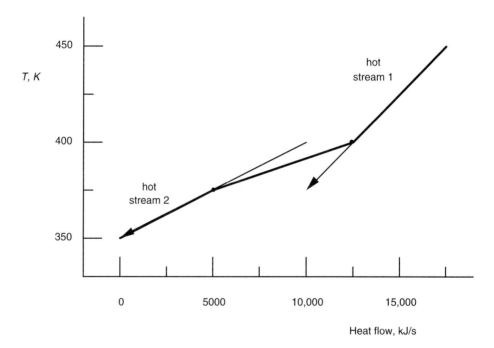

FIGURE 10.9 Merging two hot streams within a common temperature range.

Sec. 10.1   The Basic Heat Exchanger Network Synthesis (HENS) Problem   355

for the combined streams is straight, we connect these two points to merge the streams in this common temperature range. The thicker line is then a plot of temperature versus heat flow removed for the combined streams over their entire ranges. Note it has two kinks in it—at the start and at the end of the common temperature range.

We can merge this curve with a third hot stream and so forth until we have a composite curve for all the hot streams. It will be a line having several straight segments. We can merge all the cold streams for a problem in a similar manner.

To see how to use these ideas to compute the minimum utility requirements for a problem, let us examine the following example problem.

EXAMPLE 10.2 HENS Problem 4SP1

The literature contains several test problems for testing the effectiveness of heat exchanger network synthesis algorithms. Problem 4SP1 (four stream problem number 1) is one of them. We shall use it to illustrate how to use Hohmann/Lockhart composite curves to compute minimum utility use for a heat exchanger network synthesis problem. Table 10.6 gives the data for this problem.

TABLE 10.6   Stream Data for Problem 4SP1

Stream	FCp, kW/°C	T_{in}, °C	T_{out}, °C	Heat flow out, kW
C1	7.62	60	160	−762.0
C2	6.08	116	260	−875.5
H1	8.79	160	93	588.9
H2	10.55	249	138	1171.1

There are two cold streams and two hot streams in this problem. As we did for Example 10.1, we analyze this problem in Table 10.7 by partitioning it into temperature intervals. The columns labeled Hot and Cold under Temperatures (columns 3 and 4) show this partitioning. The hottest temperature in the data is a cold temperature of 260°C. We list it at the top of the cold temperature column and its corresponding hot temperature (270), which is $\Delta T_{min} = 10$°C hotter, alongside under the hot temperature column. We find there are seven temperature intervals in this problem. Each is numbered from the bottom between the two temperature columns.

We next tabulate the amount of heat that the composite of the hot streams has available for each interval. This tabulation is column 1. H2 enters the problem at 249°C, which is the upper temperature for interval 6; it is the hotter of the two hot streams and the only hot stream in this interval. In interval 6 it contributes 833.5 kW = $(FCp)_{H1}(T_{6,up} - T_{6,low})$ = 7.62 kW/°C (249 − 170)°C. H1 is also the only hot stream in interval 5, contributing another 105.5 kW. Interval 4 has both hot streams present; the amount of heat flow contributed is the sum for both over this interval, 425.5 kW = $((FCp)_{H1}+(FCp)_{H2})(T_{4,up} - T_{4,low})$.

We do the same for the cold streams, tabulating the amount of heat the composite cold streams need in each interval. This we do in the fifth column labeled "Req'd Heat" under the heading "Composite Cold Streams." Stream C2 only is present in interval 7 and requires 127.7 kW to heat it in that interval.

TABLE 10.7 Extended Problem Table for 4SP1 for $\Delta T_{min} = 10°C$

Composite Hot Streams		Temperatures		Composite Cold Streams		Grand Composite Hot and Cold Streams		
Avail Heat	Cascaded Heat	Hot	Cold	Req'd Heat	Casc'd Heat	Net Heat	Casc'd Heat	Adj Casc'd Heat
		(270)	260	0.0		0.0	127.7	
		249	7	127.7	127.7	−127.7		
—	0.0		(239)	127.7		−127.7	0.0	
833.5		(170)	6	480.3	608.0	353.1	225.4	353.1
	833.5		160					
105.5		160	5	137.0	745.0	−31.5	193.9	321.6
	939.0		(150)					
425.5		138	4	301.4	1046.4	124.1	318.0	445.7
	1364.5		(128)					
105.5		(126)	3	164.4	1210.8	−58.9	259.1	386.8
	1470.0		116					
290.1		93	2	251.5	1462.3	38.6	297.7	425.4
	1760.1		(83)					
—		(70)	1	175.3	1637.6	−175.3	122.4	250.1
	1760.1		60					

We can now establish the composite curves giving temperature versus heat flow for the hot streams and then for the cold streams. We accumulate ("cascade") the heats to develop the needed numbers. In the second column of numbers, we accumulate the heat produced by the hot streams as we move down the intervals from interval 7 to interval 1. We place a zero the top of this column. We then add the amount of heat produced by the composite hot streams in interval 6, getting 833.5 kW at the bottom of this interval. Adding the 105.5 kW from interval 5 brings the number to 939 kW. Another 425.5 kW brings the total to 1364.5 kW. We continue accumulated these heats until we reach the bottom interval, where we find that the hot streams produce 1760.1 kW total. We cascade the heats needed for the cold streams in a similar fashion. Starting at the top, we place 0 kW at the top of the sixth column. Adding in 127.7 for interval 7, we get 127.7 at the top of interval 6. Adding another 480.3 brings the total to 608 kW. We again continue until we reach the bottom interval, where we find that the cold streams need a total of 1637.6 kW.

In Figure 10.10 we plot both these cascaded heat flow columns from Table 10.7. We show both the hot and the cold temperature scales on the right ordinate. We plot the hot cascaded data versus the hot temperature to get the hot composite curve and the cold cascaded data versus the cold temperature to get the cold composite curve. We plot both by starting in the upper right with the hot end of both streams. The hot is the solid line starting at the lower temperature and against the right vertical axis. The cold is the solid line starting at the higher temperature, also against the right vertical axis. Note that the heat flows start at zero on the right and increase as we move to the left.

Sec. 10.1 The Basic Heat Exchanger Network Synthesis (HENS) Problem 357

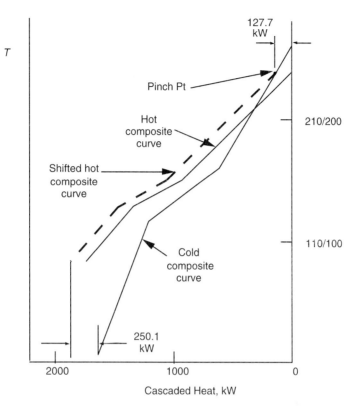

FIGURE 10.10 Cascaded composite hot and cold heat flows. Hohmann/Lockhart diagram obtained by plotting temperature versus composite hot and composite cold cascaded heat flow data for problem.

These curves are a plot of temperature versus heat flow. They could represent the temperature profiles we would see within a heat exchanger. However, we know that the temperature of the hot stream in an exchanger must be everywhere at least $\Delta T_{min} = 10°C$ hotter than the temperature of the cold stream. We see these curve cross. Crossing violates this requirement of a 10°C driving force. To make the profiles feasible within an exchanger, we can shift one of the two curves right or left until the cold curve is everywhere below the hot curve. It can touch because we plotted the two curves with a 10°C offset between them (remember we used two different temperature scales to plot). We shift the hot curve to the left as shown by the dashed line. In the position shown it just touches the cold curve at the very top of it. From that point on the hot curve is at least 10°C hotter than the cold as we move to the left.

Where the curves are one above the other, we can interpret them as profiles within a counter current heat exchanger. The hot curve is hotter by at least 10°C. They are in heat balance as they are both plotted against heat flow. If we move the hot curve even further to the left, the two curves would overlap less than they do in the position shown. Thus, they would exchange less heat between them. We cannot move it to the right as the curves would then not have the required minimum approach temperature between them everywhere. This position is where they exchange the maximum amount of heat possible.

The portions of the cold curve where heat is not transferred from the hot curve (that is, there is no hot curve directly above it as on the far right in Figure 10.10) must be added using hot utilities. We see we must add 127.7 kW. Similarly the portion of the hot curve where heat is not transferred to the cold streams (to the far left) must be removed using cold utilities; we must remove 250.1 kW using cold utilities. These are the minimum amounts of heat we must add and remove for this problem. We also see that the heat exchanger network we invent to give these results will have a point in it at 249°C (hot)/ 239°C (cold) where the two curves just touch in this plot and thus where the minimum driving force of 10°C will occur. This temperature is called the pinch point for the problem. If a pinch point exists on this plot and, therefore, within the heat exchanger network, we will in general need both to add and to remove heat from the problem.

10.1.2 The Grand Composite Curve (GCC)

We can carry our calculations in Table 10.7 one step further and generate the data representing the overall net heat flow for the problem. The resulting plot is called the *grand composite curve* or GCC. This curve is one of the most important to understand for the HENS problem (Umeda et al., 1979). First, we do the mechanics needed and then we interpret what the resulting curve means.

We need to produce the last three columns in Table 10.7. The first of these columns is the net heat expelled from an interval. We obtain it by subtracting the heat required by the cold streams from the heat produced by the hot streams; that is, we subtract the numbers in column 5 (Req'd Heat) from those in column 1 (Avail Heat). The number we compute for interval 7 is $0 - 127.7 = -127.7$ kW, for example. We then cascade these numbers, getting column 8, which we label "Casc'd Heat." We start column 8 with zero at the hot end of interval 7. We add the Net Heat for interval 7, getting -127.7 at the bottom of interval 7. We then add the 353.1 kW of net heat from interval 6, getting $+225.4$ kW at the bottom of interval 6. By the time we reach the bottom we have an entry of $+122.4$ kW, the net amount of heat the problem must expel over what it must take in (i.e., 1760.1 kW − 1637.6 kW—off by one digit in the last place due to rounding by the spreadsheet program used to create this table).

Cascaded heat is the amount of heat the problem has available from the hot streams over that required by the cold streams as we move from the higher temperatures to the lower ones. Anywhere we see a negative number in this cascaded heat column, we know that the hot streams have not produced enough heat to satisfy the needs of the cold streams above this entry. We look for the most negative number in this column, here a negative 127.7 kW. This amount of heat must be supplied by hot utilities. We can accomplish this addition by putting 127.7 kW of heat into the top interval, which we do to create the last column in the table. Cascading the heat again, but starting with this heat input will make the -127.7 entry of the previous column exactly zero. No entry in the cascaded column is now negative. The point where we find this zero entry is the pinch point for our problem.

The top number in this last column, 127.7 kW, is the minimum amount of heat we must put into the problem from hot utilities; the bottom number, 250.1 kW, is the mini-

Sec. 10.1 The Basic Heat Exchanger Network Synthesis (HENS) Problem   359

mum amount of heat we must remove from the problem using cold utilities These numbers agree with those we found on Figure 10.10. In Figure 10.11 we can plot this last column on a temperature versus heat flow diagram. This curve is the grand composite curve (GCC). It is rich with meaning.

As one moves down this curve, one can see, for every temperature, if the process is producing heat or consuming heat. To see this, compare the data in the last column of Table 10.7 to the form for the plot in Figure 10.11. Interval 7 covers the hot temperature range from 270°C down to 249°C, a range of only 21°C. It shows up on the plot as a line segment moving down and to the left. Table 10.7 indicates that this interval requires 127.7 kW of heat input; it is acting locally as a *heat sink*. Interval 6 covers the range from 249°C to 170°C, a range of 79°C. Table 10.7 indicates that it produces 353.1 kW of excess heat; it is acting locally as a *heat source*. The line segment for it in Figure 10.11 moves down and to the right. If we continue looking at the intervals and the plot, we note that every interval that acts as a heat sink has a line segment that moves down and to the left, and every interval that acts as a heat source has a line segment that moves down and to the right. Thinking about this curve, we see that this must be the case.

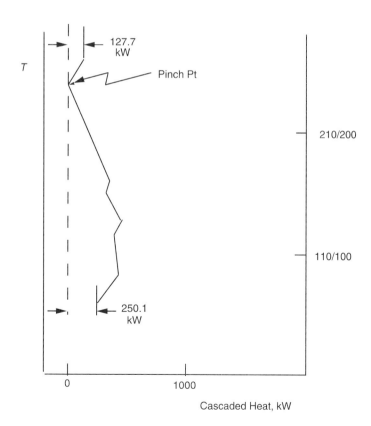

FIGURE 10.11 The grand composite curve for 4SP1.

Whenever there is a heat source segment just above a heat sink segment, we get what we can call a "right-facing nose," as we illustrate in Figure 10.12. We use this figure to prove that we can always heat integrate right-facing noses. We reverse the direction of the temperature curve for the heat source part where it is just above the heat sink part, as we show on the right of Figure 10.12. If we put these streams into a countercurrent heat exchanger, this reversed temperature profile just above the heat sink portion of the nose corresponds to a feasible heat exchange. First, as the horizontal distance is the same, the segments are in heat balance; the heat source produces exactly the heat needed by the sink. Second, the heat source temperature must be everywhere equal to or above the sink temperature by construction. As we have constructed the data with heat sources always ΔT_{min} hotter than sinks, just touching indicates that the minimum driving force is present. Being strictly above indicates an even larger driving force. Therefore, we can provide the heat needed by the sink using the heat from the source that is just above it.

We can cancel the right-facing noses on the GCC, which we do in Figure 10.13. Wherever there is a heat source above a sink, we can "slice" off the nose. Parts of the GCC to the right of the dashed lines have been sliced off in this manner. Here we do it with one slice overall but could have sliced recursively to get the same result. We next assume we integrate such a nose locally. What remains of the GCC, shown in bold lines, is the part of the problem we have not figured out how to heat integrate. If the problem has a pinch in it, there will be heat sink segments only above the pinch and heat source segments only below the pinch. The pinch will be the left-most point on this plot, which is where the vertical solid line is just touching the curve on its left side.

The bold segment at the top requires heat from hot utilities. The bold segment below the pinch must expell its heat using cold utilities. For this problem, we can dump the heat from interval 6, where the cold temperature ranges from 160°C to 239°C. This temperature range is much hotter than the coldest temperature in the problem. Such a cold utility, if one is available, will generally be less expensive per kW expelled to it. (If the temperatures are hot enough, one could use the heat to raise steam for use elsewhere on

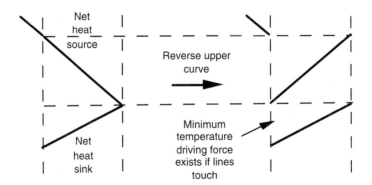

FIGURE 10.12 Illustrating that a right-facing nose on a GCC can always be heat integrated locally.

Sec. 10.1   The Basic Heat Exchanger Network Synthesis (HENS) Problem   361

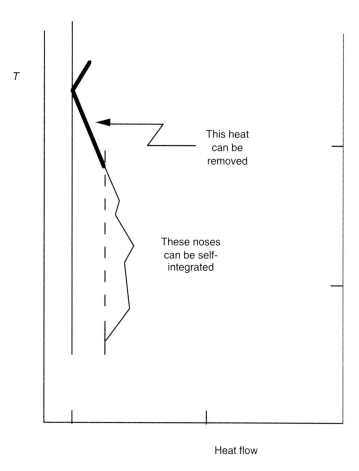

FIGURE 10.13   Cancelling right-facing noses on the GCC.

the plant site. Not only would the cold utility be less expensive, it would actually allow us to make money from this process heat.)

The bold part of the curve that is left after slicing off the right-facing noses provides us with the coldest temperatures at which we can provide needed steam and the hottest temperature at which we can provide needed cold utilities for a problem. This result is extremely valuable. It allows us to pick among the available utilities and select the least expensive ones to supply and remove heat. We can establish *how much* of *which kind* of utilities we need without inventing a heat exchange network.

10.1.3   No Heat Passes Across the Pinch for a Minimum Utility Solution

In Figure 10.14 we partition our HENS problem at the pinch point. Using arguments based on the self-integration of right-facing noses, we can prove we need only add heat

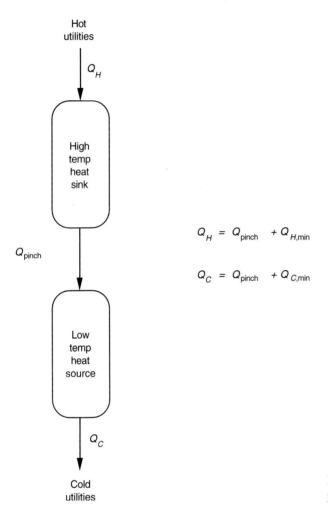

FIGURE 10.14 Pinch point breaks process into two uncoupled parts.

from utilities above the pinch and only expel heat to utilities below the pinch; that is, we never need to remove heat above the pinch nor add it below the pinch. As shown in Figure 10.14, we are adding heat into the hot end of the problem and removing it from the cold end. Let us assume that an amount of heat, Q_{pinch}, passes from the part of the network above the pinch to that below, as shown. We are taking heat from a part of the process that we already know to be deficient and passing it to a part that we know has too much. If we take heat from the part above the pinch, that heat must be supplied by utilities. Similarly, if we add any heat below the pinch, we must then remove it using cold utilities. Minimum utility use, therefore, dictates we pass no heat across the pinch point.

This observation allows us to partition our heat exchanger network synthesis problem into two parts, each of which we can solve by itself if we want only solutions that feature the minimum use of utilities. We will almost always win in a big way with such a partitioning. The number of alternative configurations possible for a problem typically grows at a rate proportional to $N!$, where N is the number of parts in the solution. If we have a heat exchanger network synthesis problem involving 10 exchanges, partitioning into two problems with six and four exchanges in each part will reduce the number of alternatives by a factor of roughly $6! \times 4!/10! = 1/210$, clearly a very significant reduction.

Above the pinch we have a problem for which need only supply hot utilities. Below the pinch we have another problem for which we need only supply cold utilities. In principle we may carry out these two designs separately. In fact, however, some of the choices made for each of these designs will guide the choices made for the other.

10.1.4 The Pinch Design Approach to Inventing a Network

Let us suppose we wish to design a heat exchanger network that uses precisely the minimum utilities computed for it and that nowhere in this network will any temperature driving force be less than ΔT_{min}. How might we proceed to get a first design? Our first seven steps can be the following.

1. Select a ΔT_{min}.
2. Compute the minimum utility use based on this value for ΔT_{min}.
3. Using the grand composite curve, pick which utilities to use and their amounts.
4. If the problem has a pinch point in it (which will occur if step 2 discovers the need for both heating and cooling), divide the problem into two parts at the pinch. We shall design the two parts separately. Remember that the part above the pinch requires only hot utilities and the part below only cold utilities.
5. Estimate the number of exchanges for each partition as $N - 1$, where N is the number of streams in that part of the problem.
6. Invent a network using all insights available. All exchangers that exist at the pinch point will have the minimum driving force at that point. A small driving force for heat transfer implies a large area. The exchangers near the pinch will tend to be large. Therefore, bad design decisions near the pinch point will tend to be more costly. We should generally make design decisions in the vicinity of the pinch first.
7. Remove heat cycles if possible.

Let us apply these ideas to problem 4SP1.

1. We chose a ΔT_{min} of 10°C. The actual value we should select can range from as low as 1°C for below-ambient processes (such as air liquefaction processes) to as high as 30°C for refineries that have to process a wide variety of crude oils. We will discuss shortly how one can be more systematic in selecting the right minimum approach temperature for a problem.

2. For this minimum approach temperature, we determined a minimum rate of heat input of 127.7 kW from hot utilities and a minimum rate of expelling heat of 250.1 kW to be passed to cold utilities.

3. For this problem, we only have steam and cooling water, so we will use these for our utilities. The GCC for this problem did suggest we could remove the heat with a cheaper utility than cooling water, were one to exist. For example, we could consider generating low pressure steam with this rejected heat.

4. The problem does have a pinch point at hot temperature of 249°C (and equivalent cold temperature 239°C). We partition the problem into a hot part and a cold part at this temperature.

5. Above the pinch, only C2 and steam exist. Therefore, we estimate we need one exchange to accomplish this part of the network. Below the pinch all four process streams exist plus cooling water. We estimate we need four exchangers for this part.

6. In Figure 10.15 we start our design by looking at the two parts of the problem near the pinch. The ordinate is temperature with hot temperatures labeled on the left and equivalent cold temperatures on the right. Above the pinch point temperature (249°C (hot)/ 239°C (cold)) there can be only one solution. We must heat C2 using steam. We show a single heat exchanger with steam supplying the heat at the required rate of 127.7 kW.

We work next on a design for below the pinch. Here only streams H2 and C2 exist in the vicinity of the pinch. We can have no hot utilities below the pinch so the top part of C2 adjacent to the pinch must be heated using H2. We also notice that the top part of C1 must also be heated by H2. H1 is not hot enough to supply either of these heating requirements. We can start then by heating the hot end of C2 below the pinch, using the hot end of H2. We might try to heat C2 all the way from its inlet temperature of 116°C, but, if we do, we find we will cool H2 to 166°C. That temperature is too cold to heat the top part of C1. We should do only about half this amount of cooling to H2. We note in Table 10.7 that we need 480.3 kW to heat C2 across interval 6—from 160°C to 239°C. That is roughly half the heat needed to heat it from its inlet temperature, so we propose to exchange 480.3 kW between H2 and C2. The temperature for H2 now drops only to 203.5°C, which is hot enough to bring C1 to its target temperature of 160°C. Now we have to decide how much heat we should supply to C1 before returning to heating C2 (H1 is not hot enough to heat the remaining part of C2).

We can use all the heat from H1 to heat the colder part of C1. This heats C1 to 137.3°C. We then need to heat C1 only from 137.3°C to its target, 160°C, using H2, which we compute requires a heating rate of 173.1 kW. We use H2, which is now at 203.5°C, and further cool it to 187.1°C. That is hot enough to supply heat to the part of C2 we still have not heated, that is, from its inlet at 116°C to 160°C. Another exchange of 276.5 kW accomplishes that heating. However, we cool H2 only to 161.7°C. A heat balance tells us we need to remove 250.1 kW from H2 to finish cooling it, exactly the amount we know we must remove with cooling water. Therefore, we finish cooling H2 with cooling water. We have a first design.

Sec. 10.1   The Basic Heat Exchanger Network Synthesis (HENS) Problem   365

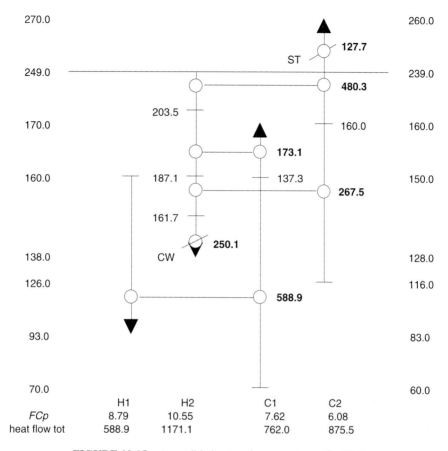

FIGURE 10.15   A possible heat exchanger network for 4SP1.

We count the number of exchanges below the pinch and find five, one more than the number we predicted we might need.

7. We can now attempt to remove any cycles in our design. Because we needed one extra exchanger below the pinch over the number we estimated, we look for a cycle in that part of the design. Here the cycle is obvious when we look at Figure 10.15. We see two exchanges between H2 and C2. We need to remove one of these exchanges. One approach we might try is to split H2 at the pinch and use it to heat C1 and C2 in parallel, which we do in Figure 10.16. Here we heat all of C2 with one branch of H2; we use the other branch to heat the top part of C1. We then cool the bottom part of this second branch, removing all of the 250.1 kW we have to expel to cooling water.

We have a design that meets all our targets. It should definitely be among those we consider for the design of this network. Its one disadvantage is that we have split H2. There are two disadvantages to splitting a stream when designing a heat exchanger net-

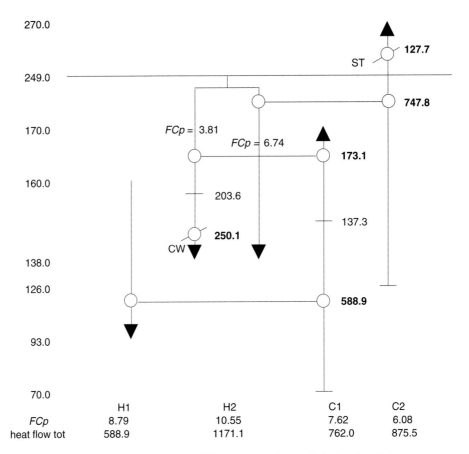

FIGURE 10.16 Design for 4SP1 after removing cycle below the pinch.

work. First, we will have to control the flows in these two branches so they split as needed. Second, splitting a stream means each branch has a lower flowrate than that for the entire stream. A lower flow means a decreased heat transfer coefficient, which means larger heat exchanger areas (unless the stream provides its heat by condensing or vaporizing). To avoid these disadvantages, designers often try to find solutions that do not split any of the streams. The pinch point offers us an interesting opportunity. As we shall now see, a simple analysis will tell us if we must split the streams at the pinch point to obtain a minimum utility use design (Linnhoff and Hindmarsh, 1983).

IS STREAM SPLITTING REQUIRED AT THE PINCH?

Suppose we have partitioned our problem at the pinch point and are looking at an exchange that is above but starts at the pinch. Figure 10.17 shows how the termperature profiles must appear in such an exchange. The pinch is at the left side; as we have seen any

Sec. 10.1   The Basic Heat Exchanger Network Synthesis (HENS) Problem   367

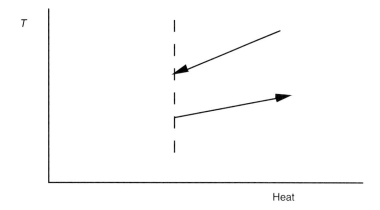

FIGURE 10.17   Temperature profile of streams in vicinity of pinch.

exchange at the pinch point occurs with the minimum driving force for a minimum utility solution. The driving force cannot become smaller as we move away from the pinch, else we would have an exchange with too small a driving force. Therefore, FCp for the hot stream must be smaller than or equal to FCp for the cold stream in the match. The composite curves also must not get closer as they move from the pinch. The composite FCp for the hot streams must also be smaller than or equal to the composite FCp for the cold streams.

Figure 10.18 represents a heat exchange problem where streams H1, H2, C1, and C2 all exist at and just above the pinch point. The nodes indicate the FCp values for each of the streams. For example, H1 has an FCp value of 5 kW/°C. The total FCp value for the hot streams is 6 kW/°C while that for the cold is 7 kW/°C. Thus, the composite streams will have their temperature profiles move apart as the temperature increases above the pinch.

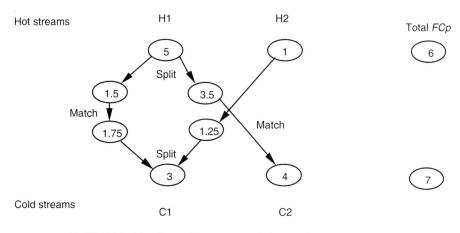

FIGURE 10.18   Case where stream splitting required at pinch point.

We next would like to propose matches between individual stream pairs starting at the pinch. If we match H1 with either C1 or C2, we would find the temperature profiles in either match moving closer as we moved away from the pinch end of the exchange because FCp for H1 (5 kW/°C) is larger than FCp for either C1 (3 kW/°C) or C2 (4 kW/°C). We must split stream H1 into parts whose FCp values are small enough for a match. For example and as illustrated in Figure 10.18, we can split H1 into two streams, one with an FCp of 1.5 kW/°C and the other 3.5 kW/°C. We then match the 3.5 kW/°C part against C2, which has an FCp of 4 kW/°C. As there can be no cold utilities above the pinch to cool H1 or H2, we must match C1 against H2 and the rest of H1. We split C1 into one part with an FCp of 1.75 kW/°C and match that part against the rest of H1 (FCp of 1.5 kW/°C). The remaining part of C1 with an FCp of 1.25 kW/°C can then match against H2 (FCp of 1 kW/°C).

Which streams we split is not necessarily unique. For example, we leave it to the reader to find a solution that splits H1 and C2 for this example.

With H1 having an FCp larger than any of the cold streams, we found we were forced to split that stream. This type of analysis tells us if we need to use stream splitting and aids us to enumerate the alternatives.

10.1.5 Picking the Right Minimum Temperature Driving Force, ΔT_{min}

We have now seen how to compute the minimum utilities required and how to estimate the fewest exchanges we will need before we configure a heat exchanger network that can solve our problem. We then developed a strategy to find a network that features the minimum use of utilities and that either has or comes close to having the fewest exchanges. The strategy required us to pick the minimum temperature driving force we will allow in our solution. We now need a method to select the right minimum driving force.

As the minimum driving force decreases, so will the minimum utilities required. However, with smaller driving forces, heat exchanger areas increase. Smaller utility costs imply larger investment costs and vice versa. There is a trade-off. When we are selecting the minimum allowed temperature driving force, we are attempting to make the right trade-off between utility costs and investment costs. Processes operating below ambient temperatures and requiring refrigeration have very expensive utilities. The proper trade-off for these processes is to reduce utility use and pay more for the exchangers; air liquefaction plants run with driving forces of only one to two degrees centigrade. Plants with very inexpensive utilities will run with large driving forces to reduce equipment costs. Some operate with minimum driving forces of 30°C.

Given a minimum driving force we can estimate the amount and kinds of utilities we need. What we are missing is a way to estimate the cost of the equipment. This section presents a simple approach to enable us to do just this. The method results from the form of the equation we use to estimate the area needed for heat exchange. We can partition the equation into the sum of two terms. One term computes the contribution to the total area needed for exchange by the hot stream and the other by the cold stream. We will show how to make a reasonable assumption based on the Hohmann/Lockhart composite curves that will allow us to compute each term so it is not a function of the stream against which

Sec. 10.1　The Basic Heat Exchanger Network Synthesis (HENS) Problem

it is matched. Thus, we shall be able to estimate the contribution to the total area for each stream independently.

We next divide the total area by the number of exchanges we have previously estimated to estimate the area per exchanger. We then buy that many exchangers of that size to estimate the network investment cost.

Finally, to find the right minimum temperature driving force, we compute annual utility costs and *annualized* area costs for a range of ΔT_{min} values, selecting the one that minimizes the sum of these costs. We do all this before we attempt any design for the network. In the next subsection, we shall show a very approximate way to do the area estimates. The astute reader may immediately see ways to improve these estimates.

EXAMPLE 10.3　Estimating Total Heat Exchanger Areas

Let us estimate the areas for the problem whose stream data we give in Table 10.8. We have added a column in which we estimate the film heat transfer coefficients for each of the streams. We include the utilities we have available.

TABLE 10.8　Stream Data for Example 10.3

Stream	FCp kW/K	T_{in} K	T_{out} K	Q_{avail} kW	h W/m² K
H1	10,000	600	450	1,500,000	800
H2	10,000	500	400	1,000,000	700
ST		650	650		5000
C1	15,000	450	590	−2,100,000	600
CW		300	325		600

We next generate the extended problem table, Table 10.9, based on whatever value of ΔT_{min} we are investigating next. Let us assume we are now investigating the value of 20 K. For this value for the minimum temperature driving force, this table tells us that we need 650,000 kW of heat from steam and we shall expell 1,050,000 kW to cooling water (the top and bottom numbers in the last column). The pinch point for this problem occurs at 500 K (hot)/480 K (cold).

If there were more than one each of the hot and cold utilities, we should next plot the grand composite curve to see which of the utilities and how much of each to use. Here there is only steam for heating, and it is hotter than all other temperatures in the problem. We shall add utility heat from steam at the highest temperatures possible. Similarly, we have only one cold utility, and it is colder than all the other temperatures in the problem. We shall expell heat into cooling water at the coldest temperatures.

Figure 10.19 shows the Hohmann/Lockhart composite curves (as bold curves) for integrating two hot streams and one cold stream. We also show the utilities explicitly. Each region has

- The same streams present
- Composite curves that are straight line segments when plotting heat transferred versus temperature

TABLE 10.9 Extended Problem Table for Example 10.3

Composite Hot Streams		Temperatures		Composite Cold Streams		Grand Composite Hot and Cold Streams		
Avail Heat (000)	Casc'd Avail Heat (000)	Hot	Cold	Req'd Heat (000)	Casc'd Req'd Heat (000)	Net Heat (000)	Casc'd Net Heat (000)	Adj Casc'd Heat (000)
		(610)	590		0		0	650
				150		−150		
	0	600	(580)		150		−150	500
1000				1500		−500		
	1000	500	(480)		1650		−650	0 pinch
600				450		150		
	1600	(470)	450		2100		−500	150
400						400		
	2000	450	(430)				−100	550
500						500		
	2500	400	(380)				400	1050

If we have a straight line segment and a constant heat transfer coefficient in a counter current heat exchanger, we can compute its area using the familiar equation based on the log mean temperature difference for the exchange.

We propose using the following equation to estimate heat exchanger areas for this problem without first inventing any heat exchanger network for the problem. To do our estimate we write the following equation for computing area.

$$A = \int_1^2 dA = \int_1^2 \frac{1}{U \times \Delta T(T)} dQ(T) = \int_1^2 \left(\frac{1}{h_{cold}} + \frac{1}{h_{hot}} \right) \times \frac{1}{\Delta T(T)} dQ(T) \quad (10.3)$$

where A is the area; U the overall heat transfer coefficient; h_{cold} and h_{hot} the individual heat transfer coefficients for the cold and hot side of the transfer respectively; $\Delta T(T)$ is the driving force in the exchange; T the temperature; $dQ(T)$ the incremental amount of heat provided by the hot stream to the cold stream between the temperature T and $T + dT$; and points 1 and 2 are at the opposite ends of the counter current heat exchanger.

We note that $dQ(T)$ is either $-(FCp)_{cold} dT$ or $+(FCp)_{hot} dT$ in such a match. We partition this computation into the sum of two integrals, one that computes a contribution for the cold stream heat transfer coefficient and one for the hot stream heat transfer coefficient. We call these two contributions A_{cold} and A_{hot}. Remember each stream sees an area equal to the *sum* of these two contributions.

$$A = \int_1^2 \frac{1}{h_{cold}} \times \frac{1}{\Delta T(T)} (FCp)_{cold} dT + \int_1^2 \frac{1}{h_{hot}} \times \frac{1}{\Delta T(T)} (FCp)_{hot} dT = A_{cold} + A_{hot} \quad (10.4)$$

Sec. 10.1   The Basic Heat Exchanger Network Synthesis (HENS) Problem   371

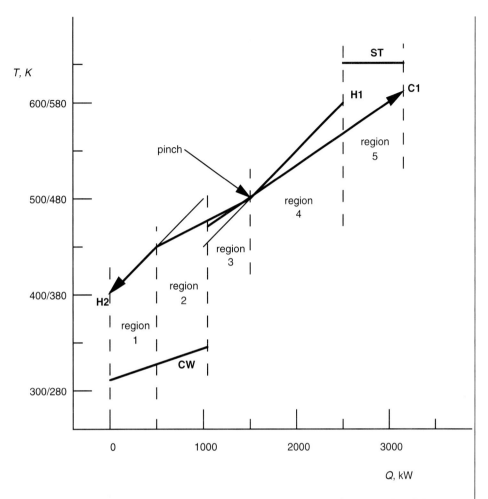

FIGURE 10.19 Hohmann/Lockhart composite curves for Example 10.3, including utilities.

The driving force is a function of the two streams we match in each exchange. However, suppose we assume that the network we invent will have temperature driving forces very close to those of the Hohmann/Lockhart composite curves. In the vicinity of the pinch point, the temperatures are close so it is here that we require a disproportionate share of the area needed by the network. Away from the pinch point, the temperature driving forces are larger giving us smaller areas. Thus, it will not matter too much if we fail to estimate these areas accurately. As we integrate versus the temperature of a stream, we can look on these composite curves for the appropriate $\Delta T(T)$. Assuming this $\Delta T(T)$ as the temperature driving force, the two terms may be computed independently. We do the computation region by region using the following equation for each stream in that region.

$$A = \frac{Q}{h_i \times \Delta T_{LM}} \tag{10.5}$$

where

$$\Delta T_{LM} = \frac{(T_{hot,1} - T_{cold,1}) - (T_{hot,2} - T_{cold,2})}{\ln\left(\dfrac{T_{hot,1} - T_{cold,1}}{T_{hot,2} - T_{cold,2}}\right)} \tag{10.6}$$

For every exchange in a region, the same log mean temperature driving force results so we compute it once per region. Table 10.10 shows the area calculations for our example. For instance, in region 5 we compute

$$\Delta T_{LM} = \frac{(650 - 590) - (650 - 546.7)}{\ln\left(\dfrac{650 - 590}{650 - 546.7}\right)} = 79.7 \tag{10.7}$$

which then results in the area contribution for the steam side of

$$A_{\text{steam side}} = \frac{650,000}{5000 \times 79.7} = 1.6 \tag{10.8}$$

TABLE 10.10 Area Calculations for Example 10.3

			Heat Exchanger					
			hot end		cold end			
Stream	Q	h	$T_{hot,1}$	$T_{cold,1}$	$T_{hot,2}$	$T_{cold,2}$	ΔT_{LM}	Area
region 5								
ST	650,000	5000	650	590	650	546.7	79.7	1.6
C1	650,000	600						13.6
region 4								
H1	1.00E + 06	800	600	546.7	500	480	34.0	36.8
C1	1.00E + 06	600						49.0
region 3								
H1	225,000	800	500	480	477.5	450.0	23.6	11.9
H2	225,000	700						13.6
C1	450,000	600						31.8
region 2								
H2	275,000	700	477.5	325	450	311.9	145.2	2.7
H1	275,000	800						2.4
CW	550,000	600						6.3
region 1								
H2	500,000	700	450	311.9	400	300	118.0	6.1
CW	500,000	600						7.1

Sec. 10.2   Refrigeration Cycles   373

The total area above the pinch is the total of the area contributions in regions 4 and 5 in Table 10.10: 1.6 + 13.6 + 36.8 + 49.0 = 101.0 m^2. Since there are three streams involved above the pinch (ST, H1, and C1), we estimate this area is distributed across $N - 1 = 2$ exchangers. We buy two exchangers, each with an area of half this total: 50.5 m^2. Similarly, the total area below the pinch is the sum for regions 1 through 3: 81.9 m^2. There are four streams present (H1, H2, C1, and CW), suggesting we need three exchangers. We buy three, each with one-third this area: 27.3 m^2. The total cost to purchase these exchangers is our estimated investment cost for choosing $\Delta T_{min} = 20$ K. We also know we need steam heat at the rate of 650,000 kW and heat removal using cooling water at the rate of 1,050,000 kW. We can use a cash flow analysis to determine the present worth of these two cash flows as a way to evaluate the worth (actually a negative worth results—i.e., a cost) of this design.

We repeat these computations for a range of minimum temperature driving force values and choose the value for ΔT_{min} that gives us the best value for its worth (i.e., least cost). Again note that we have done all these computations without inventing a network. We can use the investment and operating costs for that value of minimum driving force as an estimate for the cost of heat integration within the context of the larger problem of designing an entire flowsheet.

10.2 REFRIGERATION CYCLES

A refrigerator uses a heat pump to move heat from a low temperature to a high temperature. A heat pump is the reverse of a power cycle. For example, a home refrigerator removes heat from food that is just above freezing, say 5°C, and ejects that heat into the room, which is at ambient temperature, say 25°C. The work we put into the pump to move the heat to the higher temperature degrades to heat. It must be expelled along with the heat we remove from the food. In contrast, a power cycle (such as a Carnot cycle or a Rankine cycle) degrades high temperature heat, converting part to work and expelling the rest at low temperature. Figure 10.20 illustrates the comparison. Degrading heat from a high temperature to a low temperature allows us to create work; using work allows us to elevate the temperature of heat.

Figure 10.21 shows the component parts for a typical refrigeration cycle. We start examining the cycle at the exit of the condenser, point 1. Here the refrigerant is a high pressure liquid, very near to saturation (i.e., about ready to boil). We reduce the pressure on the liquid by passing it through an adiabatic valve. It partially vaporizes, point 2. The heat required for vaporization comes from the fluid itself, cooling it. We next pass this fluid through the refrigeration coils where the rest of the liquid evaporates. In doing so, it takes heat from its surroundings (from the food). We now have a low pressure fluid, point 3, which is all vapor and very near saturation (just ready to condense). We increase the pressure on the fluid by compressing it. An ideal compressor operates isentropically (i.e., at constant entropy), arriving at point 4. It will heat up, becoming a superheated vapor well above saturation. We then cool it by expelling heat to the surroundings (i.e., from the coils in the back of the refrigerator to the room), returning ultimately to being a liquid at high pressure, point 1.

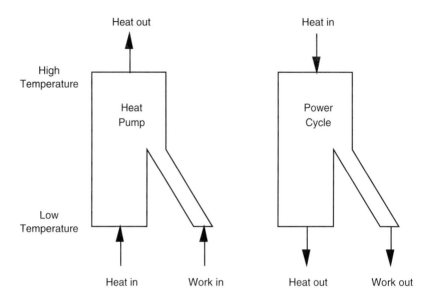

FIGURE 10.20 Comparison of a heat pump to a power cycle.

In Figure 10.22, we show this cycle on a plot of temperature versus entropy. Mechanical engineers typically view refrigeration cycles on such a plot (while chemical engineers often view it on a pressure versus enthalpy diagram). The advantage of viewing such a cycle on a temperature versus entropy diagram is that the area enclosed in the cycle represents the ideal work needed to run the cycle. Improvements to the cycle will show up

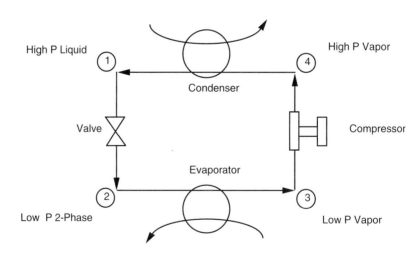

FIGURE 10.21 A typical refrigeration cycle.

Sec. 10.2   Refrigeration Cycles

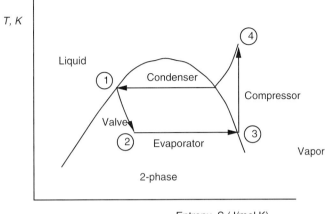

FIGURE 10.22   Temperature-entropy diagram for refrigeration cycle.

as reductions in this area, provided the we pick up the same amount of heat in the evaporator both before and after the improvement.

We illustrate two improvements in Figure 10.23. The first is to use a multistage compressor as shown on the right. We compress only part way and then cool the vapor back to its saturation temperature. We compress again to the final pressure. We point at the area saved—on the right side. The second is to use a let down turbine rather than a valve to drop the pressure of the high pressure liquid, as shown on the left side of this fig-

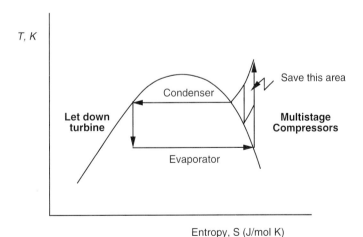

FIGURE 10.23   Using multistage compression and let down turbines to save on work required for refrigeration cycle.

ure. This step appears to increase the area, but it also increases the length of the line representing the heat we pick up in the evaporator. It really is an improvement, because the area per unit of heat we pick in the evaporator is actually reduced when we use the let down turbine.

We should normally use one cycle to elevate the low temperature heat by no more than about 30°C. If we need to increase the temperature of the heat more than that, it pays to use multiple cycles where a lower temperature cycle passes heat to the cycle above it, which in turn passes it to the cycle above it, repeating until the top cycle, which passes the heat to ambient conditions. We show a double cycle in Figure 10.24. Refrigeration cycles are expensive to purchase and very expensive to operate as they involve the use of compressors. They should be run with much smaller driving forces than are typical for above ambient processes. Smaller driving forces mean we will pay much more for the equipment but less for the operating costs as the processes operate nearer to reversible conditions.

The evaporator/condenser that connects the two cycles in Figure 10.24 requires a temperature driving force for the heat to transfer. The lower cycle must raise its heat to a temperature just above the temperature of the fluid in the upper cycle so it can transfer heat to it. If it is reasonable to use the same refrigerant in both cycles, we can eliminate this loss of temperature driving force by exchanging heat between the two cycles as shown in Figure 10.25. Here we replace the evaporator/condenser unit with a flash unit. The two cycles trade fluid rather than just heat. The lower cycle puts vapor into the flash

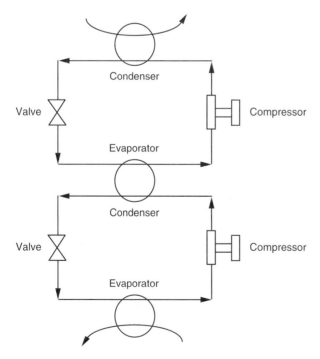

FIGURE 10.24 Two-stage refrigeration cycle.

Sec. 10.2 Refrigeration Cycles   377

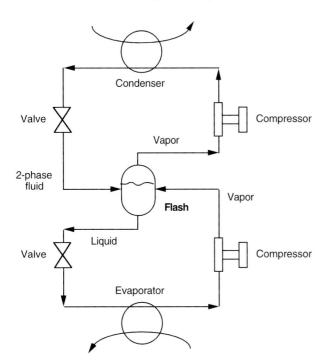

FIGURE 10.25 Replacing evaporator/condenser with flash to save loss of temperature driving force.

unit while the upper cycle feeds in 2-phase fluid. The lower cycle takes away the liquid, while the upper cycle takes the vapor from the flash unit. Material balance requires each cycle to remove the same amount of refrigerant as it put into the flash unit. The lower cycle trades vapor for liquid, while the upper trades vapor and liquid for vapor alone. It is as if they have traded heat. This trade is done with no temperature driving force and makes it an attractive alternative to improve a cascaded refrigeration cycle.

10.2.1 Using Grand Composite Curves to Design Refrigeration Cycles

There are many other ways to improve refrigeration cycles, and they are described extensively in the mechanical engineering literature. Our interest at this point is to design a good refrigeration cycle for a given process using the types of insights we developed earlier for heat exchanger networks. In Figure 10.26 we show a heat pump on a plot of temperature versus the heat transferred (i.e., T versus Q), the same axes we used when we analyzed heat exchanger networks. At the right side we show a shaded area, the width of which respresents the work we have to put into the process to elevate the temperature. The higher we raise the temperature, the more work we will have to use to run the process, so the width grows as we move up in temperature.

The second law places a constraint on the amount of work it will take for us to raise the temperature of the heat we pick up at the lower temperature and expel at the higher temperature, namely:

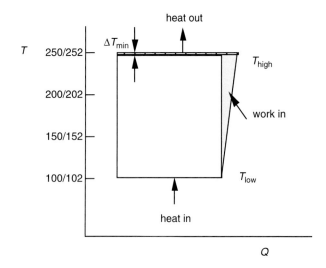

FIGURE 10.26 A heat pump on a T versus Q diagram.

$$W \geq Q\left(1 - \frac{T_{low}}{T_{high}}\right) = Q\left(\frac{T_{high} - T_{low}}{T_{high}}\right) = \frac{Q\Delta T}{T_{high}} = \frac{Area}{T_{high}} \quad (10.8)$$

where W is the amount of work the cycle requires and Q the total of the heat the cycle rejects at the higher temperature. Let us assume the amount of work is small relative to the amount of heat, so the heat picked up is approximately that rejected by the cycle. The term $Q\Delta T$ will then be approximately the area of the square unshaded box shown on this diagram. Our goal for designing heat pumps for a process is to make this area as small as possible.

We have to make an adjustment to area because of the dual temperature scale. The area has to have a height from T_{low} to T_{high}. T_{low} is the temperature at which the pump picks up the heat. This temperature is a "cold" temperature on the temperature scale. On the other end of the box is the temperature at which the pump must eject its heat to an ambient heat sink, which must be a "hot" temperature. Thus, the height of the box is $\Delta Tmin$ taller than one might at first think. We show this extra height with the strip across the top of the box in Figure 10.26. If it is narrow enough, we can ignore it in the approximate analysis we use in what follows.

We shall use the grand composite curve (GCC) to aid our design process for such cycles. In Figure 10.27 we plot of the grand composite curve for the part of our process that is operating at temperatures that are below ambient. Remember that there are two temperature scales on this plot—the hot and the cold, which are ΔT_{min} different and shown as the same value on the vertical axis. We use the hot temperature scale to give the temperature for a hot stream and the cold for a cold stream. The GCC is the zigzag line that starts just below ambient on the left and moves downward and to the right, then to the left and then to the right again. We remember that we can self-integrate the heat in the right-facing noses. Thus, we must pump only the "uncovered" heat below the grand composite curve to ambient conditions and reject it.

Sec. 10.2 Refrigeration Cycles

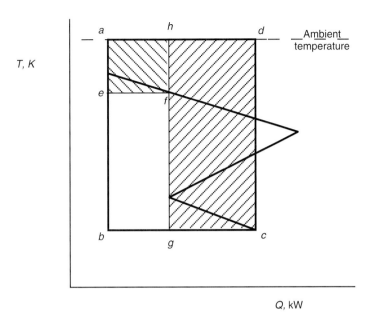

FIGURE 10.27 Use of refrigeration cycles (heat pumps) to transfer heat from low temperatures to ambient temperatures.

We could propose to use a single heat pump that will pick up all the process heat at one temperature. This temperature has to be below the coldest temperature of heat we have to recover—that is, it has to just touch the GCC at the coldest point where we pick up heat. The area for the heat pump is approximated by the single large box abcd. It has a width that covers all the heat that we have to pump.

We could also propose to use two heat pumps, whose areas are marked by the hatched boxes. We reduce the area of the heat pumping required by the difference in the larger box abcd and these two hatched boxes, aefh and hgcd—that is, the unhatched part of the larger box in the lower left. Should we use two pumps that have a smaller area? The two-pump option will have a lower operating cost, but it may have a larger investment. Our decision will have to depend on how these numbers work out when we analyze them.

As we discussed in Chapter 4, we can approximate the investment costs for equipment with an equation of the form:

$$\text{Investment cost} = A \text{ size}^M, \text{ where } 0 < M < 1 \qquad (10.9)$$

M often takes the value 0.6. We can use the work required by the heat pump to characterize its size. Thus, if we plot cost versus the work W required to run the heat pump, we would get a plot as shown in Figure 10.28. The marginal cost to buy a piece of equipment reduces as the equipment gets larger.

We can approximate this cost curve charging just for having the equipment and then

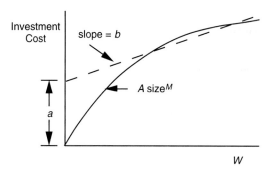

FIGURE 10.28 Investment cost versus equipment size.

adding a term that is linear in size, as shown by the dashed line in Figure 10.28. The equation for such a cost curve is:

$$\text{equipment cost} \approx a + b\,W \qquad (10.10)$$

where a is known as a "fixed charge" term in this equation. The operating cost for a heat pump will be proportional to the work done:

$$\text{operating cost} = c\,W \qquad (10.11)$$

and thus the total cost is of the form:

$$\text{cost} \approx \text{equipment cost} + \text{operating cost} = a + (b+c)W = a + \beta W \qquad (10.12)$$

Substituting the work done into this equation from earlier, we get

$$\text{cost} \approx a + \beta\left(\frac{\text{Area}}{T_{\text{high}}}\right) \qquad (10.13)$$

where area is approximately that for the box representing the heat pump on a T versus Q diagram. We call this form of approximate cost equation a linear fixed charge model.

We are now ready to decide if we should use two heat pumps to replace one. The cost for two heat pumps versus one would be

$$\text{cost}(2) \approx 2a + \beta\left(\frac{\text{Area}(2)}{T_{\text{high}}}\right) \text{ vs. } \text{cost}(1) \approx a + \beta\left(\frac{\text{Area}(1)}{T_{\text{high}}}\right) \qquad (10.14)$$

where Area(2) is the total area for the two heat pumps and Area(1) for the single heat pump. The difference in cost is

$$\text{cost}(2) - \text{cost}(1) = a + \frac{\beta}{T_{\text{high}}}(\text{Area}(2) - \text{Area}(1)) \qquad (10.15)$$

If cost(2) is smaller than cost(1), we would choose two heat pumps, else we would choose one. The point where the two are equal is where

Sec. 10.2   Refrigeration Cycles   381

$$\text{Area}(1) - \text{Area}(2) = \frac{aT_{\text{high}}}{\beta} \qquad (10.16)$$

which is an area difference we compute once we know a, β and T_{high}. We can place a box with this area on our T versus Q diagram, picking any convenient height and its corresponding width for the box. We should introduce another heat pump any time we can cause a saving in area greater than this amount by doing so.

Let us return to our example. In Figure 10.29 we show a shaded box with the area = aT_{high}/β in the lower left. Any heat pump saving more than this area on this diagram is worth introducing. We can readily justify the use of at least the two heat pumps in Figure 10.29. The area saved is much more than this shaded box.

Can we justify introducing even more heat pumps? In fact, we can. We can replace the large heat pump on the right of Figure 10.27 by using two pumps having different temperatures at the bottom (i.e., different temperatures at which they pick up heat from the process). We show this as a stepping of the low end of the box.

We suggested earlier that we would heat integrate the right-facing nose for the process. Let us not do this integration. That means the heat at the top part of the nose where the process is a net producer of heat (sloping down as we move to the right) is no longer consumed by the bottom part where the process is a net consumer of heat (sloping down as we move to the left). We can dump the heat from the two heat pumps 3 and 4 that we just introduced into the bottom part of the nose where it is a heat sink. A bit of the nose to the far right is left untouched by this process. We can self-integrate this small part. We must then pick up the heat from the upper part of the nose

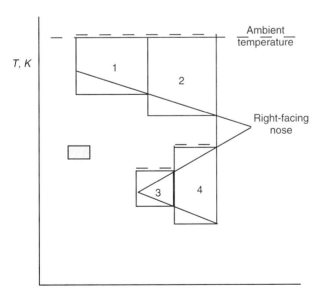

FIGURE 10.29 Adding cycles to save on work required for cycle.

that is no longer integrated with the bottom. We can use heat pump 2 and part of 1 to do this. We can see savings if we use two temperatures at least in picking up this heat and that to the left that we picked up before in Figure 10.27 with the smaller heat pump. We use four heat pumps here. Comparing this figure to the earlier one shows the areas we save. The two coldest temperatures save the corner we notched out, which is to the left of heat pump 4 and below heat pump 3. Ejecting the heat into the bottom of the right-facing nose and then pumping the heat from the top of that nose saves the area between part of pump 1 and all of pump 2 and pumps 3 and 4. Using two temperatures for pumps 1 and 2 saves the area of the notch below pump 1 to the left of pump 2. Each of these savings is larger than the box in the lower left. Thus, each would save us money.

In this section we looked at an design problem that uses insights the grand composite curve can provide to us. We discovered that we can visualize a very good but quite complex solution to the design of a below ambient heat recovery process. In the next chapter, Chapter 11, we are going to look at synthesis methods for separating relatively ideal liquid mixtures using distillation. With the background we gain in Chapter 11, we return in Chapter 12 to heat-integrating processes involving distillation columns. We shall find that the representations we developed in this chapter also help in this synthesis activity.

REFERENCES

Gunderson, T., & Naess, L. (1988). The synthesis of cost optimal heat exchanger networks. An industrial review of the state of the art. *Comput. Chem. Engng.,* **12**, 503.

Hohmann, E. C. (1971). *Optimum Networks for Heat Exchange*. Ph.D. Thesis, University of So. Cal.

Linnhoff, B. (1993). Pinch analysis—A state-of-the-art overview. *Trans. IChemE.,* **71**(A), 503.

Linnhoff, B. et al. (1982). *User Guide on Process Integration for the Efficient Use of Energy*. Inst. Chem. Engrs.: Rugby.

Linnhoff, B., & Hindmarsh, E. (1983). The pinch design method of heat exchanger networks. *Chem. Engng. Sci.,* **38**, 745.

Umeda, T., Harada, T., & Shiroko, K. (1979). A thermodynamic approach to the synthesis of heat integration systems in chemical processes. *Comput. Chem. Engng.,* **3**, 273.

EXERCISES

1. In this exercise, you are shown a shortcut method to construct composite curves if the FCp is constant for the various streams in the problem.

Exercises

Plot, as follows, the temperature (ordinate) against the heat required (abscissa) for the first two cold streams. Temperature should increase as you move from left to right for each stream. First plot the line for stream 1. Create the plot for stream 2 just to the right of stream 1, with the starting heat value for stream 2 being the ending heat value for stream 1. Then, where the two streams share the same temperature range (from 300 to 350 K), connect with a straight line the point where stream 1 is at 300 K to the point where stream 2 is at 350 K. Argue that this part of plot you have created is the composite heating curve for the two streams where their temperatures overlap. Plot the third stream and using this geometric approach, construct the composite curve for all three streams.

Stream no.	Inlet temperature K	Outlet temperature K	FC_p kW/K
1	250	350	10
2	300	400	20
3	270	370	15

The following data are to be used for problems 2 through 22.

Available Utilities

Utility	Inlet temperature K	Outlet temperature K	Cost per million kJ
Steam, Hi P	500	500	$5.50
Steam, Lo P	350	350	$2.00
Cooling Water	305	≤ 325	$0.80

Heat Transfer Coefficients When Sizing Heat Exchangers

Phase	Film coefficient W/(m^2 K)
Vapor	200
Liquid	1000
Condensing vapor	9000
Evaporating liquid	9000

Annualized installed heat exchanger cost:

$$\text{annualized cost} = 7000 \ \$/\text{yr} \ (A/100)^{0.65}$$

where area is in square meters.

HENS I

Stream	T_{in}, K	T_{out}, K	FCp, kW/K	Comment
H1	430	340	15	Liquid
C1	310	395	7	liquid
C2	370	460	32	Vapor

HENS II

Stream	T_{in}, K	T_{out}, K	FCp, kW/K	Comment
H1	450	325	5	Liquid
H2	400	375	10	Vapor
	375	374	1000	Condensing vapor
	374	330	18	Liquid
C1	310	350	8	liquid
C2	370	460	15	Vapor

HENS III

Stream	T_{in}, K	T_{out}, K	FCp, kW/K	Comment
H1	460	330	5	Liquid
H2	405	366	12	Vapor
	366	365	600	Condensing vapor
	365	330	15	Liquid
C1	310	345	40	liquid
C2	370	470	10	Vapor

Do the following for HENS I.

2. For a ΔT_{min} of 10 K, develop the problem table for this problem. (Hint: You should use a spreadsheet program here.)
3. Draw the Hohmann/Lockhart composite curves.
4. Draw the grand composite curve. Estimate the minimum utility costs that should occur for this problem if ΔT_{min} is 10 K.
5. Estimate the fewest number of heat exchangers needed above and below the pinch IF no heat can be exchanged across it. Estimate the fewest if heat can be transferred across the pinch.
6. What is the minimum utility requirement for this problem, as a function of the minimum allowed temperature driving force? In other words, develop a plot of minimum utility cost vs ΔT_{min}. Range ΔT_{min} from 2 to 50 K. (This part of the problem

Exercises

demands that you use a spreadsheeting program to solve it. Otherwise, it is far too much effort.)

7. On this same plot, indicate the area costs as a function of temperature driving force. Pick the "best" driving force for this problem.
8. For this "best" driving force, develop a heat exchanger network and compare the area costs to those estimated in question 7.

9–15. Repeat homework problems 2 to 8 for HENS II.
16–22. Repeat homework problems 2 to 8 for HENS III.
23. How many refrigeration cycles should you use for the following subambient process? The grand composite curve is based on a driving force of 2 K. The temperatures shown on the ordinate are cold-side temperatures (i.e., hot-side temperatures are 2 K hotter). Indicate clearly why you have arrived at the answer you have.

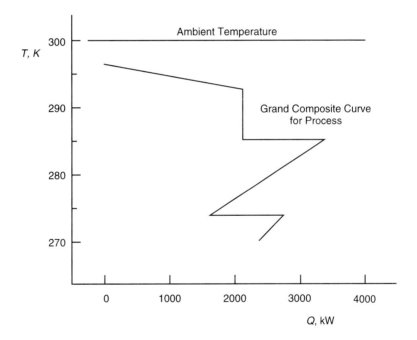

FIGURE 10.30 Grand composite curve for subambient process.

The cost for a cycle is given by

$$\text{Cost}\{\$/\text{yr}\} = 20{,}000\{\$/\text{yr}\} + 3000\{\$/\text{yr}/\text{kW}\}W\{\text{kW}\}$$

where W is the work required to operate a cycle.

24. The following streams exist at and just above the pinch point for a heat exchanger network synthesis problem. Propose all possible configurations which correspond to matches that split the fewest streams. Split a stream into at most two branches.

Stream	FC_p
H1	10
H2	6
H3	1
C1	9
C2	7
C3	2

IDEAL DISTILLATION SYSTEMS   11

In this chapter we shall look at the synthesis of distillation-based separation systems. A separation system is a collection of devices to separate a multicomponent mixture in two or more desired final products. We shall start this chapter by designing a process to separate a mixture of three normal alkanes. We shall next look at separating a mixture of five alcohols, using insights from the first problem but adding a few as the problem has many more design alternatives. These mixtures display fairly ideal behavior and are much easier to consider than mixtures that display highly nonideal behavior. The heat integration of distillation processes is the subject of the next chapter while the separation of nonideal mixtures is the subject of Chapter 14.

11.1 SEPARATING A MIXTURE OF n-PENTANE, n-HEXANE, AND n-HEPTANE

In this example we assume we have an equimolar mixture flowing at 10 mol/s that is 20 mole % n-pentane, 30% n-hexane, and 50% n-heptane. Our goal is to separate this mixture into three products: 99% pure n-pentane, 99% pure n-hexane, and 99% pure n-heptane. Let us assume the feed and the products will all be liquids at their bubble points—that is, each is just ready to boil. If we were to decide to use distillation to accomplish this separation, Figure 11.1 shows two process alternatives that we should consider. In the direct sequence, we remove the most volatile species, pentane, in the first column and then separate the hexane and heptane in the second, while in the indirect sequence, we remove the heaviest species, heptane, first and then separate pentane from hexane. We might be interested in discovering which is less expensive to buy and operate. When we

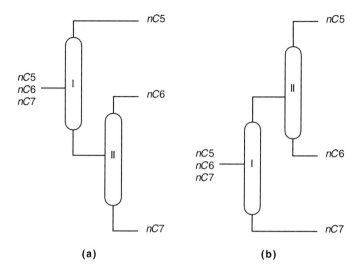

FIGURE 11.1 Two alternatives to separate $nC5$, $nC6$, and $nC7$ using distillation: (a) the direct sequence, and (b) the indirect sequence.

consider heat integrating columns—as we shall do in the next chapter—we can readily propose several other distillation-based separation schemes.

11.1.1 Do the Species Behave Ideally for Distillation?

We must first decide if these species display fairly ideal behavior during distillation. It does little good to design a system assuming ideal behavior if the mixture does not display it. For example, suppose we wish to separate toluene from water. We could assume ideal behavior and propose using distillation. However, these two species do not like each other at all. They will spontaneously separate into two fairly pure liquid phases: a toluene-rich phase and a water-rich phase. If the separation is complete enough for our needs, then the cost of separating is the cost of a decanter. A decanter will likely be much less costly than the column we would have designed assuming ideal behavior. Another possibility is that some of the species form azeotropes, as ethanol and water do. If any do, then we must design a very different process even if we can use distillation to accomplish the final separation. We will look at how to check for nonideal behavior in more detail in Chapter 14.

For species that are very similar—as are the n-alkanes we are considering here—we should expect close to ideal behavior. Table 11.1 contains some preliminary physical property data for these species. From this data we see there is quite a difference in normal boiling points, which should make the separation easier. All normal boiling points are above room temperature, although n-pentane is only just above. We include the critical properties so we have an idea of the extreme conditions we would dare to consider.

One of the first steps we might take is to compute several flash simulations for these species to see the volatility behavior they will display in a distillation column. In particu-

Sec. 11.1   Separating a Mixture of *n*-Pentane, *n*-Hexane, and *n*-Heptane   **389**

TABLE 11.1 Property Data for Alkane Example

Property	*n*-Pentane	*n*-Hexane	*n*-Heptane
MW	72.151	86.176	100.205
$T_{boiling}$, K	309.187	341.887	371.6
T_C, K	469.8	507.9	540.2
P_C, K	33.3	29.3	27.0

lar, we might wish to see what their relative volatilities are and how much they vary as composition varies. Table 11.2 shows the relative volatilities when we perform three flash calculations for the feed composition using a simulator: a bubble point flash, one where 50% of the feed exits as vapor, and a dewpoint flash. We used the Unifac method to evaluate liquid activity coefficients (as a precaution against surprising nonideal behavior). We see that the relative volatilities do not change too much. When we consider the behavior at infinite dilution (we did a bubble point calculation for each of three mixtures, each having a composition of a part per million for two of the species in the third), the relative volatilities range from 4.99 to 9.03 for $nC5$ relative to $nC7$ and from 2.25 to 3.02 for $nC6$ relative to $nC7$. These variations should not be ignored, but they do not indicate particularly nonideal behavior either.

11.1.2 Goals for Our System Design

What might be the goals for our system design? One goal is to create the system having the least cost, but what do we mean by "cost"? As we saw in Chapter 5, we can measure the cost by modeling the cash flow caused by our design. In this case there will be an initial investment in purchasing and installing the equipment and then there will be annual costs in operating it. Operating costs will include utility and labor costs. The present worth of these cash flows can be the cost we then choose to minimize.

We also want our process to be safe. It should not needlessly employ hazardous chemicals. Indeed, if the species are sufficiently hazardous, we may choose not to build the process. It should not operate at extreme conditions of temperature and pressure if we can

TABLE 11.2 Example Relative Volatilities

	Percent of Feed Vaporized in Flash	Temperature (K)	Relative Volatility for $nC5$ Relative to $nC7$	Relative Volatility for $nC6$ Relative to $nC7$
Bubble Point	0	341.6	6.24	2.46
	50	351.3	5.51	2.32
Dewpoint	100	357.4	5.76	2.36

avoid it. It should also be environmentally benign. It should be flexible enough to operate at expected levels of production. From both a safety and an environmental point of view, not introducing any other species to carry out the separation would have its advantages.

For our original screening, we shall concentrate on minimizing costs, but we shall always watch out for safety and environmental issues as they arise.

11.1.3 Evaluating Cost

It will take us some effort to compute the costs for a column. We are trying at this point just to screen among alternatives; perhaps we can use a simpler evaluation. One we might consider is the vapor flow predicted within the column. The larger this flow, the larger the column diameter must be to accomodate it. Also, for a given feed, a larger vapor flow indicates a more difficult separation, which suggests there are more trays. Finally, the utilities consumed in a column create vapor in the reboiler and condense it in the condenser. Thus, the vapor flow directly reflects the utility use in a column. For this reason several authors have suggested its use in preliminary screening of design alternatives for separation systems consisting only of distillation columns. They suggest choosing the separation process that minimizes the sum of the vapor flows in its columns.

MINIMUM VAPOR FLOWS

How can we estimate vapor flow in a column? For nearly ideal behavior where we are willing to assume constant relative volatilities and constant molar overflow throughout a column, we can use Underwood's method to estimate minimum internal vapor and liquid flows. For preliminary design purposes, we may set the actual vapor flow to be a multiple, say 1.2, times the minimum vapor flow estimated for each column. If we do, then the total of the actual vapor flows in a column sequence will be 1.2 times the total of the minimum vapor flows for it. Thus, we can search for the better sequences using the total of their minimum vapor flows.

Underwood's method uses the following three equations:

$$\sum_i \frac{\alpha_{ik}}{\alpha_{ik} - \phi} f_i = (1-q)F \qquad (11.1)$$

$$(R_{\min} + 1)D = \sum_i \frac{\alpha_{ik}}{\alpha_{ik} - \phi} d_i = V_{\min} \qquad (11.2)$$

$$\overline{R}_{\min} B = -\sum_i \frac{\alpha_{ik}}{\alpha_{ik} - \phi} b_i = \overline{V}_{\min} \qquad (11.3)$$

where α_{ik} is the relative volatility of species i to k, f_i the molar flow of species i in the feed, q the fraction of the feed that joins the liquid stream at the feed tray, F the total molar flow of the feed, D the molar flow of the distillate, $R_{\min}$ the minimum reflux ratio (= $L_{\min}/D$), d_i the molar flow of species i in the distillate, $V_{\min}$ the minimum vapor flow possible in the top section of the column to accomplish the desired separation, $\overline{R}_{\min}$ the

Sec. 11.1   Separating a Mixture of *n*-Pentane, *n*-Hexane, and *n*-Heptane

minimum reboil ratio (= $\overline{V}_{min}/B$), b_i the molar flow for species i in the bottoms product, and $\overline{V}_{min}$ the minimum vapor flow in the bottom section of the column). The final variable in these equations is ϕ, which we shall define through its use in the next subsections.

Estimating Product Compositions.   We wish to estimate the minimum vapor flows needed to separate our given feed mixture of 20% *n*-pentane, 30% *n*-hexane, and 50% *n*-heptane into one 99% pure product for each of the three species. To use Underwood's equations we must estimate the compositions for the feeds and products to a column. To make these estimates, we need to make some assumptions about what exactly is contaminating each product. Let us assume that a product is contaminated only by species immediately adjacent to it in volatility. If there are two adjacent species—one more volatile and one less—let us further assume that they each supply half the allowed contamination. We assume, therefore, that the pentane product is contaminated only with hexane, that the hexane will be contaminated equally with both pentane and heptane, and the heptane is contaminated only with hexane. Thus, we start our problem by assuming that the product compositions are as shown in Table 11.3, where product I is the one rich in pentane, product II in hexane, and III in heptane. These product specifications are to hold no matter the distillation sequence we select.

We can write equations based on molar flows, μ, for our process as follows.

$$\mu_I(nC5) + \mu_{II}(nC5) = 2 \text{ mol/s}$$

$$\mu_I(nC6) + \mu_{II}(nC6) + \mu_{III}(nC6) = 3 \text{ mol/s}$$

$$\mu_{II}(nC7) + \mu_{III}(nC7) = 5 \text{ mol/s}$$

We note, from the initial product specifications, that we can also write:

Product I: $\mu_I(nC5) = 99\, \mu_I(nC6)$

Product II: $\mu_{II}(nC5) = \dfrac{5}{990}\mu_{II}(nC6)$, $\mu_{II}(nC7) = \dfrac{5}{990}\mu_{II}(nC6)$

Product III: $\mu_{III}(nC7) = 99\, \mu_{III}(nC6)$

Substituing these latter four into the first three gives us three equations in the three flows for hexane that we can readily solve. Therefore, we can quickly compute the flows shown in Table 11.4.

TABLE 11.3   **First Guess at Product Molar Percentages**

	Feed	Product I *n*C5 rich	Product II *n*C6 rich	Product III *n*C7 rich
*n*C5	20	99	0.5	0
*n*C6	30	1	99	1
*n*C7	50	0	0.5	99

TABLE 11.4 Flows for Process in Figure 11.1a that Satisfy Composition Specifications Given in Table 11.3

Species	Product I mol/s	Product I mol%	Product II mol/s	Product II mol%	Product III mol/s	Product III mol%
$nC5$	1.985	0.99	0.015	0.005	0	0
$nC6$	0.020	0.01	2.930	0.99	0.050	0.01
$nC7$	0	0	0.015	0.005	4.985	0.99
total	2.005	1	2.960	1	5.035	1

Note that for high purity products (as here), one can readily estimate these flows using approximate computations. The contaminant flow for Product I is approximately 1% of the flow of pentane, i.e., 1% of 2 mol/s or 0.02 mol/s of hexane. The contaminants for Product II are each 0.5% of the flow of the heptane: 0.015 mol/s each of pentane and heptane. Finally, the contaminant flow for product III is 1% of 5 mol/s or 0.05 mol/s. We then correct the flow of pentane leaving in product I by reducing it by 0.015 mol/s, for hexane in Product II by removing 0.015 + 0.05 mol/s and for heptane in Product III by removing 0.015 mol/s.

Estimating Minimum Vapor Flows. For Underwood's method we start by using Eq. (11.1) to estimate the unknown variable ϕ. This equation involves only relative volatilities and information on the overall feed to the process. Thus, its value does not depend on the sequence we select to carry out the separation. We know from earlier that the relative volatilities are not constant, but they are nearly so. We need to use reasonable values; let us pick those we obtained when flashing 50% of the feed, as given in Table 11.2. For a bubble point feed, feed quality as indicated by q is equal to unity. Thus, we write:

$$\frac{5.51}{5.51-\phi} \times 2 \text{ mol/s} + \frac{2.32}{2.32-\phi} \times 3 \text{ mol/s} + \frac{1}{1-\phi} \times 5 \text{ mol/s} = (1-1) \times 10 \text{ mol/s} = 0$$

Inspecting this equation, we will discover that it has three values for ϕ that satisfy it, one between $\alpha_{1,3} = 5.51$ and $\alpha_{2,3} = 2.32$, one between $\alpha_{2,3} = 2.32$, and $\alpha_{3,3} = 1.0$ and one at infinity. To see this behavior, let ϕ take a value just below 5.51, say 5.5099999999. The first term on the left-hand side will be very large and positive; it will dominate the left-hand side terms. As ϕ decreases and approaches 2.32 from above, the second term starts to dominate and move to negative infinity. The left-hand side thus decreases from plus infinity to negative infinity as ϕ moves from 5.51 to 2.32. At the same time, the right hand side remains at zero. Thus, there must be a solution between 5.51 and 2.32 where the left hand side crosses zero. The second and third terms on the left-hand side display the same behavior as ϕ moves from just below 2.32 to just above 1. Finally, the left-hand side asymptotically approaches zero as ϕ approaches either plus or minus infinity. We can use a root finder, for example, the goal seeking tool in Excel©, to find the two finite roots, which are 3.806 and 1.462.

Sec. 11.1 Separating a Mixture of n-Pentane, n-Hexane, and n-Heptane

At this point we must select which of the two sequences we wish to analyze. For the direct sequence, the first column separates pentane from the other two species; its light key is pentane and its heavy key hexane. Its distillate product is product I, and its bottom product is everything else: the sum of products II and III. Underwood's method requires us to select the value for ϕ that lies between the volatilities for the key components for the column. Therefore, we select $\phi = 3.806$ and substitute this value into Eq. (11.2) to compute V_{min}, getting:

$$V_{min} = \frac{5.51}{5.51 - 3.806} \times 1.985 + \frac{2.32}{2.32 - 3.806} \times 0.020 = 6.4 \text{ mol/s}$$

Note we have used the distillate product flows for this column in this equation.

To compute the minimum vapor flow for the second column in the direct sequence, we must first establish its feed, which, as we noted above, is the sum of products II and III in Table 11.4: 0.015, 2.98, and 5 mol/s respectively for species $nC5$, $nC6$, and $nC7$ respectively. The light and heavy key components for this column are $nC6$ and $nC7$ respectively.

For this column Underwood's Eq. (11.1) becomes:

$$\frac{5.51}{5.51 - \phi} \times 0.015 + \frac{2.32}{2.32 - \phi} \times 2.98 + \frac{1}{1 - \phi} \times 5 = 0$$

and the root between the volatilities for the key components is 1.553. The minimum vapor rate is given by Underwood's equation II to be:

$$V_{min} = \frac{5.51}{5.51 - 1.553} \times 0.015 + \frac{2.32}{2.32 - 1.553} \times 2.93 + \frac{1}{1 - 1.553} \times 0.015 = 8.9 \text{ mol/s}$$

The total of the minimum vapor flows is, therefore, 15.3 mol/s for the direct sequence.

The two columns for the indirect sequence, as shown in Figure 11.1b, give minimum vapor flow of 10.7 and 5.5, respectively, for a total of 16.2 mol/s. According to the heuristic we should select the direct sequence.

MARGINAL VAPOR FLOWS

We introduce here an even less complicated evaluation function to compare sequences. Both of the sequences to separate $nC5$, $nC6$, and $nC7$ split $nC5$ from $nC6$ and $nC6$ from $nC7$. In the direct sequence, we carry out the $nC5/nC6$ split in the presence of all of the $nC7$ in the original feed, while the $nC6/nC7$ split is without any $nC5$ present. In the indirect sequence the reverse is true: The $nC6/nC7$ split has all of the $nC5$ present while the $nC5/nC6$ has no $nC7$ present.

Let us compare sequences by looking at how each is impacted by the presence of other species in carrying out a split between the key components for the column. Underwood's equations give us a possible way to make this estimate. Let us rewrite Eq. (11.1) in the form:

$$\sum_i \frac{\alpha_{ik}}{\alpha_{ik}-\phi} f_i = \sum_i \frac{\alpha_{ik}}{\alpha_{ik}-\phi} d_i + \sum_i \frac{\alpha_{ik}}{\alpha_{ik}-\phi} b_i = (1-q)F$$

Rearranging and using Eq. (11.2) gives:

$$V_{min} = \sum_i \frac{\alpha_{ik}}{\alpha_{ik}-\phi} d_i = (1-q)F - \sum_i \frac{\alpha_{ik}}{\alpha_{ik}-\phi} b_i \qquad (11.4)$$

This equation relates V_{min} to a sum of terms for the presence of the species that exit in the distillate and to those that exit in the bottoms. Let us assume that the value of ϕ does not move very much whether the species other than the key species are in the feed or not for a column. Then the marginal contribution we might expect to V_{min} in the first column of the direct sequence caused by the presence of $nC7$ is approximately:

$$\Delta V_{min}(nC5/nC6, nC7) = -\frac{\alpha_{nC7,nC7}}{\alpha_{nC7/nC7}-\phi} = -\frac{1}{1-3.806} \times 5 \text{ mol/s} = 1.8 \text{ mol/s}$$

We note further that ϕ has a value somewhere between the relative volatilities of the two key species, $nC5$ and $nC6$. Let us assume it takes a value that is the average: 3.915. We would then estimate the extra vapor flow to be

$$-\frac{1}{1-3.915} \times 5 \text{ mol/s} = 1.7 \text{ mol/s}$$

For the indirect sequence we estimate the marginal vapor flow in the first column using the same type of argument to be

$$\frac{5.51}{5.51 - \frac{2.32+1}{2}} \times 2 \text{ mol/s} = 2.9 \text{ mol/s}$$

The indirect sequence shows a marginal flow that is 1.2 mol/s larger than the direct sequence. Our more accurate analysis above using Underwood's method gave a difference in total minimum flows of 16.2 − 15.3 or 0.9 mol/s. Both are estimates for the same differences, and both are telling us the direct sequence is better.

A SIMPLE MEASURE TO COMPARE SEQUENCES

We appear to have a very simple measure we can use to compare distillation sequences for separating relatively ideal mixtures using conventional distillation. It says to form the term

$$\left| \frac{\alpha_{i,k}}{\alpha_{i,k} - \frac{\alpha_{lk,k} + \alpha_{hk,k}}{2}} \times f_i \right| \qquad (11.5)$$

for each species i that is not a key component for a column but is present in the feed to a column. The sum of such terms will indicate the increase in the minimum vapor flow caused by the presence of these nonkey species for that column. We would prefer those sequences having the lowest total of marginal flows for all columns in them.

Looking at the form of this term we see that, the more the relative volatility differs from the volatilities of the key components, the larger the denominator and thus the lower the marginal flow. That is intuitively appealing. We also see that the marginal flowrate is directly proportional to the flowrate of the species in the feed, also intuitively appealing. A bit less obvious is that, the higher the volatility of the nonkey species present, the more it increases the marginal flowrate. It appears that the presence of the more volatile species is bad news. This suggests we should find ourselves preferring the direct sequence more often than the indirect one. The extra species for the direct sequence are always the less volatile ones in the mixture.

Reexamining our results for choosing between the direct sequence and the indirect for separating $nC5$, $nC6$, and $nC7$, we see that the lesser amount of $nC5$ favors the indirect sequence (it would be the better extra species present based on its flowrate of 2 mol/s versus 5 mol/s for $nC7$), but the higher volatility of $nC5$ (5.51 versus 1) favors the indirect sequence. The denominators are $3.9 - 1 = 2.9$ for the direct versus $5.5 - 2.2 = 3.3$ for the indirect suggesting their difference is not too important here in deciding. The higher volatility consideration dominates, and we choose the direct sequence.

11.2 SEPARATING A FIVE-COMPONENT ALCOHOL MIXTURE

We learned a lot from our previous example that will make this example much easier to analyze. Suppose we have a mixture of five alcohols that we shall label A, B, C, D, and E with flows in the feed of 1, 0.5, 1, 7, and 10 mol/s respectively, for a total of 19.5 mol/s. Suppose further that their relative volatilities are 4.3, 4, 3, 2, and 1 respectively. We note there is a lot of the heaviest species, which suggests we might prefer to remove it early in the best sequences.

We would like to find the preferred separation sequence based on the use of "simple" distillation columns. We use our approximate measure that estimates marginal vapor flows to choose among them. Table 11.5 gives the estimated marginal vapor flows we evaluate for each species over all possible key component pairs. For example, for a column to split D from E, having C present will increase the minimum vapor flow by 2.000

TABLE 11.5 Marginal Vapor Flows Estimated for Nonkey Species for Alcohol Example

	A	B	C	D	E
A/B	—	—	2.6	6.5	3.2
B/C	5.3	—	—	9.3	4.0
C/D	2.4	1.3	—	—	6.7
D/E	1.5	0.8	2.0	—	—

mol/s, having B present by 0.800 mol/s, and so on. Having both C and B present will add $2.000 + 0.800 = 2.800$ mol/s to the minimum vapor flow.

In Figure 11.2 we tabulate the total marginal vapor flows for all the columns that can exist in any of the separation processes possible based on simple distillation columns. They are placed in such a way that we can more easily see the total flows for each of the different sequences we can construct. For example, suppose we select the direct sequence. From this figure, the marginal flows should be for *A/BCDE*, *B/CDE*, and *C/DE* for a total of $12.3 + 13.3 + 6.7 = 32.2$ mol/s.

We wish to find the sequence with the minimum sum of marginal flows. We can readily do this from this figure by performing a *branch and bound* search. We start by comparing all the first separations we might make for the original feed: *A/BCDE*, *AB/CDE*, *ABC/DE* and *ABCD/E*. The one with the lowest marginal vapor flow is the split *ABCD/E* at 4.3 mol/s. With this split made, we next compare flows for *A/BCD*, *AB/CD* and *ABC/D*, choosing the split *ABC/D* for a total of $4.3 + 3.7 = 8.0$ mol/s. We have the mixture *ABC* to separate and compare A/BC and *AB/C*; we select *A/BC* to add another 2.6 mol/s for a total of 10.6 mol/s.

We now have a complete solution. We need to examine only solutions that can be less that 10.6 mol/s. Backing up to the decision among the alternatives *A/BCD*, *AB/CD* and *ABC/D*, we see that the second best decision, *A/BCD* with a flow total marginal flow of $4.3 + 9.1 = 13.4$ mol/s, will lead to a partial solution that exceeds 10.6 mol/s. Thus we

		13.3	6.7
		B/CDE	C/DE
12.3		8.0	2.0
A/BCDE		BC/DE	CD/E
18.6		2.8	9.3
AB/CDE		BCD/E	B/CD
10.4		9.1	1.3
ABC/DE		A/BCD	BC/D
4.3		14.6	2.6
ABCD/E		AB/CD	A/BC
		3.7	5.4
		ABC/D	AB/C

FIGURE 11.2 Total marginal flows for each of the columns making up all separation sequences for five components.

Sec. 11.2   Separating a Five-Component Alcohol Mixture

back up to our first decision. Only the decision *ABC/DE* could be less expensive but it has a marginal flow already of 10.4 mol/s. To complete this sequence we must add in the flows for separating *ABC*, a decision we already examined. The lowest marginal cost comes from using *A/BC* with a flow of 2.6 mol/s; it leads to too high a final marginal flow. Thus we now know that our solution—*ABCD/E, ABC/D*, and *A/BC*—must be the best solution based on the marginal flow estimates we have made to carry out our search.

We can easily enumerate the marginal vapor flows for the fourteen possible sequences for this example; we do so in Table 11.6. We see that marginal vapor flows range from a minimum of 10.6 mol/s to maximum of 32.3 mol/s.

11.2.1 Discussion

Selecting the best distillation-based separation sequence among those possible for separating relatively ideally behaving species has been the subject of many publications over the past quarter century. The emphases in these publications have been many: how to reduce the effort to search among the alternatives, the posing and testing of heuristics to select among the alternatives, how to evaluate alternatives. We shall start this section by exposing the size of the search problem.

NUMBER OF POSSIBLE SEQUENCES

As we have seen in the alcohol example above, we can readily generate many different separation sequences to separate a given mixture into desired products. A formula exists to estimate the number of sequences for separating *n* species into *n* pure component prod-

TABLE 11.6 Total Marginal Vapor Flows for all Fourteen Possible Sequences for Alcohol Example

Seq. No.	Separations in Sequence	Marginal Vapor Cost	Rank
1	A/BCDE, B/CDE, C/DE, D/E	32.3	14
2	A/BCDE, B/CDE, CD/E, C/E	27.6	13
3	A/BCDE, BC/DE, B/C, D/E	20.3	8
4	A/BCDE, BCD/E, B/CD, C/D	24.4	11
5	A/BCDE, BCD/E, BC/D, B/C	16.4	6
6	AB/CDE, A/B, C/DE, D/E	25.3	12
7	AB/CDE, A/B, CD/E, C/D	20.6	9
8	ABC/DE, A/BC, B/C, D/E	13.0	2
9	ABC/DE, AB/C, A/B, D/E	15.8	5
10	ABCD/E, A/BCD, B/CD, C/D	22.7	10
11	ABCD/E, A/BCD, BC/D, B/C	14.7	4
12	ABCD/E, AB/CD, A/B, C/D	18.9	7
13	ABCD/E, ABC/D, A/BC, B/C	10.6	1
14	ABCD/E, ABC/D, AB/C, A/B	13.4	3

ucts using simple sharp separators. In this section we shall first define and illustrate what a simple sharp separator is and then present the formula.

Simple Sharp Separators. A simple sharp separator splits its feed into two products, each having no species in common with the other. A simple distillation column that splits its feed containing species A, B, C, and D into the two products A and BCD is an example of a simple sharp separator.

There are other separation processes that act as simple sharp separators. For example, an extractive distillation column immediately followed by a column to recover the extractive agent is a simple sharp separator. Consider, for example, using an extractive agent is to separate propylene from propane, as illustrated in Figure 11.3. We feed propane and propylene into this two-column process and remove a pure propane and a pure propylene product from it. Thus, the two columns together act like a sharp separator. The extractive agent simply recycles. (Of course some of the agent is lost with the products and must be made up using a small makeup solvent stream.)

The relative volatility between propylene and propane varies from about 1.06 to 1.09, with propylene being the more volatile. Using distillation to separate propylene from propane requires a very large column, 150 or more stages, and a reflux ratio of 20 or

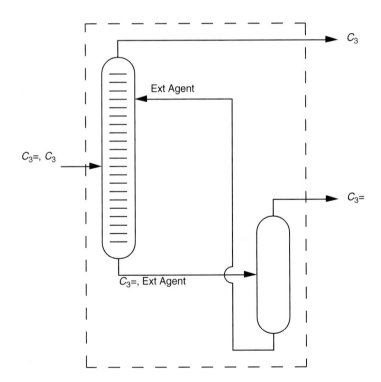

FIGURE 11.3 Separating propylene and propane using an extractive agent.

more. This reflux ratio says we must condense 20 moles of top product (propylene) and reflux it for every mole of propylene product we remove from the column. Thus, it requires the expenditure of a lot of utilities for each mole of product.

An extractive agent is typically a heavy species that preferentially "likes" one of the two species. Here acrylonitrile with its double bonds is a candidate. The extractive agent is fed into the column a few trays below the top so it will be present in the liquid phase on all stages below where it is fed. The propylene/propane feed enters the column well below the extractive agent. The agent alters the activity coefficients for propylene and propane in such a way that propylene becomes much less volatile than propane, thus the stages between the two feeds remove the propylene from the propane in the presence of the extractive agent. Only propane makes it to the tray where we feed the extractive agent. Being much more volatile that the extractive agent, a few additional trays above that feed allows us to separate the propane from the agent. Propylene and agent become the bottoms product. We then have to separate the propylene and agent in a second column, recycling the agent back to the first column.

The Thompson and King Formula to Compute the Number of Sequences.
Thompson and King (1972) developed the following formula to compute the number of sequences that can be developed based on simple sharp separators to separate a mixture containing n components into n pure component products:

$$\text{no. sequences} = \frac{(2(n-1))!}{n!(n-1)!} S^{n-1}$$

Table 11.7 lists the number of sequences for different numbers of species in the mixture and for up to three separation methods. While the numbers of sequences grow large quite quickly as a function of the number of species, they grow almost explosively when one allows different types of separators to carry out each task. Thus, many efforts in the synthesis of separation processes have emphasized how one can search these large spaces and/or how one can quickly find good solutions among the large number of alternatives.

TABLE 11.7 Number of Sequences to Separate n Components into n Single Component Products Using S Different Separation Methods

$n\backslash S$	1	2	3
2	1	2	3
3	2	8	18
4	5	40	135
5	14	224	1134
6	42	1344	10,206
7	132	8448	96,228
10	4862	2,489,344	95,698,746

HEURISTICS

One approach to finding good separation processes quickly is to use heuristics. These are guidelines based on experience that aid a designer to find the better solutions for the type of problem at hand. If we have a good solution to our separation problem, we know we need not look further at any other solution that we can prove will cost more. We used such a bounding idea in the branch and bound search we carried out in the alcohol example above. We can also use heuristics in a negative way where we eliminate any part of a solution that we believe will be much too expensive to be in any solution. Not allowing certain separation steps in any solution can often dramatically reduce the size of a search.

We list in Table 11.8 a set of commonly used heuristics for designing separation sequences (for example, see Seader and Westerberg, 1977). Note that the last heuristic states that we have listed these heuristics in order of importance in our decision making.

Let us apply these heuristics to find a separation process for the example in section 11.2, the example to separate five alcohols. To remind ourselves, we have a mixture of five alcohols that we labeled A, B, C, D, and E with flows in the feed of 1, 0.5, 1, 7, and 10 mol/s respectively, for a total of 19.5 mol/s. These species have relative volatilities of 4.3, 4, 3, 2, and 1 respectively.

Heuristic 1 is not applicable as we are treating none of these alcohols as dangerous or corrosive. For heuristics 2 and 3, we need first to compute relative volatilities for each possible pair of key components. These relative volatilities are simply the ratio or the relative volatility for the light key divided by that for the heavy key: 4.3/4 = 1.075 for A/B, 4/3 = 1.333 for B/C, 3/2 = 1.5 for C/D and 2/1 = 2 for D/E. While one is only just larger than 1.05, none is less so we skip to heuristic 4. Heuristic 4 tells us to make the easiest split first, suggesting we make the split between D and E where the relative volatility between the key components is the largest with a value of 2. Heuristic 5 also suggests we make a split that leads to the removal of species E. Heuristic 6 proposes we remove species A first (the direct sequence). For heuristic 7, we note that all species are desired

TABLE 11.8 Heuristics for Designing Separation Processes

Heuristic 1:	Remove dangerous and/or corrosive species first.
Heuristic 2:	Do not use distillation when the relative volatility between the key components is less than 1.05.
Heuristic 3:	Use extractive distillation only if the relative volatility between the key components is much better than for regular distillation—say 6 times better.
Heuristic 4:	Do the easy splits (i.e., those having the largest relative volatilities) first in the sequence.
Heuristic 5:	Place the next split to lead to the removal of the major component.
Heuristic 6:	Remove the most volatile component next (i.e., choose the direct sequence).
Heuristic 7:	The species leading to desired products should appear in a distillate product somewhere in the sequence if at all possible.
Heuristic 8:	These heuristics are listed in order of importance.

products. The direct sequence would maximize the number of them that would appear in a distillate somewhere in the sequence.

The last heuristic says to carry out the decision supported by the heuristic with the lowest number. So we elect to remove species E, as supported by both heuristics 4 and 5. A similar set of arguments leads us to remove species D next. The B/C split is much easier than the A/B split so we elect it next, leaving us with the A/B split last. This solution is the third best among the fourteen possible based on marginal vapor flows (see Table 11.6). It is only slightly worse than the second best. Using these heuristics, the effort we took to find it was minimal.

With a little thought it is possible to develop a variety of different search strategies using just these heuristics. For example, one might enumerate all sequences where at least one heuristic supports each decision leading to it. We will not examine any of the others.

The next chapter (Chapter 12) will look at heat integrating distillation columns. Chapter 14 looks at the synthesis of separation processes for species that behave highly nonideally. In Chapter 17 and part of Chapter 18 we shall look again at the search problem for distillation sequences for relatively ideally behaving species, but this time we shall propose search algorithms that use mixed integer programming.

REFERENCES

Perry, J. H. (Ed.). (1950). *Chemical Engineers' Handbook,* 3rd ed. New York: McGraw-Hill.

Seader, J. D., & Westerberg, A. W. (1977). A combined heuristic and evolutionary strategy for synthesis of simple separation sequences. *AIChEJ,* **23,** 951.

Thompson, R. W., & King, C. J. (1972). Systematic synthesis of separation systems. *AIChEJ,* **18,** 941.

EXERCISES

The first four problems are a review of undergraduate distillation concepts. Students who cannot do these should review appropriate undergraduate textbook material on distillation.

1. Consider a column to separate acetone from ethanol. The equilibrium data for acetone in ethanol at one atm are in Table 11.9 (Perry, 1950).

    The feed has a flowrate of 0.1 kgmol/s. It is 50 (mole)% acetone and is liquid at its bubble point ($q = 1$). Products are liquids at their respective bubble points. Assume 99% of the ethanol and 96% of the acetone are recovered in their respective products. The column operates at one atm.

TABLE 11.9 Acetone Vapor/Liquid Equilibrium Compositions for Acetone/Ethanol Mixtures

x	y	x	y
0	0	40	60.5
5	15.5	50	67.4
10	26.2	60	73.9
15	34.8	70	80.2
20	41.7	80	86.5
25	47.8	90	92.9
30	52.4	100	100
35	56.6		

    a. Using a McCabe-Thiele diagram, determine the number of stages to separate acetone from ethanol.

    b. Should the column have been designed for one atm? If not, how would you choose the pressure? Explain your answer.

    c. Compute the condenser and reboiler duties for the column. How close are they to being equal? Can you guess why they are this close?

    d. Should you preheat the feed to the column when it is running at one atm? Explain. You can answer this question without doing any computations. Look at the impact of preheating the feed on the construction of the McCabe-Thiele diagram to make your argument.

2. A column is a passive piece of equipment once it is designed and built. Assuming it is properly designed, how is it that one can "make" a column carry out the separation desired? For example, consider separating a mixture of ABC into two products A and BC. Explain how to operate a column so it gives one 99% of species A in the distillate product (top product) while forcing 99% of species B and virtually all of species C to the bottom product. What would you control? Assume A is most volatile and C least.

3. Using all that you know about the use of the McCabe-Thiele diagram for analyzing binary distillation columns, demonstrate that the number of degrees of freedom is five plus the total of those associated with completely specifying the feed.

4. Show that the mole fraction averaged relative volatility

$$\bar{\alpha}_k \equiv \sum_j \alpha_{jk} x_j$$

is equal to $1/K_k$, the reciprocal of the K-value for the selected key component.

5. You are to separate the following relatively ideally behaving mixture of A, B, and C. The feed is at its bubble point of 345.8 K at 1 bar.

Exercises

Component	feed, kmol/hr	VPA, unitless	VPB, K	VPC, K
A	50	11.1	3000	−70
B	100	10.2	2800	−70
C	30	10	3000	−70

The last three columns are the Antoine constants for evaluating vapor pressure, using the following formula:

$$P_i^{sat} \{bars\} = \exp\left(VPA_i - \frac{VPB_i}{T\{K\} + VPC_i}\right)$$

a. Show that the bubble point temperature for the feed is 345.8 K when pressure is 1 bar.

b. The Underwood roots for the original feed are 1.116 and 2.826. Show that the minimum vapor flow in the top of the column for the A/BC column should be approximately 828 kmol/hr. What assumptions do you need to make to do this computation?

c. The minimum vapor flows for the following columns are similarly computed to be:

$$V_{min}(AB/C) = 254 \text{ kmol/hr}$$
$$V_{min}(A/B) = 830$$
$$V_{min}(B/C) = 183$$

Which sequence is to be preferred: A/BC, B/C or AB/C, A/B? Why?

d. Compute marginal vapor flows using the very approximate method developed in this chapter. Are they in rough agreement with numbers that can be computed from the information given above? Do they predict the same sequence?

6. You have a mixture of 35 mole % n-heptane, 30% n-hexane, 10% isobutane, and 25% n-pentane.

a. Determine the bubble and dewpoint temperatures for the above mixture. Pressure is one atmosphere. Assume Raoult's law for expressing vapor-liquid equilibrium.

b. You want to run a flash unit for the above mixture in which 50% of the n-hexane leaves in the vapor product. Determine the fraction of the other species that leave in the vapor product. The pressure is one atmosphere. Repeat this computation for a pressure of two atmospheres. Do you notice anything interesting here? (Hint: Note first that

$$\frac{v_i/l_i}{v_k/l_k} = \frac{y_i V / x_i L}{y_k V / x_k L} = \frac{y_i / x_i}{y_k / x_k} = \frac{K_i}{K_k} = \alpha_{ik}$$

$$y_i = K_i x_i = \frac{\alpha_i}{\overline{\alpha}} x_i \approx \frac{P_i^{sat}(T)}{P} x_i \Rightarrow \frac{P_i^{sat}(T)}{P} \approx \frac{\alpha_i}{\overline{\alpha}}$$

If 50% of the *n*-hexane leaves in the vapor product, what is the ratio $v_{n\text{-hexane}}/l_{n\text{-hexane}}$? If you know P, can you estimate T and vice versa? You should note that, for each guess of the relative volatility, the flash computation asked for here does not require iteration.)

c. Assume that you wish to design a column to separate the *n*-heptane from the remaining three species as the first column in the sequence selected to carry out the complete separation. Assume the feed and both products are bubble point liquids. Estimate the minimum reflux ratio for this column. Is the method you used justified for computing this minimum reflux? Explain.

d. Develop the condenser and reboiler heat duties for the column for pressures of 1, 5, 10, and 20 atm. Plot heat duties versus the condenser temperature for this column. Do you notice anything special about this plot?

e. Would you use this method for a column to separate acetone from ethanol (see exercise 1)? Explain.

7. Enumerate all the simple sharp separation sequences possible for separating a mixture of *ABCDE* into products *AC*, *BE*, and *D* given the following three separation methods:
   - Method m1: Component volatility order *ABCDE*
   - Method m2: Component volatility order *CBADE*
   - Method m3: Component volatility order *BCED*

   For you to use method 3, species *A* may not be present.

8. Estimate the minimum reflux using Underwood's method for separating the following mixture into the products indicated.

Species	Feed	Recovery in Distillate
n-pentane	20%	100%
n-hexane	50%	99.5%
n-heptane	30%	0.2%

9. Consider the mixture in Table 11.10. Using Underwood's equations, compute the minimum reflux to recover 90, 95, 99, and 99.9% of the key components in their respective products for the following separation problem. Species *C* and *D* are the key components.

TABLE 11.10 Mixture for HW Problems

Species	Relative Volatility	Feed Flow, mol/s
A	2.7	10
B	2	5
C	1.5	40
D	1	15

10. Again consider the mixture in Table 11.10. Underwood's equations can be used for computing the minimum reflux when the key components are not adjacent in the

separation. Let the light key be species A and recover 99.5% of it in the distillate. Let the heavy key be species C and recover 99% of it in the bottoms product. Find two roots for the first Underwood equation: the one that lies between A and B and the one that lies between B and C. Write the second of the Underwood equations twice, once using the AB root and once using the BC root. You should have two linear equations in two unknowns: the flow d_B and in the minimum vapor flow, V_{min}. Solve these two equations for these flows.

11. Let the light key remain the same as in the previous problem. Let the heavy key be species D. Recover 98% of it in the bottoms product. What is the minimum vapor flow for this column? Problem 10 describes how to solve this problem.

12. Discover the best sequence among those possible for the following problem based on minimizing the total of the estimated vapor flows in the columns.

Species	Relative Volatility	Amount kmol/hr
A	2	10
B	1.5	20
C	1.2	10
D	1	60

Is the answer consistent with any of the heuristics in Table 11.8? Explain.
    Suppose that species B is very corrosive. Estimate the extra cost in terms of added vapor flow for following the "dangerous or corrosive species" heuristic.

13. Consider again the mixture consisting of 35 mole % *n*-heptane, 30% *n*-hexane, 10% isobutane, and 25% *n*-pentane. Using Eq. (11.5), estimate marginal vapor rates and determine which of the possible sequences constructed from simple two product columns are likely to be the best. Would you expect this heuristic to give the right answer here? Explain.

14. Find the best distillation-based separation sequence if the following data hold for marginal vapor flows using a branch and bound search. The components behave relatively ideally.

	A	B	C	D	E
A/B	—	—	100	1	1
B/C	1	—	—	1	1
C/D	1	100	—	—	1
D/E	1	1	100	—	—

Prove that you have the best answer by listing the total marginal vapor flows for all sequences.

15. You wish to separate a mixture of species A, B, and C using distillation. These species have fairly ideal vapor/liquid equilibrium behavior, having relative volatili-

ties of 4.0, 2.0, and 1.0 respectively. The flowrate of species C in the mixture is 1 kmol/hr. Estimate the flowrates of A and B in the feed such that you would be indifferent to choosing between the direct (A/BC, B/C) and the indirect (AB/C, A/B) sequences for separating them.

16. Consider separating the mixture in Table 11.11 into four pure component products.

    TABLE 11.11 Feed Flow for Exercise 16

Species	Feed Flow, mol/s
n-pentanol	10
isobutanol	5
n-hexanol	40
n-heptanol	15

    a. Using Underwood's equations, find the sequence having the lowest total for the minimum vapor flows in each of the columns in it.
    b. Use the marginal flow estimator given by Eq. (11.5) and find the sequence having the lowest total for the minimum vapor flows in the columns in it.
    c. Compute the marginal flows using the results from part a and compare them to part b.

17. Using the heuristics in Table 11.8, find a reasonable separation sequence for the feed in Table 11.11. If you have done the previous problem, how does this answer compare?

18. Using the heuristics in Table 11.8, propose separation sequences for the following problem.

    Separate a mixture of six components $ABCDEF$ into products A, BDE, C, and F.

    Use either of two methods in developing your sequences

    - Distillation, method I   Component volatility order $ABCDEF$
    - Extractive distillation, method II   Component volatility order $ACBDEF$

    Component amounts

    - A: 4.55 kmols/hr, B: 45.5, C: 155.0, D: 48.2, E: 36.8 and F: 18.2.

    Relative volatilities of the key species

    - Method m1: A/B 2.45, B/C 1.55, C/D 1.03, E/F 2.50
    - Method m2: C/B 1.17, C/D 1.70

19. Show that the direct sequence is the correct one for the following problem. Note that all the volatility ratios for adjacent species, $\alpha_{i,i+1} = r$, are equal to 1.2 here.

Species	Relative Volatility	Amount kmol/hr
A	$1.2^3 = 1.728$	1
B	$1.2^2 = 1.44$	1
C	$1.2^1 = 1.2$	1
D	$1.2^0 = 1$	1

20. Show that the result for the previous problem is general for any ratio $\alpha_{i,i+1} = r$ and not just for $r = 1.2$.

21. List the total number of instances of *extra species* present for each of the possible sequences when splitting an 8-component feed mixture into 8 relatively pure component products. Which sequence has the fewest number of extra species overall? Discuss the implications of having the fewest total number of extra species on the marginal vapor flow.

There is an heuristic that says that a column should attempt to split each mixture in a separation process into roughly equal parts. Explain how the above observation on extra species may support this heuristic.

HEAT INTEGRATED DISTILLATION PROCESSES

12

In this chapter we combine the topics of the last two chapters to look at the heat integration of systems of distillation columns. We shall also look at special column configurations that feature intercooling and interheating as well as columns that have side strippers and enrichers.

12.1 HEAT FLOWS IN DISTILLATION

12.1.1 A Base Case (Andrecovich and Westerberg, 1985)

Distillation columns require heating for the reboiler and cooling for the condenser. Unfortunately, but, not surprisingly, the reboiler, always hotter than the condenser, cannot directly use the condenser heat. Columns are *heat integrated* if heat removed from one is used to provide heat for another. Often, we have to adjust the temperature levels for the columns involved so they can be integrated, but, fortunately, we can increase or decrease the operating temperatures for a column by simply increasing or decreasing its operating pressure.

Columns can be viewed as devices that degrade heat to carry out separation. They receive higher temperature heat into their reboilers and expel lower temperature heat from their condensers. Higher temperature heat should, and had better, cost more per unit of heat than lower temperature heat. In an ideal world, we would buy utilities at just the temperature needed, paying a price for them that reflects their temperature. In such a situation, passing heat from one column to another would probably not be economic.

However, most utility systems for processes provide heat at only a few fixed temperature levels—for example, from high, medium, and low pressure steam at 350, 275,

Sec. 12.1 Heat Flows in Distillation

and 200°C, respectively. Suppose we have a column that has a condenser temperature of 50°C at one atmosphere (hot enough to pass the heat into cooling water) and a reboiler at 90°C. We would like to use 100°C utility heat, except there is none. We find we must use 200°C steam. It could prove economical to use this same heat to run one or more other columns before it passes through this column. This column will degrade the heat passing through it by only about 40°C (90°C less 50°C) plus the sum of the temperature differences used as driving forces in its reboiler and condenser, say another 20 to 30°C.

It is also possible that we could exchange heat with other streams in the process.

When heat is degraded and passed to another part of the process to degrade it further, there is a cost. The temperature driving forces for heat exchange will become smaller. If small enough, the heat exchangers for a column can cost more to purchase than the column itself. (Nothing is free.) The following ideas illustrate those instances when heat integration might be attractive because of the potential utility savings. Only these ideas need to be investigated as they are the only ones that could produce a savings that can pay for the extra exchanger area required.

Both the first and second laws are at work here. We would like to reduce the use of utilities by reusing heat (first law savings). However, the heat is degraded each time we use it (second law cost). Because of the large temperature drops available when using only a few temperature levels for utilities, we are often forced to pay for the large temperature drops whether we use them or not. Forced to have them, we should try to use them.

In order to explore these possibilities, we need to understand and be able to compute the heat flows in columns. That is the purpose of this section.

We start by considering a *base case* column, one that we shall use to compare the operation of all others. Assumptions for this base case column are:

- Feed and products are all liquids at their respective bubble points (i.e., they are liquids at their boiling point).
- Internal reflux and reboil flow rates are large relative to feed and product flow rates.

A heat balance around the column gives

$$h_F(T_{f,\text{bub}})F + Q_{\text{reb}} = h_D(T_{D,\text{bub}})D + h_B(T_{B,\text{bub}})B + Q_{\text{cond}}$$

With the above assumptions, the terms Q_{reb} and Q_{cond}, which involve latent heats, are very large compared to the remaining terms which involve only differences in sensible heats. Thus, we can write

$$Q_{\text{reb}} \approx Q_{\text{cond}}$$

A column for the base case degrades approximately $Q \approx Q_{\text{reb}} \approx Q_{\text{cond}}$ units of heat from T_{reb} to T_{cond}. In Figure 12.1 we sketch this base case as horizontal heat source and heat sink lines of width Q on a plot of T versus *Heat*. We can think of the horizontal lines being joined top to bottom to form a box *for this case*. While tempting and something we have done often, we will not show columns as boxes because the duties are often not equal for a column, for example, when the feed is dewpoint vapor.

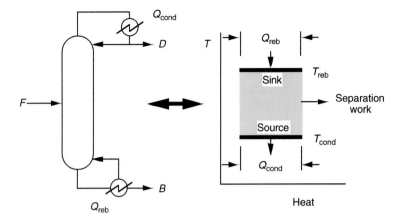

FIGURE 12.1 Base case heat balance for column—the T-Q diagram for distillation.

OBSERVATIONS ON T-Q DIAGRAM

The following observations for a column come from having carried out computations for many different examples. Most experience is with relatively ideally behaving species.

- Higher pressure → higher temperature operation → *both* more heat required and a larger temperature drop across column—that is, the box gets larger in both dimensions.

Intuition would suggest that more heat should be needed as higher pressures generally lead to smaller relative volatilities between the species; at least that is the experience with normal hydrocarbons. Thus, more reflux would be required. One's intuition probably would not suggest that the temperature drop should also increase, but it does.

- Having other species present typically increases both the heat duties and the temperature drop across the column.

We saw in the previous chapter that there is an added vapor flow when other species are present. The temperature drop increase is also expected as having D present for the B/C split will increase the bubble point for the reboiler (CD rather than for C alone).

COMPUTING REBOILER AND CONDENSER DUTIES

The following is a recipe to estimate condenser and reboiler duties for a column. Because of the effects of composition on enthalpies, it cannot be exact.

Sec. 12.1   Heat Flows in Distillation   411

- Estimate the minimum reflux/reboil ratio required for column.
- Select a reflux/reboil that is, say, 1.2 times as large as the minimum needed.
- Multiply the heat of vaporization for the distillate/bottoms times the reflux/reboil used.

SYSTEMS OF HEAT INTEGRATED COLUMNS

To indicate the type of thinking involved in heat integrating columns, we consider the following example where we shall use NO numbers. The T versus Q representation for heat flows in columns will allow us to gain insights into the design for this problem none-the-less.

EXAMPLE 12.1

Split the following mixture of components.

Species	Amount	Ease of Separation
A	lots	
		difficult
B	moderate amount	
		very easy
C	moderate amount	
		very very difficult
D	lots	

Figure 12.2 sketches the T-Q flows for each of the separations for this example. Separating C from D is difficult, indicating they have close boiling points. The temperature drop across the column is, therefore, small, but the amount of heat required is very large, as shown. On the other hand, separating B from C is easy. Here the normal boiling points will be very different—that is, there is a large temperature drop, but the heat needed is very little. Finally, splitting A from B is somewhere in between.

We make the following observations based on our understanding of how distillation processes work.

- *C/D* should be done without other species present—other species will enlarge the amount of heat required for a column that has a large heat requirement already.
- *B/C* should be done without other species present. This preference conflicts with the previous one. With a large temperature drop, it is difficult to heat integrate this column with others and still be within the allowed utility temperatures. The potential benefits of reusing heat passing through this column are greatly reduced.
- The *C/D* split could conceivably be done in two columns that are heat integrated to reduce the utility consumption (carrying out the same separation in two columns and heat integrating them is termed multi-effect distillation—for reasons that hopefully are obvious.

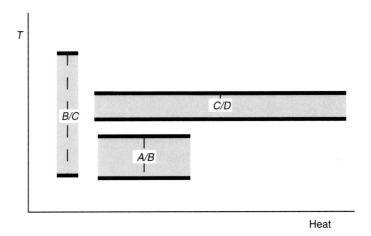

FIGURE 12.2 *T-Q* heat flows for example splits.

We select the candidate design in Figure 12.3 based on these assumptions for the process.

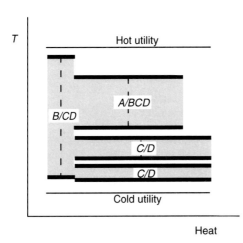

FIGURE 12.3 Heat integrated design for separation problem. A box placed vertically above another implies heat passes from the condenser of the column corresponding to the upper box into the reboiler of the column for the lower box.

The following give the reasons for this design.

- *C/D* is done without other components present.
- The box for *C/D* was split in a manner that both parts will have same width in the final design. Thus, the heat required by one is exactly the heat given up by the other.
- The two boxes for the *C/D* split are operated at the coldest temperature possible to reduce the dimensions for them. Their width impacts directly the amount of the utilities that are consumed.
- The split *B/C* is done with fewest other components possible; if others have to be present, we choose to have heavy species as they have a smaller effect on added heat duties; here we must have *D* present if the *C/D* is split is done with no other species present.

Sec. 12.1   Heat Flows in Distillation   413

> Note that the dimensions for the heat flows and temperature drops reflect that the columns are operating at different conditions than in the previous figure (different temperature levels, different components present).

12.1.2 Intercooling/Heating

An interheated and/or intercooled column is one in which heat is added and/or removed from trays within the column (the following analysis is from Terranova and Westerberg, 1989). In our previous columns, all heat was added to the reboiler and removed from the condenser. Questions we might ask are:

- Why use intercooling or interheating?
- Is more or less heat required?
- What are the costs?

We start by examining a binary separation for which we can construct a McCabe-Thiele diagram. The column in Figure 12.4 has two envelopes for which we might write component material balances at the top of the column, one above the intercooler and one below.

The operating lines for each are a result of writing component material balances:

$$y = \frac{L^I}{V^I} x + \frac{L^I}{D} x_D$$

$$y = \frac{L^{II}}{V^{II}} x + \frac{L^{II}}{D} x_D$$

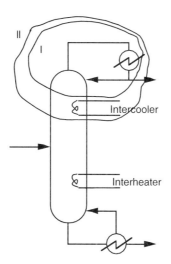

FIGURE 12.4 Material balance line for intercooling.

Since the top product is the same for both envelopes, both operating lines must go through the same point $[x_D, x_D]$ on the 45-degree line. The only thing that can vary is the slope for each of them, which can be written in the following form for both.

$$\text{slope} = \frac{L}{L+D} = \frac{1}{1+\dfrac{D}{L}}$$

Intercooling will cause L to be larger for envelope II, and therefore its slope, by the above, will be larger (i.e., larger L implies a smaller denominator implies a larger quotient). As a point of interest, we also note that since $V = L + D$ for both cases, V must also be larger for envelope II.

Figure 12.5 illustrates the McCabe-Thiele plot for a binary separation with intercooling and interheating. In the top part of the column, not removing enough heat from the condenser to run the column leads to an operating line with too small a slope to reach the bottom operating line before it crosses the equilibrium curve. Removing heat partway down pivots the operating line downward to give it a steep enough slope. We see similar behavior for the bottom of the column, where not placing enough heat into the reboiler leads to an operating line that is too steep to reach the upper line before it crosses the equilibrium curve. Shown also are the stages required for this column. We note that the temperature for a column increases as we march down it, so $T_1 < T_2 < \cdots T_9$.

We step along the first operating line (stages 1 and 2), getting warmer as we go, until we step over the intercooler, where we move to the lower operating line. With an intercooler we can remove less heat in the condenser than needed for an ordinary column because we can "rescue" the operating line and move it down by intercooling before it

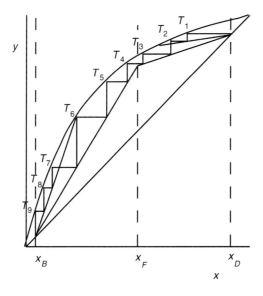

FIGURE 12.5 McCabe-Thiele plot for a column with intercooling and interheating.

Sec. 12.1   Heat Flows in Distillation   **415**

pinches with the equilibrium line. The intercooler removes heat partway down the column and, therefore, at a higher temperature than the condenser.

We can also observe that the minimum reflux requirement for the column dictates the slope for the *second* operating line only, irrespective of whether we have an intercooler. We can therefore argue that the total heat removed, which dictates the slope for the second operating line, has not changed. We have only altered the conditions at which some of the heat has been removed.

Answers to questions about intercooling that we asked earlier are now more evident.

1. Intercooling allows us to remove only part of the heat in the condenser. At a warmer temperature (between T_2 and T_3 in our example), we then remove the remaining heat. By a similar set of arguments, interheating allows us to inject only a part of the heat into the reboiler where the column is hottest. At a lower temperature we then inject the remaining heat needed to run the column.

2. If we do not move the operating line for envelope II and insist on producing the same products, then the same total amount of heat is removed and injected as for a normal column, and we find that we require more trays (as the steps along the operating line for envelope I are smaller). We also need to purchase the heat exchanger equipment, and, if we use the same utilities, it will have a smaller temperature driving force and thus require more heat transfer area. The heat exchanger equipment will almost certainly be more expensive.

3. If we have a column with a fixed number of trays (as we would for the retrofit case) and we leave the operating lines to have the same slope for envelope II, then the column will give a poorer separation. To accomplish the same separation, we have to increase the reflux we use in the column, moving the operating line for envelope II, and possibly for envelope I, closer to the 45-degree line. We would almost certainly need more heat exchanger equipment.

For 2 above, one gains on the second law—i.e., one can remove heat at hotter temperatures and inject it at lower temperatures, stays even on the first law—i.e., the column uses the same amount of heat, and finally one has to spend more on equipment as more trays and exchanger equipment are needed. For 3, one again gains on the second law but either loses on the separation accomplished or loses on the first law.

HEAT FLOWS FOR INTERCOOLED/INTERHEATED COLUMNS

The *T* versus *heat* diagram should have the shape shown in Figure 12.6, where the darkened lines are the heat in and out lines. The outer box is the *T* versus *heat* diagram for a column without interheating and intercooling. The impact of interheating and intercooling is to notch the box for the same separation task without interheating or intercooling, moving part of the heating duty to lower temperatures and part of the cooling duty to higher temperatures. The same total heat is degraded.

We would like to establish the dimensions for this diagram. We can accomplish this by performing an analysis for the pinch point. Assuming both operating (material balance

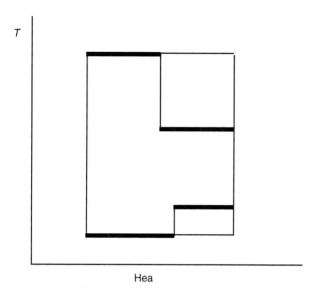

FIGURE 12.6 Expected notched structure for heat cascade diagram for intercooling and interheating.

around top of column) and equilibrium equations ($y_i = \alpha_{ik} x_i / \overline{\alpha}_k$) hold at a pinch and solving for compositions x_i, we get:

$$x_i = \frac{Dx_{D,i}}{\dfrac{\alpha_{ik} V}{\overline{\alpha}_k} - L} \tag{12.1}$$

and

$$\sum_i \alpha_{ik} x_i = \overline{\alpha}_k = \sum_i \frac{Dx_{D,i}}{\dfrac{\alpha_{ik} V}{\overline{\alpha}_k} - L} \tag{12.2}$$

We proceed as follows, given the flow and composition for the top product.

1. Set reflux ratio R to zero.
2. Compute $L = RD$ and $V = (R + 1)D$.
3. Guess all relative volatilities.
4. Iteratively solve Eq. (12.2) for $\overline{\alpha}_k$. Solve Eq. (12.1) for all x_i.
5. Using a rigorous analysis package, determine the bubble point temperature, $T_{\text{bub}}(x_i)$, for this pinch point composition. New (composition, temperature, and pressure dependent) α_i are automatically computed as a part of this calculation.
6. Iterate from step 4 until no changes occur in the variable values. (This computation is rigorous and works even for nonideal physical property behavior.)

Sec. 12.1   Heat Flows in Distillation   417

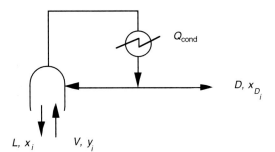

FIGURE 12.7   Top of column.

7. Compute, as follows, Q_{cond} using the heat balance around top of column (see Figure 12.7).

$$Q_{cond} = H_V V - h_L L - h_D D$$

Again, this calculation is also an exact one; no approximations are needed to do it. It will require using a rigorous physical properties package. The values for R and the corresponding values for L and V are *at the pinch point*.

8. Plot the point representing T_{bub} versus Q_{cond} for this value of the reflux ratio, R, on a plot.
9. Increment R by a small amount and repeat until R equals the value required for the normal column.

You will obtain the lower curve shown in Figure 12.8. Note that when the amount of reflux is zero, enough heat must still be removed to condense the top product; thus the heat removal value at the lowest temperature is not zero if the top product is bubble point

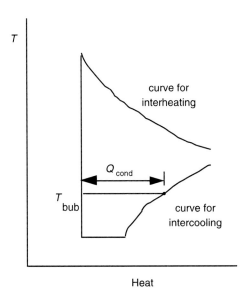

FIGURE 12.8   Intercooling and interheating temperature curves.

liquid. Repeating a similar analysis, stepping the reboil ratio from zero to its value in a normal column, allows one to plot the top curve for interheating.

The bottom curve plotted gives the amount of heat to be removed in the condenser to make the operating line intersect the equilibrium surface at the temperature shown. This is *least* amount of heat that can be removed to get down to this temperature before removing any added heat.

You could remove heat at every stage and keep the steps exactly on the equilibrium surface. The column almost carries out a reversible separation. It fails to be totally reversible (see Fonyó, 1974a, b and Koehler et al., 1992) because the feed is not required to have the same composition as the liquid on the feed tray. The enclosed area for the heating and cooling curves for this case is as small as it can be for the given feed and products; it is a limiting diagram. Of course, one would require an infinite number of stages and infinite area in the exchangers to obtain this performance for the column. Thus, if this limiting diagram is used to formulate heat integration alternatives, one could expect it to yield the best that could be done with the column.

This plot, once completed, allows one to determine the size of the "notches" in the box for the base case that corresponds to intercooling. Figure 12.9 illustrates.

We select a temperature for intercooling, a temperature that is hotter than the condenser temperature. Locate this temperature on the lower curve above and draw a vertical line to the base line shown for the base case (i.e., to the box). We must remove at least the amount of heat to the left of this line from the condenser. We should really remove more so the column does not pinch at the chosen temperature. The amount of heat not removed by the condenser must then be removed in the intercooler.

A similar construction accounts for interheating.

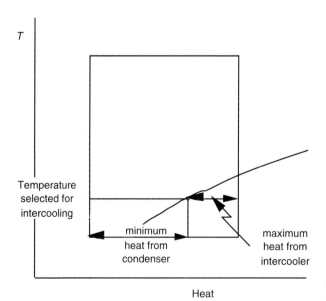

FIGURE 12.9 Discovering the amount of heat to remove from the condenser and the intercooler.

Sec. 12.1   Heat Flows in Distillation   419

EFFECT OF CHANGING THERMAL CONDITITION OF FEED

We argue here that the curves for intercooling and interheating are valid regardless of the thermal condition of the feed. Examining the method to obtain the intercooling and interheating curves, we see that they are a trajectory of pinch points whose T and Q values are determined by stating the top and bottom product compositions and thermal conditions only. Nothing in the analysis involves the thermal condition of the feed; therefore, these curves must be valid whether the feed is a bubble point liquid, dewpoint vapor, two phase, superheated, or subcooled.

We argue that the thermal condition of the feed only changes how far along these curves we proceed before reaching the reflux ratio needed for the top or the corresponding reboil ratio for the bottom. Given the thermal condition of the feed, we can find the reflux and reboil ratios needed by using whatever analysis is appropriate, for example, by using Underwood's method. If the feed is bubble point $(q = 1)$, the heat duties are nearly equal, as argued earlier. If the feed is preheated, the condenser duty will exceed the condenser duty, as shown in Figure 12.10. Here we feed bubble point liquid into a feed preheater that changes its thermal condition. Arguing as before that the sensible heats are small, the heat removed from the condenser, Q_C, has to equal approximately the heat used to preheat the feed, Q_F, plus the heat into the column reboiler, Q_R—that is,

$$Q_C \approx Q_F + Q_R$$

In Figure 12.11, we can parameterize the pinch point curves for determining interheating and intercooling with values of q to reflect the thermal condition of the feed into the column. We see that, for the base case of bubble point feed where $q = 1$, we have the box-shaped figure as before. For $q = 0$ (dewpoint vapor), the reboiler heat is less as expected, while the condenser heat *is more* than for the base case. In other words, preheating

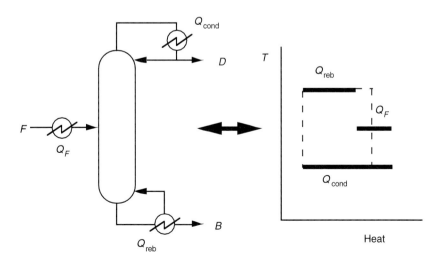

FIGURE 12.10   Preheating the column feed.

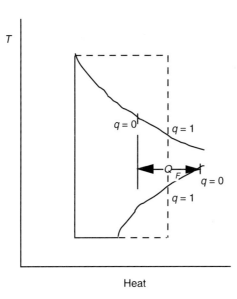

FIGURE 12.11 Changing thermal condition of feed. Both duties change when preheating or precooling the feed.

the feed simultaneously *increases* the condenser duty and *decreases* the reboiler duty. By similar arguments, when one precools the feed, the condenser duty reduces and the reboiler duty increases.

The difference in the duties is approximately the amount of heat to change the feed from being a bubble point liquid to the condition being fed to the column.

EXAMPLE FOR USING INTERHEATING/COOLING

Suppose we would like to reduce the utilities required to run two columns that are separating heat-sensitive fatty alcohols. As sketched in Figure 12.12, the column temperatures cannot be increased very much or the alcohols will rapidly decompose in the columns. These temperature limitations preclude the "stacking" of either column on top of the other. We can still get some integration by interheating in one column while intercooling in the other as illustrated in the right-hand side of the figure, carrying out a partial integration for both.

12.1.3 Heat Flows in Side Strippers and Side Enrichers (Carlberg and Westerberg, 1989)

SIDE STRIPPERS

Consider the column configuration shown on the left-hand side of Figure 12.13. This configuration is called a side stripper. As illustrated, such a configuration is capable of separating three ideally behaving species that would normally require the use of two columns. We do see two column shells, each with a reboiler here, but there is only one condenser. We have saved a piece of equipment.

Sec. 12.1　Heat Flows in Distillation

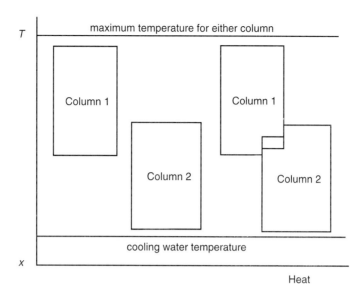

FIGURE 12.12 Example process for which intercooling/interheating is a candidate to improve integration.

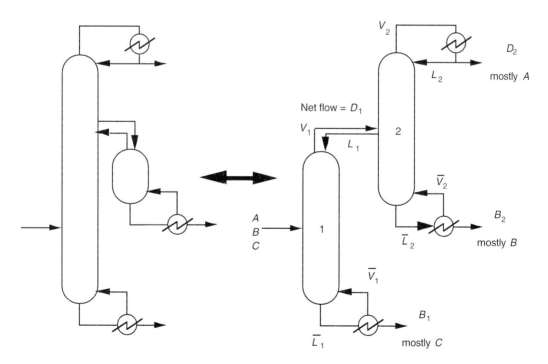

FIGURE 12.13 A side stripper with a topologically equivalent structure to it.

Column simulations show that this configuration requires less heating and cooling than would two separate columns, often as much as 25 to 40% less, so it appears to have a second very interesting advantage. It must have a cost or else it would be more widely used. The disadvantage is, in part, that the temperature drop across it ranges from the bubble point of the top to the bubble point of the bottom, which for this example is from the boiling point of A to the boiling point of C. The top of the column is pressure coupled to the bottom; indeed, since pressure must decrease as one moves up the column (so the vapor will flow up the column), the temperature drop is more than if the column could operate at a fixed pressure throughout (the lowest pressure occurs where the lowest temperature occurs making it even lower relative to the highest). Heat must degrade over this entire range to run this column. With two columns, one can decouple their pressures, adjusting them to reduce the temperature drop over which their heat is degraded.

In summary, then, we buy one less exchanger and gain on the first law—often substantially less heat is degraded—but we lose on the second law—it must be degraded over what is often a much larger temperature drop.

Another point to make for side strippers (and enrichers) is that they are really like heat integrated columns. Therefore, it is inappropriate to consider their use against columns run using only utilities. We should only compare them from a utility consumption point of view to columns where we allow the conventional columns to be heat integrated.

Let us first learn how this column performs by developing an approach to analyze it. We shall start by seeing how to compute the minimum reflux for it.

Examination of the second configuration in Figure 12.13 should make it evident it is topologically equivalent to the first, but we shall find it is easier to analyze than the first. One can think of the side stripper as two columns. We illustrate it with a 3-component feed—A, B, and C—to make it clearer what the configuration is really doing. The first column splits AB from C while the second splits A from B. The side stripper has the separation capability of two columns, but it has only one condenser.

Let us develop the following equations for the second configuration.

$$D_1 = V_1 - L_1$$

We can then write the following:

$$\overline{L}_2 = L_2 + q_2 D_1 = L_2 + q_2(V_1 - L_1)$$

However, we are taking liquid from the second column to provide reflux for the first, giving

$$L_1 = L_2 - \overline{L}_2$$

Solving for q_2 using these two equations, we get

$$q_2 = \frac{-L_1}{V_1 - L_1}$$

We can relate the reflux ratio in the first column to its internal flows, getting

$$R_1 = \frac{L_1}{D_1} = \frac{L_1}{V_1 - L_1}$$

Sec. 12.1   Heat Flows in Distillation   423

which gives the remarkable result that

$$q_2 = -R_1$$

that is, the thermal condition for the feed to the second column is the negative of the reflux ratio for the first. Since R_1 is strictly positive, q_2 is strictly negative, which corresponds to the net feed to the second column being superheated. One explanation for this is that one is passing vapor to the second column and getting back a part of that vapor as liquid. The net flow to the second column can be thought of as the net material flow as vapor plus the heat obtained by cooling the rest from vapor to liquid.

A way to analyze this configuration, then, is the following:

- Establish the bottom and then the top products for the first column.
- Determine the minimum reflux ratio for the first column, using Underwood's method if it is applicable.
- Set the reflux ratio for the first column to some factor (like 1.2) times the minimum reflux ratio for the first column.
- The thermal condition for the feed to the second column is then the negative of this reflux ratio. Determine the minimum reflux ratio for the second column. Set its value to something like $1.2 * R_{2,\min}$.

SIDE ENRICHERS

The side enricher in Figure 12.14 is also shown in a topologically equivalent form that is easier to analyze.

The analysis here is similar to that for a side enricher. Here we find the thermal condition for the feed to the second column is given as:

$$q_2 = \overline{R}_1 + 1$$

that is, it is equal to the reboil ratio for the first column plus one. It will always exceed one, a value that occurs for subcooled liquid feed. The design procedure is precisely the same, except the above should be used to set the thermal condition of the feed for the second column.

T VERSUS Q DIAGRAMS FOR SIDE STRIPPERS AND ENRICHERS

Let us consider the side stripper just analyzed. We see that it has two reboilers and one condenser. The feed to column 2 is acting like superheated vapor, as we discussed before.

As argued earlier in the section on interheating/cooling, feeding a column with superheated vapor simultaneously decreases the reboiler duty, but it also increases the condenser duty. The heat flows for the side stripper configuration act as if the column $T\text{-}Q$ diagrams have overlapped, as shown in Figure 12.15. The second column reboiler duty has decreased while its condenser duty has increased, consistent with our above observation about preheating its feed.

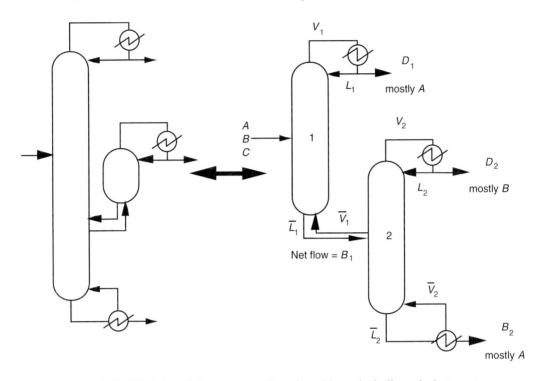

FIGURE 12.14 Side enricher configuration with topologically equivalent structure.

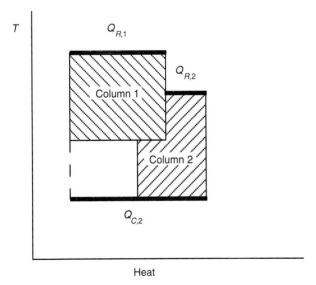

FIGURE 12.15 Heat duties on a T versus heat diagram for side stripper configuration.

This diagram suggests that there could be an advantage to placing a condenser at the top of column 1, allowing its duty to be removed at a higher temperature than from the condenser at the top of the second column.

If the advantage of the <u>higher</u> temperature for <u>heat removal</u> is needed for heat integration, then it may be a good idea. Or, if the extra driving force reduces the heat exchanger area enough, then it may be a good idea.

For a side-enricher configuration, we will simultaneously decrease the condenser duty and increase the reboiler duty, getting a diagram that is the same as the above except that it is flipped vertically, having one reboiler temperature and two condenser temperatures.

For further reading on heat flows in columns, see Dhole and Linnhoff (1992).

REFERENCES

Andrecovich, M. J., & Westerberg, A. W. (1985). A simple synthesis method based on utility bounding for heat-integrated distillation sequences. *AIChEJ.,* **31**, 363.

Carlberg, N., & Westerberg, A. W. (1989). Temperature-heat diagrams for complex columns: 2. Underwood's method for side strippers and enrichers. *I&EC Res.,* 28, 1379–1386.

Dhole, V. R., & Linnhoff, B. (1992). Distillation column targets. In *Proceedings from the European Symposium on Computer Aided Process Design-1,* 97.

Fonyó, Z. (1974a). Thermodynamic analysis of rectification I. Reversible model of rectification. *Intern. Chem. Eng.,* **14**, 18.

Fonyó, Z., (1974b). Thermodynamic analysis of rectification II. Finite cascade models. *Intern. Chem. Eng.,* **14**, 203.

Koehler, J., Aguirre P., & Blass, E. (1992). Evolutionary thermodynamic synthesis of zeotropic distillation sequences. *Gas Sep. Purif.,* **6**, 4153.

Terranova, B., & Westerberg, A. W. (1989). Temperature-heat diagrams for complex columns: 1. Intercooled/interheated distillation columns. *I&EC Res.,* **28,** 1374–1379.

EXERCISES

A flowsheet simulation program be used to aid in solving the following problems. However, they can also be done using Raoult's law within a spreadsheeting program.

1. Consider a mixture of 35 (mole) % *n*-heptane, 30% *n*-hexane, 10% isobutane, and 25% *n*-pentane. Using Raoult's law, develop the condenser and reboiler heat duties and temperatures for the column separating into two products of two species each for pressures of 1, 5, 10, and 20 atm when running the column at 1.2 times the minimum reflux ratio for it. 99 mole % of the key components should be recovered in their respective products. Use a partial reboiler and a total condenser. The feed to

the column is bubble point liquid. Plot the temperature drop across the column and the average of the two heat duties versus the condenser temperature. Can you notice anything special about this plot? Note, also how close to equal the reboiler and condenser heat duties are for each of the columns (the base case assumption).

2. Repeat the previous exercise but this time do the computation using a commercial flowsheeting system.

3. Repeat the analysis of exercise 1 for the column that produces a distillate that is a single species.

4. Consider the mixture in exercise 1 again. Desired products are all the single component products, each of which is to be 99% pure. Discover the 3-column sequence that requires the least amount of total heat for this separation problem, after the best heat integration you can discover is done between condensers and reboilers. Hot utility is available so heating of a stream up to 425 K is possible; cooling of a stream down to 305 K is possible with cooling water.

5. Repeat the previous exercise, but this time you are allowed to use a maximum of five columns. With five columns, you can propose solutions that involve multi-effecting.

6. You have been asked by another engineer to check over the flowsheet shown in Figure 12.16. Note that the condenser for the first column is a partial condenser. (a) List any obvious design errors. (b) Is the other engineer's analysis believable? (c) Should the feed to column 1 be preheated? Explain your answers.

7. Compute the intercooling/interheating diagram for separating the isobutane from the remaining components at one bar for exercise 1. Assume the feed and products are at their bubble points.

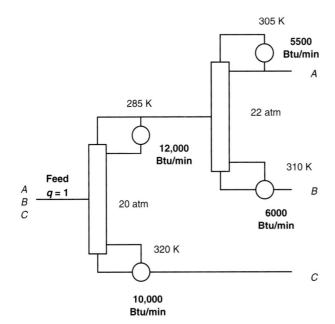

FIGURE 12.16 Separation flowsheet.

Exercises

8. You are grading a senior chemical engineering design project. The flowsheet the student group proposes contains the separation scheme shown in Figure 12.18. What comments (in red pen) would you make on it? What suggestions would you make for the group to improve this part of their flowsheet?

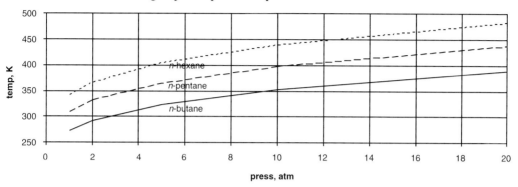

FIGURE 12.17 Temperature vs. vapor pressure for components in Exercise 8.

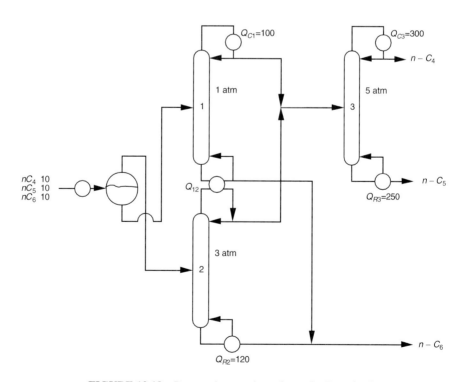

FIGURE 12.18 Proposed separation scheme for Exercise 8.

9. You are given the following mixture. Propose the better heat integrated distillation-based separation sequences to produce five relatively pure single component products. Explain your answer.

Species	Relative Volatility	Amount, kg/s
A		2
	$\alpha_{AB} = 3$	
B		2
	$\alpha_{BC} = 1.15$	
C		2
	$\alpha_{CD} = 3$	
D		1
	$\alpha_{DE} = 1.15$	
E		0.25

10. This problem is a major effort, taking perhaps several tens of hours. Do not casually choose to do it or assign others to do it. Repeat exercise 4—or for the more hearty, exercise 5—but this time worry about the cost of the equipment to carry out the separations and heat exchange. Transfer coefficients for all heat exchangers can be assumed to be 1000 W/m² K (assuming both sides are condensing/vaporizing fluids with some fouling having occurred). This problem will require you to consult information not provided here, such as cost estimation correlations for equipment. Remember that the product from a column, if withdrawn as a bubble point liquid and fed to another column, will not be bubble point liquid unless the next column is at the same pressure. To do this problem right, you will have to adjust column pressures to alter the column temperatures and thus reduce the cost for the heat exchangers.

GEOMETRIC TECHNIQUES FOR THE SYNTHESIS OF REACTOR NETWORKS

13

In the previous chapters we saw the development of synthesis strategies for energy integration and separation systems. However, virtually all process and flowsheet development begins with the reaction chemistry. Up to this point we have assumed that these reactions along with the reactor network and its performance were specified *before* the design stage. Nevertheless, the reactor network strongly influences the character of the entire flowsheet and consideration of the reactor network has a dominant effect in improving the process. On the other hand, except for simple systems that can often be designed with qualitative arguments, relatively little development has gone into the systematic synthesis of reactor networks. This is due to the complex and nonlinear behavior of the reacting system, coupled with combinatorial aspects of the network structure that are inherent in all synthesis problems. This chapter introduces the synthesis problem and provides a brief description of some simple geometric techniques for reactor network synthesis.

As with energy integration in Chapter 10, we will consider a reactor network *targeting strategy,* which seeks to describe the performance of the network without its explicit construction. Once obtained, a network is then determined that is guaranteed to match this target. To achieve these properties, we introduce a new approach based on recently developed geometric concepts. These concepts are used to construct a region in concentration space that describes the performance of a complete family of reactor networks. This region is known as the *attainable region,* and with this approach, performance targets for the network can be synthesized, in principle, for isothermal and nonisothermal systems with arbitrarily complex kinetics. Moreover, in Chapter 19, we will extend these concepts further by combining them with optimization formulations in order to solve larger and more difficult problems. We will also show how reactor network synthesis problems can be integrated into the overall flowsheet synthesis problem.

13.1 INTRODUCTION

In Chapter 2, the problem statement was defined by first specifying the reaction chemistry and describing performance characteristics of the reactor. We assumed that these were made available from an experimental study and were fixed for the flowsheet development. Key characteristics of the reactor are the *conversion* of reactant, based on the reactor feed stream, and the *selectivity* of the converted feed to desired product. Following the interactions sketched in Figure 13.1, we see that these variables determine the entire nature of the flowsheet. In particular, the reactant conversion determines the recycle structure of the flowsheet as the reactants are separated and sent back. A high conversion leads to a small recycle stream and lower equipment costs for this section. The selectivity, on the other hand, determines the downstream separation sequence in order to recover desired product from the by-products and waste products. A high selectivity to desired product reduces the need for by-product separation and significantly lowers these capital and energy costs. Reactor performance is especially important in order to avoid the generation of environmentally hazardous by-products, as this has direct savings in waste treatment costs.

These two variables combine to determine the *overall conversion* of raw material to desired product in the flowsheet. For the optimization of the flowsheet, Douglas (1988) notes that frequently the reaction kinetics cause these variables to conflict with each other; a low selectivity (high separation costs and low overall conversion) is achieved with high reactor conversion (and small recycle costs) and vice versa. Consequently, the optimum flowsheet consists of a trade-off of these two variables, which needs to be as-

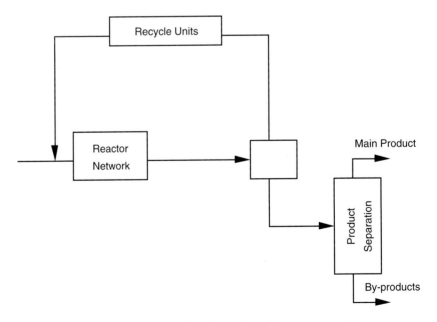

FIGURE 13.1 Flowsheet interactions for reactor network.

Sec. 13.1 Introduction

sessed quantitatively. As a starting point for this analysis we consider the synthesis of reactor networks that maximize reactor conversion, selectivity, or an economic objective derived from both variables.

As seen in Figure 13.1, the reactor system is therefore the heart of the chemical process, as it dictates the downstream processes (e.g., separation and waste treatment) and strongly influences the recycle and flowsheet structures, as well as the energy network. Despite this, the general approach is to design the reactor system in isolation and then design the remaining subsystems. As we will see in later chapters, this approach is often suboptimal and large improvements in the overall process can be made through process integration. In this chapter we will develop the concepts that will be exploited later for this integrated approach.

Synthesis of chemical reactor networks can be defined by the following problem statement:

> *Given the reaction stoichiometry and rate laws, initial feeds, a desired objective, and system constraints, what is the optimal reactor network structure? In particular:*
> - What is the flow pattern of this network?
> - Where should mixing occur in this network?
> - Where should heating and cooling be applied in this network?

Despite significant research in reactor modeling and analysis and in the design of specific reactors, relatively little work has been reported in reactor network synthesis, while other areas of process synthesis, including heat integration and separation synthesis, have advanced much more. This is due to several reasons. First, reacting systems are typically more difficult to model and generally have more diverse elements than energy or separation systems. This is typified by an important (and expensive) experimental component. Moreover, given the resource constraints in process development, there is often little opportunity to develop a detailed kinetic model or to investigate the many alternatives to find an optimal reactor network.

Previous work in reactor network synthesis can be classified into three categories: *heuristics* for reactor selection that apply to simple, well-understood reaction mechanisms and are generalized to more complex ones, *structural optimization* of a candidate reactor network, and construction of *attainable regions* in concentration space, for instance, that contain all of the candidate reactor networks. In the first category, heuristics can be derived from graphical results and rules that emphasize the effects of mixing for various reaction orders, and heating for exothermic and endothermic reactions. These results usually apply to single reactions or for series and parallel reaction cases, and are used to guide the selection of ideal reactors (e.g., plug flow (PFR) and continuous stirred tank (CSTR) reactors). Extending these heuristics beyond simple reaction cases is not always easy and these approaches have limitations when applied to more complex problems. Instead, quantitative approaches are required to establish proper trade-offs for such systems. In the next section we will outline some of these simple approaches and briefly review some basic reactor types.

The structural optimization of the reactor network is a direct and natural way to assess and improve these quantitative trade-offs. A survey of these approaches is given in Chapter 19 along with a brief description of the optimization formulations. However, formulation of the optimization problem is complicated and introduces a number of difficulties for solution. First, equations describing reactor systems are fraught with nonlinearities and nonconvexities that lead to local solutions. Given the likelihood of extreme nonlinear behavior, such as bifurcation and multiple steady states, even locally optimal solutions can be quite poor. In addition, optimization of a reactor network superstructure is plagued by the question of completeness of the network, and the possibility that a better network may have been overlooked by posing an incomplete family of solutions (or superstructure). This is exacerbated by reaction systems with many networks that have identical performance characteristics for a given objective. (For instance, a single PFR can be approximated by a large train of CSTRs.) In most cases, the simpler network is clearly more desirable. A review of optimization studies for reactor network synthesis will be highlighted in Chapter 19.

To deal with the question of a "complete" superstructure, we consider geometric concepts for the reactor network synthesis problem. Instead of postulating a family of solutions for the best reactor network, we turn the problem around to consider the characteristics of the particular reaction and mixing processes, and we use these to define the complete family of reactor networks. The approach developed in this chapter is based on geometric concepts for *attainable regions* (AR) in concentration space, for example, wherein all possible reactor structures must lie. Construction of this region is based on identifying the conditions that the attainable region must satisfy and then successively constructing regions and testing these conditions. Once we have this region, we are assured that a complete family of reactor networks has been considered that contains the optimal solution. This approach was initially suggested by Horn (1964) and developed by Glasser et al. (1987). A more complete literature summary of this area is given at the end of this chapter.

In the next section we consider some basic reactor types and summarize some simple methods for selecting among them. In section 13.3 we describe and summarize the geometric properties that relate to the attainable region and present a reactor network synthesis method through construction of attainable regions. We illustrate this approach on examples whose regions can be plotted in two dimensions. In section 13.4, we apply the method of reaction invariants that can extend the two dimensional AR approach to a class of larger problems. The concepts in both sections will be illustrated with numerous examples. Finally, section 13.5 summarizes the chapter and outlines areas for further reading.

13.2 GRAPHICAL TECHNIQUES FOR SIMPLE REACTING SYSTEMS

In Chapter 3, we assumed the reactor conditions were specified prior to flowsheet development. Here we consider the possibility of selecting a reactor network to improve overall profitability of the flowsheet. For simple reacting systems, such as for single reactions and series or parallel reactions, this topic is discussed in many standard texts on reactor

Sec. 13.2 Graphical Techniques for Simple Reacting Systems

design (see section 13.5 for a summary) and selection of steady state reactors from basic reactor types is based on qualitative behaviors and monotonic trends in the rate laws. Here we consider a selection among three basic types of reactors—the tubular reactor (idealized through plug flow (PFR)), the mixed flow reactor, also known as the continuous stirred tank reactor (CSTR), and the recycle reactor. In Chapter 19, we also consider a more complex reactor, the differential sidestream reactor (DSR). In this section, we briefly summarize some general concepts described in Levenspiel (1972) in order to provide some background for an alternative reactor design strategy in the next section.

Consider the ideal reactor types illustrated in Figure 13.2. For an isothermal system, plug flow reactors (PFR) are modeled as:

$$d(Fc)/dV = r(c), \quad c(0) = c_0 \tag{13.1}$$

where c is the vector of molar concentrations, F is the volumetric flowrate, V is the reactor volume, and $r(c)$ is the reaction rate. Continuous stirred tank reactors (CSTR), on the other hand, are expressed as:

$$F c - F_0 c_0 = V r(c) \tag{13.2}$$

Finally, recycle reactors (RR) can be written as

$$d(F c)/dV = r(c), \quad c(0) = (R F(V) c(V) + F_0 c_0)/ (R F(V) + F_0) \tag{13.3}$$

where $F(V)$ and $c(V)$ are the outlet volume flowrates and concentrations, respectively, and R is the recycle ratio. For the case of constant density systems, these expressions simplify to:

$$dc/d\tau = r(c), \quad c(0) = c_0 \tag{13.4}$$

for the PFR case, where τ is residence time, V/F. Similarly, we have:

$$c - c_0 = \tau\, r(c) \tag{13.5}$$

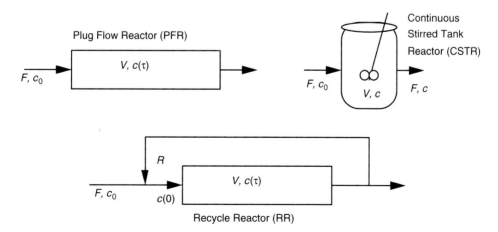

FIGURE 13.2 Ideal reactor types.

for the CSTR case, and

$$(R + 1) \, dc/d\tau = r(c), \qquad c(0) = (R \, c(V) + c_0)/(R + 1) \qquad (13.6)$$

for the recycle reactor case.

A common strategy for reactor selection arises in the single reaction case. Here we choose the limiting component (say, component A) and plot, $-1/r_A$ versus c_A. Rearrangement of the design equations for each reactor type leads to the following evaluations of residence time in each reactor. In Figure 13.3, we note that residence time for a PFR can be obtained from Eq. (13.4) and is represented as the area under this curve, while for a CSTR the residence time is obtained from Eq. (13.5) and is represented as a rectangle with reaction rate evaluated at the exit. For recycle reactors, the residence time is evaluated through mixing of the feed followed by reaction in the PFR case as described by Eq. (13.6). Here we consider integration over the PFR portion and subsequently represent

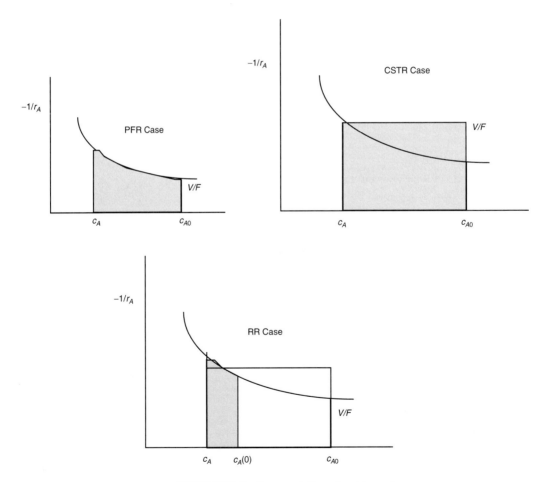

FIGURE 13.3 Representation of residence times.

Sec. 13.2 Graphical Techniques for Simple Reacting Systems

the residence time as a rectangle with the reaction rate evaluated at an intermediate point between the recycled feed and exit. From the graph and the design equations it is therefore easy to visualize that the limits of operation for recycle reactors are the PFR ($R = 0$) and the CSTR ($R = \infty$).

We note also that as long as $-1/r_A$ is monotonically decreasing with c_A, the PFR reactor leads to the smallest residence time. This is particularly true when power law kinetics are applied and the reaction rate has a positive order with respect to c_A. This property can be summarized as:

> *For reaction rates* $-r_A = k\, c_A^n$, *the PFR reactor always requires a smaller residence time than the CSTR or recycle reactor (RR) reactors. Analogously, for a given residence time, the conversion with a PFR is always greater for power law kinetics with* $n > 0$.

In addition, we can also consider the bimolecular reaction, $A + B \to C$, where separate feeds for A and B are available. Here, the yield can be exploited by varying the feed ratios of reactants. For instance, Levenspiel observed that in isothermal systems, an excess of one reactant is often exploited to lead to better reactor networks, with the PFR again requiring a smaller volume.

More generally, one can analyze simple reactions where the best reactor (e.g., smallest residence time) may not be a PFR. For instance, for a single reaction where $-1/r_A$ does not decrease monotonically with c_A, CSTRs and recycle reactors can have more desirable characteristics. This is especially the case for the autocatalytic reaction considered in the example below.

EXAMPLE 13.1   Design of Reactor with Autocatalytic Reaction

Consider the liquid phase (constant density) isothermal reaction, $A + B \to 2B$, where the rate expression is

$$r_A = -k\, c_A^n\, c_B^m, \tag{13.7}$$

where $n = m = 1$, $k = 2$ l/mol-sec and the initial concentration is 0.99 mol/l A and 0.01 mol/l B. If we desire an exit concentration of $c_B = 0.95$, which reactor gives the lowest residence time?

From the mass balance we have: $c_A + c_B = 1.0$ and thus the rate expression can be written as

$$r_A = -2\, c_A\, (1 - c_A). \tag{13.8}$$

For the CSTR case we have for $c_{A0} = 0.99$ and $c_A = 0.05$:

$$c_A - c_{A0} = r_A\, (V/F) \tag{13.9}$$

$$V/F = (c_{A0} - c_A)/2(c_A(1-c_A)) = 9.895\ \text{sec}$$

For the PFR case we have:

$$V/F = \int_{0.99}^{0.05} \frac{dc_A}{r_A} = -\int_{0.99}^{0.05} \frac{dc_A}{2c_A(1 - c_A)}$$

$$= 1/2\, [\ln(1/c_A - 1) - \ln(1/c_{A0} - 1)] \tag{13.10}$$

$$= 3.77\ \text{sec}$$

Finally, for the recycle reactor case we have:

$$V/F = (1+R) \int_{(0.05R+0.99)/(1+R)}^{0.05} \frac{dc_A}{r_A} = -(1+R) \int_{(0.05R+0.99)/(1+R)}^{0.05} \frac{dc_A}{2 c_A (1-c_A)} \quad (13.11)$$

and for a recycle ratio of 1, $V/F = 3.0244$ sec. We can further reduce the residence time by optimizing with respect to R. Setting the derivative of (V/F) with respect to R to zero gives:

$$d(V/F)/dR = 1/2[ln(19) - ln((0.95R + 0.01)/(0.05R + 0.99))] -$$
$$0.47(1 + R)/((0.01 + 0.95R)(0.05R + 0.99)) = 0 \quad (13.12)$$

Solving for R gives an optimal recycle ratio of 0.2934 with a minimum residence time of $V/F = 2.7105$ sec. Thus, for this autocatalytic reaction, the recycle reactor is the best of the three.

13.2.1 Multiple Reactions: Series and Parallel Cases

For multiple reactions we can generalize the behavior of power law kinetics by considering relative reaction rates. For process design the preferred objective is frequently reactor selectivity not reactant conversion, and here the relative reaction order determines which reactor type is preferred. Levenspiel summarizes the following concepts for multiple reaction systems:

- For reactions in parallel the concentration level of reactants strongly influences the product distribution. Higher reaction concentration favors reaction of higher order and a low concentration favors the reaction of lower order. Otherwise, there is no effect of mixing.
- For reactions in series, mixing of fluid of different composition strongly influences formation of intermediate. The maximum possible amount of intermediates is obtained if fluids of different compositions are not allowed to mix within the reactor network.
- Series-parallel reactions can be analyzed in terms of their constituent series and parallel reaction components or optimum contacting.

Finally, heat and pressure effects play an important role in the decision of reactor types, as well as ratios of reactants and the type of operation. These can be summarized by the following statements:

- For irreversible reactions, maximum yield is obtained by maintaining the profile at the highest allowable temperature within a PFR.
- For reversible reactions in gas phase, an increase in pressure increases conversion if the moles of products are fewer than the moles of reactants, and vice versa.
- For reversible reactions equilibrium concentration rises with increasing temperature for endothermic reactions and falls for exothermic reactions.
- A high temperature favors the reaction of higher activation energy, a low temperature favors the reaction of lower activation energy.

Sec. 13.2 Graphical Techniques for Simple Reacting Systems

These qualitative statements can easily be verified with quantitative examples. Moreover, there are many specific instances that can be further abstracted from these general concepts. As a result, they can be quite useful for the selection of reactor networks with simple reaction mechanisms. A more detailed explanation and demonstration of these concepts can be also found in reactor design textbooks which are listed at the end of this chapter.

However, in the context of reactor design where more complicated trade-offs exist, applying these heuristics can often lead to conflicting results. Consider, for example, the series reaction $A \rightarrow B \rightarrow C$. Qualitatively, the reaction curves have the behavior in Figure 13.4, where we note the following influence of the reactor on the flowsheet. If component C is the valuable component, then clearly region III is the desired region of operation. On the other hand, if component B is desired and either A is not valuable or C can also be sold as valuable by-product, then region II is of interest. On the other hand, if A and B are valuable and C is a useless or harmful by-product, then region I is the preferred region of operation. Of course, the exact operating points depend on the prices and costs of products and raw materials. Also to be considered is the cost effect on reactor size as well as the cost of recycling the reactants (as well as purge losses of A). Benefits of a given network can be argued qualitatively, but the best decision often requires consideration of a detailed optimization problem.

In the case of reactor design, however, the optimal choice of regions is further complicated by the appropriate choice of reactor network and the corresponding operating conditions. To make this problem tractable, we therefore need a complete representation of the family of reactor networks in order to achieve the desired operating point. Here a heuristic strategy can lead to good networks, but evaluation of trade-offs can only be done quantitatively with a rich enough family of alternatives. In Chapter 19 we will review superstructure optimization approaches to evaluate these trade-offs quantitatively.

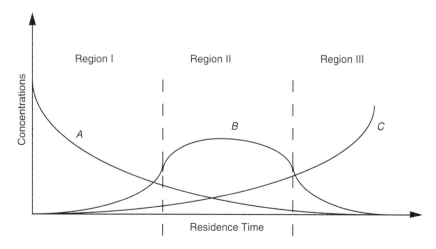

FIGURE 13.4 Choice of residence time for process flowsheet.

A key concern to superstructure optimization is to assure that it is sufficiently general to contain the optimal network. In the next section we turn this problem around and consider the conditions that describe *all* possible reactor networks. In particular, we use these conditions to construct an attainable region that describes the performance of all of these networks. Once we have this region, we then examine the reactor networks that make up its boundary and this gives us a complete set of reactors that can be considered for selection. In the next section we therefore consider the concept of *attainable regions* that define all networks and capture the processes of reaction and mixing. This region (say, in concentration space) is closed to any further addition or refinement of the family of reactor networks that make up the attainable region. Once this region is created we will see that construction of the family of optimal solutions can be obtained directly from the boundary of the attainable region.

13.3 GEOMETRIC CONCEPTS FOR ATTAINABLE REGIONS

For chemical reactor networks, the attainable region concept was first presented by Horn (1964), who noted that:

> ... *variables such as recycle flow rate and composition of the product form a space which in general can be divided into an attainable region and a non-attainable region. The attainable region corresponds to the totality of physically possible reactors* ... *Once the border is known the optimum reactor corresponding to a certain environment can be found by simple geometric considerations.*

To illustrate this concept, consider the attainable region for the series reaction:

$$A \rightarrow B \rightarrow C$$

which can be defined in the space of concentrations for A and B as shown in Figure 13.5. At each point on this graph we can evaluate the rate vectors r_A, r_B, and r_C; these are all unique functions of c_A and c_B. The generation of the components B and C from A can therefore be calculated from r_A and r_B; the slopes, dc_B/dc_A, at all points in Figure 13.5 are given by r_B/r_A. To consider different reactor types in the attainable region, we now can plot reactor trajectories. By mixing points on these trajectories we can also create the shaded regions that represent concentrations attainable by mixing all points that are generated by the particular reactor. Note that from the definitions in Appendix A, this region is convex because any nonextreme point c^* in the region can be given by a *convex combination* of two other points (say, c_1 and c_2) in that region, that is:

$$c^* = (1 - \lambda) c_1 + \lambda c_2, \quad 0 \leq \lambda \leq 1 \tag{13.13}$$

and c^* is a point that can be generated by mixing compositions of c_1 and c_2.

To observe the path of a PFR with a variable residence time and a fixed feed c_{A0} and c_{B0}, one can solve the ordinary differential equations from the feed point:

Sec. 13.3   Geometric Concepts for Attainable Regions

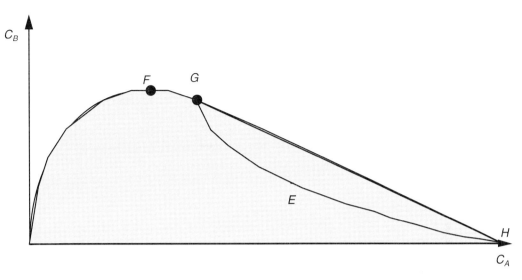

FIGURE 13.5   Attainable region in concentration space.

$$dc_A/d\tau = r_A$$
$$dc_B/d\tau = r_B \qquad (13.14)$$

or, more directly:

$$dc_B/dc_A = r_B/r_A \qquad (13.15)$$

From this differential equation, we plot the trajectory HEGF in Figure 13.5.

On the other hand, the path of a CSTR from the feed is generated from the equations:

$$c_A - c_{A0} = \tau\, r_A$$
$$c_B - c_{B0} = \tau\, r_B \qquad (13.16)$$

and the concentration trajectory of CSTR reactors is obtained from solving these equations for increasing values of τ. Note that for each point of the CSTR trajectory we have:

$$(c_A - c_{A0})/(c_B - c_{B0}) = r_A/r_B \qquad (13.17)$$

For instance, in Figure 13.5 a particular CSTR is represented by the line segment GH starting from the feed (c_{A0}, c_{B0}) that is collinear with the slope at (c_A, c_B) on the PFR trajectory HEGF.

Note that in Figure 13.5 we assume a fixed feed, an initial temperature, and trajectories that are determined entirely by the state equations for concentration Eqs. (13.14–13.17). This is true in steady state for isothermal or adiabatic systems. Does the region in Figure 13.5 represent the performance of the complete family of reactor networks? We answer this question by checking the shaded region to see if there are any additional reactors that can increase its size (by testing the conditions given below).

If the region cannot be increased, we consider this region the *attainable region* for our particular reacting system. With this attainable region, we clearly see that point F and the line segment GH represent the maximum concentration of B and maximum selectivity of B to C, respectively. Moreover, the maximum concentration and selectivity points can be achieved by the reactor networks that make up the attainable region boundary. In addition, if a more complex objective has an optimum represented in terms of c_A and c_B that yields an interior point, then this point can be achieved by any linear combination of the boundary structures. Using the attainable region is an especially powerful concept, because, once it is known, performance of the network can be determined without the network itself.

To construct the attainable region, we note that the concentration space is a vector field with a rate vector (e.g., in Figure 13.5, $dc_B/dc_A = r_B/r_A$) defined at each point. Moreover, we are not restricted to concentration space for the attainable region. We could also consider any other variable that satisfies a linear conservation law (e.g., mass fractions, residence time, energy, and temperature—for constant heat capacity and density). Recently, Glasser, Crowe, and Hildebrandt (1987) developed geometric properties of the attainable region along with a constructive approach for determining this region. They defined the necessary conditions for the attainable region as follows:

- The attainable region (AR) must be convex. Any point that is created by a convex combination of two points in the AR (13.13) must be in the AR, as it can be created by mixing these two points. Moreover, this property ensures that the AR cannot be extended by further mixing.
- Reaction vectors on the AR boundary cannot point out of the AR. If this were the case, then the AR could be extended further by PFR reactors, which have trajectories that are always tangent to the rate vectors, Eq. (13.15).
- Reversed reaction vectors in the complement of the AR cannot point back into the AR. This condition ensures that the AR cannot be extended further by a CSTR, because a CSTR is represented in the AR by a line with ends at the feed and outlet concentrations, and the rate vector at the CSTR outlet is collinear with this line, Eq. (13.17).

These properties hold for all dimensions and, in fact, are stronger than the simple exclusion of CSTRs, PFRs, and mixing. Hildebrandt (1989) proved that an AR closed to further extension by PFRs and CSTRs is also closed to extension by recycle PFRs, as long as the AR is not constrained in concentration. Hildebrandt et al. (1990) also showed how these properties could be applied to systems with nonconstant densities and heat capacities.

Sec. 13.3 Geometric Concepts for Attainable Regions

EXAMPLE 13.2 van de Vusse Reaction

Consider the isothermal van de Vusse (1964) reaction, which involves four species. The objective is the maximization of the yield of intermediate species B, given a feed of pure A. The reaction mechanism is given by

$$A \xrightarrow{k_1} B \xrightarrow{k_2} C, \quad 2A \xrightarrow{k_3} D$$

Here the reaction from A to D is second order. The feed concentration is $c_{A0} = 0.50$ mol/l and the reaction rates are $k_1 = 1 \ s^{-1}$, $k_2 = 1 \ s^{-1}$ and $k_3 = 1 \ l/(\text{mol s})$. The reaction rate vector for components A, B, C, D respectively is given in dimensionless form by:

$$r(c) = [-c_A - c_A^2, c_A - c_B, c_B, c_A^2] \tag{13.18}$$

where we also define $X_A = c_A/c_{A0}$, $X_B = c_B/c_{A0}$. As seen in Figure 13.6, by tracing out a PFR in the space of X_A and X_B we see that the attainable region is convex and the relative rate vectors (r_A/r_B) on the boundary are only tangent to this region and cannot point out of the region. Finally, by examining the vector field in Figure 13.6, it can be verified that no relative rate vectors outside of the attainable region can be reflected back into the region. Therefore, the above properties are satisfied and the PFR trajectory describes the *complete attainable region* for this example. Here the maximum yield is given by $X_B^{\text{exit}} = 0.3394$ and this is the globally optimal solution. In Figure 13.6 we also plot the PFR residence time with the conversion X_A, and from this curve we see that the optimal PFR has a residence time of 0.94 seconds.

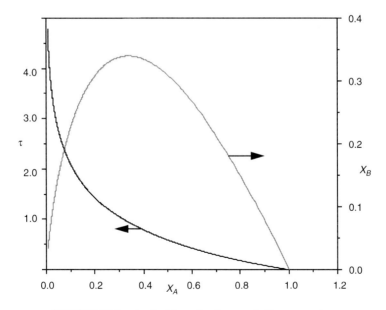

FIGURE 13.6 Attainable region for van de Vusse reaction.

For systems that can be represented in two dimensions, as in Example 13.2, the construction of the AR is particularly easy. Here the attainable region can be constructed by alternately constructing PFR and CSTR trajectories (using Eqs. 13.14–13.17) and checking the attainable region properties given above. Hildebrandt and Biegler (1995) formalized this procedure with the following algorithm.

1. Start from the feed point and work towards the equilibrium or endpoint by drawing a PFR from the feed point.

2. If the PFR forms a convex region, then we have found a candidate attainable region. We then check that no rate vectors external to this region can be reflected back into the region. If there are none, we stop.

3. Otherwise, if the PFR trajectory is not a convex region, we then find the convex hull of the PFR trajectory by drawing straight lines to fill in the nonconvex parts of the trajectory. We then check along the straight line sections of the convex hull to see if reaction vectors point outwards. If no reaction vectors point outwards, then we have a candidate attainable region and we repeat the procedure in step 2.

4. Otherwise, if reaction vectors point outwards, then we can find a CSTR trajectory, starting from the PFR trajectory, that intersects the straight line section at the point where the reaction vector becomes tangent. We then draw in the CSTR trajectory, with feed on the PFR trajectory, that increases the region most, and then find the convex hull of the new extended region by filling in nonconvex parts in the CSTR trajectory.

5. Next, draw a PFR trajectory from the end of the straight line that fills in the nonconvex part of the CSTR trajectory. If this PFR trajectory is convex, then we have a candidate for the attainable region and we return to step 2. Otherwise, we repeat from step 3 until all the nonconvex portions are filled in and we have reached the equilibrium point or endpoint.

To illustrate this approach we consider an extension of Example 13.2, by making the first reaction reversible and slightly changing the rate law, as shown in Hildebrandt and Biegler (1995).

EXAMPLE 13.3 Reversible van de Vusse Reactions

The reactions below are a slight extension of the reactions in Example 13.2

$$A \underset{k_{1r}}{\overset{k_{1f}}{\Leftrightarrow}} B \xrightarrow{k_2} C \text{ and } 2A \xrightarrow{k_3} D$$

with the following rate constants: $k_{1f} = 0.01$, $k_{1r} = 5$, $k_2 = 10$, and $k_3 = 100$. We assume that the feed is pure A where $c_A^0 = 1$ and we define $c = (c_A, c_B)$ where:

$$r(c) = [-0.01 c_A + 5 c_B - 100\, c_A^2, 0.01 c_A - 5 c_B - 10\, c_B]. \tag{13.19}$$

Sec. 13.3 Geometric Concepts for Attainable Regions   443

What is the attainable region for this reaction system?

By applying the above procedure, we can show the following construction of the attainable region.

Step 1. Construct the PFR profile using the rate expressions in Eq. (13.19). This yields the profile ABD shown in Figure 13.7a.

Step 2. This profile is not convex so we need to construct the convex hull of this trajectory and this is shown by the dashed line segment AEB in Figure 13.7a.

Step 3. We can fill in the nonconvex portions with straight lines, for example, starting from the feed point, A, we get AEB. This forms the candidate attainable region. By evaluating the rate vectors (r_B/r_A) along this line, we see that there are rate vectors at point E that point out of the candidate attainable region. This requires us to consider additional CSTR trajectories.

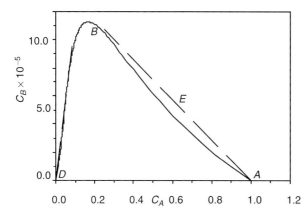

FIGURE 13.7a Initial PFR profile with convex hull for Example 13.3.

Step 4. We draw in the CSTR trajectory starting from the point on the convex hull that extends the region the most. In this case, this is the feed point. Drawing in the CSTR trajectory at point A leads to Figure 13.7b. Note that this trajectory overlaps the PFR region and a convex hull can be formed from the two. However, at point F it is clear that there is a rate vector pointing out of this region.

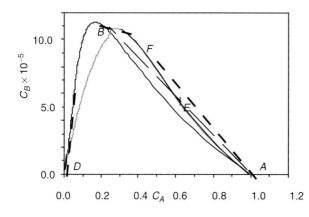

FIGURE 13.7b PFR and CSTR profiles with convex hull for Example 13.3.

Step 5. From point *F* we continue with a PFR trajectory to equilibrium (the trajectory FGD) and we obtain the trajectories shown in Figure 13.7c. Here we note that there still is a small nonconvex region starting at point *H,* toward the equilibrium point, *D*. As a result, we return to step 3 and repeat the process of generating CSTR and PFR trajectories. This leads to filling in the nonconvex portion with the line segment HD.

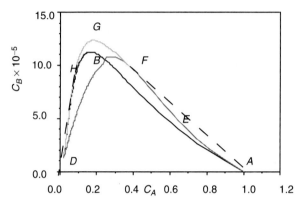

FIGURE 13.7c   PFR/CSTR/PFR profiles and convex hull, Example 13.3.

Finally, we obtain the attainable region shown in Figure 13.7d. Here we can see four different reactor structures lying on the attainable region boundary and the individual structures are simple combinations of CSTRs and PFRs. The line segment AF represents a CSTR with bypass and point F represents a CSTR. The trajectory AGH represents a CSTR followed by a PFR and the segment HD is the CSTR/PFR series in parallel with any reactor that gives an equilibrium

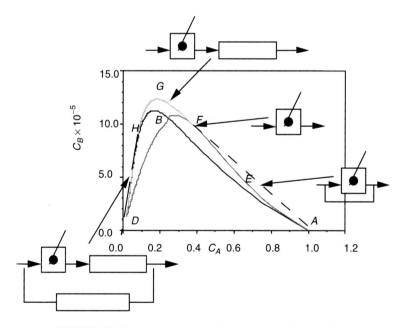

FIGURE 13.7d   Complete attainable region for Example 13.3.

Sec. 13.3 Geometric Concepts for Attainable Regions

product. Note that points F and H define important parts of this attainable region. Point F occurs at the point where the reaction vector, the tangent vector to the CSTR trajectory with feed A and the line AF are all collinear. Point H occurs where the reaction vector on the PFR trajectory with feed F is collinear with the line from the equilibrium point, D. Once we have determined the attainable region we can now solve any optimization problem where the objective function is a function of c_A and c_B only. Thus, for example, if we wanted to maximize the concentration c_B, we could read the answer off from Figure 13.7d at point G and we also know the optimal reactor structure. It is just a CSTR followed by a PFR, following the trajectory AFG.

13.3.1 A Remark on Recycle Reactors

Note from the construction of the attainable region that recycle reactors were not included in the synthesis procedure. The reason for this can be seen from the example sketched in Figure 13.8. Here we note that the recycle reactor represented by line ABD and the PFR trajectory BCD can be included within the convex hull ACD. If the trajectory is smooth and does not violate any imposed constraints (e.g., mass balance), this convex hull is itself represented by the CSTR given at point C (reaction rate collinear with segment AC) followed by the PFR given by CD. Thus, since the recycle reactor trajectory is itself not a

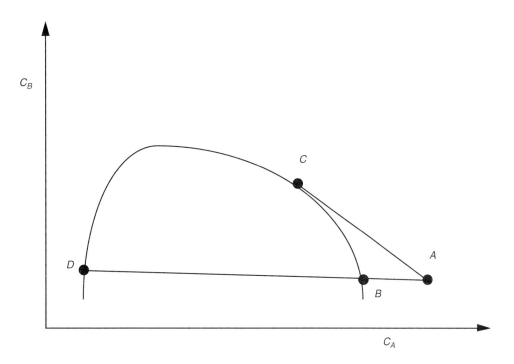

FIGURE 13.8 Convex hull of recycle reactor.

convex region, other reactors (e.g., the CSTR, PFR sequence) form the boundary of the attainable region instead. As a result of this argument we see that recycle reactors do not occur on the AR boundary and need not be considered in the construction of the attainable region.

EXAMPLE 13.4 Autocatalytic Reaction Revisited

In section 13.2 we saw that in the case of autocatalytic reactions, the optimal recycle reactor led to a lower residence time than either a PFR or a CSTR. Now if we consider an attainable region in V/F and c_A, it would appear that the recycle reactor should lie on the lower boundary of the attainable region. However, since recycle reactors cannot form the boundary of an attainable region (whether the upper or lower boundary), it appears at first glance that Example 13.1 is a counterexample to this property. What is the attainable region for this problem?

To address this anomaly, we consider the construction of the attainable region using the rate laws and reactor trajectories derived in Example 13.1. PFR trajectories are given by:

$$\tau = V/F = 1/2\,[ln\,(1/c_A - 1) - ln(1/c_{A0} - 1)] \qquad (13.20)$$

while CSTR trajectories are given by:

$$\tau = V/F = (c_{A0} - c_A)/2(c_A(1 - c_A)) \qquad (13.21)$$

and we consider the construction of an attainable region in the dimensions of c_A and the residence time, τ. This is allowed as the residence time is additive and also follows a linear mixing rule.

Using the algorithm given above we construct the attainable region in Figure 13.9. Starting from the feed point $c_{A0} = 0.99$ (point A) and using Eqs. (13.20) and (13.21), we trace both PFR and CSTR trajectories for τ vs. c_A in Figure 13.9. Note that both the CSTR and PFR trajectories have infinite residence times for a total conversion of A. Thus, the upper part of the region is obtained by filling in the nonconvex portion in the PFR trajectory with a vertical line at point A

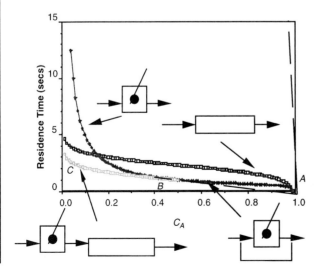

FIGURE 13.9 Attainable region for Example 13.4: Autocatalytic reactions.

Sec. 13.4 Reaction Invariants and Reactor Network Synthesis

and this line goes to infinity. If we concentrate on the lower portion of the attainable region we see that until a concentration of 0.15, the CSTR trajectory lies below the PFR trajectory. It then rises steeply and becomes unbounded as c_A goes to zero. At this point the PFR trajectory is lower. Note that at this crossover point (as well as at earlier points) on the CSTR trajectory, there are rate vectors that point out of the attainable region.

To construct the convex hull of the lower section of the attainable region, we fill in the concave portion in the CSTR trajectory with the line AB. We note that the rate vector points out of this region at point B with $c_A = 0.5$. Therefore, we extend the lower portion of the attainable region with a PFR at point B (curve BC). As all of the conditions are now satisfied, curve ABC therefore forms the lower boundary of the attainable region. Moreover, for the exit concentration of $c_A = 0.05$, specified in Example 13.1, we observe that for the CSTR/PFR serial combination the residence time is $V/F = 2.4522$ sec, which is considerably less than the optimal recycle reactor residence time in Example 13.1 (2.7105 sec). The attainable region for this problem becomes the *entire region above ABC*. Thus, for two-dimensional problems, this region is still formed by PFR/CSTR combinations.

For problems with more than three dimensions, however, geometric constructions become more complex and reactor networks can require more complicated reactors than PFRs and CSTRs. We defer discussion of the properties and methods for higher dimensional problems to Chapter 19. Nevertheless, many higher dimensional problems can be reduced to two dimensions through the application of dimension reduction techniques. In the next section we consider the concept of reaction invariants that allows us to reduce the number of dimensions in these problems.

13.4 REACTION INVARIANTS AND REACTOR NETWORK SYNTHESIS

In the previous section, we constructed attainable regions for two-dimensional problems. Before considering methods for more difficult, higher-dimensional cases, we extend the application of these two-dimensional concepts. Omtveit et al. (1994) enhanced this strategy to deal with higher-dimensional problems, through projections in concentration space that allow a complete two-dimensional representation. These projections were accomplished through the principle of reaction invariants (Fjeld et al., 1974) and have also been extended to include the imposition of additional system specific constraints.

The principle of reaction invariants follows by imposing atomic balances on the reacting species. As these balances always hold, concentrations during reaction can be projected into the reduced space of "independent" components and the complete system can be represented as a lower-dimensional problem. If this representation is then only in the space of two dimensions, we can apply the attainable region constructions mentioned above. To develop this strategy, consider the moles n_i of species i in the reacting system where each component i contains a_{ij} atoms of element j. Since the number of atoms for each element in the reacting system remains constant, we combine the changes in the number of component moles into vector Δn and the coefficients a_{ij} into a matrix A. We then express the atom balances as: $A \, \Delta n = 0$. Partitioning Δn and A into:

$$A = [\,A_d \mid A_f\,] \tag{13.22}$$

$$\Delta n^T = [\Delta n_d^T \mid \Delta n_f^T]$$

with components that are dependent and independent, and ensuring that A_d is square and nonsingular, we substitute this partition into the atom balances and with minor rearrangement we obtain:

$$\Delta n_d = -A_d^{-1} A_f \Delta n_f \tag{13.23}$$

Now for cases where the dimension of n_f is no more than two (this is the number of components minus the number of elements in these components), we can apply the attainable region algorithm given in the previous section. To illustrate these concepts we briefly consider a steam reforming example based on the study of Omtveit et al. (1994).

EXAMPLE 13.5 Attainable Region for Steam Reforming

Steam reforming reactions can be written as:

$$CH_4 + 2\,H_2O \leftrightarrow CO_2 + 4\,H_2$$

$$CH_4 + H_2O \leftrightarrow CO + 3\,H_2$$

$$CO + H_2O \leftrightarrow CO_2 + H_2$$

This system has five components and three elements, so it can be reduced to a two-dimensional system. The atom balances for C, H, and O can be written as:

C balance: $\quad \Delta n(CH_4) + \Delta n(CO_2) + \Delta n(CO) = 0$

H balance: $\quad 4\,\Delta n(CH_4) + 2\,\Delta n(H_2O) + 2\,\Delta n(H_2) = 0$

O balance: $\quad \Delta n(CO) + \Delta n(H_2O) + 2\,\Delta n(CO_2) = 0$

Defining the vector of mole changes as:

$$\Delta n^T = [\,\Delta n(H_2O),\, \Delta n(H_2),\, \Delta n(CO_2),\, \Delta n(CH_4),\, \Delta n(CO)\,]$$

and assembling the coefficients into matrix A leaves us with:

$$A = \begin{bmatrix} 0 & 0 & 1 & 1 & 1 \\ 2 & 2 & 0 & 4 & 0 \\ 1 & 0 & 2 & 0 & 1 \end{bmatrix}$$

Now, selecting CH_4 and CO as independent components allows us to partition the matrix and establish the following dependence according to Eqs. (13.22) and (13.23).

$$A = [\,A_d \mid A_f\,] = \begin{bmatrix} 0 & 0 & 1 & 1 & 1 \\ 1 & 0 & 0 & 4 & 0 \\ 1 & 0 & 2 & 0 & 1 \end{bmatrix} \quad \text{and} \quad -A_d^{-1} A_f = \begin{bmatrix} 2 & 1 \\ -4 & -1 \\ -1 & -1 \end{bmatrix}$$

and the dependent components can be written as:

$$\Delta n(H_2O) = 2\,\Delta n(CH_4) + \Delta n(CO)$$

$$\Delta n(H_2) = -4\,\Delta n(CH_4) - \Delta n(CO)$$

$$\Delta n(CO_2) = -\,\Delta n(CH_4) - \Delta n(CO)$$

We now consider the construction of an attainable region using results of Omtveit et al. (1994) and reaction kinetics from Xu and Froment (1989). Now it can be shown that the total number of moles in the system is given by:

$$n_T = [n(H_2O) + n(H_2) + n(CO_2) + n(CH_4) + n(CO)]_0 - 2\,\Delta n(CH_4)$$

and the rate expressions can therefore be rewritten by substituting $P\,n(i)/n_T$ for p_i and all of these partial pressures are functions of the independent components CO and CH_4. The attainable region for this system is shown in Figure 13.10 below.

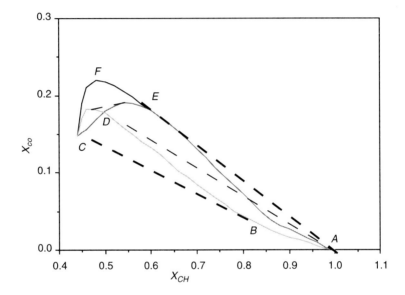

FIGURE 13.10 Attainable region for Example 13.5.

These are plotted in the space of normalized concentrations, $X_{CO} = c_{CO}/c_{CH4,0}$ and $X_{CH4} = c_{CH4}/c_{CH4,0}$. To construct the attainable region, we repeat the steps described in section 13.3. This construction is very similar to the one described for Example 13.3.

Step 1. We trace a PFR trajectory (*ABDC*) from the feed point (*A*) up to the equilibrium point (*C*).

Steps 2, 3. Filling in the concavities with line segments above and below the PFR trajectory leads to a convex region. Along line *AD* we have rate vectors pointing out of the attainable region but not along line *BC*. Thus, the curve *AB* and line segment *BC* forms the lower boundary of the attainable region.

> *Step 4.* We now extend this region by plotting a CSTR trajectory (AEDC) from the feed point (where the attainable region is extended the most). We fill in the concavity for the CSTR trajectory with line AE.
>
> *Step 5.* Note that the region can be extended from this point with a PFR trajectory starting at point E. This trajectory has a maximum at point F.
>
> Checking the attainable region properties, we see that region *AEFCBA* is convex, has no rate vectors pointing out of the region and no rate vectors in the complement of this region that can be reversed into region. Thus, *AEFCBA* forms the attainable region for the steam reforming problem. From here we see that the maximum yield of CO is at point *F*, the maximum conversion of CH_4 is at point C and the maximum selectivity of CO to CH_4 is at point *E*.

The application of the reaction invariance principle to the two-dimensional construction of attainable regions is possible when the (number of reacting species) − (number of elements in species) ≤ 2. Otherwise, higher-dimensional constructions are required. Nevertheless, this example illustrates the application of component reduction techniques for the simple construction of attainable regions.

13.5 CHAPTER SUMMARY AND GUIDE TO FURTHER READING

This chapter summarizes concepts for the synthesis of reactor networks. As noted by Nishida et al. (1980), reactor network synthesis has seen far less development than strategies for separation and heat exchanger systems. A key reason for this is the highly nonlinear behavior of reacting systems, which leads to difficulties for both heuristic and optimization-based approaches. To deal with these issues, we introduce a recently developed concept for this problem: construction and analysis of attainable regions for the synthesis of reactor networks. With this approach we have a general tool to construct a region in concentration space (with extensions to residence time and temperature) that is closed for all mixing and reaction operations. This approach also serves to extend well known heuristics for reactor design to more complex reaction systems.

In section 13.2, we briefly reviewed some reactor selection criteria for simple reaction systems. These are developed and discussed in some detail in many standard textbooks in kinetics and reactor design (e.g., Fogler, 1992; Froment and Bischoff, 1979; Kramers and Westerterp, 1963; Levenspiel, 1972). These criteria can be generalized to heuristics for more complex systems; when applied systematically, they can yield reasonably good reactor networks. In fact, the READPERT expert system that embodies these heuristics was recently developed and demonstrated successfully on several real-world design problems (Schembecker et al., 1994). A summary of these heuristics is also given in Chitra and Govind (1985) and Hartmann and Kaplick (1990). On the other hand, when conflicting terms arise in the design objective or the reactions have multiple characteristics, trade-offs in the design problem need to be evaluated directly. Therefore, for the design problem, a quantitative search strategy is necessary and candidate solutions for the reactor network need to be selected and optimized.

Another way to approach this problem is to turn the problem around and consider the conditions that define a complete set of reactor networks for a given reacting system.

Sec. 13.5 Chapter Summary and Guide to Further Reading

The *attainable region* approach is a systematic strategy for postulating this complete family of solutions. Moreover, the reactors that make up the boundary of the attainable region are sufficient to decide the reactor network, as any interior point in the attainable region can be realized by mixing the boundary points. In section 13.3, we develop and applied the principles of the attainable region to reacting systems that could be represented in two dimensions. These concepts were developed by Glasser, Hildebrandt, and coworkers (1987, 1990, 1992). From these concepts, we know that the attainable region:

- Is convex.
- Has no rate vectors pointing out of the region.
- Has no rate vectors in the AR complement that can be reversed into the region.

In two dimensions the reactor system only needs to consist of PFRs and CSTRs and it was also shown that recycle reactors are not needed to form the boundary of the attainable region. This boundary (and consequently the family of network solutions) was then constructed through a systematic algorithm where PFR and CSTR trajectories were constructed and nonconvex portions were filled in with line segments. The above conditions were also checked at each iteration to decide on termination of the algorithm. This approach was illustrated through three small example problems.

In section 13.4, we consider a further extension by projecting the reacting species into a smaller subspace. The projection of species was performed by exploiting the concept of reaction invariants introduced by Fjeld et al. (1974). If this projected subspace of concentration has only two dimensions, then the approach of section 13.3 could be applied readily. By introducing constraints that enforce atom balances, dependent component behavior can be described entirely through the reaction paths of selected independent components. Omtveit et al. (1994) applied this projection to two independent components in order to construct the attainable region, consisting only of CSTRs and PFRs. This approach was illustrated on a steam reforming problem.

When the reacting system cannot be represented entirely in two dimensions, construction of the attainable region becomes more difficult. First, in higher dimensions, more complicated reactor types can arise, such as the differential sidestream reactor (DSR). Also, while the attainable region has been applied to several interesting three-dimensional problems (Hildebrandt et al., 1990), it is very difficult to extend these geometric constructions beyond three dimensions. Instead, optimization problems can be formulated that apply the steps of the geometric algorithm and allow the strategy to "see" and construct the attainable region in higher dimensions. These formulations will be developed in detail in Chapter 19.

Finally, this chapter has not applied attainable region concepts to reactor networks that are embedded within flowsheets. A desirable synthesis strategy should also seek to exploit flowsheet interactions among the reaction, energy and separation subsystems. As an illustration, consider the flowsheet in Figure 13.11 with van de Vusse kinetics:

$$A \rightarrow B \rightarrow C \qquad 2A \rightarrow D$$

and B as the desired product.

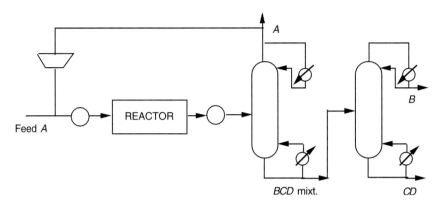

FIGURE 13.11 Reactor-based flowsheet example.

To synthesize the reactor network, we need to specify the feed to the reactor system but the recycle flowrate, composition and temperature still need to be determined from the flowsheet. Moreover, the effluent of the reactor network influences the character of the downstream (and upstream) separation systems. And the energy supplies and demands for reaction and separation systems need to be handled through the synthesis of an efficient heat exchanger network. These interactions are hard to integrate through the heuristic or geometric approaches in this chapter unless severe restrictions are imposed on the synthesis problem (e.g., only pure A is recycled). On the other hand, the concept of attainable regions can be embodied into optimization formulations that integrate models for these separate subsystems and address the trade-offs that result from the integration. This topic will also be developed further in Chapter 19.

REFERENCES

Chitra, S. P., & Govind, R. (1985). Synthesis of optimal serial reactor structure for homogenous reactions, part II: Nonisothermal reactors. *AIChE J.*, **31**(2), 185.

Douglas, J. M. (1988). *Conceptual Design of Chemical Processes*. New York: McGraw-Hill.

Fjeld, M., Asbjornsen, O. A., & Astrom, K. J. (1974). Reaction invariants and the importance of in the analysis of eigenvectors, stability and controllability of CSTRs. *Chem. Eng. Science*, **30**, 1917.

Fogler, H. S. (1992). *Elements of Chemical Reaction Engineering*. Englewood Cliffs, NJ: Prentice-Hall.

Froment, G. F., & Bischoff, K. B. (1979). *Chemical Reactor Analysis and Design*. New York: Wiley.

Glasser, D., Crowe, C., and Hildebrandt, D. (1987). A geometric approach to steady flow reactors: The attainable region and optimization in concentration space. *I & EC Research*, **26**(9), 1803.

Glasser, B., Hildebrandt, D., & Glasser, D. (1992). Optimal mixing for exothermic reversible reactions. *I & EC Research,* **31**(6), 1541.

Hartmann, K., & Kaplick, K. (1990). *Analysis and Synthesis of Chemical Process Systems.* Amsterdam: Elsevier.

Hildebrandt, D. (1989). PhD Thesis, Chemical Engineering, University of Witwatersrand, Johannesburg, South Africa.

Hildebrandt, D., & Biegler, L. T. (1995). Synthesis of reactor networks. In L. T. Biegler & M. F. Doherty (Eds.), *Foundations of Computer Aided Process Design '94* (p. 52). AIChE Symposium Series, **91**.

Hildebrandt, D., Glasser, D., & Crowe, C. (1990). The geometry of the attainable region generated by reaction and mixing: With and without constraints. *I & EC Research,* **29**(1), 49.

Horn, F. (1964). Attainable regions in chemical reaction technique. In *The Third European Symposium on Chemical Reaction Engg.* London: Pergamon.

Kramers, H., & Westerterp, K. R. (1963). *Elements of Chemical Reactor Design and Operation.* New York: Academic Press.

Levenspiel, O. (1972). *Chemical Reaction Engineering,* 2nd ed. New York: Wiley.

Nishida, N., Stephanopoulos, G., & Westerberg, A. W. (1981). Review of process synthesis. *AIChE J.,* **27,** 321.

Omtveit, T., Tanskanen, J., & Lien, K. (1994). Graphical targeting procedures for reactor systems. *Comp. and Chem. Engr.,* **18,** S113.

Schembecker, G., Droge, T., Westhaus, U., & Simmrock, K. (1995). A heuristic-numeric consulting system for the choice of chemical reactors. In L. T. Biegler & M. F. Doherty (Eds.), *Foundations of Computer Aided Process Design '94,* AIChE Symposium Series, 91.

Trambouze, P.J., & Piret, E. L. (1959). Continuous stirred tank reactors: Designs for maximum conversions of raw material to desired product. *AIChE J.,* **5,** 384.

van de Vusse, J. G. (1964). Plug flow vs. tank reactor. *Chem. Eng. Sci.,* **19,** 994.

Xu, J., & Froment, G. (1989). Methane steam reforming: Diffusional limitations and reactor simulation. *AIChE J.,* **35**(1), 88.

EXERCISES

1. Consider the autocatalytic reaction in Example 13.1 but with the rate law: $r_A = -10\, c_A^2 c_B$. Which reactor type is optimal if the feed is pure A with a concentration of 5 mol/l?

2. Resolve problem 1 with the same rate law but with initial feed concentration of $c_{A0} = 0.5$ mol/l.

3. Derive the representation of the residence time for the recycle reactor in Figure 13.3.

4. Consider the isothermal parallel reaction $A \to B$, $A \to C$, where $r_B = 4c_A$ and $r_C = 2c_A^2$.
   a. Using the reactor selection criteria in section 13.2, choose the best reactor for this system to maximize the yield of component B.
   b. Construct the attainable region for this system and find the reactor network that maximizes the yield of B.

5. The isothermal van de Vusse (1964) reaction involves four species for which the objective is the maximization of the yield of intermediate species B, given a feed of pure A. The reaction network is given by

$$A \xrightarrow{k_1} B \xrightarrow{k_2} C$$
$$\downarrow k_3$$
$$D$$

Here the reaction from A to D is second order. The feed concentration is $c_{A0} = 0.58$ mol/l and the reaction rates are $k_1 = 10 \text{ s}^{-1}$, $k_2 = 1 \text{ s}^{-1}$ and $k_3 = 1 \text{ l/(mol s)}$. The reaction rate vector for components A,B,C,D respectively is given in dimensionless form by:

$$r(X) = [-10X_A - 0.29X_A^2, 10X_A - X_B, X_B, 0.29X_A^2],$$

where $X_A = c_A/c_{A0}$, $X_B = c_B/c_{A0}$, and c_A, c_B are the molar concentrations of A and B respectively.
   a. Synthesize the optimal reactor network using the attainable region approach if the objective function is yield of component B.
   b. Synthesize the optimal reactor network using the attainable region approach if the objective function is the selectivity of B to A.

6. The Trambouze reaction (Trambouze & Piret, 1959) involves four components and has the following reaction scheme:

$$A \xrightarrow{k_1} B \quad A \xrightarrow{k_2} C \quad A \xrightarrow{k_3} D$$

where the reactions are zero order, first order and second order, respectively, with $k_1 = 0.025 \text{ mol/(l min)}$, $k_2 = 0.2 \text{ min}^{-1}$, $k_3 = 0.4 \text{ l/(mol min)}$ and an initial concentration of $c_A = 1 \text{ mol/l}$. Using the attainable region algorithm, find the reactor network that maximizes the selectivity of C to A.

7. Resolve the Trambouze example where the first two reactions are first order and last is second order, with $k_1 = 0.02 \text{ min}^{-1}$, $k_2 = 0.2 \text{ min}^{-1}$, $k_3 = 2.0 \text{ l/(mol min)}$ and an initial concentration of $c_A = 1 \text{ mol/l}$. Using the attainable region algorithm, find the reactor network that maximizes the selectivity of C to A.

8. Consider the steam reforming system in Example 13.5. Choose methane and carbon dioxide as independent components. How do the remaining components depend on methane and carbon dioxide?

SEPARATING AZEOTROPIC MIXTURES

14

In Chapter 11 we examined the synthesis of distillation-based processes to separate mixtures that behave fairly ideally. In this chapter we shall look at the synthesis of processes to separate mixtures that display highly nonideal phase equilibrium behavior. We shall look in particular at the separation of mixtures that display azeotropic behavior and possibly heterogeneous behavior. An azeotrope occurs for a boiling mixture of two or more species when the vapor and liquid phases in equilibrium have the same composition. As a consequence, we cannot separate such a mixture by boiling or condensing it. Heterogeneous behavior means a liquid mixture partitions into two or more liquid phases at equilibrium.

We all know that when we try to separate water from ethyl alcohol using distillation, the mixture forms an azeotrope. At a pressure of one atmosphere, this azeotrope occurs at 85.4 mole % ethanol. We would find that this mixture boils 78.1°C, which is lower than the normal boiling points for both ethanol (78.4°C) and water (100°C). We say that ethanol and water form a minimum boiling azeotrope. We find that a mixture of acetone and chloroform at a composition of 64.1 mole % acetone forms a maximum boiling azeotrope at one atmosphere. Acetone boils at 56.5°C, chloroform 61.2°C while the azeotrope boils at 64.43°C.

Mixtures of ethyl alcohol, water, and toluene display complex azeotropic behavior. Each of the three possible binary pairs (ethyl alcohol/water, ethyl alcohol/toluene, and water/toluene) form a binary azeotrope. Also water and toluene form two liquid phases.

At one atmosphere n-butanol and water will break into two liquid phases for n-butanol compositions less than about 40 mole % at temperatures below 94°C. We can separate such mixtures by allowing them to settle into two liquid layers and decanting them. We also find that at 94°C n-butanol and water form an azeotrope at about 24% n-butanol. The behavior of this mixture is obviously very complex.

14.1 SEPARATING A MIXTURE OF *n*-BUTANOL AND WATER

In this first example let us synthesize a process to separate a 15 mole % mixture of *n*-butanol and water into its pure components.

The first activity we must undertake when devising separation processes for this mixture is to determine if mixtures of these species display azeotropic and/or heterogeneous behavior. If they do, the separation systems we must consider will be very different from the systems we designed in Chapter 11. How might we check for such behavior?

14.1.1 Detecting Azeotropic Behavior

First of all, we can attempt to find if experimental data exists for the phase behavior of *n*-butanol and water mixtures. In this case, we would be successful in finding such data. Another way we might proceed is to use an available physical property estimation package to see the type of behavior it predicts for this mixture. Many of these packages contain estimation techniques for phase behavior that are very good for several types of mixtures, and our experts would tell us that these packages will perform very well for mixtures of *n*-butanol and water.

Figure 14.1 is a plot of the phase behavior of *n*-butanol and water versus temperature at one atmosphere. We base it on data appearing in an older edition of Perry's *Chemi-*

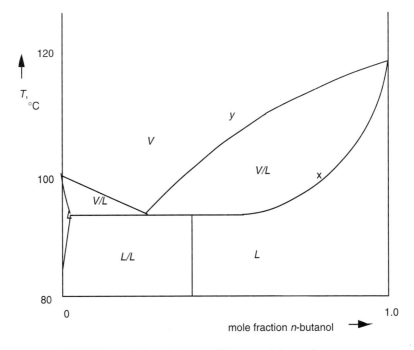

FIGURE 14.1 Phase behavior of the water/*n*-butanol system.

Sec. 14.1 Separating a Mixture of *n*-Butanol and Water

TABLE 14.1 Infinite Dilution K-values for *n*-Butanol/Water System. (For example, the K-value for a drop of water in *n*-butanol is 21.0.)

Trace\Plentiful	Water	*n*-Butanol	Temperature, K
water	1.0	21.0	373.3
n-butanol	2.4	1.0	390.7

cal Engineering Handbook (3rd edition, 1950). We see immediately the complex behavior we described above.

We can also discover this behavior if we can accurately compute the vapor/liquid behavior for *n*-butanol and water at the two extreme conditions of a drop of water in *n*-butanol and a drop of *n*-butanol in water, that is, at infinite dilution of each species in the other. We perform two flash calculations using the Unifac method with the trace species having a molar amount 10^{-4} times that of the plentiful species. Table 14.1 gives the results we obtain at one atmosphere. We see that the infinite dilution K-value for a drop of water in *n*-butanol is 21.0 and for a drop of *n*-butanol is water is 2.4. Both are greater than unity, the importance of which we shall now discuss.

To interpret these results, consider Figure 14.2, which illustrates a *T* versus composition diagram at a constant pressure for two well-behaved species. The upper line gives the vapor composition and the lower line gives the liquid composition when the mixture partitions into two phases. They both start and end at the boiling point temperatures for the two species. Drawing a horizontal line at temperature T_1, we find the vapor composition $y_B(T_1)$ on the upper line, which is in equilibrium with the liquid composition $x_B(T_1)$ on the lower line at that temperature.

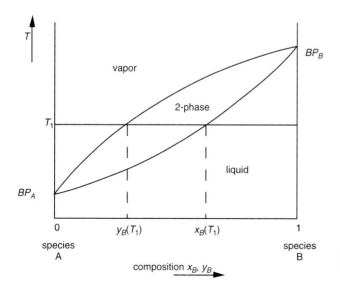

FIGURE 14.2 Typical binary vapor/liquid equilibrium boundaries.

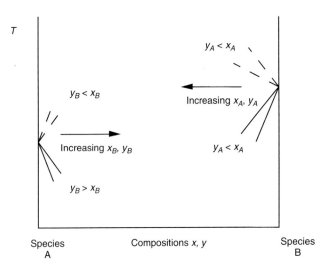

FIGURE 14.3 Behavior of compositions at infinite dilution.

Figure 14.3 illustrates our approach to discover if the two species form an azeotrope. Let us start at the left side, which is pure species A. The two equilibrium phase boundary lines start at the temperature equal to the boiling point of pure species A and will either both point upward or both point downward. As is evident from the previous figure, the upper line gives vapor compositions while the lower line gives liquid compositions versus temperature. Thus, either the two dashed lines or the two nondashed lines could indicate the behavior. When the two lines point upward on the far left (the dashed lines), we know that y_B is less than x_B at that point. When they both point downward (the two solid lines), we know that y_B is greater than x_B at that point. A similar argument tells us that the dashed lines on the right occur when y_A (vapor composition of the trace species) has a composition less than its corresponding liquid composition while the two solid lines indicate the reverse.

If both lines for both species point upward, there must be a maximum boiling azeotrope at some intermediate composition. If both lines for both species point downward, there must be a minimum boiling azeotrope. If they point up on the left (the lower boiling species) and down on the right, nonazeotropic behavior is indicated (though not assured as there could be two azeotropes—one maximum and one minimum—occurring between). If we deem having two azeotropes to be a rare event, then we will consider this situation to be one without azeotropes. The one remaining option, namely the right pair points upward while the left points downward, would require and even number and at least two azeotropes between. Again this situation rarely occurs.

The infinite dilution K-values in Table 14.1 tell us what we need to know. K-values are ratios of y to x. For n-butanol and water, both indicate y is greater than x for the trace species; the phase equilibrium lines must both point downward in Figure 14.3. There must be a minimum boiling azeotrope between.

14.1.2 Detecting Liquid/Liquid Behavior

Detecting the likelihood of liquid/liquid behavior is more complex but is also based on carrying out these same two flash calculations. This time, however, we need to extract infinite dilution activity coefficients from the results. Thus, we need to do no added work over that we have done already. If the flash program does not report activity coefficients, we can estimate them by noting

$$\gamma_{i \text{ in } j}^\infty = \frac{\phi_i P}{f_i^0} K_{i \text{ in } j}^\infty \approx \frac{P_j^{\text{sat}}(T)}{P_i^{\text{sat}}(T)} K_{i \text{ in } j}^\infty \tag{14.1}$$

If the mixture is at one atmosphere, T is the normal boiling point for the plentiful species j, and its saturation pressure $P_j^{\text{sat}}(T)$ will be one atmosphere. Then the activity coefficient for i in j will be the infinite dilution K-value for species i in j divided by the vapor pressure of species i at the normal boiling point of species j. Table 14.2 lists our estimates for the infinite dilution activity coefficients for water and n-butanol at one atmosphere. We see that a drop of water in lots of n-butanol has an activity coefficient of 40.5.

As a word of caution, physical property packages compute liquid/liquid activity coefficients using different physical property parameters (e.g., different Unifac parameters) than they would use to calculate vapor/liquid phase behavior. We are "cheating" somewhat when we use the same parameters for vapor/liquid behavior to assess liquid/liquid behavior. However, we are attempting here only to assess if we need to worry about liquid/liquid behavior so we will "cheat.")

To proceed, we need to understand why a mixture forms two liquid phases at equilibrium, and we shall use Figure 14.4 to aid us in this explanation. At constant temperature and pressure, an equilibrium mixture will minimize its total molar Gibbs free energy. We frequently compute the molar Gibbs free energy for a mixture as the sum of three contributions. The first contribution simply mixes the pure component molar Gibbs free energies:

$$\frac{G_{\text{ave}}}{RT} = x_A \frac{G_A}{RT} + x_B \frac{G_B}{RT}$$

where G_A and G_B are the molar Gibbs free energies for pure liquid species A and B respectively, x_A and x_B are their corresponding mole fractions in the mixture, R is the uni-

TABLE 14.2 $\gamma_{i \text{ in } j}^\infty$ for Water/ n-Butanol Mixtures

$i \setminus j$	Water	n-Butanol
Water	1.0	40.5
n-butanol	1.3	1.0

versal gas constant, and T the absolute temperature. Note we can specify only one of the mole fractions independently as they add to unity.

The second contribution reflects the effect of the entropy of ideal mixing on the Gibbs free energy:

$$\frac{\Delta G_{mix}}{RT} = x_A \ln(x_A) + x_B \ln(x_B)$$

while a third term corresponds to its nonideal behavior during mixing, $\Delta G_{excess}/RT$. The mixing and excess contributions are zero for pure components, becoming nonzero only as we mix species A and B. The excess term is estimated by any one of a number of different models, such as a Margules, an NRTL, or a Unifac model (the Wilson equation cannot predict liquid/liquid behavior and should not be used here). Figure 14.4 shows a plot of these terms. As we just stated above, the total molar Gibbs free energy for the mixture is the sum of these three contributions.

For convenience we have placed G_A/RT at the origin for pure species A in this plot. The average term is along the straight line connecting G_A/RT to G_B/RT. The ideal mixing term is exactly as shown no matter the species as it depends only on the mole fractions of the two species. It is the excess term that can have a variety of different shapes starting with a quadratic shape for the simplest Margules models. In Figure 14.4 we show it making a fairly large positive contribution. It is the excess Gibbs free energy that relates directly to the activity coefficients for the mixture. If it were zero everywhere, activity coefficients would be unity everywhere.

We have labeled the total molar Gibbs free energy in this figure. Because of the shape of the excess contribution, we see the total curve starts out downward from G_A/RT

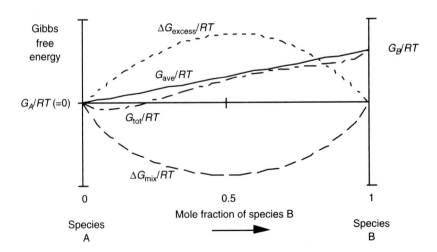

FIGURE 14.4 Terms contributing to total molar Gibbs free energy for a mixture.

Sec. 14.1 Separating a Mixture of n-Butanol and Water

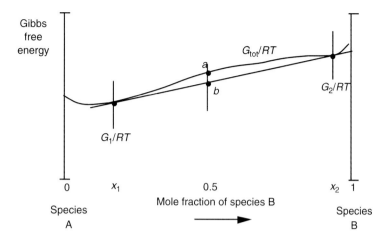

FIGURE 14.5 Case where liquid mixture will break into two liquid phases.

but is curving upward. It passes through an inflection point, after which it curves downward again, through another inflection point, and then curves back upward, finally reaching G_B/RT. In the middle portion of this curve the total molar Gibbs free energy for the mixture is "concave downward."

We emphasize this shape in Figure 14.5 by drawing a straight line that supports the curve from below in more than one place, here at two points labeled G_1/RT and G_2/RT. It is a support line because the entire total Gibbs free energy curve lies above it. The concave downward shape is required for us to draw a support line that touches in more than one point. For this diagram, let us consider having one mole of a 50/50 mixture of A and B. The computation for its total molar Gibbs free energy as given above would lie at point a, which is on the curve for G_{tot}/RT at $x_B = 0.5$.

It turns out we can lower the molar Gibbs free energy for this mixture if we partition it into two liquid phases, one corresponding to the composition x_1 at G_1/RT and one to the composition x_2 at G_2/RT. The mixture splits according to the lever rule, which says

$$\frac{m_1}{m_2} = \frac{x_2 - 0.5}{0.5 - x_1}$$

where m_1 and m_2 sum to one mole (the amount of the original mixture) and are the molar amounts in each of the two phases. The total molar Gibbs free energy for these two phases is then $m_1 G_1/RT + m_2 G_2/RT$, which we would find to be the value at point b, the point on the straight line connecting the two support support points that lies directly below point a.

We can generalize these results for any number of species. We plot the total molar Gibbs free energy divided by RT versus composition. Think of it as a surface above the composition space. Place a "support" plane below this surface at the composition of the mixture. If the plane does not touch the surface at this composition, then the system can lower its total molar Gibbs free energy to the value on the plane by partitioning into those liquid phases whose compositions correspond to the points where this support surface just touches the total molar Gibbs free energy surface.

For our *n*-butanol and water mixture, we need to ascertain if the free energy surface can have the shape required for the system to break into two liquid phases. The infinite dilution activity coefficients can provide us with a clue. We (Westerberg and Wahnschafft, 1996) have carried out computations using a Margules equation to predict activity coefficients and found that it predicts the onset of liquid/liquid behavior if either of the following is (approximately) true:

- If either infinite dilution activity coefficient is greater than 9
- If the larger of the two activity coefficients is larger than 9 times the cube root of the smaller

To illustrate the use of the second condition, if the larger activity coefficient $\gamma^\infty_{A \text{ in } B}$ is 1.8, then we need to worry about liquid/liquid behavior if the smaller activity coefficient $\gamma^\infty_{B \text{ in } A}$ is less than about $(1/9 \times 1.8)^3 = 0.008$.

We can propose to use these tests to alert us to the potential for liquid/liquid behavior. For example, we might consider the need to check more thoroughly for liquid/liquid behavior if we replace the 9 by a 6 and either of these tests passes.

From Table 14.2 we see that the infinite dilution activity coefficient for water in *n*-butanol is 40.5. That is well above 9 and the first of the above tests strongly suggests this system displays liquid/liquid behavior. That being the case, we need to spend time to find or develop the phase diagram for this system. Fortunately, we already have its phase behavior, as shown in Figure 14.1, given as a plot of temperature versus vapor/liquid composition. We now look at how we can synthesize a separation process to split our feed, which is 15 mole% *n*-butanol into relatively pure water and *n*-butanol.

14.1.3 Synthesizing a Separation Process

Effective design procedures often depend on our devising a good way to represent the problem we are trying to solve. As we shall see, such will certainly be the case here. We need a representation that highlights the important features related to distillation and decanting of liquid phases on which we make our design decisions. The representation should hide or suppress the other features. In the top part of Figure 14.6 we map the essential features from the phase diagram on the composition axis. We show the compositions for the liquid/liquid boundaries and for the azeotrope. We also show the feed composition.

With this feed we can either cool the mixture to the two-liquid phase region and allow the phases to separate, or we can distill off the water. We develop the first alternative in the middle portion of Figure 14.6. We show the feed entering a horizontal line

Sec. 14.1   Separating a Mixture of *n*-Butanol and Water

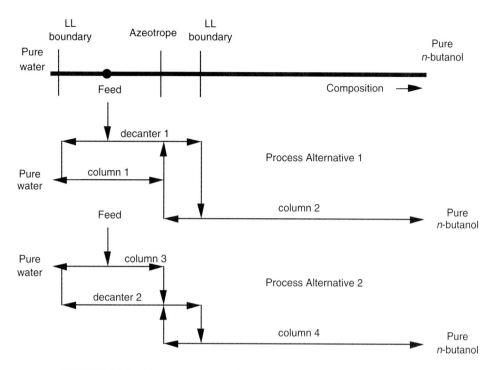

FIGURE 14.6 Abstract representation for synthesizing separation processes for water/*n*-butanol mixtures.

labeled decanter 1. The ends of this line indicate approximately the compositions we can reach with a decanter. The left side of the decanter is the water-rich phase, and it still has a few percent of *n*-butanol in it. We feed it to a distillation column, which we label column 1. This column can give us relatively pure water as a bottoms product. The distillate can be arbitrarily close to but always below the azeotrope composition. We can also distill the phase rich in *n*-butanol from the decanter (right side above), producing *n*-butanol (product) as the bottom stream and azeotrope as the top. We now have an azeotrope that we must separate. We can cool it and feed it to a second decanter. However, we could also feed it back to the first decanter as it can accept any feed between its two products.

If we decide to distill the original feed (column 3), then we produce water product and azeotrope as shown in the lower part of Figure 14.6. We can cool the azeotrope and decant it in decanter 2. We recycle the water rich phase from the decanter back to column 3 and send the *n*-butanol rich phase to column 4. Column 4 produces *n*-butanol product and azeotrope. Again we recycle the azeotrope back to the decanter.

These two designs are minor variations of each other, and we show both in Figure 14.7.

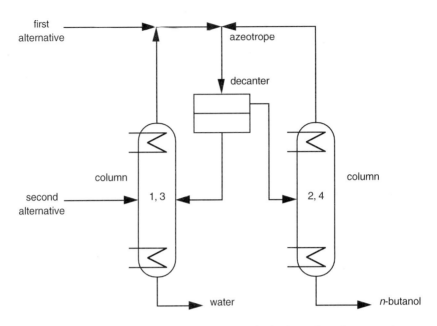

FIGURE 14.7 Flowsheet corresponding to both alternatives for separating *n*-butanol and water mixture.

14.2 SEPARATING A MIXTURE OF ACETONE, CHLOROFORM, AND BENZENE

We start by computing infinite dilution K-values and activity coefficients to assess possible nonideal behavior. Table 14.3 gives the resulting K-values and Table 14.4 the activity coefficients. We need to carry out three flash calculations, one for each species in the mixture. The first has one mole of acetone and 0.0001 moles each of chloroform and benzene in the feed, the second one mole of chloroform and 0.0001 moles each of acetone and benzene, and so on. The ratio of y_i to x_i for each species i in the vapor and liquid product streams provides us with the K-values. Either extracting them directly from the simulation output if they are provided or using Eq. (14.1), we estimate the infinite dilution activity coefficients.

TABLE 14.3 Infinite Dilution K-values for Trace of Species *j* in *i*

Trace of *j* in *i*	*j* = Acetone	Chloroform	Benzene
i = Acetone	K: 1.00	0.45 max	0.77 normal
Chloroform	0.60	1.00	0.43 normal
Benzene	3.08	1.54	1.00

Sec. 14.2   Separating a Mixture of Acetone, Chloroform, and Benzene   465

TABLE 14.4   Infinite Dilution Activity Coefficients for Trace Species j in i

Trace of j in i	j = Acetone	Chloroform	Benzene
i = Acetone	γ: 1.00	0.52 (7.25)	1.73 (10.8)
Chloroform	0.51 (7.2)	1.00	0.81 (8.4)
Benzene	1.45 (10.2)	0.85 (8.5)	1.00

We find that the infinite dilution K-values for acetone in chloroform (0.60) and chloroform in acetone (0.45) are both less than 1.0, indicating the existence of a maximum boiling azeotrope for this binary pair. The other two pairs, acetone/benzene and chloroform/benzene, have infinite dilution K-values on both sides of one, indicating normal behavior—that is, no azeotropes. The activity coefficients range in value from 0.51 to 1.73. All are substantially less than 9 in value so none by itself suggests liquid/liquid behavior (using the first test given earlier). The second earlier test says the larger activity coefficient of the pair must exceed 9 times the cube root of the smaller for us to worry if there is liquid/liquid behavior. We enclose in parentheses 9 times the cube root of each activity coefficient next to the value of the activity coefficient. For the acetone/benzene pair, the activity coefficients are 1.73 and 1.45. Nine time the cube root of the smaller is 10.2; the larger is nowhere this size, so again the numbers suggest no liquid/liquid behavior for this pair. None of the other pairs suggest liquid/liquid behavior.

We do have azeotropic behavior that tells us we should not design a distillation-based separation process using the approaches of Chapter 11 where we assumed ideal behavior.

14.2.1   Representing Phase Behavior for Three Species

We need a means to represent the phase behavior for three species that will aid us in designing separation processes. Humans have a difficult time seeing things in more than two dimensions—that is, as a diagram on a sheet of paper. Can we create a way to look at the vapor/liquid phase behavior for our three species on a two-dimensional diagram that aids us to then design a separation process? Fortunately we can. Figure 14.8 is such a representation. On a triangular composition diagram, we superimpose "distillation" curves. Each point on this diagram represents the three mole fractions of a mixture of these species. A point exactly in the middle is an equimolar mixture with mole fraction of 0.3333 for each species. Points at the corners are pure species. This plot is a two-dimensional plot because there are only two independent mole fractions; the third must be such that the mole fractions add to one.

To understand what a distillation curve is, consider the distillation column section in Figure 14.9. It shows the trays at the top of a column operating at total reflux and thus with no top product. Because there is no top product, material balances show that the total flows and the compositions of the opposing liquid/vapor streams between any pair of trays are the same, that is, what goes in comes out. If we assume each tray is an equilib-

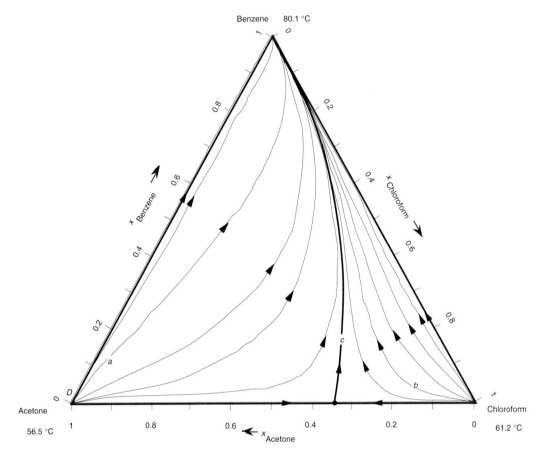

FIGURE 14.8 Distillation curves for acetone, chloroform, benzene mixtures.

rium tray, then the vapor leaving from the top of tray k is in equilibrium with the liquid leaving from the bottom of that same tray. Suppose we know the liquid compositions $x_{i,k}$ for all species i leaving a tray. Then a bubble point calculation will give us the vapor compositions, $y_{i,k}$, for all species i in equilibrium with that liquid composition. The compositions of the liquid and vapor stream pair between two trays must be equal, that is, $x_{i,k-1} = y_{i,k}$ for all species i. A bubble point computation gives us the compositions $y_{i,k-1}$ for all species i. In this manner we can march up the column tray by tray by doing a series of bubble point computations. To march down a column requires we do a series of dew-point calculations; that is, we know the vapor composition and compute the liquid composition in equilibrium with it. A *distillation curve* is defined to be a smooth curve that passes through these compositions for a column.

We can construct the map in Figure 14.8 by picking any arbitrary composition and generating points on the distillation curve emanating both up and down a column from it;

Sec. 14.2 Separating a Mixture of Acetone, Chloroform, and Benzene

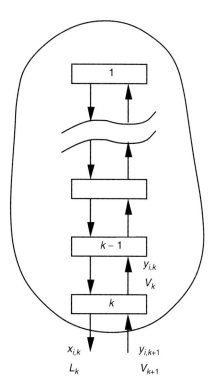

FIGURE 14.9 Top section of a distillation column operating at total reflux.

we then pick another composition near the curve just generated and repeat. Each curve is thus the result of doing a number of bubble/dewpoint calculations starting at some arbitrary composition on it.

Let us now look at the behavior of these curves. Suppose we start at the composition marked a near to the pure acetone corner in Figure 14.8. As we compute successive bubble points to move up the column, we would expect the compositions to move toward the most volatile species in the mixture, here acetone. Indeed, we follow this curve and find it moves downward, asymptotically approaching the pure acetone corner in the lower left. If we were to compute a series of dewpoints to move down the column, we would expect the compositions to move toward the least volatile component, here benzene, and, following the curve upward, we find exactly that behavior.

Remembering that acetone and chloroform form a maximum boiling azeotrope, we pick the composition b nearer to the chloroform corner and try again. We find that the trajectory moves up the column by moving toward chloroform rather than acetone as before. It still moves down the column by moving to benzene. Plotting a number of such trajectories, we find some reach acetone while others reach chloroform for the top of the column. They all end at benzene in the bottom of the column. Indeed, we discover there is a particular distillation curve that separates the composition diagram into those distillation curves reaching acetone from those reaching chloroform. Marked c, we find it reaches the lower edge at exactly the maximum boiling azeotrope that we knew had to exist between acetone and chloroform. We call this particular distillation curve a distillation boundary (for what we hope are obvious reasons).

Suppose we superimpose bubble point temperatures on these distillation curves. It was once thought that a distillation boundary corresponded to a ridge in this temperature surface; however, this conjecture is not true. There appears to be no clear relationship between distillation curves and the shape of the temperature surface except to note that, as one moves down a column, the temperature always increases.

As a final step in constructing a distillation curve, we place an arrow showing the direction of increasing temperature on it.

From this figure which shows several of the distillation curves for this system, we note that columns operating at total reflux cannot operate across the distillation boundary we have labeled c. Our intuition suggests that columns operating at total reflux should give us the most separation possible for a column; thus, we are likely to conclude this boundary is a firm one. Our intuition fails us again but not by much, as we shall see later in this chapter. It turns out we can cross the boundary curve c with a column operating at less than total reflux, but we cannot operate very far across it. Thus, these boundaries are soft but strongly indicate where we can and cannot operate columns separating these species.

We should remember that to create this figure from which we are developing our insights, we have had to compute distillation curves, each of which requires us to do a series of bubble and dewpoint calculations.

14.2.2 Designing Alternative Separation Sequences

Let us use a diagram like this to design alternatives based on distillation for separating a mixture of acetone, chloroform, and bezene. We shall find that it matters in which region we place the feed. Let us place the feed as shown in Figure 14.10 at 36 mole % acetone, 24% chloroform, and 40% benzene.

Suppose we simulate a distillation column having this feed using a large number of trays—say 50 of them—and a high reflux ratio—say about 10. Our goal is to have a column that carries out what we guess will be the maximum separation possible for the way we choose to operate it. We first operate it by asking it to produce a distillate whose flow is 1% of the flow of the feed. We should expect and will find that the column produces relatively pure acetone as the distillate; however, the distillate will remove only 1/36 or just under 3% of the acetone that is in the feed; the rest together with all the chloroform and benzene will leave in the bottoms product.

On a composition diagram, the feed composition must lie on a straight line between the distillate product composition and the bottom product composition. The distillate is pure acetone. Its composition is in the lower left corner. The bottoms product composition must lie on the line from the acetone corner to the feed composition and then just past it. Since the distillate is 1% of the feed, the lever rule says the distance from the acetone to the feed is 99 times the distance from the feed to the bottoms composition.

Now let us carry out a simulation that removes 2% of the feed. Again, the distillate will be pure acetone; the bottoms will be the rest of the acetone and all the chloroform and

Sec. 14.2   Separating a Mixture of Acetone, Chloroform, and Benzene   469

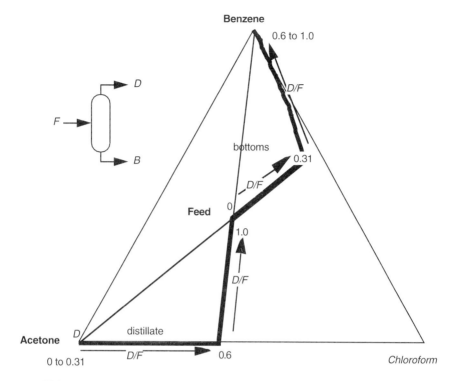

FIGURE 14.10 All products reachable by a column for a given feed for the acetone, chloroform, benzene system.

benzene. The composition of the bottoms product will move a little farther away from the feed, again in a direction directly away from the acetone corner.

We keep increasing the amount of the distillate. If there were no irregularities in the VLE behavior of these species, we would expect the distillate to remain relatively pure acetone until we have removed all of the acetone in the feed, that is, until the distillate flow is 36% of the feed flow. However, we find the top product starts to contain noticeable amounts of chloroform when we try to remove more than 31% of the feed as distillate. Adding more trays to the column and increasing the reflux ratio does not help. The distillate starts to move to the right along the bottom edge of the composition triangle toward the chloroform corner. The bottoms product moves along what we can now readily recognize is the distillation boundary we saw in Figure 14.8, always at the other end of the straight line passing from the distillate composition through the feed composition.

The distillate composition will continue to move along the lower edge until we are removing 60% of the feed as distillate. At this point the distillate is all of the acetone and chloroform in the feed. The bottoms product is essentially all the benzene in the feed. We thus have a point where we have sharply separated acetone and chloroform from benzene. If we remove more than 60% of the feed as distillate, we must withdraw benzene, too.

The bottoms product will be pure benzene, but it will not be all the benzene in the feed. The distillate trajectory moves on a straight line toward the feed composition until we withdraw 100% of the feed as distillate, in which case it is precisely the feed.

The compositions we have just mapped out are those we can reach for this feed. We should now have become aware that, if we had the distillation curves and boundaries plotted as we do in Figure 14.8, we could have drawn these product trajectories without carrying out all these column simulations. Thus, we might be well advised to create this diagram, at least when we are trying to separate a three-species mixture.

Now how do we invent different separation schemes? For species displaying relatively ideal behavior, we started by enumerating two obvious alternative schemes: *A/BC* followed by *B/C* or *AB/C* followed by *A/B*. Here, however, we cannot separate acetone completely from benzene and chloroform. While the distillate can be pure acetone, the bottoms product will contain acetone no matter how we design and operate the column. We can, however, sharply separate benzene from acetone and chloroform.

In this type of problem we must include the step of identifying the "interesting" products we can reach with our feed in a column. We see three interesting products: (1) pure acetone (but unfortunately not 100% of it), (2) pure benzene, and (3) acetone and chloroform with no benzene. The last two we produce in one column.

DESIGN ALTERNATIVE 1

Let us propose to carry out the separation that gets us two interesting products right away: the column that produces acetone and chloroform as the distillate and benzene as the bottoms. We label this separation step as col 1 in Figure 14.11. We have produced one of our desired products, all the benzene. If we now use the distillate as a feed to a second column, col 2, we find the distillate is acetone but, unfortunately, the bottoms product is at best the maximum boiling azeotrope between acetone and chloroform.

We now need to devise a way to separate this azeotrope. None is obvious here. We will likely need some third species or a separation method not based on distillation. It might occur to us that we just removed a third species that could have helped: benzene. Perhaps we should not remove benzene first.

DESIGN ALTERNATIVE 2

We start again. This time we elect to produce the first interesting product, pure acetone. We sketch the steps in Figure 14.12. First, we separate as far as we can in col 1. The distillate is pure acetone; the bottoms is near to the distillation boundary and contains all three species. We now propose to split this bottoms product into pure benzene and a mixture of acetone and chloroform in col 2. We now have to ask if we can accomplish this separation. The distillate product, a mixture of acetone and chloroform, is on the other side of the distillation boundary from the column feed. We abided by the rule that the distillate, bottoms, and feed compositions must all lie on a straight line to satisfy the material balance relationships for a column. We managed to get the distillate end on the other side of the boundary only because the boundary is curved such that it bulges to the right. The

Sec. 14.2   Separating a Mixture of Acetone, Chloroform, and Benzene   471

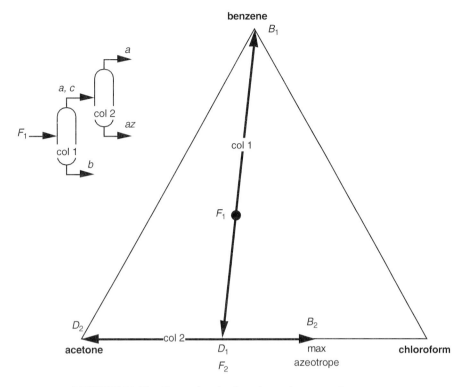

FIGURE 14.11   Generating the first alternative separation process.

feed to the second column is tucked inside this bulge, allowing us to draw a straight line through it to a point on the right of the maximum azeotrope between acetone and chloroform. We note that benzene is in either region so we should have no trouble reaching it with our second column. But can we reach the distillate shown?

It turns out that there is no requirement for the liquid compositions on the trays for a column to equal the feed composition anywhere in the column. What is required is that the liquid compositions on the trays should generally stay in one region for the entire column. Can that happen here? Start at the distillate, D_2. The liquid composition will move away from the distillate D_2 in the right-hand side region toward the feed, "curtsy" toward the feed but stay in the right-hand side region, and then proceed in the right-hand side region to the bottoms product, benzene. In this manner the trajectory can stay on one side of the boundary throughout the column. Simulation shows that this column can indeed exist as we have sketched it.

We see that the bulge in the distillation boundary is important for us to develop this separation process. It is the feature that allows us to use distillation and get across the boundary. Note we step over the maximum boiling azeotrope between acetone and chloroform in this manner, but we had to have benzene present to do it.

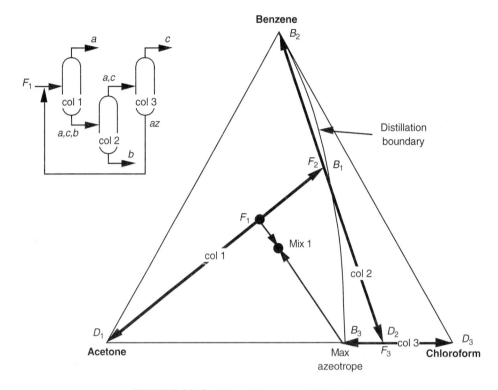

FIGURE 14.12 Generating a second alternative.

In col 3 we separate the acetone/chloroform mixture into pure chloroform and azeotrope. Again we seem to be in trouble. What can we do with the azeotrope we seem destined to produce? There is a significant difference this time from the first alternative we generated above. This time the process we have developed already produces all three products: pure acetone, pure benzene, and pure chloroform. Thus, we can propose to feed the azeotrope into this process, letting this partially complete process separate it. We are getting a "recursion" in the design, just as we did for the water/n-butanol process earlier. To feed the azeotrope back, we can mix it with the original feed, moving the actual feed for column 1 (col 1) on the straight line connecting the azeotrope to the feed and toward the azeotropic composition. Moving the feed to column 1 does not change the topology of the separation problem, only the details. It should work. Simulation of the total process at steady state verifies that the final feed to column 1 settles onto a composition between the original feed and the azeotrope, about where we show it here.

Thus, we have successfully completed a design for this process. We did it by examining the structure of the distillation curves plotted on a triangular composition diagram. This representation seems to be suited for inventing such a process.

DESIGN ALTERNATIVE 3

Are there any other alternatives? What if we would accept a chloroform product containing 2% acetone in it? Then an "interesting" product shows up along the distillation boundary where the ratio of acetone to chloroform is 2 to 98. There will be lots of benzene present, but, as we have already demonstrated, separating out the benzene is not the problem here. Our first column could produce this special product as shown in Figure 14.13. We separate out the benzene in col 2, getting a chloroform product that is 2% acetone directly. Col 3 separates the distillate from col 1 into acetone and azeotrope. Since the process created already produces all the desired final products, we recycle the azeotrope back, mixing it with the original feed.

DESIGN ALTERNATIVE 4

There is even another option, again provided we will accept a chloroform product with 2% acetone in it. In the second and third alternatives (Figures 14.12 and 14.13) we pro-

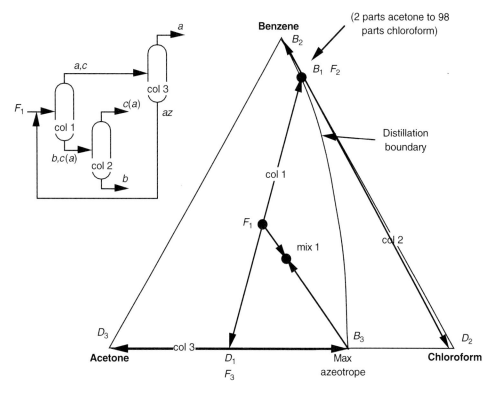

FIGURE 14.13 Third alternative based on producing a first bottoms product containing benzene and a mixture of 2 parts acetone to 98 parts chloroform.

474 Separating Azeotropic Mixtures Chap. 14

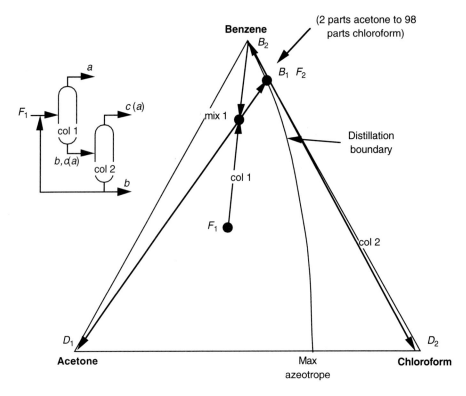

FIGURE 14.14 Fourth alternative, which eliminates the need for a third column by first mixing benzene with the feed.

duced only one interesting product in the first column. The former produced pure acetone while the second produced the benzene mixture that had acetone and chloroform in it in the ratio of 2 to 98. We could produce both in the first column if we moved the feed to that column so its composition lies on a straight line between these products. Figure 14.14 illustrates. We add benzene to the feed to move it so it lies directly between our two interesting products. A second column produces benzene (some of which we recycle to the feed) and the chloroform product with 2% acetone in it. This solution has only two columns in it.

14.2.3 Discussion

We have now designed separation processes for two nonideal mixtures, water/n-butanol and acetone/chloroform/benzene. Do the ideas generalize? We suggest that to a large extent they do. The first step we took in each case was to discover if the mixture will display nonideal behavior. One flash calculation per species in the mixture provided us with the clues needed. The next step was to find a representation that could aid us to see the design alternatives. For the binary mixture, we first looked at the phase behavior on a plot of temperature versus vapor/liquid compositions. For the ternary mixture, we used a composition triangle in which we plotted distillation curves, each of which is found by doing a series of bubble and dewpoint calculations. We then found "interesting" compositions on these plots and reduced our problem largely to looking for separation schemes that could create these interesting compositions. Unlike separating ideal mixtures, we discovered that we typically create azeotropes as products from a distillation column somewhere in the process. If we encounter them after we have enough of a process to produce all the species in them as products, we found we could simply recycle them back into the process, letting it separate the azeotrope. We noted the design represented a "recursive" solution. Recycle of material is not needed in separating ideally behaving species which makes this problem qualitatively quite different.

We also discovered that we need to be careful to observe all the interesting products. Some might be well disguised, as the mixture with lots of benzene but with an acetone to chloroform ratio of 2 to 98 proved interesting to us in the second example. We needed to look ahead to understand this mixture might be interesting. In this case we already knew we could remove benzene from any mixture so having it in that mixture was not a problem.

14.3 SKETCHING DISTILLATION AND THE CLOSELY RELATED RESIDUE CURVES

While the McCabe-Thiele diagram is a very useful design tool to determine reflux flows and number of stages for binary distillation, it may well be that its most important role for chemical engineers is to provide the qualitative insights they can gain from examining it for different situations. For example, in Chapter 12 we used it to motivate how we should think about intercoolers and interheaters for columns. It is also an excellent tool to argue why one would or would not preheat the feed to a column. Often the insights gained hold or generalize in straightforward ways for multicomponent distillation. For ternary distillation, the plot of distillation curves and the closely related residue curves on a composition diagram turn out to be excellent tools for gaining insights into the complex behavior of nonideal ternary mixtures. For this reason we shall devote this section to showing you how to think about them and in particular how to sketch them—with more details on them then you might expect you could put there.

Closely related and looking very similar to a distillation curve map is a plot called the "residue" curve map. It has some very useful geometry for understanding distillation, so we shall develop here the analysis to construct such a diagram. This plot contains a number of trajectories tracing the composition of the liquid residue in a pot that we are slowly boiling away with time, in contrast to a distillation curve plot that maps the trajectories passing through the composition of the liquid on the trays in a column operating at total reflux as we move down a column. As the more volatile species boil off, the pot becomes richer and richer in the less volatile species. If the operating pressure remains fixed, the pot becomes hotter with time as the less volatile species have higher boiling points. Residue curves move in the direction of higher temperatures and higher concentrations of the less volatile species, which is the same direction distillation curves move as we progress down a column. This is the reason we chose that as the direction for distillation curves. The following analysis supports the construction of residue plots.

Suppose we boil a pot of liquid, always removing vapor that is in equilibrium with the liquid in the pot. What would be the trajectory of the composition of the liquid in the pot versus time on a composition diagram? Figure 14.15 illustrates. The overall material balance for this unit is

$$\frac{dM}{dt} = -V$$

where M is the molar holdup in the pot in mols, V the vapor flowrate in mols/time leaving the pot, and t the time. The component material balance for species i in the pot is

$$\frac{dx_i M}{dt} = x_i \frac{dM}{dt} + M \frac{dx_i}{dt} = x_i(-V) + M \frac{dx_i}{dt} = -y_i V$$

Rearranging the terms and letting τ be dimensionless time $t/(M/V)$, we get

$$\frac{dx_i}{d\tau} = x_i - y_i \tag{14.2}$$

We can integrate these differential equations and plot the trajectory for the compositions x_i versus τ on the triangular composition plot for a ternary mixture. As we just stated, the curves we get are very similar to distillation curves. Their direction in time is to higher temperatures and less volatile species.

These curves all start at composition points that represent the lowest temperature in a region and end up at the highest temperature in the same region. These points correspond to the pure component and the azeotropes in the mixture. We term the lowest temperature nodes *unstable nodes,* as all trajectories leave from them. We term the highest temperature points in a region *stable nodes,* as all trajectories ultimately reach them. Finally, there are points that the trajectories approach from one direction and leave in the other, and we call these *saddle points.* The maximum boiling azeotrope in Figure 14.8 is such a point. The trajectories along the lower binary acetone/chloroform edge approach this point while those that are interior to the composition triangle on the distillation curve labeled c move away from the azeotrope and toward benzene.

There are geometric implications to the Eqs. (14.2) that define the residue curve trajectories. These equations say that the direction in which the liquid composition moves at

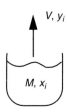

FIGURE 14.15 A boiling pot.

any instant in time is along the vector x-y, which is the vector pointing from the vapor composition toward the liquid composition. Thus, the trajectory moves directly away from the vapor composition. This observation makes sense. If we distill off a drop of vapor that has a composition y, then the pot composition, the starting liquid composition, and the final liquid composition must lie on a straight line, with the starting liquid composition falling between the other two as it represents the mixing of the other two. We assume the vapor composition is in equilibrium with the liquid as we boil it off. For any point on a residue curve, we know then that the vapor composition is along the line tangent to the curve at that point—in the opposite direction the curve is moving with time.

If we were to plot a residue plot for the acetone/chloroform/benzene mixture, it would look very similar to the distillation curve plot in Figure 14.8. Trajectories on the left side start at the lowest temperature point in that region, pure acetone, and move to the benzene corner. Those on the right start at chloroform and also end at benzene. There is a residue curve boundary the separates those trajectories on the right from those on the left. Each region on either plot has a corresponding region on the other.

What we wish to impart here is how to sketch these diagrams to discover the regions and even their general shape. Often one can sketch such a diagram for a ternary mixture knowing just the existence of and the type (maximum or minimum) of the binary azeotropes. Sometimes we need to know the temperatures of the azeotropes to get a unique plot, and in rare situations we also need to know if the points are stable nodes, unstable nodes, or saddle points.

Zharov and Serafimov (1975) and independently Doherty and Perkins (1979) developed an equation that relates the number of nodes (stable and unstable) and saddle points one can have in a legitimately drawn ternary residue plot. The equation is based on topological arguments. One form for this equation is

$$4(N_3 - S_3) + 2(N_2 - S_2) + (N_1 - S_1) = 1 \tag{14.3}$$

where N_i is the number of nodes (stable and unstable) involving i species and S_i the number of saddles involving i species. To illustrate the use of this equation, consider Figure 14.8 for the acetone/chloroform/benzene system. It has three pure components points and one maximum boiling azeotrope between acetone and chloroform. The corner points for acetone and chloroform are single species points and both are unstable nodes—all residue curves leave. The corner point for benzene is a single species point which is a stable node—all residue curves enter. All three are nodes; none are saddles, thus $N_1 = 3$ and $S_1 = 0$. The binary azeotrope involves two species. Trajectories along the lower edge enter this point while trajectories along the residue curve boundary internal to the composition space leave. It is a saddle point, and it is the last point we need to consider. Thus $S_2 = 1$ as a result and $N_2 = N_3 = S_3 = 0$. Substituting these numbers into the left hand side of Eq. (14.3) yields

$$4(0 - 0) + 2(0 - 1) + (3 - 0) = 0 - 2 + 3 = 1$$

which satisfies the equation and indicates this plot has a valid topology.

To see the usefulness of this equation, suppose we wish to construct a plot for three species having boiling points of 160, 170, and 180°C. There is an azeotrope that boils at 175°C between the two more volatile species. We start our sketch of a residue (distilla-

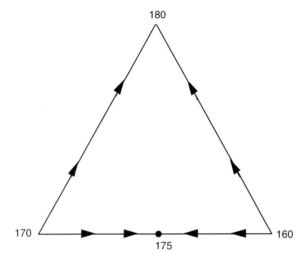

FIGURE 14.16 Starting sketch for residue curve map for three species having boiling points of 160, 170, and 180°C with a maximum binary azeotrope between the two more volatile species.

tion) curve map by sketching the triangular diagram in Figure 14.16, placing arrows pointing from lower to higher temperatures around the edges as shown.

We see that the species along the lower edge are unstable nodes, while the species at the upper edge is a stable node. This figure is very similar to that for the acetone/chloroform/benzene system. We quickly sketch the residue curve map in Figure 14.17, which we know is a valid topology. We should now wonder if there might be any other topologies that could be consistent with this same information.

Let us assume there is at most one ternary azeotrope in any of these diagrams. Let us further assume there will be at most one binary azeotrope between any pair of binary components. From the information in Figure 14.16, we know the nature of the three corner points: two are unstable nodes and one is a stable node. There is only one binary azeotrope so either N_2 or S_2 in Eq. (14.3) will be one while the other will be zero. We write Eq. (14.3) for the binary azeotrope being a saddle, getting

$$4(N_3 - S_3) + 2(0 - 1) + (3 - 0) = 1$$

or

$$N_3 - S_3 = 0$$

Either N_3 is one and S_3 is zero or the reverse or both are zero. The only way we can satisfy this equation is for both to be zero.

We next write Eq. (14.3) for the binary azeotrope being a node, getting

$$4(N_3 - S_3) + 2(1 - 0) + (3 - 0) = 1$$

or

$$4(N_3 - S_3) = -4$$

which we can satisfy for the assumption that at most one ternary azeotrope exists if S_3 is one and N_3 is 0. So there is another topology that is legal. It has the binary azeotrope we

Sec. 14.3 Sketching Distillation and the Closely Related Residue Curves

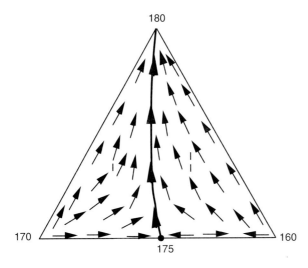

FIGURE 14.17 Sketch of the residue curve map consistent with information in Figure 14.16.

know exists being a node and has a ternary azeotrope that is a saddle. To be a node, the maximum binary azeotrope must also have trajectories from the interior entering it—that is, it must be a stable node. Figure 14.18 illustrates such a diagram. We show a temperature on the ternary azeotrope that is consistent with this diagram to show that it makes sense.

Generally, therefore, we cannot construct a unique diagram if we know only the existence of the azeotropes and their temperatures. If we know the temperature and also the nature of all the pure component and azeotrope points (type of saddle, stable node, unstable node), then we can draw a unique diagram. If we know there is a ternary saddle

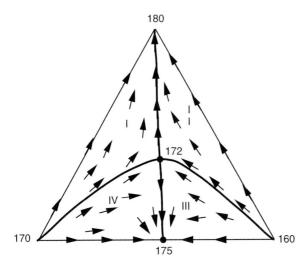

FIGURE 14.18 Another residue curve map consistent with information in Figure 14.16.

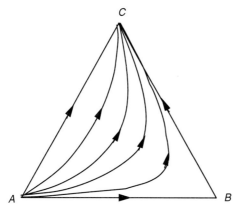

FIGURE 14.19 Symmetric distillation curves for ideal components having nearly equal "adjacent" relative volatilities.

azeotrope and that the binary azeotrope is a node, we could directly sketch the second diagram. There are a few more topological insights we could draw on, but we now want to look at the shape of these maps and not just the topology.

Figure 14.19 is a sketch of the residue (or distillation) curves for a constant relative volatility (ideal) mixture of species A, B, and C. This figure corresponds to the "adjacent" relative volatility of A to B being about the same as the "adjacent" relative volatility of B to C. An example would be if the volatilities were $\alpha_{AB} = \alpha_{BC} = 1.5$. Note that $\alpha_{AC} = \alpha_{AB} \alpha_{BC} = 2.25$. We start by noting that A, being the most volatile, will have the lowest temperature. C will have the highest temperature. The trajectories will all start at the corner for pure A and end at the corner for pure C. They will move towards the corner for the intermediate component, B, before bending back to the corner for C. The map will be symmetric as shown.

Let us next assume that the adjacent volatility between A and B is much larger than between B and C. An example would be $\alpha_{AB} = 6$, $\alpha_{BC} = 1.5$. A will preferentially leave the mixture without much of either B or C until the concentration of A becomes quite small. We would expect that, when A is present, the vapor composition will contain more A than for the previous case. The consequence of these observations, shown in Figure 14.20, is that the residue curves emanating from the corner for A will be straight, indicating the ratio of B to C does not change much, until most of the A is gone. Only then will the curves bend toward C. If C is markedly less volatile than A and B, then the lines emanating from the C corner will be straight until one approaches the AB edge, when they will bend toward A.

Next, let us do a rough sketch in Figure 14.21 of the curves for a mixture of water, ethyl alcohol, and ethylene glycol. Ethylene glycol is much less volatile than either water or ethyl alcohol. Water and ethyl alcohol form a minimum boiling azeotrope at about 85% ethyl alcohol. The other two binary pairs do not form any azeotropes. We see that the topology looks very similar to the previous one except the minimum boiling azeotrope is the unstable node from which all trajectories emanate. We show the residue curves almost as straight lines from the EG node until one gets close to the lower edge.

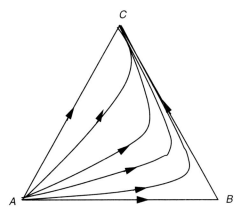

FIGURE 14.20 Residue curves for species *A* being very volatile.

Consider the data shown in Table 14.4 for the infinite dilution K-values for acetone, chloroform, and benzene. Let us see how much of the shape we can predict for the distillation curve map in Figure 14.8.

- The temperature for the maximum boiling azeotrope is slightly more than that for the boiling point of chloroform, which is higher than the boiling point for acetone. The closeness of the boiling point for the azeotropic point to the boiling point of chloroform suggests that the azeotrope composition will be nearer to chloroform than to acetone.
- Assuming we know that the azeotrope is a saddle (points approach along the lower edge and leave it into the interior of the composition diagram) and we know the temperatures for all the points, we would then know there is a residue curve boundary from benzene to the azeotrope.
- We look in Table 14.3 to see the behavior of acetone and chloroform in lots of benzene. We see that the infinite dilution K-values for acetone and chloroform in benzene

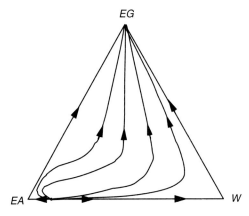

FIGURE 14.21 Rough sketch of distillation curves for ethyl alcohol, water, and ethylene glycol.

are 3.08 and 1.54 respectively. When we are near to pure benzene, the system thinks that chloroform is the intermediate species and acetone the most volatile. Starting at benzene, the residue curves will head toward the intermediate, chloroform, before bending back toward acetone. We would expect the distillation boundary to be bowed toward chloroform while heading to the azeotrope as a result. In other words, we have anticipated the curvature of the residue curve boundary. We would have expected straight lines had the infinite dilution K-values both been about equal.

- The adjacent relative volatilities are 3.08/1.54 = 2.0 and 1.54 for acetone and chloroform in benzene. These are not too different so the residue curves start out looking more like the curves in Figure 14.19 than those in Figure 14.20.

14.4 SEPARATING A MIXTURE OF *n*-PENTANE, WATER, ACETONE, AND METHANOL

Our third example is a difficult one. We will find that this mixture displays highly nonideal behavior. We should expect heterogeneous behavior when we see *n*-pentane and water in the same mixture. To design separation alternatives for this mixture, we will use distillation, liquid/liquid extraction, and extractive distillation. To make this problem particularly difficult, we have taken an equimolar mixture of these species and let it decant into a pentane-rich phase and a water-rich phase. We define as our problem here to separate the pentane-rich phase into four 99.9% pure species. The composition of the *n*-pentane-rich phase is 75.11 mole % *n*-pentane, 12.13% acetone, 11.34% methanol, and 1.42% water.

Table 14.5 gives the infinite dilution K-values and Table 14.6 the activity coefficients we compute for these species using the Unifac method assuming an ideal vapor phase. We see highly nonideal behavior predicted here. If either activity coefficient is larger than 9, we labeled the pair as possibly displaying heterogenous behavior. Where this first test does not suggest heterogeneous behavior, then the number in parentheses next to the smaller activity coefficient of a pair is the value (9 times its cube root) the other activity coefficient should be for the two together to predict heterogeneous behavior. Only the water and methanol pair behave in a somewhat ideal manner.

The suspicious prediction is that water and acetone will form two liquid phases. If we look up experimental data for these two species, we find they do not. They also do not

TABLE 14.5 Infinite Dilution K-values for a Trace of Species *j* in Species *i*

Trace of *j* in *i*	*j* = Pentane	Acetone	Methanol	Water
i = Pentane	1.0	3.0 (min)	5.9 (min)	71.4 (min)
Acetone	7.9	1.0	1.3 (min)	1.05 (min)
Methanol	29.6	2.4	1.0	0.4 (normal)
Water	8106	38.5	7.8	1.0

Sec. 14.4 Separating a Mixture of n-Pentane, Water, Acetone, and Methanol

TABLE 14.6 Infinite Dilution Activity Coefficients for a Trace of Species j in Species i

Trace of j in i	j = Pentane	Acetone	Methanol	Water
i = Pentane	1.0	6.6	23.1 (het)	1537 (het)
Acetone	4.7 (15.1)	1.0	2.0	7.4 (het)
Methanol	14.4	2.0	1.0	1.6 (10.5)
Water	3213	11.5	2.2	1.0

form a minimum boiling azeotrope. However, expermental data shows that the equilibrium curve nearly pinches for these two when there is a small amount of water in the acetone, which means they come very close to forming an azeotrope.

With water and pentane in the mixture, and with them disliking each other as they do (activity coefficients predicted to be 3213 and 1537), we would normally first suggest decanting this mixture. However, we know this is the pentane rich phase from decanting an equimolar mixture of these species so this mixture will not partition into two liquid phases, at least not at the temperature and pressure we decanted the equimolar mixture.

We could propose distillation, but virtually every pair of species forms an azeotrope. We have not looked for ternary azeotropes yet, but we should be suspicious that there could be some. Distillation is problematic.

Another separation method that suggests itself is liquid/liquid extraction. We already have in this mixture at least two species that "hate" each other. Perhaps we could wash the methanol and acetone from the pentane using water. Or perhaps we could wash the acetone and water out using methanol. Our data suggests that methanol also forms two liquid phases with pentane.

How do we assess if liquid/liquid extraction will be of use? Liquid/liquid extraction is indicated if the two species we wish to separate distribute very differently between the two liquid phases we propose to use for the extraction process. For example, would methanol distribute very differently than pentane between a water-rich phase and a pentane-rich phase? To find out, we proceed as follows. When we express equilibrium between two liquid phases, we write that the fugacities for each species i in the two phases, I and II, are equal to each other:

$$\hat{f}_i^I = \gamma_i^I x_i^I f_i^o = \hat{f}_i^{II} = \gamma_i^{II} x_i^{II} f_i^o$$

where f is a fugacity, γ an activity coefficient, and x a mole fraction. Superscripts I and II represent liquid phases I and II respectively, and o the standard state for pure species i in the liquid phase. The standard states cancel. Thus, this equation says that the ratio of the mole fractions of a species i in the two phases is the inverse of the ratio of its activity coefficients in those two phases, that is,

$$\frac{x_i^I}{x_i^{II}} = \frac{\gamma_i^{II}}{\gamma_i^I}$$

TABLE 14.7 Separability Factors for Species i and j using k-rich and l-rich Phases

		trace	
k/l ↓ \ i/j →	a/p	m/p	w/p
a/p	31	54	980
rich m/p	48	330	14,000
w/p	1800	34,000	4,900,000

To check if methanol will separate from pentane when using a water-rich and a pentane-rich phase, we can form the ratio

$$S_{methanol/pentane}^{water\text{-}rich/pentane\text{-}rich} = \frac{x_{methanol}^{water\text{-}rich} / x_{methanol}^{pentane\text{-}rich}}{x_{pentane}^{water\text{-}rich} / x_{pentane}^{pentane\text{-}rich}} = \frac{\gamma_{methanol}^{pentane\text{-}rich} / \gamma_{methanol}^{water\text{-}rich}}{\gamma_{pentane}^{pentane\text{-}rich} / \gamma_{pentane}^{water\text{-}rich}} = \frac{23.1/2.2}{1/3213} = 34{,}000$$

which is called a separability factor. A number markedly different from unity as we have here indicates that we can readily separate methanol from pentane using water. We can check out all the separability factors for having an acetone-rich, a methanol-rich, or a water-rich phase together with a pentane-rich phase. Table 14.7 lists all these factors. As we already surmised when looking at the activity coefficients, the largest separability factor of 4.9 million is between water and pentane, splitting between water- and pentane-rich phases. Water and pentane make the best two phases to use. We see that acetone and methanol have separability factors with pentane of 1800 and 34,000 respectively when we use water-rich and pentane-rich phases (the last row of the table).

We could also consider using methanol-rich and pentane-rich phases. Water will split easily from pentane with a separability factor of 14,000, but acetone has only a modest separability factor of 48 with pentane in this case.

Let us propose, therefore, to extract the methanol from the pentane by using water as the extraction agent. We can simulate this process, adjusting the water flow until we remove enough of the methanol to meet product specifications. We could also propose to remove both the methanol and acetone using water, but, when we simulate, we fail to get enough of the acetone away from the pentane, no matter the amount of water we use. (Note that these simulations will require significant effort to set up and solve even using commercial flowsheeting packages.) We place this liquid/liquid extraction unit as the first in our process (unit 1 on the left side of Figure 14.22).

We look next at the pentane-rich product from the extraction unit. From the simulation we find it to be most of the n-pentane and about a third of the acetone. It has virtually no methanol in it (by design) and only a trace of water. The infinite dilution K-values in Table 14.5 indicate that pentane and acetone form a minimum boiling azeotrope. With all the pentane in this mixture, we expect to be on the n-pentane side of the azeotrope. If so, distilling it would recover relatively pure pentane as the bottoms product and the pen-

Sec. 14.4 Separating a Mixture of *n*-Pentane, Water, Acetone, and Methanol   485

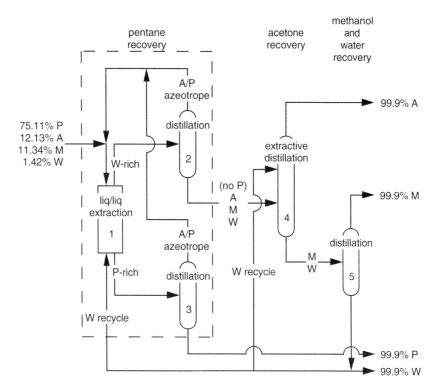

FIGURE 14.22 Synthesized flowsheet to separate a mixture of *n*-pentane, acetone, methanol, and water. Note that no other species are introduced to effect this separation process.

tane/acetone azeotrope as the distillate. Simulation verifies this behavior and shows that the trace of water exits with the azeotrope, as we might well have expected. The amount of pentane/acetone azeotrope is small, with a total flow about one-fifth that of the original feed; we propose to recycle it back to join the feed to the liquid/liquid extraction unit. Relatively small changes occur in the overall composition to that unit when we carry out material balances involving the recycle so recycling is not a problem.

The water-rich phase leaving the liquid/liquid extraction unit has virtually all the methanol, about two-thirds of the acetone, and a small amount of pentane, along with the water. We had to use about three parts of water for every two of methanol to extract all the methanol. The water is about 40% of this stream as a result. We propose to distill this mixture to recover all the pentane in the distillate. We do; the distillate is mostly the pentane/acetone azeotrope with a small amount of methanol and virtually no water. We propose to recycle the distillate back to join the feed to the liquid/liquid extraction unit.

When we look at the three units we have now proposed—left side of Figure 14.22—we find that together they have provided a means to remove all the pentane as a 99.9% pure product. We remove a pure pentane product while the stream we pass to the rest of the process contains no pentane. We draw a dashed box around these units and label them as the pentane removal section. Simulation verifies that when we include the two pentane/acetone azeotrope recycles, these three units function as proposed.

We now have a mixture of acetone, methanol, and water to separate. While the data in Table 14.5 suggest that acetone and water form a minimum boiling azeotrope and may also form two liquid phases, experimental data indicate that they do not, but, as we mentioned before, they do form a near pinch at the acetone-rich end during distillation. Methanol and acetone do, however, form a minimum azeotrope. Thus, if we separate out the water first, we will then have to break this azeotrope afterwards.

We look to see if we can break the azeotrope with water present (as we broke the acetone/chloroform azeotrope with benzene present in the section 14.2). Looking at the infinite dilution K-values for acetone and methanol in lots of water, we find them to be 38.5 and 7.8 respectively. Acetone is over four times more volatile than methanol with lots of water present. Water is less volatile than both of these species. One way to separate methanol and acetone with lots of water present is to use extractive distillation.

One typically feeds an extractive agent, here water, on a tray near the top of the extractive column. Being the least volatile it will move down the column and will therefore be present in the liquid on all the stages below where we have fed it. We then feed the acetone, methanol, and water mixture onto a tray partway down the column. In the presence of lots of water, the section of trays above where we have fed the acetone, methanol, and water feed will remove the methanol and water from this mixture, leaving only acetone to migrate up the column to the point where we are feeding the water being used as the extractive agent.

Above the water feed, only acetone and water are present. The top of the column will act like the top of an acetone/water distillation column. We can separate the acetone from the water, albeit with lots of trays and high reflux as there is the acetone/water near pinch we discussed earlier at high acetone concentrations. The extractive column in Figure 14.22 accomplishes this step. We simulate this column and discover that it functions as proposed here. We are left to separate methanol and water. They do not form an azeotrope; we accomplish this separation easily using a conventional column, the last column in Figure 14.22. We recycle some of the water back to the liquid/liquid extraction unit and to the extractive distillation column to be used in both cases as the extractive agent.

14.4.1 Discussion

ARE THERE OTHER ALTERNATIVES?

If we distill the original feed, we produce both distillate and bottoms products having all the species in them. There are no really "interesting" products produced. Our liquid/liquid extraction unit directly removes methanol from *n*-pentane, which is interesting.

N-pentane is also the most plentiful species in the feed. Separation heuristics strongly suggest we remove it first, which we have done here.

If we allow ourselves to introduce other species, we could look for other extractive agents in the liquid/liquid extraction unit. However, we will seldom wish to introduce other species as we then have to handle them in addition to those already there. Water is hard to beat as an extractive agent. We could look at using methanol or acetone as the extractive agent in this unit. Water is so superior in terms of its separability factor that it is unlikely either would be a better alternative to use. We also mentioned using water in the liquid/liquid extraction unit to also remove the acetone, in addition to the methanol, from the n-pentane. However, when we simulate this unit, we find we cannot remove enough of the acetone to meet the n-pentane product specification of 99.9% purity. If we were willing to back off on the purity specification for the n-pentane to that we could reach, then this would be an alternative.

We should look for alternatives to separate the water-rich product from the liquid/liquid extraction unit. The obvious interesting product when applying distillation is the one that removes all the pentane, leading the process we chose.

If we were to use simple distillation to separate the acetone, methanol, and water mixture, we would remove the water first from the material passing up the column, leaving ourselves with an acetone/methanol mixture where we know there is an azeotrope. Thus that will not work.

THE GENERAL APPROACH

The general approach is to assess if one can distill the mixture easily, based on the very powerful heuristic: "Distill if at all possible." If not, then look for simple measures that suggest other separation methods might work. For most separation methods that we propose, we cannot readily tell exactly what they will do when applied to the mixture we are attempting to separate. Here we resorted to a number of simulations to find out, always looking for "interesting" products. At one extreme, the separation method may be simple and allow us to predict the products without effort. At the other we may need to carry out experiments, something we would like to avoid because of the expense and time involved. We then propose alternatives based on producing at least one of the interesting products, often at the cost of producing a second product that we know will be very difficult to separate. However, we often have a partial separation process available. We may be able to recycle the difficult product back to it.

With the first and second examples, we were able to show how to predict performance of distillation processes without carrying out detailed simulations. In the first, which was for separating two species, we needed to produce a T versus composition diagram; in the second, for three species, we sketched distillation curves within a triangular composition diagram. We could imagine developing such a sketch for four components, but our result would be distillation curves in a three-dimensional tetrahedron that we would find difficult but not impossible to examine and understand.

We also illustrated three ways we can break azeotropes. The first, water and n-butanol, used another method, decantation. The second and third examples used distillation. In the second case we used the curvature of the distillation boundary appearing in the acetone, chloroform, and benzene composition space. In the third we used the difference in volatility of the two species involved in the azeotrope, acetone and methanol, when in the presence of a third species, water, and used extractive distillation.

One other way to break azeotropes is to distill at a two different pressures. In some cases the composition of the azeotrope moves sufficiently that one can get an economic process. Generally, however, one has to look for other separation methods: membranes, adsorption, absorption, forming intermediate chemical complexes that easily separate and then decompose when heated, and so on.

14.5 MORE ADVANCED WORK

The literature on distillation is extensive. Recent review articles will lead the reader to this literature [Poellmann and Blass, 1994; Fein and Liu, 1994; Widagdo and Seider, 1996; Westerberg and Wahnschafft, 1996]. We outline here briefly some of the concepts covered.

14.5.1 Assessing Nonideal Component Behavior

We need to assess component behavior to give us a simple means to predict the interesting products when we distill.

- Many articles exist in the thermodynamics area on predicting phase behavior. A good start into this literature are the chemical engineering thermodynamics textbooks.
- Articles exist on how to compute all the azeotropes predicted for a mixture given a good thermodynamic model for the mixture. These articles also show how to compute the eigenvalues and eigenvectors for these azeotrope and pure components assuming that one is boiling a mixture with a composition near one of these points. A positive eigenvalue says that any composition along the corresponding eigenvector computed near the point moves away from the point if we were to boil the mixture while a negative eigenvalue says the reverse. As the compositions must add to unity, there are $n - 1$ eigenvalue/eigenvector pairs for each point in a space of n species. A point with only positive eigenvalues is called an *unstable node* (all trajectories will move away from it), with all negative eigenvalues a *stable node* (it attracts all trajectories), or a *saddle* (some trajectories move away while others are attracted).
- An extensive literature exists on finding all the distillation regions for a mixture of n species. This is a very complex issue. Many of these articles present topological arguments to tell which combinations of stable nodes, unstable nodes, and saddles can be present in a topologically correct map. Most of the work is for ternary systems. However, the generalizations for n component systems exist.
- A literature exists on how all these concepts extend to batch distillation.
- Much work exists on how to predict liquid/liquid behavior for mixtures. Much of it involves developing thermodynamic models as we mentioned above, while other literature assumes good thermodynamic models exist and describes how to solve

Sec. 14.5 More Advanced Work

them when one does not know the number of phases that might be present. This problem is a nonconvex optimization problem requiring considerable care to assure one does not discover local optima.

14.5.2 Insights into Column Operation

We just discussed methods that allow us to predict interesting products when distilling a mixture that did not require us to simulate the column at a large number of different operating conditions. These insights are to aid in doing these predictions and are largely based on plotting distillation and the closely related "residue" curves for distillation in composition space.

- There are articles on how to assess column operation given the distillation regions. Most limit these insights to mixture involving three species. Contrary to our intuition, total (infinite) reflux conditions do not always lead to maximum separations. Some of the literature demonstrates that one can cross distillation boundaries by using finite reflux. Recent literature shows how to compute exactly how far one can cross these boundaries for ternary systems. One of the goals for this work is to map out all possible products one can reach using a distillation column on a given feed.
- Work exists to discover the reachable products for ternary extractive distillation columns, both continuous and batch, when viewed on a triangular composition diagram. The extra degree of freedom is the solvent feed rate. One finds that these columns have not only a minimum reflux ratio but also a maximum one.
- There are articles that show how single columns and systems of columns may display multiple operating states. Here one can fix entirely values for all the degrees of freedom for the column or system of columns and find that it will operate in multiple ways. Imagine the control issues this suggests.

14.5.3 Designing Columns to Perform Given Tasks

When we have a proposed flowsheet, we still must size the equipment.

- There is an extensive literature on simulating the performance of columns.
- Another body of literature talks about how to compute minumum reflux conditions for a column, including extractive distillation (where there is a maximum rate too) and heat integrated columns such as side strippers and enrichers.
- To design a column, one needs to determine the number of trays, the feed tray location, and the column diameter. The more recent literature tells how to do this for nonideal mixtures. Trading off the number of trays versus the reflux ratio is often turned into an optimization problem. Some of this work is based on doing tray-by-tray computations in a stable manner. Other suggests using collocation models.

REFERENCES

Doherty M. F., & Perkins, J. D. (1979). On the dynamics of distillation processes-III (The topological structure of ternary residue curve maps). *Chem. Eng. Sci.,* 34, 1401–1414.

Fein, G. A. F., & Liu, Y. A. (1994). Heuristic synthesis and shortcut design of separation processes using residue curve maps: A review. *Ind. Eng. Chem. Res.,* 33, 2505–2522.

Perry, J. H. (Ed.). (1950). *Chemical Engineers' Handbook*, 3rd ed. New York: McGraw-Hill.

Poellmann, P., & Blass, E. (1994). Best products of homogeneous azeotropic distillations. *Gas Sepn & Purification,* 8(4), 194–228.

Widagdo, S., & Seider, W. D. (1996). Journal review: Azeotropic distillation. *AIChE J,* 42(1), 96–130.

Westerberg, A. W., & Wahnschafft, O. (1996). Synthesis of distillation-based separation processes. In J. L. Anderson (Ed.), *Advances in Chemical Engineering, Vol. 23, Process Synthesis* (pp. 63–170). New York: Academic Press.

Zharov, W., & Serafimov, L. A. (1975). *Physicochemical Fundamentals of Distillations and Rectifiations* (in Russian). Leningrad: Khimiya.

EXERCISES

1. Synthesize a process to separate a 70 mole% mixture of *n*-butanol and water. Is there a second readily apparent process or not? Explain. How does it compare to the process we designed for a 15% feed?

2. Sketch the products that one can obtain when separating an equal molar mixture of *A*, *B*, and *C* on a triangular diagram vs *D/F*. The vapor/liquid equilibrium behavior of these species is ideal. *A* is the most volatile and *C* the least.

3. Figure 14.23 characterizes the phase behavior of the toluene, water, pyridine system on a ternary composition diagram. The two-phase behavior does provide pure

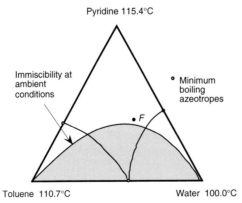

FIGURE 14.23 Ternary composition diagram for toluene, water, and pyridine.

enough water and toluene given a mixture of just these two substances. (They really do not like each other.)

a. Develop all the alternative processes you can for the feed that is shown. There should be at least three that you can sketch; one has only one column.

b. Sketch the reachable products on both this triangular diagram versus the fraction of the feed flow taken off as distillate.

4. Figure 14.24 characterizes the behavior for ethanol, water, and toluene. Devise alternative separation schemes for a mixture of these species.

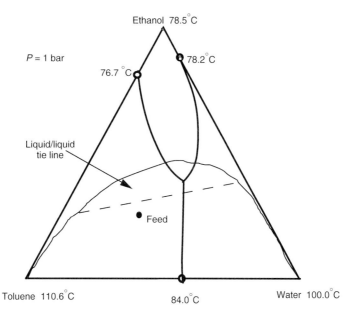

FIGURE 14.24 Ternary composition diagram for ethanol, water, and acetone.

5. We can represent the composition space for a three-component mixture as a triangular diagram that lies in a plane. The total molar Gibbs free energy function will then be a surface plotted above this triangle. Suppose that a support plane from below at the composition 0.333, 0.333, 0.333 does not touch this free energy surface at that point. This support plane will then touch the surface at three points, suggesting that the mixture will break into three liquid phases at equilibrium. Discuss how it could be that the mixture will only break into two liquid phases without invalidating this geometrical view of the problem.

6. We separated methanol from acetone in the third example in this chapter by using water as an extraction agent. Argue why similar reasoning would not suggest that we use benzene as an extractive agent to separate acetone from chloroform.

7. You need to separate a mixture of 10 mole % A in B. Your design group has proposed species C, D, E, F, and G as candidate extractive agents for use in an extractive distillation column. Following are the infinite dilution K-values for these components.

Trace of j in i	$j = A$	B	C	D	E	F	G	NBP, K
$i = A$	1	1.8	0.15	4	0.2	0.25	3	370
B	2	1	0.05	7	0.08	0.07	1.1	390
C	4	1.7						450
D	0.6	0.1						330
E	1.3	3.2						430
F	2.2	2.1						430
G	0.3	0.96						375

a. For each candidate extractive agent, sketch the residue curve maps for A, B, and the agent. Put in as much detail as you can.
b. Sketch the extractive column, indicating where the agent should be fed into the column and in which product the major portion of each species will exit. Clearly indicate where trays should exist in any section of the column.
c. Which of the candidate agents would be good extractive agents? Which would be poor? Explain your answers.

8. Synthesize a separation process to separate a mixture of 10% acetone, 80% chloroform, and 10% benzene. Note that this mixture is in the lower right-hand corner (near to chloroform), to the right of the curved distillation boundary in Figure 14.8. Create a process based entirely on distillation. (Hint: What if you create two intermediate products you mix?)

9. Sketch all triangular diagrams that are compatible with the following infinite dilution K-values.

inf dil K-val of in	A	B	C	NBP, C
A	1	50	0.8	100
B	20	1	0.72	120
C	5	0.2	1	150

Describe exactly how these infinite dilution K-values are to be computed. Assume you have a commercially available physical property package.

10. Sketch all composition triangular diagrams that are consistent with the following data. This table indicates the temperature of any azeotropes between the binary pairs and the boiling points for the pure components at one atmosphere.

	B	C	NBP
A	84.1°C	93.0°C	100.0°C
B		109.9°C	110.7°C
C			115.4°C

11. Consider the ternary diagram shown in Figure 14.25.
    a. If you are given no added information than what appears on the figure, sketch all the different topologies possible for this problem.
    b. Choose a topology that has no ternary azeotrope. Can a column operating as shown work? You have to decide which would be the distillate product and which the bottom product if you think these two products are possible. The numbers shown are temperatures.

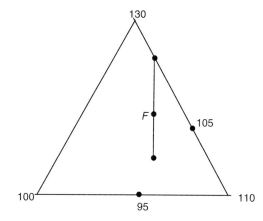

FIGURE 14.25 Ternary composition diagram for Exercise 11.

12. Consider the triangular composition diagram shown in Figure 14.26. Shown are all binary azeotropes.

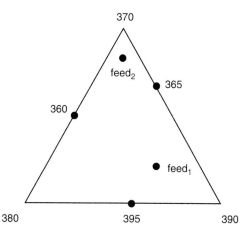

FIGURE 14.26 Ternary composition diagram for Exercise 12.

a. Sketch the possible topologies for the triangular diagram in the figure. Can there be more than one? Explain.
   b. Sketch the possible products that one can produce for feed$_1$ shown for the case that there is a ternary node.
   c. Sketch the possible products that one can produce for feed$_2$ for the case that there is no ternary azeotrope.
13. Consider any three species that form two liquid phases at ambient conditions. Argue that neither liquid phase can be completely pure no matter how much these species dislike each other. Describe clearly what it means for them to "dislike" each other.

   Hint: Look closely at the behavior of ΔG_{mix} at compositions of 0 and 1.

PART IV

OPTIMIZATION APPROACHES TO PROCESS SYNTHESIS AND DESIGN

BASIC CONCEPTS FOR ALGORITHMIC METHODS 15

15.1 INTRODUCTION

Some fundamental insights were presented in Part III that can greatly reduce the large combinatorial problem in process synthesis. These insights have the advantage of providing a basic understanding of the nature of these problems. However, as the reader may have noted, most of these insights come from analyzing particular subproblems, for example, heat exchanger networks, heat and power systems, distillation sequences, reactor networks. While these are clearly essential for successfully tackling synthesis problems, it is also clear that they have the following limitations:

1. The possible interactions between material flow and energy flow are generally complex and not taken into account. A major question is, then, how to determine the trade-offs between raw material utilization and energy consumption when selecting the flows of the process streams?

2. With few exceptions, insights for heat integration, separation, and reaction tend to rely on physical principles without explicitly considering capital costs. Therefore, we will also need to consider the question of how to develop synthesis procedures where trade-offs between raw material, capital, and energy costs are explicitly accounted for so as to produce cost-effective systems.

3. Finally, while insights do reduce very significantly the combinatorial problem, they do not always provide all the information that is required to synthesize an optimal or near optimal system. In general, one may still be left with the problem of having to search among a relatively large number of alternatives. For example, in the heat exchanger network problem, the insight of the minimum utility target limits the combinations of matches that must be considered. However, these insights do not supply all the information on what matches are actually required nor how to interconnect them. The same limitations apply to reactor networks, distillation sequences, or heat and power systems. An important question that then arises is how to systematically determine an optimal or near optimal structure. Furthermore, can we automate to a great extent this task in the computer and take advantage of its increasing computational power?

In Part IV we will present algorithmic methods that to a great extent can address some of the questions posed in the above three points. These algorithmic methods will rely on optimization techniques, mainly mixed-integer optimization methods. That is, methods where we can model discrete and continuous decisions that are required in process synthesis. Also, we will see how a number of the insights presented in Part III can actually be incorporated effectively into these methods so as to simplify the optimization problems. The major emphasis throughout will be on modeling.

This chapter will cover three basic elements that are required in the development of algorithmic methods for process synthesis: problem representation, modeling, and solution strategies. In Part III we already saw the great importance of problem representations in the analysis of heat flows, distillation residue curves, and attainable regions in reactor networks. In this chapter we will study how different problem representations have an impact upon models and solution strategies, and how these can in fact also motivate representations for synthesis problems. We will also emphasize the modeling of constraints with 0-1 variables.

15.2 PROBLEM REPRESENTATION

In general, there are different ways in which we can develop algorithmic methods for process synthesis. The differences arise on the particular problem representation that is used. In these representations the objective is to include explicitly or implicitly a family of flowsheets, all of which are potential candidates for the optimal solution. Depending on what particular problem representation we use, we may have to resort to different search techniques as will be shown in sections 15.3, 15.4, and 15.5.

EXAMPLE 15.1   Sharp-split Separation of a Multicomponent Feed

Assume we have a mixture of four components A,B,C,D. As was shown in Chapter 11, there are five different sequences where the simplest option is to represent these with the tree shown in Figure 15.1 (Hendry and Hughes, 1972). You will note that any given sequence is determined by a path that goes from the root node to a terminal node. So, for example, the direct sequence is given by the path that starts at the root node and goes sequentially through nodes 1, 2, and 3.

In the tree representation of Figure 15.1, however, we have multiple representations for some of the separators. In particular, the binary separators *A/B, B/C, C/D* appear twice in the terminal nodes. Can we avoid this duplication? The answer is yes, as shown in Figure 15.2 (Andrecovich and Westerberg, 1985). We just simply mix the streams from those separators that lead to the same binary separator. In this representation, however, we no longer have a tree. We have instead created a network, where we only have 10 nodes for these separators, as opposed to having 13 as in Figure 15.1.

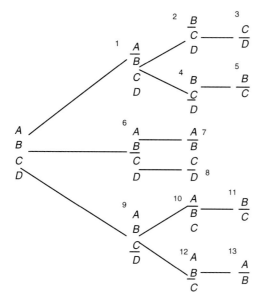

FIGURE 15.1 Tree representation of four-component separation.

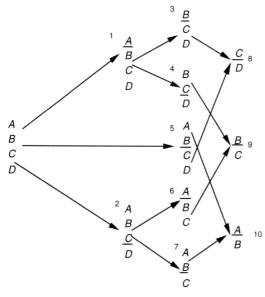

FIGURE 15.2 Network representation of four-component separation.

EXAMPLE 15.2 Flowsheet for Ammonia Process

Assume that the following alternatives are to be considered in the development of a flowsheet for manufacturing ammonia in which the major processing steps are shown in Figure 15.3:

a. Reaction with a tubular or multibed-quench reactor
b. Separation of product by flash condensation or absorption/distillation
c. Possible recovery of hydrogen with membrane separation in the purge stream

Clearly, we can represent these alternative choices through the tree in Figure 15.4, where each terminal node corresponds to one of the eight different flowsheet structures. Note that we again have duplication of some nodes. Also, you might note that any path in the tree starting at the root node has not a direct resemblance to the flowsheet structure implied by a given terminal node. This is simply because there is no recycle in the tree. How can we include the effect of the recycle in our representation? If we replace the decision blocks in Figure 15.3 by the alternative choices, we can obtain the network representation in Figure 15.5, which is also known as a "superstructure". Note that this superstructure has embedded all the flowsheets implied by Figure 15.4. As seen in Figure 15.6, we just simply obtain them by "deleting" some of the streams and units in Figure 15.5. In addition, we may even create new flowsheet structures if we do not delete all the streams as seen in Figure 15.7. You should also note that, as in Example 15.1, the network representation in Figure 15.5 has no duplication of alternative choices.

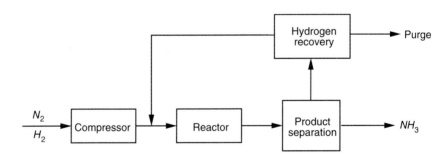

FIGURE 15.3 Major processing steps for NH_3 production.

Sec. 15.2   Problem Representation

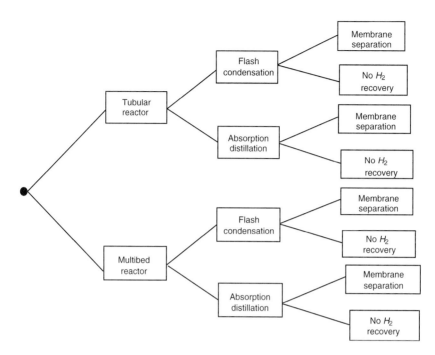

FIGURE 15.4 Tree representation for alternatives in NH_3 flowsheet.

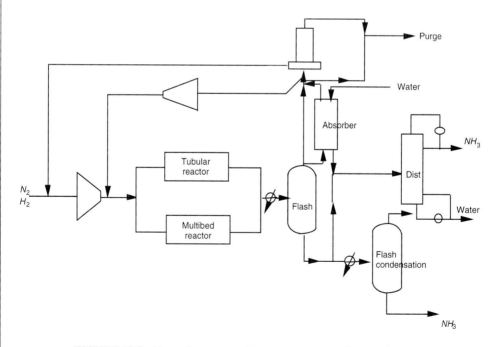

FIGURE 15.5 Network representation or superstructure for NH_3 flowsheet.

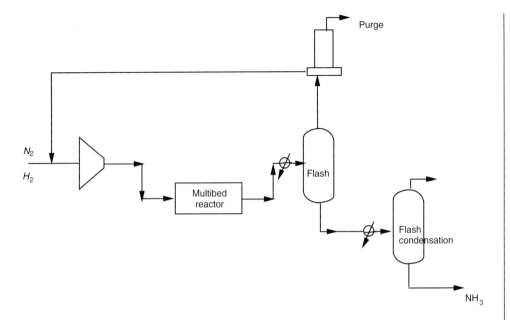

FIGURE 15.6 Alternative for multibed reactor/flash condensation/membrane separation that is contained in the network of Figure 15.5.

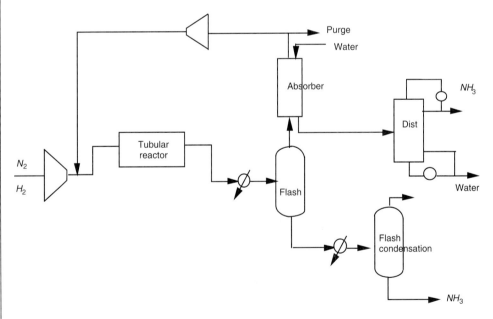

FIGURE 15.7 Alternative for tubular reactor/flash condensation and absorption distillation/membrane separation.

The next few chapters will present different problem representations that can be used for modeling various synthesis problems. It is important, however, to consider first general aspects of solution strategies and how they relate to the tree and network representations.

15.3 SOLUTION STRATEGIES FOR TREE REPRESENTATIONS

Having developed a particular problem representation, the next question that we need to consider is how to search for the optimal flowsheet structure. For the case where we have developed a tree representation, we will be able to decompose the solution of the problem by analyzing a sequence of nodes in the tree.

Each node will typically involve the sizing and costing of a process unit. We have the two following alternatives for the analysis of the nodes: exhaustive enumeration and implicit enumeration. The first is clearly only practical for trees of small size. The second is a strategy that requires the examination of a subset of nodes and is in general suitable for large trees. Therefore, we will concentrate on implicit enumeration strategies, which are often also denoted as branch and bound methods. For the sake of simplicity, we will assume that our objective is cost minimization.

When we consider a tree, we will have the root node or initial node, intermediate nodes, and terminal nodes whose path from the root node defines a complete solution. For any particular node in the tree we can obtain a partial cost that is given by the sum of costs of the previous nodes involved in the path that starts at the root node. Since the partial cost increases monotonically along any path in the tree, we have the two following properties:

1. For an intermediate node in the tree, its partial cost is a lower bound on the cost of any of the successor nodes. This is just simply because successor nodes incur in additional costs.
2. For a terminal node, its total cost is an upper bound to the cost of the original problem. This follows from the fact that a terminal node defines a particular solution to our problem that may or may not be optimal.

Based on these simple properties, we can prune any node in the tree whose partial cost is greater or equal than the current upper bound. In addition to this bounding rule, however, we also need to specify the order in which the nodes will be enumerated, or in other words a rule for selecting nodes. The two options that are most commonly used for the selection of nodes are the following:

1. Depth-first. Here we successively perform one branching on the most recently created node. When no nodes can be expanded, we backtrack to a node whose successor nodes have not been examined.

2. **Breadth-first.** Here we select the node with the lowest partial cost and expand all its successor nodes.

To illustrate more clearly these node selection rules and how they are applied within an implicit enumeration scheme where we prune nodes according to the bounding rule, let us consider Example 15.1 on distillation sequences. Figure 15.8 displays the tree structure for this problem with the associated costs at each node. These would have been obtained if we had used an exhaustive enumeration of all the nodes, which in turn would have implied sizing and costing all the columns in the tree. As you can see from Figure 15.8, the optimal separation sequence is (A/BCD)–(BC/D)–(C/D) (i.e., nodes 1,4,5) with a total cost of 16.

If we use a depth-first procedure for the implicit enumeration, this would be the order in which we would examine the nodes in the tree (see Figure 15.9):

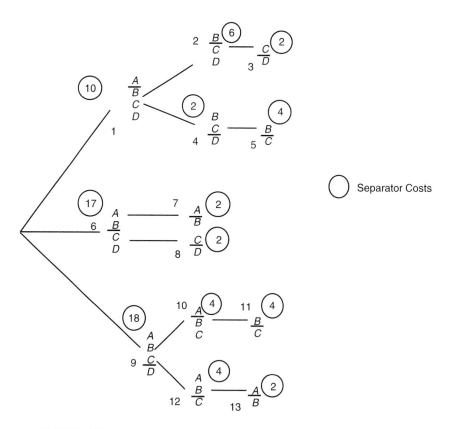

FIGURE 15.8 Tree representation for Example 15.1 with costs of separators (in 10^3 \$/yr).

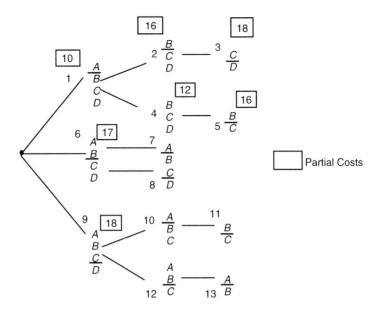

FIGURE 15.9 Depth-first search for tree in Figure 15.8.

Branch from root node to node 1; partial cost = 10
Branch from node 1 to node 2; partial cost = 10 + 6 = 16
Branch from node 2 to node 3; partial cost = 16 + 2 = 18
    Since node 3 is terminal, current upper bound = 18;
    current best sequence (1,2,3)
Backtrack to node 2
Backtrack to node 1
Branch from node 1 to node 4; partial cost = 10 + 2 = 12 < 18
Branch from node 4 to node 5; partial cost = 12 + 4 = 16
    Since node 5 is terminal and 16 < 18, current upper bound = 16;
    current best sequence (1,4,5)
Backtrack to node 4
Backtrack to node 1
Backtrack to root node
Branch from root node to node 6; partial cost = 17
    Since 17 > 16 (current upper bound), prune node 6.
Backtrack to root node
Branch from root node to node 9; partial cost = 18
    Since 18 > 16 (current upper bound), prune node 9.
Backtrack to root node.
    Since all branches from the root node have been examined, stop.
    Optimal sequence (1,4,5), cost = 16.

Note that with this depth-first strategy we examined 7 nodes out of the 13 that we have in the tree. Therefore, we only need to size and cost 7 columns.

If we use a breadth-first procedure, this would be the order in which we have to examine the nodes (see Figure 15.10):

Branch from root node to:
    node 1; partial cost = 10
    node 6; partial cost = 17
    node 9; partial cost = 18
Select node 1 since it has the lowest partial cost;
Branch from node 1 to:
    node 2; partial cost = 10 + 6 = 16
    node 4; partial cost = 10 + 2 = 12
Select node 4 since it has the lowest partial cost among nodes 6,9,2,4;
Branch from node 4 to:
    node 5; partial cost = 12 + 4 = 16
Since node 5 is terminal, current best upper bound = 16,
    current best sequence (1,4,5).
From the remaining nodes, 6,9,2, the one with lowest partial cost is
    node 2 with partial cost = 16;
Since 16 = 16 (current best upper bound); prune nodes 6,9,2, stop.
    Optimal sequence (1,4,5), cost =16

Note that with this breadth-first strategy we only had to examine 6 nodes out of the 13 nodes in the tree, one less than with the depth-first procedure.

It should be noted that in general the breadth-first strategy requires the examination of fewer nodes and no backtracking is required. However, depth-first requires less storage

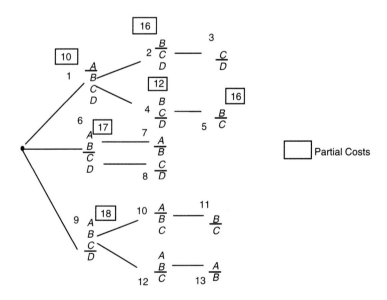

FIGURE 15.10 Breadth-first search for tree in Figure 15.8.

of nodes since the maximum nodes to be stored at any point is the number of levels in the tree. Breadth-first in general requires storing a much larger number of nodes. For this reason the depth-first strategy is commonly used. Also this strategy has the tendency of finding the optimal solution early in the enumeration procedure when compared to breath-first. The two strategies will often require the examination of a relatively small fraction of the nodes in the tree. For very large trees however, the number of nodes to be examined might still be very large, so that one may have to develop sharper bounds or else resort to heuristics to prune more effectively the nodes. Finally, the search methods outlined here are also used in the solution of MILP problems in the form of branch and bound methods (see Appendix A). The main difference is that LP subproblems are solved in each node.

Another point to be noted from this example is the fact that the optimal sequence in this example could have been obtained by successively selecting the cheapest separator; that is, node 1 is cheaper than nodes 6, 9, node 4 is cheaper than node 2 (see Figure 15.8). This procedure however, does not in general guarantee that we can find the optimal solution (see exercise 1). For this reason this procedure is called a "greedy" heuristic and is only useful for generating initial estimates.

Finally, it should also be noted that if we wanted to optimize continuous parameters at the nodes in the tree, interactions may start to take place among the different nodes. In this case an implicit enumeration scheme might no longer be valid unless the interactions among the nodes or units is small. In our separation example, optimizing pressures and reflux ratios will normally not produce large interactions.

15.4 MODELS AND SOLUTION STRATEGIES FOR NETWORK REPRESENTATIONS

For the case when a network is used as the basis of the representation, it is often not possible or even desirable to decompose the problem by analyzing a sequence of nodes as we did in the case of the tree representation. Here the basic approach will be to consider a simultaneous optimization of the network through an appropriate mathematical programming problem (Minoux, 1986; Nemhauser et al., 1989). The motivation for a simultaneous solution is that the network will often be nonserial in nature due to the presence of recycles. Even if no recycles are present, it might still be more efficient to consider the problem simultaneously, especially when both the structure and the parameters in the network are to be optimized.

In general, when we optimize a network for synthesizing a processing system, we would like to model both the discrete decisions on the nodes or units that should be included in the optimal solution as well as the continuous parameters that define flows and operating conditions (e.g., pressures and temperatures). In this way we can introduce two types of variables:

1. Binary variables y_i, that are defined for each node or unit i as:

$$y_i = \begin{cases} 1 \text{ if unit } i \text{ is selected in the optimal structure} \\ 0 \text{ if unit } i \text{ is not included in the optimal structure} \end{cases}$$

2. Continuous variables x that represent flowrates, pressures, temperatures, compositions, splits, conversions, sizes of units.

The objective function (e.g., cost), $f(x,y)$, will in general be a function of both types of variables. The continuous variables x, which for physical reasons are assumed to be non-negative, must in general obey mass and energy balances, equilibrium relationships, and sizing equations. That is, these variables must satisfy equations $h(x) = 0$, where usually $\dim(h) < \dim(x)$, since there are commonly degrees of freedom for the optimization.

Both continuous and binary variables must also satisfy design specifications (e.g., product purity, physical operating limits) as well as logical constraints (e.g., select only one reactor in the network; the flow in a column must be zero if it is not selected). We will represent these constraints as inequalities of the form $g(x,y) \leq 0$.

In this way, the optimization of a network or superstructure where we wish to "extract" the optimal flowsheet structure with its associated continuous parameters can be posed as the mathematical programming problem (P0):

$$\begin{aligned}
\min \quad & f(x,y) \\
\text{s.t.} \quad & h(x) = 0 \\
& g(x,y) \leq 0 \\
& x \geq 0, \, y \in \{0, 1\}^m
\end{aligned} \tag{P0}$$

The solution of the desired flowsheet will then be defined by the non-zero flows and units whose binary variables are equal to one in the network.

For the case when $f, h, g,$ are nonlinear functions, problem (P0) corresponds to a mixed-integer nonlinear programming (MINLP) problem. If $f, g, h,$ are linear, problem (P0) corresponds to a mixed-integer linear programming (MILP) problem. The special case when no binary variables y are present corresponds in the two cases above to a nonlinear programming (NLP) problem (see Chapter 9) and linear programming (LP) problem, respectively.

LP problems are by far the easiest to solve and very large-scale problems involving thousands of variables can be handled very effectively. MILP and NLP problems are next in difficulty. The former can be handled with reasonable expense as long as neither the number of binary variables nor the relaxation gap is very large. The latter can be handled effectively for problems with few hundred variables as long as the sparsity of the constraints is properly exploited. MINLP problems are the most difficult, although with recent advances the computational expense in solving these problems has been reduced.

LP problems are commonly solved with the well-known simplex algorithm (Hillier and Lieberman, 1986). MILP problems are solved with branch and bound methods where the search tree is given by assignments of the 0–1 variables (Nemhauser and Wolsey, 1988). As opposed to the implicit enumeration schemes, LP subproblems are solved at those nodes that have to be examined in the tree. However, the basic search strategies are similar as the ones we presented in section 15.2. Both LP and MILP problems can be solved so as to obtain the global optimum solution. NLP problems are commonly solved to obtain a local optimum solution with reduced gradient or successive quadratic programming methods (Bazaraa and Shetty, 1979). Finally, MINLP problems are solved through a sequence of NLP and MINLP problems using either Generalized Benders decomposition or outer-approximation methods (Grossmann, 1990). In this case global optimality cannot always be guaranteed, but it is often more likely than in the NLP case.

Appendix A presents a brief summary of some of these methods, and references to computer software are also given since these are actually required to solve some of the exercises. The reader is advised to read Appendix A before proceeding with the next section.

15.5 ALTERNATIVE MATHEMATICAL PROGRAMMING FORMULATIONS

In this section we would like to illustrate through a simple example the modeling of networks as mathematical programming problems. Furthermore, what we would like to stress in this section are some of the implications of modeling synthesis problems as LP, MILP, NLP, or MINLP problems. The small example problem will allow us to gain some insights into these implications.

EXAMPLE 15.3 Selection of Reactors

Assume that we have the choice of selecting the two reactors in Figure 15.11a for the reaction $A \rightarrow B$. Reactor I has a higher conversion (80%) but is more expensive, while reactor II has lower conversion (66.7%) but is cheaper. We will consider here that we need to produce 10 kmol/hr of

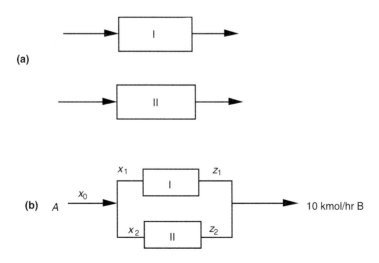

FIGURE 15.11 (a) Selection between high conversion and low conversion reactors. (b) Network representation.

product B, and that the cost of the feed A is \$5/kmol. To select the reactor that minimizes the cost of the reactor and the cost of the feed, we can develop the small network in Figure 15.11b to account for the choice of either reactor, or a combination of the two.

If we model the mass balances for the network in Figure 15.11b, by denoting with x the flows of A, and by z the flows of B, we obtain:

Mass balance initial split: $\qquad x_0 = x_1 + x_2 \qquad$ (15.1)

Mass balance reactor I: $\qquad z_1 = 0.8x_1 \qquad$ (15.2)

Mass balance reactor II: $\qquad z_2 = 0.67x_2 \qquad$ (15.3)

Mass balance mixer: $\qquad z_1 + z_2 = 10 \qquad$ (15.4)

Finally, we assume that the cost of reactors I and II is given in terms of the feed flows by the cost equations:

Reactor I: $\qquad 5.5(x_1)^{0.6} \qquad$ \$/hr $\qquad$ (15.5)

Reactor II: $\qquad 4.0\,(x_2)^{0.6} \qquad$ \$/hr $\qquad$ (15.6)

With this, our objective function becomes

$$\min C = 5.5(x_1)^{0.6} + 4.0(x_2)^{0.6} + 5.0\, x_0 \qquad (15.7)$$

The objective function in Eq. (15.7), subject to constraints in Eqs. (15.1) to (15.4) and non-negativity conditions on the x and z variables, will then define an NLP problem. To gain some geometrical insight into the nature of this optimization problem, let us eliminate the variables z_1, z_2, and x_0 from the above equations. Our problem then reduces to

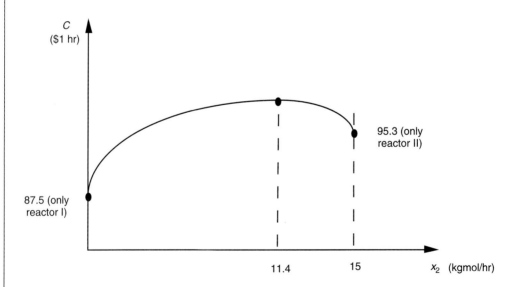

FIGURE 15.12 Cost as a function of x_2 when cost Eqs. (15.5) and (15.6) are used for the network in Figure 15.11a.

Sec. 15.5 Alternative Mathematical Programming Formulations

$$\min C = 5.5 \, (x_1)^{0.6} + 4.0 \, (x_2)^{0.6} + 5.0x_1 + 5.0x_2$$
$$\text{s.t.} \quad 0.8 \, x_1 + 0.67x_2 = 10 \tag{15.8}$$
$$x_1 \geq 0 \quad x_2 \geq 0$$

If we eliminate x_1, we can then easily plot C as a function of x_2 as seen in Figure 15.12. Note that the cost function is concave and exhibits two local minima at the extreme values 0 and 15 of x_2. At 0 we have the global optimum (\$87.5/hr), which corresponds to selecting reactor I, while at 15 we have a local optimum that corresponds to selecting reactor II (\$95.3/hr). Further, at 11.4 we actually have a global maximum. Clearly, this is an undesirable feature as it means that when using standard NLP algorithms our solution will be dependent on the starting point. It is possible to use in this case special global optimization algorithms such as the ones presented in Grossmann (1996). Since the application of these techniques is out of the scope of this book, we consider instead approximations that yield alternative problem formulations that are tractable. Since the concave cost functions in Eqs. (15.5) and (15.6) are responsible for the multiplicity of local solutions, let us assume that we replace these cost functions by linear fixed cost charge models. As shown in Figure 15.13, we will replace the nonlinear concave cost by a cost function that is linear with a fixed cost for $x > 0$, while for $x = 0$ the cost will be equal to zero. We can model such a discontinuous function with binary variables. For our example we will define the binary variables,

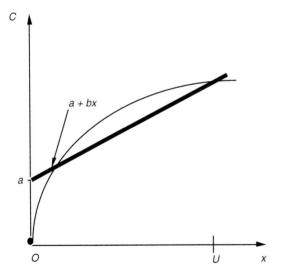

FIGURE 5.13 Fixed charge cost model.

$$y_1 = \begin{cases} 1 \text{ if reactor I is selected} \\ 0 \text{ otherwise} \end{cases} \qquad y_2 = \begin{cases} 1 \text{ if reactor II is selected} \\ 0 \text{ otherwise} \end{cases}$$

The cost functions in Eqs. (15.5) and (15.6) we will replace by linear approximations with fixed charges,

$$\text{Reactor I: } 7.5y_1 + 1.4x_1 \qquad \text{Reactor II: } 5.5y_2 + 1.0x_2 \tag{15.9}$$

Since we want the flows x to be zero when the binary variables are zero, we need to consider the logical constraints

$$x_1 - 20y_1 \leq 0 \qquad x_1 \geq 0 \qquad (15.10)$$

$$x_2 - 20y_1 \leq 0 \qquad x_2 \geq 0 \qquad (15.11)$$

where 20 has been selected as an arbitrary upper bound for x_1 and x_2.

Note, for example, that if $y_1 = 0$, Eq. (15.10) will force x_1 to zero; hence, the cost of reactor I as given in Eq. (15.9) is also zero. If, on the other hand, $y_1 = 1$, x can lie anywhere between 0 and 20, and in that case the cost equation in Eq. (15.9) for reactor I will correspond to a linear cost in the feed with the fixed charge of 7.5.

Using the new cost models in Eq. (15.9), our problem Eq. (15.8) can then be written as the MILP problem:

$$\min C = 7.5y_1 + 6.4 x_1 + 5.5 y_2 + 6.0 x_2$$

$$\text{s.t.} \quad 0.8 x_1 + 0.67 x_2 = 10 \qquad (15.12)$$

$$x_1 - 20y_1 \leq 0 \qquad x_2 - 20y_1 \leq 0$$

$$x_1, x_2 \geq 0 \qquad y_1, y_2 = 0,1$$

As mentioned above, this problem can be solved with a branch and bound enumeration procedure (see Appendix A) in which we do not require the analysis of all possible 0–1 combinations. However, since this problem is very small, let us consider an exhaustive enumeration of all the combination of the binary variables y. For each combination of fixed y values, problem (Eq. 15.12) reduces to an LP that has a unique global optimum because it is a convex optimization problem. The results are as follows:

y_1	y_2	$C(\$/hr)$
0	0	Infeasible solution-the mass balance is violated
1	0	87.5
0	1	95.5
1	1	93.0

This solution indicates that the global optimal solution is given by selecting reactor I with a cost of \$87.5/hr. We have thus been able to locate the global optimum with the MILP formulation. In essence, what we have done through this formulation is to discretize the search space so as to be able to handle the nonconvex cost functions for the reactors. If we had used no binary variables but only linear costs, we would have obtained an LP that would actually yield the same type of result for this particular problem (i.e., select reactor I). However, the limitation of the LP model is that it does not account for the effect of economies of scale (see Figure 15.13). Therefore, when we deal with larger networks, the solutions will tend to exhibit more units and streams than is actually practical (see exercise 4). Further, by not having binary variables we cannot impose other logical constraints to our problem. For instance, we may want to specify in our example that at least one of the reactors be selected. This we can easily specify in an MILP with the constraint,

$$y_1 + y_2 \geq 1 \tag{15.13}$$

Finally, if, instead of using fixed conversions for the reactors, we had nonlinear equations for the conversions, the problem would correspond to an MINLP problem.

15.6 SUMMARY OF MATHEMATICAL MODELS

From the previous sections in this chapter we can conclude the following general points for modeling optimization problems for synthesis:

Given a superstructure of alternatives for a given design problem, problem (P0) corresponds in general to an MINLP. Given the fact that 0–1 variables normally appear linearly in the objective and constraints, the more specific form of the mathematical programming model is

$$\begin{aligned}
\text{Min} \quad & Z = c^T y + f(x) \\
\text{st} \quad & h(x) = 0 \\
& g(x) + My \leq 0 \\
& x \in X, y \in Y
\end{aligned} \tag{P1}$$

where x is the vector of continuous variables involved in design, such as pressures, temperatures, and flowrates; while y is the vector of binary decision variables, such as existence of a particular stream or unit. Integer variables might also be involved but these are often expressed in terms of 0–1 variables. Also, model (P1) may contain among the inequalities pure integer constraints for logical specifications (e.g., select only one reactor type).

We can either solve this problem directly or reduce it to the following problems:

- NLP if we remove the binary variables.
- MILP if we use linear approximations for the cost and performance equations while keeping the binary variables.
- LP as the above but binary variables are excluded.

Global optimum solutions can be determined with LP and MILP formulations. The former, however, may lead to systems with many units and streams as it ignores effects of economies of scale. With NLP and MINLP formulations, unless special algorithms for global optimization are used, there is a significant risk of not obtaining the global optimum solution if the problem is nonconvex (e.g., due to the concave cost functions). Global optimum solutions are guaranteed if the problem is convex.

With binary variables in the MINLP or MILP we can handle logical constraints that are often very useful in synthesis problems. In the next section, we will show how propositional logic can be used to help us to model these constraints. In the next chapters, we will actually make use of LP, MILP, NLP, and MINLP formulations for modeling synthesis problems. However, we will keep in mind the above guidelines when developing these models.

15.7 MODELING OF LOGIC CONSTRAINTS AND LOGIC INFERENCE

Because a large part of the next chapters will deal with the development of mixed-integer optimization methods, we will present in this section a framework that should be helpful for deriving constraints involving 0–1 variables. Some of these constraints are quite straightforward, but some are not. For instance, specifying that exactly only one reactor be selected among a set of candidate reactors $i \in R$ is simply expressed as,

$$\sum_{i \in R} y_i = 1 \qquad (15.14)$$

On the other hand, consider representing the constraint: "if the absorber to recover the product is selected or the membrane separator is selected, then do not use cryogenic separation". We could by intuition and trial and error arrive at the following constraint,

$$y_A + y_M + 2y_{CS} \le 2 \qquad (15.15)$$

where y_A, y_M, and y_{CS} represent 0–1 variables for selecting the corresponding units (absorber, membrane, cryogenic separation). Note that if $y_A = 1$ and/or $y_M = 1$ (Eq. 15.15) forces $y_{CS} = 0$. We will see, however, that we can systematically arrive at the alternative constraints,

$$y_A + y_{CS} \le 1 \qquad (15.16)$$
$$y_M + y_{CS} \le 1$$

which are not only equivalent to Eq. (15.15) but also more efficient in the sense that they are "tighter" because they constrain more the feasible region (see exercise 7).

In order to systematically derive constraints involving 0–1 variables, it is useful to first think of the corresponding propositional logic expression that we are trying to model as described in Raman and Grossmann (1991). For this we first must consider basic logical operators to determine how each can be transformed into an equivalent representation in the form of an equation or inequality. These transformations are then used to convert general logical expressions into an equivalent mathematical representation (Cavalier and Soyster, 1987; Williams, 1985).

To each literal P_i that represents a selection or action, a binary variable y_i is assigned. Then the negation or complement of P_i ($\neg P_i$) is given by $1 - y_i$. The logical value of true corresponds to the binary value of 1 and false corresponds to the binary value of 0. The basic operators used in propositional logic and the representation of their relationships are shown in Table 15.1. From this table, it is easy to verify, for instance, that the logical proposition in $y_1 \vee y_2$ reduces to the inequality in Eq. (15.13).

Sec. 15.7 Modeling of Logic Constraints and Logic Inference

TABLE 15.1 Constraint Representation of Logic Propositions and Operators

Logical Relation	Comments	Boolean Expression	Representation as Linear Inequalities
Logical OR		$P_1 \vee P_2 \vee .. \vee P_r$	$y_1 + y_2 + .. + y_r \geq 1$
Logical AND		$P_1 \wedge P_2 \wedge .. \wedge P_r$	$y_1 \geq 1$ $y_2 \geq 1$ .. $y_r \geq 1$
Implication	$P_1 \Rightarrow P_2$	$\neg P_1 \vee P_2$	$1 - y_1 + y_2 \geq 1$
Equivalence	P_1 if and only if P_2 $(P_1 \Rightarrow P_2) \wedge (P_2 \Rightarrow P_1)$	$(\neg P_1 \vee P_2) \wedge (\neg P_2 \vee P_1)$	$y_1 = y_2$
Exclusive OR	Exactly one of the variables is true	$P_1 \veebar P_2 \veebar .. \veebar P_r$	$y_1 + y_2 + .. + y_r = 1$

With the basic equivalent relations given in Table 15.1 (e.g., see Williams, 1985), one can systematically model an arbitrary propositional logic expression that is given in terms of OR, AND, IMPLICATION operators, as a set of linear equality and inequality constraints. One approach is to systematically convert the logical expression into its equivalent *conjunctive normal form* representation, which involves the application of pure logical operations (Raman and Grossmann, 1991). The conjunctive normal form is a conjunction of clauses, $Q_1 \wedge Q_2 \wedge ... \wedge Q_s$ (i.e., connected by AND operators $\wedge$). Hence, for the conjunctive normal form to be true, each clause Q_i must be true independent of the others. Also, since a clause Q_i is just a disjunction of literals, $P_1 \vee P_2 \vee ... \vee P_r$ (i.e., connected by OR operators $\vee$), it can be expressed in the linear mathematical form as the inequality.

$$y_1 + y_2 + ... + y_r \geq 1 \qquad (15.17)$$

The procedure to convert a logical expression into its corresponding conjunctive normal form was formalized by Clocksin and Mellish (1981). The systematic procedure consists of applying the following three steps to each logical proposition:

1. Replace the implication by its equivalent disjunction,

$$P_1 \Rightarrow P_2 \quad \Leftrightarrow \quad \neg P_1 \vee P_2 \qquad (15.18)$$

2. Move the negation inward by applying DeMorgan's Theorem:

$$\neg (P_1 \wedge P_2) \quad \Leftrightarrow \quad \neg P_1 \vee \neg P_2 \qquad (15.19)$$

$$\neg (P_1 \vee P_2) \quad \Leftrightarrow \quad \neg P_1 \wedge \neg P_2 \qquad (15.20)$$

3. Recursively distribute the "OR" over the "AND", by using the following equivalence:

$$(P_1 \wedge P_2) \vee P_3 \Leftrightarrow (P_1 \vee P_3) \wedge (P_2 \vee P_3) \quad (15.21)$$

Having converted each logical proposition into its conjunctive normal form representation, $Q_1 \wedge Q_2 \wedge ... \wedge Q_s$, it can then be easily expressed as a set of linear equality and inequality constraints.

The following two examples illustrate the procedure for converting logical expressions into inequalities.

EXAMPLE 15.4

Consider the logic condition we gave above "if the absorber to recover the product is selected or the membrane separator is selected, then do not use cryogenic separation". Assigning the boolean literals to each action P_A = select absorber, P_M = select membrane separator, P_{CS} = select cryogenic separation, the logic expression is given by:

$$P_A \vee P_M \Rightarrow \neg P_{CS} \quad (15.22)$$

Removing the implication, as in (15.18), yields,

$$\neg (P_A \vee P_M) \vee \neg P_{CS} \quad (15.23)$$

Applying De Morgan's Theorem, as in Eq. (15.20), leads to,

$$(\neg P_A \wedge \neg P_M) \vee \neg P_{CS} \quad (15.24)$$

Distributing the OR over the AND gives,

$$(\neg P_A \vee \neg P_{CS}) \wedge (\neg P_M \vee \neg P_{CS}) \quad (15.25)$$

Assigning the corresponding 0–1 variables to each term in the above conjunction, and using Eq. (15.17),

$$\begin{aligned} 1 - y_A + 1 - y_{CS} &\geq 1 \\ 1 - y_M + 1 - y_{CS} &\geq 1 \end{aligned} \quad (15.26)$$

which can be rearranged to the two inequalities in Eq. (15.16),

$$\begin{aligned} y_A + y_{CS} &\leq 1 \\ y_M + y_{CS} &\leq 1 \end{aligned} \quad (15.27)$$

EXAMPLE 15.5

Consider the proposition

$$(P_1 \wedge P_2) \vee P_3 \Rightarrow (P_4 \vee P_5) \quad (15.28)$$

By removing the implication, the above proposition yields from Eq. (15.18),

$$\neg [(P_1 \wedge P_2) \vee P_3] \vee P_4 \vee P_5 \quad (15.29)$$

Further, from Eqs. (15.19) and (15.20), moving the negation inwards leads to the following two steps,

Sec. 15.7   Modeling of Logic Constraints and Logic Inference

$$[\neg(P_1 \wedge P_2) \wedge \neg P_3] \vee P_4 \vee P_5 \qquad (15.30)$$

$$[(\neg P_1 \vee \neg P_2) \wedge \neg P_3] \vee P_4 \vee P_5 \qquad (15.31)$$

Recursively distributing the "OR" over the "AND" as in Eq. (15.21) the expression becomes

$$(\neg P_1 \vee \neg P_2 \vee P_4 \vee P_5) \wedge (\neg P_3 \vee P_4 \vee P_5) \qquad (15.32)$$

which is the conjunctive normal form of the proposition involving two clauses. Translating each clause into its equivalent mathematical linear form, the proposition is then equivalent to the two constraints,

$$\begin{aligned} y_1 + y_2 - y_4 - y_5 &\leq 1 \\ y_3 \quad\quad - y_4 - y_5 &\leq 0 \end{aligned} \qquad (15.33)$$

From the above example it can be seen that logical expressions can be represented by a set of inequalities. An integer solution that satisfies all the constraints will then determine a set of values for all the literals that make the logical system consistent. This is a logical inference problem where given a set of n logical propositions, one would like to prove whether a certain clause is always true.

It should be noted that the one exception where applying the above procedure becomes cumbersome is when dealing with constraints that limit choices, for example, select no more than one reactor. In that case it is easier to directly write the constraint and not go through the above formalism.

As an application of the material above, let us consider logic inference problems in which given the validity of a set of propositions, we have to prove the truth or the validity of a conclusion that may be either a literal or a proposition. The logic inference problem can be expressed as:

$$\begin{aligned} \text{Prove} \quad & Q_u \\ \text{st} \quad & B(Q_1, Q_2 ... Q_s) \end{aligned} \qquad (15.34)$$

where Q_u is the clause or proposition expressing the conclusion to be proved and B is the set of clauses Q_i, $i = 1, 2, ..., s$.

Given that all the logical propositions have been converted to a set of linear inequalities, the inference problem in Eq. (15.34) can be formulated as the following MILP (Cavalier and Soyster, 1987):

$$\begin{aligned} \text{Min} \quad & Z = \sum_{i \in I(u)} c_i y_i \\ \text{st} \quad & A \, y \geq a \\ & y \in \{0,1\}^n \end{aligned} \qquad (15.35)$$

where $A \, y \geq a$ is the set of inequalities obtained by translating $B(Q_1, Q_2, .., Q_s)$ into their linear mathematical form, and the objective function is obtained by also converting the clause Q_u that is to be proved into its equivalent mathematical form. Here, $I(u)$ corresponds to the index set of the binary variables associated with the clause Q_u. This clause is always true if $Z = 1$ on minimizing the objective function as an integer programming problem. If $Z = 0$ for the optimal integer solution, this establishes an instance where the clause is false. Therefore, in this case, the clause is not always true.

In many instances, the optimal integer solution to problem (15.35) will be obtained by solving its linear programming relaxation (Hooker, 1988). Even if no integer solution is obtained, it may be possible to reach conclusions from the relaxed LP problem if the solution is one of the following types (Cavalier and Soyster, 1987):

1. $Z_{relaxed} > 0$: The clause is always true even if $Z_{relaxed} < 1$. Since Z is a lower bound to the solution of the integer programming problem, this implies that no integer solution with $Z = 0$ exists. Thus, the integer solution will be $Z = 1$.
2. $Z_{relaxed} = 0$, and the solution is fractional and unique: The clause is always true because there is no integer solution with $Z = 0$.

For the case when $Z_{relaxed} = 0$ and the solution is fractional but not unique, one cannot reach any conclusions from the solution of the relaxed *LP*. The reason is that there may be other integer-valued solutions to the same problem with $Z_{relaxed} = 0$. In this way, just by solving the relaxed linear programming problem in Eq. (15.35), one might be able to make inferences. The following example will illustrate a simple application in process synthesis.

EXAMPLE 15.6

Reaction Path Synthesis involves the selection of a route for the production of the required products starting from the available raw materials. All chemical reactions can be expressed in the form of clauses in propositional logic and can therefore be represented by linear mathematical relations. The specific example problem is to investigate the possibility of producing H_2CO_3 given that certain raw materials are available and the possible reactions.

The chemical reactions are given by

$$\begin{aligned} H_2O + CO_2 &\longrightarrow H_2CO_3 \\ C + O_2 &\longrightarrow CO_2 \end{aligned} \quad (15.36)$$

assuming that H_2O, C, and O_2 are available. Expressing the reactions in logical form yields

$$\begin{aligned} H_2O \wedge CO_2 &\Rightarrow H_2CO_3 \\ C \wedge O_2 &\Rightarrow CO_2 \end{aligned} \quad (15.37)$$

The objective is to prove whether H_2CO_3 can be formed given that H_2O, C, and O_2 are available. Define binary variables corresponding to each of C, O_2, CO_2, H_2O, and H_2CO_3. Translating the above logical expressions into linear inequalities, the inference problem in Eq. (15.35) becomes the following MILP problem,

$$\begin{aligned} Z = \text{Min} \quad & y_{H2CO3} \\ \text{st} \quad & y_{H2O} + y_{CO2} - y_{H2CO3} \leq 1 \\ & y_C + y_{O2} - y_{CO2} \leq 1 \\ & y_{H2O} = 1 \\ & y_C = 1 \\ & y_{O2} = 1 \\ & y_C, y_{O2}, y_{CO2}, y_{H2O}, y_{H2CO3} \in \{0,1\} \end{aligned} \quad (15.38)$$

Sec. 15.8   Modeling of Disjunctions

> The objective involves the minimization of y_{H2CO3} because the objective is to prove whether H_2CO_3 can be found. Solving the relaxed LP problem yields an integer solution with $Z = 1$ and $y_{H2CO3} = y_{CO2} = 1$. This solution is then interpreted as "H2CO3 can always be produced from H_2O, C, and O_2 given the above reactions".

Finally, it should be noted that the MILP in Eq. (15.35) can easily be extended for handling heuristic rules that may be violated (Raman and Grossmann, 1991). To model the potential violation of heuristics, the following logic relation is considerded,

$$\text{Clause OR } v \qquad (15.39)$$

where either the clause is true or it is being violated (v). In order to discriminate between weak and strong rules, penalties are associated with the violation v_i of each heuristic rule, $i = 1,...,m$. The penalty w_i is a non-negative number that reflects the uncertainty of the corresponding logical expression. The more uncertain the rule, the lower the penalty for its violation. In this way, the logical inference problem with uncertain knowledge can be formulated as an MILP problem where the objective is to obtain a solution that satisfies all the logical relationships (i.e., $Z = 0$), and if that is not possible, to obtain a solution with the least total penalty for violation of the heuristics:

$$\begin{aligned} \text{Min} \quad & Z = w^T v \\ \text{st} \quad & A y \geq a \quad : \text{ Logical facts} \\ & B y + v \geq b \quad : \text{ Heuristics} \\ & y \in \{0,1\}^n, v \geq 0 \end{aligned} \qquad (15.40)$$

Note that no violations are assigned to the inequalities $Ay \geq a$ since these correspond to hard logical facts that always have to be satisfied. In this way Eq. (15.40) can be used to solve inference problems involving logic relations and heuristics. Clearly, if the solution is $Z = 0$, it means that it is possible to find a solution without violating heuristics. In general, the solution to Eq. (15.40) will determine a design that best satisfies the possibly conflicting qualitative knowledge about the system.

15.8 MODELING OF DISJUNCTIONS

In the previous section we presented a systematic framework based on logic for modeling constraints involving 0–1 variables. In a number of cases, however, we will have to deal with logic constraints that involve continuous variables. A good example is the following condition when selecting among two reactors:

If select reactor 1, then pressure P must lie between 5 and 10 atmospheres.
If select reactor 2, then pressure P must lie between 20 and 30 atmospheres.

To represent logic with continuous variables we will consider linear disjunctions of the form:

$$\underset{i \in D}{\vee}[A_i x \leq b_i] \qquad (15.41)$$

where $\vee$ is the OR operator that applies to a set of disjunctive terms D. In the above example, Eq. (15.41) reduces to:

$$\begin{bmatrix} P \leq 10 \\ -P \leq -5 \end{bmatrix} \vee \begin{bmatrix} P \leq 30 \\ -P \leq -20 \end{bmatrix} \qquad (15.42)$$

where the first term is associated to reactor 1 (y_1) and the second term to reactor 2 (y_2).

The simplest way to convert Eq. (15.41) into mixed-integer constraints is by using "big-M" constraints, which are given as follows:

$$A_i x \leq b_i + M_i(1 - y_i) \quad i \in D$$
$$\sum_{i \in D} y_i = 1 \qquad (15.43)$$
$$y_i = 0, 1 \quad i \in D$$

Note that the 0–1 variable y_i is introduced to denote which disjunction i in D is true ($y_i = 1$). The second constraint in Eq. (15.43) only allows one choice of y_i. The first set of inequalities, $i \in D$, introduce on the right-hand side a big parameter M_i, which renders the inequality redundant if $y_i = 0$. Note that if $y_i = 1$, the inequality is enforced.

As applied to Eq. (15.42) the big-M constraints yield:

$$P \leq 10 + M_1 (1 - y_1)$$
$$-P \leq -5 + M_1 (1 - y_1) \qquad (15.44)$$
$$P \leq 30 + M_2 (1 - y_2)$$
$$-P \leq 20 + M_2 (1 - y_2)$$
$$y_1 + y_1 = 1$$

Large values, such as $M_1 = 100$, $M_2 = 100$, are valid choices but produce weak "relaxations" or bounds for the objective function when the y's are treated as continuous variables. This would be, for instance, the first step in the LP branch and bound method.

An alternative for avoiding the use of big-M parameters in Eq. (15.43) is the use of the convex hull formulation, which requires disaggregating continuous variables. As shown in Balas (1985) and discussed in Turkay and Grossmann (1996), the convex hull model of Eq. (15.41) is given by:

$$x = \sum_{i \in D} z_i$$
$$A_i z_i \leq b_i y_i \qquad i \in D$$
$$\sum_{i \in D} y_i = 1$$
$$0 \leq z_i \leq U y_i \qquad i \in D \qquad (15.45)$$
$$y_i = 0, 1$$

In the above z_i are continuous variables disaggregated into as many new variables as there are terms for the disjunctions. The first equation simply equates the original variable x to the disaggregated variables z_i. The second constraint corresponds to inequalities written in terms of the disaggregated variables z_i and a 0–1 variable y_i. The third simply states that only one y_i can be set to one. The fourth constraint is optional in that it is only included if $y_i = 0$ in the second inequality does not imply $z_i = 0$. The importance of the constraints in Eq. (15.45) is that they do not require the introduction of the big-M parameter yielding a tight LP relaxation. The disadvantage is that it requires a larger number of variables and constraints.

Applied to Eq. (15.42), Eq. (15.45) yields,

$$P = P_1 + P_2$$
$$P_1 \leq 10 \, y_1 \qquad P_2 \leq 30 \, y_2 \qquad (15.46)$$
$$-P_1 \leq -5 \, y_1 \qquad -P_2 \leq -20 \, y_2$$
$$y_1 + y_2 = 1$$

It is important to note that often the convex hull formulation will simplify if there are only two terms in the disjunction and one requires the variable to take a value at zero. For instance, consider a flow $F \geq 0$ for which

$$[F \leq 20] \vee [F = 0] \qquad (15.47)$$

It can easily be shown that applying Eq. (15.45) to Eq. (15.47), since $F_2 = 0.y_2$, $F = F_1$, and hence the convex hull at Eq. (15.47) is given by

$$F \leq 20 \, y_1 \qquad (15.48)$$

In practice, the big-M constraints as in Eq. (15.43) are easiest to use and will not cause major difficulties if the problem is small. For larger problems the convex hull formulation is often the superior one.

15.9 NOTES AND FURTHER READING

A recent review on optimization approaches to process synthesis can be found in Grossmann and Daichendt (1996). Modeling is largely an art that has a large impact in mixed-integer programming. Good practices can be learned from examples. The book by Williams (1985) is perhaps the most useful. Similarly, the book by Schrage (1984) has a good number of examples for LP and MILP problems. Nemhauser and Wolsey (1988) also present some interesting examples. Finally, the papers by Raman and Grossmann (1991, 1994) provide logic-based formalisms for the modeling of the 0–1 and disjunctive constraints.

REFERENCES

Andrecovich, M. J., & Westerberg, A. W. (1985). MILP formulation for heat-integrated distillation sequence synthesis. *AIChE J.*, 31, 1461.

Balas, E. (1985). Disjunctive programming and a hierarchy at relaxations for discrete optimization problems. *SIAM J. Alg. Disc. Metn.*, **6,** 466.

Bazaraa, M. S., & Shetty, C. M. (1979). *Nonlinear Programming.* New York: Wiley.

Cavalier, T. M., & Soyster, A. L. (1987). *Logical Deduction via Linear Programming.* IMSE Working Paper 87-147, Department of Industrial and Management Systems Engineering, Pennsylvania State University.

Clocksin, W. F., & Mellish, C. S. (1981). *Programming in Prolog.* New York: Springer-Verlag.

Grossmann, I. E. (1996). *Global Optimization in Engineering Design.* Amsterdam: Kluwer.

Grossmann, I. E. (1990). MINLP Optimization strategies and algorithms for process synthesis. In J. J. Siirola, I. E. Grossmann, & G. Stephanopoulos (Eds.), *Foundations of Computer-Aided Design,* Amsterdam: Cache-Elsevier.

Grossmann, I. E., & M. M. Daichendt. (1996). New trends in optimization-based approaches to process synthesis. *Computers and Chemical Engineering,* **20,** 665–683.

Hendry, J. E., & Hughes, R. R. (1972). Generating separation process flowsheets. *Chem. Eng. Progress,* **68,** 69.

Hillier, F. S., & Lieberman, G. J. (1986). *Introduction to Operations Research.* San Francisco: Holden Day.

Hooker, J. N. (1988). Resolution vs cutting plane solution of inference problems: Some computational experience. *Operations Research Letters,* **7**(1), 1.

Minoux, M. (1986). *Mathematical Programming: Theory and Algorithms.* New York: Wiley.

Nemhauser, G. L., & Wolsey, L. A. (1988). *Integer and Combinatorial Optimization.* New York: Wiley-Interscience.

Nemhauser, G. L., Rinnoy Kan, A. H. G., & Todd, M. J. (Eds.). (1989). *Optimization. Handbook in Operations Research and Management Science, Vol. 1.* North Holland, Amsterdam: Elsevier.

Raman, R., & Grossmann, I. E. (1991). Relation between MILP modelling and logical inference for chemical process synthesis. *Computers and Chemical Engineering,* **15,** 73.

Raman, R., & Grossmann, I. E. (1994). Modeling and computational techniques for logic based integer programming. *Computers and Chemical Engineering,* **18,** 563.

Schrage, L. (1984). *Linear, Integer and Quadratic Programming with LINDO.* Redwood City: The Scientific Press.

Turkay, M., & Grossmann, I. E. (1996). Disjunctive programming techniques for the optimization of process systems with discontinuous investment costs—multiple size regions. *Ind. Eng. Chem. Research,* **35,** 2611–2623.

Williams, H. P. (1985). *Model Building in Mathematical Programming.* New York: Wiley-Interscience.

EXERCISES

1. Given a mixture of four components A, B, C, D, (A-most volatile, D-heaviest) for which two separation technologies (I and II) are to be considered:
   a. Determine the tree representation and the network representation for all the alternative sequences.
   b. Find the optimal sequence with depth-first and breadth-first given the costs below for each separator.
   c. Compare the optimal solution with the heuristic design that is obtained by determining the cheapest separator at each level of the tree.

Cost of separators ($/yr)

Separator	Technology I	Technology II
A/BCD	55,000	44,000
AB/CD	37,000	56,000
ABC/D	29,000	19,000
A/BC	42,000	34,000
AB/C	27,000	32,000
B/CD	38,000	45,000
BC/D	25,000	18,000
A/B	35,000	39,000
B/C	23,000	44,000
C/D	21,000	18,000

2. a. Show that the number of nodes in a tree where all possible combinations of m 0–1 binary variables are represented as
$$2^{m+1} - 1$$
   b. If a complete enumeration of all the nodes in the tree were required, by what factor would this enumeration increase with respect to the direct enumeration of all 0–1 combinations?

3. Suppose we would like to extend the fixed charge cost model given in section 15.5 and Figure 15.13 to handle the following condition:
$$\text{If } L \leq x \leq U \text{ then cost } C = a + b\,x$$
$$\text{If } x = 0 \text{ then cost } C = 0$$
   What would be the form of the cost function and the required constraints if L, U are positive lower and upper bounds?

4. Given are three candidate reactors for the reaction $A \rightarrow B$, where we would like to produce 10 kmol/hr of B. Up to 15 kmol/hr of reactant A are available at a price of $2/kmol. The data on the three reactors is as follows:

	Conversion	Linear cost	Fixed-charge cost
Reactor I	0.8	2.2xfeed	8.0 + 1.5xfeed
Reactor II	0.667	1.5xfeed	5.4 + 1.0xfeed
Reactor III	0.555	0.73xfeed	2.7 + 0.5xfeed

    a. Develop a network representation for this problem.
    b. Determine an LP formulation for linear reactor costs and solve.
    c. Determine an MILP formulation using the fixed-charge cost models and solve.
    d. Compare the solutions in b and c and explain any qualitative differences that might exist in the two solutions.

5. A company is considering producing a chemical C that can be manufactured with either process II or process III, both of which use as raw material chemical B. B can be purchased from another company or else manufactured with process I, which uses A as a raw material. Given the specifications below, draw the corresponding superstructure of alternatives and formulate an MILP model and solve it to decide:

    a. Which process to build (II and III are exclusive)?
    b. How to obtain chemical B?
    c. How much should be produced of product C?

    The objective is to maximize profit.

    Consider the two following cases:

    i. Maximum demand of C is 10 tons/hr with a selling price of $1800/ton.
    ii. Maximum demand of C is 15 tons/hr; the selling price for the first 10 tons/hr is $1800/ton, and $1500/ton for the excess.

Data	Investment and Operating Costs	
	Fixed ($/hr)	Variable ($/ton raw mat.)
Process I	1000	250
Process II	1500	400
Process III	2000	550

Prices: A: $500/ton
               B: $950/ton

Conversions: Process I   90% of A to B
                    Process II   82% of B to C
                    Process III   95% of B to C

Maximum supply of A: 16 tons/hr

NOTE: You may want to scale your cost coefficients (e.g., divide them by 100).

6. Repeat problem 5 for the case of an MINLP model in which the input/output relations in processes II and III are given by the nonlinear equations:

$$\text{Process II:} \quad C = 6.5 \, \ell n \, (1 + B)$$
$$\text{Process III:} \quad C = 7.2 \, \ell n \, (1 + B)$$

where B and C are the corresponding amounts of B and C in tons/hr.

7. Plot the constraints in Eqs. (15.15) and (15.16) in the unit hypercube in terms of the variables y_A, y_M, and y_{CS} to show that the constraints in Eq. (15.16) are tighter in the sense that the size of their feasible region is smaller than with Eq. (15.15). Also, which are the extreme points in the hypercube for the two alternatives?

8. Apply the procedure given in section 15.7 to convert the logic expression below into a system of inequalities with 0–1 variables:

$$\neg P1 \vee P2 \Rightarrow P3 \vee \neg P4$$

9. Formulate linear constraints in terms of binary variables for the four following cases:
   a. At least K out of M inequalities $f_j(x) \leq 0 \; j = 1,...M$ must be satisfied ($K<M$).
   b. If A is true and B is true, then C is true or D is true (inclusive OR).
   c. The choice of all 0–1 combinations for y_j, $j \in J$ is feasible, except the one for which $y_j = 0, j \in N$, $y_j = 1, j \in B$, where N and B are specified partitions of J.
   d. Given are two binary variables, x and y. Define a third binary variable z to be one if $x = y$, and $z = 0$ if x and y are different.

10. Formulate linear constraints in terms of the binary variables that are assigned for given units in the following logical conditions:
    a. Among three candidate reactors only one should be selected.
    b. Among two candidate processes for a chemical complex at most one process can be selected.
    c. If the absorber is selected, then the distillation column must be included. However, if the distillation column is selected, the absorber may or may not be included.
    d. The temperature approach constraint for a heat exchanger

    $$T_{in} - t_{out} \geq DT_{min}$$

    should only hold if the exchanger is actually selected.
    e. If reactor $R1$ and the distillation column are selected, set the minimum reactor pressure to 50 atm. Otherwise, set the minimum pressure to 65 atm.

11. Assume that it is desired to manufacture acetone. The raw materials available are ethyl alcohol (CH_3CH_2OH) and methane (CH_4). The candidate chemical reactions are listed below. Assuming that the catalysts required for all reactions and all the inorganic chemicals required are available except for CrO_3 and O_3, determine with the MILP in Eq. (15.35) if it is feasible to manufacture acetone from the given raw materials and if so, specify a reaction path.

Chemical Reactions

$$CH_3CO_2C_2H_5 \xrightarrow{NaOC_2H_5 / C_2H_5OH} CH_3COCH_2CO_2C_2H_5$$

$$CH_3COCH_2CO_2C_2H_5 \xrightarrow{H_3O^+} CH_3COCH_3 + C_2H_5OH + CO_2$$

$$CH_3CN + CH_3MgI \xrightarrow{Et_2O} CH_3C(NMgI)CH_3 \xrightarrow{H_2O / HCl} CH_3COCH_3$$

$$CH_3CHO + CH_3MgI \xrightarrow{Et_2O / H_3O^+} CH_3CHOHCH_3$$

$$CH_3CHOHCH_3 \xrightarrow{CrO_3 / H_2SO_4} CH_3COCH_3$$

$$CH_2=C(CH_3)_2 \xrightarrow{O_3 / H_2O / H_2O_2} CH_3COCH_3 + HCO_2H$$

$$CH_3I \xrightarrow{Mg / Et_2O} CH_3MgI \xrightarrow{CH_3CO_2CH_3} (CH_3)_3COH$$

$$(CH_3)_3COH \longrightarrow CH_2=C(CH_3)_2$$

$$CH_4 + I_2 \longrightarrow CH_3I + HI$$

$$CH_4 + Cl_2 \longrightarrow CH_3Cl + HCl$$

$$CH_3CH_2OH + O_2 \xrightarrow{Cr_2O_3 / Cu} CH_3CHO$$

$$CH_3Cl + NaCN \xrightarrow{H_2O} NaCl + CH_3CN$$

$$CH_3COOH + C_2H_5OH \longrightarrow CH_3CO_2C_2H_5$$

$$CH_3CHO + O_2 \longrightarrow CH_3COOH$$

12. Determine a convex hull formulation for the two following disjunctions and determine in each case whether it is possible to eliminate the use of disaggregated variables:
    a. [Cost = 10] OR [Cost = 0]
    b. [$T1 - T2 \geq 5$] OR [$T1 - T2 \leq -5$]

SYNTHESIS OF HEAT EXCHANGER NETWORKS 16

16.1 INTRODUCTION

In Chapter 10, a number of powerful insights were presented that can greatly simplify the problem of synthesizing heat exchanger networks. These insights can be summarized as follows:

- Given a minimum temperature approach, the exact amount for minimum utility consumption can be predicted prior to developing the network structure.
- Based on the pinch temperatures for minimum utility consumption, the synthesis of the network can be decomposed into subnetworks.
- The fewest number of units in each subnetwork is often equal to the number of process and utility streams minus one.
- It is possible to develop good a priori estimates of the minimum total area of heat exchange in a network.

While these insights narrow down the alternative designs for a network very considerably, by themselves they do not provide an explicit procedure for deriving the configuration of a heat exchanger network. In other words, the user has to examine by trial and error matches and stream interconnections that will hopefully come close to satisfying the targets for utility consumption, number of units, and total area. Quite often, this might not be a trivial task, especially when one is faced with a rather large number of process streams, and when splitting of streams is required. Furthermore, if we were to rely only on these insights, it is rather difficult to develop a computer program that can automatically synthesize heat exchanger networks of arbitrary structure (e.g., with stream splitting, by-passing of streams). Moreover, networks satisfying the targets may not necessarily correspond to designs with minimum cost.

In this chapter we will present algorithmic optimization models for the synthesis of heat exchanger networks that illustrate two major synthesis strategies: sequential optimization and simultaneous optimization. First, we consider sequential optimization models that exploit the above insights, and at the same time provide systematic procedures that allow the automation of this synthesis problem in the computer. The models (LP, MILP, NLP) will also allow us to expand the type of problems that we can consider (e.g., multiple utilities, constraints on the matches, stream splitting). Secondly, we will present an MINLP model in which the energy recovery, selection of matches, and areas are all optimized simultaneously.

Three basic heuristic rules that are motivated by the insights of Chapter 10 will be used in the development of algorithmic methods based on sequential optimization. In particular, it will be assumed that an optimal or near optimal network exhibits the following characteristics:

Rule 1. Minimum utility cost
Rule 2. Minimum number of units
Rule 3. Minimum investment cost

Clearly, it is possible in general to have conflicts among these rules. Therefore, we will assume that Rule 1 has precedence over Rule 2, and Rule 2 over Rule 3. In this way, our objective will be to consider first candidate networks that exhibit minimum utility cost, among these the ones that have the fewest number of units, and among these the one that has the minimum investment cost. We will show in this chapter how for each of these three steps we can develop appropriate optimization models to generate networks with all possible options for sequencing, stream splitting, mixing and bypassing. We can consider the optimization of the minimum heat recovery approach temperature (HRAT) either in an outer loop of this procedure or else through the approximate procedure presented in Part III. Also, the precedence order of the heuristics can be indirectly challenged through constraints on matches. In section 16.3 we will present a simultaneous MINLP model in which the above rules do not have to be applied.

16.2 SEQUENTIAL SYNTHESIS

16.2.1 Minimum Utility Cost

Let us consider the following example to motivate a useful problem representation for the prediction of the minimum utility cost.

Sec. 16.2 Sequential Synthesis   529

EXAMPLE 16.1

Determine the minimum utility consumption for the two hot and two cold streams given below:

	Fcp (MW/C)	Tin (C)	Tout (C)
H1	1	400	120
H2	2	340	120
C1	1.5	160	400
C2	1.3	100	250

Steam : 500°C
Cooling water: 20–30°C
Minimum recovery approach temperature (HRAT): 20°C

The data for this problem are displayed in Table 16.1, where heat contents of the hot and cold processing streams are shown at each of the temperature intervals, which are based on the inlet and highest and lowest temperatures. The flows of the heat contents we can represent in the heat cascade diagram of Figure 16.1. Here the heat contents of the hot streams are introduced in the corresponding intervals, while the heat contents of the cold streams are extracted also from their corresponding intervals. The variables R_1, R_2, R_3, represent heat residuals, while Q_s, Q_w represent the heating and cooling loads respectively.

TABLE 16.1 Temperature Intervals and Heat Contents (MW) for Example 16.1

Temperature Intervals (K)			C1	H1	H2	C1	C2
	420 — 400		↑				
	int 1						
H1	400 — 380					30	
	int 2						
H2	340 — 320			60		90	
	int 3		250				
	180 — 160		↑	160	320	240	117
	int 4						
↓ ↓	120 — 100			60	120		78
			C2	280	440	360	195

The usefulness of the heat cascade diagram in Figure 16.1 is that it can be regarded as a transshipment problem that we can formulate as a linear programming problem (Papoulias and Grossmann, 1983). In terms of the transshipment model, hot streams are treated as source nodes, and cold streams as destination nodes. Heat can then be regarded as a commodity that must be transferred from the sources to the destinations through some intermediate "warehouses" that correspond to the temperature intervals that guarantee feasible heat exchange. When not all of

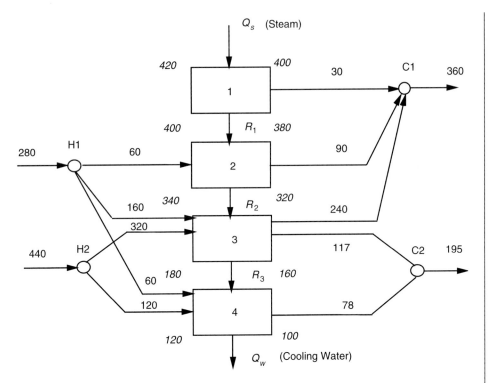

FIGURE 16.1 Heat cascade diagram.

the heat can be allocated to the destinations (cold streams) at a given temperature interval, the excess is cascaded down to lower temperature intervals through the heat residuals.

To show how we can formulate the minimum utility consumption in Table 16.1 as an LP transshipment problem, let us consider first the heat balances around each temperature level in Figure 16.1. These are given by:

$$\begin{aligned} R_1 + 30 &= Q_s \\ R_2 + 90 &= R_1 + 60 \\ R_3 + 357 &= R_2 + 480 \\ Q_w + 78 &= R_3 + 180 \end{aligned} \quad (16.1)$$

From Eq. (16.1) it is clear that we have a system of 4 equations in 5 unknowns: R_1, R_2, R_3, Q_s, Q_w. Thus, there is one degree of freedom, which in turn implies that we have an optimization problem.

By considering the objective of minimization of utility loads, rearranging Eq. (16.1) and introducing nonnegativity constraints on the variables, our problem can be formulated as the LP:

Sec. 16.2 Sequential Synthesis

$$\min Z = Q_s + Q_w$$
$$\text{s.t.} \quad R_1 - Q_s = -30$$
$$R_2 - R_1 = -30$$
$$R_3 - R_2 = 123 \quad (16.2)$$
$$Q_w - R_3 = 102$$
$$Q_s, Q_w, R_1, R_2, R_3 \geq 0$$

If we solve this problem with a standard LP package (e.g., LINDO), we obtain for the utilities $Q_s = 60$ MW, $Q_w = 225$ MW, and for the residuals $R_1 = 30$ MW, $R_2 = 0$, $R_3 = 123$ MW. Since $R_2 = 0$ this means that we have a pinch point at the temperature level 340°–320°C, which lies between intervals 2 and 3 (see Figure 16.1).

The above example then shows that we can formulate the minimum utility consumption problem as an LP. This model is actually equivalent to the calculation of the problem table that was given in Part III. This can be shown if we rearrange the constraints in Eq. (16.2) by successively substituting for the heat residuals so as to leave the right-hand sides as a function of Q_s; that is,

$$\min Z = Q_s + Q_w$$
$$\text{s.t.} \quad R_1 = Q_s - 30$$
$$R_2 = R_1 - 30 = Q_s - 60$$
$$R_3 = R_2 + 123 = Q_s + 63 \quad (16.3)$$
$$Q_w = R_3 + 102 = Q_s + 165$$
$$R_1, R_2, R_3, Q_s, Q_w \geq 0$$

Suppose we now want to determine the smallest Q_s such that all the variables in the left-hand side are nonnegative. Clearly if $Q_s = 0$, the largest violation of the nonnegativity constraints will be –60 in the second equation of Eq. (16.3). Therefore, if we set $Q_s = 60$ MW, this will be the smallest value for which we can satisfy all nonnegativity constraints. By then substituting for this value in Eq. (16.3), we get $R_1 = 30$, $R_2 = 0$, $R_3 = 123$, $Q_W = 225$, which is the same result that we obtained for the LP in Eq. (16.2).

Thus, we have shown that the LP for minimum utility consumption leads to equivalent results as the problem table given in Chapter 10. We may then wonder what the advantages are of having such a model. As we will see, the transshipment model can be easily generalized to the case of multiple utilities, and where the objective function corresponds to minimizing the utility cost. Furthermore, we will show in the next sections how this model can be expanded so as to handle constraints on the matches, and so as to predict the matches for minimizing the number of units. In Chapters 17 and 18 we will also see how we can embed the equations of the transsshipment model within an optimization model for synthesizing a process system (e.g. separation sequences, process flowsheets) where the flows of the process streams are unknown.

The transshipment model for predicting the minimum utility cost given an arbitrary number of hot and cold utilities can be formulated as follows. First, we consider that we have K temperature intervals that are based on the inlet temperatures of the process streams, highest and lowest stream temperatures, and of the intermediate utilitites whose inlet temperatures fall within the range of the temperatures of the process streams (see Chapter 10). We assume as in the above example that the intervals are numbered from the top to the bottom. We can then define the following index sets:

$$H_k = \{\, i \mid \text{hot stream } i \text{ supplies heat to interval } k\,\}$$
$$C_k = \{\, j \mid \text{cold stream } j \text{ demands heat from interval } k\,\} \quad (16.4)$$
$$S_k = \{\, m \mid \text{hot utility } m \text{ supplies heat to interval } k\,\}$$
$$W_k = \{\, n \mid \text{cold utility } n \text{ extracts heat from interval } k\,\}$$

When we consider a given temperature interval k, we will have the following known parameters and variables (see Figure 16.2):

Known parameters: Q^H_{ik}, Q^c_{jk} heat content of hot stream i and cold stream j in interval k

                       c_m, c_n unit cost of hot utility m and cold utility n

Variables: Q^S_m, Q^W_n heat load of hot utility m and cold utility n

           R_k heat residual exiting interval k

The minimum utility cost for a given set of hot and cold processing streams can then be formulated as the LP (Papoulias and Grossmann, 1983):

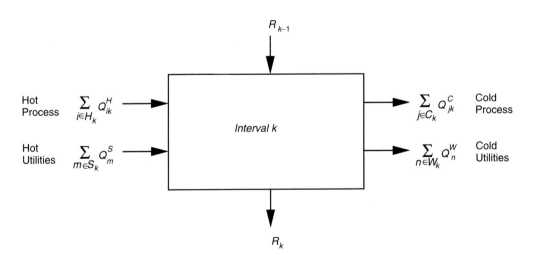

FIGURE 16.2 Heat flows in interval k.

Sec. 16.2 Sequential Synthesis

$$\min Z = \sum_{m \in S} c_m Q_m^S + \sum_{n \in W} c_n Q_n^W \quad (16.5)$$

$$\text{s.t. } R_k - R_{k-1} - \sum_{m \in S_k} Q_m^S + \sum_{n \in W_k} Q_n^W = \sum_{i \in H_k} Q_{ik}^H - \sum_{j \in C_k} Q_{jk}^C \quad k = 1, \ldots K$$

$$Q_m^S \geq 0 \quad Q_n^W \geq 0 \quad R_k \geq 0 \quad k = 1, \ldots K-1$$

$$R_0 = 0, R_K = 0$$

In the above, the objective function represents the total utility cost, while the K equations are heat balances around each temperature interval k. Note that this LP will in general be rather small as it will have K rows and $n_H + n_c + K - 1$ variables. The model in Eq. (16.5) we will denote as the condensed LP transshipment model to differentiate it from the LP that will be given in section 16.3 for constrained matches. It should also be noted that in the above formulation it would be very easy to impose upper limits on the heat loads that are available from some of the utilities (e.g., maximum heat from low pressure steam).

EXAMPLE 16.2

Given the data in Table 16.2 for two hot and two cold processing streams and two hot and one cold utility, determine the minimum utility cost with the LP transshipment model in Eq. (16.5). By considering the temperature intervals in Table 16.3, and calculating the heat contents of the process streams at each interval, the *LP* for this example is:

$$\min Z = 80000 \, Q_{HP} + 50000 \, Q_{LP} + 20000 \, Q_{CW}$$

$$\text{s.t.} \quad R_1 - Q_{HP} = -60$$

$$R_2 - R_1 = 10 \quad (16.6)$$

$$R_3 - R_2 - Q_{LP} = -15$$

$$-R_3 + Q_{CW} = 75$$

$$R_1, R_2, R_3, Q_{HP}, Q_{LP}, Q_{CW} \geq 0$$

TABLE 16.2 Data for Example 16.2

	FCp (MW/K)	T_{in}(K)	T_{out}(K)
H1	2.5	400	320
H2	3.8	370	320
C1	2	300	420
C2	2	300	370

HP Steam: 500K $80/kWyr
LP Steam: 380K $50/kWyr
Cooling Water: 300K $20/kWyr
Minimum Recovery Approach Temperature (HRAT): 10K

TABLE 16.3 Temperature Intervals of Example 16.2

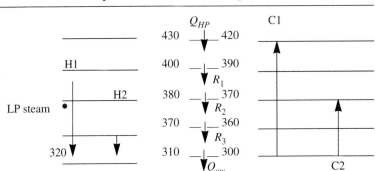

The solution to this LP yields the following results:

Utility cost: $Z = 6{,}550{,}000$ \$/yr.
Heat load high pressure steam: $Q_{HP} = 60$ MW
Heat load low pressure steam: $Q_{LP} = 5$ MW
Heat load cooling water: $Q_{CW} = 75$ MW
Residuals: $R_1 = 0$, $R_2 = 10$ MW, $R_3 = 0$.

The two above zero residuals imply that there are two pinch points for this problem: at 400–390 K, and at 370–360 K. This means that the temperature intervals in this problem can be partitioned into three subnetworks:

Subnetwork 1: above 400–390 K
Subnetwork 2: between 400–390 K and 370–360 K
Subnetwork 3: below 370–360 K

16.2.2 Minimum Utility Cost with Constrained Matches

In practice it might not always be desirable or possible to exchange heat between any given pair of hot and cold streams. This could be due to the fact that the streams are too far apart or because of other operational considerations such as control, safety or startup. Therefore, it would be clearly desirable to extend our LP transshipment formulation to the case when we impose certain constraints on the matches. The most common would simply be to forbid the heat exchange between certain pairs of streams. We could also think of requiring that a minimum or maximum amount of heat be exchanged between certain pairs of streams (e.g. forcing the use of utilities on some of the streams).

The LP transshipment model in Eq. (16.5) implicitly assumes that any given pair of hot and cold streams can exchange heat since there was no information as to which pairs

Sec. 16.2 Sequential Synthesis 535

of streams actually exchange heat. In order to develop an LP formulation where we do have that information, we can consider the two following alternative models:

1. Transportation model where we consider directly all the feasible links for heat exchange between each pair of hot and cold streams over their corresponding temperature intervals (Cerda and Westerberg, 1983). Figure 16.3 illustrates this representation for Example 16.1.
2. Expanded transshipment model (Papoulias and Grossmann, 1983) where we consider within each temperature interval a link for the heat exchange between a given pair of hot and cold streams, where the cold stream is present at that interval and the hot stream is either also present, or else it is present in a higher temperature interval. Figure 16.4 illustrates this representation for Example 16.1.

In principle we could use either of the two representations. However, we will concentrate on the second one for continuity with the previous section, and also because it leads to LP problems of smaller size. So let us now try to explain in greater detail on how the representation in Figure 16.4 is obtained.

The basic idea in the expanded transshipment model is as follows. First, instead of assigning a single overall heat residual R_k exiting at each temperature level k, we will assign individual heat residuals R_{ik}, R_{mk} for each hot stream i and each hot utility m that are present at or above that temperature interval k. Secondly, within that interval k we will define the variable Q_{ijk} to denote the heat exchange between hot stream i and a cold stream j. Likewise, we can define similar variables for the exchange between process streams and

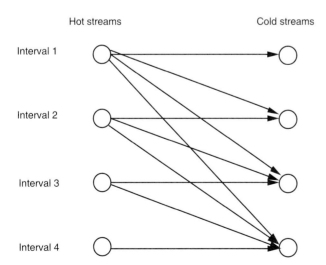

FIGURE 16.3 Representation of heat flows for transportation model.

536 Synthesis of Heat Exchanger Networks   Chap. 16

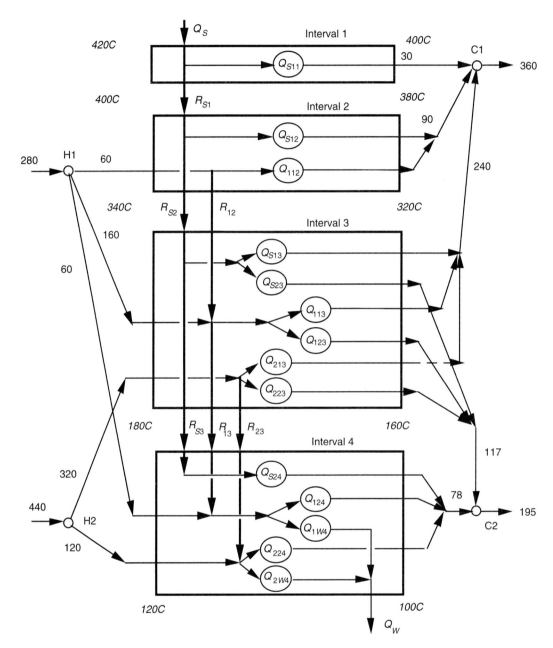

FIGURE 16.4   Representation of expanded transshipment model for Example 16.1.

Sec. 16.2 Sequential Synthesis

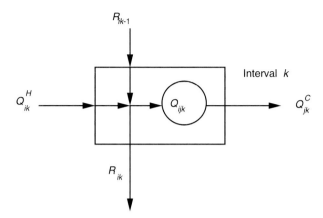

FIGURE 16.5 Interval for expanded transshipment model.

utilities. Figure 16.5 illustrates the above ideas for an interval k where we consider a hot stream i and a cold stream j.

We should note that in general a given pair of streams can exchange heat within a given temperature interval k if either of the two following conditions hold:

1. Hot stream i and cold stream j are present in interval k. This case is obvious as seen in Figure 16.5.
2. Cold stream j is present in interval k, but hot stream i is only present at a higher temperature interval. An example of this case is shown in Figure 16.6, where hot stream i can exchange heat at interval 3, although it is not present there. The reason the heat exchange can take place is simply because hot stream i is transferring heat to interval 3 through the residual R_{i2} that is coming from interval 2. Another example is shown in Figure 16.4 where steam can exchange heat with cold stream C1 at interval 2.

Based on the above observations we can then formulate an expanded LP transshipment model where we do include the information on the exchange of heat between any given pair of streams. Let us define first the following index sets:

$$H'_k = \{i \mid \text{hot stream } i \text{ is present at interval } k \text{ or at a higher interval}\} \quad (16.7)$$
$$S'_k = \{m \mid \text{hot utility } m \text{ is present at interval } k \text{ or at a higher interval}\}$$

The index sets C_k, W_k are defined the same as in Eq. (16.4).

As for the parameters and variables, we will have the following (see Figure 16.7):

Q_{ijk}: Exchange of heat of hot stream i and cold stream j at interval k
Q_{mjk}: Exchange of heat of hot utility m and cold stream j at interval k
Q_{ink}: Exchange of heat of hot stream i and cold utility n at interval k \quad (16.8)
R_{ik}: Heat residual of hot stream i exiting interval k
R_{mk}: Heat residual of hot utility m exiting interval k

538   Synthesis of Heat Exchanger Networks   Chap. 16

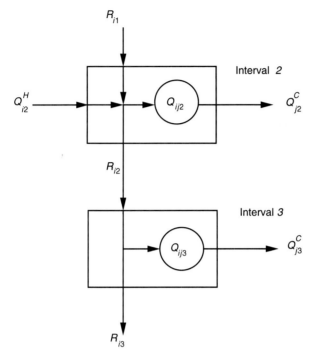

FIGURE 16.6 Example of heat flows in case a hot stream does not provide heat to all intervals.

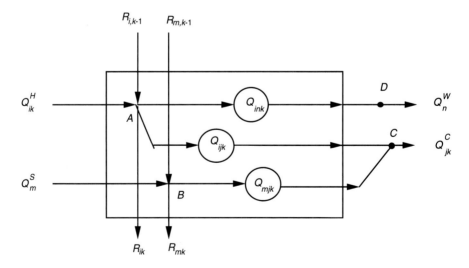

FIGURE 16.7 Heat flows in expanded transshipment model.

Sec. 16.2 Sequential Synthesis

The variables Q_m^S, Q_n^W and the parameters Q_{ik}^H, Q_{jk}^C, c_m, c_n are identical to those of the previous section.

In contrast to the compact LP transshipment model Eq. (16.5) where we simply did an overall heat balance around each temperature level, in this case we have to perform balances at the following points within each temperature interval:

1. For the hot process and utility streams at the internal nodes that relate the heat content, residuals, and heat exchanges (i.e., nodes A and B in Figure 16.7).
2. For the cold process and utility streams at the destination nodes that relate the heat content and heat exchanges (i.e., nodes C and D in Figure 16.7).

In this way the expanded LP transshipment model by Papoulias and Grossmann (1983) can be formulated as:

$$\min \ Z = \sum_{m \in S} c_m Q_m^S + \sum_{n \in W} c_n Q_n^W$$

$$\text{s.t.} \quad R_{ik} - R_{i,k-1} + \sum_{j \in C_k} Q_{ijk} + \sum_{n \in W_k} Q_{ink} = Q_{ik}^H \quad i \in H_k'$$

$$R_{mk} - R_{m,k-1} + \sum_{j \in C_k} Q_{mjk} - Q_m^S = 0 \quad m \in S_k'$$

$$\sum_{i \in H_k} Q_{ijk} + \sum_{m \in S_k} Q_{mjk} = Q_{jk}^C \quad j \in C_k \quad\quad (16.9)$$

$$\sum_{i \in H_k} Q_{ink} - Q_n^W = 0 \quad\quad n \in W_k \quad k = 1, \ldots K$$

$$R_{ik}, \ R_{mk}, \ Q_{ijk}, \ Q_{mjk}, \ Q_{ink}, \ Q_m^S, \ Q_n^W \geq 0$$

$$R_{i0} = R_{iK} = 0$$

Note that the size of this LP is obviously larger than the one in Eq. (16.5). The importance of the formulation in Eq. (16.9) is the fact that we can very easily specify constraints on the matches. For example, if we want to forbid a match between hot i and cold j all we need to do is to set $Q_{ijk} = 0$ for all intervals k. Or, alternatively, we just simply delete these variables from our formulation. For the case when we want to impose a given match we can do this by specifying that its total heat exchange, which is the sum of Q_{ijk} over all intervals, must lie within some specified lower and upper bounds. That is,

$$Q_{ij}^L \leq \sum_{k=1}^K Q_{ijk} \leq Q_{ij}^U \quad\quad (16.10)$$

Obviously we can also simply specify a fixed value for the sum in Eq. (16.10).

EXAMPLE 16.3

Let us consider the example in Table 16.1 that we examined in section 16.2. For that example we found that by not imposing any restriction on the matches, the minimum heating is 60 MW, and the minimum cooling is 225 MW. If the cost of the heating and cooling utilities is $80/kWyr and $20/kWyr, respectively, this would mean an annual cost of $9,300,000/yr. In addition, we found a pinch point at 340–320°C. Let us assume now that we were to impose as a constraint that the match for stream H1 and C1 is forbidden. Referring to Figure 16.4, the formulation in Eq. (16.9) leads to the LP problem shown in Table 16.4. The solution to this LP is as follows:

Minimum utility cost Z = $15,300,000/yr
Heating utility load Q_S = 120 MW
Cooling utility load Q_W = 285 MW

TABLE 16.4 Expanded LP for Restricted Match in Example 16.3

Utility Cost:	min $Z = 80000\, Q_S + 20000 Q_W$
Interval 1:	s.t. $R_{S1} + Q_{S11} - Q_S = 0$
	$Q_{S11} = 30$
Interval 2:	$R_{12} + Q_{112} = 60$
	$R_{S2} - R_{S1} + Q_{S12} = 0$
	$Q_{S12} + Q_{112} = 90$
Interval 3:	$R_{13} - R_{12} + Q_{113} + Q_{123} = 160$
	$R_{23} + Q_{213} + Q_{223} = 320$
	$R_{S3} - R_{S2} + Q_{S13} + Q_{S23} = 0$
	$Q_{113} + Q_{213} + Q_{S13} = 240$
	$Q_{123} + Q_{223} + Q_{S23} = 117$
Interval 4:	$-R_{13} + Q_{124} + Q_{1W4} = 60$
	$-R_{23} + Q_{224} + Q_{2W4} = 120$
	$-R_{S3} + Q_{S24} = 0$
	$Q_{124} + Q_{224} + Q_{S24} = 78$
	$Q_{1W4} + Q_{2W4} - Q_W = 0$
Forbidden match:	$Q_{112} = Q_{113} = 0$ (H1–C1 do not exchange heat)

In other words, the heating utility consumption has doubled, while the utility cost has increased by $6,000,000/yr with respect to the case when no matches are forbidden. In addition, there is no longer a pinch point since the sum of heat residuals exiting each interval is greater than zero. It is interesting to note that if we specify the match H2–C2 as a forbidden match, the utility cost will be identical to the case when no constraints are imposed. This example, then, shows that by imposing constraints on the matches the minimum utility cost may or may not increase.

Sec. 16.2   Sequential Synthesis   541

16.2.3   Prediction of Matches for Minimizing the Number of Units

As was shown in Chapter 10, the fewest number of units in a network is very often equal to the number of process streams and utilities minus one. This estimate applies either to each subnetwork when we partition the problem by pinch points or to the overall network when we do not perform the partitioning. In this section we will show how we can extend the expanded transshipment model Eq. (16.9) to rigorously predict the actual number of fewest units, as well as the stream matches that are involved in each unit, and the amount of heat that they must exchange.

Our first reaction might be to think that the expanded LP in Eq. (16.9) is already giving us the information on the stream matches, and that therefore we can work from there the required number of units. The reason why this is not true in general, is because the objective function in Eq. (16.9) does not have the information that we want to minimize the number of units. In fact, it is quite possible to have solutions of the expanded LP that have the same minimum cost but involve different number of matches. Therefore, it is clear that we require a formulation where we explicitly include the objective of minimizing number of matches.

Since at this point we would have performed the minimum utility cost calculation with or without match constraints, we would know the heat loads of the heating and cooling utilities. Therefore, at this point hot process streams and hot utilities can be treated simply as additional hot streams i, while cold process streams and cold utilities can be treated as cold streams j.

Assume we partition our problem into subnetworks. Each subnetwork q will then have an associated set of K_q temperature intervals. In addition, to represent the potential match of a given pair of hot and cold streams, we will define the following binary variables at the subnetwork q:

$$y^q_{ij} = \begin{cases} 1 & \text{hot stream } i, \text{ cold stream } j \text{ exchange heat} \\ 0 & \text{hot stream } i, \text{ cold stream } j \text{ do not exchange heat} \end{cases} \quad (16.11)$$

It should be noted that for each of the predicted matches as given by the above binary variables with a value of one, we will be able to associate it to a single exchanger unit. Therefore, the sum of units in the subnetwork will be simply given by the sum of the binary variables in Eq. (16.11). Since our objective is to minimize the number of units, it can be expressed as:

$$\min \sum_{i \in H} \sum_{j \in C} y^q_{ij} \quad (16.12)$$

As for the constraints, we will use the heat balances in Eq. (16.9) since they contain the information on the heat exchange between pairs of streams. However, we can simplify these equations for the two following reasons. One is that we know the heat contents of the utility streams, the other is that we use a common index i for hot process and utility streams, and the common index j for cold process and utility streams. In this way, the equations for the heat balances can be written for each interval k as:

$$R_{ik} - R_{i,k-1} + \sum_{j \in C_k} Q_{ijk} = Q_{ik}^H \qquad i \in H_k' \qquad k = 1,\ldots K_q \tag{16.13}$$

$$\sum_{i \in H_k} Q_{ijk} = Q_{jk}^C \qquad j \in C_k$$

$$R_{ik}, Q_{ijk} \geq 0$$

Finally, in a similar way as in the fixed cost charge model that we considered in Chapter 15, we need a logical constraint that states that if the binary variable is zero, the associated continuous variable must also be zero. In this case, we want to express the fact that if the match is not selected (i.e., $y_{ij}^q = 0$), then the heat exchanged for that match should also be zero. For any pair of hot i and cold j, this constraint can be written as:

$$\sum_{k=1}^{K_q} Q_{ijk} - U_{ij} y_{ij}^q \leq 0 \tag{16.14}$$

In this case, the upper bound U_{ij} will be given by the smallest of the heat contents of the two streams. For example, if hot i has 100 MW and cold j has 200 MW, then we can set U_{ij} to 100 MW as this is the maximum amount of heat that the two streams can exchange.

In this way, the problem defined by the objective function in Eq. (16.12), subject to the heat balances in Eq. (16.13), the logical constraints in Eq. (16.14), zero-one constraints in Eq. (16.11), and non-negativity constraints for the heat residuals and heat exchanges in Eq. (16.13), corresponds to an MILP transshipment problem (Papoulias and Grossmann, 1983). This problem we can solve independently for each subnetwork q (as implied by the above equations) or simultaneously over all the subnetworks. We can, of course, also develop a virtually identical formulation when we do not partition the problem into subnetworks.

The solution of the MILP transshipment problem will then indicate the following:

- Matches that take place $\left(y_{ij}^q = 1 \right)$

- Heat exchanged at each match $\sum_{k=1}^{K_q} Q_{ijk}$

This information can then be used to derive a network structure, either manually or automatically, as will be shown in the next section.

An important point to be noted here is the fact that the solution of this MILP is not necessarily unique. This follows from the fact that there might be several network configurations for the same number of units and utility cost. Furthermore, a given network configuration may not necessarily have its heat loads defined in a unique way due to the presence of heat loops.

Sec. 16.2 Sequential Synthesis

EXAMPLE 16.4

Let us consider again the problem in Table 16.1. We will assume that no constraints are imposed on the matches, so that 60 MW will be required for the heating and 225 MW for the cooling. Referring to Figure 16.8, which follows from Figure 16.4, Eqs. (16.12) to (16.14) lead to the problem shown in Table 16.5. If we solve the MILP, the solution that we obtain involves the six following matches:

Above pinch:
| Match Steam–C1 | 60 MW | $(y_{S1A} = 1, Q_{S11} = 30, Q_{S12} = 30)$ |
| Match H1–C1 | 60 MW | $(y_{11A} = 1, Q_{112} = 60)$ |

Below pinch:
Match H1–C1	25 MW	$(y_{11B} = 1, Q_{113} = 25)$
Match H1–C2	195 MW	$(y_{12B} = 1, Q_{123} = 117, Q_{124} = 78)$
Match H2–C1	215 MW	$(y_{21B} = 1, Q_{123} = 215)$
Match H2–W	225 MW	$(y_{2WB} = 1, Q_{2W4} = 225)$

TABLE 16.5 MILP Model for Example 16.4

Number of units:	$\min Z = y_{S1}^A + y_{11}^A + y_{11}^B + y_{12}^B + y_{1W}^B + y_{21}^B + y_{22}^B + y_{2W}^B$
Interval 1: s.t.	$R_{S1} + Q_{S11} = 60$
	$Q_{S11} = 30$
Interval 2:	$R_{12} + Q_{112} = 60$
	$R_{S2} - R_{S1} + Q_{S12} = 0$
	$Q_{S12} + Q_{112} = 90$
Interval 3:	$R_{13} - R_{12} + Q_{113} + Q_{123} = 160$
	$R_{23} + Q_{213} + Q_{223} = 320$
	$Q_{113} + Q_{213} + Q_{S13} = 240$
	$Q_{123} + Q_{223} + Q_{S23} = 117$
Interval 4:	$-R_{13} + Q_{124} + Q_{1W4} = 60$
	$-R_{23} + Q_{224} + Q_{2W4} = 120$
	$Q_{124} + Q_{224} + Q_{S24} = 78$
	$Q_{1W4} + Q_{2W4} = 225$
Matches above pinch:	$Q_{S11} + Q_{S12} - 60\, y_{S1}^A \leq 0$
	$Q_{112} - 60\, y_{11}^A \leq 0$
Matches below pinch:	$Q_{113} - 220\, y_{11}^B \leq 0$
	$Q_{123} + Q_{124} - 195\, y_{12}^B \leq 0$
	$Q_{1W4} - 220\, y_{1W}^B \leq 0$
	$Q_{213} - 240\, y_{21}^B \leq 0$
	$Q_{223} + Q_{224} - 60\, y_{22}^B \leq 0$
	$Q_{2W4} - 225\, y_{2W}^B \leq 0$

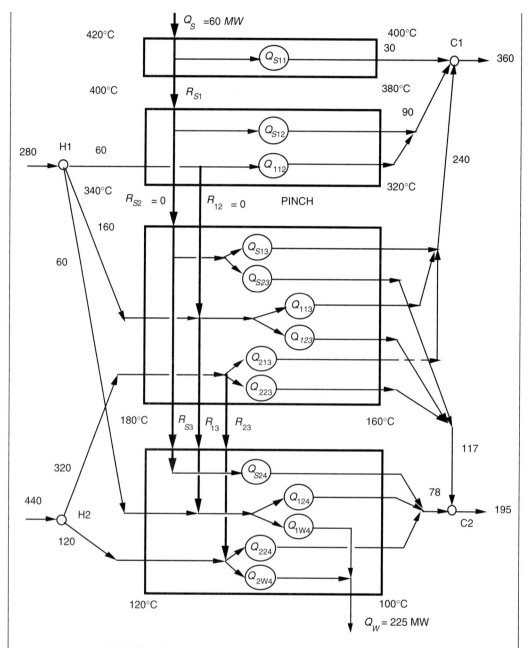

FIGURE 16.8 Representation of heat flows in MILP transshipment.

Sec. 16.2 Sequential Synthesis

Based on the above information of matches and heat loads, we can manually derive the network configuration, shown in Figure 16.9, with six units. The solution of the MILP, however, is not unique. If we set the binary variable $y_{11}^B = 0$ for the match H1–C1 below the pinch, we obtain a different set of six matches:

Above pinch:

| Match Steam–C1 | 60 MW | $(y_{s1A} = 1, Q_{s11} = 30, Q_{s12} = 30)$ |
| Match H1–C1 | 60 MW | $(y_{11A} = 1, Q_{112} = 60)$ |

Below pinch:

Match H1–C2	195 MW	$(y_{12B} = 1, Q_{123} = 117, Q_{124} = 78)$
Match H2–C1	240 MW	$(y_{21B} = 1, Q_{213} = 240)$
Match H1–W	25 MW	$(y_{1WB} = 1, Q_{1WB} = 25)$
Match H2–W	200 MW	$(y_{2WB} = 1, Q_{2WB} = 200)$

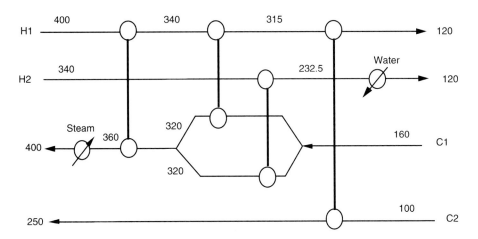

FIGURE 16.9 Network configuration for matches predicted from MILP in Example 16.4.

Thus, there are different matches and changes in the heat loads below the pinch. The above matches can be translated into the network configuration shown in Figure 16.10.

Finally, we could also solve the above MILP problem without partitioning into subnetworks. In this case, the only change required in the formulation of Table 16.5 is that for each potential match only one binary variable is defined, and the logical conditions are written also for each potential match. For example, the match H1–C1 is denoted by the binary y_{11}, and its logical condition is given by (see Figure 16.8):

$$Q_{112} + Q_{113} - 220 y_{11} \leq 0$$

If we solve the MILP with no pinch partitioning, we obtain the following five matches:

Match Steam–C1	60 MW
Match H1–C1	85 MW
Match H1–C2	195 MW

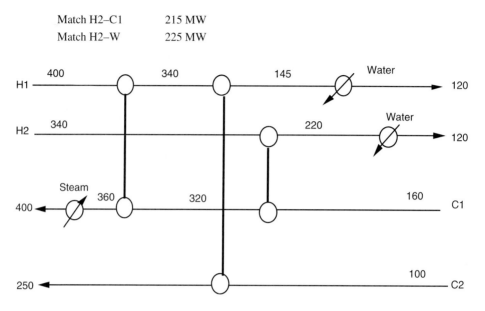

FIGURE 16.10 Alternative network configuration for Example 16.4.

These results would suggest that we should be able to derive a network with only five units. This is, in fact, possible if the match H1–C1 is placed across the pinch, has a driving force equal to the temperature approach (20°C), and if we introduce bypass streams in the network (see Wood et al., 1985). The configuration that has been derived manually for the above five matches is shown in Figure 16.11. Note that the match H1–C1 would require a large area due to its small driving force. It is of course not that trivial to derive manually a network like the one in Figure 16.11. Can we possibly automate this procedure?

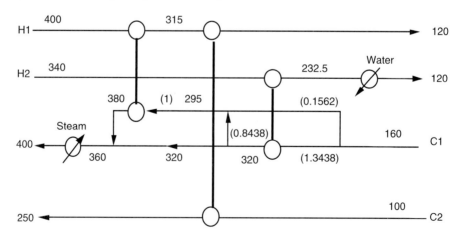

FIGURE 16.11 Five-unit network for Example 16.4.

16.2.4 Automatic Derivation of Network Structures

In this section we will show how we can make use of the information provided by the MILP transshipment model to automatically derive heat exchanger network configurations (Floudas, Ciric, and Grossmann, 1986).

The basic idea here will be to postulate a superstructure for each stream that has the following characteristics:

- Each exchanger unit in the superstructure corresponds to a match predicted by the MILP transshipment model (with or without pinch partitioning). Each exchanger will also have as heat load the one predicted by the MILP.
- The superstructure will contain those stream interconnections among the units that can potentially define all configurations with no stream splitting, with stream splitting and mixing, and with possible bypass streams. The stream interconnections will be treated as unknowns that must be determined.

An example of such a superstructure is given in Figure 16.12 for the case of one hot and two cold streams in which the two predicted matches are H1–C1 and H1–C2. Note that in this superstructure stream H1 is split initially into two streams that are directed to the two units. The outlets of these units are then also split into two streams: one that is directed to the inlet of the other unit, and one that is directed to the final mixing point.

By "deleting" some of the streams in the superstructure of Figure 16.12, we can easily verify that it has embedded all possible network configurations for the two matches. As shown in Figure 16.13, we have embedded the following alternatives:

1. Units H1–C1, H1–C2 in series
2. Units H1–C2, H1–C1 in series

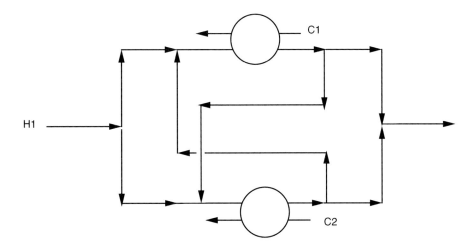

FIGURE 16.12 Superstructure for matches H1–C1, H1–C2.

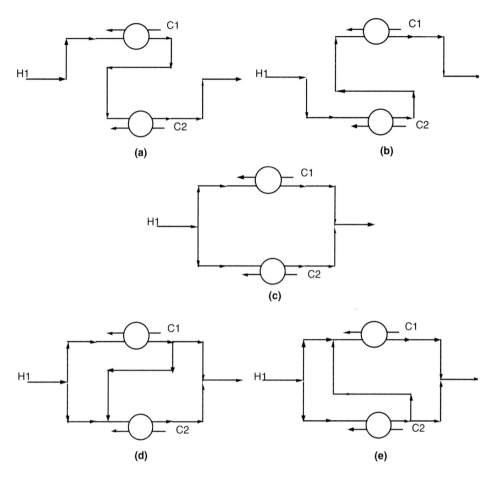

FIGURE 16.13 Alternatives embedded in the superstructure of Figure 16.12.

3. Units H1–C1, H1–C2 in parallel
4. Units H1–C1, H1–C2 in parallel with bypass to H1–C2
5. Units H1–C1, H1–C2 in parallel with bypass to H1–C1

Thus, in the network superstructure of Figure 16.12 we have embedded all possible configurations for a two-unit network.

Before we consider the extension of the superstructure to an arbitrary number of stream matches, let us see how we can model the superstructure in Figure 16.12 in order to determine the network structure with minimum investment cost. First, we assign the variables representing heat capacity flowrates (F, f), temperatures (T, t), heat loads (Q), and areas as shown in Figure 16.14. Note that the following variables are known:

Sec. 16.2 Sequential Synthesis

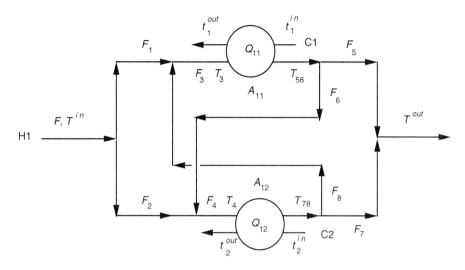

FIGURE 16.14 Variables for superstructure with two matches.

- For stream H1, the heat capacity flowrate F, and the inlet and outlet temperatures T^{in}, T^{out}.
- For stream C1, the heat capacity flowrate f_1 and the inlet and outlet temperatures t_1^{in}, t_1^{out}.
- For stream C2, the heat capacity flowrate f_2, and the inlet and ouelt temperatures t_2^{in}, t_2^{out}.
- The heat loads Q_{11}, Q_{12} as predicted by the MILP transshipment model.

The objective function representing the minimization of the investment cost will be given by:

$$\min \; C = c_1 A_{11}^{\beta} + c_2 A_{12}^{\beta} \quad (16.15)$$

where c_1, c_2, β are cost parameters. We can express this objective function in terms of temperatures by replacing the areas through the design equation $Q = UA\,\text{LMTD}$ for countercurrent heat exchangers. However, the LMTD function can lead to numerical difficulties when the temperature differences θ_1, θ_2, at both ends are the same. Therefore, we replace the definition of the LMTD

$$\text{LMTD} = \frac{\theta_2 - \theta_1}{\ln \dfrac{\theta_2}{\theta_1}} \quad (16.16)$$

by the Chen (1987) approximation $\text{LMTD} \cong [\theta_1 \theta_2 (\theta_2 + \theta_1)/2]^{1/3}$

That is,

$$\min C = C_1 \left[\frac{Q_{11}}{U_{11}\left[\theta_1^1 \theta_2^1 (\theta_1^1 + \theta_2^1)/2\right]^{1/3}} \right]^\beta + C_2 \left[\frac{Q_{12}}{U_{12}\left[\theta_1^2 \theta_2^2 (\theta_1^2 + \theta_2^2)/2\right]^{1/3}} \right]^\beta \quad (16.17)$$

where U_{11}, U_{12} are the overall heat transfer coefficients for the two exchangers.

Thus, the constraints that apply to the superstructure are as follows (see Figure 16.13):

1. Mass balance for initial splitter

$$F_1 + F_2 = F \quad (16.18)$$

2. Mass and heat balances for mixers at inlet of two units

$$F_1 + F_8 - F_3 = 0$$
$$F_1 T^{in} + F_8 T_{78} - F_3 T_3 = 0 \quad (16.19)$$
$$F_2 + F_6 - F_4 = 0$$
$$F_2 T^{in} + F_6 T_{56} - F_4 T_4 = 0$$

3. Mass balance for splitters at outlet of exchangers

$$F_3 - F_6 - F_5 = 0 \quad (16.20)$$
$$F_4 - F_7 - F_8 = 0$$

4. Heat balances in exchangers

$$Q_{11} - F_3 (T_3 - T_{56}) = 0 \quad (16.21)$$
$$Q_{12} - F_4 (T_4 - T_{78}) = 0$$

5. Definition temperature differences

$$\theta_1^1 = T_3 - t_1^{out}$$
$$\theta_2^1 = T_{56} - t_1^{in}$$
$$\theta_1^2 = T_4 - t_2^{out} \quad (16.22)$$
$$\theta_2^2 = T_{78} - t_2^{in}$$

6. Feasibility constraints for temperatures

$$\theta_1^1 \geq \Delta T_{min}$$
$$\theta_2^1 \geq \Delta T_{min}$$
$$\theta_1^2 \geq \Delta T_{min} \quad (16.23)$$
$$\theta_2^2 \geq \Delta T_{min}$$

Sec. 16.3   Simultaneous MINLP Model

7. Nonnegativity conditions on the heat capacity flowrates

$$Fj \geq 0 \qquad j = 1, 2, \ldots 8 \qquad (16.24)$$

The optimization problem defined by the objective function in Eq. (16.17) subject to the constraints in Eqs. (16.18) to (16.24) corresponds to a nonlinear programming problem that has as variables the flows F_j, $j = 1,2,\ldots 8$, and the temperatures T_3, T_4, T_{56}, T_{78}. Those flowrates that take a value of zero will then "delete" the streams that are not required in the superstructure.

It should be noted that the likelihood of multiple local optima in this problem is somewhat reduced because the areas of the units cannot take a value of zero due to the fixed heat loads. We may recall the example on selection of reactors in section 15.5 of Chapter 15, where local solutions were mainly due to the deletion of the reactors.

The superstructure and its nonlinear programming formulation can be readily extended to the case of an arbitrary number of stream matches with the following procedure:

1. Develop a superstructure for any stream involving two or more matches according to the following scheme:
   a. Initial split where the streams are directed to all the units in that superstructure.
   b. Outlet of units is split and mixed with the inlets of other units and with the final mixing point.
2. All stream superstructures are joined through an NLP formulation similar to Eqs. (16.17) to (16.23), having the heat loads predicted by the MILP transshipment model Eqs. (16.12) to (16.14).
3. The resulting NLP is solved to obtain the optimal network configuration. This NLP can be solved with a large-scale reduced gradient method (e.g., MINOS).

This strategy for automatic network synthesis has been implemented in the interactive computer program MAGNETS, developed by Amy Ciric, as described by Floudas, Ciric, and Grossmann (1986). The optimization of the minimum temperature approach can be performed in an outer loop, and constraints on matches can be easily handled as discussed in section 16.3. Figure 16.15 shows an example of a network configuration that was automatically synthesized with MAGNETS for the data given in Table 16.6.

16.3   SIMULTANEOUS MINLP MODEL

While the sequential targeting and optimization approach presented in the previous sections has the advantage of decomposing the synthesis problem, it has the disadvantage that the trade-offs between energy, number of units and area are not rigorously taken into account. The reason for this is that the optimization problem:

$$\min \text{Total Cost} = \text{Area Cost} + \text{Fixed Cost Units} + \text{Utility Cost} \qquad (16.25)$$

is being approximated by a problem that conceptually can be stated as follows:

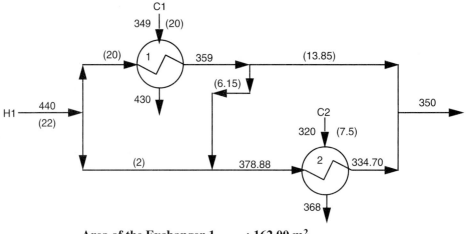

Area of the Exchanger 1	: 162.00 m²
Area of the Exchanger 2	: 56.50 m²
Total Area	: 218.50 m²
ΔT_{min}	: 10.00 K
Total Cost	: $77,972.00/yr

FIGURE 16.15 Network structure obtained from NLP superstructure approach.

$$\begin{aligned} \min \quad & \text{Area Cost} \\ \text{st.} \quad \min \quad & \text{Number Units} \\ \text{s.t} \quad & \text{Minimum Utility Cost} \end{aligned} \qquad (16.26)$$

In this section we will show that the simultaneous optimization as implied in Eq. (16.25) can be performed with an MINLP optimization model on a somewhat different superstructure in which we will be able to express the constraints in linear form. The MINLP model is based on the stage-wise superstructure representation proposed by Yee

TABLE 16.6 Data for One Hot/Two Cold Stream Problem

Stream	TIN (K)	TOUT (K)	Fcp (kW/K)	h (kW/m²K)	Cost ($/kW-yr)
H1	440	350	22	2.0	—
C1	349	430	20	2.0	—
C2	320	368	7.5	0.67	—
S1	500	500	—	1.0	120
W1	300	320	—	1.0	20

Minimum Approach of Temperatures (EMAT) = 1 K
Exchanger Cost = 6,600 + 670 (Area)$^{0.83}$

Sec. 16.3 Simultaneous MINLP Model

et al. (1990) (see Ciric and Floudas, 1991, for an alternative model). The superstructure for the problem is shown in Figure 16.16. Within each stage of the superstructure, potential exchanges between any pair of hot and cold streams can occur. In each stage, the corresponding process stream is split and directed to an exchanger for a potential match between each hot stream and each cold stream. It is assumed that the outlets of the exchangers are isothermally mixed, which simplifies the calculation of the stream temperature for the next stage, since no information of flows is needed in the model. The outlet temperatures of each stage are treated as variables in the optimization. The number of stages should in general coincide with the number of temperature intervals to ensure maximum energy recovery. However, in most cases selecting the number of stages as the maximum of hot and cold streams suffices.

As shown in Figure 16.16, the two stage representation for the problem involves eight exchangers, with four possible matches in each stage. Note that alternative parallel and series configurations are embedded as well as possible rematching of streams. However, the use of by-passes and split streams with two or more matches in each branch is not included. A heater or cooler is placed at the outlet of the superstructure for each process stream. Optimization of the MINLP model identifies the least cost network embedded within the superstructure by identifying which exchangers are needed and the flow configuration of the streams. A major advantage of this model is its capability of easily handling constraints for forbidding stream splits.

With the superstructure in Figure 16.16, the formulation can now be presented. The notation follows the ones used in Yee and Grossmann (1990). Process streams are divided into two sets, set HP for hot streams, represented by index i, and set CP for cold streams, represented by index j. Index k is used to denote the superstructure stages given by the set

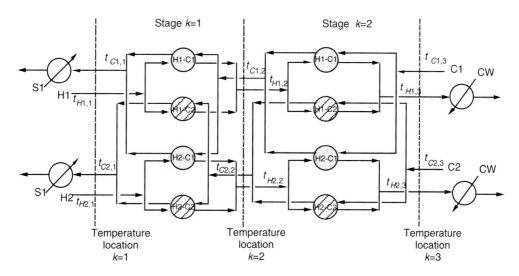

FIGURE 16.16 Two-stage superstructure.

ST. Indices HU and CU correspond to the heating and cooling utilities respectively. Also, the following parameters and variables are used in the formulation:

Parameters

TIN = inlet temperature of stream
$TOUT$ = outlet temperature of stream
F = heat capacity flow rate
U = overall heat transfer coefficient
CCU = unit cost for cold utility
CHU = unit cost of hot utility
CF = fixed charge for exchangers
C = area cost coefficient
β = exponent for area cost
NOK = total number of stages
Ω = upper bound for heat exchange
Γ = upper bound for temperature difference

Variables

dt_{ijk} = temperature approach for match (i,j) at temperature location k
$dtcu_i$ = temperature approach for the match of hot stream i and cold utility
$dthu_j$ = temperature approach for the match of cold stream j and hot utility
q_{ijk} = heat exchanged between hot process stream i and cold process stream j in stage k
qcu_i = heat exchanged between hot stream i and cold utility
qhu_j = heat exchanged between hot utility and cold stream j
$t_{i,k}$ = temperature of hot stream i at hot end of stage k
$t_{j,k}$ = temperature of cold stream j at hot end of stage k
z_{ijk} = binary variable to denote existence of match (i,j) in stage k
zcu_i = binary variable to denote that cold utility exchanges heat with stream i
zhu_j = binary variable to denote that hot utility exchanges heat with stream j

With the above definitions, the formulation can now be presented.

1. *Overall heat balance for each stream.* An overall heat balance is needed to ensure sufficient heating or cooling of each process stream. The constraints specify that the overall heat transfer requirement of each stream must equal the sum of the heat it exchanges with the other process streams at each stage plus the exchange with the utility streams,

$$(TIN_i - TOUT_i) F_i = \sum_{k \in ST} \sum_{j \in CP} q_{ijk} + qcu_i \quad i \in HP$$

$$(TOUT_j - TIN_j) F_j = \sum_{k \in ST} \sum_{i \in HP} q_{ijk} + qhu_j \quad j \in CP$$

(16.27)

2. *Heat balance at each stage.* An energy balance is also needed at each stage of the superstructure to determine the temperatures. Note that for the two-stage superstructure as shown in Figure 16.16, three temperatures, t, are required. Temperatures for the streams are highest at temperature location $k = 1$ and lowest at $k = 3$. Also, due to the isothermal mixing assumption, no variables are required for the flows.

$$\begin{aligned}(t_{i,k} - t_{i,k+1})F_i &= \sum_{j \in CP} q_{ijk} \qquad k \in ST,\ i \in HP \\ (t_{j,k} - t_{j,k+1})F_j &= \sum_{i \in HP} q_{ijk} \qquad k \in ST,\ j \in CP\end{aligned} \qquad (16.28)$$

3. *Assignment of superstructure inlet temperatures.* Fixed supply temperatures (TIN) of the process streams are assigned as the inlet temperatures to the superstructure. In Figure 16.16, for hot streams the superstructure inlet corresponds to temperature location $k=1$, while for cold streams, the inlet corresponds to location $k = 3$.

$$\begin{aligned}TIN_i &= t_{i,1} \\ TIN_j &= t_{j,NOK+1}\end{aligned} \qquad (16.29)$$

4. *Feasibility of temperatures.* Constraints are also needed to specify a monotonic decrease of temperature at each successive stage. In addition, a bound is set for the outlet temperatures of the superstructure at the respective stream's target temperature. Note that the outlet temperature of each stream at its last stage does not necessarily correspond to the stream's target temperature since utility exchanges can occur at the outlet of the superstructure.

$$\begin{aligned}t_{i,k} &\geq t_{i,k+1} & k &\in ST,\ i \in HP \\ t_{j,k} &\geq t_{j,k+1} & k &\in ST,\ j \in CP \\ TOUT_i &\leq t_{i,NOK+1} & i &\in HP \\ TOUT_j &\geq t_{j,1} & j &\in CP\end{aligned} \qquad (16.30)$$

5. *Hot and cold utility load.* Hot and cold utility requirements are determined for each process stream in terms of the outlet temperature in the last stage and the target temperature for that stream. The utility heat load requirements are determined as follows:

$$\begin{aligned}(t_{i,NOK+1} - TOUT_i)F_i &= qcu_i & i &\in HP \\ (TOUT_j - t_{j,1})F_j &= qhu_j & j &\in CP\end{aligned} \qquad (16.31)$$

6. *Logical constraints.* Logical constraints and binary variables are needed to determine the existence of process match (i,j) in stage k and also any match involving utility streams. The 0–1 binary variables are represented by z_{ijk} for process stream matches, zcu_i for matches involving cold utility, and zhu_j for matches involving hot utility. An integer value of one for any binary variable designates that the match is present in the optimal network. The constraints, then, are as follows:

$$\begin{aligned}q_{ijk} - \Omega z_{ijk} &\leq 0 & i &\in HP,\ j \in CP,\ k \in ST \\ qcu_i - \Omega zcu_i &\leq 0 & i &\in HP \\ qhu_j - \Omega zhu_j &\leq 0 & j &\in CP \\ z_{ijk},\ zcu_i,\ zhu_j &= 0,1\end{aligned} \qquad (16.32)$$

7. *Calculation of approach temperatures.* The area requirement of each match will be incorporated in the objective function. Calculation of these areas requires that approach temperatures be determined. To ensure feasible driving forces for exchangers that are selected in the optimization procedure, the binary variables are used to activate or deactivate the following constraints for approach temperatures:

$$dt_{ijk} \leq t_{i,k} - t_{j,k} + \Gamma(1 - z_{ijk}) \quad\quad k \in ST, i \in HP, j \in CP$$

$$dt_{ijk+1} \leq t_{i,k+1} - t_{j,k+1} + \Gamma(1 - z_{ijk}) \quad\quad k \in ST, i \in HP, j \in CP$$

$$dtcu_i \leq t_{i,NOK+1} - TOUT_{CU} + \Gamma(1 - zcu_i) \quad\quad i \in HP \quad\quad (16.33)$$

$$dthu_j \leq TOUT_{HU} - t_{j,1} + \Gamma(1 - zhu_j) \quad\quad j \in CP$$

Note that these constraints can be expressed as inequalities because the cost of the exchangers decreases with higher values for the temperature approaches *dt*. Also, the role of the binary variables in the constraints is to ensure that non-negative driving forces exist for a selected match. When a match (*i,j*) occurs in stage *k*, z_{ijk} equals one and the constraint becomes active so that the approach temperature is properly calculated. However, when the match does not occur, z_{ijk} equals zero, and the contribution of the upper bound Γ on the right-hand side deems the constraint inactive. Note that the upper bounds can be set to zero for the utility exchangers since for the data given, all the temperature differences are always positive. Also, one can specify a minimum approach temperature so that in the network, the temperature between the hot and cold streams at any point of any exchanger will be at least EMAT:

$$dt_{ijk} \geq EMAT \quad\quad (16.34)$$

8. *Objective function.* Finally, the objective function can be defined as the annual cost for the network. The annual cost involves the combination of the utility cost, the fixed charges for the exchangers, and the area cost for each exchanger. LMTD, which is the driving force for a countercurrent heat exchanger, is approximated using the Chen approximation (1987).

$$LMTD \approx [(dt1*dt2)*(dt1+dt2)/2]^{1/3} \quad\quad (16.35)$$

This approximation is used to avoid the numerical difficulties of the LMTD equation when the approach temperature (*dt*1, *dt*2) for both sides of the exchanger are equal. Furthermore, when the driving force on either side of the exchanger equals zero, the driving force will be approximated to zero. The objective function is defined as follows:

$$\min \quad \sum_{i \in HP} CCU\, qcu_i + \sum_{j \in CP} CHU\, qhu_j$$

$$+ \sum_{i \in HP} \sum_{j \in CP} \sum_{k \in ST} CF_{ij} z_{ijk} + \sum_{i \in HP} CF_{i,CU} zcu_i + \sum_{j \in CP} CF_{j,HU} zhu_j \quad\quad (16.36)$$

Sec. 16.3 Simultaneous MINLP Model

$$+ \sum_{i \in HP} \sum_{j \in CP} \sum_{k \in ST} C_{ij} [q_{ijk} / (U_{ij} [dt_{ijk} dt_{ijk+1}) (dt_{ijk} + dt_{ijk+1})/2]^{1/3})]^{\beta_{ij}}$$

$$+ \sum_{i \in HP} C_{i,CU} [qcu_i / (U_{i,CU} [(dtcu_i) (TOUT_i - TIN_{CU}) \{dtcu_i + (TOUT_i - TIN_{CU})\}/2]^{1/3})]^{\beta_{i,CU}}$$

$$+ \sum_{j \in CP} C_{HU,j} [qhu_j / (U_{HU,j} [(dthu_j) (TIN_{HU} - TOUT_j) \{dthu_j + (TIN_{HU} - TOUT_j)\}/2]^{1/3})]^{\beta_{j,HU}}$$

where $\quad \dfrac{1}{U_{ij}} = \dfrac{1}{h_i} + \dfrac{1}{h_j}; \quad \dfrac{1}{U_{i,CU}} = \dfrac{1}{h_i} + \dfrac{1}{h_{cu}}; \quad \dfrac{1}{U_{HU,j}} = \dfrac{1}{h_j} + \dfrac{1}{h_{HU}}.$

The proposed MINLP model for the synthesis problem consists of minimizing the objective function in Eq. (16.36) subject to the feasible space defined by Eqs. (16.27) to (16.35). The continuous variables (t, q, qhu, qcu, dt, $dtcu$, $dthu$) are non-negative and the discrete variables z, zcu, zhu are 0–1. Although Eqs. (16.27) to (16.34) are all linear, the nonlinearities in the objective function Eq. (16.36) may lead to more than one local optimal solution due to their nonconvex nature.

It should be noted that the simplifying assumption of isothermal mixing at the stage outlets for the stream splits is rigorous for the case when the network to be synthesized does not involve stream splits. For structures where splits are present, the assumption may lead to an overestimation of the area cost since it will restrict trade-offs of area between the exchangers involved with the splits stream. In this case one possibility is to refine the temperatures by introducing flow variables in the selected network structure and perform the corresponding optimization through an NLP model similar to the one in section 16.5.

An interesting feature of the MINLP model is that it is possible to add constraints to avoid generating structures with no stream splits. This is simply accomplished by requiring that not more than one match be selected for every stream at each stage; that is,

$$\sum_{i \in HP} z_{ijk} \leq 1 \quad j \in CP \quad k \in ST, \quad \sum_{j \in CP} z_{ijk} \leq 1 \quad i \in HP \quad k \in ST \qquad (16.37)$$

Finally, an important point in the application of the proposed model is the selection of number of stages. A simple alternative is to set the number of stages equal to the maximum of the number of hot or cold process streams. This choice is often adequate but may exclude networks with maximum heat recovery. As discussed in Daichendt and Grossmann (1994), a rigorous choice is to set the number of stages equal to the number of temperature intervals with EMAT as the minimum approach. These authors proposed a procedure by which matches can be eliminated from the superstructure thus greatly reducing the size of the MINLP.

EXAMPLE 16.5

Consider the synthesis of one hot and two cold streams given in Table 16.6.

If we solve the MINLP model with two stages and with a code such as DICOPT++ (Viswanathan and Grossmann, 1990) we obtain the design given in Figure 16.17. Note that the design requires neither heating nor cooling, and it is somewhat cheaper than the design obtained

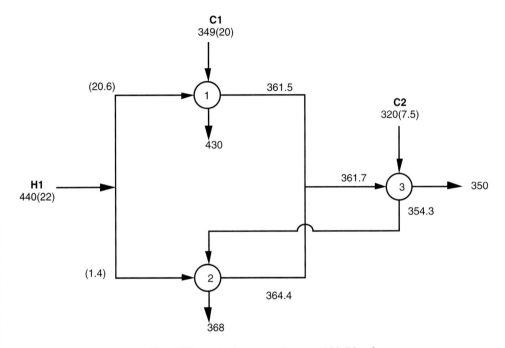

Total Heat Exchangers Area = 182.78 m^2
Utilities:
Heaters heat load = 0 KW
Coolers heat load = 0 KW
Costs: Investment = $ 76, 445.00 per year
Total = $ 76, 445.00 per year

FIGURE 16.17 Optimal network with no constraints on split streams.

with the sequential approach in Figure 16.15 ($76,445 vs. $77,972/year), although it involves one more unit. However, its structure is simpler. On the other hand, the network still requires stream splitting, which from a practical point of view is not always attractive, as this requires the additional investment of a control valve and a potentially more complex operation. We can easily generate a network structure with no stream splitting by adding the inequalities in Eq. (16.37). The resulting solution is shown in Figure 16.18. Note that the new structure does require heating and cooling, although in small amounts. Also, the network consists now of four instead

Sec. 16.4   Comparison of Sequential and Simultaneous Synthesis

of three units. In fact, the investment penalty for not having stream splits is rather modest ($78,944/yr vs. $76,445/yr), although the total cost is increased rather substantially to $86,222/yr due to the use of utilities.

This example, then, shows the versatility of the simultaneous MINLP model.

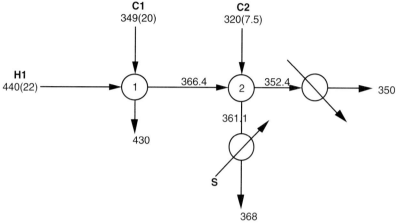

Total Heat Exchangers Area = 165.32 m²
Utilities:
        Heaters heat load = 51.98 KW
        Coolers heat load = 51.98 KW
Costs:
Utilities = $ 7, 277.59 per year
        Investment = $ 78, 944.00 per year
        Total = $ 86, 222.00 per year

FIGURE 16.18   Network structure with no stream splits.

16.4 COMPARISON OF SEQUENTIAL AND SIMULTANEOUS SYNTHESIS

The main advantage in the sequential synthesis approach is that the problem is made more managable by solving a sequence of smaller problems. Clearly, targets are essential for setting up these smaller problems as was the case of the minimum utility cost, minimum number of units, and minimum area targets. On the other hand, the advantage of the simultaneous approach is that the trade-offs are all taken simultaneously into account, thus increasing the possibility of finding improved solutions. However, the computational requirements are greatly increased; for this reason, this motivates simplifications like the one that was presented on isothermal mixing for the MINLP model.

One important aspect, though, that is offered by simultaneous optimization models is that they do not rely on heuristics. To illustrate this point, consider the two networks in Figure 16.19. The one in Figure 16.19.b was synthesized with the simultaneous MINLP model using an EMAT = 1K. Having obtained the solution to that problem, the heat recovery approach temperature, HRAT, that would correspond to that problem was deter-

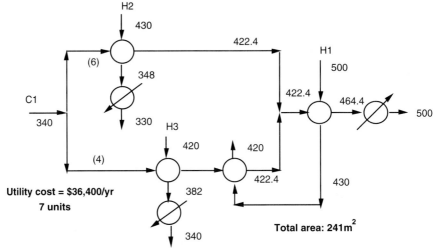

(a) Sequential Design: $72,257.30/yr

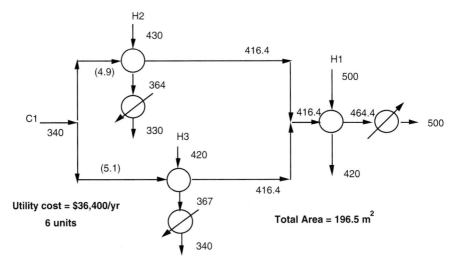

(b) Simultaneous Design: $67,762.80/yr

FIGURE 16.19 Synthesis designs obtained with (a) sequential and (b) simultaneous optimization.

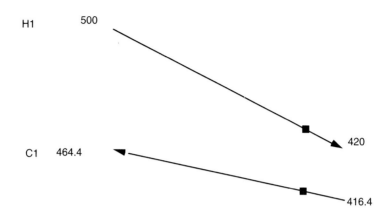

FIGURE 16.20 Match H1–C1 from simultaneous model placed across the pinch.

mined to be 7.6K. The sequential synthesis strategy was applied for that value of HRAT yielding the network in Figure 16.19.a. Note that the design obtained with the sequential strategy is more expensive and involves one more unit, although it does meet the units target of 7, above the pinch, $N_{min} = 2 + 1 + 1 - 1 = 3$, and below the pinch, $N_{min} = 3 + 1 + 1 - 1 = 4$. In contrast, the network in Figure 16.19.b requires only 6 units. Note that both networks have the same energy requirements ($36,400/yr). The reason for the improved design by the simultaneous synthesis strategy is that it violates the heuristic guideline of partitioning the network above and below the pinch points (430K–422.4K). It can be seen in Figure 16.20 that the match H1–C1 is in fact placed across the pinch, with the actual approach temperature being as low as 3.6K. What this example shows is that the guideline of not placing matches across the pinch is a heuristic that ought to be challenged.

16.5 NOTES AND FURTHER READING

For a review of the state-of-the-art up to the late 1980s, see the excellent survey paper by Gundersen and Naess (1988). The LP transshipment model predicts the exact target for minimum utility cost for the cases of unrestricted and restricted matches. The MILP transshipment predicts an exact target for the minimum number of matches but its solution may not be unique. Gundersen and Grossmann (1990) proposed a "vertical" transshipment model that will tend to favor the selection of matches that exhibit vertical heat transfer.

It is interesting to note that El-Halwagi and Maniousiouthakis (1989) have shown that the problem of synthesizing mass exchanger networks can be formulated with LP and MILP transshipment models similar to the ones for heat exchanger networks.

In addition to the program MAGNETS by Floudas, Ciric, and Grossmann (1986), which implements the sequential synthesis strategy, the program RESHEX by Saboo, Morari, and Colberg (1986a,b) implements the LP and MILP transshipment models by Papoulias and Grossmann (1983). The program SYNHEAT (Bolio et al. 1994) implements the simultaneous MINLP model.

Global optimization of the MINLP model by Yee and Grossmann (1990) has been addressed with a rigorous deterministic method by Quesada and Grossmann (1993) for the case of fixed network configurations, linear costs and arithmetic mean temperature differences.

REFERENCES

Bolio, B., Turkay, A., Yee, T. F., & Grossmann, I. E. (1994). *Manual SYNHEAT*. Pittsburgh: Computer Aided Process Design Laboratory, Carnegie Mellon University.

Cerda, J., & Westerberg, A. W. (1983). Synthesizing heat exchanger networks having restricted stream/stream match using transportation problem formulations. *Chem. Engng. Sci.*, **38**, 1723.

Cerda, J., Westerberg, A. W., Mason, D., & Linnhoff, B. (1983). Minimum utility usage in heat exchanger network synthesis—A transportation problem. *Chem. Engng Sci.*, **38**, 373.

Chen, J. J. J. (1987). Letter to the Editor: Comments on improvement on a replacement for the logarithmic mean. *Chem. Engng. Sci.*, **42**, 2488.

Ciric, A. R., & Floudas, C. A. (1991). Heat exchanger network synthesis without decomposition. *Computers Chem. Eng.*, **15**, 385.

Daichendt, M. M., & Grossmann, I. E. (1994). Preliminary screening procedure for the MINLP synthesis of process systems. II. Heat exchanger networks. *Comp. and Chem. Engng.*, **18**, 679.

El-Halwagi, M., & Maniousiouthakis, V. (1989). Synthesis of mass exchange networks. *AIChE J.*, **35**, 1233.

Floudas, C. A., & Ciric, A. R. (1989). Strategies for overcoming uncertainties in heat exchanger network synthesis. *Comp. and Chem. Engng.*, **13**(10), 1117.

Floudas, C. A., Ciric, A. R., & Grossmann, I. E. (1986). Automatic synthesis of optimum heat exchanger network configurations. *AIChEJ.*, 32, 276.

Gundersen, T., & Grossmann, I. E. (1990). Improved optimization strategies for automated heat exchanger network synthesis through physical insights. *Comp. and Chem. Engng.*, **14**(9), 925.

Gundersen, T., & Naess, L. (1988). The synthesis of cost optimal heat exchanger networks. An industrial review of the state of the art. *Comp. and Chem. Engng.*, **12**(6), 503.

Papoulias, S. A., & Grossmann, I. E. (1983). A structural optimization approach to process synthesis—II. Heat recovery networks. *Comp. and Chem. Engng.*, **7**, 707.

Quesada, I., & Grossmann, I. E. (1993). Global optimization algorithm for heat exchanger networks. *Ind. Eng. Chem. Res.*, **32**, 487.

Saboo, A. K., Morari, M., & Colberg, R. D. (1986a). RESHEX—an interactive software package for the synthesis and analysis of resilient heat exchanger networks—I. Program description and application. *Comput. Chem. Engng.*, **10**, 577.

Saboo, A. K., Morari, M., & Colberg, R. D. (1986b). RESHEX—an interactive software package for the synthesis and analysis of resilient heat exchanger networks—II. Discussion of area targeting and network synthesis algorithms. *Comput. Chem. Engng.*, **10**, 591.

Viswanathan, J., & Grossmann, I. E. (1990). A combined penalty function and outer-approximation method for MINLP optimization. *Comp. and Chem. Eng.*, **14**, 769.

Wood, R. M., Wilcox R. J., & Grossmann, I. E. (1985). A note on the minimum number of units for heat exchanger network synthesis. *Chemical Eng. Communications*, **39**, 371.

Yee, T. F., Grossmann, I. E., & Kravanja, Z. (1990). Simultaneous optimization models for heat integration—I. Area and energy targeting and modeling of multistream exchangers. *Comp. and Chem. Engng.*, **14**(10), 1165.

Yee, T. F., & Grossmann, I. E. (1990). Simultaneous optimization models for heat integration—II. Heat exchanger network synthesis. *Comp. and Chem. Engng.*, **14**(10), 1165.

EXERCISES

1. Formulate the LP transshipment problem for minimum utility cost for the process streams and utilities given below:

	FCp(KW/K)	T_{in}(K)	T_{out}(K)
H1	10	450	270
C1	5	360	480
C2	5	300	400
C3	4	300	400

   HP Steam 500K, $80/KWyr   LP Steam 420K, $60/KWyr
   CW 300K, $20/KWyr   Refrigerant 260K, $100/KWyr
   HRAT = 10K

2. Show that the expanded form of the LP transshipment model in Eq. (16.9) can be reduced to the compact LP transshipment model in Eq. (16.5) if there are no constraints on the heat loads of the individual matches.

3. Assume that a consulting company tells you that for a given set of hot and cold streams with fixed flows and inlet and outlet temperatures, the minimum utility cost is $120,000/yr, requiring a minimum of 8 exchanger units. An engineer working for

you reports a utility cost of $110,000/yr using only 7 exchanger units. If both used exactly the same data and there are no arithmetic mistakes, what might be the reasons for the discrepancies in the results?

4. In the stream data below apart from having a heating and a cooling utility, there is a stream of saturated water that can be used to generate steam. This steam will produce a revenue to the network. Formulate the LP transshipment model that will maximize the annual profit of the network.

Stream data

	Fcp(KW/h)	T_{in}(K)	T_{out}(K)
H1	20	600	350
C1	8	400	560
C2	10	340	420

Utilities: Steam 610 K cost = 150 ($/KWyr), Cooling water 300–320 K cost = 20 ($/KWyr)

Saturated water for steam generation: Temperature 440 K net profit = 50 ($/kW/y) HRAT = 10K

5. Given is a process that involves the following set of hot and cold streams:

Stream	Fcp(KW/K)	T_{in}(K)	T_{out}(K)
H1	20	700	420
H2	40	600	310
H3	70	460	310
H4	94	360	310
C1	50	350	650
C2	180	300	400

The following utilities are available for satisfying heating and cooling requirements:

		Maximum available
Fuel	@ 750K , $5 × 10^{-6}/kJ	
HP steam	@ 510K , $3 × 10^{-6}/kJ	1000 KW
LP steam	@ 410K , $1.8 × 10^{-6}/kJ	500 KW
Cooling water	300–325K , $7 × 10^{-7}/kJ	

a. Formulate the LP transshipment that will predict the minimum annual utility cost and solve it with a computer code.

b. Indicate the loads predicted for the different utilities (in KW) and the location of pinch points.

c. Derive a configuration for a network with minimum utility cost (either by hand or with the MILP transshipment model).

NOTE: Assume operating time 8000 hrs/yr, and consider a minimum temperature approach of 10K.

6. Given the two hot and two cold streams below, determine:
   a. Minimum utility consumption
   b. Minimum number of units
   c. Network configuration that satisfies two above targets.

   Use the LP and MILP transshipment formulations for a and b.

	Fcp(MW/K)	T_{in}(K)	T_{out}(K)
H1	1	450	350
H2	1.2	450	350
C3	1	320	400
C4	2	350	420

   $\Delta T_{min} = 10K$
   Heating utility at 500K
   Cooling utility at 300K

7. Given the two hot and two cold streams below, determine a feasible heat exchanger network configuration with minimum utility consumption, fewest number of units, and which *does not* involve a match between hot stream H1 and cold stream C1. Formulate the corresponding LP and MILP transshipment models and solve.

	Fcp(MW/°C)	T_{in}(°C)	T_{out}(°C)
H1	1	400	120
H2	2	340	120
C1	1.5	160	400
C2	1.3	100	250

   Heating utility at 500°C. Cooling utility at 30°C, = ΔT_{min} 20°C.

8. a. Discuss why the inequalities Eq. (16.33) of the *MINLP* model for simultaneous synthesis will be active (i.e., behave as equations) when the corresponding exchangers are selected (i.e., variable z set to one).
   b. Assume that the inequalities in Eq. (16.33) are simplified by setting $\Gamma = 0$, which effectively enforces the constraint regardless of the choice of the 0–1 variables z. Discuss the difficulties that can arise in the MINLP model.

9. Given are a set of two hot and two cold process streams and steam and cooling water as utilities. The objective is to determine a heat exchanger network that exhibits least annual cost yet satisfies the heating and cooling requirements of the process streams. The table below shows the supply and target temperatures for the

streams, the heat capacity flow rates, the heat transfer coefficients, and the cost of utilites and exchangers. Costs to be considered include the utility cost and the annualized capital cost for the countercurrent heat exchangers.

Solve the problem as follows:

a. Sequential synthesis strategy with HRAT = 5K, solving the LP, MILP transshipment, and NLP models.
b. Simultaneous synthesis strategy with TMAPP = 5K solving the MINLP model.

Problem Data for Example

Stream	TIN (K)	TOUT (K)	FCp (kW/K)	h (KW/m^{2}K)	Cost ($/KW-yr.)
H1	650	370	10.0	1.0	—
H2	590	370	20.0	1.0	—
C1	410	650	15.0	1.0	—
C2	353	500	13.0	1.0	—
S1	680	680	—	5.0	80
W1	300	320	—	1.0	10

Exchanger cost = $5500 + $150 * Area ($m^2$)
Minimum approach temperature = 10K

SYNTHESIS OF DISTILLATION SEQUENCES   17

17.1   INTRODUCTION

In Chapter 11 a number of heuristic rules and physical insights were presented for synthesizing distillation sequences for ideal systems. Also for the heat integration case the use of T-Q diagrams was illustrated. In this chapter we will examine how one can develop MILP models for distillation sequences based on the network representation that was given in Chapter 15. Also, for the case of heat integration we will present two alternative models, one based on continuous temperatures and the other one on discrete temperatures. For simplicity in the presentation, we will concentrate mainly on the problem where a single multicomponent feed is given that must be separated into essentially pure components through the use of simple sharp split separators.

In order to reduce the size and complexity of the MILP models, we will rely on a number of simplifying assumptions in this chapter. These assumptions can be relaxed but at the expense of increasing the problem size or by introducing nonlinearities, as will be shown later in this chapter when considering rigorous MINLP models.

17.2   LINEAR MODELS FOR SHARP SPLIT COLUMNS

Firstly, as shown in Figure 17.1, we will consider single-feed distillation columns in which sharp splits are performed for light and heavy key components that are adjacent to each other. If we consider a fixed pressure and reflux ratio, then by performing short-cut calculations with any of the methods presented in Part II, we can obtain linear mass balance relationships in terms of the feed flowrates as given by (see Figure 17.2):

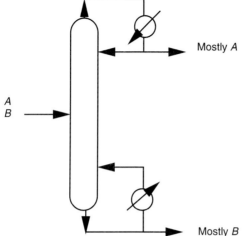

Mostly A

A
B

Mostly B **FIGURE 17.1** Sharp split separation.

$$d_i = \gamma_i f_i$$
$$b_i = (1-\gamma_i) f_i \qquad (17.1)$$

where d_i and b_i represent the mass flowrates of component in the distillate and bottoms, and γ_i are the corresponding recovery fractions that are typically obtained from the mass balance in the short-cut model for a selected feed composition. By assuming the fractions γ_i to be constant, it is clear that Eq. (17.1) reduces to linear equations.

Although in principle we can use the mass balance equations as given in Eq. (17.1), we will consider a further simplification with which we can pose our model only in terms

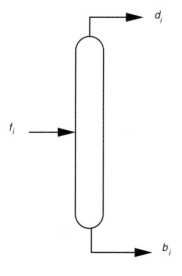

FIGURE 17.2 Mass balance for multicomponent column.

Sec. 17.2 Linear Models for Sharp Split Columns

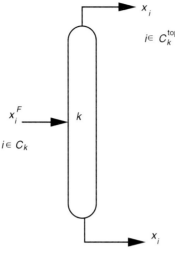

FIGURE 17.3 Module for total flow with sharp split.

of total feed flowrates for each column. If we assume 100% recoveries, then for each column k we can determine a priori the fractions of the total feed that are recovered at the top and at the bottoms by the following equations (see Figure 17.3):

$$\xi_k^{top} = \sum_{i \in C_k^{top}} x_i^F / \sum_{i \in C_k} x_i^F \qquad \xi_k^{bot} = \sum_{i \in C_k^{bot}} x_i^F / \sum_{i \in C_k} x_i^F \qquad (17.2)$$

where x_i^F is the mole fraction of component i in the initial mixture, C_k, C_k^{top}, and C_k^{bot}, are the sets of components that are involved in the feed, overhead, and bottoms of column k.

As an example, consider the column in Figure 17.4, which has as feed the initial multicomponent mixture. Applying Eq. (17.2), it is clear that $\xi^{top} = 0.2 + 0.4 = 0.6$, and $\xi^{bot} = 0.3 + 0.1 = 0.4$. For the column in Figure 17.5 that only has components C and D in the feed, it follows from Eq. (17.2) that $\xi^{top} = 0.3/(0.1 + 0.3) = 0.75$, $\xi^{bot} = 0.1/(0.1 + 0.3) = 0.25$. With these fractions we can then express the flows of the two product streams in the column in terms of the total feed flowrate, F, into the column as seen in Figures 17.4 and 17.5.

Since from the above assumptions we can model the mass balances through total feed flowrates for each column, it is convenient to model the heat duties of the condenser and reboiler and the capital cost in terms of these variables. Assuming the same loads in the condenser and reboiler, the heat duties for column k can be expressed as the linear functions:

$$Q_k = K_k F_k \qquad (17.3)$$

where K_k is a constant derived from a short-cut calculation. Finally, the annualized cost of the column, that includes the fixed-charge cost model for investment and the utility costs, will be given by:

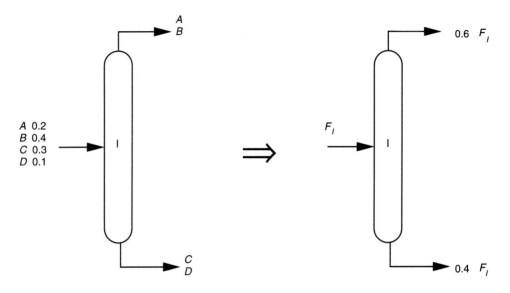

FIGURE 17.4 Example of initial split of four-component mixture.

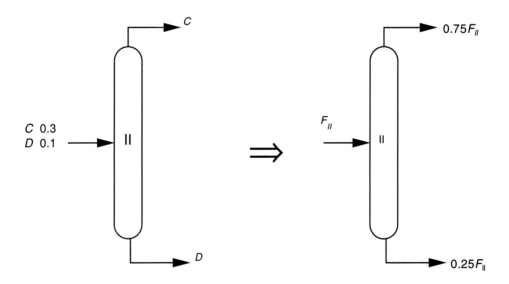

FIGURE 17.5 Example of intermediate split for two-component mixture.

$$C_k = \alpha_k y_k + \beta_k F_k + (c_H + c_C) Q_k \qquad (17.4)$$

where α_k is the annualized fixed-charge cost in terms of the 0–1 variable y_k, β_k is the size-factor for the column in terms of the total flow F_k, and c_H, c_C are unit costs for the heating and cooling in the reboiler and condenser, respectively.

17.3 EXAMPLE OF MILP MODEL FOR FOUR-COMPONENT MIXTURE

Before presenting the general form of the MILP model that is based on the linear models of the previous section, let us consider as an example the case where we have a mixture of 4 components A,B,C,D, that we want to separate into essentially pure products. The data on the composition of this mixture, the constants for heat balance, and the cost data are given in Table 17.1.

Firstly, we need to develop the superstructure for this problem. The corresponding network representation by Andrecovich and Westerberg (1985) that we discussed in Chap-

TABLE 17.1 Data for Example Problem

a) Initial field

F_{TOT} = 1000 kmol/hr

Composition (mole fraction)
- A   0.15
- B   0.3
- C   0.35
- D   0.2

b) Economic data and heat duty coefficients

| | | Investment cost | | Heat duty |
| | | | | coefficients, K_k, |
k	Separator	α_k, fixed (10^3 \$/yr)	β_k, variable (10^3 \$hr/kmol yr)	(10^6 kJ/kgmol)
1	A/BCD	145	0.42	0.028
2	AB/CD	52	0.12	0.042
3	ABC/D	76	0.25	0.054
6	A/BC	125	0.78	0.024
7	AB/C	44	0.11	0.039
4	B/CD	38	0.14	0.040
5	BC/D	66	0.21	0.047
10	A/B	112	0.39	0.022
9	B/C	37	0.08	0.036
8	C/D	58	0.19	0.044

Cost of utilities:

Cooling water C_C = 1.3 (10^3\$/10^6kJyr)
Steam C_H = 34 (10^3\$/10^6kJyr)

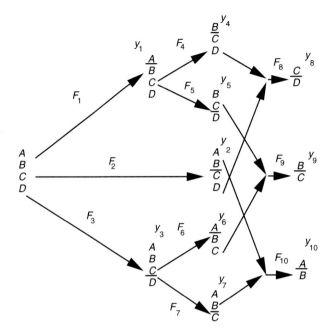

FIGURE 17.6 Network for four-component example.

ter 12, is shown in Figure 17.6. To each of the 10 columns in this network we can assign a 0–1 variable y to denote its potential existence and a variable F for its feed flowrate.

In order, to derive the mass balance equations, we need to compute first the split fractions as given by Eq. (17.2). Based on the feed composition in Table 17.1, and assuming sharp splits with 100% recoveries, the corresponding split fractions are shown in Table 17.2. The mass balances are then as follows.

TABLE 17.2 Split Fractions in Superstructure of Figure 17.6

$\xi_1^A = 0.15$	$\xi_6^A = 0.188$
$\xi_1^{BCD} = 0.85$	$\xi_6^{BC} = 0.812$
$\xi_2^{AB} = 0.45$	$\xi_7^{AB} = 0.5625$
$\xi_2^{CD} = 0.55$	$\xi_7^C = 0.437$
$\xi_3^{ABC} = 0.8$	$\xi_8^C = 0.636$
$\xi_3^D = 0.2$	$\xi_8^D = 0.364$
$\xi_4^B = 0.353$	$\xi_9^B = 0.462$
$\xi_4^{CD} = 0.647$	$\xi_9^C = 0.538$
$\xi_4^{BC} = 0.765$	$\xi_{10}^A = 0.333$
$\xi_5^D = 0.235$	$\xi_{10}^B = 0.667$

Sec. 17.3 Example of MILP Model for Four-Component Mixture

For the initial node in the network we have,

$$F_1 + F_2 + F_3 = 1000 \tag{17.5}$$

For the remaining nodes in the network, instead of considering mass balances around each column, we will consider mass balances for each intermediate product. The reason for this is that in the superstructure of Figure 17.6 we have associated flows only to the feed to each column, so that product streams do not necessarily have associated a flow as is the case of columns 2,4,5,6 and 7.

Based on the recovery fractions given in Table 17.2, the mass balance for each intermediate product is as follows:

1. Intermediate (BCD) which is produced in column 1, and directed to columns 4 and 5,

$$F_4 + F_5 - 0.85 F_1 = 0 \tag{17.6}$$

2. Intermediate (ABC), which is produced in column 3 and directed to columns 6 and 7,

$$F_6 + F_7 - 0.8 F_3 = 0 \tag{17.7}$$

3. Intermediate (AB), which is produced in columns 2 and 7 and directed to column 10,

$$F_{10} - 0.45 F_2 - 0.563 F_7 = 0 \tag{17.8}$$

4. Intermediate (BC), which is produced in columns 5 and 6 and directed to column 9,

$$F_9 - 0.765 F_5 - 0.812 F_6 = 0 \tag{17.9}$$

5. Intermediate (CD), which is produced in columns 2 and 4 and directed to column 8,

$$F_8 - 0.55 F_2 - 0.647 F_4 = 0 \tag{17.10}$$

The 10 flows in Eqs. (17.5) to (17.10) are related to the binary variables y through the following inequalities for each column (see Chapter 15):

$$F_k - 1000 y_k \leq 0, \quad F_k \geq 0, \quad y_k = 0,1, \quad k = 1,...10 \tag{17.11}$$

where we have selected 1000 as an upper bound because it corresponds to the feed flow rate of the initial mixture. Recall that the inequalities in Eq. (17.11) have the effect of setting a flow to zero if its corresponding binary variable is set to zero. If, on the other hand, the binary variable is set to 1, the flow has an upper bound of 1000.

The heat duties of condensers and reboilers, we can represent by the continuous variables Q_k, $k = 1, ...10$, and from Eq. (17.3), they are given through the equations

$$Q_k = K_k F_k, \quad k = 1,...10 \tag{17.12}$$

where the parameters K_k are given in Table 17.1. Note that the above equation assumes the loads in the condensers and reboilers to be the same. In practice, these values are often close.

FIGURE 17.7 Optimal separation sequence.

Finally, the objective function will be given by the minimization of the sum of the costs given in Eq. (17.4) for the 10 columns. That is,

$$\min C = \sum_{k=1}^{10}(\alpha_k y_k + \beta_k F_k) + (34+1.3)\sum_{k=1}^{10} Q_k \qquad (17.13)$$

where the cost coefficients α_k, β_k, are given in Table 17.1.

The objective function in Eq. (17.13), subject to the constraints in Eqs. (17.5) to (17.12), corresponds, then, to the MILP model for determining the optimum distillation sequence in the superstructure of Figure 17.6. Note that we have 20 continuous variables ($F_k, Q_k, k = 1,...10$) and 16 equations: Eqs. (17.5) to (17.10) and the ten in Eq. (17.12). Hence, this problem has 4 degrees of freedom. Also, we have ten 0–1 variables, and the ten logical inequalities in Eq. (17.11) that relate the flows and the binary variables.

If we solve the above MILP problem (e.g., with LINDO), we obtain the optimal sequence shown in Figure 17.7, which has an annualized cost of $3,308 × 10³ /yr. We can also obtain the second, and the third best solutions from the MILP by resolving it with the use of integer cuts (see Appendix)

Since the optimal solution in Figure 17.7 is given by $y_2 = y_8 = y_{10} = 1$, we can make this choice of binaries infeasible by adding the inequality

$$y_2 + y_8 + y_{10} \leq 2 \qquad (17.14)$$

By resolving the MILP with the additional inequality above, we obtain the second best solution, which as shown in Figure 17.8, corresponding to the direct sequence that has an annualized cost of $3,927 × 10³/yr. To obtain the third best solution we make the selection of this configuration infeasible by adding Eq. (17.14) and the inequality

$$y_1 + y_4 + y_8 \leq 2 \qquad (17.15)$$

Resolving the MILP we obtain the indirect sequence which is shown in Figure 17.9 with an annualized cost of 4,102 × 10³ $/yr. It is interesting to note from Figures 17.7, 17.8, and 17.9 that the optimal sequence in Figure 17.7 is the one that has the lowest total mass flow (2000 kmol/hr), which is consistent with the heuristic of selecting the sequence

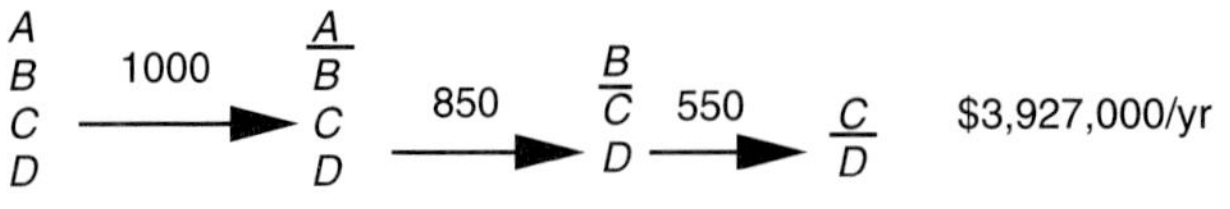

FIGURE 17.8 Second best sequence.

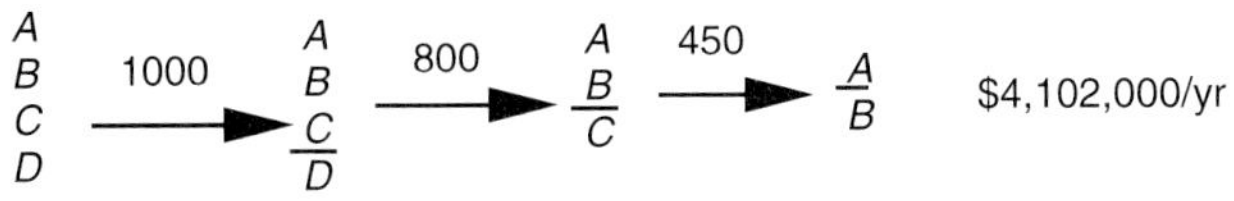

$$\begin{array}{c}A\\B\\C\\D\end{array} \xrightarrow{1000} \begin{array}{c}A\\B\\\hline C\\D\end{array} \xrightarrow{800} \begin{array}{c}A\\B\\\hline C\end{array} \xrightarrow{450} \begin{array}{c}A\\\hline B\end{array} \quad \$4{,}102{,}000/\text{yr}$$

FIGURE 17.9 Third best solution.

with minimum total mass flow. Note, however, that the third best solution has a lower total mass flow (2250 kmol/hr) than the second best solution (2400 kmol/hr).

17.4 MILP MODEL FOR DISTILLATION SEQUENCES

Based on the example in the previous section, we can now easily generalize the MILP model for synthesizing distillation sequences for any mixture of n components that is to be separated into pure components.

First, we will need to define the following index sets, which we will illustrate with the example of the previous section:

1. $IP = \{m \mid m \text{ is an intermediate product}\}$
   e.g., $IP = \{(ABC), (BCD), (AB), (BC), (CD)\}$
2. $COL = \{k \mid k \text{ is a column in the superstructure}\}$
   e.g., $COL = \{1, 2, \ldots, 9, 10\}$
3. $FS_F = \{\text{columns } k \text{ that have as feed the initial mixture}\}$
   e.g., $FS_F = \{1, 2, 3\}$
4. $FS_m = \{\text{columns } k \text{ that have as feed intermediate } m\}$
   e.g., for $m = (BCD)$, $FS_m = \{4, 5\}$
5. $PS_m = \{\text{columns } k \text{ that produce intermediate } m\}$
   e.g., for $m = (CD)$, $PS_m = \{2, 4\}$

Through these sets, the objective function in Eq. (17.13) and the constraints in Eqs. (17.5) to (17.12) can then be written as the MILP (Andrecovich and Westerberg, 1985):

$$\min C = \sum_{k \in COL} \left[\alpha_k y_k + \beta_k F_k + (C_H + C_C) Q_k \right]$$

$$\text{s.t.} \sum_{k \in FS_F} F_k = F_{TOT}$$

$$\sum_{k \in FS_m} F_k - \sum_{k \in PS_m} \xi_k^m F_k = 0 \quad m \in IP \quad (17.16)$$

$$Q_k - K_k F_k = 0 \qquad k \in COL$$

$$F_k - U y_k < 0 \qquad k \in COL$$

$$F_k, Q_k \geq 0, \; y_k = 0, 1 \qquad k \in COL$$

where F_{TOT} is the flowrate of the initial mixture, ξ_k^m are the recoveries of intermediate m in column k, and U is an upper bound for the flowrates, which for simplicity we can select as F_{TOT}.

Note that the size of the above MILP is a function of the number of separators in the network representation of the superstructure and not a function of the number of sequences. It should also be noted that the above model can be easily extended so as to handle flowrates of individual components with individual split fractions as given by Eq. (17.1). For reasons of problem size, however, it is convenient to keep the form of the MILP model as in Eq. (17.16), especially for the next section where heat integration is considered as part of the synthesis problem.

17.5 HEAT INTEGRATION AND PRESSURE EFFECTS

In the MILP model that was presented in the previous section, it was assumed that the cooling in the condensers and the heating in the reboilers would be performed with utilities (e.g., cooling water and steam respectively). However, as was shown in Chapter 10, it is often desirable to perform heat integration in distillation sequences, because energy, more than capital, tends to be the dominant cost.

The two major alternatives that we can consider for heat integration in distillation columns are shown in Figures 17.10 and 17.11. In Figure 17.10 we have an indirect se-

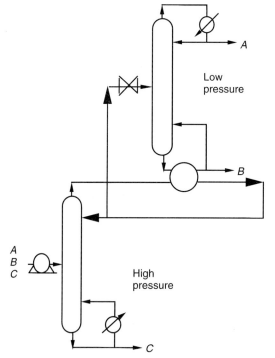

FIGURE 17.10 Heat integration between tasks.

Sec. 17.5 Heat Integration and Pressure Effects

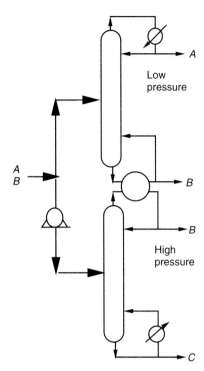

FIGURE 17.11 Multieffect heat integration.

quence for the separation of (*ABC*). Here the first column operates at a high pressure so that its condenser can be used as a heat source for the reboiler in the second column, which operates at low pressure. In Figure 17.11 the separation of *A* and *B* is performed with two columns; one at low pressure and the other at high pressure, so that the condenser of the latter can be used as a source of heat for the reboiler of the former. In other words, Figure 17.11 represents an alternative for heat integration through multieffect distillation for the same separation task, while Figure 17.10 represents an alternative of heat exchange between columns that perform different separation tasks. In both cases, it is clear that the selection of column pressures is of critical importance.

Treating the pressure of the columns in our models explicitly will introduce nonlinearities because in order to consider the temperature effects, we need to compute bubble and dewpoint temperatures as shown in Figure 17.12. Although it is possible to explicitly include these equations in an MINLP model, we will make the assumption that ΔT_{RC}, the difference between the reboiler temperature (dew point) and condenser temperature (bubble point), is a constant that is independent of the column pressure (see Chapter 10, Part III). This constant would be typically computed from a short-cut model at nominal pressure. We will assume throughout this chapter that the columns consist of total condensers and total reboilers as in Figure 17.12. Furthermore, we will assume constant temperatures for the distillate and bottoms streams.

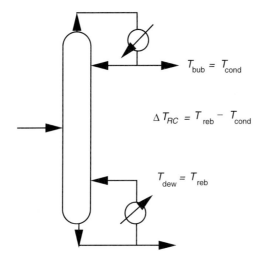

FIGURE 17.12 Modelling pressure changes through ΔT_{RC}.

In addition, we assume the heat duty coefficients K_k to be constant and independent of temperatures. In this way it is possible to model the problem of synthesizing heat integrated distillation sequences by augmenting the MILP model in Eq. (17.16) with additional constraints that only depend on the temperatures of condensers and reboilers. We will see in the next two sections that we can accomplish this with two different model types: (a) continuous temperatures and (b) discretized temperatures.

17.6 MILP MODEL WITH CONTINUOUS TEMPERATURES

We will assume for the model in this section that heat integration will only be considered between different separation tasks. Thus the possibility of synthesizing multi-effect columns will be excluded, and therefore the superstructure in terms of columns will remain the same as in sections 17.3 and 17.4. Also, we will assume only one heating and one cooling utility (e.g., steam and cooling water). Finally, in order to retain linearity in the model we will assume that only the fixed cost of the column varies with pressure, or equivalently, with the temperature in the condenser (Raman and Grossmann, 1993). The general form that will be considered is as follows:

$$\text{Fixed cost} = \alpha \left[1 + \gamma \left(\frac{T_C - T_{CW} - EMAT}{T_{CW} + EMAT} \right) \right] \quad (17.17)$$

where T_C is the condenser temperature, T_{CW} the temperature of cooling water, $EMAT$ the minimum exchanger approach temperature, and α and γ cost coefficients. Note that $T_C \geq T_{CW} + EMAT$. Also, if the column is not selected the fixed cost has to be set to zero. This can be accomplished by introducing the new variable μ^k to represent fixed charges that are to be minimized in the objective function and that satisfy the following inequalities,

Sec. 17.6 MILP Model with Continuous Temperatures

$$\mu_k \geq \alpha \left[1 + \gamma \left(\frac{T_C - T_{CW} - EMAT}{T_{CW} + EMAT} \right) \right] - U_k(1 - y_k) \quad (17.18)$$

$$\mu_k \geq 0 \qquad y_k = 0, 1$$

where U_k represents a valid upper bound on the fixed cost of column k. In this way, if the column is selected, $y_k = 1$, μ_k takes the value of the right hand side as in Eq. (17.17). If, on the other hand, the column is not selected, $y_k = 0$, the inequality becomes redundant; since μ_k is restricted to be non-negative, it will take the value of zero.

In addition to the mass balance equations in (17.16), we need constraints that represent the heat exchange. If T_R^k and T_C^k are the reboiler and condenser temperatures of column k, ΔT_{RC} is the temperature difference in the column, $EMAT$ is the minimum exchanger approach temperature, and T_S and T_{CW} are the temperatures of steam and cooling water, the three following constraints apply:

$$\left. \begin{aligned} T_R^k &= T_C^k + \Delta T_{RC} \\ T_R^k &\leq T_S - EMAT \\ T_C^k &\geq T_{CW} + EMAT \end{aligned} \right\} k \in COL \quad (17.19)$$

The first simply establishes the relation between the condenser and reboiler temperatures, while the two inequalities provide the limits of these temperatures in terms of the steam and cooling water temperatures.

To consider the potential exchanges of heat we define the variable QEX_{kj} to denote the amount of heat exchanged between the condenser of column k and the reboiler of column j (see Figure 17.13), and we also define the binary variable z_{kj}:

$$z_{kj} = \begin{cases} 1 & \text{condenser column } k \text{ supplies} \\ & \text{heat to reboiler column } j \\ 0 & \text{otherwise} \end{cases} \quad (17.20)$$

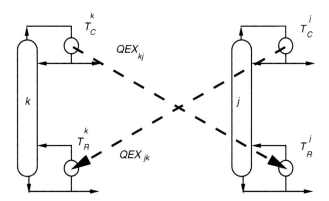

FIGURE 17.13 Definition of variables for heat integration between different separation tasks.

Then the two following conditional constraints apply where Ω_k, Λ_k, are valid bounds:

$$\left.\begin{array}{l} QEX_{kj} - \Omega_k z_{kj} \leq 0 \\ T_C^k \geq T_R^j + EMAT - \Lambda_{kj}(1 - z_{kj}) \end{array}\right\} \quad k, j \in \text{COL } k \neq j \qquad (17.21)$$

Note that if $z_{kj} = 1$, the temperature of the condenser of column k is forced to be larger than the temperature of the reboiler in column j; on the other hand, if $z_{kj} = 0$, the first inequality forces $QEX_{kj} = 0$ and the inequality for the temperature becomes redundant. Finally, heat balance must hold for the exchanges of heat QEX_{kj} and the cooling and heating duties, QW_k and QS_k, supplied to satisfy the load Q_k of each column. The following equations then apply:

$$\left.\begin{array}{l} \sum_{j \in \text{COL} \backslash k} QEX_{kj} + QW_k = Q_k \\ \sum_{j \in \text{COL} \backslash k} QEX_{jk} + QS_k = Q_k \end{array}\right\} \quad k \in \text{COL} \qquad (17.22)$$

Defining the objective function in terms of the investment cost of the column as in Eq. (17.16) and the utility cost and considering the constraints in Eqs. (17.16) and (17.18) to (17.22), the MILP model is given as follows:

$$\min C = \sum_{k \in \text{COL}} [\mu_k + \beta_k F_k + C_H QS_k + C_C QW_k]$$

s.t. $\mu_k \geq \alpha_k \left[1 + \gamma \left(\dfrac{T_C - T_{CW} - EMAT}{T_{CW} + EMAT}\right)\right] - U_k(1 - y_k) \quad k \in \text{COL}$

st.

$$\sum_{k \in FS_F} F_k = F_{\text{TOT}}$$

$$\sum_{k \in FS_m} F_k - \sum_{k \in PS_m} \xi_k^m F_k = 0 \qquad m \in \text{IP} \qquad (17.23)$$

$$\left.\begin{array}{l} Q_k - K_k F_k = 0 \\ F_k - U y_k \leq 0 \end{array}\right\} \quad k \in \text{COL}$$

$$\left.\begin{array}{l} \sum_{j \in \text{COL} \backslash k} QEX_{kj} + QW_k = Q_k \\ \sum_{j \in \text{COL} \backslash k} QEX_{jk} + QS_k = Q_k \end{array}\right\} \quad k \in \text{COL}$$

Sec. 17.7   MILP Model with Discrete Temperatures

$$\left.\begin{array}{l} T_R^k = T_C^k + \Delta T_{RC} \\ T_R^K \leq T_S - EMAT \\ T_C^k \geq T_{CW} + EMAT \end{array}\right\} \; k \in COL$$

$$\left.\begin{array}{l} QEX_{kj} - \Omega_k z_{kj} \leq 0 \\ T_C^k \geq T_R^j + EMAT - \Lambda_{kj}(1 - z_{kj}) \end{array}\right\} \; k, j \in COL \; k \neq j$$

$$F_k, Q_k, \mu_k \geq 0, y_k = 0,1 \quad k \in COL; \quad QEX_{kj} \geq 0, z_{kj} = 0,1 \quad j,k \in COL, j \neq k$$

Note that in the above formulation the cost of the exchangers is not directly accounted for. The simplest option would be to add a fixed charge that would be associated to each binary variable z_{kj}. The other option would be to add the nonlinear equation of the area with which Eq. (17.23) would become an MINLP. Assuming we solve the model as in Eq. (17.23), we will find that the computational cost can be rather high, mainly due to the large number of binary variables (y_k, z_{kj}). We can expedite the solution of this MILP with three additional types of constraints:

1. Number of columns cannot exceed number of components minus one:

$$\sum_{k \in COL} y_k \leq \text{No. Comp.} - 1 \tag{17.24}$$

2. If a column is not selected, the corresponding matches to that column cannot take place:

$$\left.\begin{array}{l} z_{jk} \leq y_k \\ z_{kj} \leq y_k \end{array}\right\} \; \begin{array}{l} k, j \in COL \\ k \neq j \end{array} \tag{17.25}$$

3. Either column j supplies heat to column k, or vice versa:

$$z_{jk} + z_{kj} \leq 1 \quad k, j \in COL, k \neq j \tag{17.26}$$

It should be noted that although the three above constraints are redundant, they help to limit the search space in the branch and bound enumeration of the MILP problem.

Finally, it is clear that once the optimum solution has been found with the MILP model in Eq. (17.23), the operating pressures of the columns can simply be back-calculated from the temperatures of the condensers. An MINLP version of this model has been developed by Floudas and Paules (1988).

17.7 MILP MODEL WITH DISCRETE TEMPERATURES

In this section we will present an alternate MILP model for synthesizing heat integrated sequences (Andrecovich and Westerberg, 1985) that is based on discretizing the temperatures. Although in principle this may appear to be more restrictive, we will see how this facilitates the consideration of multi-effect integration as well as the aggregation of the heat integration through the transshipment equations, thus eliminating the need of introducing 0–1 variables for the matches.

In principle, we could approach the heat integration problem in distillation sequences as follows. First, in the network of Figure 17.6 we could postulate a different number of candidate columns for each separation task. This number would be typically the maximum number of columns we are willing to have for multi-effect separation. So, for example, for the task A/BCD we might postulate two different columns, each operating at a different fixed pressure. The condensers could then be treated as hot streams and the reboilers as cold streams, both with fixed temperatures. Only their flowrates would be unknown. Based on this discretization scheme for the temperatures we can simply add to our basic MILP problem in Eq. (17.16) the equations of the LP transshipment model of Chapter 16. We will see that the fact that the flows are unknown poses no problem to preserve linearity.

In order to postulate the superstructure we consider here a simplified version of the procedure suggested by Andrecovich and Westerberg (1985). Having determined ΔT_{RC} for each separation task, a procedure for selecting candidate columns operating at discrete temperature levels, is as follows:

1. Define the allowable range of temperatures for heat integration:

   Highest temperature = Hottest hot utility temperature – *EMAT*

   Lowest temperature = Coldest cold utility outlet temperature + *EMAT*

2. Within the allowable range, for each separation task create a stack of columns from bottom to top with temperature change of ΔT_{RC} and with *EMAT* difference between successive columns; if the stack misses the top by more than *EMAT*, create a second stack of columns from top to bottom.

To illustrate more clearly this procedure, consider the three-component example in Table 17.3. As seen in Figure 17.12, the allowable temperature range is given from 330K to 540K. The 330K is obtained by adding 10K to the outlet of the cooling water temperature, and the 540K by subtracting 10K from the high pressure steam temperature.

TABLE 17.3 Data for Three-Component Mixture

	Task		ΔT_{RC} (K)
	A/BC		100
	AB/C		80
	A/B		50
	B/C		50
Utilities:		High pressure steam:	550 K
		Low pressure steam:	460 K
		Cooling water:	300–320 K
EMAT = 10 K			

Sec. 17.7 MILP Model with Discrete Temperatures

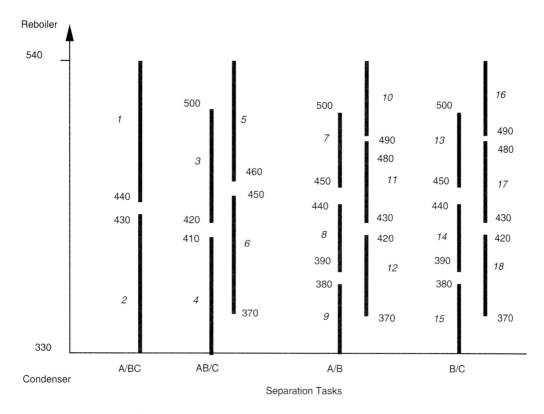

FIGURE 17.14 Potential columns for multieffect heat integration.

Let us consider the separation task, *A/BC*, which has ΔT_{RC} of 100K as seen in Figure 17.14. By starting at 330K we first consider a column with condenser temperature at 330K and reboiler temperature at 430K. With an *EMAT* of 10K, we stack on top of this column one whose condenser is at 440K and whose reboiler is at 540K. In this case there are then two columns that we can exactly fit within the range 330 to 540K, as seen in Figure 17.14. In these two columns the condenser of column 1 at 440K can potentially exchange heat with the reboiler of column 2 at 430K. For the separation task *AB/C*, we have a ΔT_{RC} of 80K. In this case, the first column has the condenser at 330K and the reboiler at 410K. With the *EMAT* of 10K, we stack on top of this column one whose condenser is at 420K and its reboiler at 500K. Since we miss the top of the range by 40K, we now create a second stack from top to bottom. The first column in the second stack has the reboiler at 540K and the condenser at 460K; the second column has the reboiler at 450K and the condenser at 370K. Thus, we will postulate four columns for this separation task as seen in Figure 17.14.

For the separation tasks *A/B*, *B/C*, both of which have ΔT_{RC} of 50K, the procedure is entirely analogous as above. In each case we obtain two stacks with six columns as seen in Figure 17.14.

From Figure 17.14 we can then see that through the discretization scheme we will consider a total of 18 potential columns. The operating pressure of these columns could be obtained, for instance, by doing a bubble point calculation for the condenser temperatures.

Assuming that we have determined the discrete set of potential columns as in Figure 17.14, let us consider how we can represent the heat integration among these columns. Rather than considering all individual heat exchanges that are possible between all condensers and reboilers as we did in the previous section, we can embed the heat integration into the MILP through a heat cascade. That is, by treating the condensers as hot streams and the reboilers as cold streams, we can construct a heat cascade that is based on the temperatures of these streams. Since these temperatures can be assumed to be a constant, it is convenient to represent the temperature intervals at constant temperatures. In this way, based on the temperatures of the reboilers and condensers in Figure 17.14 and the different utilities, we can construct the heat cascade shown in Figure 17.15. On the left, we have as inputs the heat of the condensers, Q_k, and the heat of the low pressure steam, Q_{LP}. On the right, we have as outputs the heats of the reboilers, Q_k. At the top of the cascade we have as input the heat of the high pressure steam, Q_{HP}, and at the bottom the output of the heat of the cooling water, Q_{CW}.

Note that all the heat loads in Figure 17.15 are unknown. However, this poses no difficulty since we can perform heat balances around each temperature interval in a similar way as we did with the transshipment model. That is, for each interval $\ell = 1,2,...L$ we can write the equation,

$$R_\ell - R_{\ell-1} - \sum_{i \in HU^\ell} Q_{HU}^i + \sum_{j \in CU^\ell} Q_{CU}^j - \sum_{k \in I_C^\ell} Q_k + \sum_{k \in I_R^\ell} Q_k = 0 \qquad (17.27)$$

where R^ℓ is the heat residual exiting from interval ℓ; Q_{HU}^i, Q_{CU}^j are the heat loads of the hot utilities $i \in HU^\ell$ and cold utilities $j \in CU^\ell$ in interval ℓ, and Q_k are the heat loads of the columns. I_C^ℓ, I_R^ℓ are the set of columns whose condenser and reboiler temperatures coincide with the ones of temperature interval ℓ (e.g., $I_C^\ell = \{10,16\}$, $I_R^\ell = \{11,17\}$ for interval (490–480K) in Figure 17.15). In this way, by including Eq. (17.27) in the MILP model Eq. (17.16), we can ensure maximum heat integration in the columns since the heating and cooling utility loads will be included as operating costs in the objective function.

Based on the discretization scheme of the previous section, we can develop a network superstructure that is similar to the one in Figure 17.6. As an example, assume we have the ternary mixture (ABC), for which the discretization would yield two columns for each task: (A/BC), (AB/C), (A/B), (B/C). We would then simply duplicate columns in the superstructure as seen in Figure 17.16. Note that here we are still assigning to each column a feed flow and a corresponding 0–1 variable. To such a superstructure we can assign similar index sets as we did in section 17.4.

By replacing the utility loads in the objective function in Eq. (17.16), and by adding the transshipment equations for heat integration in (17.27), the MILP model yields:

$$\min C = \sum_{k \in COL} (\alpha_k y_k + \beta_k F_k) + \sum_{i \in HU} C_{HU}^i Q_{HU}^i + \sum_{j \in CU} C_{CU}^j Q_{CU}^j \qquad (17.28)$$

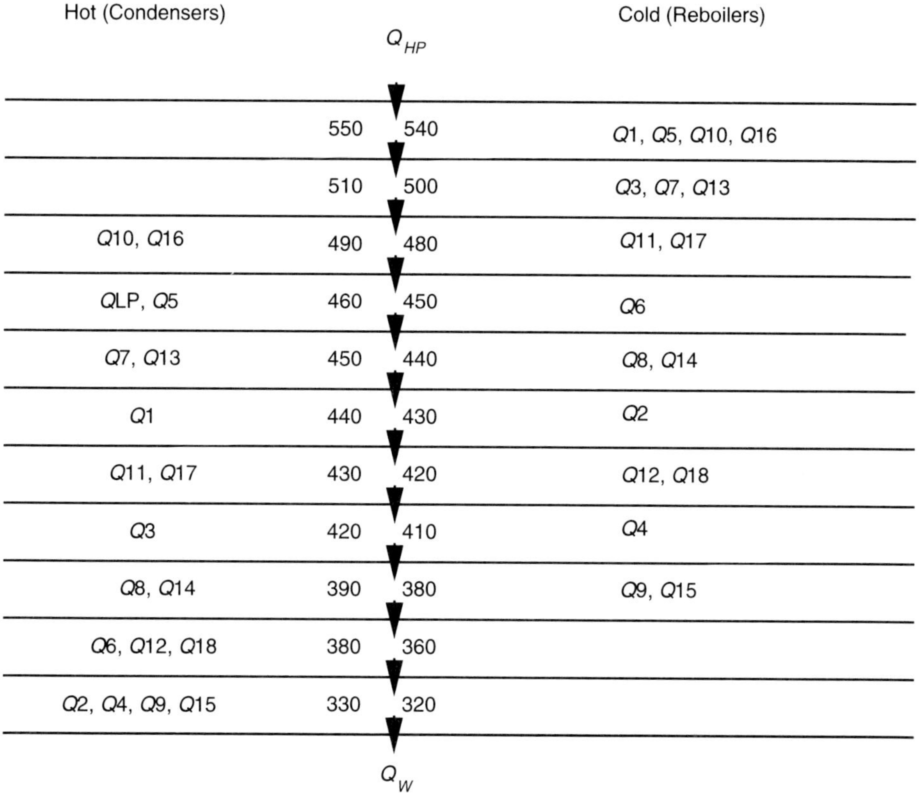

Sec. 17.7 MILP Model with Discrete Temperatures

Hot (Condensers)			Cold (Reboilers)
		Q_{HP} ↓	
	550	540	Q1, Q5, Q10, Q16
	510	500	Q3, Q7, Q13
Q10, Q16	490	480	Q11, Q17
QLP, Q5	460	450	Q6
Q7, Q13	450	440	Q8, Q14
Q1	440	430	Q2
Q11, Q17	430	420	Q12, Q18
Q3	420	410	Q4
Q8, Q14	390	380	Q9, Q15
Q6, Q12, Q18	380	360	
Q2, Q4, Q9, Q15	330	320	
		↓ Q_W	

FIGURE 17.15 Heat flows for potential columns in Figure 17.14.

$$s.t. \sum_{k \in FS_F} F_k = F_{TOT}$$

$$\sum_{k \in FS_m} F_k - \sum_{k \in PS_m} \xi_k F_k = 0 \quad m \in IP$$

$$Q_k - K_k F_k = 0 \quad k \in COL$$

$$F_k - U y_k \leq 0 \quad k \in COL$$

$$R_\ell - R_{\ell-1} - \sum_{i \in HU^\ell} Q_{HU}^i + \sum_{j \in CU^\ell} Q_{CU}^j - \sum_{k \in I_C^\ell} Q_k + \sum_{k \in I_R^\ell} Q_k = 0 \quad \ell = 1, 2, \dots L$$

$$F_k, Q_k, \geq 0, y_k = 0, 1 \quad k \in COL$$

$$Q_{HU}^i \geq 0 \quad i \in HU, \quad Q_{CU}^j \geq 0 \quad j \in CU, \quad R_\ell \geq 0 \quad \ell = 1, \dots L$$

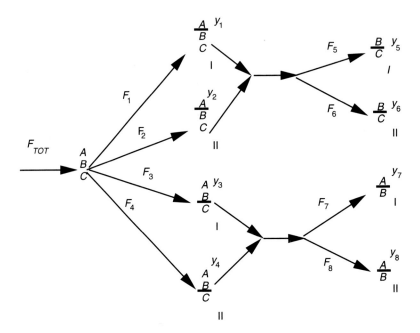

FIGURE 17.16 Superstructure and variables for three-component mixture with two columns per separation task.

The solution of this model would then indicate the columns that are selected from the superstructure and the heat loads of the condensers, reboilers, and utilities. These loads will feature maximum heat integration due to the inclusion of the transshipment model equations. Hence, once the MILP solution is obtained, the detailed heat recovery network structure can be derived either manually or through the models that were given in Chapter 16.

In a similar way as in the previous section, we can determine the second or third solution by resolving the MILP with integer cuts. Also, if no multi-effect columns are allowed, one can simply exclude this option by specifying the constraint that no more than one column be selected for each separation task. For example, for the separation task (A/BC) in Figure 17.16, we would specify the constraint

$$y_1 + y_2 \leq 1 \tag{17.29}$$

As a final point, it is apparent that while the discretization scheme for heat integration has the advantage of keeping our synthesis problem as an MILP, it has two limitations. First, the temperatures cannot be treated as continuous variables for the optimization as was the case with the MILP model in Eq. (17.23). Secondly, although no 0–1 variables are used for the matches, the number of these variables is often increased due to the different columns that must be included in the superstructure. Nevertheless, the model in Eq. (17.28) can be solved with reasonable computational expense because it usually has a much smaller relaxation gap than the model in Eq. (17.23).

17.8 DESIGN AND SYNTHESIS WITH RIGOROUS MODELS

In the previous sections of this chapter we considered highly simplified models for synthesizing separation sequences. While in principle their simplified nature is a major limitation, they can still serve as useful tools for examining alternatives for preliminary design. On the other hand, at one point of the synthesis one has to consider design models that are more rigorous in nature. In this section we will present such a model by Viswanathan and Grossmann (1990) for determining the optimal feed tray location in columns with specified number of trays. The extension for optimizing the number of trays will also be discussed.

Consider the superstructure for distillation columns in Figure 17.17 with N stages, including the condenser and the (kettle-type) reboiler (we consider the total condenser case; the other cases are dealt with similarly). To model the optimal feed tray location, it is assumed that the feed is split into streams that in principle can each be fed into every tray, although clearly one can easily restrict the candidate trays as discussed below. Let

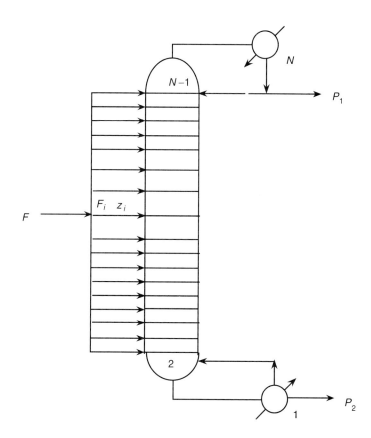

FIGURE 17.17 Superstructure for optimal feed tray location.

the trays be numbered bottom upwards so that the reboiler is the first tray and the condenser is the last (Nth) tray. Let $I = \{1,2,...N\}$ denote the set of trays and let $R = \{1\}$, $C = \{N\}$, COL $= \{2,3,...N-1\}$ denote the subsets corresponding to the trays in the reboiler, the condenser, and those within the column respectively.

Let c be the number of components in the feed and let F, T_f, P_f, z_f, h_f, denote respectively, the molar flowrate, temperature, pressure, the vector of mole-fractions (with components z_{fj}, $j = 1,2,...c$), and the molar specific enthalpy of the feed. The pressure prevailing on tray i is denoted by p_i. Let $p_{reb} = p_1$, $p_{bot} = p_2$, $p_{top} = p_{N-1}$, $p_{con} = p_N$ be given. We have $p_1 \geq p_2 \geq ... \geq p_{N-1} \geq p_N$ and for simplicity we assume $p_f \geq p_{bot}$.

Let L_i, x_i, h_i^L, and f_{ij}^L denote the molar flowrate, the vector of mole-fractions, the molar specific enthalpy, and the fugacity of component j, respectively, of the liquid leaving tray i. Similarly, let V_i, y_i, h_i^V, and f_{ij}^V denote the corresponding quantities of the vapor leaving tray i. Denoting the temperature of tray i by T_i, we have

$$\begin{aligned} f_{ij}^L &= f_{ij}^L(T_i, p_i, x_{i1}, x_{i2},...x_{ic}) \\ f_{ij}^V &= f_{ij}^V(T_i, p_i, y_{i1}, y_{i2},...y_{ic}) \\ h_i^L &= h_i^L(T_i, p_i, x_{i1}, x_{i2},...x_{ic}) \\ h_i^V &= h_i^V(T_i, p_i, y_{i1}, y_{i2},...y_{ic}) \end{aligned} \quad (17.30)$$

where the functions on the right-hand side depend on the thermodynamic model used.

Let P_1 and P_2 denote the top and bottom product rates, respectively. The subset of (contiguous) candidate tray locations for the feed are specified by the index set LOC, where LOC $\subset$ COL $\subset I$. Let z_i, $i \in$ LOC, denote the binary variable associated with the selection of i as the feed tray; i.e., $z_i = 1$ iff i is the feed tray. Let F_i, $i \in$ LOC denote the amount of feed entering tray i.

The modeling equations are then as follows:

a. Phase equilibrium: $\quad f_{ij}^L = f_{ij}^V \quad j = 1,...c, \quad i \in I \quad (17.31)$

b. Phase equilibrium error: $\quad \sum_j x_{ij} - \sum_j y_{ij} = 0 \quad i \in I \quad (17.32)$

c. Total material balances:

$$\begin{aligned} V_{i-1} - (L_i + P_1) &= 0 \quad i \in C \\ L_i + V_i - L_{i+1} - V_{i-1} &= 0 \quad i \in \text{COL}\backslash\text{LOC} \\ L_i + V_i - L_{i+1} - V_{i-1} - F_i &= 0 \quad i \in \text{LOC} \\ V_i + P_2 - L_{i+1} &= 0 \quad i \in R \end{aligned} \quad (17.33)$$

d. Component material balances:

$$\begin{aligned} V_{i-1} y_{i-1,j} - (L_i + P_1) x_{ij} &= 0 \quad j = 1,...c, i \in C \\ L_i x_{ij} + V_i y_{ij} - L_{i+1} x_{i+1,j} - V_{i-1} y_{i-1,j} &= 0 \quad j = 1,...c, i \in \text{COL}\backslash\text{LOC} \\ L_i x_{ij} + V_i y_{ij} - L_{i+1} x_{i+1,j} - V_{i-1} y_{i-1,j} - F_i z_{fj} &= 0 \quad j = 1,...c, i \in \text{LOC} \\ V_i y_{i,j} + P_2 x_{ij} - L_{i+1} x_{i+1,j} &= 0 \quad j = 1,...c, i \in R \end{aligned} \quad (17.34)$$

Sec. 17.8 Design and Synthesis with Rigorous Models

e. Enthalpy balances:

$$L_i h_i^L + V_i h_i^V - L_{i+1} h_{i+1}^L - V_{i-1} h_{i-1}^V - F_i h_f = 0 \quad i \in \text{LOC}$$
$$L_i h_i^L + V_i h_i^V - L_{i+1} h_{i+1}^L - V_{i-1} h_{i-1}^V = 0 \quad i \in \text{COL\textbackslash LOC}$$
(17.35)

f. Constraints on feed location :

$$\sum_{i \in LOC} z_i = 1$$
$$\sum_{i \in LOC} F_i = F$$
$$F_i - F z_i \leq 0, \ i \in \text{LOC}$$
(17.36)

The last constraint in Eq. (17.36) expresses the fact that if tray $i \in$ LOC is selected as the feed tray, then the amount of feed entering other candidate locations is zero. This follows from the fact $z_j = 0, j \neq 1, j \in$ LOC. In addition, there may be constraints on purity, recovery, reflux ratio, and so on. The MINLP problem, then, is to minimize (or maximize) a given objective function subject to the equality and inequality constraints Eqs. (17.30) to (17.36). Note that in this model, the variables z_i are binary, while all other variables are continuous.

An example of the application of the above MINLP model reported in the GAMS Optimization Case Study of CACHE (Viswanathan and Grossmann, 1991) will be considered next.

Given is a distillation column with seven ideal stages, a total condenser, and a kettle-type reboiler. The feed consists of a mixture of 70% mole benzene and 30% mole toluene entering at its bubble point at 1.12 bar. The top product must have a purity of at least 95% mole benzene. The objective is to maximize profit, which is proportional to the top product rate minus the cost of the energy used, expressed in terms of the reflux ratio. Additional data and specifications are given in Table 17.4. The problem is to determine the optimum location of the feed plate; i.e find the best location for introducing the feed to maximize profit.

The objective function chosen ($P1 - 50 \ast r$) is indicative of the trade-offs between increasing the throughput (primary objective) and the corresponding increase in reboiler duty—measured roughly by the value of the reflux ratio.

The optimal solution obtained with DICOPT++ is given by:

Obj. function	=	13.144
r	=	0.925
P_1	=	59.396
P_2	=	40.604
feed plate	=	tray no. 4

TABLE 17.4 Data for Optimal Feed Tray Problem

System	Benzene-toluene
Thermodynamic model	liquid - ideal
	vapor - ideal
Source of thermodynamic data	Reid et al. (1987)
Condenser type	Total
No. of trays (N)	9
(including condenser and reboiler)	
Candidates for feed tray LOC =	$\{2, 3, ... 8\}$

Specifications:
$F = 100$, $p_f = 1.12$ bar, $T_f = 359.6$ K, $z_f = (0.70, 0.30)$
$p_{reb} = 1.20$, $p_{bot} = 1.12$, $p_{top} = 1.08$, $p_{con} = 1.05$ bar

Constraints : r = reflux ratio ≤ 0.95
                 purity of benzene in the distillate : $x_{9,1} \geq 0.95$
             Objective function : $P\ 1 - 50 * r$

Note that the solution was found in the first step in the relaxed NLP where the binary variables are continuous variables with values between zero and one. The CPU time required on a HP-UX 9000/835 was 6.7 seconds.

Finally, it is worth to note that the above MINLP model can be extended to optimize the number of trays for single or multiple feeds (see Viswanathan and Grossmann, 1993a,b). The main idea is to extend the superstructure representation in Figure 17.17 to the one in Figure 17.18. Note that the reflux is potentially returned to every tray i with the variable r_i and its associated binary variable w_i. Selecting one of these return points will determine the "redundant" trays at the top of the column that only handle vapor flow and perform no separation. Multiple feeds could be handled by defining the variables F_i^k, z_i^k, for the flows and 0–1 variables for each feed k at each tray i.

17.9 NOTES AND FURTHER READING

A review of algorithmic methods up to the late 1980s for the systems of distillation sequences can be found in Floquet et al. (1988).

The first comprehensive approach for design and synthesis (Sargent and Gaminibandara, 1976) proposed the optimization of a superstructure with rigorous models. The first simplified MILP model was proposed by Andrecovich and Westerberg (1985). This work was subsequently extended to incorporate the use of bypasses for non-sharp splits (Wehe and Westerberg, 1987). The Andrecovich and Westerberg (1985) work was also extended as an MINLP model by Floudas and Paules (1988) who modeled the nonlinearities in the heat exchangers. More complex cases like multiple feeds and multiple products and sharp splits have been modeled as an NLP and as an MINLP by Floudas (1987) and

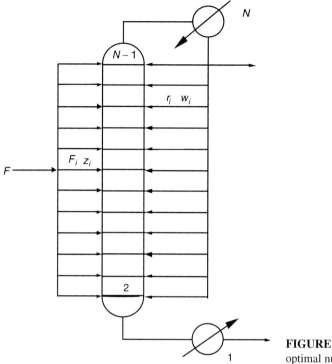

FIGURE 17.18 Superstructure for optimal number of trays.

by Aggarwal and Floudas (1990), respectively. Also, Quesada, and Grossmann (1995) have developed a rigorous global optimization method for the NLP model. The use of rigorous models for column design (feed trays, number of trays) within MINLP techniques has been addressed by Viswanathan and Grossmann (1990, 1993a,b), including the use of multiple feeds.

REFERENCES

Aggarwal, A., & Floudas, C. A. (1990). Synthesis of general distillation sequences. *Comput. Chem. Engng.*, **14**, 631.

Andrecovich, M. J., & Westerberg, A. W. (1985). An MILP formulation for heat-integrated distillation sequence synthesis. *AIChE J.*, **31**, 1461.

Eliceche, A. M., & Sargent, R. W. H. (1986). Synthesis and design of distillation sequences. *I.Chem.E. Symposium Series No. 61*, 1–22.

Floudas, C. A. (1987). Separation synthesis of multicomponent feed streams into multicomponent product streams. *AIChE J.*, **33**, 540.

Floudas, C. A., & Paules, G. E. IV. (1988). A mixed-integer nonlinear programming formulation for the synthesis of heat-integrated distillation sequences. *Comput. Chem. Engng.*, **12**, 531.

Floquet, P., Pibouleau, L., & Domenach, S. (1988). Mathematical programming tools for chemical engineering process design synthesis, *Chem. Eng. Process,* **23**, 1.

Kakhu, A. I., & Flower, J. R. (1988). Synthesising heat-integrated distillation sequences using mixed integer programming. *Chem. Eng. Res. Des.,* **66**, 241.

Quesada, I., & Grossmann, I. E. (1995). Global optimization of bilinear process networks with multicomponent streams. *Comput. Chem. Engng.*, **19**, 1219.

Raman, R., & Grossmann, I. E. (1993). Symbolic integration of logic in mixed integer linear programming techniques for process synthesis. *Computers and Chemical Engineering*, **17**, 909.

Reid, R. C., Prausnitz, J. M., & Poling, B. E. (1987). *The Properties of Gases and Liquids*, 4th ed. New York: McGraw-Hill.

Sargent, R. W. H., & Gaminibandara, K. (1976). Introduction: Approaches to chemical process synthesis. In L.C.W. Dixon (Ed.), *Optimization in Action*. London: Academic Press.

Viswanathan, J., & Grossmann, I. E. (1990). A combined outer approximation and penalty function method for MINLP optimization. *Comput. Chem. Engng.*, **14**, 769.

Viswanathan, J., & Grossmann, I. E. (1991). Optimal feed tray location. In M. Morari & I. E. Grossman (Eds.), *Chemical Engineering Optimization Problems with GAMS,* Vol. 6. CACHE Design Case Studies. Austin: CACHE.

Viswanathan, J., & Grossmann, I. E. (1993a). An alternate MINLP model for finding the number of trays required for a specified separation objective. *Comput. Chem. Engng.*, 17, 949.

Viswanathan, J., & Grossmann, I. E. (1993b). Optimal feed locations and number of trays for distillation columns with multiple feeds. *Ind. Eng. Chem.*, **32**, 2942.

Wehe, R. R., & Westerberg, A. W. (1987). An algorithmic procedure for the synthesis of distillation sequences with bypass. *Comput. Chem. Engng.*, **11**, 619.

EXERCISES

1. Solve the MILP model Eq. (17.16) for the four-component example in section 17.3 to determine the optimal separation sequence. Also, obtain the second and third best solutions. Repeat the calculations for the case when the investment cost data in Table 17.1 are such that the separator (*AB/CD*) has a fixed cost and variable cost coefficient that is three times larger.

2. To obtain the second best solution in the example of section 17.3 we used the integer cut in Eq. (17.14).
   a. Show that instead of using Eq. (17.14) we could have used the inequality

Exercises

$$y_2 + y_8 + y_{10} - y_1 - y_3 - y_4 - y_5 - y_6 - y_7 - y_9 \leq 2$$

to exclude the point $y_2 = y_8 = y_{10} = 1$, and $y_1 = y_3 = y_4 = y_5 = y_6 = y_7 = y_9 = 0$.

   b. What is the potential disadvantage of using the above inequality compared to Eq. (17.14)?

3. Develop the network superstructure for the case of a six-component mixture (*ABCDEF*) that is to be separated into pure components. How many 0–1 and continuous variables, equations, and inequalities would be involved in the MILP formulation Eq. (17.16) for this problem?

4. Repeat problem 1 but solving model MILP Eq. (17.23) for synthesizing a heat integrated sequence. Assume that steam is available at 490 K, cooling water at 320 K and EMAT = 10 K. The temperature differences between reboiler and condenser ΔT_{RC} for each column are as follows:

A/BCD	25 K	A/BC	20 K	A/B	15 K
AB/CD	20 K	AB/C	15 K	B/C	10 K
ABC/D	35 K	B/CD	15 K	C/D	25 K
		BC/D	30 K		

Finally, for the cost function in Eq. (17.18) assume the same value of α as in Table 17.1 and set $\gamma = 0.2$.

5. Extend the MILP model in Eq. (17.23) for the case of multiple utilities.

6. Show that the MILP formulation in Eq. (17.28) for heat integrated distillation sequences reduces to the LP transshipment model for minimum utility cost if the flowrates F_k and the binary variables y_k have a fixed value corresponding to a particular structure for separation.

7. Given the ternary mixture below, determine an optimal heat integrated distillation sequence using the MILP model Eq. (17.23) for continuous temperatures and the MILP Eq. (17.28) for discretized temperatures.

Feed = 250 kmol/hr A: 0.6
                               B: 0.3
                               C: 0.1

Desired products: pure *A, B, C*

Utilities

Cooling water	300–320 K	$20/kWhr
LP Steam	420 K	$55/kWhr
MP Steam	460 K	$95/kWhr
HP Steam	490 K	$120/kWhr

EMAT = 10 K

Temperature differences reboiler-condenser

A/BC: 70 K   AB/C: 60 K
A/B: 43 K   B/C: 38 K

Investment data, heat duties

	Fixed** (10^3 \$/yr)	Variable* (10^3 \$hr/kmol yr)	Heat duty coefficients* (10^6 kJ/kmol)
A/BC	32	0.27	0.048
AB/C	120	1.15	0.095
A/B	30	0.29	0.052
B/C	98	2.32	0.225

*Based on feed flowrate
**Apply the following correction factor to account for the effect of column pressure:

$$[1+ (T_c - 320)/320]$$

where T_c is the temperature of the condenser.

NOTE: Show the column configuration with the associated heat recovery network

8. Using the GAMS Optimization Case Study by CACHE (see Appendix C), solve the optimal feed tray problem in section 17.8 with the file FEEDTRAY. In addition, solve the problem with the feed composition corresponding to 75% mole benzene and 25% mole toluene. You may find it interesting to analyze the X profile (i.e., composition of the liquid leaving) of the feedtray and in neighboring trays in both the cases. Do you think this may have some thermodynamic significance?

9. Suppose there are two feeds to the column in Figure 17.17. For definiteness, assume that the first feed stream has a larger proportion of the most volatile component. Formulate the problem for the following cases:

   a. Exactly two (optimum) locations are to be determined.

   b. At most two (optimum) locations are to be determined (i.e the blending of feed streams is allowed).

Generalize the above. Consider a column with M feeds with different distributions of the components. First, state the problems precisely, making all your assumptions explicit. Then, proceed for the modelling of the general case.

SIMULTANEOUS OPTIMIZATION AND HEAT INTEGRATION 18

18.1 INTRODUCTION

In Chapter 17 the problem of heat integration was considered simultaneously with the synthesis of separation sequences. The basic idea was to design these systems so that they would be better heat integrated. In this way one can often achieve substantial savings in energy, which will then translate to lower operating costs.

When we consider a process flowsheet, however, energy is not the only item for the operating costs. In fact, the dominant cost item is usually raw materials. If we consider a typical process flowsheet involving a recycle, we can anticipate that higher recycles will increase the overall conversion, and thus reduce the expenses for the raw material. However, we would then have higher flows in our process, which will then presumably increase the energy requirements. A natural question that then arises is how to determine the proper trade-off between raw material costs and energy expenses? Or, more generally, how can we establish the optimal trade-off by also including the capital investment?

In this chapter we will show that this question can be answered if the optimization of the process is performed *simultaneously* with the heat integration of the process. Or, in other words, the idea will be to anticipate in the optimization that the process will be heat integrated.

We will first examine, through a simplified model of a recycle process, the nature of the trade-offs when heat integration is anticipated or not at the optimization stage. We will then show how to simultaneously perform the optimization and heat integration in processes that are modeled by linear and nonlinear equations. We will restrict ourselves here to fixed process configurations, since the structural optimization of flowsheets will be considered in Chapter 20. Also, Chapter 19 will consider the simultaneous optimization and heat integration in reactor networks.

18.2 SEQUENTIAL VERSUS SIMULTANEOUS OPTIMIZATION AND HEAT INTEGRATION

When designing a chemical process, we can consider basically two types of strategies for handling the heat integration (Duran and Grossmann, 1986; Lang et al. 1988; Papoulias and Grossmann, 1983a). In the sequential strategy we optimize the process at a first stage by assuming that all the heating and cooling loads will be supplied by utilities. In the second stage, having established all the stream conditions (flows, pressures, temperatures), we then perform the heat integration of the streams with any of the techniques presented in Chapter 16.

In the simultaneous strategy, on the other hand, we will perform the heat integration of the streams while we optimize the process. In order to avoid the problem of synthesizing a heat exchanger network for each process condition generated throughout the optimization (Yee et al., 1990), we will consider only the utility cost for maximum heat integration.

In order to analyze the effect of using the sequential or simultaneous strategy let us consider the processing system shown in Figure 18.1. This system consists of the following steps: (FP) feed preparation (e.g., compression); ($R1$) reaction (e.g., preheat, reaction, cooling); ($S1$) recovery of liquid product and by-products (e.g., flash separation); ($S2$) split for purge stream; ($R2$) recycle (e.g., recompression); (PR) recovery of final product (e.g., distillation). This processing scheme is representative of many chemical and petrochemical processes in which the feedstock contains some inerts, and the conversion per pass in the reactor is not very high.

In order to develop a simplified model for this process the following assumptions will be made:

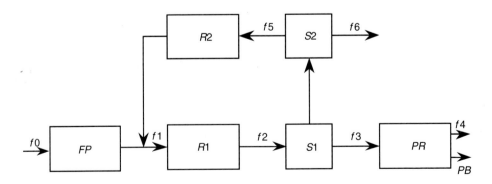

FIGURE 18.1 Processing system.

Sec. 18.2 Sequential Versus Simultaneous Optimization

- Single reaction A → B with fixed conversion per pass r.
- Feedstock contains inert C with composition y^c.
- The production rate of B, P_B is fixed.
- Fixed pressure and temperature levels throughout the flowsheet.
- Feed preparation (*FP*) and recycle (*R2*) involve only electricity demands.
- Reaction step (*R1*) and product recovery (*PR*) involve heating and cooling demands.
- Perfect split between *AC/B* in splitter *S1*.
- Fixed recovery fraction of B (β) in *PR*.
- Cost models are assumed to be linear functions of the flows f_i in Figure 18.1. The cost of the heat recovery network is neglected.

Based on the above, the cost models for the different items are as follows:

Net cost feedstock: $C_{NF} = C_F - I_P$, where
Feedstock: $C_F = c_F(f_0^A + f_0^C)$
Purge income: $I_p = c_p(f_6^A + f_6^C)$
Capital and operating expenses $= C_{FP} + C_{R1} + C_{R2} + C_{PR}$

where

Feed preparation: $C_{FP} = c_{FP}(f_0^A + f_0^C)$
Reaction step: $C_{R1} = c_{R1}(f_1^A + f_1^C)$
Recycle step: $C_{R2} = c_{R2}(f_5^A + f_5^C)$
Product recovery: $C_{PR} = c_{PR} P_B/\beta$

The unit costs c_F, c_P, c_{FP}, c_{R1}, c_{R2}, and c_{PR} are for the case of no heat integration. For the case of heat integration $c'_{R1} < c_{R1}$, $c'_{PR} < c_{PR}$ to reflect the savings in utility costs in the reaction and product recovery sections. The total cost of the flowsheet with no heat integration is then given by,

$$C = C_{NF} + C_{FP} + C_{R1} + C_{R2} + C_{PR} \qquad (18.1)$$

Given the conversion per pass in the reactor, r, the inert composition in the feed y^c, each of the terms in this cost function can be expressed as a function of x, the overall conversion of A in the feedstock to B in the amount of product P_B. By performing the appropriate mass balances in Figure 18.1 (see exercise 2) it can be shown that,

$$C_{NF}(x) = \frac{P_B}{\beta(1-y^c)}\left[\frac{c_F - c_P}{x} + c_p\left(1 - y^C\right)\right] \qquad (18.2)$$

$$C_{FP}(x) = \frac{P_B c_{FP}}{\beta(1-y^C)x} \qquad (18.3)$$

$$C_{R1}(x) = \frac{P_B c_{R1}}{\beta r}\left[1+\left(\frac{y^C}{1-y^C}\right)\left(\frac{1-r}{1-x}\right)\right] \qquad (18.4)$$

$$C_{R2}(x) = \frac{P_B c_{R2}}{\beta}\left(\frac{1}{r}-\frac{1}{x}\right)\left[1+\left(\frac{y^C}{1-y^C}\right)\left(\frac{1}{1-x}\right)\right] \qquad (18.5)$$

$$C_{PR} = \frac{P_B c_{PR}}{\beta} \qquad (18.6)$$

Based on the above equations, we can identify two major terms:

Net cost of feedstock: $C_{NF}(x)$
Operating and capital costs: $C_{OC}(x) = C_{FP}(x) + C_{R1}(x) + C_{R2}(x) + C_{PR}$

In order to determine the overall conversion that minimizes the total cost, the problem reduces to the one-dimensional optimization problem:

$$\min C = C_{NF}(x) + C_{OC}(x) \qquad (18.7)$$
$$\text{s.t. } r \le x \le 1$$

In order to illustrate how the overall conversion is affected by using the sequential and simultaneous strategies, consider the data given in Table 18.1. In this case, since it is assumed that heat integration can only be performed in the reaction step, there is only a difference in the cost coefficient c_{R1} between the sequential and simultaneous strategies. The respective values of 5 and 1 imply that 80% of the energy can be recovered in the reaction section.

The plot of the two cost terms in Eq. (18.7) as a function of the overall conversion x of the raw material, and for the data in Table 18.1, is shown in Figure 18.2. As expected,

TABLE 18.1 Data for Optimization and Heat Integration with Simplified Model

Production rate of B:	$P_B = 100$ tons/day
Recovery of B in PR:	$\beta = 0.95$
Conversion per pass:	$r = 0.1$
Cost coefficients ($/ton.day)	
• Feed	$c_F = 30$
• Purge	$c_P = 12$
• Feed preparation	$c_{FP} = 10$
• Reaction step	$c_{R1} = 5$ (no heat integration)
	$c_{R1} = 1$ (with heat integration)
• Recycle step	$c_{R2} = 1$
• Product recovery	$c_{PR} = 1$

Sec. 18.2 Sequential Versus Simultaneous Optimization and Heat Integration

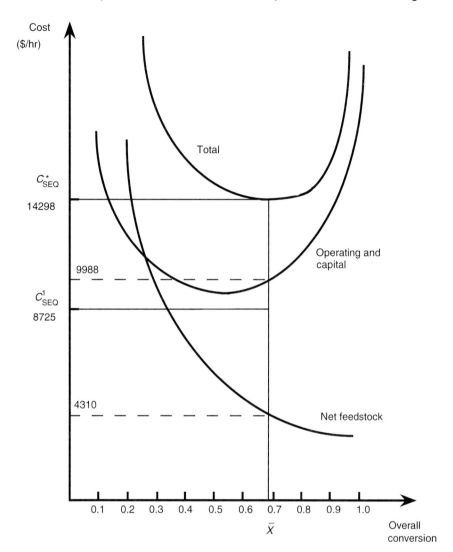

FIGURE 18.2 Plot of objective for sequential optimization.

the curve for the net cost of the feedstock is convex and decreases monotonically with the overall conversion. On the other hand, the curve for capital and operating expenses is convex, goes through a minimum, and tends to infinity for 100% overall conversion. Qualitatively, the reason is that at low overall conversion, the cost of feed preparation is high due to the large flow in the feed, while at high overall conversion the cost of the reaction and recycle is very high due to the large flow in the recycle loop.

From Figure 18.2 it can be seen that the minimum cost $C^*_{SEQ} = \$14{,}298/\text{day}$ is attained at the overall conversion $\bar{x} = 0.69$. The net cost of the feedstock is \$4,310/day and the operating and capital expenses with no heat integration are $9,988/day. If heat integration is now performed at the conversion of $\bar{x} = 0.69$, the operating and capital expenses can be reduced by $4,415/day yielding a total cost $C^1_{SEQ} = \$8,725/day$ shown in Figure 18.2.

Let us consider now the case when heat integration is considered simultaneously for determining the optimal conversion. Since in this case $c_{R1} = 1$, we obtain a lower curve for capital and operating expenses as seen in Figure 18.3. This, then, has the effect of shifting

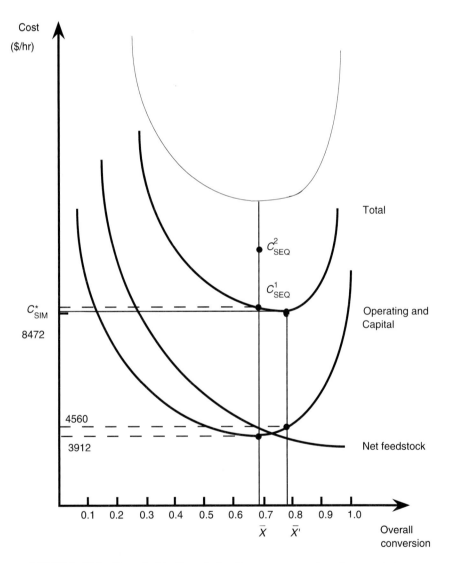

FIGURE 18.3 Plot of objective of simultaneous approach and comparison with sequential.

the optimal overall conversion towards the higher value $\bar{x}' = 0.79$, which is 10% higher than the one of the sequential strategy. Also, the minimum cost is $C^*_{SIM} = \$8,472/\text{day}$, which is lower than the cost $C^1_{SEQ} = \$8,725/\text{day}$ in the sequential strategy. The net cost of the feedstock at $\bar{x}' = 0.79$ is \$3,912/day and the operating and capital expenses are \$4,560/day.

Thus, from the above example we can conclude that the simultaneous strategy when compared to the sequential approach exhibits:

- Higher overall conversion of the raw material.
- Lower total cost.

Another point of interest in this example is that the operating and capital cost in the simultaneous strategy are greater than the one in the sequential approach (\$4,560/day vs. \$4,415/day). This, however, is compensated by the lower net cost of the feedstock (\$3,912/day vs. \$4,310/day) in the simultaneous optimization.

An important assumption in the above example is that operating conditions such as pressures and temperatures have been assumed to be constant. However, very often some of these variables will be degrees of freedom for the optimization. This implies that since fixed pressures and temperatures are considered for the heat integration of the process streams, the final cost C^2_{SEQ} in the sequential approach will typically lie above C^1_{SEQ} (see Figure 18.3), and thus will have an even greater difference with C^*_{SIM}.

Also, in this case one will often achieve savings in both the net cost of the feedstock and in the operating and capital expenses as will be shown in section 18.4.3.

In the next sections we will examine how to consider the simultaneous optimization and heat integration in processes that are modeled with linear and nonlinear equations.

18.3 LINEAR MODELS

In the previous section we considered a very simplified model of a process to show the advantages of the simultaneous optimization and heat integration. In this section we will consider the case when the units in a process flowsheet are described by linear equations given that fixed pressure and temperature levels are assumed. The only nonlinearities that will be considered are the split fractions for the recycle streams.

Let x denote the variables corresponding to the total and individual component flowrates in each stream and the sizes or capacities of the units (e.g., reactor volume, power of compressors). From among the variables x we will denote the heat capacity flowrates of the hot and cold streams by F_i, $i = 1...n_H$, f_j, $j = 1...n_C$, respectively. Each of these streams is assumed to undergo constant temperature changes ΔT_i, and Δt_j respectively.

To simplify the presentation we will assume one single hot utility and a single cold utility. The case of multiple utilities can be easily extended (see exercise 4). The load of the hot utility will be denoted by Q_S, and the load of the cold utility by Q_W.

When we consider the optimization of the process with no heat integration, the problem can be formulated as follows:

$$\min C = c^T x + c_s Q_s + c_w Q_w \qquad (18.8a)$$

$$\text{s.t.} \quad A x = a \qquad (18.8b)$$

$$B x \leq a \qquad (18.8c)$$

$$s(x) = 0 \qquad (18.8d)$$

$$Q_S = \sum_{j=1}^{n_c} f_j \Delta T_j \qquad (18.8e)$$

$$Q_W = \sum_{i=1}^{n_H} F_i \Delta t_i \qquad (18.8f)$$

$$x, Q_S, Q_W \geq 0, \quad F_i \geq 0 \; i = 1...n_H, \quad F_j \geq 0 \; j = 1...n_c$$

The objective function in (18.8a) involves the linear cost $c^T x$ in terms of sizes and flows, and the cost of the heating and cooling utility. Equations (18.8b) are linear mass balances and design equations that are constrained by the linear inequalities in Eq. (18.8c). Equations (18.8d) are nonlinear equations for the splitters in the recycle, and Eqs. (18.8e) and (18.8f) are the heat balances to determine the loads of the utility streams. In this way problem (18.8a) assumes that all the heat loads of the process streams are satisfied by utilities.

Problem (18.8a) can actually be solved as an NLP or as an MILP depending on how we treat the equations $s(x) = 0$ for the splitters in the recycle. As an example, consider the splitter in Figure 18.4. If α is a variable denoting the split fraction of the recycle stream, the mass balance equations are as follows:

$$\left\{ \begin{array}{l} x_R^c = \alpha x_V^c \\ x_P^c = x_V^c - x_R^c \end{array} \right\} \; c \in \text{COMP} \qquad (18.9)$$

$$0 \leq \alpha \leq 1$$

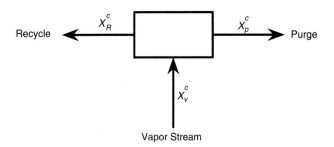

FIGURE 18.4 Splitter for recycle in a process.

Sec. 18.3   Linear Models

where x_R^c, x_P^c, x_V^c are the flowrates for component c in the recycle, purge, and vapor stream, respectively. The latter stream will be commonly the vapor overhead of a flash unit or the vapor exit stream in an absorber.

Since the first equation in (18.9) involves the split fraction α times the flowrate x_V^c, it is nonlinear. Therefore, if we treat the equations as in (18.9) the problem in (18.8a) corresponds to an NLP. However, we can also formulate the problem as an MILP as follows. Consider L discrete values for the split α: $\alpha_1, \alpha_2, \ldots \alpha_L$. If we assign to each of these splits a binary variable y_ℓ, we can approximate the equations in (18.9) by:

$$\begin{aligned}
x_R^c &= \sum_{\ell=1}^{L} x_{R\ell}^c \\
x_V^c &= \sum_{\ell=1}^{L} x_{V\ell}^c \qquad c \in \text{COMP} \\
x_P^c &= x_V^c - x_R^c \\
x_{R\ell}^c &= \alpha_\ell x_{V\ell}^c \qquad c \in \text{COMP} \\
x_{V\ell}^c - U y_\ell &\leq 0 \quad \ell = 1,\ldots L \\
\sum_{\ell=1}^{L} y_\ell &= 1
\end{aligned} \qquad (18.10)$$

where U is a valid upper bound and all the x variables are nonnegative. The reader can easily verify that the selection of a given split fraction a_1 is performed by activating only one binary variable y_1 to one, which then yields the corresponding mass balances for that split.

In order to perform the heat integration simultaneously with the optimization in problem (18.8a), this can be done by replacing equations (18.8e) and (18.8f) for the heat balances of the utilities by constraints that ensure the maximum heat integration of the process streams for any given values of the flowrates of the streams. Since in this case we are assuming fixed temperature levels in the process, this can simply be accomplished by incorporating the heat integration constraints of the transshipment model in Chapter 16. That is, let K be the temperature intervals that arise from the different temperatures of the process streams for a given value of ΔTmin (HRAT). Also, let us assume that no constraints are imposed on the matches. If we recall from Chapter 16, the constraints for minimum utility cost or consumption for the transshipment model (Papoulias and Grossmann, 1983a,b) have the form

$$R_k - R_{k-1} - Q_s + Q_W = \sum_{i \in H_k} Q_{ik}^H - \sum_{j \in C_k} Q_{jk}^C \qquad k = 1\ldots K \qquad (18.11)$$

where R_k, R_{k-1}, are heat residuals, and Q_{ik}^H, Q_{jk}^C are the heat contents of hot and cold streams in the interval k. These heat contents, however, are not constant when the flowrates are unknown. They are given by the linear equations,

$$Q_{ik}^H = F_i \Delta T_i^k \quad i = 1...n_H$$
$$Q_{jk}^C = f_j \Delta t_j^k \quad j = 1...n_C \tag{18.12}$$

where ΔT_i^k, Δt_j^k are the fixed changes of temperature of hot stream i and cold stream j in interval k.

If we substitute Eq. (18.12) in Eq. (18.11) and incorporate these equations in place of constraints, Eqs. (18.8e) and (18.8f), the problem of simultaneous optimization and heat integration can be posed as follows:

$$\min C = c^T x + c_S Q_S + c_W Q_W$$
$$\text{s.t.} \quad Ax = a$$
$$Bx \leq a$$
$$s(x) = 0 \tag{18.13}$$
$$R_k - R_{k-1} - Q_S + Q_W - \sum_{i \in H_k} F_i \Delta T_i^k + \sum_{j \in C_k} f_j \Delta t_j^k = 0 \quad k-1,...K$$
$$x, Q_S, Q_W \geq 0, \quad R_k \geq 0 \quad k = 1,...K-1, \quad R_0, R_K = 0$$
$$F_i \geq 0 \quad i = 1...n_H, \quad f_j \geq 0 \quad j = 1...n_C$$

In this way, this formulation will consider for the optimization the fact that the required utility loads Q_S and Q_W correspond to the maximum heat integration. Using a similar line of reasoning, we can easily extend problem (18.13) to the case of multiple utilities and restricted matches (see exercise 4). The reader should try to apply the formulation (18.13) in exercise 5.

18.4 NONLINEAR MODELS

In general, it will be desirable to model a process with nonlinear performance equations where pressures and temperatures are also variables. The main difficulty that arises is that we can no longer apply the equations of the transshipment model directly as we did in the previous section, since the temperature intervals will now be variable.

A simple-minded approach to circumvent this problem would be to use a "black-box" approach. Here the utility loads are computed at each iteration of the nonlinear optimization for the corresponding flows and temperatures with a subroutine for minimum utility cost. This strategy might be suitable for a process simulator (Lang et al., 1988). However, given that discrete decisions are made in the selection of intervals, nondifferentiabilities will be introduced that can commonly cause numerical difficulties with NLP solvers. Therefore, it is desirable to develop equivalent expressions to the ones of the transshipment equations but which can handle both variable flowrates and temperatures. In order to devise such a model (Duran and Grossmann, 1986), let us consider first the

Sec. 18.4 Nonlinear Models

nonlinear optimization problem with no heat integration. Here we will denote by x all the variables in the process among which are included the heat capacity flowrates and inlet and outlet temperatures, F_i, T_i^{in}, T_i^{out}, $i = 1...n_H$, f_j, t_j^{in}, t_j^{out}, $j = 1...n_C$, of hot and cold streams respectively. The loads of the hot and cold utilities are denoted by Q_S, Q_W. The optimization problem corresponds then to:

$$\min C = f(x) + c_S Q_S + c_W Q_W \qquad (18.14a)$$

$$\text{s.t. } h(x) = 0 \qquad (18.14b)$$

$$g(x) \leq 0 \qquad (18.14c)$$

$$Q_S = \sum_{j=1}^{n_C} f_j \left(t_j^{out} - t_j^{in} \right) \qquad (18.14d)$$

$$Q_W = \sum_{i=1}^{n_H} F_i \left(T_i^{in} - t_i^{out} \right) \qquad (18.14e)$$

Q_S, $Q_W \geq 0$, F_i, T_i^{in}, $T_i^{out} \geq 0$ $i = 1...n_H$, f_j, t_j^{in}, $t_j^{out} \geq 0$ $j = 1...n_C$

$$x \in R^n$$

In this formulation, the objective term $f(x)$, the equations $h(x) = 0$, and the constraints $g(x) \leq 0$ are in general nonlinear. Also note that in this model the flowrates F_i, f_j and the temperatures T_i^{in}, T_i^{out}, T_j^{in}, T_j^{out} are *variables* for the optimization. In order to replace Eqs. (18.14d) and (18.14e) by heat integration constraints, it is essential to remove the definition of temperature intervals since they are not fixed for problem (18.14a). Hence, we will need a new representation for the heat integration problem.

18.4.1 Pinch Location Method

Let us assume in this section that the flowrates and inlet and outlet temperatures of the streams are fixed. We will show how to perform the minimum utility calculation with a pinch location method that does not require the definition of temperature intervals. We will then incorporate the appropriate equations in problem (18.14a) in section 18.4.2.

To illustrate the idea behind the pinch location method (Duran and Grossmann, 1986), consider the problem data in Table 18.2. Using the problem table or the transshipment model we can determine that the minimum utility consumption is $Q_S = 35$ KW, $Q_W = 145$ KW, and that the pinch occurs at 450–430 K. However, in this calculation we required the definition of temperature intervals.

Let us consider the following procedure. In Figure 18.5, we have plotted the T-Q curves at a value ΔT_{min} (HRAT) greater than 20 K. Suppose we now were to pinch each of the inlet of the streams as shown in Figure 18.6 and determine the corresponding heating and cooling requirements. Clearly the pinch at 450–430 K which is defined by hot

TABLE 18.2 Stream Data for Example Problem

Hot 1:	$F_1 = 1$ kW/K,	$T_1^{in} = 450$, $T_1^{out} = 350$ K
Hot 2:	$F_2 = 4$ kW/K,	$T_2^{in} = 400$, $T_2^{out} = 350$ K
Cold 1:	$f_1 = 2$ kW/K,	$t_1^{in} = 300$, $t_1^{out} = 360$ K
Cold 2:	$f_2 = 0.5$ kW/K,	$t_2^{in} = 360$, $t_2^{out} = 500$ K
	$\Delta T_{min} = 20$ K	

stream H1 is the correct one (Figure 18.6a). Note that all the others (Figures 18.6b, 18.6c, 18.6d) exhibit temperatures crossings, and hence *lower* utility consumptions. Therefore, what this figure would suggest is that the criteria for selecting the correct pinch to define the minimum heating and cooling that is feasible is to select the one that exhibits largest heating and cooling among all the pinch candidates.

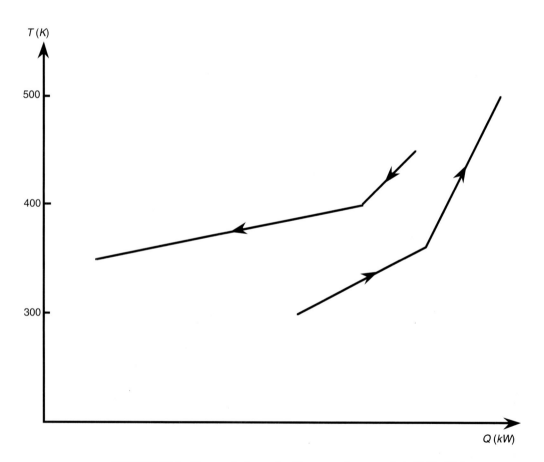

FIGURE 18.5 Composite hot and cold streams for example in Table 18.2.

Sec. 18.4 Nonlinear Models

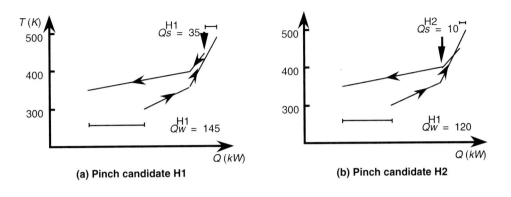

(a) Pinch candidate H1

(b) Pinch candidate H2

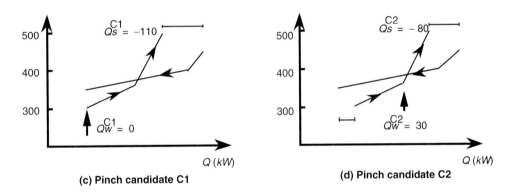

(c) Pinch candidate C1

(d) Pinch candidate C2

FIGURE 18.6 Utility requirements for different pinch candidates.

Mathematically, this condition can be expressed as follows:

$$Q_S = \max_{p \in P} \{Q_S^p\}$$
$$Q_W = \max_{p \in P} \{Q_W^p\} \tag{18.15}$$

where P is the index set of all the hot and cold streams, $i = 1\ldots n_H$, $j = 1\ldots n_C$, and Q_S^p, Q_W^p, are the heating and cooling loads that result from each pinch candidate.

We can simplify Eq. (18.15) if we consider the overall heat balance

$$Q_W = \Omega + Q_S \tag{18.16}$$

where

$$\Omega = \sum_{i=1}^{n_H} F_i\left(T_i^{in} - T_i^{out}\right) - \sum_{j=1}^{n_C} f_j\left(t_j^{out} - t_j^{in}\right) \tag{18.17}$$

is the total heat surplus.

We can then replace the second equation in (18.15) so that our basic criterion for the pinch location reduces to:

$$Q_S = \max_{p \in P} \{Q_S^p\} \tag{18.18}$$
$$Q_W = \Omega + Q_S$$

The only remaining point is then how to develop an explicit expression for the terms Q_S^p in Eq. (18.18) in terms of flows and temperatures. From Figure 18.6 it is clear that these terms are obtained from the heat balance

$$Q_S^p = QA_C^p - QA_H^p \tag{18.19}$$

where QA_C^p and QA_H^p are the total heat content above the candidate pinch p of the cold and of the hot streams, respectively. Or, in other words, $QA_C^p - QA_H^p$ represents the heat deficit that exists above the candidate pinch $p \in P$.

To develop explicit expressions of QA_C^p and QA_H^p, let us consider as an example the hot stream i in Figure 18.7. We can clearly see that the heat content of this stream above the pinch depends on whether the stream is entirely above the pinch, whether it crosses the pinch, or whether it is below the pinch. In each case, we get different algebraic expressions for the heat content above the pinch. An equation that however, can capture the three cases is given below:

$$\text{Heat content above pinch } p \text{ for hot stream } i = F_i[\max\{0, T_i^{in} - T_i^p\} - \max\{0, T_i^{out} - T_i^p\}] \tag{18.20}$$

We can verify the three cases as follows:

1. Stream lies above pinch, $T_i^{in} > T_i^{out} > T^p$, which implies that Eq. (18.20) reduces to

$$F_i[\{T_i^{in} - T^p{}_i\} - \{T_i^{out} - T^p{}_i\}] = F_i[T_i^{in} - T_i^{out}]$$

2. Stream crosses the pinch, $T_i^{in} > T^p > T_i^{out}$, which implies that Eq. (18.20) reduces to

$$F_i[\{T_i^{in} - T_i^p\} - \{0\}] = F_i[T_i^{in} - T_i^p]$$

3. Stream lies below the pinch, $T^p > T_i^{in} > T_i^{out}$, which implies that Eq. (18.20) reduces to

$$F_i[\{0\} - \{0\}] = 0$$

Or in other words, Eq. (18.20) provides an explicit equation for the heat content above the pinch for all cases. In this way, QA_H^p will be given by

Sec. 18.4 Nonlinear Models

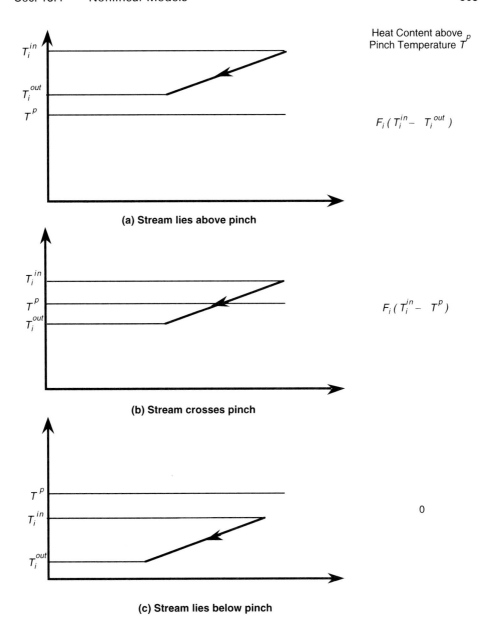

FIGURE 18.7 Heat content above pinch of hot stream i for different cases.

$$QA_H^P = \sum_{i=1}^{n_H} F_i \left[\max\{0, T_i^{in} - T^p\} - \max\{0, T_i^{out} - T^p\} \right] \quad (18.21)$$

and using a similar reasoning, QA_C^P will be given by

$$QA_C^P = \sum_{j=1}^{n_C} f_j \left[\max\{0, t_j^{out} - (T^p - \Delta T_{min})\} - \max\{0, t_j^{in} - (T^p - \Delta T_{min})\} \right] \quad (18.22)$$

where the pinch temperatures, T^p are defined as follows:

$$T^p = \begin{cases} T_i^{in} & \text{if candidate } p \text{ is hot stream } i \\ t_j^{in} + \Delta T_{min} & \text{if candidate } p \text{ is cold stream } j \end{cases} \quad (18.23)$$

Table 18.3 presents the calculations involved in Eq. (18.18) using Eqs. (18.19), (18.21), (18.22), and (18.23) to perform the minimum utility calculation for the example in Table 18.2. Note in Figure 18.6 that the utility requirements for the different pinch candidates are the same as the ones displayed in Table 18.3.

18.4.2 Nonlinear Optimization with Heat Integration

Based on the equations developed in the previous section where we obtained explicit expressions of the heat integration in terms of flowrates and temperatures, we can easily modify the formulation in Eq. (18.14) so as to perform simultaneous optimization and heat integration. By expressing the first equation in (18.18) as a set of inequalities, and substituting Eqs. (18.21) and (18.22) in Eq. (18.19), and Eq. (18.19) and (18.17) in Eq. (18.18), the formulation is as follows:

$$\min C = f(x) + c_S Q_S + c_W Q_W$$
$$\text{s.t.} \quad h(x) = 0 \quad (18.24)$$
$$g(x) \leq 0$$

$$Q_S \geq \sum_{j=1}^{n_C} f_j \left[\max\{0, t_j^{out} - (T^p - \Delta T_{min})\} - \max\{0, t_j^{in} - (T^p - \Delta T_{min})\} \right]$$

$$- \sum_{i=1}^{n_H} F_i \left[\max\{0, T_i^{in} - T^p\} - \max\{0, T_i^{out} - T^p\} \right] \quad p \in P$$

$$Q_W = Q_S + \sum_{i=1}^{n_H} F_i (T_i^{in} - T_i^{out}) - \sum_{j=1}^{n_C} f_j (t_j^{out} - t_j^{in})$$

$$Q_S, Q_W \geq 0, \ F_i, T_i^{in}, T_i^{out} \geq 0 \ i = 1...n_H, \ f_j, t_j^{in}, t_j^{out} \geq 0 \ j = 1...n_C \ x \in R^n$$

where $T^p, p \in P$, is given by Eq. (18.23).

Sec. 18.4   Nonlinear Models   611

TABLE 18.3   Calculation with Pinch Location Method

Pinch p	T^p(K)	QA^p_H	QA^p_C	Q^p_S	Q^p_W
H1	450	0	35	35	145
H2	400	50	60	10	120
C1	320	300	190	−110	0
C2	380	150	70	−80	30

$\Omega = 1(450 - 350) + 4(400 - 350) - 2(360 - 300) - 0.5(500 - 360) = 110$ kW
$\Delta T_{min} = 20$ K
$Q_S = \max\{35, 10, -110, -80\} = 35$ kW
$Q_W = 110 + 35 = 145$ kW

Note that the above formulation can treat the flows and the temperatures as variables for the optimization and the heat integration. The difficulty with Eq. (18.24) is the presence of max operators that are nondifferentiable. However, as shown in Appendix B, a smooth approximation procedure can be used that avoids difficulties with the use of NLP solvers (Balakrishna and Biegler, 1992; Duran and Grossmann, 1986). This formulation can also be extended to the case of multiple utilities (see exercise 8). For the case of streams with constant temperatures, the above model requires that a finite temperature change be specified for all the streams. In this case, however, an approach that models directly the matches in Section 17.6 of Chapter 17 might be more suitable (see exercise 11).

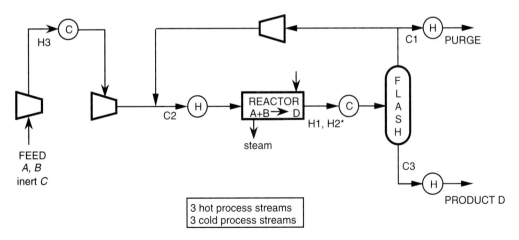

*H1 superheat to dewpoint
H2 dewpoint to supercool

FIGURE 18.8   Flowsheet example for simultaneous optimization and heat integration.

18.4.3 Numerical Example

It is out of scope for this book to present a detailed example with the formulation in Eq. (18.24). Therefore, we will simply quote the results of Duran and Grossmann (1986) for the nonlinear optimization of the flowsheet in Figure 18.8. This flowsheet involves three hot and three cold streams. Streams H1 and H2 are physically the same one, but they have been treated separately, since the former has to be cooled from superheated vapor to the dewpoint, and the latter from the dewpoint to the two-phase region.

As can be seen in Table 18.4, a very substantial difference in the profit is obtained between the simultaneous and the sequential strategy ($19 million/yr vs. $10 million/yr). This big difference was not only due to the higher overall conversion of the simultaneous strategy (82% vs. 75%), but also to the much lower heating requirements ($2.8 million/yr

TABLE 18.4 Results Flowsheet Optimization and Heat Integration

	Simultaneous	Sequential
Economic		
Expenses ($\times \$10^6$/yr):		
Feedstock	22.6717	26.4166
Capital investment	3.7596	3.9108
Electricity compression	2.3774	2.4871
Heating utility	2.8244	14.4586
Cooling utility	0.7900	0.7247
Earnings ($\times \$10^6$/yr):		
Product	41.5300	41.5300
Purge	4.5169	6.8242
Generated steam	5.6407	9.7441
Annual Profit	19.2645 90% HIGHER!	10.1005
Technical		
Overall conversion A [%]	81.68	75.13
Pressure reactor [atm]	12.10	13.87
Conversion per pass [%]	30.43	37.53
Temp. inlet reactor [°K]	450.00	450.00
Temp. outlet reactor [°K]	502.65	450.00
Steam generated [kW]	10119.12	17479.60
Pressure in flash [atm]	9.10	10.87
Temperature flash [°K]	320.00	339.88
Purge rate [%]	9.66	19.66
Power compressors [kW]	1353.60	11877.44
Heating utility [kW]	1684.27	8622.04
Cooling utility [kW]	10632.04	9752.77
Total heat exchanged [kW]	31962.20	28720.61

Note: Simultaneous has higher overall conversion (i.e., less feedstock) and lower heating requirements.

TABLE 18.5 Resulting Flowrates and Temperatures of Process Streams

		SIMULTANEOUS			
Stream	F kmol/sec	Cp_e [KJ/(kmol°K)]	T^{in} [K]	T^{out} [K]	Q [kW]
H1	3.1826	35.1442	502.65	347.41	17363.58
H2	3.1826	115.4992	347.41	320.00	10075.58
H3	1.0025	29.6588	405.48	310.00	2838.90
C1	0.2724	33.9081	320.00	670.00	3232.80
C2	3.5510	31.8211	368.72	450.00	9184.37
C3	0.3617	297.7657	320.00	402.76	8913.40
		SEQUENTIAL			
Stream	F [kmol/sec]	Cp_e [KJ/kmolK]	T^{in} [K]	T^{out} [K]	Q [kW]
H1	2.4545	35.1438	450.00	363.08	7497.76
H2	2.4545	158.6957	363.08	39.88	9036.83
H3	1.1681	29.6596	412.87	310.00	3563.97
C1	0.4115	33.9116	339.88	670.00	4606.69
C2	2.8494	31.8188	387.33	450.00	5681.95
C3	0.3617	340.8035	339.88	410.30	8680.58

vs. $14 million/yr). This was accomplished because the flows and temperatures selected by the simultaneous strategy (see Table 18.5) lead to a much better integration than the one of the sequential strategy. This is clearly displayed in the T-Q curves of Figure 18.9. Note that the simultaneous strategy led to two pinch points due to streams H1 and C2, while the sequential had only one due to stream H2.

Similar results for simultaneous optimization and heat integration have been reported for an ammonia and a methanol process by Lang et al. (1988).

18.5 NOTES AND FURTHER READING

As has been shown in this chapter, in the case of process flowsheets the main advantage of performing simultaneous optimization and heat integration is to improve the overall conversion of raw material with which the economics can be significantly improved. However, we have restricted ourselves in this chapter to the simplest models: transshipment and pinch location, which rely on the assumption of a fixed ΔT_{min} or HRAT. This implies that these models do not take into account the areas of the heat recovery network, thereby underestimating the real cost. Also, the network is derived in a second phase that may yield suboptimal designs. Kravanja and Grossmann (1990) have developed an iterative strategy that ex-

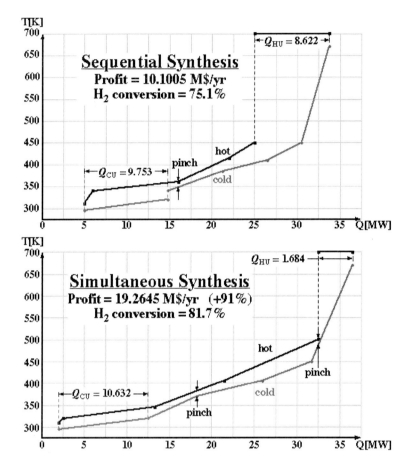

FIGURE 18.9 T-Q curves obtained with the sequential and simultaneous strategies.

tends the model of Duran and Grossmann (1986) to take into account the area cost. Also, Yee et al. (1990) have proposed to integrate the staged superstructure given in Chapter 16 in order to explicitly derive the network structures as part of the optimization.

REFERENCES

Balakrishna, S., & Biegler, L. T. (1992). Targeting strategies for the synthesis and energy integration of nonisothermal reactor networks. *IE&C Research*, **31,** 2152.

Duran, M. A., & Grossmann, I. E. (1986). Simultaneous optimization and heat integration of chemical processes. *AIChE J.*, **32,** 123.

Kravanja, Z., & Grossmann, I. E. (1990). PROSYN—An MINLP process synthesizer. *Computers and Chemical Engineering*, **14**, 1363.

Lang, Y. D., Biegler, L. T., & Grossmann, I. E. (1988). Simultaneous optimization and heat integration with process simulators. *Computers and Chemical Engineering*, **12**, 311.

Papoulias, S. A., & Grossmann, I. E. (1983a). A structural optimization approach in process synthesis. Part II: Heat recovery networks. *Comput. Chem. Engng.*, **7**, 707.

Papoulias, S. A., & Grossmann, I. E. (1983c). A structural optimization approach in process synthesis. Part III: Total processing systems. *Comput. Chem. Engng.*, **7**, 723.

Yee, T. F., Grossmann, I. E., & Kravanja, Z. (1990). Simultaneous optimization models for heat integration. III. Optimization of process flowsheets and heat exchanger networks. *Computers and Chemical Engineering*, **14**, 1185.

EXERCISES

1. Using the simplified model in section 18.2, determine the optimal overall conversion for the sequential and simultaneous optimization with the following data:

    $P_B = 500$ tons/day

    $\beta = 0.98$, $\gamma = 0.1$

    Cost coefficients ($/ton/day):

    $c_F = 40$, $c_P = 25$, $c_{FP} = 10$, $c_{R2} = 1$

    $c_{R1} = 4$, $c_{PR} = 2$ (no heat integration)

    $c_{R1} = 1$, $c_{PR} = 1$ (heat integration)

2. Derive Eqs. (18.2) to (18.6) in section 18.2.

3. Consider the case of a process where the cost of the raw material is much smaller than the capital and operating expenses. Using the simplified model in section 18.2, determine whether higher overall conversions are always achievable with the simultaneous strategy.

4. Extend the formulation in Eq. (18.13) for the two following cases:
    a. Multiple utilities, unrestricted matches.
    b. Multiple utilities, restricted matches.

5. Given the flowsheet in Figure 18.10, optimize it using formulations (18.8) and (18.13) to compare the sequential and simultaneous strategies:

    Data:
    Conversion per pass in reactor: 10% of A
    Recoveries in overhead of flash:
    95% A, 100% C, 5% B.

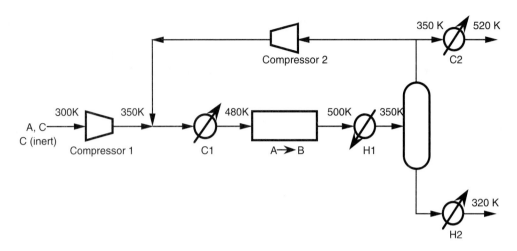

FIGURE 8.10

Purity specification product: min of 90% mole of B
Production rate: 150 tons/day

Heat capacities (cal/g °C)

$$c_{pH1} = 0.5 \quad c_{pH2} = 1.8 \quad c_{pc1} = 0.5 \quad c_{pc2} = 0.9$$

Molecular weights (g/mol)

$$M_A = M_B = 80, \quad M_C = 14$$

Cost of compressors:
  Compressor 1: $8.23/kg/day
  Compressor 2: $1.75/kg/day

Cost of reactor: $1.35 /kg/day
Cost of steam (550K): $95/kWhr
Cost of cooling water (300-320 K): $18/kWhr
$\Delta T_{min} = 10K$

6. Given the following stream data, determine the minimum utility consumption with the pinch location method of section 4.1.

	F_{cp}(kW/K)	T_{in}(K)	T_{out} (K)
H1	1.5	480	340
H2	2	420	330
C1	1	320	410
C2	2	350	460

Exercises

7. Consider a single cold stream j and use a figure similar to Figure 18.7 to verify Eq. (18.22).

8. Extend the formulation in (18.24) to the case of multiple utilities. Consider that intermediate utilities can give rise to pinch points, and that these streams are available at constant temperature.

9. Repeat problem 6 by specifying the inlet and outlet temperatures within ± 10K of the values given above, and by treating the heat capacity flows as variables through two multiplicative factors, $R1$ and $R2$, for both hot and cold streams so as to allow ± 20% variations; i.e.

$$F_{cpH1} = 1.5R1, \quad F_{cpH2} = 2R1$$
$$F_{cpC1} = 1R2, \quad F_{cpC2} = 2R2$$

Also, consider the cost function to be:

$$\text{Cost} = -2500 F_{cpH1} + 3200 F_{cpC2} + 80 Q_H + 20 Q_C$$

Formulate the corresponding NLP optimization model, and solve it with a code such as GAMS/MINOS.

10. Suppose the nonlinear simultaneous optimization and heat integration were applied to a sequence of distillation columns. What differences is one likely to encounter when compared to the sequential strategy?

11. Assume the optimization model in Eq. (18.24) is applied to a refrigeration system in which all the hot and cold streams have the same inlet and outlet temperatures since they are pure components undergoing vaporization and condensation. What difficulties can arise in the model?

OPTIMIZATION TECHNIQUES FOR REACTOR NETWORK SYNTHESIS

19

Earlier in this text, synthesis strategies were developed using optimization formulations. The advantage of these strategies is that they describe a rich problem space within an optimization framework. This approach is continued here with the synthesis of reactor networks. As was described in Chapter 13, complex and nonlinear behavior of the reacting system, coupled with combinatorial aspects inherent in all synthesis problems, makes reactor network problems difficult. Consequently, synthesis approaches for these problems are less developed than for the systems considered in previous chapters. This chapter summarizes current optimization-based studies for reactor network synthesis and outlines some directions for future research.

As in Chapter 13, we will concentrate on a reactor network *targeting strategy,* which seeks to describe the performance of the network without its explicit construction. Once obtained, a network is then determined that is guaranteed to match this target. To achieve these properties, we rely on recent geometric concepts based on *attainable regions.* Moreover, we will show how they can be combined with optimization formulations in order to solve larger and more difficult problems, and how reactor network synthesis problems can be integrated into the overall flowsheet synthesis problem.

19.1 INTRODUCTION

In Chapter 13, the reactor synthesis problem was stated as:

> For given reaction stoichiometry, rate laws, a desired objective, and system constraints, what is the optimal reactor network structure and its flow pattern? Where should mixing, heating, and cooling be introduced into the network?

Sec. 19.1 Introduction

In addition to synthesis of the reactor network itself, we also need to consider interactions with other units in the flowsheet, especially those pertaining to energy and separation subsystems. In Chapter 13, heuristic and geometric strategies for selecting reactor types and generalizations to reactor networks were outlined and illustrated on several small examples. These strategies allow the designer a clear understanding of the trade-offs in the reactor system. Moreover, explicit construction of the *attainable region* (*AR*) leads to a complete space of the performance behavior for the reacting system with fixed external specifications (e.g., feeds, heat input, output requirements).

However, for reaction systems that must be represented in three or more independent dimensions (see Chapter 13), the attainable region becomes difficult to construct and interpret geometrically. Moreover, if the feed conditions or other external problem parameters change due to evaluation of more complicated trade-offs in an overall process, the AR approach may need to be performed repeatedly; this leads to a tedious design procedure. In this chapter we explore the incorporation of attainable region concepts within NLP and MINLP formulations for process synthesis. Here we take advantage of powerful methods to solve nonlinear and mixed integer nonlinear programming problems developed in previous chapters. The resulting optimization formulations have a number of advantages. First, conceptual limitations due to system dimensionality are avoided. Also, trade-offs due to different mechanisms or competing terms in the objective function are handled in a straightforward manner. Finally, interactions from other flowsheet subsystems can be incorporated directly and naturally. While this leads to larger optimization problems, current methods for NLP and MINLP, discussed in Chapters 9 and 15, respectively, can readily handle these formulations.

Most structural optimization strategies for reactor network synthesis start by postulating a network of idealized reactors and performing a structural optimization on this enlarged network or "superstructure." These optimization-based approaches can lead to very useful results for reactor networks, but they have a number of limitations. First, the reactor superstructure often leads to nonconvex optimization problems, usually with local optimization tools used to solve them. As a result, only locally optimal solutions can be guaranteed from the network superstructure. Moreover, because reacting systems often have extreme nonlinear behavior, such as bifurcations and multiple steady states, even locally optimal solutions can be quite poor. In addition, superstructure approaches are usually plagued by the question of completeness of the network, and the possibility that a better network may have been overlooked by a limited superstructure (e.g., not enough reactors in the formulation). Finally, many reactor networks can have identical performance characteristics. (For instance, a single PFR can be approximated by a large train of CSTRs.) As a result, secondary characteristics, such as a simpler network would need to be considered.

In this chapter, we will see that the integration of AR concepts with optimization-based synthesis strategies leads to superior problem formulations, because they consider the richness of the solution space and lead to valuable insights in formulating and initializing the optimization problem. Moreover, the attainable region properties often lead to simpler optimization problem formulations than with superstructure approaches. In the next section, attainable region concepts from Chapter 13 are introduced and applied to develop optimization formulations for isothermal systems. This section also extends these formulations to nonisothermal systems. Section 19.3 then describes the integration of reactor targeting optimization problems to process flowsheets and heat exchanger networks. Finally, section 19.4 summarizes the chapter and provides a guide to further reading.

19.2 REACTOR NETWORK SYNTHESIS WITH TARGETING FORMULATIONS

In this section, we apply the concepts of attainable regions to develop simple and efficient optimization formulations for reactor synthesis. In this development, we confine ourselves to homogeneous, constant density reacting systems, although the concepts can be extended to more general cases. The motivation for this approach is that both superstructure and geometric approaches to reactor network synthesis have several limitations. In superstructure-based approaches, the optimal reactor network is limited by the richness of the superstructure, and the synthesis strategy can suffer from convergence to local or nonunique solutions that are characteristic of reactor networks. On the other hand, geometric approaches, considered in Chapter 13, have limitations in treating problems with more than three dimensions. By combining AR concepts and optimization formulations, we instead create performance targets for the optimal reactor network through the solution of small optimization problems. This is applied first to isothermal systems in the next subsection.

19.2.1 Isothermal Reactor Networks

Once the reaction stoichiometry and rate laws are established for an isothermal system, a simple, but incomplete, representation of the reactor network is the segregated flow model, illustrated in Figure 19.1. Here, we assume that only the system molecules of the same age, t, can be perfectly mixed and that molecules of different ages will mix only at the reactor exit. As a result, the behavior of this reactor model is completely determined by its residence time distribution function (RTD), $f(t)$. By finding the optimal $f(t)$ for a specified reactor network objective, one can solve the synthesis problem in the absence of mixing.

Since mixing is not allowed for molecules of different ages, they react according to the following differential equation:

$$\frac{dX_{\text{seg}}}{dt} = R(X_{\text{seg}}) \tag{19.1}$$

$$X_{\text{seg}}(0) = X_0$$

where X_{seg} is the concentration vector (e.g., normalized by a feed concentration) and $R(X)$ is the corresponding rate vector. From the definitions of the residence time distribution we have:

Sec. 19.2   Reactor Network Synthesis with Targeting Formulations

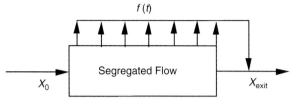

FIGURE 19.1 Segregated flow model.

$$X_{exit} = \int_0^{t_{max}} f(t) X_{seg}(t)\, dt$$

$$\int_0^{t_{max}} t f(t)\, dt = \tau \qquad (19.2)$$

$$\int_0^{t_{max}} f(t)\, dt = 1$$

where X_{exit} is the dimensionless output concentration of the segregated flow system and this system has a residence time τ. The isothermal formulation for maximizing the performance index in segregated flow is given by:

$$\begin{aligned}
&\underset{f(t)}{\text{Max}} \quad J(X_{exit}, \tau) \\
&\frac{dX_{seg}}{dt} = R(X_{seg}) \\
&X_{seg}(0) = X_0 \\
&X_{exit} = \int_0^{t_{max}} f(t) X_{seg}(t)\, dt \\
&\int_0^{t_{max}} t f(t)\, dt = \tau \\
&\int_0^{t_{max}} f(t)\, dt = 1
\end{aligned} \qquad \text{(P1)}$$

The objective function, J, can be specified by the designer as any function of X_{exit} and τ. Moreover, if the dimensionless feed concentration, X_0, is prespecified, we know that $X_{seg}(t)$ is independent of $f(t)$ and the differential equation system (19.1) can be uncoupled from the rest of the model and solved offline. Once X_{seg} is determined, we then find $f(t)$, which satisfies a set of linear constraints.

Problem (P1) can be simplified to an NLP if Gaussian quadrature on finite elements is applied to the integrals over the domain [0, t_{max}], where t_{max} is some large final time. This leads to the following linearly constrained problem:

$$\begin{array}{c} \underset{f_{ij}}{\text{Max}} \quad J(X_{exit}, \tau) \\ \\ \sum_i \sum_j w_j f_{ij} \Delta \alpha_i = 1 \\ \tau = \sum_i \sum_j w_j f_{ij} t_{ij} \Delta \alpha_i \\ X_{exit} = \sum_i \sum_j w_j f_{ij} X_{seg\ ij} \Delta \alpha_i \end{array} \quad \text{(P2)}$$

where

i	=	Index set of finite elements
j	=	Index set of Gauss quadrature (or collocation) points
f_{ij}	=	RTD function at j^{th} quadrature point in i^{th} element (point $[i,j]$)
$X_{seg\ ij}$	=	Dimensionless concentration at point $[i,j]$
w_j	=	Weights of Gaussian quadrature
$\Delta \alpha_i$	=	Length of ith finite element (fixed)

If J is a concave objective function, solution of (P2) gives us a globally optimal network that is restricted to segregated flow. Moreover, for both yield and selectivity objective functions we can reduce the above problem to a linear program by applying suitable transformations (see exercise 6). As a result, (P2) can often be solved as a linear program.

Solution of (P2) provides a good lower bound for the best reactor network. Moreover, in some cases, the segregated flow model is sufficient to describe the attainable region. For instance, a two-dimensional attainable region is complete under segregated flow if the PFR trajectory encloses a convex region (Hildebrandt, 1989). For higher dimensional attainable regions, two-dimensional projections of the PFR trajectory in the space of the reactants and products can be analyzed for convexity (Balakrishna and Biegler, 1992a), and this leads to sufficient conditions for the attainable region. Moreover, the segregated flow model can be optimal even if these convexity conditions are not satisfied. However, if the segregated flow region (for P2) is not sufficient, we need to generate optimization formulations that extend the region described by (P2). The main idea for this approach is:

> *Given a candidate region for the AR, can reactors be generated that extend this region? If yes, then create this reactor extension and, on the convex hull of the extended region, check for further extensions that improve the objective function. Continue this procedure until no further reactor extensions improve the objective function.*

A key point to this approach is that the residence time distributions, $f(t)$, act as convex combinations of the segregated flow profile. As a result, the region in X_{seg} enclosed by the segregated flow model is always convex, as are the feasible regions in (P1) and

Sec. 19.2 Reactor Network Synthesis with Targeting Formulations

(P2). Given a candidate region for the AR, we now aim to develop an algorithm where we can check and, if possible, extend this region. For simplicity of presentation, we first consider constructions with PFR and CSTR extensions only.

The first candidate for the attainable region is the feasible region formed by (P2). Each combination of the RTD, $f(t)$, and X_{seg} gives a unique point in the feasible region. In order to check whether another reactor provides an extension to the region defined by (P2), we consider problem (P3). Here, we combine PFR and CSTR extensions into a single, concise formulation as a recycle reactor (RR) extension. The model for this extension is given by:

$$\frac{dX_{rr}}{dt} = R(X_{rr}), \qquad X_{rr}(t=0) = \frac{R_e X_{exit} + X_{P2}}{R_e + 1} \tag{19.3}$$

where the feed to the recycle reactor, X_{P2}, is found from the solution of (P2), and R_e is the recycle ratio for the recycle reactor. If $R_e = 0$ Eq. (19.3) reduces to an equation for the PFR (19.1); if $R_e \to \infty$, then the reactor becomes a CSTR. Note however, that from Chapter 13 we know that recycle reactors themselves do not form the boundary of an attainable region, as any AR extended by an RR can also be extended by a CSTR. Consequently, formulations for CSTR or PFR extensions can also be developed along the same lines.

For (P3) we see that if $J_{rr} > J_{P2}$, then the recycle reactor provides an extension to the AR that improves the objective function.

$$\begin{aligned}
\text{Max} \quad & J_{rr}(X_{exit}) \\
& X_{P2} = \sum_i \sum_j w_j f_{ij} X_{seg\,ij} \Delta\alpha_i \\
& \frac{dX_{rr}}{dt} = R(X_{rr}) \\
& X_{rr}(t=0) = \frac{R_e X_{exit} + X_{P2}}{R_e + 1} \\
& X_{exit} = \sum_i \sum_j w_j f_{rij} X_{rr\,ij} \Delta\alpha_i \\
& \sum_i \sum_j w_j f_{ij} \Delta\alpha_i = 1.0 \\
& \sum_i \sum_j w_j f_{rij} \Delta\alpha_i = 1.0 \\
& \tau < \tau_{max} \\
& \ell \le X_{exit} \le u
\end{aligned} \tag{P3}$$

where

J_{rr} = Objective function at the exit of the recycle reactor extension.
X_{rr} = Dimensionless concentrations within the RR
X_{exit} = Vector of reactor exit concentrations
f_r = Linear combiner of all the concentrations from the plug flow section of the recycle reactor.

In (P3) the first equation describes the concentrations available from the segregated flow model and this leads to X_{P2}. The model equations for a recycle reactor Eq. (19.3) have a feed that starts from any feasible point described by the first equation. The fourth equation gives the concentration at the exit of the recycle reactor. Here the vectors l and u are lower and upper bounds, respectively, on the exit concentration vector. The RR model (P3) provides an extension over (P2) if $J_{rr} > J_{P2}$.

Note that problem (P3) requires a differential equation constraint for the recycle reactor. Unlike the segregated flow formulation (P2), this equation has a variable initial condition and cannot be solved in advance. Instead, the differential equation can be converted to an algebraic relation in order to solve (P3) as a nonlinear program. To do this, we apply the method of collocation on finite elements, and this will be illustrated in Example 19.2 below.

From (P3), CSTR, PFR, and RR extensions can be applied to any convex candidate region, not just the one defined by (P2). (Linear combinations of these convex candidates are described by optimization formulations that contain these convex regions.) As a result, a sequence of convex hulls of the attainable region can be generated until the conditions for completeness are satisfied (i.e., there are no further extensions). Figure 19.2 presents a synthesis flowchart that illustrates these ideas. In the algorithm, we first check the possibility of a complete attainable region for (P2). If this solution is suboptimal, then a more complex model can be solved to update the solution. Thus, a new or updated convex hull based on the new concentrations is generated, and the following subproblem, which represents the third box in Figure 19.2, is solved.

$$\text{Max } J(X_{exit})$$

$$\frac{dX_{rr}}{dt} = R(X_{rr})$$

$$X_{rr}(t=0) = \frac{R_e X_{exit} + X_{updat}}{R_e + 1} \qquad (P4)$$

$$X_{update} = \sum_i \sum_j f_{ij} X_{seg\ ij} + \sum_k f_{model(k)} X_{model(k)}$$

$$X_{exit} = \sum_i \sum_j f_{rij} X_{rr\ ij}$$

$$\sum_i \sum_j f_{ij} + \sum_k f_{model(k)} = 1.0$$

In problem (P4), $X_{model(k)}$ is a constant vector and reflects the concentration at the exit in the models chosen from (P2), (P3), or previous instances of (P4). A convex combination of $X_{model(k)}$ with the segregated flow region described by (P2) gives the fresh feed point for the recycle reactor in (P4), X_{update}. The exit concentration of the RR is X_{exit}, and if $J(X_{exit}) > J(X_{model(k)})$, the previous model chosen is insufficient. The problem variables are R_e, f_{ij}, and $f_{model(k)}$ and these describe the linear combinations for the convex candidates in (P4). Note that the last equation in this formulation checks for completeness of the convex hull of the region found by (P4).

Sec. 19.2 Reactor Network Synthesis with Targeting Formulations

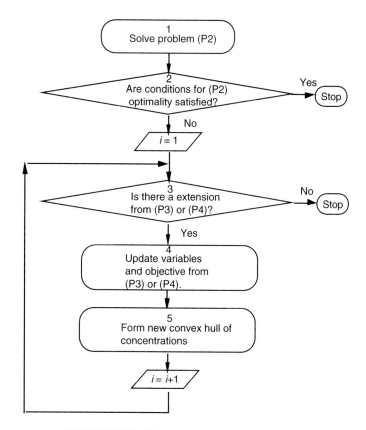

FIGURE 19.2 Flowchart for stagewise synthesis.

A geometric interpretation to the solution of (P4) is shown in Figure 19.3. If the solution of (P4) indicates that the objective function can be improved by extending the *AR* (say, that was generated by (P2)), we consider a more complex model. Thus, the expression for X_{update} automatically includes all the points in the convex hulls generated from (P2) in addition to favorable recycle reactor extensions from (P4). We continue to check for extensions by augmenting (P4) with additional models and terminate when there are no further extensions that improve the objective function. Note that with the solution from this sequential approach, the reactor network can be synthesized easily and retains the flavor of the algorithm developed in Chapter 13; An important difference, though, is that the approach in Chapter 13 searches for *all* possible extensions of candidate ARs, not just the ones that improve the objective function. On the other hand, this requires checking an infinite number of points on the convex hull of the candidate region.

Because of this difference with Chapter 13, one disadvantage to the algorithm in Figure 19.2 is that it may not find the entire attainable region. For instance, there could be an extension that does not improve the objective function but still enlarges the AR. From this enlargement, we may be able to find a further extension that *does* improve the objec-

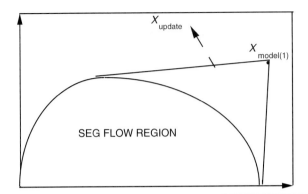

FIGURE 19.3 Illustration for extension of the convex hull (P4).

tive function beyond what we started with. This *non-monotonic* increase in the objective is a limitation of the algorithm in Figure 19.2; in section 19.2.4 we will present an MINLP formulation that overcomes this approach. Moreover, we note that even though the attainable space of concentrations is always convex, (P4) is not always a convex nonlinear program, and therefore we may not find the global optimum to (P4). Therefore, with local NLP solvers, multiple starting points need to be tried to improve the likelihood of finding a global optimum for P4; good initial points are often obtained from the solution to (P2).

We conclude this subsection with two example problems to illustrate our approach. Both examples illustrate the problem formulations (P2) and (P3) in detail. The first example satisfies the sufficiency conditions for segregated flow and is relatively easy to solve. The second example, on the other hand, does not satisfy these properties but is readily solved by the algorithm of Figure 19.2. Several additional problems are considered in Balakrishna and Biegler (1992a) and Lakshmanan and Biegler (1995).

EXAMPLE 19.1

The isothermal van de Vusse (1964) reaction shown below involves four species. However, if we wish to maximize the yield of the intermediate species B from a feed of pure A, then only the species A and B need to be considered. This problem is similar to Example 13.2 in Chapter 13, but uses different rate vectors and initial concentrations. The reaction network is given by

$$A \xrightarrow{k_1} B \xrightarrow{k_2} C$$
$$k_3 \downarrow$$
$$D$$

Here the reaction from A to D is second order. The feed concentration is $c_{A0} = 0.58$ mol/l and the reaction rates are $k_1 = 10\ s^{-1}$, $k_2 = 1\ s^{-1}$ and $k_3 = 1\ l/(gmol\ s)$. The reaction rate vector for components A, B, C, D respectively is given in dimensionless form by:

$$R(X) = [-10X_A - 0.29X_A^2,\ 10X_A - X_B,\ X_B,\ 0.29X_A^2], \tag{19.4}$$

where $X_A = c_A/c_{A0}$, $X_B = c_B/c_{A0}$, and c_A, c_B are the molar concentrations of A and B respectively. For this problem, the differential equations:

$$\frac{dX_{seg}}{dt} = R(X_{seg}),\ X_{seg}(0) = X_0 \tag{19.1}$$

become:

$$dX_{seg,A}/dt = -10 X_{seg,A} - 0.29 X_{seg,A}^2 \quad X_{seg,A}(0) = 1.0$$
$$dX_{seg,B}/dt = 10 X_{seg,A} - X_{seg,B} \quad X_{seg,B}(0) = 0. \tag{19.5}$$

These equations are solved first and the profiles are shown in Figure 19.4.

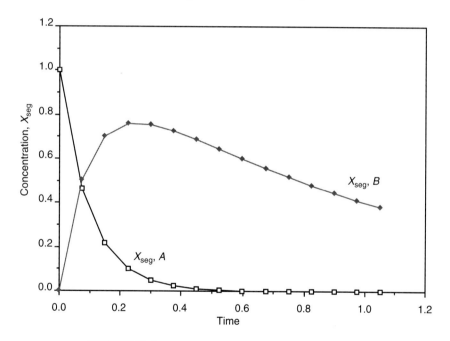

FIGURE 19.4 Concentration profiles for Example 19.1.

We now discretize these profiles and form the problem (P2). Here we set $\Delta\alpha_i = 0.075$ and we choose fourteen finite elements so that $t \in [0, t_{max}]$ and $t_{max} \geq 1$. The quadrature points in each element are chosen to be roots of orthogonal polynomials, τ_j, and the quadrature weights, w_j, in (P2) are calculated to correspond to the integration of these polynomials. Values for τ_j and w_j are tabulated in a number of references (see, e.g., Carnahan et al., 1969). Now if we choose three quadrature points, then we have:

$\tau_j = [0.1127, 0.5, 0.8873]$ and $w_j = [0.5555, 0.8888, 0.5555], j = 1,...3$

For the finite elements i and quadrature points, j, we define the quadrature points in time as:

$$t_{ij} = \sum_{k=1}^{i-1} \Delta\alpha_{(k)} + \Delta\alpha_i \, \tau_j \qquad (19.6)$$

and we evaluate the profiles in Figure 19.4 at these points, so that $X_{\text{seg } ij} = X_{\text{seg}}(t_{ij})$. Similarly the profile for the residence time distribution, $f(t)$, is also evaluated at these points so that $f_{ij} = f(t_{ij})$. Substituting this information into (P2) leads to the following optimization problem:

TABLE 19.1   Linear Program for Example 19.1: Optimal Reactor Network in Segregated Flow

Max $X_{\text{exit},B}$
subject to

$0.075[(0.5555)f_{1,1} + (0.8888)f_{1,2} + (0.5555)f_{1,3}$
$+ (0.5555)f_{2,1} + (0.8888)f_{2,2} + (0.5555)f_{2,3}$
$+ (0.5555)f_{3,1} + (0.8888)f_{3,2} + (0.5555)f_{3,3}$
$+ (0.5555)f_{4,1} + (0.8888)f_{4,2} + (0.5555)f_{4,3}$
$+ (0.5555)f_{5,1} + (0.8888)f_{5,2} + (0.5555)f_{5,3}$
$+ (0.5555)f_{6,1} + (0.8888)f_{6,2} + (0.5555)f_{6,3}$
$+ (0.5555)f_{7,1} + (0.8888)f_{7,2} + (0.5555)f_{7,3}$
$+ (0.5555)f_{8,1} + (0.8888)f_{8,2} + (0.5555)f_{8,3}$
$+ (0.5555)f_{9,1} + (0.8888)f_{9,2} + (0.5555)f_{9,3}$
$+ (0.5555)f_{10,1} + (0.8888)f_{10,2} + (0.5555)f_{10,3}$
$+ (0.5555)f_{11,1} + (0.8888)f_{11,2} + (0.5555)f_{11,3}$
$+ (0.5555)f_{12,1} + (0.8888)f_{12,2} + (0.5555)f_{12,3}$
$+ (0.5555)f_{13,1} + (0.8888)f_{13,2} + (0.5555)f_{13,3}$
$+ (0.5555)f_{14,1} + (0.8888)f_{14,2} + (0.5555)f_{14,3}] = 1$

$0.075[(0.5555)\,0.0080\,f_{1,1} + (0.8888)\,0.0375\,f_{1,2} + (0.5555)\,0.0665\,f_{1,3}$
$+ (0.5555)\,0.0834\,f_{2,1} + (0.8888)\,0.1125\,f_{2,2} + (0.5555)\,0.1415\,f_{2,3}$
$+ (0.5555)\,0.1584\,f_{3,1} + (0.8888)\,0.1875\,f_{3,2} + (0.5555)\,0.2165\,f_{3,3}$
$+ (0.5555)\,0.2334\,f_{4,1} + (0.8888)\,0.2625\,f_{4,2} + (0.5555)\,0.2915\,f_{4,3}$
$+ (0.5555)\,0.3084\,f_{5,1} + (0.8888)\,0.3375\,f_{5,2} + (0.5555)\,0.3665\,f_{5,3}$
$+ (0.5555)\,0.3834\,f_{6,1} + (0.8888)\,0.4125\,f_{6,2} + (0.5555)\,0.4415\,f_{6,3}$
$+ (0.5555)\,0.4584\,f_{7,1} + (0.8888)\,0.4875\,f_{7,2} + (0.5555)\,0.5165\,f_{7,3}$
$+ (0.5555)\,0.5334\,f_{8,1} + (0.8888)\,0.5625\,f_{8,2} + (0.5555)\,0.5915\,f_{8,3}$
$+ (0.5555)\,0.6084\,f_{9,1} + (0.8888)\,0.6375\,f_{9,2} + (0.5555)\,0.6665\,f_{9,3}$
$+ (0.5555)\,0.6834\,f_{10,1} + (0.8888)\,0.7125\,f_{10,2} + (0.5555)\,0.7415\,f_{10,3}$
$+ (0.5555)\,0.7584\,f_{11,1} + (0.8888)\,0.7875\,f_{11,2} + (0.5555)\,0.8165\,f_{11,3}$
$+ (0.5555)\,0.8334\,f_{12,1} + (0.8888)\,0.8625\,f_{12,2} + (0.5555)\,0.8915\,f_{12,3}$
$+ (0.5555)\,0.9084\,f_{13,1} + (0.8888)\,0.9375\,f_{13,2} + (0.5555)\,0.9665\,f_{13,3}$
$+ (0.5555)\,0.9834\,f_{14,1} + (0.8888)\,1.0125\,f_{14,2} + (0.5555)\,1.0415\,f_{14,3}] = \tau$

$X_{\text{exit},A} = 0.075[(0.5555)\,0.9210\,f_{1,1} + (0.8888)\,0.6810\,f_{1,2} + (0.5555)\,0.5070\,f_{1,3}$
$+ (0.5555)\,0.4271\,f_{2,1} + (0.8888)\,0.3182\,f_{2,2} + (0.5555)\,0.2375\,f_{2,3}$
$+ (0.5555)\,0.2004\,f_{3,1} + (0.8888)\,0.1495\,f_{3,2} + (0.5555)\,0.1117\,f_{3,3}$

Sec. 19.2 Reactor Network Synthesis with Targeting Formulations

TABLE 19.1 *Continued*

$$+ (0.5555)\ 0.0943\ f_{4,1} + (0.8888)\ 0.0705\ f_{4,2} + (0.5555)\ 0.0527\ f_{4,3}$$
$$+ (0.5555)\ 0.0445\ f_{5,1} + (0.8888)\ 0.0332\ f_{5,2} + (0.5555)\ 0.0248\ f_{5,3}$$
$$+ (0.5555)\ 0.0210\ f_{6,1} + (0.8888)\ 0.0157\ f_{6,2} + (0.5555)\ 0.0117\ f_{6,3}$$
$$+ (0.5555)\ 0.0099\ f_{7,1} + (0.8888)\ 0.0074\ f_{7,2} + (0.5555)\ 0.0055\ f_{7,3}$$
$$+ (0.5555)\ 0.0047\ f_{8,1} + (0.8888)\ 0.0035\ f_{8,2} + (0.5555)\ 0.0026\ f_{8,3}$$
$$+ (0.5555)\ 0.0022\ f_{9,1} + (0.8888)\ 0.0016\ f_{9,2} + (0.5555)\ 0.0012\ f_{9,3}$$
$$+ (0.5555)\ 0.0010\ f_{10,1} + (0.8888)\ 0.0008\ f_{10,2} + (0.5555)\ 0.0006\ f_{10,3}$$
$$+ (0.5555)\ 0.0005\ f_{11,1} + (0.8888)\ 0.0004\ f_{11,2} + (0.5555)\ 0.0003\ f_{11,3}$$
$$+ (0.5555)\ 0.0002\ f_{12,1} + (0.8888)\ 0.0002\ f_{12,2} + (0.5555)\ 0.0001\ f_{12,3}$$
$$+ (0.5555)\ 0.0001\ f_{13,1} + (0.8888)\ 0.0001\ f_{13,2} + (0.5555)\ 0.0001\ f_{13,3}]$$

$$X_{\text{exit},B} = 0.075[(0.5555)\ 0.0765\ f_{1,1} + (0.8888)\ 0.3053\ f_{1,2} + (0.5555)\ 0.4651\ f_{1,3}$$
$$+ (0.5555)\ 0.5354\ f_{2,1} + (0.8888)\ 0.6261\ f_{2,2} + (0.5555)\ 0.6871\ f_{2,3}$$
$$+ (0.5555)\ 0.7122\ f_{3,1} + (0.8888)\ 0.7416\ f_{3,2} + (0.5555)\ 0.7574\ f_{3,3}$$
$$+ (0.5555)\ 0.7620\ f_{4,1} + (0.8888)\ 0.7636\ f_{4,2} + (0.5555)\ 0.7592\ f_{4,3}$$
$$+ (0.5555)\ 0.7546\ f_{5,1} + (0.8888)\ 0.7441\ f_{5,2} + (0.5555)\ 0.7310\ f_{5,3}$$
$$+ (0.5555)\ 0.7226\ f_{6,1} + (0.8888)\ 0.7071\ f_{6,2} + (0.5555)\ 0.6908\ f_{6,3}$$
$$+ (0.5555)\ 0.6810\ f_{7,1} + (0.8888)\ 0.6639\ f_{7,2} + (0.5555)\ 0.6468\ f_{7,3}$$
$$+ (0.5555)\ 0.6368\ f_{8,1} + (0.8888)\ 0.6197\ f_{8,2} + (0.5555)\ 0.6029\ f_{8,3}$$
$$+ (0.5555)\ 0.5932\ f_{9,1} + (0.8888)\ 0.5767\ f_{9,2} + (0.5555)\ 0.5606\ f_{9,3}$$
$$+ (0.5555)\ 0.5514\ f_{10,1} + (0.8888)\ 0.5359\ f_{10,2} + (0.5555)\ 0.5207\ f_{10,3}$$
$$+ (0.5555)\ 0.5121\ f_{11,1} + (0.8888)\ 0.4975\ f_{11,2} + (0.5555)\ 0.4834\ f_{11,3}$$
$$+ (0.5555)\ 0.4753\ f_{12,1} + (0.8888)\ 0.4618\ f_{12,2} + (0.5555)\ 0.4486\ f_{12,3}$$
$$+ (0.5555)\ 0.4411\ f_{13,1} + (0.8888)\ 0.4285\ f_{13,2} + (0.5555)\ 0.4163\ f_{13,3}$$
$$+ (0.5555)\ 0.4093\ f_{14,1} + (0.8888)\ 0.3976\ f_{14,2} + (0.5555)\ 0.3862\ f_{14,3}]$$

$$\underset{f_{ij}}{\text{Max}}\quad X_{\text{exit},B}$$

$$\sum_i \sum_j w_j f_{ij} \Delta\alpha_i = 1$$

$$\tau = \sum_i \sum_j w_j f_{ij} t_{ij} \Delta\alpha_i \tag{19.7}$$

$$X_{\text{exit},A} = \sum_i \sum_j w_j f_{ij} X_{\text{seg},A\ ij} \Delta\alpha_i$$

$$X_{\text{exit},B} = \sum_i \sum_j w_j f_{ij} X_{\text{seg},B\ ij} \Delta\alpha_i$$

and the variables in this problem (f_{ij}, τ, $X_{\text{exit},A}$ and $X_{\text{exit},B}$) appear linearly in the constraint and objective functions. If we substitute the numerical values for the constants t_{ij}, w_{ij}, $X_{\text{seg},A\ ij}$, $X_{\text{seg},B\ ij}$ and $\Delta\alpha_i$ into (19.7) we obtain the linear program given in Table 19.1. From the algorithm in Figure 19.2, we find that the solution of Eq. (19.7) is sufficient to obtain a globally optimal reactor network for this system. This follows because the profiles for X_A and X_B form a convex candidate AR, as can be seen in Figure 19.5. Moreover, it can be shown, by using the information in Figure 19.5, that there are no CSTRs that further extend the attainable region. This linear programming problem Eq. (19.7) was modeled in GAMS and its solution required only 0.58 CPU secs. on a Sun 3 workstation.

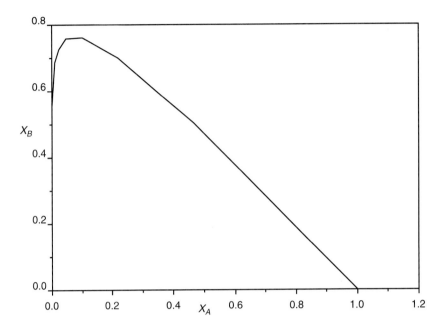

FIGURE 19.5 Attainable Region for Example 19.1.

Here the linear program in Table 19.1 has the solution $X_{exit,A} = 0.0705$, $X_{exit,B} = 0.7636$, $\tau = 0.2625$ and $f_{4,2} = 1$, with the optimal value of the objective function given by $X_{exit,B} = 0.7636$. As seen from the attainable region in Figure 19.5, this (globally) optimal solution is realized by a PFR with a residence time of 0.263 seconds. Previous literature values with superstructure approaches (Chitra and Govind, 1985; Kokossis and Floudas, 1990) report optimal yields of 0.752 with residence times around 0.25 s. Their results are only slightly lower and differences could be attributed to numerical approximations of the differential equations. In fact, solving equations (19.5) off-line for (P2) helps to improve the solution accuracy.

EXAMPLE 19.2

The Trambouze reaction (Trambouze and Piret, 1959) has the following reaction scheme and also involves four components:

$$A \xrightarrow{k_1} B \qquad A \xrightarrow{k_2} C \qquad A \xrightarrow{k_3} D$$

The three reactions are zero order, first order, and second order, respectively, with $k_1 = 0.025$ mol/(lit min), $k_2 = 0.2$ min^{-1}, $k_3 = 0.4$ l/(mol min) and pure feed with $c_{A0} = 1$ gmol/l. Again, we define $X_A = c_A/c_{A0}$ and $X_C = c_C/c_{A0}$, but here we maximize the selectivity of C to A defined by $X_C/(1 - X_A)$. This problem is solved in two stages.

CANDIDATE REGION FOR SEGREGATED FLOW

Following the algorithm in Figure 19.2, we first integrate the differential equations from (P2):

$$dX_{seg,A}/dt = -0.025 - 0.2 X_{seg,A} - 0.4 X_{seg,A}^2 \quad X_{seg,A}(0) = 1.0$$
$$dX_{seg,C}/dt = 0.2 X_{seg,A} \quad X_{seg,C}(0) = 0. \quad (19.8)$$

and this leads to the concentration profiles in Figure 19.6a. Moreover, for comparison with the graphical method of Chapter 13, the attainable region is shown in Figure 19.6b.

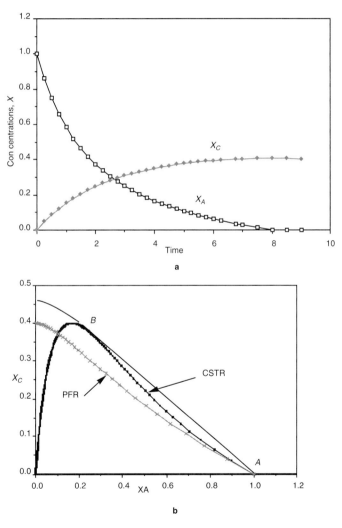

FIGURE 19.6 (a) Concentration profiles, and (b) Attainable region for Example 19.2.

We now discretize these profiles and form problem (P2). Here we set $\Delta\alpha_i = 0.25$ for $0 \leq t < 7$ and $\Delta\alpha_i = 0.5$ for $7 \leq t \leq 9$, $t_{max} = 9.0$ and we choose 32 finite elements. The quadrature points and weights are determined as in Example 19.1. If we choose two quadrature points in each element, then we have:

$$\tau_j = [0.2113, 0.7887] \quad \text{and} \quad w_j = [1.0, 1.0] \quad j = 1,2$$

For the finite elements i and quadrature points, j, we obtain the profiles in Figure 19.6a so that $X_{seg\ ij} = X_{seg}(t_{ij})$ and $f_{ij} = f(t_{ij})$. Substituting this information into (P2) leads to the following optimization problem:

$$\text{Max}_{f_{ij}} \quad (X_{exit,C}) / (1 - X_{exit,A})$$

$$\sum_i \sum_j w_j f_{ij} \Delta\alpha_i = 1$$

$$\tau = \sum_i \sum_j w_j f_{ij} t_{ij} \Delta\alpha_i \tag{19.9}$$

$$X_{exit,A} = \sum_i \sum_j w_j f_{ij} X_{seg,A\ ij} \Delta\alpha_i$$

$$X_{exit,C} = \sum_i \sum_j w_j f_{ij} X_{seg,C\ ij} \Delta\alpha_i$$

and the only variables in this problem are f_{ij}, τ, $X_{exit,A}$, and $X_{exit,C}$. Problem (19.9) can be simplified to a linear program by defining new variables. First, we define the variable $S = 1/(1 - X_{exit,A})$ and we assume that it is always positive. We then define:

$$g_{ij} = S f_{ij} \qquad \gamma = S\tau$$

$$Y_{exit,A} = S X_{exit,A} \qquad Y_{exit,C} = S X_{exit,C}$$

and substitute into the above problem as:

$$\text{Max}_{f_{ij}} \quad Y_{exit,C}$$

$$\sum_i \sum_j w_j g_{ij} \Delta\alpha_i = S$$

$$\gamma = \sum_i \sum_j w_j g_{ij} t_{ij} \Delta\alpha_i \tag{19.10}$$

$$Y_{exit,A} = \sum_i \sum_j w_j g_{ij} X_{seg,A\ ij} \Delta\alpha_i$$

$$Y_{exit,C} = \sum_i \sum_j w_j g_{ij} X_{seg,C\ ij} \Delta\alpha_i$$

which is a linear program and leads to globally optimal reactor network *for segregated flow*. A generalization of this property is discussed in exercise 6. The solution to Eq. (19.10) is $Y_{exit,C} = X_{exit,C}/(1 - X_{exit,A}) = 0.422$, with $S = 0.893$, $X_{exit,C} = 0.377$ and $X_{exit,A} = 0.107$. This corresponds to a single PFR with a residence time, $\tau = 5.01$ minutes. However, from Figure 19.6b we see that the PFR profile is not convex and, as a result, the reactor network can be further improved. This will now be verified by the algorithm of Figure 19.2.

SOLUTION OF (P3) BY COLLOCATION ON FINITE ELEMENTS

As in the discretization of problem (P1) to (P2), we represent the problem profiles with the subscript i denoting the ith finite element, and the subscript j (or k) denoting the jth (or k^{th}) collocation point in any finite element. There are a total of N finite elements and K collocation points

Sec. 19.2 Reactor Network Synthesis with Targeting Formulations

($i = 1, N$; $j = 1, K$). The normalized concentrations, X, are approximated over each finite element by a polynomial written in Lagrange form. Here Lagrange interpolation basis functions ($L_k(\alpha)$) are given by:

$$X(t) = \sum_{k=0}^{K} X_{ik} L_k(t) \quad \text{for } t_{i0} \leq t \leq t_{i+1,0}, \quad \text{and}, \quad L_k(t) = \prod_{l=0; \neq k}^{K} \left[\frac{t - t_{il}}{t_{ik} - t_{il}} \right]$$

At each of the quadrature points (which we also use as collocation points) we note that the basis functions have the property that $L_k(t_{ij}) = 0$ for $j \neq k$ and $L_k(t_{ij}) = 1$ for $j = k$. This leads to the nice property that Lagrange polynomial coefficients are equal to the value of the polynomial at the collocation point, that is, $X(t_{ij}) = \sum_{k=0}^{K} X_{ik} L_k(t_{ij}) = X_{ij}$. Also notice that from: $t_{ij} = \sum_{k=1}^{i-1} \Delta \alpha_k + \Delta \alpha_i \, \tau_j$, we have, $L_k(t_{ij}) = L_k(\tau_j)$ and

$$L'_k(\tau_j) = d\, L_k(\tau_j)/d\tau = d\, L_k(\tau_j)/dt\, (dt/d\tau) = \Delta \alpha_i\, d\, L_k(\tau_j)/dt$$

Over each finite element we now substitute the polynomial approximation for X into the differential equations for the recycle reactor in (P3):

$$dX_{rr,A}/dt = -0.025 - 0.2\, X_{rr,A} - 0.4\, X_{rr,A}^2, \quad (19.11)$$

$$X_{rr,A}(0) = (R_e\, X_{\text{exit},A} + X_{P2,A})/(R_e + 1)$$

$$dX_{rr,C}/dt = 0.2\, X_{rr,A},$$

$$X_{rr,C}(0) = (R_e\, X_{\text{exit},C} + X_{P2,C})/(R_e + 1)$$

After some rearrangement (see Exercise 7), this leads to the following algebraic equations at each of the collocation points.

$$\sum_{k=0}^{K} X_{ik,A} L'_k(\tau_j) = \Delta \alpha_i (-0.025 - 0.2\, X_{ij,A} - 0.4\, X_{ij,A}^2) \quad i = 1, N,\ j = 1, K$$

$$X_{10,A} = (R_e\, X_{\text{exit},A} + X_{P2,A})/(R_e + 1)$$

$$\sum_{k=0}^{K} X_{ik,C} L'_k(\tau_j) = \Delta \alpha_i (0.2\, X_{ij,A}) \quad\quad\quad i = 1, N,\ j = 1, K$$

$$X_{10,C} = (R_e\, X_{\text{exit},C} + X_{P2,C})/(R_e + 1)$$

In addition to these collocation equations, we also add an additional set of constraints that ensure continuity of the concentration profiles at the limits of the finite elements. For this example, they are given by:

$$\sum_{k=0}^{K} X_{ik,A} L_k(1.0) = X_{(i+1,0)A} \quad \sum_{k=0}^{K} X_{ik,C} L_k(1.0) = X_{(i+1,0)C} \quad (19.12)$$

Note that the coefficients $L_k(1.0)$ and $L'_k(\tau_j)$ are constants that can be calculated and tabulated in advance. Substitution of the collocation and continuity equations for the differential equations in (P3) leads to the following nonlinear program, the solution of which gives us the optimal recycle ratio, values of $f(t)$ and $f_r(t)$, and $X_{\text{exit},A}$ and $X_{\text{exit},C}$.

Max $(X_{exit,C}) / (1 - X_{exit,A})$ (19.13)

s.t. $\sum_i \sum_j w_j f_{ij} \Delta \alpha_i = 1$

$\tau = \sum_i \sum_j w_j f_{ij} t_{ij} \Delta \alpha_i$

$X_{P2,A} = \sum_i \sum_j w_j f_{ij} X_{seg,A\,ij} \Delta \alpha_i$

$X_{P2,C} = \sum_i \sum_j w_j f_{ij} X_{seg,C\,ij} \Delta \alpha_i$

$\sum_{k=0}^{K} X_{ik,A} L'_k(\tau_j) = \Delta \alpha_i (-0.025 - 0.2 X_{ij,A} - 0.4 X_{ij,A}^2)$ $i = 1, N, j = 1, K$

$X_{10,A} = (R_e X_{exit,A} + X_{P2,A})/(R_e + 1)$

$\sum_{k=0}^{K} X_{ik,C} L'_k(\tau_j) = \Delta \alpha_i (0.2 X_{ij,A})$ $i = 1, N, j = 1, K$

$X_{10,C} = (R_e X_{exit,C} + X_{P2,C})/(R_e + 1)$

$\sum_{k=0}^{K} X_{ik,A} L_k(1.0) = X_{(i+1,0)A}$ $\sum_{k=0}^{K} X_{ik,C} L_k(1.0) = X_{(i+1,0)C}$

$i = 1, N$

$X_{exit,A} = \sum_i \sum_j w_j f_{rij} X_{ij,A} \Delta \alpha_i$

$X_{exit,C} = \sum_i \sum_j w_j f_{rij} X_{ij,C} \Delta \alpha_i$

$\sum_i \sum_j w_j f_{rij} \Delta \alpha_i = 1.0$

From the solution of Eq. (19.13) we obtain a CSTR extension (R_e becomes unbounded in the recycle reactor) from the feed point of the segregated flow model. The optimal reactor network is therefore a single CSTR with an exit stream of $X_A = 0.25$, $X_C = 0.375$, a selectivity of 0.5, and residence time of 7.5 sec. Following the stagewise approach in Figure 19.2, we observe no further recycle reactor extensions by solving problem (P4) with this collocation approach.

For this example problem, Achenie and Biegler (1990) observe a selectivity of 0.4999 in a two CSTR combination. Kokossis and Floudas (1990) report many optimal networks to this problem with the same objective function of 0.5. Using the graphical approach from Chapter 13, Glasser et al. (1987) also observe that this problem has an infinite number of optimal solutions (a CSTR with variable bypass) with a selectivity of 0.5. This solution can be seen from the attainable region in Figure 19.6b where the selectivity is the slope of line segment *AB*. Consequently, for this example we see that the algorithm of Figure 19.2 yields the optimal reactor network as well.

19.2.2 Nonisothermal Systems

In this subsection, we extend the formulation for the synthesis of isothermal reactor networks to nonisothermal systems. Here, the optimization formulation also deals with optimal temperature profiles and, as with (P3) and (P4), we require the solution of dynamic optimization problems. With these formulations, we again consider the sequential solution of small nonlinear programming problems as in Figure 19.2; the solution to each NLP generates an additional component of the reactor network. This provides a constructive technique for the synthesis of nonisothermal reactor networks, using any general objective function and process constraints.

For nonisothermal systems, temperature is an additional profile that often needs to be maintained at added cost. However, an inexpensive technique for temperature manipulation in exothermic reactions is cold shot cooling, even though mixing may not always be optimal in the space of concentrations. To address this, we consider as our basic targeting model a different reactor flow model that can address temperature manipulation both by feed mixing as well as by external heating or cooling. The model consists of a particular differential sidestream reactor (DSR), shown in Figure 19.7, which has a sidestream concentration set to the feed concentration. It also includes a general exit flow distribution function.

Feinberg and Hildebrandt (1997) showed that for higher dimensional (≥ 3) problems the DSR is an essential element for the boundary of an AR. In our optimization formulation, we consider a (more restricted) DSR by considering a sidestream given by the feed concentration as our basic model. This model allows the manipulation of reactor temperature by feed mixing. From Figure 19.7, we define X_0 as the dimensionless concentration of the feed that is entering the reactor network, t is the independent variable denoting length (normalized by residence time) along the reactor, and $T(t)$ denotes the temperature as a function of the reactor length. We define $f(t)\Delta t$ as the fraction of molecules in the reactor exit that leave between points t and $t + \Delta t$ of the reactor (an exit flow distribution function), and $q(t)$ is the distribution function for a molecule entering the system at point t in the reactor. Thus, the number of molecules entering between points t and $t + \Delta t$ is given by $q(t)Q_0 \Delta t$, where Q_0 is the flow rate entering the reactor network. Finally, we will assume instantaneous mixing between the feed and the mixture in the reactor and will only

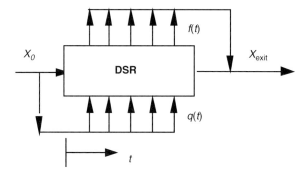

FIGURE 19.7 A particular differential sidestream reactor (DSR) model.

consider constant density systems here, although the formulation can easily be extended to variable density systems.

As seen from Figure 19.7, our DSR model allows for a number of special cases. For instance, when $q(t)$ is zero throughout the reactor and we have a nonzero $f(t)$, we recover the equations for a segregated flow model. On the other hand, when $f(t)$ is a Dirac delta exactly at one point, and we have a general non-zero $q(t)$, this model reduces to the Zwietering (1959) model of maximum mixedness. Based on this nomenclature, a differential mass balance on an element Δt leads to:

$$\frac{dX}{dt} = R(T(t), X) + \frac{q(t)Q_0}{Q(t)}(X_0 - X(t)) \tag{19.14}$$

where $Q(t)$ is the volumetric flowrate at point t. With this governing equation (19.14), the mathematical model for maximizing the performance index in with the extended DSR can be derived as shown below:

$$\underset{q(t), f(t), T(t)}{\text{Max}} \quad J(X_{\text{exit}}, \tau)$$

$$\frac{dX}{dt} = R(T(t), X) + \frac{q(t)Q_0}{Q(t)}(X_0 - X(t))$$

$$X(0) = X_0$$

$$X_{\text{exit}} = \int_0^\infty f(t) X(t)\, dt$$

$$\int_0^\infty f(t)\, dt = 1 \tag{P5}$$

$$\int_0^\infty q(t)\, dt = 1$$

$$Q(t)/Q_0 = \int_0^t [q(t') - f(t')]\, dt'$$

$$\int_0^\infty \int_0^t [q(t') - f(t')]\, dt'\, dt = \tau$$

Here, the last two equations define the flow rate and the mean residence time, respectively. This formulation is an optimal control problem, where the control profiles are $q(t)$, $f(t)$, and $T(t)$. The solution to (P5) gives us a lower bound on the objective function for the nonisothermal reactor network along with the optimal temperature and mixing profiles and we could use this formulation to construct an algorithm similar to the one in Figure 19.2.

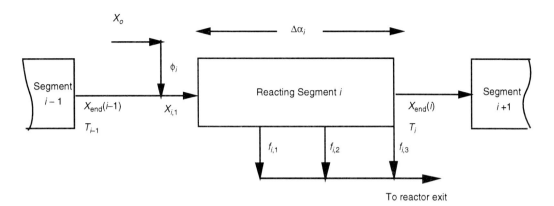

FIGURE 19.8 Reactor representation for discretized extended DSR model. [Reprinted with permission from Balakrishna., S., & Biegler, L. T., *Ind. Eng. Chem. Research*, **31**, p. 2152 (1992). Copyright 1992, American Chemical Society]

A simple modification of this problem can also be considered by discretizing the feed distribution profile, $q(t)$, as shown in Figure 19.8. This leads to an approximate DSR formulation where mixing occurs before each element and reaction occurs within an element. From Figure 19.8, we discretize (P5) based on collocation on finite elements, as the differential equations can no longer be solved in advance. Application of this discretization leads to the following nonlinear program, the solution of which gives us the optimal control variables at the collocation points.

$$\text{Max} \quad J(X_{\text{exit}}, \tau) \tag{P6}$$
$$\phi_i, f_{ij}, T_i$$

$$\sum_k X_{ik} L_k'(\tau_j) - R(X_{ij}, T_{ij}) \Delta \alpha_i = 0 \quad j = 1, K, , i = 1,...N \tag{a}$$
$$X(0) = X_0$$

$$X_{i\text{end}} = \sum_k X_{ik} L_k(1.0), i = 1,...N \tag{b}$$

$$X_{i,0} = \phi_i X_0 + (1-\phi_i) X_{(i-1)\text{end}}, \; i = 1,...N \tag{c}$$

$$X_{\text{exit}} = \sum_i \sum_j \Delta \alpha_i \, w_j X_{ij} f_{ij} \tag{d}$$

$$\sum_i \sum_j w_j \Delta \alpha_{ij} \, Q_{ij}/Q_0 = \tau \tag{e}$$

$$\sum_i \sum_j \Delta \alpha_i \, w_j f_{ij} = 1 \tag{f}$$

$$\phi_i \, Q_{i,1} = q_{i,1} Q_0 \, , \, i = 1,...N \tag{g}$$

$$Q_0 [\sum_{i<i'} \sum_j w_j \Delta \alpha_{ij} (q_{ij} - f_{ij})] = Q_{i',1}, \, i' = 1,...N \tag{h}$$

$$0 \le \phi_i \le 1, \, i = 1,...N$$

where

- ϕ_i = Ratio of the side inlet flow rate to the bulk flow rate within the reactor after mixing before element i.
- $\Delta\alpha_i = t_{i+1,0} - t_{i,0}$, is the length of each finite element i.
- f_{ij} = Exit flow distribution at collocation point j in element i at t_{ij}.
- q_{ij} = Fraction of inlet flow entering at t_{ij}.
- T_{ij} = Temperature at t_{ij}.
- X_{ij} = Dimensionless concentration at t_{ij}.
- $X_{i\text{end}}$ = Concentration at end of ith finite element.

In this formulation, Eqs. (P6a) and (P6b) are the differential equations for the reacting elements, approximated with orthogonal collocation. The equations (P6d), (P6e), (P6f), and (P6h) represent Gaussian quadrature applied to the integrals in (P5). These approximations are illustrated in Examples 19.1 and 19.2. Note also that ϕ_i in (P6) is a pointwise approximation to $q(\alpha)Q_0/Q(\alpha^+)$. Equation (P6g) follows from the pointwise discretization of $q(t)$ and (P6c) represents the feed mixing point in Figure 19.8.

It can be shown that if the finite elements ($\Delta\alpha_i$) are chosen sufficiently small, then (P6) simply reduces to a numerical scheme for solving (P5). Thus, (P5) can be approximated and solved as a nonlinear program, to obtain the optimal set of f, T, and ϕ over each element. Also, note that even though the temperature along the reactor is a control variable, part of the temperature manipulation can be readily accomplished by feed mixing if this is optimal for the reactor.

The solution to (P6) provides a lower bound to the performance index of the reactor network. By applying the optimization formulations detailed in section 19.3, we now develop techniques for extending the reactor network provided by (P6). Note that the constraints of (P6) define the feasible region for any achievable DSR and a convex combination of the concentrations in this region provides the entire region attainable by the DSR and mixing. This corresponds to the first candidate for the AR. Based on the convex hull extensions illustrated in section 19.2.1, we now consider an NLP subproblem to check whether a reactor can provide an extension to the candidate AR. Here, we can again consider a recycle reactor extension, since it includes the PFR and CSTR extensions as special cases. Also, in this nonisothermal recycle reactor we assume that the temperature is a control profile along the length of the plug flow section of the recycle reactor. The inlet temperature to this reactor, will also follow a convex combination rule, if intermediate heating or cooling is not permitted. The resulting formulation is similar to the isothermal extension in (P3).

$$\text{Max} \quad J_{rr}(X_{\text{exit}}, \tau_R)$$

$$X_{P6} = \sum_i \sum_j \lambda_{ij} X_{DSRij}$$

$$\frac{dX_{rr}}{dt} = R(X_{rr}, T_{rr}) \tag{P7}$$

$$X_{rr}(t=0) = \frac{R_e X_{\text{exit}} + X_{P6}}{R_e + 1}$$

Sec. 19.2 Reactor Network Synthesis with Targeting Formulations

$$X_{exit} = \Sigma_i \Sigma_j \Delta\alpha_i\, w_j f_{rij} X_{rrij}$$

$$\Sigma_i \Sigma_j \lambda_{ij} = 1.0$$

$$\Sigma_i \Sigma_j f_{r\,ij} = 1.0$$

$$\tau_R < \tau_{max}$$

$$l \le X_{exit} \le u$$

Here, J_{rr} is the value of the objective function at the exit of the recycle reactor; X_{P6} is the concentration vector obtained from the solution of (P6) and λ_{ij} is the convex combiner of all points available from the *DSR* model. The variables T_{rr}, X_{rr}, and R_e represent the temperatures, concentrations, and the recycle ratio, respectively, in the recycle reactor extension. X_{exit} is the vector of exit concentrations from the *RR* reactor and f_r is a linear combiner of all the concentrations from the plug flow section of the recycle reactor.

The nonisothermal synthesis algorithm follows the same scheme as in Figure 19.2, except that (P6) is substituted for (P2), and (P7) is substituted for (P3). Similarly, the next iteration of the nonisothermal algorithm consists of creating the new convex hull of concentrations, which includes the concentrations obtained from (P7) and checking for favorable recycle reactor extensions from this point. Continuing at iteration (p), we substitute (P8) for (P4) and consider the following nonlinear programming problem:

$$\underset{R_e,\, \lambda_{ij},\, f_{model(p)},\, T_{rr}(t)}{\text{Max}} \quad J^{(P+1)}$$

$$\frac{dX_{rr}}{dt} = R(X_{rr}, T_{rr}(t))$$

$$X_{rr}(t=0) = \frac{R_e X_{exit} + X_{update}}{R_e + 1} \qquad (P8)$$

$$X_{update} = \Sigma_i \Sigma_j \lambda_{ij} X_{DSRij} + \sum_{p=1}^{P} f_{model(p)} X_{model(p)}$$

$$X_{exit} = \Sigma_i \Sigma_j f_{rij} X_{rr\,ij}$$

$$\Sigma_i \Sigma_j \lambda_{ij} + \sum_{p=1}^{P} f_{model(p)} = 1.0,\ \lambda_{ij} \ge 0,\ f_{model(p)} \ge 0$$

In (P8), $X_{model(p)}$ is a constant vector representing the concentration at the exit at iteration (p) in the models previously chosen. A convex combination of this vector with the model described by (P6) gives the fresh feed point for the recycle reactor we are looking for, X_{update}. X_{exit} then represents the concentration at the exit of the recycle reactor; and if $J^{(P+1)} > J^{(P)}$, then the earlier model chosen is insufficient and we have found an extension to the candidate AR. The control profiles are $[f, f_{model(p)}]$ and T_{rr}, which are the linear combiners used to provide a convex candidate and the temperature profile in the recycle reactor, respectively. This procedure is repeated at each iteration (p) until no further improvement in the objective function is observed. Finally, it is easy to see that with this approach, the reactor network is synthesized readily from the extensions generated at each iteration. This approach is illustrated in the next example.

EXAMPLE 19.3

Here we maximize the conversion in the catalytic oxidation of sulfur dioxide in fixed bed reactors, which has been investigated by Lee and Aris (1963). Assuming pseudohomogeneous reaction kinetics, we can use the following information:

$$SO_2 + \frac{1}{2}O_2 = SO_3$$

$$R(g, \theta) = 3.6 \cdot 10^6 \left[\exp\left\{12.07 - \frac{50}{1+0.311\theta}\right\} \frac{\{2.5-g\}^{0.5}\{3.46-0.5g\}}{\{32.01-0.5g\}^{1.5}} \right.$$

$$\left. - \exp\left\{22.75 - \frac{86.45}{1+0.311\theta}\right\} \frac{g\{3.46-0.5g\}^{0.5}}{\{32.01-0.5g\}\{2.5-g\}^{0.5}} \right] \quad (19.15)$$

where g is defined as the number of moles of SO_3 formed per unit mass of mixture, θ is defined as $(T - T_0)/J$, T is the temperature, T_0 is 310 K (fresh feed temperature), and $J = 96.5$ K kg/mol. The rate of reaction, $R(g, \theta)$, is defined as in terms of (kgmol of SO_3 produced)/(h-kg catalyst). The extent of reaction for moles/(total mass) of SO_3 formed is limited by the inlet mass flow of SO_2, which is fixed at 2.5 moles/(total mass) SO_2. Lee and Aris assumed adiabatic reactor sections, with cold shot cooling in their optimization. Instead, we maximize the yield of SO_3 without restrictions on the reactor network or the temperature profile.

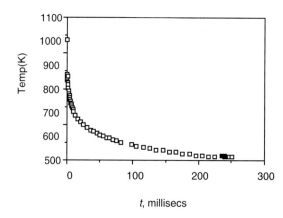

FIGURE 19.9 Temperature profile for Lee-Aris example. [Reprinted with permission from Balakrishna., S., and Biegler, L. T., *Ind. Eng. Chem. Research*, **31**, p. 2152 (1992). Copyright 1992, American Chemical Society]

Solving this example with (P6), we first constrain the residence time to 0.25 secs. The maximum reaction extent of 2.42 for this formulation is obtained in a PFR with the temperature profile shown in Figure 19.9. The resulting optimization problem (P6) required 555 equations and 753 variables and took 1503 CPU secs on a VAX 3200 workstation. Moreover, if the constraint on the residence time is removed, the extent of reaction (as defined by g) asymptotically approaches the upper bound of 2.5 in a PFR with a sufficiently large residence time. For instance, with a residence time bound of 2.2 secs, we obtain an extent of reaction of 2.48. Additional nonisothermal examples have also been considered in Balakrishna and Biegler (1992b).

19.2.3 Improvements to the Targeting Algorithm

The reactor network targeting algorithms described above generally lead to superior networks when applied to literature examples. However, because the algorithm generates only those extensions to the attainable region that improve the objective, it can terminate prematurely. This problem occurs when the extension to a candidate attainable region offers no improvement to the objective function. However, once this extension is added, the candidate attainable region is expanded so that further extensions may improve the objective. To overcome this *nonmonotonic behavior*, we could consider a superstructure of reactor networks. From this superstructure we can develop an MINLP formulation which would then pick the best alternative.

However, as discussed in section 19.1, superstructure approaches by themselves have some drawbacks and it is important to consider *AR* concepts in their development. To motivate this approach, consider the superstructure of Kokossis and Floudas (1990). This superstructure consists of CSTRs or a series of subCSTRs that represent PFRs; the resulting MINLP problem is able to handle complex kinetics for both isothermal and nonisothermal cases. A particular representation of their superstructure for two PFRs and two CSTRs is given in Figure 19.10. The optimization formulation is derived from mass and energy balance equations for the splitters, mixers, recycle streams, and bypass streams. In addition, CSTR equations are introduced for each CSTR or subCSTR. Integer variables are introduced to select both the flow pattern and the number of reactors in the network. Note that the superstructure in Figure 19.10 is particularly rich in that it allows both local and global recycles and bypasses in the optimal network. On the other hand, this formulation leads to a large, complex MINLP. In addition, the authors also demonstrate the interaction of the reactor network with other parts of the process flowsheet. Finally, they also incorporated stability constraints within the MINLP problem in order to avoid the selection of unstable network structures.

Using AR concepts as well as representation of PFRs through collocation on finite elements, we can develop a simpler superstructure and MINLP formulation. Like the targeting algorithm, this MINLP approach still relies on a stagewise construction procedure, but now retains all of the previous solutions in order to allow for nonmonotonic behavior. Here we exploit two properties of the attainable region (Feinberg and Hildebrandt, 1997; Hildebrandt, 1989):

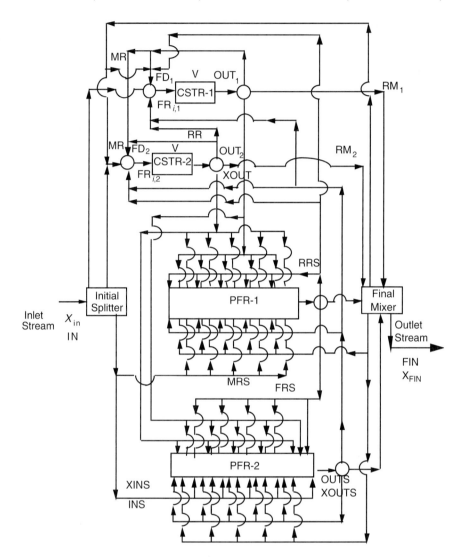

FIGURE 19.10 MINLP superstructure for reactor network synthesis (Kokossis and Floudas, 1990).

- Recycle reactors and networks with recycles across several reactors are not required to form the boundary of the attainable region.
- The attainable region is made up of PFRs, CSTRs, and straight line segments for two-dimensional problems. For higher dimensional problems, DSRs can also form the boundary of the AR.

Sec. 19.2 Reactor Network Synthesis with Targeting Formulations

By incorporating these concepts, we directly generate a superstructure of DSRs and CSTRs. Note that the DSR itself becomes a PFR if the sidestream flow, $q(t)$, is set to zero. This superstructure has several common features with the Kokossis-Floudas superstructure shown in Figure 19.10, but also three important simplifications. First, the PFR models can be represented more concisely and accurately through collocation on finite elements, rather than as subCSTRs. Second, as recycles are unnecessary, the stages or modules in the superstructure require only a series-parallel structure. Third, DSRs with feeds starting from other network points are represented directly in the superstructure. This greatly simplifies the network as it can now be constructed by linking reactor modules. For example, a two-reactor module linkage is shown in Figure 19.11. This pair includes a CSTR and a DSR, and the modules are augmented by splitters, mixers, and bypass streams, also shown in Figure 19.11.

In a similar manner to Figure 19.10, the MINLP formulation can be derived from balance equations for all of the streams as well as the reactor equations. Integer variables are introduced to indicate the presence and types of reactors. Based on these variables, the superstructure allows a full set of bypasses as well as series and parallel reactor structures. A key advantage of this superstructure is that it avoids the *nonmonotonic behavior* that leads to premature termination in the targeting algorithm. Here the algorithm simultaneously considers all reactor networks in the superstructure instead of the sequential strategy in Figure 19.2. As a result, the MINLP formulation retains all of the candidate solutions within the superstructure, even if they do not improve the candidate attainable region. As solution of the MINLP problem (e.g., by the OA algorithm described in Chapter 15 and Appendix A) proceeds and additional reactor modules are considered, these candidates are retrieved as needed.

Based on the structure in Figure 19.11, the MINLP formulation (P9) uses the modules k that are linked together. The number of modules, N, is increased successively and an improving sequence of MINLPs is solved until no further improvement is obtained. The isothermal MINLP formulation is given by:

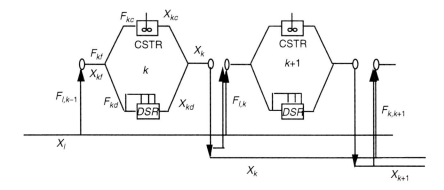

FIGURE 19.11 Overall structure for MINLP targeting model.

Max $J(X_{exit}, \tau_{kc}, \tau_{kd})$

s.t.

$$X_{kc} = R(X_{kc})\,\tau_{kc} + X_{kf}$$

$$dX_{kd}/dt = R(X_{kd}) + (Q_{side,k}\,q_k(t)/Q_k(t))\,(X_{side,k} - X_{kd})$$

$$X_{kd0} = X_{kf}$$

$$X_{kde} = \int_0^{tmax} f_k(t)\,X_{kd}(t)\,dt$$

$$1 = \int_0^{tmax} f_k(t)\,dt$$

$$1 = \int_0^{tmax} q_k(t)\,dt$$

$$\tau_{kd} = \int_0^{tmax}\int_0^t [Q_{side,k}\,q_k(t')/Q_k(t) - f_k(t')]\,dt'$$

$$F_{kf} = \sum_{l=0}^{k-1} F_{l,k-1} \tag{P9}$$

$$F_{kf}X_{kf} = \sum_{l=0}^{k-1} F_{l,k-1}X_l$$

$$F_{kf} = F_{kc} + F_{kd}$$

$$F_{kf}X_k = F_{kc}X_{kc} + F_{kd}X_{kde}$$

$$1 = Y_{kc} + Y_{kd}$$

$$F_{kf}X_k = \sum_{j=k}^{N} F_{kj}X_k,\ k = 1,\ldots N$$

$$X_{exit} = X_N \quad 0 \le F_{kc} \le U\,Y_{kc}$$

$$0 \le F_{kd} \le U\,Y_{kd} \quad Y_{kc} \in \{0,1\} \quad Y_{kd} \in \{0,1\}$$

where the variables

F_{kf} = Flowrate at the inlet of the kth reactor module

$F_{l,k-1}, X_{l,k-1}$ = Flowrate and concentration from the exit of the lth stage which is an inlet stream to the kth stage $l = 0, k-1$

X_{kf} = Concentration at the inlet to the kth reactor module

F_{kc}, F_{kd} = Flowrate of stream passing through the CSTR and DSR in the kth reactor module

$X_{kc\,in}, X_{kc}, X_{kd\,in}, X_{kde}$ = Concentration at the inlet and exit of the CSTR and DSR respectively in the kth reactor module

X_{side}	= Sidestream composition for DSR taken from any network point
Q_{side}	= DSR sidestream flowrate
X_k	= Concentration at the exit of the kth reactor module
Y_{kc}, Y_{kd}	= Binary variable associated with the CSTR and DSR in the ith reactor module

The differential equations and the integrals in (P9) are discretized using collocation and quadrature on finite elements, as shown in Examples 19.1 and 19.2 and this leads to potentially large optimization problems. However, by successively increasing N in the MINLP formulation, we ensure that the problem size remains only as small as needed. To illustrate this approach with (P9) we consider a modification of the van de Vusse problem that exhibits nonmonotonic behavior and achieves a suboptimal network with the targeting algorithm of Figure 19.2. Here we demonstrate how the MINLP (P9) formulation overcomes this problem.

EXAMPLE 19.4

We revisit the van de Vusse reaction of Example 19.1 with altered rate constants. The objective function again is the yield of intermediate species B. The rate vector is given by $R(X) = [-X_A - 20X_A^2, X_A - 2X_B, 2X_B, 20X_A^2]$. In this case, the segregated flow model (P2) gives a yield of 0.061. However, the sufficiency conditions for this formulation are not satisfied as the PFR trajectory is nonconvex. Here the algorithm of Figure 19.2, with recycle reactor extensions (P4), leads to a recycle reactor (recycle ratio = 0.772, τ = 0.1005 sec) in series with a PFR (τ = 0.09 sec) with a yield of 0.069. This is solved using GAMS and CONOPT with a computational time of 0.038 sec on a HP-UX 9000-720 workstation.

Glasser et al. (1987), on the other hand, report a yield of approximately 0.071 with a graphical approach. This solution is given by a CSTR followed by a PFR. The lower yield obtained with the targeting formulation is attributed to *nonmonotonic behavior* of the algorithm. Instead, if we consider the two modules shown in Figure 19.11 for problem (P9), the MINLP problem is represented by 294 continuous variables, 218 constraints and 4 integer variables. Upon solution, a yield of 0.0703 is obtained and the reactor network matches the one obtained by Glasser et al (residence times for CSTR and PFR are 0.302 s and 0.161 s, respectively). Solution of the MINLP problem requires only 0.041 CPU secs on an HP-UX 9000/720 workstation.

19.3 REACTOR NETWORK SYNTHESIS IN PROCESS FLOWSHEETS

In this section we extend the targeting algorithm considered in the previous section to deal with a more general process synthesis problem. Reactor networks are rarely designed in isolation, but rather form an important part of an overall flowsheet. Moreover, since feed preparation, product recovery, and recycle steps in a process are directly influenced by the reactor network, the synergy among these subsystems is a key factor in establishing an optimum process. Because of reactant recycling, overall conversion to product is influenced by selectivity to desired products rather than reactor yield, as noted by Conti and Paterson (1985). Douglas (1988) extends this notion of process and reactor interactions by establishing trade-offs among conversion of raw materials, capital costs, and operating costs. Here, although selectivity maximization leads to optimum overall conversion to product, capital and operating costs affected by high recycles can improve if reactor yield is increased instead. Hence, to balance these trade-offs, Douglas suggests a reactor network that operates between maximum yield and maximum selectivity.

A geometric approach to reactor/flowsheet integration was developed by Omtveit and Lien (1993) where separations and recycles were incorporated into the construction of the attainable region. Here, geometric constructions need to be performed iteratively as the reactor feed is unknown in the optimum flowsheet. Omtveit and Lien (1994) therefore construct a family of attainable regions and use constraints due to reaction limitations to represent this problem in only two dimensions. This approach was demonstrated on the HDA process (Douglas, 1988) as well as methanol synthesis. In both problems the optimal reactor turned out to be a plug flow reactor and quantitative trade-offs were established between the purge fraction, reactor yield, and economic potential.

While the qualitative concepts mentioned above yield useful insights for process integration, many quantitative evaluations, along with discrete and continuous decisions, still have to be made. A natural way to account quantitatively for process trade-offs and to represent the interactions of process subsystems is to develop targeting models based on NLP and MINLP formulations. Again, as with reactor network targeting, the goal of these formulations is to predict process performance without explicitly developing the network itself. Consequently, AR concepts are extremely useful here and dimensionality limitations can be overcome through the NLP formulations. In this section we first consider an NLP formulation for flowsheet integration on the Williams-Otto process. Following this, a more comprehensive nonisothermal example is considered that involves flowsheet integration and the synthesis of heat exchanger networks.

19.3.1 Targeting Strategy Integrated with Process Flowsheet

The targeting approach, coupled with the simultaneous solution strategy presented before, allows for integration of the reactor with the flowsheet. Though this integrated approach is independent of a particular reactor network, it is effective because the capital cost of the reactor is generally low compared to raw material and downstream processing costs. Here, we replace the reactor within the flowsheet by our targeting model. For integration within a process flowsheet, the reactor feed concentration usually cannot be specified but is defined by the flowsheet constraints. Therefore, the differential equations in (P1) cannot be solved offline, but have to be treated simultaneously with the optimization problem. As with the recycle reactor extension problem (P4), this is done through discretization using collocation on finite elements.

For the objective function, the capital cost for the reactor is approximated as a function of the residence time. The initial and final conditions for this model may be related through the variables of the flowsheet such as the feed rate, recycle ratio, and so on. Using the stagewise approach with formulations (P4) or (P8) augmented by the flowsheet equations and the algorithm sketched in Figure 19.2, we can find the best reactor network for all initial concentrations dictated by the constraints imposed by the flowsheet.

EXAMPLE 19.5

Consider the Williams and Otto (Williams and Otto, 1960) flowsheet problem, which has already been presented in Chapters 8 and 9. The flowsheet for this problem is shown in Figure 19.12.

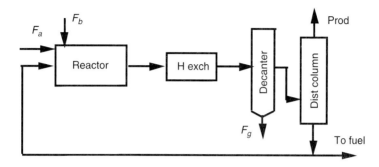

FIGURE 19.12 Williams and Otto flowsheet [Reprinted with permission from Balakrishna., S., and Biegler, L. T., *Ind. Eng. Chem. Research*, **31**, p. 300 (1992). Copyright 1992, American Chemical Society].

The plant consists of a reactor, a heat exchanger to cool the reactor effluent, a decanter to separate a waste product G, and a distillation column to separate product P. A portion of the bottom product is recycled to the reactor, and the rest is used as fuel. The plant model can be defined without an energy balance and we further simplify this problem to consider only isothermal reactions for the manufacture of compound P. These are given by:

$$A + B \to C$$
$$C + B \to P + E$$
$$P + C \to G$$

The rate vector for components A, B, C, P, E, G, respectively, is given by

$$R(X) = [-k_1 X_A X_B; -(k_1 X_A + k_2 X_C) X_B; 2k_1 X_A X_B - 2k_2 X_B X_C - k_3 X_P X_C; \\ k_2 X_B X_C - 0.5 k_3 X_P X_C; 2k_2 X_B X_C; 1.5 k_3 X_P X_C] \tag{19.16}$$

where

$k_1 = 6.1074 \ h^{-1}$ wt fraction^{-1}.
$k_2 = 15.0034 \ h^{-1}$ wt fraction^{-1}.
$k_3 = 9.9851 \ h^{-1}$ wt fraction^{-1}.

Here the X's denote the weight fractions of the components. F_A and F_B are the flowrates of fresh A and B, respectively; F_G is the flowrate of waste G; and F_P is the fixed exit flowrate of pure P out of the plant. Previous researchers have solved this problem by assuming the reactor to be a CSTR and maximizing the rate of return on investment. Here we replace the CSTR by the segregated flow targeting model embedded within the flowsheet. The objective function, the return on investment (ROI), includes all raw material and separation costs for the entire plant and an optimal ROI value of 130% is typically obtained for this problem with the fixed CSTR model. With a segregated flow model integrated within the flowsheet, an ROI of 278% is obtained. We now look for CSTR extensions from the one-compartment model by solving (P4) for a CSTR extension by including all the constraints imposed by the flowsheet. No CSTR extensions that improve the ROI are observed. Therefore, the optimal network is just a PFR with a residence time of 0.0111 hr. Modeled in GAMS, the overall formulation requires 153 variables and 133 constraints. It solves on a Sun 3 workstation in 397 CPU secs. Moreover, optimality of this network was verified by the MINLP targeting approach in section 19.2.3. These results indicate that significant savings can be obtained by integrating the reactor with the flowsheet, even with very simple targeting models.

19.3.2 Energy Integration of Reactor Networks

As discussed in Chapter 16, algorithms for the "isolated" construction of heat exchanger networks (HENs) are well known. However, the synergy among process subsystems is a key area for the exploitation of energy integration. Reactor networks, in particular, are associated with significant heat effects and strongly influence the behavior of other subsystems. In this subsection, we address integration of the heat effects within the reactor with the rest of the process and demonstrate the effectiveness of the optimization formulations in section 19.2.

We consider two approaches for the integration of the reactor and energy network, the *sequential* and the *simultaneous* formulations. In the conventional sequential approach, the reactor and separator schemes appear at a higher level compared to energy integration. In other words, once the "optimal" flowsheet parameters have been determined for the reactor target and the separation system, the reactor network is realized, and the heat exchanger network is derived in a straightforward manner. However, it is well known that this approach can be suboptimal with respect to the overall flowsheet (see Chapter 18).

For the simultaneous approach, we consider both reactor network synthesis and energy integration at the same level. This approach considers the strong interaction between the chemical process and the heat exchanger network, but it is not a trivial problem. Here, unlike the approach in Chapter 15, the flowrates and the temperatures for heat integration are not known in advance. Moreover, we consider general nonisothermal reacting systems and a general temperature profile within the reactor. As a result, the streams within the reactor cannot be classified as hot or cold streams a priori, because the optimal temperature trajectory within the reactor is unknown. Instead, we discretize the temperature trajectory in the extended DSR model (P6), and introduce the concept of candidate streams within the reactor network. Here, we approximate the optimal temperature trajectory with piecewise constant segments. Temperature changes occur between these segments as shown in Figure 19.13. Here the curve represents the actual temperature profile. The piecewise constant segments represent the approximation; the horizontal lines represent isothermal reacting segments, while the vertical lines represent the temperature changes needed to follow an optimal trajectory.

Sec. 19.3 Reactor Network Synthesis in Process Flowsheets

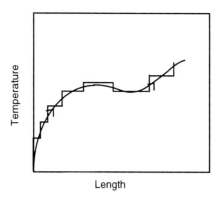

FIGURE 19.13 Piecewise constant approximation of optimal temperature profile [Reprinted with permission from Balakrishna., S., and Biegler, L. T., *Ind. Eng. Chem. Research*, **31**, p. 2152 (1992). Copyright 1992, American Chemical Society]

In the heat integration, the isothermal horizontal segments correspond either to hot streams or cold streams, depending on whether the reaction is exothermic or endothermic. The vertical sections require heating or cooling in the reactor; therefore, we assume the presence of both heaters and coolers between the reacting segments. Also, we term these hot or cold streams *candidate* streams, because they may or may not be present in the heat exchanger network. This will depend on the number of reacting segments and hence the corresponding temperature profiles. Figure 19.14 shows the reactor representation corresponding to the above approximation. Note that this is a straightforward extension of the extended DSR model in Figure 19.8 but also includes heat exchangers between elements.

Again the subscript i refers to the ith finite element corresponding to the discretization and T^i_{mix} corresponds to the temperature after mixing the reacting stream with the feed. T^i_{hin}, T^i_{hout} are the temperatures of the streams entering and leaving the cooler, and t^i_{cin}, t^i_{cout} are the temperatures of the streams entering and leaving the heater. In the optimal heat exchanger network, at most one of these two heat exchangers will be chosen since only cooling or heating will be needed. Also, $\Delta\alpha_i$ corresponds to the length of the finite element, which may also be variable in the optimization problem (subject to con-

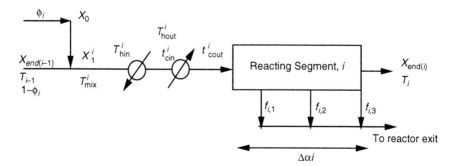

FIGURE 19.14 Reacting segment for heat integration [Reprinted with permission from Balakrishna., S., and Biegler, L. T., *Ind. Eng. Chem. Research*, **31**, p. 2152 (1992). Copyright 1992, American Chemical Society]

straints on approximation error). Thus, our heat integration problem is now defined since we know the hot and cold streams a priori, even if the flow rates and the temperatures are not known. Also, some amount of temperature control can be achieved by mixing of process streams in Figure 19.14. Otherwise, the temperature profile is determined by the utilities or the heat flows within the network. Using the framework for reactor targeting from above, we now integrate this within a suitable energy targeting framework. In our optimization formulation we assume that utility costs will dominate capital costs in the HEN and this solution will be adequate for preliminary design. However, overall capital cost and area estimates for the HEN can also be included into the objective function if desired.

In Chapter 18, analytical expressions were derived for minimum utility consumption as a function of flow rates and temperatures of the heat exchange streams. From a set of hot and cold streams we consider pinch point candidates as the inlets of these streams and, as in Chapter 10, we define an approach temperature, ΔT_m, for heat integration. Now the minimum heating utility consumption is given by $Q_H = \max(z_H^p)$, where, z_H^p is the difference between the heat sources and sinks above the pinch point for each pinch candidate p. Therefore, for hot and cold streams with inlet temperatures given by T_h^{in} and t_c^{in}; and outlet temperatures T_h^{out} and t_c^{out} respectively, $z_H^p(y)$ is given by

$$z_H^p(y) = \sum_{c \in C} (FCp)_c [\max\{0; t_c^{out} - \{T_p - \Delta T_m\}\} - \max\{0; t_c^{in} - \{T_p - \Delta T_m\}\}] - \sum_{h \in H} (FCp)_h [\max\{0; T_h^{in} - T_p\} - \max\{0; T_h^{out} - T_p\}] \quad (19.17)$$

for $p = 1, N_p$, where N_p is the total number of heat exchange streams. Here, the temperatures T_p in Eq. (19.17) correspond to all the candidate pinch point temperatures, which are the inlet temperatures for all hot streams and the inlet temperatures ($+ \Delta T_m$) for the cold streams. $(FCp)_c$ and $(FCp)_h$ are the heat capacity flows for the hot and cold streams, and the vector y represents all of the variables in the reactor and energy network. Finally, the minimum cooling utility is given by a simple energy balance as $Q_C = Q_H + \Omega(y)$, where $\Omega(y)$ is the difference in heat content between the hot and the cold process streams. It is defined by:

$$\Omega(y) = \Sigma_h (FCp)_h (T_h^{in} - T_h^{out}) - \Sigma_c (FCp)_c (t_c^{out} - t_c^{in}) \quad (19.18)$$

The above concepts for reactor and energy network synthesis now lead to a simultaneous reactor-energy synthesis formulation. We first classify the process streams into four sets; the sets H_R and C_R represent the hot and cold streams, respectively, associated with the reactor network and H_P, C_P represent the hot and cold streams, respectively, in the process flowsheet. These sets have the elements $h \in H = H_R \cup H_P$, and $c \in C = C_R \cup C_P$. If the reactor network is modeled by NE isothermal reacting segments, and if the reaction is exothermic, then we have NE hot reacting streams in H_R from which the heat of reaction is to be removed in order to maintain a desired temperature in each segment. Conversely, for an endothermic system, we have NE cold reacting streams in C_R from which the heat of reaction needs to be added to maintain the desired temperature.

Sec. 19.3 Reactor Network Synthesis in Process Flowsheets

Also, between the elements, there are hot and cold streams corresponding to the discretization shown in Figure 19.14 (the vertical distances). Hence, for exothermic systems, H_R is a set of cardinality 2NE, while the set C_R has NE elements. For an endothermic system, C_R and H_R have cardinalities in the reverse order, as the reacting segments now correspond to cold streams. Therefore, we always have 3NE candidate streams. We further define F_h and F_c as the mass flow rates of the hot and cold streams respectively and F_i denotes the mass flow at the entry point of reacting segment i. F_0 is the total inlet flow into the reactor and the heat capacities, Cp, in our formulation are allowed to be temperature dependent. In addition, the vector ω constitutes the remaining variables in the flowsheet. Based on these assumptions, a simultaneous reactor-energy synthesis can be obtained by extending (P7) to incorporate the heat integration model in Chapter 18 for the reactor network and flowsheet. This leads to the following nonlinear programming problem:

Max $\quad \Phi(\omega, y, Q_H, Q_C) = J(\omega, y) - c_H Q_H - c_C Q_C$

s. t. $\quad \sum_k X_{ik} L_k'(\alpha_j) - R(X_{ij}, T_{ij}) \Delta\alpha_i = 0 \quad j = 1, K$

$X(0) = X_0(\omega, y)$

$X_{iend} = \sum_k X_{ik} L_k(t_{end})$

$X_{i,0} = \phi_i X_0 + (1-\phi_i) X_{(i-1)end}$

$X_{exit} = \sum_i \sum_j X_{ij} f_{ij}$

$\sum_{ij} (f_{ij} - q_{ij}) \alpha_{ij} = \tau$

$\sum_{ij} f_{ij} = 1, f_{ij} \geq 0, q_{ij} \geq 0$ \hfill (P10)

$F_{(0)} = \phi_0 F_0$

$F_{(ij)} = \sum_i \phi_i F_{(i,0)} - \sum_{ij} f_{ij} F_0$

$Q_C = Q_H + \sum_{h \in H} (FCp)_h [T_h^{in} - T_h^{out}] - \sum_{c \in C} (FCp)_c [t_c^{out} - t_c^{in}]$

$Q_H \geq z_H p(y)$

$h(\omega, y) = 0$

$g(\omega, y) \leq 0$

Here, T_p corresponds to the pinch candidates that are derived from T_h^{in} for the hot streams, and $t_c^{in} + \Delta T_m$ for the cold streams. The heats of reaction are directly accounted for by the definition of (FCp) of the reacting streams, as follows. If Q_R is the heat of reaction to be removed (or added, for endothermic reactions) to maintain an isothermal reacting segment, the equivalent $(FCp)_h$ (or $(FCp)_c$) for this reacting stream is equated to Q_R, and we assume a 1 K temperature difference for this reacting stream. Finally, the constraints $h(\omega, y)$ and $g(\omega, y)$ are derived from interactions of the flowsheet with the heat integration and reactor networks.

In (P10) it should be noted that the max(0,Z) functions, which make up the $z_H{}^p(y)$ relations and have a nondifferentiability at the origin—this can lead to failure of the NLP solver. Here we approximate max (0,Z) as shown in Appendix B, using:

$$f(Z) = \max(0, Z) = \frac{(Z^2 + \varepsilon^2)^{0.5}}{2} + Z/2 \tag{19.19}$$

With a value of $\varepsilon = 0.01$, we obtain a good approximation to the max function for (P10).

If we use the algorithm in Figure 19.2 for the reactor network, then solution of problem (P10) gives us only a lower bound on the best objective function for the flowsheet. This is because the DSR model may not be sufficient for the network, and we need to check if there are reactor extensions that improve our objective function beyond (P10). As in formulations (P4) and (P7), we can therefore check for CSTR (or RR) extensions from the convex hull of the DSR model. This algorithm is similar to Figure 19.2, except that now all the flowsheet constraints must be included. As a result, for this simultaneous reactor energy synthesis, the dimensionality of the problem increases with each extension of the network, because the heat effects in the reactor affect the heat integration of the process streams. In order to keep the problem formulation simple, we consider CSTR extensions only. The CSTR extension to the convex hull of the DSR leads to the addition of the following relations to (P10) and we now maximize $\Phi^{(2)}$ instead of Φ:

$$\begin{aligned} \text{Max} \quad & \Phi^{(2)}(\omega, y^{(2)}, Q_H, Q_C) = J(\omega, y^{(2)}) - c_H Q_H - c_C Q_C \\ \text{s.t.} \quad & X_{cstr} = X_{exit} + R(X_{cstr}, T_{cstr}) \tau_{cstr} \\ & \tau \geq 0, X_{cstr} \geq 0 \end{aligned} \tag{P11}$$

Here, X_{cstr} corresponds to the concentration from the new reactor extension and $y^{(2)}$ is the vector of new variables in the reactor and energy network. In addition to the variables ω and y in (P10), we include the variables corresponding to the new CSTR extension, namely, X_{cstr}, T_{cstr}, τ_{cstr}, as well as three more candidate streams for heat exchange. This is because we add two heat exchangers that will either cool or heat the feed to the CSTR (only one of these will exist in the optimal network) and an additional exchanger within the CSTR to maintain a desired temperature. As in Figure 19.2, if $\Phi^{(2)*} \geq \Phi^*$, we have a reactor extension that improves the objective function. We continue this procedure with a new convex hull of concentrations, and then check for extensions that improve our objective function within the flowsheet constraints. As in section 19.2, we terminate this procedure when there are no extensions that improve the objective function.

Finally, as seen from section 19.2.3, this approach can also be improved (and nonmonotonic behavior due to Figure 19.2 can be avoided) by using the direct MINLP formulation based on an extension of (P9).

EXAMPLE 19.6

To illustrate the simultaneous synthesis of reactor and energy networks, we consider the process flowsheet shown in Figure 19.15. Here, we consider a van de Vusse reaction mechanism but with nonisothermal kinetic expressions different from those used in Examples 19.1 or 19.4.

Sec. 19.3 Reactor Network Synthesis in Process Flowsheets

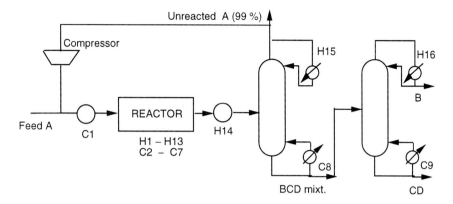

FIGURE 19.15 Flowsheet for reactor-energy network synthesis [Reprinted with permission from Balakrishna., S., and Biegler, L. T., *Ind. Eng. Chem. Research*, **31**, p. 2152 (1992). Copyright 1992, American Chemical Society]

The process feed consists of pure A and this is mixed with the recycle gas stream consisting of almost pure A. The combined stream is preheated (C1) before entering the reactor and after reaction, the mixture of A, B, C, and D passes through an aftercooler prior to separation of the raw material from products. In the first distillation column, A is recovered and recycled overhead, while in the second column, the desired product B is separated from C and D, which are used as fuel. The distillation columns are modeled to operate with a constant temperature difference between reboiler and condenser temperatures (Andrecovich and Westerberg, 1985). The reflux ratios in the column models are fixed and the column temperatures are functions of the column pressures, which are allowed to vary so that efficient heat integration can be attained between the distillation columns and the rest of the process. The reactions involved in this process are given by:

$$A \xrightarrow{k_1} B \xrightarrow{k_2} C$$
$$A \xrightarrow{k_3} D$$

where

$k_i = k_{i0} \exp(-E_i/RT)$
$k_{10} = 8.86 \times 10^6 \ h^{-1}$
$k_{20} = 9.7 \times 10^9 \ h^{-1}$
$k_{30} = 9.83 \times 10^3 \ lit\text{-mol}^{-1} \ h^{-1}$
$E_1 = 15.00$ kcal/gmol
$E_2 = 22.70$ kcal/gmol
$E_3 = 6.920$ kcal/gmol
$\Delta H_{A \to B} = -0.4802$ kcal/gmol
$\Delta H_{B \to C} = -0.918$ kcal/gmol
$\Delta H_{A \to D} = -0.792$ kcal/gmol of A.

The extended DSR reactor is represented by the discretization shown in Figure 19.14. This model has seven reactor segments (NE = 7) with uniform segment lengths, $\Delta\alpha_i$. Since the reaction is exothermic, we obtain 14 hot streams and 7 cold streams. Thus, the streams in the reactor may be enumerated as hot streams H1–H13 (2NE − 1 segments are required since the entry point is fixed by a preheater), and cold streams C1–C7. Also, the streams H15–H16 and C8–C9 correspond to the condensers and reboilers of the distillation columns. As described in (P10), the heat capacities Cp for these hot and cold streams are assumed to be linear with temperature. Finally, the objective function for this example is the total plant profit given in simplified form by:

$$J = 1.7F_B + 0.8F_{CD} - 6.95 \times 10^{-5}\tau F_0 - 0.4566F_B(1 + 0.01(T^{in}_{H15} - 320)) \\ - 0.7(F_B + F_{CD}) - 0.2F_{A0} - 0.007Q_C - 0.08Q_H \quad (19.20)$$

In this process model, F_B and F_{CD} represent the production rates of product B and the by-products C and D, while F_{A0} is the flow rate of fresh feed. A target production rate of 40000 lb/hr is assumed for the desired product B. The third term in the objective Eq. (19.20) corresponds to the reactor capital cost, which is assumed proportional to τ, the residence time, and F_0, the total reactor feed. We assume that the cost of the reactor is independent of the reactor type and this assumption can be justified because the capital cost of the reactor itself is usually an order of magnitude or more smaller than capital costs of the other major units. The fourth and the fifth terms in Eq. (19.20) correspond to the capital cost of the distillation columns and the operating costs of the columns are directly incorporated into the energy network in terms of condenser and reboiler heat loads. Further details of this process can be found in Balakrishna and Biegler (1992b).

For this process we now consider two cases. First, we consider a sequential approach, where the reactor network is optimized first and we then determine the heat exchanger network that maintains this optimal profile, integrated with the energy flows in the rest of the flowsheet. In the second case, we consider the simultaneous formulation proposed in (P10).

The optimization model for the sequential case has 342 equations and 362 variables for the reactor and flowsheet optimization. Finding the optimal reactor network requires 96 CPU secs on VAX 6320. Using the formulation in Chapter 18, the energy integration was modeled with 200 equations and 161 variables for the energy integration; 170 CPU secs were required for solution. On the other hand, the simultaneous optimization model (542 equations, 523 variables) was solved in 1455 CPU secs after it was initialized with the solution from the sequential model. Table 19.2 provides a brief comparison between the results for sequential and simultaneous cases for reactor and heat exchanger network synthesis.

From Table 19.2 it is clear that the simultaneous formulation leads to a significant improvement in the overall profit. This is accompanied by an increased conversion due to the correct anticipation of the energy costs in the reactor design. Note that the shape of the temperature profiles is not markedly different. However, the temperatures in the simultaneous case are lower, as seen in Figure 19.16. This lower temperature leads to a reduction in the degradation of product B to by-product C, as seen from Table 19.2. Since the B–C reaction is the most exothermic, a lower reaction rate leads to less heat evolved and less cold utility consumed. Furthermore, more efficient conversion to B leads to less consumption of raw material A, and higher overall conversion for the simultaneous case. Coincidentally, the optimal reactor in both sequential and simultaneous cases (a nonisothermal PFR) has the same residence time of 0.59 secs. However, note that since the temperatures are lower in the simultaneous case, the conversion per pass of A is also lower, thus leading to higher recycles in the simultaneous case. Finally, of the 20 candidate streams for heat integration, only 12 are actually used in the optimal network. This is because the strictly falling temperature profile in the reactor avoids the use of any cold streams (C2–C7) within the reactor network.

Sec. 19.3 Reactor Network Synthesis in Process Flowsheets

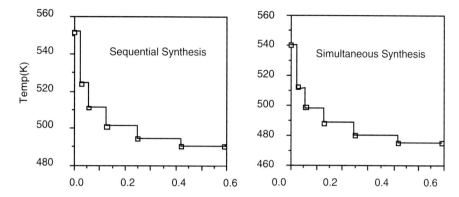

FIGURE 19.16 Reactor temperature profiles [Reprinted with permission from Balakrishna., S., and Biegler, L. T., *Ind. Eng. Chem. Research,* **31,** p. 2152 (1992). Copyright 1992, American Chemical Society]

TABLE 19.2 Comparison between Sequential and Simultaneous Formulations

	Sequential	Simultaneous
Overall Profit	$38.98 \times \$10^5$/yr	$74.02 \times \$10^5$/yr
Overall Conversion	49.6 %	61.55%
Hot utility load	3.101×10^5 BTU/hr	2.801×10^5 BTU/hr
Cold utility load	252.2×10^6 BTU/hr	168.5×10^6 BTU/hr
Fresh Feed A	8.057×10^4 lb/hr	6.466×10^4 lb/hr
Degraded Product C	3.112×10^4 lb/hr	1.44×10^4 lb/hr
By-Product D	0.933×10^4 lb/hr	1.00×10^4 lb/hr
Unreacted (Recycled) A	1.22×10^4 lb/hr	1.963×10^4 lb/hr

Also, no further extensions were observed to this reactor network by solving (P11), for either the sequential or the simultaneous cases, within the constraints on the residence time ($\tau^{up} = 1.00$ s). The final network is therefore just a PFR, and cold shot cooling, allowed in formulation of (P10), was not used at all. However, this decision is directly influenced by the ratio of the raw material to energy costs. If the energy cost is high, it may lead to the use of cold shots in the reactor in order to reduce utility consumption, even if mixing lowers the product yield.

Finally, the pinch points correspond to 546.5 K and 535.1 K for the sequential and simultaneous schemes, respectively, and the heat contents for the hot streams are significantly higher than those for the cold streams. Thus, the T–Q curves for the hot streams will be nearly horizontal in the process. In addition, the pinch point corresponds to the inlet temperature of the hottest hot stream and in either case, no part of the T–Q curve for the hot streams will extend beyond the pinch. Therefore, the matches below the pinch are easy to make because of the large temperature difference between hot and cold streams. Here, streams C1, C8, and C9 can be matched with any of the streams from H1 to H11, without any alteration in the utility consumption. The resulting network is thus innately flexible, and this is due to the large heat effects in the reactor. One possible set of matches for the heat exchanger network corresponds to the cold streams C1, C8, and C9 diverted to suitable jacketed reactor compartments, as shown in Figure 19.17.

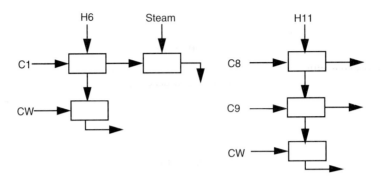

FIGURE 19.17 Heat exchanger network substructure [Reprinted with permission from Balakrishna., S., and Biegler, L. T., *Ind. Eng. Chem. Research*, **31**, p. 2152 (1992). Copyright 1992, American Chemical Society]

The remaining hot streams from the reactor are not shown in the above network as they are matched directly with cooling water (CW). Also the amount of steam used in this process is very small. The network in Figure 19.17 requires the same minimum utility consumption predicted by the solution of (P10). This network is equally suitable for both simultaneous and sequential solutions. In fact, if we have an exothermic reacting system where the reactor temperature is the highest process temperature, the pinch point is often known a priori as the highest reactor temperature (in this case, the feed temperature) and the inequality constraints in (P10), $Q_H \geq z_H p(y)$, $p \in P$, can be replaced by a simple energy balance constraint. This simplification greatly reduces the computational effort to solve (P10).

19.4 SUMMARY AND FURTHER READING

In this chapter we extend the mathematical programming approach developed in previous chapters to the synthesis of chemical reactor networks. Previous reactor network synthesis approaches based on mathematical programming rely on general superstructure optimization formulations. However, the limitations of these stem from solutions that may be local or nonunique and that are only as complete as the superstructure itself. To address these issues, geometric approaches based on attainable region (AR) concepts have been developed and were discussed in Chapter 13. There an attainable region in concentration space was constructed that cannot be extended with further mixing and/or reaction. This geometric approach leads to important insights into the structure of the optimal network, but its construction is currently based on graphical tools and two- or three-dimensional problem representations. Nevertheless, these AR concepts are quite useful when incorporated within a mathematical programming framework.

Consequently, the reactor network synthesis approach in this chapter addresses the drawbacks of the superstructure and graphical AR techniques through a constructive, *optimization-based* targeting strategy. This approach proceeds by considering simplified reactor models and applies the concept of attainable regions to verify the sufficiency of these models. The main idea in this targeting approach is that we develop optimization problem formulations that allow us to explore the attainable region in higher dimensions. We first start with the segregated flow limit to this model, which can often be solved through a simple linear program. The example problems in section 19.2 show that the segregated flow model can be sufficient to describe the network. When the segregated flow model is not sufficient, simple nonlinear programs can be solved to enhance the target. These include the extension of the attainable region with additional CSTR or recycle reactors. Alternatively, an MINLP formulation is postulated that combines these DSR and CSTR models within a compact superstructure. Based on the properties of Feinberg and Hildebrandt, this superstructure does not require recycle streams in the reactor network. Most of these optimization formulations require the discretization of differential equations using collocation on finite elements. This was illustrated in Example 19.2 and more information on this method can be found in Ascher et al. (1988).

The extension of this approach to nonisothermal systems follows simply by considering temperature as an additional control profile. Here we extend our optimization formulations to maintain this optimal temperature profile. We accomplish this by postulating a differential sidestream reactor (DSR) model as the initial targeting model, since this allows for temperature control through feed mixing as well. In contrast to isothermal synthesis, the variable temperature profile in the initial DSR representation itself encompasses a larger choice for the AR. In fact, we often observe that without temperature constraints, the conversion asymptotically can approach a stoichiometric upper bound for some systems. In section 19.2.4, this was illustrated by the Lee-Aris sulfur-dioxide oxidation, where the extent of reaction asymptotically approaches the upper bound through manipulation of the temperature profile.

The optimization formulations for reactor network synthesis also allow us to address the interaction of the reactor design on the other process subsystems within the flowsheet. In section 19.3 we consider the integration of the reactor network synthesis algorithm with other parts of the process including the process recycle and the heat exchanger network. In particular, reactors with significant heat effects allow for very efficient integration of with energy networks. Here we provide a general formulation for the integration of the reactor targeting formulation with an energy targeting scheme, based on minimum utility costs. The results for a small process flowsheet with van de Vusse kinetics indicate that significant increases in profit can be obtained by considering the reactor and energy subsystems within a unified framework. Also, for this example high reaction exothermicities lead to a very flexible heat exchanger network, as described in section 19.3.2.

19.4.1 Guide to Further Reading

The optimization-based approach for reactor network synthesis can be traced to Aris (1961) where dynamic programming was applied to a series of reactors. This concept was further investigated by Horn and Tsai (1967), Jackson (1968), and Ravimohan (1971) through the analysis of optimal control policies. More recently, Waghmere and Lim (1981) exploited analogies between the synthesis of optimal reactor networks and the optimization of feeding policies in batch reactors.

At an algorithmic level, the superstructure approach was also advanced and summarized by Hartmann and Kaplick (1990) as well as through direct search optimization of serial recycle reactors (Chitra and Govind, (1985). More efficient uses of nonlinear programming to solve superstructure problems were also developed by Pibouleau, Floquet, and Domenech (1988) and Achenie and Biegler (1990). Finally, the most comprehensive superstructure approach is described in several studies by Kokossis and Floudas (1990), where sophisticated mixed integer nonlinear programming (MINLP) strategies were applied to a large reactor network. These authors also extended this approach to a number of interesting cases including interactions with the separation and recycle system (Kokossis and Floudas, 1991), nonisothermal systems (Kokossis and Floudas, 1994a) and ensuring stability of the optimal reactor network (Kokossis and Floudas, 1994b).

Concepts for the construction of two-dimensional and three-dimensional regions have been firmly established by the work of Glasser, Hildebrandt, and coworkers. For higher dimensional systems, Feinberg and Hildebrandt (1997) have rigorously established a number of properties that lead to useful insights for processes with reaction and mixing. In particular, they showed that the boundary of the attainable region is made up of PFR trajectories and straight line segments. As a result, all points on this boundary can be found by a combination of PFRs, CSTRs, and differential sidestream reactors (DSRs). However, constructive procedures for higher dimensional attainable regions that incorporate these properties still need to be developed.

Finally, further research for integrated reactor network synthesis includes the design of reactive separation processes. Exploiting the strong integration of reaction and separation processes can lead to significant improvements and savings in the design of new processes and this has led to dramatic industrial successes (Agreda et al., 1990). A preliminary approach for identifying the potential for coupling these reaction and separation processes is developed in Balakrishna and Biegler (1993) and Lakshmanan and Biegler (1996), but detailed phenomena still need to be modeled carefully with this approach. Often the nonlinearity and complexity of the reaction and phase equilibrium models make this problem very difficult. Nevertheless, as with reactor networks, geometric insights can lead to simplification of the synthesis procedure as well as refinement of the optimization formulation. Finally, reactor network synthesis can be applied to a number of design problems in the synthesis of waste minimizing flowsheets. In fact, the approaches described in this chapter have been applied directly to these problems, simply by considering waste minimization as part of the objective function. Lakshmanan and Biegler (1995) recently considered this problem and established trade-offs in reactor targeting between profitability and waste generation in the overall process.

REFERENCES

Achenie, L. E. K., & Biegler, L. T. (1986). Algorithmic synthesis of chemical reactor networks using mathematical programming. *I & EC Fund*, **25,** 621.

Achenie, L. E. K., & Biegler, L. T. (1988). Developing targets for the performance index of a chemical reactor network. *I & EC Research*, **27,** 1811.

Achenie, L. E. K., & Biegler, L. T. (1990). A superstructure based approach to chemical reactor network synthesis. *Comput. Chem. Engg.*, **14**(1), 23.

Agreda, V. H., Partin, L. R., & Heise, W. H. (1990). High purity methyl acetate via reactive distillation. *Chem. Eng. Prog.*, **86** (2).

Andrecovich, M. J., & Westerberg, A. W. (1985). An MILP formulation for heat integrated distillation sequence synthesis. *AIChE J.*, **31**, 363.

Aris, R. (1961). *The Optimal Design of Chemical Reactors.* New York: Academic Press.

Ascher, U., Mattheiij, R., & Russell, R. (1988). *Numerical Methods for the Solution of Boundary Value Problems for Ordinary Differential Equations.* Englewood Cliffs, NJ: Prentice Hall.

Balakrishna, S. (1992). PhD Thesis, Carnegie Mellon University, Pittsburgh, PA.

Balakrishna, S., & Biegler, L. T. (1992a). A constructive targeting approach for the synthesis of isothermal reactor networks. *Ind. Eng. Chem. Research,* **31**, 300.

Balakrishna, S., & Biegler, L. T. (1992b). Targeting strategies for the synthesis and heat integration of nonisothermal reactor networks. *Ind. Eng. Chem. Research,* **31**, 2152.

Balakrishna, S., & Biegler, L. T. (1993). A unified approach for the simultaneous synthesis of reaction, energy and separation systems. *Ind. Eng. Chem. Research,* **32**, 1372.

Carnahan, B., Luther, C., & Wilkes, J. (1969). *Applied Numerical Methods.* New York: Wiley.

Chitra, S. P., & Govind, R. (1985). Synthesis of optimal serial reactor structure for homogenous reactions, Part II: Nonisothermal reactors. *AIChE J.*, **31**(2), 185.

Conti, G. A. P., & Paterson, W. (1985). Chemical reactors in process synthesis. *Process Systems Engineering '85.* I ChemE Symp. Ser. # 92, 391.

Douglas, J. M. (1988). *Conceptual Design of Chemical Processes.* New York: McGraw-Hill.

Duran, M. A., & Grossmann, I. E. (1986). Simultaneous optimization and heat integration of chemical processes. *AIChE J.,* **32**, 123.

Feinberg, M., & Hildebrandt, D. (1997). Optimal reactor design from a geometric viewpoint: I. Universal properties of the attainable region, *Chem Eng. Sci.,* to appear.

Glasser, D., Crowe, C., & Hildebrandt, D. (1987). A geometric approach to steady flow reactors: The attainable region and optimization in concentration space. *I & EC Research,* **26**(9), 1803.

Hartmann, K., & Kaplick, K. (1990). *Analysis and Synthesis of Chemical Process Systems.* Amsterdam: Elsevier.

Hildebrandt, D. (1989). PhD Thesis, University of Witwatersrand, Johannesburg, South Africa.

Horn, F. J. M., & Tsai, M. J. (1967). The use of adjoint variables in the development of improvement criteria for chemical reactors. *J. Opt. Theory and Applns.,* **1**(2), 131.

Jackson, R. (1968). Optimization of chemical reactors with respect to flow configuration. *J. Opt. Theory and Applns.*, **2**(4), 240.

Kokossis, A. C., & Floudas, C. A. (1990). Optimization of complex reactor networks—I. Isothermal operation. *Chemical Engineering Science,* **45**(3), 595.

Kokossis, A. C., & Floudas, C. A. (1991). Synthesis of isothermal reactor-separator-recycle systems. *Chemical Engineering Science*, **46**(5/6), 1361.

Kokossis, A. C., & Floudas, C. A. (1994a). Optimization of complex reactor networks—II. Nonisothermal operation. *Chemical Engineering Science*, **49**(7), 1037.

Kokossis, A. C., & Floudas, C. A. (1994b). Stability in optimal design: Synthesis of complex reactor networks. *AIChE J.*, **40**(5), 849.

Lakshmanan, A., & Biegler, L. T. (1995). Reactor network targeting for waste minimization. In M. El-Halwagi & D. Petrides (Eds.), *Pollution Prevention via Process and Product Modifications* (p. 128), AIChE Symposium Series, **90**.

Lakshmanan, A., & Biegler, L. T. (1996a). Synthesis of optimal reactor networks. *I & EC Research*, **35**(4), 1344.

Lakshmanan, A., & Biegler, L. T. (1996b). Synthesis of optimal chemical reactor networks with simultaneous mass integration. *I & EC Research,* **35**(12), 4523.

Lee, K. Y., & Aris, R. (1963). Optimal adiabatic bed reactors for sulphur dioxide with cold shot cooling. *Ind. Eng. Chem. Proc. Des. Dev.*, **2**, 300.

Omtveit, T., & Lien, K. (1993). Graphical targeting procedures for reactor systems. Proc. ESCAPE-3, Graz, Austria.

Pibouleau, L. Floquet, L., & Domenech, S. (1988). Optimal synthesis of reactor separator systems by nonlinear programming method. *AIChE Journal*, **34**, 163.

Ravimohan, A. (1971). Optimization of chemical reactor networks with respect to flow configuration. *JOTA*, **8**(3), 204.

Trambouze, P. J., & Piret, E. L. (1959). Continuous stirred tank reactors: Designs for maximum conversions of raw material to desired product. *AIChE J.*, **5**, 384.

van de Vusse, O. (1964). Plug flow type reactor vs. tank reactor. *Chemical Engineering Science*, **19**, 994.

Viswanathan, J. V., & Grossmann, I. E. (1990). A combined penalty function and outer-approximation method for MINLP optimization. *Comput. and Chem. Engg.*, **14**, 769–782.

Waghmere, R. S., & Lim, H. C. (1981). Optimal operation of isothermal reactors. *I & EC Fund.*, **20**, 361.

Williams, T. J., & Otto, R. E. (1960). A generalized chemical processing model for the investigation of computer control. *Trans. Am. Inst. Elect. Engrs.*, **79**, 458.

Zwietering, N. (1959). The degree of mixing in continuous flow systems. *Chemical Engineering Science,* **11**, 1.

EXERCISES

1. Derive the MINLP formulation for the Kokossis and Floudas superstructure shown in Figure 19.10. Write the balance equations, reactor equations, and constraints. How does the problem size increase with the number of reactors in the superstructure?

Exercises

2. Derive the MINLP formulation (P9) for the superstructure shown in Figure 19.11. Write the balance equations, reactor equations, and constraints and discretize them using collocation and quadrature on finite elements. How does the problem size increase with the number of reactors in the superstructure?

3. The α-pinene problem is a reaction network that consists of five species and has the following reaction network (Figure 19.18). The objective function here is the maximization of the selectivity of C over D given a feed of pure A.

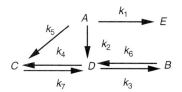

FIGURE 19.18

The reaction vector for the components A, B, C, D, E, respectively, is given by

$$R(X) = [-(k_1 + k_2)X_A - 2k_5 X_A^2, -k_6 X_B + k_3 X_D, k_5 X_A^2 + k_4 X_D^2 - k_7 X_C, k_2 X_A + k_6 X_B - k_3 X_D - 2k_4 X_D^2 + 2k_7 X_C, k_1 X_A]$$

where

$X_i = c_i / c_{A0}$ and $c_{A0} = 1$ mol/l
$k_1 = 0.33384 \ s^{-1}$
$k_2 = 0.26687 \ s^{-1}$
$k_3 = 0.14940 \ s^{-1}$
$k_4 = 0.18957 \ l\text{-mol}^{-1} s^{-1}$
$k_5 = 0.009598 \ l\text{-mol}^{-1} s^{-1}$
$k_6 = 0.29425 \ s^{-1}$
$k_7 = 0.011932 \ s^{-1}$

   a. Solve this problem by applying (P2) with a maximum residence time of 60 sec.
   b. Increase the residence time to 600 sec and resolve this problem with (P2).
   c. Based on the behaviors in parts a and b, what can you conclude about the optimal reactor network for this problem?

4. Resolve the Trambouze problem with a feed concentration of A at 10 gmol/l given in Example 2. How does the solution change?

5. Resolve the van de Vusse problem in Example 19.1 with a feed concentration 5.6 gmol/l. How does the solution change?

6. Show that if selectivity is the objective function in (P2), the problem can still be reformulated and solved as a linear program. (Hint: consider the problem:

$$\text{Min } a^T x / b^T x$$
$$\text{s.t. } A x \leq d$$
$$x \geq 0$$

with $b^Tx > 0$. Introduce the scalar variable $z = 1/b^Tx$ and vector $y = x\,z$ and reformulate this problem as an LP.)

7. Apply collocation on finite elements to the differential equations:

$$dX_{rr,A}/dt = -0.025 - 0.2\,X_{rr,A} - 0.4\,X_{rr,A}^2$$

$$X_{rr,A}(0) = (R_e X_{exit,A} + X_{P2,A})/(R_e + 1)$$

$$dX_{rr,C}/dt = 0.2\,X_{rr,A}$$

$$X_{rr,C}(0) = (R_e X_{exit,C} + X_{P2,C})/(R_e + 1)$$

and show that this yields the algebraic equations given in (19.12).

8. Show that the discretization in (P6) leads to the formulation in (P5) if the finite elements $\Delta\alpha_i$ are sufficiently small.

STRUCTURAL OPTIMIZATION OF PROCESS FLOWSHEETS   20

20.1 INTRODUCTION

The synthesis of a process flowsheet can be performed through a superstructure optimization in which the problem is formulated as an MINLP. In order to accomplish this task two major questions need to be addressed. The first one is how to develop the superstructure; the second is how to effectively model and solve the MINLP for the selected superstructure. We briefly discuss first the issue of generating superstructures for process flowsheets. The bulk of the chapter is then devoted to the modeling and solution of the MINLP.

20.2 FLOWSHEET SUPERSTRUCTURES

To systematically develop superstructures for process flowsheets is in principle a difficult task. For instance, consider a process flowsheet that is composed of reaction, separation, and heat integration subsystems. One general approach would be to develop a superstructure by combining the detailed superstructures for each subsystem, in which each unit performs a single preassigned task. Conceptually, the advantage of such an approach is that all the interactions and economic trade-offs would be taken explicitly into account. The major disadvantage, however, is that it can lead to a very large MINLP optimization problem.

Another general approach for developing a superstructure is to consider detailed models of units that can perform multiple tasks or functions, and interconnect the units with all feasible connections (Pantelides and Smith, 1995; Smith, 1996; Umeda et al., 1972). As an example, consider the diagram in Figure 20.1, which consists of a CSTR reactor, a tubular reactor, and two distillation columns for a given feedstock, a main product, and a by-product. The idea is to consider intermediate inputs and outputs for every unit, and assign potential feasible interconnection between them. In this way, the alternatives are largely determined by the selection of streams and to a lesser extent by the selection of units. Note that in Figure 20.1 no separation tasks are preassigned to columns 1 and 2 (see exercise 1). Therefore, all separation systems for a multicomponent mixture can be considered with these columns, provided tray by tray models are used.

To circumvent the problem of dimensionality, another possible approach is to use aggregate representations for the superstructures of the subsystems. In particular, the model of

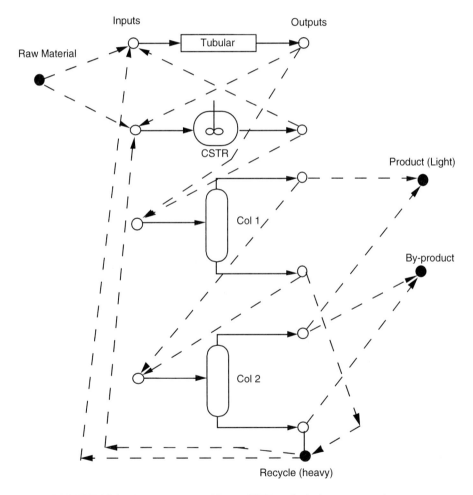

FIGURE 20.1 Superstructure with one CSTR and tubular reactor and two columns.

Sec. 20.1   Flowsheets Superstructures   665

simultaneous optimization and heat integration of Chapter 18 could be used to replace a detailed heat exchanger network superstructure such as the ones given in Chapter 16. Similarly, the targeting model for reactor networks of Chapter 19 can be used in place of a detailed superstructure for reactors. For the case of separation systems, aggregated models might also be used, although more commonly one might use a more detailed superstructure for this part of the process. While the advantage of this approach is that it greatly reduces the size of the MINLP problem, it has the disadvantage that not all economic factors are taken into account. In particular, the sizes of the individual units are often neglected or indirectly fixed with parameters such as minimum temperature approaches or maximum yields that might produce suboptimal solutions. This approach, however, is useful in preliminary design when assessing the potential of different design alternatives.

Finally, a third approach is to assume that some preliminary screening is performed (e.g., through heuristics) in order to postulate a smaller number of alternatives in the superstructure (Kocis and Grossmann, 1989). While this approach is somewhat restrictive, it does provide a systematic framework for analyzing specific alternatives at the level of tasks. As an example, consider the synthesis of an ammonia plant (see Chapter 15). A preliminary screening would indicate that the major options are as follows: for the reactor (multibed quench or tubular), for separation of product (flash condensation or absorbtion/distillation), for recovery of hydrogen in purge (membrane separation or simple purge). Figure 20.2 displays the superstructure for these alternatives. This superstructure contains eight different configurations. Figure 20.3 shows a superstructure for the HDA

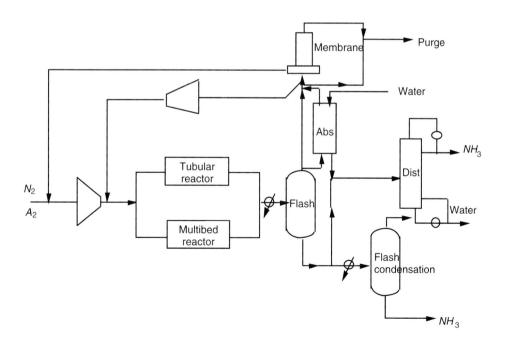

FIGURE 20.2 Superstructure for selected alternatives for ammonia production.

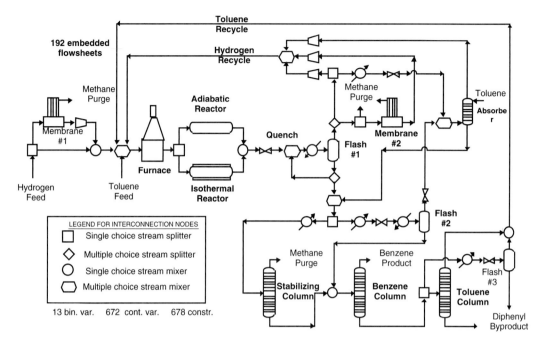

FIGURE 20.3 Superstructure for hydrodealkylation of toluene process.

process developed by Kocis and Grossmann (1989) based on alternatives that were postulated by Douglas (1988) in the hierarchical decomposition scheme. Thus, generating superstructures for process flowsheets based on specific alternatives at the level of tasks is actually not a very difficult problem. Finally, it is clear that in this approach it is possible to treat part of the problem with an aggregated model (e.g., heat integration), and the rest of the process with a detailed superstructure.

As for the modeling and solution, the MINLP models for aggregated representations are generally easier to solve, while the more detailed superstructures lead to larger MINLP problems that are more difficult to solve. One solution approach is to simply solve the MINLP problem directly without any special provisions. The other solution approach is to recognize the structure in flowsheet MINLP problems and exploit it so as to reduce the computational cost and increase its reliability. It is clear that if we want to consider the more detailed superstructures, it is of paramount importance to consider the second approach. The remainder of the chapter is devoted to this issue.

20.3 MIXED-INTEGER OPTIMIZATION MODELS

Having developed a superstructure of design alternatives, whether at a high level of abstraction or at a relatively detailed level of units, the synthesis problem can be formulated in general terms as the mixed-integer optimization model:

Sec. 20.4 MILP Approximation

$$\min_{x,y} Z = C(x, y)$$

$$\text{s.t.} \quad h(x) = 0 \qquad \qquad \text{(MIP)}$$

$$g(x,y) \leq 0$$

$$x \in X \quad y \in \{0,1\}^m$$

in which x is the vector of continuous variables representing flows, pressures, temperatures, while y is the vector of 0–1 variables to denote the potential existence of units. The equations $h(x) = 0$ are generally nonlinear and correspond to material and heat balances, while the inequalities $g(x,y) \leq 0$, represent specifications or physical limits. As we have seen in the previous chapters it should be noted that for most of the applications in process synthesis, problem (MIP) has the special structure that the 0–1 variables appear linearly in the objective function and constraints. The reason for this is that in the objective 0–1 variables are commonly used to represent fixed charges, that is,

$$C(x, y) = c^T y + f(x) \qquad (20.1)$$

while in the constraints they are used to represent logical conditions which normally can be expressed in linear form, that is,

$$g(x,y) = Cx + By - d \leq 0 \qquad (20.2)$$

Appendix A presents a brief discussion of MINLP algorithms that can be used to solve problem (MIP) given the special structure of Eqs. (20.1) and (20.2). The algorithms rely on solving a sequence of NLP subproblems and MILP master problems. The former arise when fixing the 0–1 variables in (MIP) and optimizing the continuous variables. Also, their solution provides an upper bound. The latter provide a global linear approximation to optimize the 0–1 variables, and relies on linearizations in the case of the outer-approximation algorithm, or on Lagrangian cuts in the case of Generalized Benders Decomposition. For convex problems these master problems predict a valid lower bound. As will be discussed in section 20.6, there are several reasons why it is often not advisable to solve directly the nonlinear problem (MIP) for the case of a process flowsheet, but instead use a decomposition strategy. The other option is to avoid solving the MINLP by approximating this problem as an MILP through discretization, as will be discussed in the next section. It should be noted that, in fact, in Chapter 17 we used this principle when deriving an MILP model for heat integrated distillation columns.

20.4 MILP APPROXIMATION

In order to derive an MILP approximation to problem (MIP), we will partition the continuous variables x as follows:

$$x = \begin{bmatrix} z^d \\ x^c \end{bmatrix} \qquad (20.3)$$

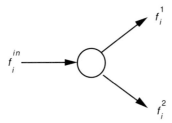

FIGURE 20.4 Stream splitter.

in which z^d is the vector of operating conditions that gives rise to the nonlinearities (e.g., pressures, temperatures, split fractions, conversions, etc.), and x^c is a vector of material, heat, and power flow variables that appear linearly (see section 18.3, Chapter 18). In this way, given a fixed value of z^d, the nonlinear equations reduce to a subset of linear equations, that is,

$$h(x) = 0 \quad \Rightarrow \quad E x^c = e \qquad (20.4)$$

in which the matrix of coefficients E and the right hand sides e are a function of z^d, $E(z^d), e(z^d)$.

Since in general we would like to consider more than one fixed value for the variables z^d, we will require the introduction of the additional 0–1 variables y^d to represent the potential selection of the discrete operating conditions. In this way, the general form of the MILP approximation will be as follows:

$$\min C = a_1^T y + a_2^T y^d + b^T x^c$$
$$\text{s.t.} \quad E_1 y^d + E_2 x^c = e \qquad \text{(MAPP)}$$
$$D_1 y + D_2 y^d + D_3 x^c \le d$$
$$y, y^d = 0,1 \quad x^c \ge 0$$

It should be noted that the derivation of the above problem generally requires the disaggregation of the vector of continuous variables x^c in terms of the discretized conditions. To illustrate this point more clearly, consider the simple splitter shown in Figure 20.4. The corresponding mass balance equations for each component i are as follows:

$$f_i^1 = \eta f_i^{in} \qquad (20.5)$$
$$f_i^2 = f_i^{in} - f_i^1 \qquad (20.6)$$

where η is the split fraction for outlet stream 1. Note that Eq. (20.5) is nonlinear (in fact, bilinear), and despite its simplicity it is a major source of nonconvexities and numerical difficulties.

Now let us assume that we consider N discrete values of η, η_k $k = 1, 2, ...N$. Then if we disagggregate the flow for the inlet stream as $f_i^{in,k}$, $k=1,2,...N$, and introduce the 0–1 variables $y^{d,k}$, $k=1,2...N$, Eqs. (20.5) and (20.6) can be replaced by the linear constraints,

Sec. 20.5   MILP Model for the Synthesis of Utility Plants   669

$$f_i^1 = \sum_{k=1}^{N} \eta_k f_i^{in,k} \tag{20.7}$$

$$f_i^{in} = \sum_{k=1}^{N} f_i^{in,k} \tag{20.8}$$

$$f_i^{in,k} - U y^{d,k} \leq 0 \qquad k=1,2...N \tag{20.9}$$

$$\sum_{k=1}^{N} y^{d,k} = 1 \tag{20.10}$$

$$f_i^2 = f_i^{in} - f_i^1 \tag{20.11}$$

While we have been able to eliminate the nonlinearities, it is clear that we have increased the number of discrete and continuous variables as well as the number of constraints. Also, in the general case the definition of the matrix of cofficients and the right-hand sides of problem (MAPP) requires an a priori evaluation or simulation of nonlinear models. The next section presents an example of an MILP approximation.

20.5 MILP MODEL FOR THE SYNTHESIS OF UTILITY PLANTS

Consider the synthesis problem in which we are given demands of electric power, mechanical power for several drivers (e.g., pumps, compressors), and steam at various levels of pressure. The problem is then to find a minimum cost configuration consisting of boilers, gas and steam turbines, electric motors, let-down valves, and waste heat boilers. The intent here is not to present a detailed model but simply to outline the nature of the model (a specific example is given in exercise 3).

Given the utility demands, it is possible to postulate a superstructure that contains the units that potentially can satisfy the demands. For example, the electricity demand can be satisfied with gas turbines (see Figure 20.5a) and steam turbines of various types (see Figure 20.5b), or one might even consider its external purchase. The power for the drivers can be satisfied with the various steam turbines or with electric motors. Finally, steam demands can be met by generating steam from boilers or by using the exhaust from steam turbines or adjusting the let-down valves.

As an example for deriving the superstructure of a utility plant, consider the case of an electricity demand that can be satisfied with a gas turbine and/or with a high pressure (HP) turbine, one power demand of a compressor that can be satisfied with backpressure, total condensation or extraction turbines operating at high pressure (HP) or medium pressure (MP), or with an electric motor. Finally, assume there are steam demands at medium and low pressure steam, and that there are waste heat boilers generating high pressure steam. Assuming three pressure headers (HP, MP, LP) for steam and that no boiler is considered for generating low pressure steam, the corresponding superstructure is shown in

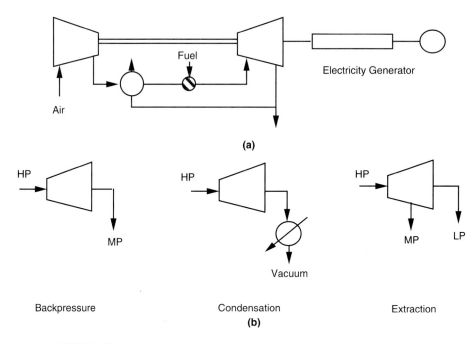

FIGURE 20.5 (a) Gas turbine for electricity generation. (b) Various types of steam turbines.

Figure 20.6. Note that rather than considering the various types of steam turbines separately, we "embed" them into a single one (e.g., backpressure + extraction + condensing). Also, note that to close the cycle a water deaerator is used for collecting the condensate. Pumps are then used to feed water to the boilers.

Having developed a superstructure like the one in Figure 20.6, the mixed-integer optimization model (Papoulias and Grossmann, 1993) has the following general form,

$$\min C = \text{Investment} + \text{Fuel Cost}$$
$$\text{s.t.} \quad \text{Material and enthalpy balances} \quad \text{(MIP)}$$
$$\text{Logical constraints}$$

The above problem can in fact be formulated as an MILP problem (see exercise 3). Without presenting a complete model, an outline is as follows.

The cost of the boilers (e.g., boiler 1) can be represented as a linear cost function in terms of the amount of steam produced F_1 and with fixed charges,

$$C_{\text{boiler}} = \alpha_1 y_{B1} + \beta_1 F_1$$
$$F_1 - U y_{B1} \leq 0 \quad (20.12)$$
$$F_1 \geq 0, \ y_{B1} = 0, 1$$

Note that in order to determine the cost coefficients α_1 and β_1 one has to perform some preliminary calculations, for instance, to relate the fuel consumption to F_1.

Sec. 20.5 MILP Model for the Synthesis of Utility Plants

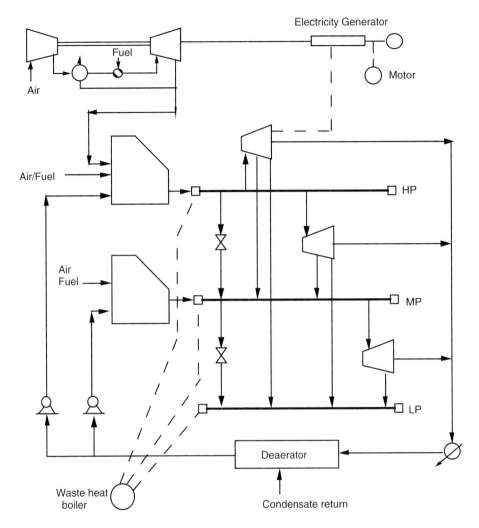

FIGURE 20.6 Superstructure for utility plant.

To see more clearly how linearity is induced by fixed operating conditions, consider a given power demand W_p that can be satisfied either with the various types of turbines operating at high pressure or at medium pressure or with an electric motor. By representing the turbines by different sections (see Figure 20.7), the equations that apply are as follows:

1. Power delivered by turbines A and B

$$W_A = F_1\eta_1(H_{HP} - H_{MP}) + F_2\eta_2(H_{MP} - H_{LP}) + F_3\eta_3(H_{LP} - H_{VAC}) \quad (20.13)$$

$$W_B = f_1\eta_1(H_{MP} - H_{LP}) + f_2\eta_2(H_{LP} - H_{VAC}) \quad (20.14)$$

where η are turbine efficiencies and H are steam enthalpies.

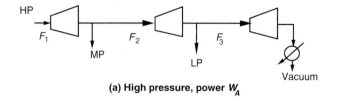

(a) High pressure, power W_A

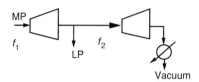

(b) Medium pressure, power W_B

(c) Electric motor, power W_e

FIGURE 20.7 Representation of alternatives for power demand.

2. Requirement for power demand W_p

$$W_A + W_B + W_e = W_p \qquad (20.15)$$

3. Select only one alternative

$$y_A + y_B + y_e = 1 \qquad (20.16)$$

$$W_A - Uy_A \le 0 \qquad W_B - Uy_B \le 0 \qquad W_e - Uy_e \le 0$$

It is clear that if the efficiencies and enthaplies in Eqs. (20.13) and (20.14) are assumed to be constant (i.e., fixed pressures and temperatures) then Eqs. (20.13) and (20.14) become linear equations. Thus, together with the remaining constraints and a cost function linear in the W variables with fixed charges, the problem reduces to an MILP. It should also be noted that one of the reasons why formulating this problem as an MILP is facilitated is because we are dealing basically with a pure component, water. Thus, fixing the operating conditions at discrete values is obviously easier than if we had multicomponent mixtures in which the enthalpies are not only a function of pressure and temperature but also of composition. In this case, applying a nonlinear model is more natural.

20.6 MODELING/DECOMPOSITION STRATEGY

In the case that nonlinearities are explicitly accounted for in problem (MIP), aside from the potentially large size of the MINLP model for the superstructure optimization of a process flowsheet, there are two other potential difficulties. The first is that when fixing the 0–1 variables for defining the corresponding NLP subproblem in a direct solution of the MINLP, one has to carry many redundant variables and equations that unnecessarily increase the dimensionality and complexity of this subproblem. The reason is that when some of the process units are not selected, the corresponding flows are fixed to zero, but yet the mass and heat balances of the "dry units" have to be converged. This usually introduces singularities that cause great difficulty in the convergence of the NLP (see Exercise 1). The second difficulty that arises from a direct solution of the MINLP is because the effects of nonconvexities are accentuated when flows take a value of zero (again effect of "dry" units). This may cause the NLP subproblem to converge to a suboptimal solution or the master problem to "cut off" the optimal 0–1 combination. It is precisely these two difficulties that motivate the modeling/decomposition (M/D) strategy described by Kocis and Grossmann (1989) in the next two sections. As will be shown, the two basic ideas are to model the MINLP so as to explicitly handle the effect of nonconvexities in the interconnection nodes, and to decompose it so as to avoid NLP subproblems with zero flows.

20.6.1 MINLP Model

It will be assumed that the superstructure of alternative flowsheets is represented in terms of interconnection nodes (splitters and mixers) and process unit nodes (reactors, compressors, distillation columns). This superstructure is then modeled as an MINLP problem in which 0–1 variables are assigned to the potential existence of units and continuous variables to the flows, pressures, temperatures, and sizes.

To define the MINLP for the network superstructure, let U and N denote the set of process units and interconnection nodes with elements u and n, respectively. Also, let S denote the set of process streams in the superstructure with elements s. Finally, let $I^{U(u)}$ and $O^{U(u)}$ represent the set of input and output streams for process unit u and $I^{N(n)}$ and $O^{N(n)}$ represent the set of input and output streams for interconnection node n. Having stated these definitions, a flowsheet superstructure can be formulated as:

$$Z = \min_{x,d,z,y} \sum_{u \in U} \{c_u y_u + f_u(d_u)\} + \sum_{s \in S} c_s$$

$$\text{s.t.} \begin{rcases} h_u(d_u, z_u, x_p, x_q) = 0 \\ g_u(d_u, z_u, x_p, x_q) \leq 0 \\ x_p^F - x_p^{F,UP} y_u \leq 0, \; x_p^F \geq 0 \\ d_u - d_u^{UP} y_u \leq 0, \; d_u \geq 0 \\ r_n(d_n, x_p, x_q) = 0 \end{rcases} \begin{array}{l} u \in U, \; p \in I^{U(u)}, \; q \in O^{U(u)} \\ \\ n \in N, \; p \in I^{N(n)}, \; q \in O^{N(n)} \end{array} \quad \text{(PF)}$$

$$x_s \in X_s = \{x_s \mid x_s^{LO} \leq x_s \leq x_s^{UP}\} \qquad s \in S$$

$$d_u \in D_u = \{d_u \mid 0 \leq d_u \leq d_n^{UP}\} \qquad u \in U$$

$$d_n \in D_n = \{d_n \mid 0 \leq d_n \leq d_n^{UP}\} \qquad n \in N$$

$$z_u \in Z_u = \{z_u \mid z_u^{LO} \leq z_u \leq z_u^{UP}\} \qquad u \in U$$

$$y \in Y = \{y \mid y \in \{0,1\}^m, Ey \leq e\}$$

The variables in problem (PF) include x_s, d_u, z_u, and $y = \{y_u, u \in U\}$. x_s is a vector of variables for each stream $s \in S$ (e.g., component flowrates, temperature, pressure, etc.) where x_s^F denotes the subvector of flowrate components. d_u denotes a vector of decision/sizing variables, z_u denotes a vector of internal/performance variables for each process unit $u \in U$, and d_n denotes a vector of decision/sizing variables for each interconnection node $n \in N$.

In the objective function of problem (PF) there is a term for each process unit u which includes a fixed-charge cost (c_u) and a cost term f_u that is a function of the decision/sizing variable d_u. The second part of the objective function represents the purchase cost or sales revenue (c_s) for the process streams. The constraints in MINLP (PF) are partitioned into two sets, which are associated with the two types of nodes, process units nodes and interconnection nodes. For each process unit $u \in U$, the model includes a vector of linear and nonlinear equality and inequality constraints, h_u, g_u, involving the continuous variables d_u, z_u, and x_s ($s \in I^{U(u)} \cup O^{U(u)}$). Also, it is necessary to have linear inequalities for each process unit to insure that the input flowrate to this unit, x_p^F and its design variables, d_u, are zero if the unit does not exist (i.e., the associated binary variable $y_u = 0$). Note that in these constraints, $x_p^{F,UP}$ and d_u^{UP} are constants that represent upper bounds on these variables when the process unit exists. Finally, for each interconnection node $n \in N$, there is a vector of equality constraints, r_n, which relates the output streams to the input streams through the decision variables d_n.

In order to maximize the occurrence of linear constraints, mass balances are expressed in terms of component flows. Finally, the interconnection nodes are modeled so as to try to remove nonconvexities whenever possible. To illustrate, the mass balances in single choice splitters—in which only one output stream is specified to exist—is modeled through linear constraints as outlined below.

Given an input stream with unknown compositions, it is possible to make use of the binary variables defined to denote the existence of the process units in each outlet of the splitter to derive a linear model for the multicomponent splitter where only one outlet stream can be selected. For a stream splitter with inlet stream F_0 and outlet streams $F_1, F_2, \ldots F_N$, of which exactly one can exist, the following linear model describes the splitter (where f_i^j denotes the flowrate of component j in stream i for $j = 1, 2, \ldots C$ and $i = 0, 1, 2, \ldots N$):

$$F_i = \sum_{j=1}^{C} f_i^j \qquad i = 0, 1, 2, \ldots N \qquad (20.17)$$

Sec. 20.6   Modeling/Decomposition Strategy

$$f^j = \sum_{i=1}^{N} f_i^j \qquad j = 1, 2, \ldots C \tag{20.18}$$

$$F_i - \rho Y_i \le 0 \qquad i = 1, 2, \ldots N$$
$$\sum_{i=1}^{N} Y_i = 1 \qquad Y_i = 0, 1 \tag{20.19}$$

where ρ is a valid upper bound on the inlet flowrate.

This model makes use of the binary variables of the process units in a way that the mass balance in the splitter is represented by a selection procedure (i.e., equating the input stream to the output stream that exists). This can be verified by observing the implication of the constraint $\sum_{i=1}^{N} Y_i = 1$. Let Y_I denote the binary variable whose value is 1, thus from Eq. (20.19) and the nonnegativity condition for this variable, $F_{i \ne I} = 0$. Eq. (20.17) in turn implies that $f_i^j = 0$ for $i \ne I$ and $j = 1, 2, \ldots C$. Finally, from Eq. (20.18), $f_0^j = f_I^j$ for $j = 1, 2, \ldots C$.

Similarly, the heat balances in single choice mixers can be modeled with linear constraints.

20.6.2   Modeling/Decomposition Algorithm

The superstructure is decomposed into the initial flowsheet and subsystems of nonexisting units. The idea here is to solve the NLP only for the existing flowsheet, while the remaining subsystems are to be suboptimized with a Lagrangian scheme in order to provide a linear approximation of the entire superstructure in the master problem.

In order to only solve the NLP of a specific flowsheet, consider a partitioning of the subset of process units, U, into a subset of existing process units, UE for which $y_u = 1$, and a set of nonexisting process units, UN, for which $y_u = 0$ ($U = UE \cup UN$). The optimization of the current flowsheet structure for a given assignment of binary variables can then be performed by solving the following reduced NLP subproblem:

$$Z = \min_{x,d,z} \sum_{u \in UE} \{c_u + f_u(d_u)\} + \sum_{s \in S} c_s x_s \tag{RP}$$

$$\left.\begin{array}{l} \text{s.t. } h_u(d_u, z_u, x_p, x_q) = 0 \\ g_u(d_u, z_u, x_p, x_q) \le 0 \\ 0 \le x_p^F \le x_p^{F,UP} \\ 0 \le d_u \le d_u^{UP} \end{array}\right\} \quad u \in UE,\ p \in I^{UE(u)},\ q \in O^{UE(u)}$$

$$x_t^F = 0 \qquad u \in UN,\ t \in I^{UN(u)}$$
$$r_n(d_n,\ x_p,\ x_q) = 0 \qquad n \in N,\ p \in I^{N(n)},\ q \in O^{N(n)}$$
$$x_s \in X_s,\ d_u \in D_u,\ d_n \in D_n,\ z_u \in Z_u \qquad s \in S,\ u \in UE,\ n \in N$$

where x_t^F corresponds to the stream flowrates in the superstructure that are inputs to the nonexisting units. The solution of the reduced NLP subproblem (RP) leads to a smaller optimization problem where the nonlinear functions of the nonexisting process units are excluded, which reduces the potential of numerical singularities.

Since subsystems with nonexisting units are connected in the superstructure through the interconnection nodes, Lagrange multipliers are available from the equations $r = 0$ in (RP). Therefore, a suboptimization problem can be formulated for the disappearing process units based on the prices of the variables x at the interconnection nodes. Also, in order for the suboptimization problem to generate nonzero conditions where nonexisting units are "likely" to operate had they existed in the current flowsheet, the input stream variables of the nonexisting subsystems can be set to the optimal values of the input variables of the interconnection nodes.

Denoting x_t the fixed inlets to the splitter nodes obtained in the solution to the NLP subproblem, the suboptimization problem for the disappearing process units is then given by:

$$Z_{sub} = \min_{x,d,z} \sum_{u \in UN} c_u + f_u(d_u) + \sum_{s \in I^{UN(u)}} \mu_s x_s - \sum_{s \in O^{UN(u)}} \mu_s x_s \qquad \text{(SP)}$$

$$\text{s.t.} \quad \left. \begin{array}{l} h_u(d_u,\ z_u,\ x_p,\ x_q) = 0 \\ g_u(d_u,\ z_u,\ x_p,\ x_q) \le 0 \end{array} \right\} \qquad u \in,\ UN\ p \in I^{UN(u)},\ q \in O^{UN(u)}$$

$$x_p = x_t \qquad u \in UN,\ p \in I^{UN(u)},\ t \in I^{N(n)}$$

$$x_s \in X_s,\ d_u \in D_u,\ d_n \in Dn,\ z_u \in Z_u \qquad u \in UN,\ s \in O^{UN(u)},\ n \in N$$

This problem provides in general a good estimation of conditions that would prevail if a nonexisting unit was included in the flowsheet structure. Hence, the solution to this NLP problem yields a good point for deriving the linearizations for the MINLP master problem (see Kocis and Grossmann, 1989). While this decomposition is obvious for the case of superstructures involving competing parallel units, it is nontrivial for more complex superstructures as will be described later in the chapter.

Having formulated the problem as an MINLP and decomposed the superstructure into the initial flowsheet and subsystems, the major steps are then as follows:

Step 1. Solve the NLP for the initial flowsheet to obtain an upper bound of the cost. In addition, the solution to this problem provides flows and Lagrange multipliers for the interconnection nodes.

Step 2. Based on the flows and Lagrange multipliers in Step 1, suboptimize each subsystem by fixing the inlet flows and by assigning the multipliers as prices for the inlet and outlet flows.

Sec. 20.6   Modeling/Decomposition Strategy   677

Step 3. Given the solution points at Steps 1 and 2, construct the first MILP master problem by incorporating the linearizations of the units of the initial flowsheet and of the subsystems. These linearizations are modified to ensure consistency at zero flows (Kocis and Grossmann, 1989). For single choice interconnection nodes, the equations are linear, so they are directly included in the master problem. For multiple choice interconnection nodes, valid linear outer-approximations as described in Kocis and Grossmann (1989) are included.

Step 4. Solve the MILP master problem to predict a new flowsheet and a lower bound. If the lower bound exceeds the current best upper bound, stop; the optimal flowsheet corresponds to the best upper bound. Otherwise go to Step 5.

Step 5. Solve the NLP for the new flowsheet structure and update the current best upper bound.

Step 6. Given the solution point at Step 5, add to the MILP master problem the linearization of the units and the valid outer-approximations to the multiple choice interconnection nodes of the flowsheet in Step 5.

Step 7. Repeat Steps 4 to 6 until the termination criterion in Step 4 is satisfied.

Note from the above that the major advantage in this strategy is that the NLP optimization is only required for the current flowsheet structure being analyzed (Steps 1 and 5). The NLP subproblems in Step 2 are required to provide information on the nonexisting units at non-zero flow conditions and usually require modest computational effort. Also, the size of the MILP master problem is kept smaller by only including the linearizations of the current flowsheet in Step 6.

20.6.3 Decomposition of Superstructure

While the attractive feature of the M/D strategy is that it avoids solving NLP subproblems with nonexisting units, an important question that must be addressed is, given the initial flowsheet, how to systematically determine the subsystems to be suboptimized (Kravanja and Grossmann, 1990). In a number of instances this is a relatively simple task, such as in the case of the superstructure shown in Figure 20.8a. Here it is clear that by selecting the initial flowsheet in Figure 20.8b, the "deleted" units 2 and 5 have the property that their interconnection nodes provide all the required information to suboptimize these units. In particular, for unit 2 node S1 provides the flow F_1 and the multiplier μ_{S1} while node M1 provides the multiplier μ_{M1}. In this way, as described in the previous section, it is possible by using problem (SP) to suboptimize unit 2 by fixing its inlet flow and by assigning prices to its inlets and outlets. The same is, of course, true with unit 5 (see Figure 20.8c).

Consider however, the superstructure in Figure 20.9a where the initial flowsheet is given by Figure 20.9b. The nonexisting units then define the subsystem in Figure 20.9c. It is clear that the difficulty that arises is that there is no information on flows and prices for the interconnection nodes S2 and M2 since they do not belong to the initial flowsheet.

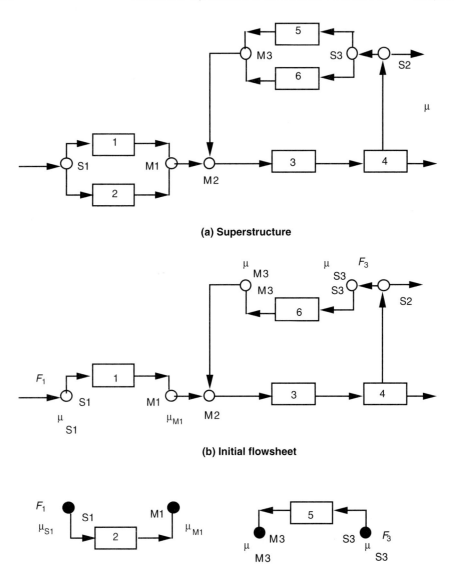

FIGURE 20.8 Superstructure decomposition for simple case.

The alternative of suboptimizing directly the subsystem in Figure 20.9c is not attractive, because there is no way to ensure that the inlet flows to units 4 and 5 will be non-zero.

To circumvent this problem, we can proceed in a recursive manner and regard the subsystem in Figure 20.9c as a "new" superstructure. In this case as shown in Figure 20.10, if we select units 3 and 4 as the "initial" flowsheet, then the nodes S2 and M2 will

Sec. 20.6 Modeling/Decomposition Strategy

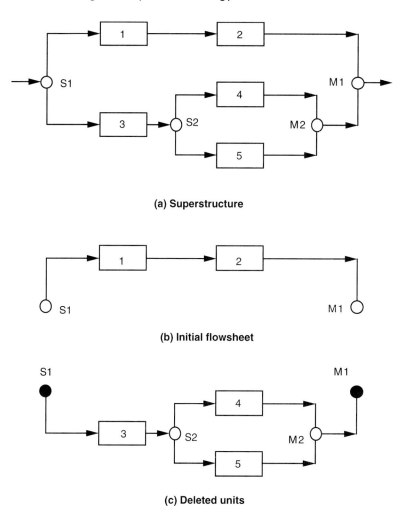

FIGURE 20.9 Decomposition of complex superstructure.

provide flows and multipliers to suboptimize unit 5 with this information. In summary, the NLP optimizations for the superstructure in Figure 20.9a would involve the optimization of the initial flowsheet in Figure 20.9b, and the suboptimization of the subsystems in Figure 20.10b (with fixed flow at node S1 and multipliers at S1 and M1) and Figure 20.10c (with fixed flow at node S2 and multipliers at nodes S2 and M2).

Based then on the observation that information of nonexisting interconnection nodes can be generated recursively, the following algorithm was proposed by Kravanja and Grossmann (1990) to systematically perform the decomposition into subsystems.

Let $U = \{u\}$ be the set of process units in the superstructure and $N = \{n\}$ be the set of interconnection nodes. The procedure is then as follows:

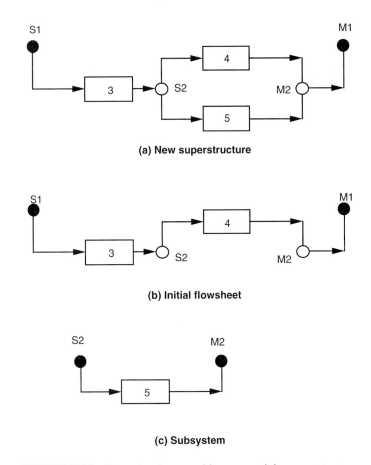

FIGURE 20.10 Recursive decomposition on remaining superstructure.

Step 0. **a.** Merge the units in the superstructure with no adjacent interconnection nodes and define the set of resulting units by U^M.
  b. Define the superstructure with merged units through the sets $U_S = U^M$, $N_S = N$.

Step 1. Select a flowsheet $U_1 \subseteq U_S$, $N_1 \subseteq N_S$, which is to be (sub) optimized.

Step 2. Determine the index sets of the current superstructure with nonexisting units, $U_R = U_S - U_1$, $N_R = N_S - N_1$.

Step 3. **a.** For (U_R, N_R) determine the sets of disjoint substructures that are not interconnected (U_l, N_l), $l = 2, N$.
  b. If $N_l = \emptyset$, the substructure U_l is to be suboptimized.

Step 4. Repeat Steps 1 to 3 by setting as the new superstructure(s) to be analyzed for the substructures in Step 3a with interconnection nodes that have not been covered; that is $U_S = U_l$, $N_S = N_l$ for $N_l \neq \emptyset$.

To illustrate the application of this procedure consider first the simple case of Figure 20.8. In this case units 3 and 4 are merged into unit 3–4. Then $U_S = \{1,2,3\text{–}4,5,6\}$, $N_S = \{S1, S2, S3, M1, M2, M3\}$. The initial flowsheet according to Figure 20.8b is $U_1 = \{1,3\text{–}4,6\}$, $N_1 = \{S1, S2, S3, M1, M2, M3\}$. The remaining flowsheet is then given by $U_R = \{2,5\}$, $N_R = \emptyset$. Since units 2 and 5 are disjoint $U_2 = \{2\}$, $U_3 = \{5\}$. Hence, the NLP optimization must be applied to the initial flowsheet U_1 and the NLP suboptimization to the subsystems U_2 and U_3.

Similarly, for the example in Figure 20.9 it can easily be determined from the above algorithm that the initial flowsheet is $U_1 = \{1\text{–}2\}$ from the first pass. In the second pass the initial flowsheet selected is $U_1 = \{3,4\}$; hence, the last subsystem is $U_2 = \{5\}$.

EXAMPLE 20.1 Structural Flowsheet Optimization

The M/D strategy has been implemented in the program PROSYN-MINLP by Kravanja and Grossmann (1990). These authors considered a modified process synthesis problem by Kocis and Grossmann (1987). The superstructure is shown in Figure 20.11a, and includes 16 flowsheet alternatives. The problem data are given in Table 20.1. The alternatives for producing product C from chemicals A and B are as follows: The chemicals are supplied to the process by either of two feedstocks, both containing reactants A and B, and inert material D. Feedstock F2 has less inert than F1 but is more expensive.

Since reaction takes place at high pressure, the feed entering the process must be compressed either in single-stage or two-stage compressors with intermediate cooling. After mixing the compressed feed with the recycle, the stream undergoes exothermic gas-phase reaction, which can be carried out in two alternative adiabatic reactors: Reactor 1 is less expensive but has lower conversion than reactor 2. The reaction is favored by high pressure, low temperature, high concentration of reactant B, and low concentration of inert D. The reactor effluent is then sent to a flash separator, where lighter reactants and inert materials are separated from the heavier product C at an unspecified pressure and temperature. The bottom stream is the product stream that must contain at least 90% of component C and must satisfy a maximum of 1 kmol/s of the market demand. Since the conversion is generally low, unconverted raw materials in the top stream are recycled. To prevent accumulation of inert D, a portion of the recycle stream must be purged and an optimal selection of purge rate stream must be determined. The recycle stream must be recompressed due to the pressure loss in the reactor and the possible lower pressure in the flash unit to achieve the desired product purity. There is a choice between a single-stage and a two-stage compressor with intermediate cooling for the recycle. In addition, the minimum temperature of both the product and the by-product streams is 400 K. Finally, the objective specified for this synthesis problem is the maximization of annual profit.

Simple, but concise PROSYN-MINLP models of process units (compressor, reactor, flash separator, heater, and cooler) and interconnection nodes (single and multiple-choice mixer and splitter) were used, as well as the proposed models for simultaneously considering heat integration and HEN costs. The resulting MINLP formulation contains 293 constraints with 279 continuous variables and 8 binary variables. The superstructure was decomposed for the optimization of the initial flowsheet (bold lines in Figure 20.11a), and for the suboptimization of three nonexisting substructures (dashed lines in Figure 20.11a).

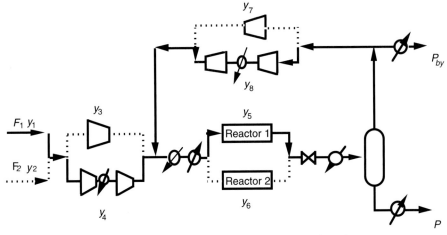

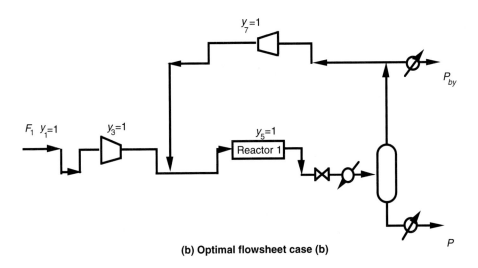

FIGURE 20.11 Superstructure and optimal flowsheet for example problem.

This example problem was solved with PROSYN-MINLP for the three following cases: (a) MINLP optimization with no heat integration, (b) simultaneous MINLP optimization and heat integration using the model by Duran and Grossmann (1986), (c) simultaneous NLP optimization and heat integration with HEN costs (see Kravanja and Grossmann, 1990) for the optimal structure obtained in case (b). For cases (a) and (b) the OA/ER algorithm was terminated based on the progress of the NLP solutions, since higher bounds on the profit were obtained from the MILP master with the proposed deactivation scheme for the linearizations of the splitter in the recycle. Results of the OA/ER algorithm are given in Table 20.2 while technical and economic results of the optimal flowsheet are given in Table 20.3. As can be seen in Table 20.2,

Sec. 20.6   Modeling/Decomposition Strategy

TABLE 20.1   Flowsheet Synthesis Problem Data

Feedstock or Product/By-product	Composition	Costs, $/kmol
F1 ≤ 10 kmol/s	60% A 25% B 15% D	0.0245
F2 ≤ 10 kmol/s	65% A 30% B 5% D	0.0294
$P \leq 1$ kmol/s	≥ 90% C	0.2614
P_{BY}		0.0163

Utilities	Costs
electricity	$0.03/(kWh)
heating (steam)	$8.0/$10^6$ kJ
cooling water	$0.7/$10^6$ kJ

Design Specifications	
	Reactor
reactor pressure, MPa	$2.5 \leq P_R \leq 15$
temp, inlet K	$300 \leq T_{in} \leq 623$
temp, outlet, K	$365 \leq T_{out} \leq 623$
	Flash Separation
pressure, MPa	$0.15 \leq P_F \leq 15$
temp, K	$300 \leq T_F \leq 500$
Operating time	8500 hrs/yr

the OA/ER algorithm requires two NLP subproblems to confirm that the initial flowsheet in case (a) is the optimum. In case (b) it requires three NLP subproblems to find the structure in Figure 20.11b. This clearly indicates that the quality of the information supplied to the MILP master problem by the M/D strategy is good.

First, consider case (a) when only the MINLP optimization of the superstructure is performed without heat integration. The optimal flowsheet is $y^k = \{1,0,0,1,1,0,0,1\}$ with annual profit of 794,000 $/yr. As seen in Figure 20.11a, it utilizes the cheaper feedstock F1, two-stage feed compression, cheap reactor R1 with low conversion and two-stage compressor for the recycle. If costs of the HEN, which are quite significant, are subsequently calculated and they are added to the profit, this leads to a loss of –$1,192,000/yr. When heat integration is simultaneously performed in the MINLP optimization of the superstructure (Duran and Grossmann, 1986), the results are at first glance much better. The optimal flowsheet in Figure 20.11b yields an annual profit of $3,403,000/yr ($2,609,000/yr more than for the nonintegrated flowsheet). The differences in the new flowsheet lies in the selection of single-stage compressors for the feed and for the recycle. Also, almost all parameters change significantly (Table 20.3) since the trade-offs between heat in-

TABLE 20.2 Results of OA/ER Algorithm for the Flowsheet Problem

Iteration	y^k	NLP[1] (CPU Time)[2]	MILP[1] (CPU Time)[2]
a) MINLP optimization, no heat integration			
1	{1,0,0,1,1,0,0,1}	794 (32)	1259 (24)
2	{1,0,1,0,1,0,1,0}	534 (20.13) and terminated	
b) MINLP heat integration			
1	{1,0,0,1,1,0,0,1}	3315 (118)	4985 (27)
2	{1,0,1,0,1,0,1,0}	3403 (39)	4208 (42)
3		3365 (81) and terminated	
c) NLP with heat integration and HEN costs			
		1679 (105) and terminated	

[1]Profit in 10^3/yr
[2]CPU time (sec) VAX-8800

TABLE 20.3 Technical and Economic Results

	MINLP only	Heat integration Duran-Grossmann	Heat integration HEN costs
Flows, kg-mol/sec			
F1	6.176	5.648	5.451
F2	0	0	0
P	1	1	1
Pby	3.027	2.682	2.618
purge rate %	14.5	14.6	19.7
Reactor			
Pin, MPa	7.048	2.500	4.377
Pout, MPa	6.343	2.250	3.939
Tout, K	378	430	419
Tin, K	332	379	356
conversion of B per pass, %	25.5	25.4	29.4
composition of reactor inlet %			
A	52.5	54.5	55.7
B	17.5	18.1	19.3
C	4.3	0.9	0.4
D	25.7	26.5	24.6
volume, m^3	55.7	49.1	67.7
Flash separation			
P, MPa	6.343	2.250	4.377
Tout, K	378	310	310

TABLE 20.3 *Continued*

	MINLP only	Heat integration Duran-Grossmann	Heat integration HEN costs
Utilities			
electricity, MW	3.718	1.798	2.78
heating, steam 10^9 MJ/year	0.114	0	0
cooling, water, 10^9 MJ/year	1.566	0.834	1.05
Other			
overall conversion of B, %	58.29	63.7	66.04
load of HEN, MW	54.9	71.5	48.0
Earnings, 10^3/yr			
Product	8000	8000	8000
By-product	1513	1341	1309
Expenses, 10^3/yr			
Feedstock	4632	4236	4088
Capital investment HEN	1986	3695	1173
other	1131	659	925
Electricity compress	948	459	709
Heating utility	912	0	0
Cooling utility	1096	584	735
Annual profit, 10^3/yr			
Without HEN costs	794	3403	2852
With HEN Costs	−1192	−292	**1679**

tegration (consumption of steam and cooling water), electricity, and consumption of feedstock are now appropriately established. Since energy is recovered within the process, no expensive heating utility is required. Note that the overall conversion of *B* is increased from 58.3% to 63.7%, and the reactor operates at 2.5 MPa instead of 7.05 MPa as in case (a).

As was mentioned previously, in the formulation for simultaneous heat integration by Duran and Grossmann, a fixed ΔT_{min} must be specified ahead of calculation (30 K in this case) and hence no area versus energy trade-offs are considered. Owing to the relatively small vertical driving forces and the gas-gas matches, the HEN costs are very high, so that annual profit when these costs are added to the expenses, reduces the profit to −$292,000/yr, which, as in case (a), also incurs in a loss (see Table 20.3).

In order to consider the HEN costs, the NLP optimization was repeated again on the flowsheet in Figure 20.11b with the stepwise procedure by Kravanja and Grossmann (1990) for simultaneous optimization and heat integration with HEN costs. The solution of the simultaneous optimization and heat integration by Duran and Grossmann was used as a starting point and the enthalpy intervals and the ordering of their temperatures were established from this solution. The new solution yielded a profit of $1,679,000/yr. As can be seen from Table 20.3, the operating conditions again undergo considerable changes. The most significant differences are a further increase in the overall conversion to 66.04%, elimination of the preheat of the reactor feed (gas-gas matches with small temperature driving forces), and selection of the reactor pressure at 4.377 MPa, which lies between the pressures of cases (a) and (b). Note that the HEN costs are significantly reduced while other capital and utility costs increased (electricity and cooling) to yield an increase in the profit of $2,871,000 /yr when compared to case (a) where no heat integration was considered, and with an increase of $1,971,000/yr compared to case (b). It should be noted that by the simultaneous stepwise procedure the load of the HEN was considerably reduced to 48 MW (versus 54.9 MW case (a) and 71.5 MW case (b)). What also gave rise to lower HEN costs was a significant increase in the vertical temperature driving forces and the elimination of one cold stream with very expensive matches. This example shows the importance of considering the heat exchanger network costs within a simultaneous optimization and heat integration scheme.

20.7 NOTES AND FURTHER READING

Pantelides and Smith (1995) have recently reported the application of global optimization techniques to superstructures such as the ones given in Figure 20.1 in which units can perform multiple functions through the use of rigorous models. Another recent publication dealing with strategies for structural flowsheet optimization is the one by Daichendt and Grossmann (1996) in which the use of aggregated models is proposed within a procedure that combines hierarchical decomposition and mathematical programming.

The use of an NLP optimization model for synthesizing utility systems has been proposed by Colmenares and Seider (1987). Kalitventzeff (1991) has developed an MINLP model that has applications in the retrofit of utility plants, while Foster (1987) developed an MINLP model for optimal operation.

An updated description of the implementation of the modeling/decomposition strategy in PROSYN-MINLP can be found in Kravanja and Grossmann (1993, 1994). Diwekar et al. (1992a, 1992b) have reported an application of the modeling/decomposition strategy in the public version of ASPEN. Finally, Turkay and Grossmann (1996) have recently shown that the modeling/decomposition strategy can be formalized within the framework of generalized disjunctive programming.

REFERENCES

Colmenares, T. R., & Seider, W. D. (1987). Heat and power integration of chemical processes. *AIChEJ*, **33**, 898.

Daichendt, M. M., & Grossmann, I. E. (1996, in press). Integration of hierarchical decomposition and mathematical programming for the synthesis of process flowsheets. *Computers and Chemical Engineering*.

Diwekar, U. M., Grossmann, I. E., & Rubin, E. S. (1992a). MINLP process synthesizer for a sequential modular simulator. *Industrial & Engineering Chemistry Research*, **31**, 313–322.

Diwekar, U. M., Frey, C. M., and Rubin, E. S. (1992b). Synthesizing optimal flowsheets.

Application to IGCC system environmental control. *Industrial & Engineering Chemistry Research*, **31**, 1927–1936.

Douglas, J. M. (1988). *Conceptual Design of Chemical Processes*. New York: McGraw-Hill.

Duran, M. A., & Grossmann, I. E. (1986). Simultaneous optimization and heat integration of chemical processes. *AIChE J.*, **32**, 123.

Foster, D. (1987). Optimal unit selection in a combined heat and power station. *I.Chem.E. Symp. Series,* **100,** 307.

Kalitventzeff, B. (1991). Mixed integer nonlinear programming and its application to the management of utility networks. *Engineering Optimization*, **18**, 183–207

Kocis, G. R., & Grossmann, I. E. (1987). Relaxation strategy for the structural optimization of process flow sheets. *Ind. Eng. Chem. Res.*, **26**, 1869.

Kocis, G. R., & Grossmann, I. E. (1989). A modeling and decomposition strategy for the MINLP optimization of process flowsheets. *Comput. Chem. Engng.*, **13**, 797.

Kravanja, Z., & Grossmann, I. E. (1990). PROSYN: An MINLP process synthesizer. *Computers and Chemical Engineering,* **14**, 1363

Kravanja, Z., & Grossmann, I. E. (1993). PROSYN—An automated topology and parameter process synthesizer. *Computers and Chemical Engineering*, **17**, S87–S94.

Kravanja, Z., & Grossmann, I. E. (1994). New developments and capabilities in PROSYN—An automated topology and parameter process synthesizer. *Computers Chem. Engng.*, **18**, 1097–1114.

Pantelides, C., & Smith, E. (1995). *A Software Tool for Structural and Parametric Design of Continuous Processes,* Paper 192b, Annual AIChE Meeting, Miami.

Papoulias, S. A., & Grossmann, I. E. (1983). A structural optimization approach in process synthesis. Part I: Utility systems. *Comput. Chem. Engng.*, **7**, 695.

Smith, E. M. B. (1996). On the optimal design of continuous processes, *Ph.D. Thesis,* London: Imperial College.

Turkay, M., & Grossmann, I. E. (1996). Logic-based MINLP algorithms for the optimal synthesis of process networks. *Computers and Chemical Engineering,* **20**, 959–978.

Umeda, T., Harada, T., & Ichikawa, A. (1972). Synthesis of optimal processing system by an integrated approach, *Chem.Eng.Sci.,* **27**, 795.

EXERCISES

1. Consider the superstructure in Figure 20.1 for which the raw material consists of two chemicals, X and Y, that react to yield product Z. Assume that the decreasing order of relative volatility is given by (Z, X, Y), and that the possibility of recycling the limiting reactant Y is considered, while X can be recovered as a by-product.

    a. Determine all possible configurations consisting of only one reactor and all separation sequences.

b. How would the superstructure be modified to include columns that each perform only one single task: (Z/XY), (ZX/Y), (Z/X), (X/Y)? Discuss advantages and disadvntages of both superstructures.

2. Develop a first order Taylor series expansion for the right-hand side of the nonlinear splitter equation in (20.5) at a given point (η^k, f_i^{ink}). Evaluate the corresponding linearization at $\eta^k = 0$, $f_i^{ink} = 0$. Discuss the potential numerical difficulties with such a linearization.

3. A utility plant must supply the following demands:
  a. Power 1 = 7500 kW
  b. Power 2 = 4500 kW
  c. Medium pressure steam = 25 ton/hr (minimum)
  d. Low pressure steam = 85 ton/hr (minimum)

Develop a superstructure that contains the alternatives described below.
Formulate and solve as an MILP to synthesize a utility system that requires minimum annual cost. Also find the second and third best solutions.

Steam: High pressure 4.83 MPa, 758 K
Medium pressure 2.07 MPa, 523 K
Low pressure 0.34 MPa, 412 K

Steam can be raised with high pressure and/or medium pressure boilers.
Let-down valves can be used.

Turbines:
Medium to low are backpressure turbines.
High pressure turbines can be expanded down to medium or to low pressure, and also have extractions to medium pressure.
Power demands can be satisfied with any of these turbines, but only one turbine can be assigned to each demand.
Efficiency turbines: 65%

Thermodynamic data:
ΔH (high to medium) = 71 kWhr/ton
ΔH (medium to low) = 112 kWhr/ton

Cost data:

	Fixed	Variable
Boiler HP	90,000 $/yr	9,600 $hr/yr ton steam
Boiler MP	40,000 $/yr	8,500 $hr/yr ton steam
MP turbine	25,000 $/yr	14.5 $/kWyr
HP turbine	45,000 $/yr	25 $/kWyr

If extraction is used in HP turbine, an additional fixed charge of 20,000 $/yr is required.

NOTE: Do not consider deaerator and return of steam condensate.

Exercises

4. Assume the superstructure in the figure below is considered with rigorous models for the separation of a mixture of four components. If the objective is to avoid the solution of the entire corresponding MINLP model, develop a decomposition into subsystems if the initial flowsheet is given by the direct sequence.

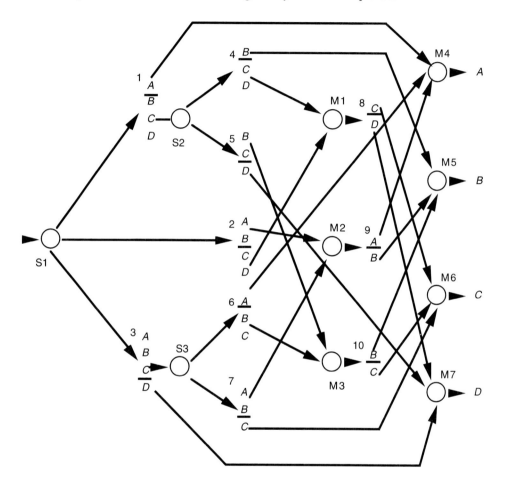

PROCESS FLEXIBILITY 21

In the previous chapters of this book we have assumed that nominal conditions are given for the specifications of a design (e.g., product demand, reaction constraints, inlet temperatures, ambient conditions). However, it is clear that these conditions will normally be different during the operation of the process. This will be due to variations that are normally encountered, as well as to uncertainties in the predicted parameters. Therefore, for a design to be useful in a practical environment it is not sufficient that it be economically optimal at the nominal conditions, but it must also exhibit good operability characteristics.

In this chapter we will address one of the important components in the operability of a chemical process, namely, flexibility (for a general review, see Grossmann et al., 1983; Grossmann and Morari, 1984; Grossmann and Straub, 1991). By flexibility we will mean the capability that a design has of having feasible steady state operation for a range of uncertain conditions that may be encountered during plant operation. Clearly, there are other aspects to the operability of a plant, such as controllability, safety, and reliability, which are equally important. However, flexibility is the first step that must be considered for the operability of a design.

In this chapter we will concentrate on two basic analysis problems for flexibility. The first problem will focus on the determination of whether a design is feasible for a fixed range of uncertainty. In the second problem we will address the question of how to actually quantify flexibility. We will present first an example to motivate the basic ideas,

21.1 MOTIVATING EXAMPLE

Let us consider the heat exchanger network structure in Figure 21.1. Note that this network only requires cooling; hence, it achieves maximum heat integration. Since this network is attractive from an economical standpoint, we would like to examine its flexibility of operation given uncertainties in the inlet temperatures T_3 and T_5 whose nominal values are 388K and 583K, respectively.

Let us assume that T_3 and T_5 can have each deviations of up to ± 10K. The question we would like to pose is whether this network, independent of area choices, has the flexibility to operate over such variations. Or, alternatively, we may want to know what are the actual temperature deviations that this network structure can tolerate. In order to address the above questions, we need to establish first the performance equations (i.e., heat balances) and the temperature specifications for the network. These are given as follows (see Figure 21.1):

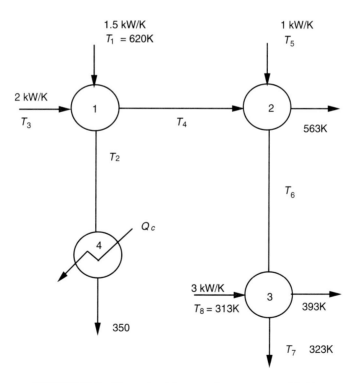

FIGURE 21.1 Network with uncertain temperatures T_3, T_5.

A. Heat balance equations:

$$\text{Exchanger 1: } 1.5(620 - T_2) = 2(T_4 - T_3) \tag{21.1}$$

$$\text{Exchanger 2: } T_5 - T_6 = 2(563 - T_4) \tag{21.2}$$

$$\text{Exchanger 3: } T_6 - T_7 = 3(393 - 313) \tag{21.3}$$

$$\text{Exchanger 4: } Q_c = 1.5(T_2 - 350) \tag{21.4}$$

B. Temperature specifications:

$$\text{Exchanger 1: } T_2 - T_3 \geq 0 \tag{21.5}$$

$$\text{Exchanger 2: } T_6 - T_4 \geq 0 \tag{21.6}$$

$$\text{Exchanger 3: } T_7 - 313 \geq 0 \tag{21.7}$$

$$\text{Exchanger 3: } T_6 - 393 \geq 0 \tag{21.8}$$

$$\text{Exchanger 3: } T_7 \leq 323 \tag{21.9}$$

Note that in the above, inequalities (21.5) to (21.8) guarantee feasible heat exchange with zero temperature approach, while Eq. (21.9) states that the outlet temperature T_7 can be delivered at any temperature equal or lower to 323K. In the equations (21.1) to (21.4), T_2, T_4, T_6, T_7 can be regarded as state variables with T_3, T_5, being uncertain parameters and Q_c, the load of the cooler, a control variable that can be adjusted in the face of changes in T_3 and T_5.

By eliminating the state variables in Eqs. (21.1) to (21.4) and substituting into Eqs. (21.5) to (21.9) yields the inequalities

$$\begin{aligned} f_1 &= T_3 - 0.666\, Q_c - 350 \leq 0 \\ f_2 &= -T_3 - T_5 + 0.5\, Q_c + 923.5 \leq 0 \\ f_3 &= -2\, T_3 - T_5 + Q_c + 1144 \leq 0 \\ f_4 &= -2\, T_3 - T_5 + Q_c + 1274 \leq 0 \\ f_5 &= 2\, T_3 + T_5 - Q_c - 1284 \leq 0 \end{aligned} \tag{21.10}$$

These inequalities will then define the feasibility of operation given a realization of T_3 and T_5 and a selection of Q_c.

If we assume that the load of the cooler Q_c remains unchanged, we can easily plot the above inequalities. Assume that Q_c is set to 75kW, which is the load at the nominal conditions $T_3 = 388K$, $T_5 = 583K$, and with T_7 at 323K—the feasible region of operation in terms of T_3 and T_5 is shown in Figure 21.2 where each of the inequalities in Eq. (21.10) has been plotted.

Note in Figure 21.2 that the nominal conditions for T_3 and T_5 lie at the boundary of constraint f_5. Clearly any increases or positive deviations in these uncertain parameters will cause infeasible operation. We may therefore be tempted to conclude that the network has very little flexibility. But is this true? Remember, we have assumed a *fixed* rate of Q_c at 75kW.

Sec. 21.1 Motivating Example

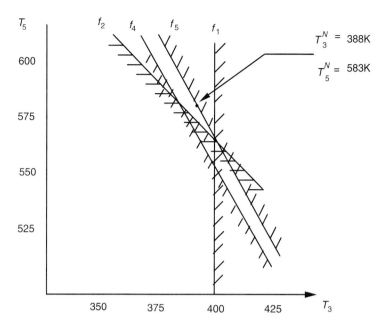

FIGURE 21.2 Feasible region for fixed $Q_c = 75$ kW.

In order to determine what happens to the network if the load Q_c is adjusted depending on the actual parameter realizations, let us consider the following flexibility test problem. At each of the four vertices or extreme values of the desired range for feasible operation $378 \leq T_3 \leq 398$, $573 \leq T_5 \leq 593$K (see Figure 21.3), we will minimize the maximum violation in the inequality constraints with respect to the heat load. That is, this problem can be formulated as the LP:

$$\psi^k = \min_{u, Q_c} u$$

s.t.
$$f_1 = T_3^k - 0.666 Q_c - 350 \leq u$$
$$f_2 = -T_3^k - T_5^k + 0.5 Q_c + 923.5 \leq u$$
$$f_3 = -2T_3^k - T_5^k + Q_c + 1144 \leq u \quad (21.11)$$
$$f_4 = -2T_3^k - T_5^k + Q_c + 1274 \leq u$$
$$f_5 = 2T_3^k + T_5^k - Q_c - 1284 \leq u$$
$$Q_c \geq 0$$

where k, $k = 1,...,4$ is an index for the vertex number, which from Figure 21.3 corresponds to:

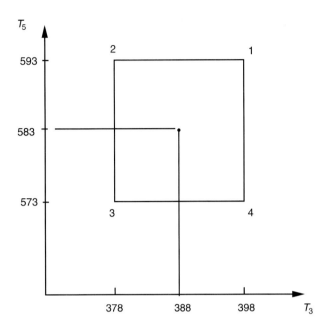

FIGURE 21.3 Desired range of feasible operation with labeled vertices.

$$\begin{aligned}
&\text{Vertex } k = 1 \; T_3^1 = 338 + 10, \; T_5^1 = 583 + 10 \\
&\text{Vertex } k = 2 \; T_3^2 = 338 - 10, \; T_5^2 = 583 + 10 \\
&\text{Vertex } k = 3 \; T_3^3 = 338 - 10, \; T_5^3 = 583 - 10 \\
&\text{Vertex } k = 4 \; T_3^4 = 338 + 10, \; T_5^4 = 583 - 10
\end{aligned} \qquad (21.12)$$

Solving Eq. (21.11) at each vertex k yields the results shown in Table 21.1. Since the maximum constraint violation ψ^k is negative in all cases the network has indeed the flexibility to operate over the assumed range of operation for the temperature variations in T_3 and T_5. But as we can see, this requires that our control variable Q_c be readjusted at each operating point and not simply set to 75 kW.

From the results in Table 21.1 it also follows that since ψ^k is strictly negative at each vertex, our network can actually tolerate variations greater than ±10K if we properly

TABLE 21.1 Results of Problem (21.11) for the Four Vertices

Vertex k	ψ^k	Q_c
1	−5	110
2	−5	—
3	−3.333	48.333
4	−3.333	88.333

Sec. 21.1 Motivating Example

adjust the load in the cooler, Q_c. We may wonder then how "flexible" our network really is.

To answer the above question, let us determine the maximum deviation that the network can tolerate along each of the four vertex directions, $k = 1,2,3,4$. This can be determined with the following LPs:

$$\delta^k = \max_{\delta, Q_c} \delta$$

s.t.
$$f_1 = T_3^k - 0.666\, Q_c - 350 \leq 0$$
$$f_2 = -T_3^k - T_5^k + 0.5\, Q_c + 923.5 \leq 0$$
$$f_3 = -2T_3^k - T_5^k + Q_c + 1144 \leq 0 \quad (21.13)$$
$$f_4 = -2T_3^k - T_5^k + Q_c + 1274 \leq 0$$
$$f_5 = 2T_3^k + T_5^k - Q_c - 1284 \leq 0$$
$$Q_c \geq 0$$

where δ is a scaled parameter deviation that for each vertex k is given as follows: (see Figure 21.3):

$$\begin{aligned}
\text{Vertex 1} \quad & T_3^1 = 338 + 10\delta,\ T_5^1 = 583 + 10\delta \\
\text{Vertex 2} \quad & T_3^2 = 338 - 10\delta,\ T_5^2 = 583 + 10\delta \\
\text{Vertex 3} \quad & T_3^3 = 338 - 10\delta,\ T_5^3 = 583 - 10\delta \\
\text{Vertex 4} \quad & T_3^4 = 338 + 10\delta,\ T_5^4 = 583 - 10\delta
\end{aligned} \quad (21.14)$$

Note that if $\delta = 1$ we get the specified expected deviation (10 K); if $\delta < 1$, it will be smaller than 10 K; if $\delta > 1$, it will be greater than 10 K.

Solving the LPs in Eq. (21.13) at each vertex yields the results shown in Table 21.2. As can be seen, the network can tolerate unbounded deviations for vertices 1 and 2. The smallest deviation is vertex 3 with $\delta^3 = 1.526$, which corresponds to the temperatures $T_3^3 = 388 - 1.53\,(10) = 372.7\,\text{K}$, $T_5^3 = 583 - 1.53(10) = 567.7$ K. Since these temperatures limit the flexibility of the network, we will denote them as the critical point. Furthermore, we can say that a quantitative measure of the flexibility of this network is 1.53. For this

TABLE 21.2 Results of Problem (21.13) for the Four Vertices

Vertex k	δ^k	Active Constraints
1	∞	—
2	∞	—
3	1.5267	(f_1, f_2)
4	2	(f_2, f_5)

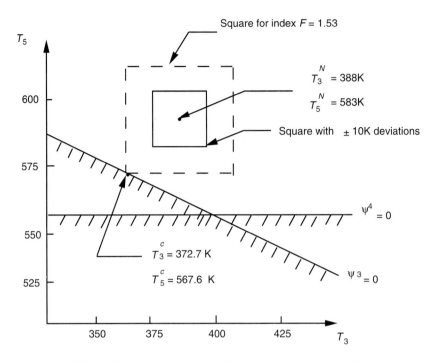

FIGURE 21.4 Feasible region with heat load Q_c as adjustable control variable.

deviation along any direction from the nominal point we will have feasible operation. We will denote the value of $\delta^3 = 1.53$ as the *index of flexibility*. As seen in Figure 21.4, this index geometrically corresponds to a square centered at the nominal point with ± 15.3 K deviations.

Finally, it is of interest to know what the actual boundary of the region of operation is when the cooler load Q_c is readjusted at each parameter point. In Table 21.2 the active constraints that were obtained in the LPs of Eq. (21.13) are given. Note that there are *two* in each case.

For vertex 3, if we equate $f_1 = f_2 = 0$, then from Eq. (21.10) algebraic manipulation and elimination of Q_c yields,

$$\psi^3 = -0.333T_3 - 1.333T_5 + 881.0255 = 0 \tag{21.15}$$

Similarly, for vertex 4, equating $f_2 = f_5 = 0$, yields

$$\psi^4 = -T_5 + 563 = 0 \tag{21.16}$$

Plotting ψ^3 and ψ^4 in terms of T_3 and T_5, and setting $\psi^3 \leq 0$, $\psi^4 \leq 0$, we obtain the region shown in Figure 21.4. As can be seen, the network has considerably more flexibility than is suggested in Figure 21.2 where Q_c was set to 75kW. Also note in Figure 21.4 that $T_3 = 372.7$, $T_5 = 567.7K$ is the critical point in that it is the closest to the nominal point lying in the boundary of the region, namely, $\psi^3 = 0$. Furthermore, the square in dashed lines corresponds to the square for the flexibility index $F = 1.53$ which is centered at the nominal point and with deviations of $\pm 15.3K$.

21.2 MATHEMATICAL FORMULATIONS FOR FLEXIBILITY ANALYSIS

In the previous section we have shown how to perform a flexibility analysis on a simple heat exchanger network. In the next two sections of this chapter we will see how we can actually generalize these ideas through mathematical formulations. We will then also consider simple vertex solution methods as well as a method that does not necessarily have to examine all the vertex points or even assume that critical points correspond to vertices.

The basic model that we will assume for the flexibility analysis will involve the following vectors of variables and parameter:

d = Design variables corresponding to the structure and equipment sizes of the plant
x = State variables that define the system (e.g., flows, temperatures)
z = Control variables that can be adjusted during operation (e.g., flows, loads utilities)
θ = Uncertain parameters (e.g., inlet conditions, reaction rate constants)

The equations that represent performance equations (e.g., heat and material balances) will be given by:

$$h(d,x,z,\theta) = 0 \qquad (21.17)$$

where by definition $\dim\{h\} = \dim\{x\}$. The constraints that represent feasible operation (e.g., physical constraints, specifications) will be given by:

$$g(d,x,z,\theta) \leq 0 \qquad (21.18)$$

Although in principle we can analyze flexibility directly in terms of Eqs. (21.17) and (21.18), for presentation purposes it is convenient to eliminate the state variables x from Eq. (21.17) as we did in section 21.1. In this way the state variables become an implicit function of d, z, and θ. That is,

$$x = x(d,z,\theta) \qquad (21.19)$$

Substituting Eq. (21.19) in Eq. (21.18) then yields the reduced inequalities

$$g(d,x(d,z,\theta),z,\theta) = f(d,z,\theta) \leq 0 \qquad (21.20)$$

Hence, the feasibility of operation of a design d operating at a given value of the uncertain parameters θ is determined by establishing whether by proper adjustment of the control variables z each inequality $f_j(d,z,\theta)$, $j \in J$ is less or equal to zero.

In the next two sections we will present mathematical formulations for both the *flexibility test problem* and the *flexibility index problem*.

21.3 FLEXIBILITY TEST PROBLEM

Let us assume that we are given a nominal value of the uncertain parameters θ^N, as well as expected deviations $\Delta\theta^+$, $\Delta\theta^-$, in the positive and negative directions. This, then, implies that the uncertain parameters θ will have the following bounds:

$$\text{Lower bound: } \theta^L = \theta^N - \Delta\theta^-$$

$$\text{Upper bound: } \theta^U = \theta^N + \Delta\theta^+$$

The flexibility test problem (Halemane and Grossmann, 1983) for a given design d will then consist of determining whether by proper adjustment of the controls z, the inequalities $f_j(d,z,\theta) \leq 0$, $j \in J$, hold for all $\theta \in T = \{\theta \,|\, \theta^L \leq \theta \leq \theta^U\}$.

In order to answer this question, we first need to consider whether for a *fixed* value of θ, the controls z can be *adjusted* to meet the constraints $f_j \leq 0$. Clearly, this can be accomplished if we select the controls z so as to minimize the largest f_j, that is,

$$\psi(d,\theta) = \min_{z} \max_{j \in J} \{f_j(d,z,\theta)\} \tag{21.21}$$

where $\psi(d,\theta)$ is defined as the feasibility function. If $\psi(d,\theta) \leq 0$, we can clearly have feasible operation; if $\psi(d,\theta) > 0$, there is infeasible operation even if we do our best in trying to adjust the control variables z. If $\psi(d,\theta) = 0$, it also means that we are on the boundary of the region of operation, since in this case $f_j = 0$ for at least one constraint j (see Figure 21.5).

Problem (21.21) can be posed as a standard optimization problem by defining a scalar variable u, such that

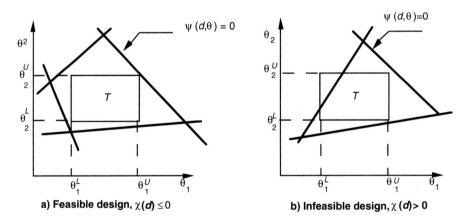

FIGURE 21.5 Regions of feasible operation for feasible and infeasible design (flexibility test problem).

$$\psi(d,\theta) = \min_{z,u} u \qquad (21.22)$$

$$s.t.\ f_j(d,z,\theta) \leq u \quad j \in J$$

This is precisely the problem we considered in Eq. (21.11), which happened to be an LP due to the linearity of f_j in z. In general, however, Eq. (21.22) will correspond to an NLP problem if f_j is nonlinear in z.

In order to determine whether we can have feasible operation in the parameter range of interest,

$$T = \{\theta \mid \theta^L \leq \theta \leq \theta^U\} \qquad (21.23)$$

we clearly need to establish whether $\psi(d,\theta) \leq 0$ for all $\theta \in T$. But this is also equivalent to stating whether the maximum value of $\psi(d,\theta)$ is less or equal than zero in the range θ. Hence, the flexibility test problem can be formulated as

$$\chi(d) = \max_{\theta \in T} \psi(d,\theta) \qquad (21.24)$$

where $\chi(d)$ corresponds to the flexibility function of design d over the range T. If $\chi(d) \leq 0$, it then clearly means that feasible operation can be attained over the parameter range T (see Figure 21.5a). If $\chi(d) > 0$, it means that at least for part of the range of T, feasible operation cannot be achieved (see Figure 21.5b). Also, the value of θ determined in Eq. (21.24) can be regarded as a critical value for the parameter range T, since at this value the feasibility of operation is the smallest ($\chi(d) \leq 0$) or where maximum constraint violation occurs ($\chi(d) > 0$).

Finally, by substituting Eq. (21.21) in Eq. (21.24), the general mathematical formulation of the flexibility test problem yields,

$$\chi(d) = \max_{\theta \in T} \min_z \max_{j \in J} f_j(d,z,\theta) \qquad (21.25)$$

The above is in general a difficult problem whose solution we will examine in sections 21.5 and 21.6.

21.4 FLEXIBILITY INDEX PROBLEM

The drawback in the flexibility test problem is that it only determines whether a design does or does not have the flexibility to operate over the specified parameter range T. It is clearly desirable to develop a quantitative measure that will indicate how much flexibility can actually be achieved in the given design. To consider this question, let us define a *variable* parameter range

$$T(\delta) = \{\theta \mid \theta^N - \delta\Delta\theta^- \leq \theta \leq \theta^N + \delta\Delta\theta^+\} \qquad (21.26)$$

where δ is a non-negative scalar variable. Note that for $\delta = 1$, $T(1) = T$; that is, in this case $T(\delta)$ becomes identical to our specified parameter range T. For $\delta < 1$, it is clear that $T(\delta) \subset T$, while for $\delta > 1$, $T(\delta) \supset T$.

We can then define as the *flexibility index*, F, the largest value of δ such that the inequalities $f_j(d,z,\theta) \leq 0$, $j \in J$, hold over the parameter range $T(F)$ (i.e., $\chi(d) \leq 0$ for $T(F)$). Mathematically, this problem can be posed as (Swaney and Grossmann, 1985b)

$$F = \max \delta$$

$$\text{s.t.} \quad \chi(d) = \max_{\theta \in T} \min_{z} \max_{j \in J} f_j(d,z,\theta) \leq 0 \tag{21.27}$$

$$T(\delta) = \{\theta \mid \theta^N - \delta\Delta\theta^- \leq \theta \leq \theta^N + \delta\Delta\theta^+\}, \delta \geq 0$$

The geometrical interpretation of this problem is shown in Figure 21.6, where it can be seen that $T(F)$ is the largest rectangle that can be inscribed within the region of operation. This rectangle is centered at the nominal point and its sides are proportional to the expected derivations, $\Delta\theta^+, \Delta\theta^-$. Note that the flexibility index also indicates the actual parameter range that can be handled by the design; this will be given by (see Figure 21.6),

$$T(F) = \{\theta \mid \theta^N - F\Delta\theta^- \leq \theta \leq \theta^N + F\Delta\theta^+\} \tag{21.28}$$

The interpretation of the flexibility index, F, is then also clarified. A value $F = 1$ implies that the design has *exactly* the flexibility to satisfy the constraints over the set T. A value $F > 1$ implies that the design *exceeds* the flexibility requirements; a value $F < 1$ indicates the *fractional* deviation that can actually be handled for any of the expected deviations.

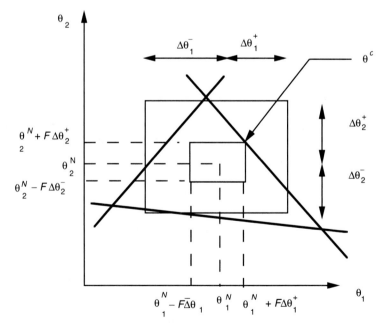

FIGURE 21.6 Geometrical representation of parameter range $T(F)$ with flexibility index F.

Finally, the value of θ determined by Eq. (21.27) corresponds to the critical parameter point, θ^c, that limits flexibility (see Figure 21.6). Thus, it is clear that the flexibility index problem can supply a great deal of useful information.

21.5 VERTEX SOLUTION METHODS

The solution of Eq. (21.25) for the flexibility test problem and of Eq. (21.27) for the flexibility index problem can be greatly simplified for the case when the critical points correspond to vertices or extreme values of the parameter sets T and $T(F)$, respectively (Halemane and Grossmann, 1983).

Consider first the flexibility test problem, and let θ^k, $k \in V$, represent the vertices of the set T. Then, Eq. (21.24) reduces to

$$\chi(d) = \max_{k \in V} \{\psi(d, \theta^k)\} \tag{21.29}$$

Note that $\psi(d, \theta^k)$ can be evaluated through the optimization problem in Eq. (21.22) at the vertex θ^k (recall section 21.1). Hence, the following simple algorithm can be applied:

Step 1: For each vertex θ^k, $k \in V$, solve the optimization problem

$$\psi(d, \theta^k) = \min_{z, u} u$$

$$\text{s.t.} \quad f_j(d, z, \theta^k) \leq u \quad j \in J$$

Step 2: Set $\chi(d) = \max_{k \in V} \{\psi(d, \theta^k)\}$.

If $\chi(d) \leq 0$, then the design is feasible to operate over the set T; otherwise, if $\chi(d) > 0$, it is not.

For the flexibility index problem a similar procedure can be applied. First, note that in Eq. (21.27), $\chi(d) = 0$ at the optimal solution, since the critical point in this case will always lie on the boundary (see Figure 21.6). Let $\Delta\theta^k$, $k \in V$, denote the vertex directions from the nominal point to the vertex points in T. Then, the maximum derivation δ^k to the boundary along $\Delta\theta^k$ will be given by the optimization problem

$$\delta^k = \max_{z, \delta} \delta$$

$$\text{s.t.} \quad f_j(d, z, \theta) \leq 0 \quad j \in J \tag{21.30}$$

$$\theta = \theta^N + \delta \Delta\theta^k$$

From among the parameter rectangles $T(\delta^k)$, $k \in V$, it is clear that only the smallest one can be totally inscribed within the feasible region. Hence,

$$F = \min_{k \in V} \{\delta^k\} \tag{21.31}$$

Thus, the following simple algorithm applies,

Step 1: Solve the optimization problem in (21.30) for each vertex $k \in V$.

Step 2: Set $F = \min_{k \in V} \{\delta^k\}$

The two above algorithms were precisely the ones that were applied to the problem in section 21.1. The question, though, is whether we can always use these procedures. The answer is no.

First, it can be shown that only under some convexity conditions (see Swaney and Grossmann, 1985a,b) for the constraint functions $f_j, j \in J$, the critical points will always correspond to vertices (e.g., linear functions). For most cases however, even when these conditions are not met, we will still have vertex critical points. The next section will show an example where we can have nonvertex critical points due to nonconvexities.

A second reason is that even if critical points are vertices, we may be faced with the problem of having to analyze far too many vertices. Say we have 10 uncertain parameters; we would have to solve $2^{10} = 1024$ optimization problems according to the above algorithms. If we have 20, we would have to solve $2^{20} = 1,048,576$ optimization problems. We will present in section 21.7 a method that can overcome these problems.

21.6 EXAMPLE WITH NONVERTEX CRITICAL POINT

Let us consider the heat exchanger network shown in Figure 21.7 (Saboo and Morari, 1984) where the heat capacity flowrate F_{H1} is an uncertain parameter. We would like to determine whether this network is feasible for the range $1 \leq F_{H1} \leq 1.8$ (kW/K).

The following inequalities are considered for feasible operation of this network:

Feasibility in exchanger 2:	$T_2 - T_1 \geq 0$	
Feasibility in exchanger 3:	$T_2 - 393 \geq 0$	(21.32)
Feasibility in exchanger 3:	$T_3 - 313 \geq 0$	
Specification in outlet temperature	$T_3 \leq 323$	

By considering the corresponding heat balances, we can solve for the above temperatures in terms of the cooling load Q_c, our control variable, and in terms of F_{H1}, the uncertain parameter. The reduced inequalities in Eq. (21.32) are then as follows:

$$f_1 = -25 + Q_c [(1/F_{H1}) - 0.5] + 10/F_{H1} \leq 0$$
$$f_2 = -190 + (10/F_{H1}) + (Q_c/F_{H1}) \leq 0 \quad (21.33)$$
$$f_3 = -270 + (250/F_{H1}) + (Q_c/F_{H1}) \leq 0$$
$$f_4 = 260 - (250/F_{H1}) - (Q_c/F_{H1}) \leq 0$$

Sec. 21.6 Example with Nonvertex Critical Point

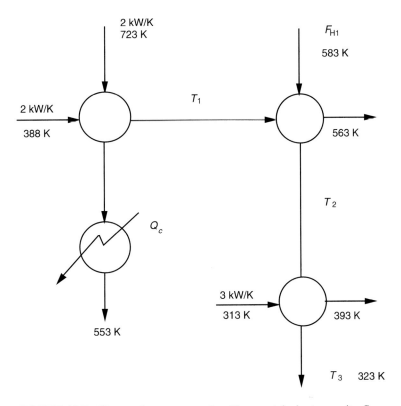

FIGURE 21.7 Heat exchanger network with uncertain heat capacity flow-rate, F_{H1}.

If we now examine the two extreme points, for F_{H1}, by solving the NLP in Eq. (21.22) for the above inequalities we get the following:

For F_{H1} = 1kW/K, $\psi^1(1) = -5$, $Q_c = 15$ kW
For F_{H1} = 1.8 kW/K, $\psi^2(1.8) = -5$, $Q_c = 227$ kW

Since $\psi^1 < 0$ and $\psi^2 < 0$, we may be tempted to conclude that the network is feasible to operate for the range $1 \leq F_{H1} \leq 1.8$ kW/K. However, let us consider an intermediate value, say F_{H1} = 1.2 kW/K for problem (21.22). We then get:

$$F_{H1} = 1.2 \text{ kW/K}, \psi(1.2) = 2.85; Q_c = 58.57 \text{ kW}$$

In other words, the network is infeasible at the *nonvertex* point F_{H1} = 1.2 kW/K. Why is that? If we plot the constraints in Eq. (21.33), as shown in Figure 21.8, we can clearly see that we have a nonconvex region where for $1.118 \leq F_{H1} \leq 1.65$ we have infeasible operation. In fact, at F_{H1} = 1.37 kW/K we have the greatest violation of constraints, since at that

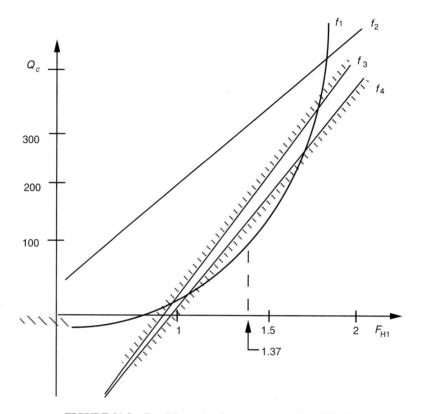

FIGURE 21.8 Feasible region for constraints in Eq. (21.33).

point $\psi(1.37) = +5.108$ attains its maximum value. Hence, $F_{H1} = 1.37$ corresponds to the critical point.

The above example, then, shows that it is possible to have nonvertex critical points, and consequently, we need an appropriate method that will be able to predict such points as we will show in the next section.

21.7 ACTIVE SET METHOD

In this section we will show how the flexibility test in problem (21.24) and the flexibility index in problem (21.27) can be formulated as mixed-integer optimization problems (Grossmann and Floudas, 1987).

Let us consider first problem (21.24), the flexibility test, which with Eq. (21.21) becomes,

Sec. 21.7 Active Set Method

$$\chi(d) = \max_{\theta \in T} \psi(d,\theta)$$
$$\text{s.t. } \psi(d,\theta) = \min_{z} \max_{j \in J} f_j(d,z,\theta) \quad (21.34)$$

The above is clearly a two-level optimization problem since it involves as a constraint the min max problem for the function ψ. In order to convert this constraint into algebraic equations, let us consider the Karush-Kuhn-Tucker conditions of the function $\psi(d,\theta)$ as defined by the problem in (21.22). These conditions yield (see Appendix A):

$$\sum_{j \in J} \lambda_j = 1 \quad (21.35a)$$

$$\sum_{j \in J} \lambda_j \frac{\partial f_i}{\partial z} = 0 \quad (21.35b)$$

$$\lambda_j \left[f_j(d,z,\theta) - u \right] = 0 \quad j \in J \quad (21.35c)$$

$$\lambda_j \geq 0, f_j(d,z,\theta) - u \leq 0 \quad j \in J \quad (21.35d)$$

where λ_j are the Lagrange multipliers for the constraints $f_j - u \leq 0$ in Eq. (21.22). Since at the optimal solution of (21.22), $\psi(d,\theta) = u$, we can reformulate Eq. (21.34) as a *single level* optimization problem.

$$\chi(d) = \max_{\theta \in T} u$$
$$\text{s.t. Contraints in (21.35)} \quad (21.36)$$

The difficulty, however, is that the complementarity conditions in Eq. (21.35c) imply making discrete choices of those constraints that become active in Eq. (21.22), that is, $f_j - u = 0$. Thus, if $\lambda_j = 0, f_j - u < 0$, constraint j is inactive. We can, however, model these discrete choices as follows.

Let $s_j \geq 0$, be the slack of constraint $f_j - u \leq 0$, such that

$$f_j(d,z,\theta) + s_j = u \quad j \in J \quad (21.37)$$

Also let y_j be a 0–1 variable defined as follows:

$$y_j = \begin{cases} 1 \text{ if constraint } f_j - u = 0 \\ 0 \text{ otherwise} \end{cases}$$

This binary variable can be related to s_j and λ_j by the logical inequalities:

$$\left. \begin{array}{l} s_j \leq U(1 - y_j) \\ \lambda_j \leq y_j \end{array} \right\} \quad j \in J \quad (21.38)$$

where U is a valid upper bound for the slacks. Note then that if $y_j = 1$, it implies $s_j = 0$, $0 \leq \lambda_j \leq 1$; if $y_j = 0$, it implies $0 \leq s_j \leq U$, $\lambda_j = 0$. In other words, the inequalities in Eq. (21.38) are equivalent to the conditions in Eq. (21.35c).

Furthermore, it can be shown that if the gradients $\partial f_j/\partial z$, $j \in J$ are linearly independent (Swaney and Grossmann, 1985a,b), then there will be $n_z + 1$ active constraints in Eq. (21.22), where n_z is the dimensionality of the control variables z. Recall that in section 21.1 we had one control variable and two active constraints. Hence, we can set

$$\sum_{j \in J} y_j \leq n_z + 1 \qquad (21.39)$$

to account for the possibility that the assumption of linear independence may not hold. By then considering Eqs. (21.37), (21.38), (21.39) in place of Eqs. (21.35c) and (21.35d), problem (21.36) can be posed as the following mixed-integer optimization problem:

$$\chi(d) = \max_{\substack{u,\theta,z \\ \lambda_j, s_j, y_j}} u$$

$$\text{s.t.} \quad f_j(d,z,\theta) + s_j = u \quad j \in J$$

$$\sum_{j \in J} \lambda_j = 1$$

$$\sum_{j \in J} \lambda_j \frac{\partial f_j}{\partial z} = 0 \qquad (21.40)$$

$$\left. \begin{array}{l} s_j - U(1-y_j) \leq 0 \\ \lambda_j - y_j \leq 0 \end{array} \right\} \quad j \in J$$

$$\sum_{j \in J} y_j \leq n_z + 1$$

$$\theta^L \leq \theta \leq \theta^U$$

$$\lambda_j, s_j \geq 0, j \in J; \; y_j = 0,1 \quad j \in J$$

Note that in the above formulation all the variables, u, θ, z, λ_j, s_j, y_j, $j \in J$ appear as variables for the optimization since these are constrained to solve the problem for $\psi(d,\theta)$ through the constraints. There are several interesting features about the formulation in Eq. (21.40):

1. If f_j is linear in z and θ, Eq. (21.40) corresponds to an MILP problem (note $\partial f_j/\partial z$ is constant for this case). Otherwise, it corresponds to an MINLP.
2. No enumeration of vertices is required, and therefore many uncertain parameters can be handled.
3. The derivation of problem (21.40) did not require the assumption of vertex critical points. Hence, we will be able to predict nonvertex critical points as will be shown in section 21.8.

We can derive a similar formulation for the flexibility index problem by reformulating Eq. (21.27) as the minimum δ to the boundary $\psi(d,\theta) = 0$. That is,

Sec. 21.8 Active Set Method for Nonvertex Example

$$F = \min \delta$$
$$\text{s.t. } \psi(d,\theta) = 0 \tag{21.41}$$

Since the constraint $\psi(d,\theta) = 0$ implies setting $u = 0$ in problem (21.40) and from the derivation of the variable parameter range in Eq. (21.26), the flexibility index problem can be posed as the following mixed-integer optimization problem:

$$F = \min_{\delta, \lambda_j, s_j, y_j} \delta$$

$$\text{s.t. } f_j(d,z,\theta) + s_j = 0 \quad j \in J$$

$$\sum_{j \in J} \lambda_j = 1$$

$$\sum_{j \in J} \lambda_j \frac{\partial f_i}{\partial z} = 0 \tag{21.42}$$

$$\left. \begin{array}{l} s_j - U(1 - y_j) \leq 0 \\ \lambda_j - y_j \leq 0 \end{array} \right\} \quad j \in J$$

$$\sum_{j \in J} y_j \leq n_z + 1$$

$$\theta^N - \delta \Delta \theta^- \leq \theta \leq \theta^N + \delta \Delta \theta^+$$

$$\delta \geq 0; s_j, \lambda_j \geq 0, j \in J; y_j = 0, 1 \quad j \in J$$

This problem has again similar features as the flexibility test problem in Eq. (21.40).

To provide some more insight behind these formulations, we will apply the flexibility test in Eq. (21.40) to the nonvertex problem in section 21.6.

21.8 ACTIVE SET METHOD FOR NONVERTEX EXAMPLE

Applying the flexibility test formulation in Eq. (21.40) to the inequalities in Eq. (21.33) for the heat exchanger network in section 21.6 yields

$$\chi(d) = \max_{\substack{u, Q_c, F_{H1} \\ s_j \lambda_j, y_j}} u \tag{21.43}$$

$$\text{s.t.} \quad -25 + Q_c[(1/F_{H1}) - 0.5] + 10/F_{H1} + s_1 = u$$

$$-190 + (10 F_{H1}) + (Q_c/F_{H1}) + s_2 = u$$

$$-270 + (250 F_{H1}) + (Q_c/F_{H1}) + s_3 = u$$

$$260 - (250 F_{H1}) - (Q_c/F_{H1}) + s_4 = u$$

$$\lambda_1 + \lambda_2 + \lambda_3 + \lambda_4 = 1$$

$$\left[\left(\frac{1}{F_{HI}}\right) - 0.5\right]\lambda_1 + \left(\frac{1}{F_{HI}}\right)\lambda_2 + \left(\frac{1}{F_{HI}}\right)\lambda_3 - \left(\frac{1}{F_{HI}}\right)\lambda_4 = 0$$

$$\left.\begin{array}{c} s_j - 1000\,(1 - y_i) \leq 0 \\ \lambda_j - y_j \leq 0 \end{array}\right\} \quad j = 1, 4$$

$$y_1 + y_2 + y_3 + y_4 = 2$$

$$1 \leq F_{H1} \leq 1.8$$

$$s_j, \lambda_j \leq 0, j = 1, 4 \,;\, y_j = 0, 1 \quad j = 1, 4$$

Problem (21.43) corresponds to an MINLP problem that can be solved with the outer approximation/equality relaxation method described in Appendix A. In fact, applying this method yields $u = 5.108$, $F_{H1} = 1.37$ kW/K, which corresponds precisely to the point of maximum constraint violation as was discussed in section 21.6. Also $y_1 = 1$, $y_4 = 1$, $y_2 = y_3 = 0$ means that constraints 1 and 4 are the active constraints responsible for the infeasibility, as in fact is the case seen in Figure 21.8.

Since the above problem in Eq. (21.43) is not too large, let us consider its analytical solution.

First, we note that two of the 0–1 variables have to be set to one; that is, we will have two active constraints. Further, from the stationary equations (21.35a) and (21.35b) in Eq. (21.43) we have,

$$\lambda_1 + \lambda_2 + \lambda_3 + \lambda_4 = 1 \tag{21.44}$$

$$\left[\left(\frac{1}{F_{H1}}\right) - 0.5\right]\lambda_1 + \left(\frac{1}{F_{H1}}\right)\lambda_2 + \left(\frac{1}{F_{H1}}\right)\lambda_3 - \left(\frac{1}{F_{H1}}\right)\lambda_4 = 0$$

Since $1 \leq F_{H1} \leq 1.8$ and two λ_j must be non-zero, there are three possible active sets that can satisfy Eq. (21.44):

Active set 1: Constraints 1,4 ($s_1 = s_4 = 0$, λ_1, λ_4 non-zero)
Active set 2: Constraints 2,4 ($s_2 = s_4 = 0$, λ_2, λ_4 non-zero)
Active set 3: Constraints 3,4 ($s_3 = s_4 = 0$, λ_3, λ_4 non-zero)

For each of the above active sets we can determine their corresponding value of u by simply setting their two constraints to u and solving the corresponding equations for u. For instance, take active set 1. By setting $f_1 = u^1, f_4 = u^1$, in Eq. (21.43) leads to:

$$u^1 = 260 - \frac{250}{F_{H1}} + \frac{520 - 570 F_{H1}}{F_{H1}(4 - F_{H1})} \tag{21.45}$$

If we now maximize u with respect to F_{H1} (e.g., with any one-dimensional optimization method) we get $F_{H1} = 1.372$ kW/K and $u^1 = +5.108$, which is precisely the nonvertex point in Figure 21.8, where it is clear that constraints 1 and 4 are responsible for the maximum infeasibility.

Let us consider now active set 2. By setting $f_2 = u^2, f_4 = u^2$, in Eq. (21.43) leads to:

$$u^2 = 35 - (120/F_{H1}) \tag{21.46}$$

The above exhibits its maximum at $F_{H1} = 1.8$, the upper bound, with $u^2 = -31.67$. As seen in Figure 21.8, at that point constraints f_2 and f_4 do not cause infeasibility.

Finally, for active set 3, we set $f_3 = u^3, f_4 = u^3$ in Eq. (21.43). This leads to $u^3 = -5$; that is, constraints f_3 and f_4 do not cause infeasibility for any value of F_{H1}, as can be seen in Figure 21.8.

Since from among the three active sets $u^1 = +5.108$ is the largest, this corresponds to the solution of problem (21.43).

The above procedure that we have outlined, which is based on individual analysis of each potential active set of constraints, can be used as an alternative to the direct solution of the MILP or MINLP in Eq. (21.40) for the flexibility test. A similar procedure can be used for the flexibility index problem in Eq. (21.42).

21.9 SPECIAL CASES FOR FLEXIBILITY ANALYSIS

In the previous sections we have made three major assumptions for the flexibility analysis problems:

1. Independent variations of the uncertain parameters θ.
2. There is always at least one control variable z.
3. The reduced inequalities are obtained by algebraically eliminating the performance equation in (21.17).

We will briefly discuss how we can handle extensions for each of these cases. First, it is quite commonly the case that we may have *correlated* uncertain parameters. For example, assume that two flowrate variations are given by

$$F_1 = 10\,(1 + \theta) \tag{21.47}$$

$$F_2 = 20\,(1 + \theta)$$

where $-0.1 \le \theta \le +0.1$. This, then, means that both flowrates increase or decrease simultaneously, but one cannot increase while the other decreases and vice versa. The simplest option is to regard only θ as an uncertain parameter and F_1 and F_2 as state variables. Alternatively, for this example, or more generally when the parameter correlations are given by algebraic equations $r(\theta) = 0$, we can simply add these as constraints in the mixed-integer optimization problems (21.40) and (21.42).

Often, we might also have problems where there are no control variables z (i.e., $n_z = 0$). We would expect our flexibility analysis problems to become simple to solve. This is indeed the case. Consider, for instance, problem (21.40) for the flexibility test. If

$n_z = 0$, the stationary conditions in Eq. (21.35) are not required. Hence, problem (21.40) reduces to:

$$\chi(d) = \max_{u,\theta,s,y_i} u$$

$$\left. \begin{array}{l} \text{s.t.} \quad f_j(d,\theta) + s_j = u \\ \phantom{\text{s.t.}} \quad s_j - U(1 - y_j) \leq 0 \end{array} \right\} \quad j \in J$$

$$\sum_{j \in J} y_j = 1 \qquad (21.48)$$

$$\theta^L \leq \theta \leq \theta^U$$

$$s_j \geq 0, \; y_j = 0,1, \quad j \in J$$

Since in the above formulation only one constraint can be active, we can easily decompose the solution to this problem by setting $s_j = 0$ and maximizing $u = f_j(d,\theta)$ for each constraint j. That is the problem reduces to:

Step 1: For each constraint $j \in J$, solve $u^j = \max\limits_{\theta^L \leq \theta \leq \theta^U} f_j(d,\theta)$.

Step 2: Set $\chi(d) = \max\limits_{j \in J} \{u^j\}$

Qualitatively, what we are doing in the above procedure is to maximize each constraint with respect to θ and setting $\chi(d)$ to that constraint with the highest value.

In a similar fashion, it can easily be shown that for $n_z = 0$ the problem for the flexibility index reduces from Eq. (21.42) to:

Step 1: For each constraint $j \in J$, solve $\delta^j = \min\limits_{\delta,\theta} \delta$

$$\text{s.t.} \; f_j(d,\theta) = 0$$

$$\theta^N - \delta\Delta\theta^- \leq \theta \leq \theta^N + \delta\Delta\theta^+$$

Step 2: Set $F = \min\limits_{j \in J} \{\delta^j\}$.

That is, for each constraint we determine the closest displacement δ^j to the boundary, $f_j(d,\theta) = 0$, and then set the index F to the smallest of all the displacements.

Finally, let us consider the case where we would like to explicitly keep the performance equations to avoid the algebraic elimination of the state variables.

The case when there are no control variables is straightforward, as we then simply have to include the equations $h_i(d,x,\theta) = 0$, $i \in I$, in the optimization problems. For example, for the flexibility test, u^j can be determined as:

Sec. 21.9 Special Cases for Flexibility Analysis

$$u^j = \max_{\theta, x} g_j(d, x, \theta)$$

$$\text{s.t. } h_i(d, x, \theta) = 0 \quad i \in I \tag{21.49}$$

$$\theta^L \leq \theta \leq \theta^U$$

For the case when $n_z \geq 1$, the feasibility function $\psi(d, \theta)$ in Eq. (21.22) must be redefined as

$$\psi(d, \theta) = \min_{u, z, x} u$$

$$\text{s.t. } h_i(d, x, z, \theta) = 0 \quad i \in I \tag{21.50}$$

$$g_j(d, x, z, \theta) \leq u \quad j \in J$$

This formulation would then be used for the vertex search method in section 21.5 for the flexibility test.

For the mixed-integer formulation in Eq. (21.40), the Karush-Kuhn-Tucker conditions of problem (21.50) must be included. Using a similar reasoning as used in section 21.7 (see exercise 8), the flexibility test problem corresponds to:

$$\chi(d) = \max_{\substack{u, \theta, z \\ \lambda_j \mu_i s_j y_j}} u$$

$$\text{s.t. } h_i(d, x, z, \theta) = 0 \quad i \in I$$

$$g_j(d, x, z, \theta) + s_j = u \quad j \in J$$

$$\sum_{j \in J} \lambda_j = 1$$

$$\sum_{i \in I} \mu_i \frac{\partial h_i}{\partial z} + \sum_{j \in J} \lambda_j \frac{\partial g_j}{\partial z} = 0 \tag{21.51}$$

$$\sum_{i \in I} \mu_i \frac{\partial h_i}{\partial x} + \sum_{j \in J} \lambda_j \frac{\partial g_j}{\partial x} = 0$$

$$\left.\begin{array}{l} \lambda_j - y_j \leq 0 \\ s_j - U(1 - y_j) \leq 0 \end{array}\right\} j \in J$$

$$\sum_{j \in J} y_j \leq n_z + 1$$

$$\theta^L \leq \theta \leq \theta^U$$

$$s_j, \lambda_j \geq 0 \ j \in J; \ y_j = 0, 1 \ j \in J$$

where μ_i are Lagrange multipliers to the equality constraints in Eq. (21.50) that are unrestricted in sign (see Appendix A). Note that in Eq. (21.51) we have the advantage of not having to eliminate equations, although we face a problem larger in size than in Eq. (21.40).

Similar extensions can be performed for the flexibility index problem in (21.42) (see exercise 8).

21.10 OPTIMAL DESIGN UNDER UNCERTAINTY

In the previous sections of this chapter we have exclusively considered the problem of analyzing the flexibility of a given design. An important question is, of course, how to systematically determine designs that can accomplish a desired degree of flexibility. In this section we will briefly address this question.

In conventional design optimization problems the design variables d must be selected so as to minimize cost at some nominal values of the uncertain parameters. When the goal of flexibility is also to be accomplished, there are basically two options: Either (a) ensure flexibilty for a fixed parameter range (i.e., satisfy the feasibility test Eq. (21.25); or (b) maximize the flexibility measure as given by Eq. (21.27), while at the same time minimizing cost. The latter problem gives rise to a multi-objective optimization problem, which in fact would normally be solved by optimizing the cost at different fixed values of the flexibility range (e.g., flexibility index). Thus, by considering the solution of case (a), one can in principle also approach the solution by option (b).

The choice of the objective for minimizing cost merits some discussion. Most of the previous work in design under uncertainty (Johns et al., 1976; Malik and Hughes, 1979) has considered the effect of the continuous uncertain parameters θ for the design optimization through the minimization of the expected value of the cost using what is normally termed a *two-stage strategy*:

$$\min_{d} E_{\theta} \left[\min_{z} C(d, z, \theta) \mid f(d, z, \theta) \leq 0 \right] \tag{21.52}$$

The reason the above is denoted as a two-stage strategy is because the problem is conceived in two stages: stage 1, which is prior to the operation (design phase), and stage 2, which is the time of operation. The design variables d are chosen in stage 1 once and for all, since they remain fixed during stage 2. At this second stage, the control variables z are adjusted during operation depending on the realizations of the parameters θ. Note that implicit in this design strategy there is the assumption of "perfect" control. That is, the control can be immediately adjusted depending on the realization of θ. No delays in the measurements, or adjustments in the control are considered.

One situation that can arise in the optimization of Eq. (21.52) is an infeasible operation at a certain value of θ. This would mean that no control z can be selected given the current selection of the design variables d in the optimization. In order to handle infeasibilities in the inner minimization, one approach is to assign penalties for the violation of constraints (e.g., $C(d,z,\theta) = \overline{C}$ if $f(d,z,\theta) > 0$. This, however, can lead to discontinuities.

The other approach is to enforce feasibility for a specified flexibility index F (e.g., see Halemane and Grossmann, 1983) through the parameter set $T(F) = \{\theta | \theta^L - F\Delta\theta^- \leq \theta \leq \theta^U + F\Delta\theta^+, r(\theta) \leq 0\}$. In this case, Eq. (21.52) is formulated as

$$\min_{d} \; \mathop{E}_{\theta \in T(F)} \left[\min_{z} \; C(d, z, \theta) \mid f(d, z, \theta) \leq 0 \right]$$

$$\text{s.t.} \; \max_{\theta \in T(F)} \psi(d, \theta) \leq 0 \tag{21.53}$$

A particular case of Eq. (21.53) is when the infinite number of points in $T(F)$ is replaced by a discrete set of points θ^k, $k = 1..K$, which are somehow specified. This gives rise to the optimal design problem,

$$\min_{d, z^1...z^k} \sum_{k=1}^{K} w_k \, C\!\left(d, z^k, \theta^k\right)$$

$$\text{s.t.} \; f\!\left(d, z^k, \theta^k\right) \leq 0 \quad k = 1..K \tag{21.54}$$

where w_k are weights that are assigned to each point θ^k, and $\sum_{k=1}^{K} w_k = 1$.

Problem (21.54) can be interpreted as a multiperiod design problem in which the weights can in fact be interpreted as probabilities, or durations, of the realization of each parameter value θ^k. As shown by Grossmann and Sargent (1978), problem (21.54) can also be used to approximate the solution of (21.53). This is accomplished by applying the following algorithm:

Step 1: Select an initial set of points θ^k.

Step 2: Solve the multiperiod optimization problem (21.54) to obtain a design.

Step 3: Check the feasibility of the proposed design over $T(F)$ by solving problem (21.25) or (21.27). If the design is feasible, the procedure terminates. Otherwise, the critical point obtained from the flexibility evaluation is included in the current set of θ points, and return to step 2.

Computational experience has shown that commonly one or two major iterations must be performed to achieve feasibility with this method.

21.11 NOTES AND FURTHER READING

General reviews on process flexibility can be found in Grossmann et al. (1983), Grossmann and Morari (1984) and Grossmann and Straub (1991). Recent methods for flexibility analysis include the branch and bound method by Kabatek and Swaney (1992), and the sensitivity based method by Varvarezos et al. (1995).

Design applications include synthesis of heat exchanger networks (Floudas and Grossmann, 1987), and retrofit design (Pistikopoulos and Grossmann, 1988, 1989). The multiperiod optimization problem is important in its own right for the design of flexible chemical plants (see Grossmann and Sargent, 1979; Varvarezos et al. 1992). Other approaches for the design problem can be found in Pistikopoulos and Grossmann (1988, 1989).

Finally, this chapter has not addressed methods that deal with a probabilistic description of the uncertain parameters. The treatment and definition of stochastic flexibility index is given in Pistikopoulos and Mazzuchi (1991) and Straub and Grossmann (1991). Issues related to design with such an index can be found in Straub and Grossmann (1993).

REFERENCES

Floudas, C. A., & Grossmann, I. E. (1987). Synthesis of flexible heat exchanger networks with uncertain flowrates and temperatures. *Comp. Chem. Eng.*, **11,** 319.

Grossmann, I. E., & Floudas, C. A. (1987). Active constraint strategy for flexibility analysis in chemical processes. *Comp. Chem. Eng.*, **11,** 675.

Grossmann, I. E., Halemane, K. P., & Swaney, R. E. (1983). Optimization strategies for flexible chemical processes. *Comp. Chem. Eng.*, **7,** 439.

Grossmann, I. E., & Morari, M. (1984). Operability, resiliency and flexibility-process design objectives for a changing world. In Westerberg & Chien. (Eds.), *Proc. 2nd Int. Conf. Foundations Computer Aided Process Design.* CACHE, 937.

Grossmann, I. E., & Sargent, R. W. H. (1978). Optimum design of chemical plants with uncertain parameters. *AIChE J.*, **24,** 1021.

Grossmann, I. E., & Sargent, R. W. H. (1979). Optimum design of multipurpose chemical plants. *Ind. Eng. Chem. Process Des. Development,* **18,** 343.

Grossmann, I. E., & Straub, D. A. (1991). Recent developments in the evaluation and optimization of flexible chemical processes. In L. Puigjaner, & A. Espuna (Eds.), *Proceedings of COPE-91.* Barcelona, Spain.

Halemane, K. P., & Grossmann, I. E. (1983). Optimal process design under uncertainty. *AIChE J.*, **29,** 425.

Johns, W. R., Marketos, G., & Rippin, D. W. T. (1976). The optimal design of chemical plant to meet time-varying demands in the presence of technical and commercial uncertainty. *Design Congress,* 76, F1.

Kabatek, U., & Swaney, R. E. (1992). Worst-case identification in structured process systems. *Comp. Chem. Eng.*, **16,** 1063.

Malik, R. K., & Hughes, R. R. (1979). Optimal design of flexible chemical processes. *Comp. Chem. Eng.*, **3,** 473.

Pistikopoulos, E. N., & Grossmann, I. E. (1988). Optimal retrofit design for improving process flexibility in linear systems. *Comp. Chem. Engng.*, **12**, 719.

Pistikopoulos, E. N., & Grossmann, I. E. (1989). Optimal retrofit design for improving process flexibility in nonlinear systems—I. Fixed degree of flexibility. *Comp. Chem. Engng.*, **13**, 1003.

Pistikopoulos, E. N., & Mazzuchi, T. A. (1990). A novel flexibility analysis approach for processes with stochastic parameters. *Comp. Chem. Eng.*, **14**(21.9), 991.

Saboo, A. K., & Morari, M. (1984). Design of resilient processing plants. IV. Some new results on heat exchanger network synthesis, *Chem. Eng. Sci.*, **39**, 579.

Straub, D. A., & Grossmann, I. E. (1990). Integrated statistical metric of flexibility for systems with discrete state and continuous parameter uncertainties. *Comp. Chem. Eng.*, **14**, 967.

Straub, D. A., & Grossmann, I. E. (1993). Design optimization of stochastic flexibility. *Comp. Chem. Eng.*, **17**, 339.

Swaney, R. E., & Grossmann, I. E. (1985a). An index for operational flexibility in chemical process design. Part 1—Formulation and theory. *AIChE J.*, **31**, 621.

Swaney, R. E., & Grossmann, I. E. (1985b). An index for operational flexibility in chemical process design. Part 2—Computational algorithms. *AIChE J.*, **31**, 631.

Varvarezos, D. K., Grossmann, I. E., & Biegler, L. T. (1992). An outer approximation method for multiperiod design optimization. *Ind. Eng. Chem. Research*, **31**, 1466.

Varvarezos, D. K., Grossmann, I. E., & Biegler, L. T. (1995). A sensitivity based approach for the flexibility analysis and design of linear process systems. *Comp. Chem. Eng.*, **19**, 1305.

EXERCISES

1. In the heat exchanger network shown in Figure 21.9 the area of exchanger 1 is 31.2 m^2, and the area of exchanger 2 is 41.2 m^2.

   a. If we have the specifications $T_{H_2} \leq 410K$ and $t_{C_2} \geq 430K$, will the network be feasible for the following range of heat transfer coefficients? Explain your answers.

   $$0.64 \leq U_1 \leq 0.96 \text{kW}/m^2 \text{ K}$$

   $$0.64 \leq U_2 \leq 0.96 \text{kW}/m^2 \text{ K}$$

   b. Repeat (a), assuming we change the areas as follows:

   Case I: Exchanger 1 from 31.2 m^2 to 37.4 m^2
               Exchanger 2 from 41.2 m^2 to 49.4 m^2

   Case II: Exchanger 1 from 31.2 m^2 to 26.0 m^2
               Exchanger 2 from 41.2 m^2 to 57.0 m^2

   Note: Use Chen approximation for LTMD (Chapter 16).

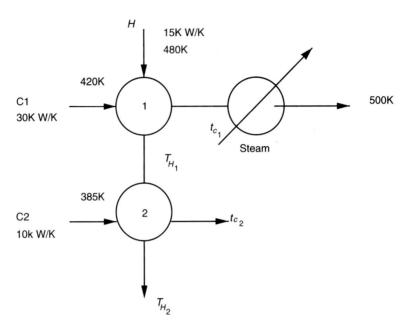

FIGURE 21.9

2. The inequality constraints for feasible operation of a design d are given by

$$f_1 = -25\theta + z\left[1 - \frac{\theta}{2}\right] + d \leq 0$$

$$f_2 = -190\theta + z + d \leq 0$$

$$f_3 = 260\theta - z - 240 - d \leq 0$$

where Θ is an uncertain parameter and z is a control variable.

For the design $d = 10$:
  a. Plot the feasible region of operation in the $z - \theta$ space.
  b. Obtain the analytical expression for the feasibility function $\psi(d,\theta)$ in the range $0.5 \leq \theta \leq 2$, and plot this function.
  c. Determine the critical point for feasible operation in this design. Explain why the critical point is a vertex or a nonvertex solution.
  d. Is this design feasible for the parameter range $0.5 \leq \theta \leq 2$?

3. Derive the mathematical formulations for the active set strategy for the following cases:
  a. Feasibility test: only inequalities, no control variables.
  b. Feasibility test: equalities and inequalities with control variables.
  c. Flexibility index for two cases above.

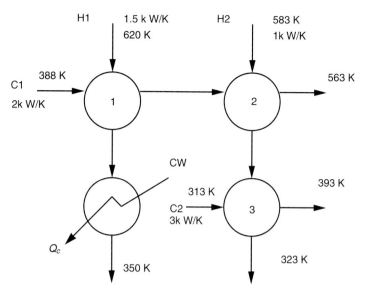

FIGURE 21.10

4. In the heat exchanger network shown In Figure 21.10 the inlet temperatures of the two hot and two cold process streams are regarded as uncertain parameters. Given the nominal values of the temperatures shown and expected deviations of ±10 K in each of these streams, determine the flexibility index for this network and its range of inlet temperatures for feasible operation.

   To solve this problem:
   a. Formulate the inequality constraints for feasible heat exchange and the specification ($T \leq 323$ K) in terms of the cooling load Q_c and the inlet temperatures using $\Delta T_{min} = O$ K.
   b. Solve for the flexibility index with a vertex enumeration scheme (i.e., 16 LPs) and with the MILP formulation.

   Note: Areas are not specified. Q_c at 300 K.

5. Show that if the feasibility function $\psi(d,\theta)$ is convex in θ, then the parametric region of feasible operation $R = \{\theta \mid \psi(d,\theta) \leq 0\}$ is convex.

6. a. Show that the three inequalities below are active in the feasibility function $\psi(d,\theta)$ for any θ_1, θ_2. Derive the explicit expression for $\psi(d,\theta)$ as a function of the two parameters.
   b. Also show that the function $\psi(d,\theta)$ has the unique critical point $\theta_1 = 2$, $\theta_2 = 2$ in the specified range

$$1 \leq \theta_1 \leq 2$$
$$1 \leq \theta_2 \leq 2$$

Sec. 21.13 Exercises

Inequalities:
$$f_1 = -z_1 + 3\theta_1 - \theta_2 \leq 0$$
$$f_2 = -z_2 - \theta_1 + 3\theta_2 \leq 0$$
$$f_3 = z_1 + z_2 - \theta_1 - \theta_2 - 4 \leq 0$$

where z_1, z_2 are control variables, θ_1, θ_2 are uncertain parameters.

7. For the case of a fixed design with one control variable and one uncertain parameter, sketch inequality constraints for which:
    a. The number of active constraints for the feasibility problem in Eq. (21.22) is two.
    b. The number of active constraints in Eq. (21.22) for same parameter values is one.

    (Hint: See Eq. (21.39).)

8. a. Derive the mixed-integer formulation for the feasibility test in Eq. (21.50) in which equations and inequalities are assumed for the process model.
    b. What would be the corresponding mixed-integer optimization model for the flexibility index?

OPTIMAL DESIGN AND SCHEDULING OF MULTIPRODUCT BATCH PLANTS

22

22.1 INTRODUCTION

In Chapter 6 we presented basic concepts related to the design and scheduling of batch processes. In this chapter we will see how some of the design and scheduling problems that we alluded to can be formulated mathematically as optimization problems. For the design problems we will restrict ourselves to the case of multiproduct or flowshop plants. At the end of the chapter we will consider the scheduling of multipurpose plants.

We will start first with the design of multiproduct batch plants for the case of single product campaigns in which no sequencing is performed among batches of different products. We will then consider the case of mixed product campaigns in which scheduling must be anticipated at the design stage. We will show that the key element for approaching this problem is the development of an aggregated scheduling model. We will consider the equipment sizing with continuous and discrete sizes. Finally, we will present the state-task-network MILP scheduling model, which can be applied to general batch plant configurations.

22.2 HORIZON CONSTRAINTS FOR FLOWSHOP PLANTS—SINGLE-PRODUCT CAMPAIGNS

As defined in Chapter 6, flowshop plants are those in which all products follow the same sequence through all the processing stages. We consider in this section the case in which the plant is operated with single-product campaigns and when no intermediate storage is available (Grossmann and Sargent, 1978; Sparrow et al., 1975). This is a relatively simple case in the sense that the production scheduling is greatly simplified, thereby facilitating the consideration of timing considerations (or horizon constraints) at the design stage.

Let us consider first the case of a plant with one unit per stage for deriving the horizon constraints. We assume that the plant consists of M stages for manufacturing N different products. Given H, the total horizon time (hrs) over which one production cycle will be considered, and given τ_{ij}, the processing time (hrs) of product i in stage j, $i = 1,...N$, $j = 1,...M$, the major variables to be determined are:

n_i = number of batches of product i that are to be produced in horizon H
T_{Li} = cycle time of product i
θ_i = time allocated to product i from time horizon H

As was shown in Chapter 6, the cycle time can be determined from the following equation:

$$T_{Li} = \max_{j = 1,M} \{\tau_{ij}\} \qquad (22.1)$$

As an example, consider the Gantt chart in Figure 22.1 of a plant with three stages for manufacturing products A and B. Clearly the cycle time for product A is 20 hours,

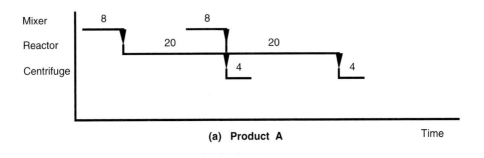

(a) Product A

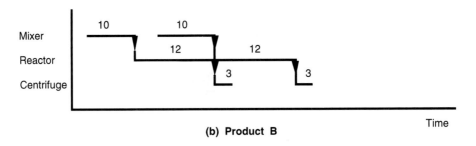

(b) Product B

FIGURE 22.1 Gantt charts with one unit per stage.

Sec. 22.2 Horizon Constraints for Flowshop Plants

while for product B it is 12 hours. Since the number of batches n_i is normally large, the "heads" and "tails" of the schedule can be neglected with which the production time θ_i devoted to each product can be approximated by

$$\theta_i = n_i T_{Li} \qquad i = 1...N \qquad (22.2)$$

$$\sum_{i=1}^{N} \theta_i \leq H \qquad (22.3)$$

Substituting Eq. (22.2), the horizon constraint for one unit per stage can be written in terms of number of batches n_i,

$$\sum_{i=1}^{N} n_i T_{Li} \leq H \qquad (22.4)$$

where the cycle time T_{Li} as given by Eq. (22.1) is a fixed parameter.

For the case when N_j parallel units might be used at each stage of the flowshop plant, the cycle time T_{Li} is expressed as follows:

$$T_{Li} = \max_{j=1,M} \{\tau_{ij}/N_j\} \qquad (22.5)$$

Assume now that in our example we have $N_{mixer} = 1$, $N_{reactor} = 2$, $N_{centrifuge} = 1$. From Eq. (22.5) and Figure 22.1, it follows that,

$$T_{LA} = \max \{8, 20/2, 4\} = 10 \text{ hrs}$$

$$T_{LB} = \max \{10, 12/2, 3\} = 10 \text{ hrs}$$

Figure 22.2 displays the operation of the plant with these cycle times.

Note that in this case for product A the bottleneck is in the reactor. However, since we can process the batches twice as fast, the cycle time is 10 hours. For the case of product B, the bottleneck is now shifted to the mixer; hence, the cycle time is 10 hours.

The horizon constraints for flowshop plants with parallel units, we can then be expressed in general form as,

$$\sum_{i=1}^{N} n_i T_{Li} \leq H \qquad (22.6)$$

$$T_{Li} = \max_{j=1,M} \{\tau_{ij}/N_j\}$$

which are a clear generalization of Eqs. (22.4) and (22.1).

722   Optimal Design and Scheduling of Multiproduct Batch Plants   Chap. 22

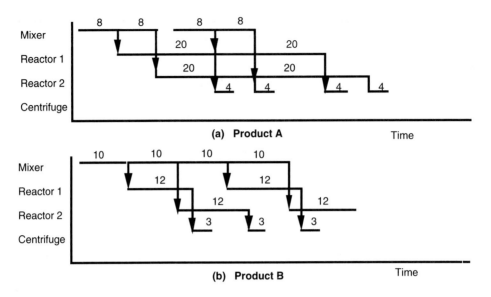

FIGURE 22.2   Gantt chart for two parallel reactors.

22.3   MINLP DESIGN MODEL FOR FLOWSHOP PLANTS—SINGLE-PRODUCT CAMPAIGNS

Having developed the appropriate horizon constraints for the case of single-product campaigns, we will present in this section an MINLP model for selecting the sizes and number of parallel units operating out of phase. The objective is to minimize the investment cost given fixed product demands.

We will present the formulation of this problem in terms of general equations and indices as reported in Kocis and Grossmann (1988). This formulation is an extension to the model proposed by Grossmann and Sargent (1978). First, we will define the following parameters:

N	= Number of products to be produced
M	= Number of stages in the batch plant
τ_{ij}	= Processing time of product i in stage j (hrs)
S_{ij}	= Size factor of product i in stage j (ℓ/kg)
H	= Horizon time (hrs)
Q_i	= Demand of product i (kg)
α_i, β_j	= Cost coefficient and cost exponent for unit j
V_j^L, V_j^U	= Lower and upper bounds of volumes

Let V_j be the variable that represents the required volume of a unit in stage j and B_i

Sec. 22.3 MINLP Design Model for Flowshop Plants

the variable that represents the size of the batch of product i at the end of the M stages. Since the volume V_j has to be able to process all the products i, we have the constraint

$$V_j \geq S_{ij} B_i \qquad i = 1\ldots N, j = 1\ldots M \tag{22.7}$$

where the right-hand side represents the actual volume needed by each product. The number of batches n_i for each product i is given by,

$$n_i = Q_i / B_i \tag{22.8}$$

Finally, the investment cost is given by

$$C = \sum_{j=1}^{M} N_j \alpha_j V_j^{\beta_j} \tag{22.9}$$

Using the horizon constraints in Eq. (22.6) as inequalities to avoid nondifferentiable functions and eliminating the variables n_i and θ_i, using Eqs. (22.2) and (22.8), yields the optimization problem

$$\min C = \sum_{j=1}^{M} N_j \alpha_j V_j^{\beta_j}$$

$$\text{s.t. } V_j \geq S_{ij} B_i \quad i = 1, N, J = 1, M$$

$$T_{Li} \geq \tau_{ij}/N_j \quad i = 1, N, j = 1, M$$

$$\sum_{i=1}^{N} \frac{Q_i}{B_i} T_{Li} \leq H \tag{22.10}$$

$$V_j^L \leq V_j \leq V_j^U, N_j = 1,2,\ldots N_j^U \quad j = 1,\ldots M$$

$$B_i \geq 0, T_{Li} \geq 0 \quad i = 1,\ldots N$$

Rather than specifying zero lower bounds for B_i and T_{Li}, it is easy to show that for a maximum of K parallel units these variables are bounded as follows:

$$T_{L_i}^L \leq T_{Li} \leq T_{L_i}^U$$
$$B_i^L \leq B_i \leq B_i^U \tag{22.11}$$

where

$$T_{Li}^L = \max_{j=1,M} \{\tau_{ij} / N_j^K\}$$

$$T_{Li}^U = \max_{j=1,M} \{\tau_{ij}\} \tag{22.12}$$

$$B_i^L = \frac{Q_i T_{Li}^L}{H}$$

$$B_i^U = \min_{j=1,M} \{V_j^U / S_{ij}\}$$

The formulation in Eqs. (22.10) and (22.11) corresponds to an MINLP problem where the variables N_j, are restricted to take integer values. Note also that the objective function is nonconvex as it involves concave terms of the form $V_j^{\beta j}$ ($0 < \beta_j < 1$). Also, except for the volume constraints, all the other constraints are nonlinear. Below we will show how we can convexify the MINLP problem in a way where we are left with only one nonlinear inequality, and where the N_j variables are expressed in terms of 0-1 variables.

For this let us define the following exponential transformations:

$$V_j = e^{v_j},\ N_j = e^{n_j},\ B_i = e^{b_i},\ T_{Li} = e^{t_{Li}} \tag{22.13}$$

where v_j, n_j, b_i, t_{Li} are the new transformed variables.

If we substitute into the objective function in Eq. (22.9) this yields:

$$C = \sum_{i=1}^{M} \alpha_j \exp(n_j + \beta_j v_j) \tag{22.14}$$

which is a convex function.

Substituting Eq. (22.13) in Eq. (22.7) yields

$$e^{v_j} \geq S_{ij} e^{b_i} \quad i = 1, N\ j = 1, M \tag{22.15}$$

which is nonlinear. However, taking logarithms on both sides yields,

$$v_j \geq \ln S_{ij} + b_i \quad i = 1, N, j = 1, M \tag{22.16}$$

Similarly, it can be shown that the second constraint in Eq. (22.10) reduces to

$$t_{Li} \geq \ln \tau_{ij} - n_j \tag{22.17}$$

which is linear, while the last constraint in Eq. (22.10) reduces to

$$\sum_{i=1}^{N} Q_i \exp(t_{Li} - b_i) \leq H \tag{22.18}$$

which is a convex constraint.

Finally, we can relate the n_j variables to 0-1 variables as follows. From Eq. (22.13) we have

$$n_j = \ln N_j \quad j = 1, M \tag{22.19}$$

Since N_j is integer (1,2,...K), we can express n_j as

$$n_j = \sum_{k=1}^{K} \ln k\, y_{jk}, \quad \sum_{k=1}^{K} y_{jk} = 1 \quad j = 1, M \tag{22.20}$$

where $y_{jk} = 1$ if k parallel units are selected and 0 otherwise. Note that the summation on y_{jk} is imposed so that only one alternative is chosen for parallel units in each stage j.

In this way, by gathering equations (22.14), (22.16) to (22.18) and (22.20), the final MINLP problem for selecting the optimal sizes and number of parallel units in flowshop plants with single-product campaigns is given by

$$\min C = \sum_{i=1}^{M} \alpha_j \exp(n_j + \beta_j v_j)$$

$$\text{s.t. } v_j \geq \ln S_{ij} + b_i \quad i = 1, N \quad j = 1, M$$

$$t_{Li} \geq \ln \tau_{ij} - n_j \quad i = 1, N \quad j = 1, M$$

$$\sum_{i=1}^{N} Q_i \exp(t_{Li} - b_i) \leq H \qquad (22.21)$$

$$n_j = \sum_{k=1}^{K} \ln k\, y_{jk}, \quad \sum_{k=1}^{K} y_{jk} = 1 \quad j = 1, M$$

$$\ln V_j^L \leq v_j \leq \ln V_j^U, \quad n_j \geq 0 \quad j = 1, M$$

$$\ln B_i^L \leq b_i \leq \ln B_i^U, \quad \ln T_{Li}^L \leq t_{Li} \leq \ln T_{Li}^U \quad i = 1, N$$

$$y_{jk} = 0, 1 \quad j = 1, M \quad k = 1..N_j^U$$

It should be noted that since the nonlinear functions involved in the above MINLP are convex, algorithms such as the outer-approximation method and Generalized Benders decomposition (see Appendix A) are guaranteed to obtain the global optimum. Also, note that if there is only one unit per stage, the above model reduces to an NLP involving only the transformed volume, v_j, and transformed batch size, b_j, as variables.

22.4 MILP REFORMULATION FOR DISCRETE SIZES

In the previous section we assumed that the equipment is available in continuous sizes being restricted only by specified lower and upper bounds. In practice, however, it is often the case that only standard sizes are available. More specifically, let us assume that the equipment size in stage j, $j = 1,...M$, is given by the set, $SV_j = \{dv_{js}, s = 1, NS(j)\}$. The most straightforward approach would be to solve problem (22.21) with continuous sizes using as lower and upper bounds the smallest and largest discrete sizes. The sizes of the continuous solution would then simply be rounded up to corresponding discrete sizes. Although this procedure might seem attractive because of its obvious simplicity, it has the drawback that the resulting solution might be suboptimal, particularly if successive increases in the discrete sizes are rather significant.

A second approach that might be used, which is rigorous, is to introduce the following 0-1 variables,

$$z_{js} = \begin{cases} 1 \text{ if size } s \text{ is selected for stage } j \\ 0 \text{ otherwise} \end{cases}$$

The selection of discrete sizes can then be enforced by adding the following constraints to the MINLP problem (22.21):

$$v_j = \sum_{s=1}^{NS(j)} \ell n\,(dv_{js})\,z_{js} \quad j=1,\,M$$

$$\sum_{s=1}^{NS(j)} z_{js} = 1 \quad j=1,\,M \tag{22.22}$$

While the above approach is rigorous, it has the disadvantage that it complicates the original MINLP model by increasing the number of variables and constraints. It turns out, however, that one can take advantage of the requirement of discrete sizes and reformulate problem (22.21) for that case as an MILP problem (Voudouris and Grossmann, 1992). We will show this for the case of one unit per stage, and leave the case of parallel units as an exercise to the reader (see exercise 12).

Eliminating the batch size from Eq. (22.7) using equations (22.8) and (22.2) the capacity constraint can be written as,

$$V_j \geq \frac{S_{ij} Q_i T_{Li}}{\theta_i} \quad i=1,\,N \quad j=1,\,M \tag{22.23}$$

or equivalently as,

$$\theta_i \geq \frac{S_{ij} Q_i T_{Li}}{V_j} \quad i=1,\,N \quad j=1,\,M \tag{22.24}$$

Let the inverse of the volume V_j be expressed as a linear combination of the inverse of the discrete sizes.

$$\frac{1}{V_j} = \sum_{s=1}^{NS(j)} \frac{z_{js}}{dv_{js}} \quad j=1,\,M \tag{22.25}$$

Substituting Eq. (22.25) into Eq. (22.24) yields the following linear inequality.

$$\theta_i \geq S_{ij} Q_i T_{Li} \sum_{s=1}^{NS(j)} \frac{z_{js}}{dv_{js}} \quad i=1,\,N \quad j=1,\,M \tag{22.26}$$

Thus, if we set the cost coefficients for each unit j at every size s as $c_{js} = \alpha_j\,(dv_{js})^{\beta_j}$ and gathering the constraints (22.26), (22.22), and (22.3), the optimal sizing of a flowshop plant with one unit per stage and operating with single-product campaigns can be formulated as the MILP,

$$\min C = \sum_{j=1}^{M} \sum_{s=1}^{NS(j)} c_{js} z_{js}$$

Sec. 22.4 MILP Reformulation for Discrete Sizes

$$\theta_i \geq S_{ij} Q_i T_{Li} \sum_{s=1}^{NS(j)} \frac{z_{js}}{dv_{js}} \quad i = 1, N \quad j = 1, M$$

$$\sum_{s=1}^{NS(j)} z_{js} = 1 \quad j = 1, M \tag{22.27}$$

$$\sum_{i=1}^{N} \theta_i \leq H$$

$$\theta_i \geq 0 \quad i = 1, N; \quad z_{js} = 0.1 \quad s = 1, NS(j) \quad j = 1, M$$

where T_{Li} is a fixed parameter as defined by Eq. (22.1). Note that the interesting feature in the above problem is that it involves fewer variables and constraints than the MINLP in Eq. (22.21) with the constraints in Eq. (22.22).

EXAMPLE 22.1

Consider the case of a multiproduct plant with one unit per stage operating under the SPC/ZW policy. The plant consists of 6 stages and is dedicated to the production of 5 products $A, B, C, D,$ and E. Data for this problem are given in Table 22.1. One way to solve the problem is using the NLP model (22.21) for continuous sizes (with $n_j = 0$). The optimal solution of the corresponding NLP has a cost of \$2,314,896. The optimal sizes of the vessels predicted are $V_1 = 6017.59$, $V_2 = 3483.6$, $V_3 = 3960.9$, $V_4 = 4823.5$, $V_5 = 4646.5$, $V_6 = 3885.55$ (in liters). Assume, however, that the vessels are only available in the following set of discrete values $SV = \{3000, 3750, 4688, 5860, 7325\}$ liters. Note that the ratio of two consecutive sizes is constant and in this case this ratio is 1.25. A simple approach to determine a design with discrete sizes would be to round up

TABLE 22.1 Data for Example 22.1 (SPC with One Unit per Stage)

	Size factor S_{ij} (l/kg)					Proc. time τ_{ij} (h)					Cost coeff.	Cost exp.
	A	B	C	D	E	A	B	C	D	E	α_j (\$)	β_j
Stage 1	7.9	0.7	0.7	4.7	1.2	6.4	6.8	1	3.2	2.1	2500	0.6
Stage 2	2	0.8	2.6	2.3	3.6	4.7	6.4	6.3	3	2.5	2500	0.6
Stage 3	5.2	0.9	1.6	1.6	2.4	8.3	6.5	5.4	3.5	4.2	2500	0.6
Stage 4	4.9	3.4	3.6	2.7	4.5	3.9	4.4	11.9	3.3	3.6	2500	0.6
Stage 5	6.1	2.1	3.2	1.2	1.6	2.1	2.3	5.7	2.8	3.7	2500	0.6
Stage 6	4.2	2.5	2.9	2.5	2.1	1.2	3.2	6.2	3.4	2.2	2500	0.6

$Q(A) = 250000$, $Q(B) = 150000$, $Q(C) = 180000$ $Q(D) = 160000$, $Q(E) = 120000$ (Kg)
$H = 6200$ hrs

the sizes predicted by the continuous model. By rounding up the NLP solution, we get $V_1 = 7325$, $V_2 = 3750$, $V_3 = 4688$, $V_4 = 5860$, $V_5 = 4688$, $V_6 = 4688$ liters and a cost of \$ 2,521,097.

Using the MILP model in Eq. (22.27), the availability of discrete sizes is taken explicitly into account. The solution in this case is $V_1 = 5860$, $V_2 = 3750$, $V_3 = 3750$, $V_4 = 5860$, $V_5 = 4688$, $V_6 = 4688$ liters with a cost of \$2,405,840, which is \$115,257 cheaper or 4.6% lower than the rounded values. It is clear that the rounding scheme can fail to predict the global optimal design when discrete sizes are involved.

22.5 NLP DESIGN MODEL—MIXED-PRODUCT CAMPAIGNS (UIS)

As discussed in Chapter 6, mixed-product campaigns, as opposed to single-product campaigns, involve sequencing of individual batches of the different products. The main motivation is to reduce idle times so as to increase equipment utilization. This can often be accomplished if the cleanup times are small. To simplify the development of models for these type of plants, we will assume only one unit per stage. Also, we will first assume that the transfer between stages is with unlimited intermediate storage (UIS) although for the optimization we will neglect the costing of the storage tanks.

From Eq. (6.3) in Chapter 6, the cycle time for a plant with one unit per stage operating with UIS policy is given by,

$$CT = \max_{j=1..M} \left\{ \sum_{i=1}^{N} n_i \tau_{ij} \right\} \qquad (22.28)$$

where n_i is the number of batches for each product i. Given H as the horizon time for satisfying the specified demands Q_i, $i = 1, ..N$, the horizon constraint is simply given by,

$$\max_{j=1..M} \left\{ \sum_{i=1}^{N} n_i \tau_{ij} \right\} \leq H \qquad (22.29)$$

Thus, considering the objective function in Eq. (22.9) for one unit per stage ($N_j = 1$), the capacity constraint in Eq. (22.7), and the horizon constraint in Eq. (22.29), eliminating the number of batches with the equation in Eq. (22.8), and by expressing Eq. (22.29) as a system of inequalities, the resulting NLP model for optimizing continuous sizes is given by (Birewar and Grossmann, 1989b),

$$\min C = \sum_{j=1}^{M} \alpha_j V_j^{\beta_j} \qquad (22.30)$$

$$\text{s.t. } V_j \geq S_{ij} B_i \quad i = 1, N, j = 1, M$$

$$\sum_{i=1}^{N} \frac{Q_i}{B_i} \tau_{ij} \leq H \quad j = 1, M$$

$$V_j^L \leq V_j \leq V_j^U, \quad j = 1,...M$$

$$B_i \geq 0, \quad i = 1,...N$$

This NLP problem can be convexified as the MINLP in Eq. (22.21). Also, if discrete sizes are involved, the problem can be reformulated as an MILP (see exercise 11). Finally, one can also show that for flowshops with one unit per stage the above NLP will provide a lower bound of the equipment cost to plants that implement transfer policies other than UIS (e.g., zero-wait).

22.6 CYCLIC SCHEDULING IN FLOWSHOP PLANTS

In the previous section the development of a design model of a flowshop plant with mixed product campaigns proved to be rather easy because we had the nice closed form expression for the cycle time in UIS plants Eq. (22.28). If the plant, however, operates with zero-wait transfer and/or the cleanup times are significant, this task becomes considerably more complex. The basic reason for this is that determining the cycle does require determining a given sequence of production that we have so far avoided with the simpler cases in the previous sections. The objective of this section is to show that for fixed number of batches n_i, $i = 1,...N$, calculating the cycle time can be reduced to solving an LP model (Birewar and Grossmann, 1989a).

The first important point in cyclic scheduling with ZW policy is to realize that forced idle times arise at the different stages, and that these idle times are sequence dependent. Fortunately, however, these idle times can easily be computed a priori. As shown in Figure 22.3a, consider two products A and B over three production stages. Since zero-wait transfer is imposed, if the timing curves of the two products is placed as close as possible, there is at least one stage where the two curves will touch, giving rise to a bottleneck with zero slack (i.e., stage 2). Thus, as seen in Figure 22.3b the forced idle times or "slacks" are 1 hour, 0 hour, and 2 hours, respectively, for each stage. Similarly, if cleanup times are needed, the procedure is similar. As shown in Figure 22.3c, if 1 hour of cleanup time is required at stage 2 and 2 hours at stages 1 and 3, the forced idle times are 0 hours, 0 hour, and 1 hour, respectively.

In fact, a simple algorithm to compute the slacks without resorting to plots for the case of a product i followed by product k and with cleanup times CL_{ikj} is as follows:

1. Define start times for product k assuming the bottleneck occurs in stage 1:

$$\begin{aligned} T_1 &= \tau_{i1} + CL_{ik1} \\ T_j &= T_{j-1} + \tau_{kj-1} \quad j = 2,3...M \end{aligned} \qquad (22.31)$$

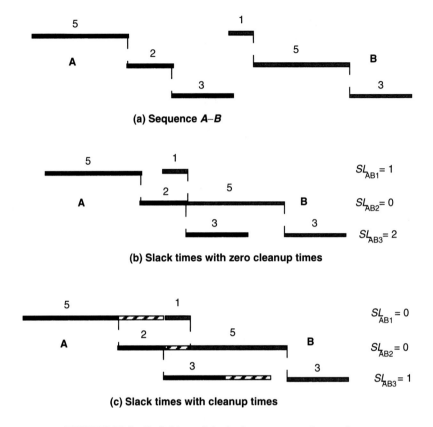

FIGURE 22.3 Definition of slacks for two successive products.

2. Calculate slacks d_j corresponding to assumption in step 1:

$$d_j = T_j - \sum_{\ell=1}^{j} \tau_{i\ell} - CL_{ikj} \quad j = 1,...M \tag{22.32}$$

and the smallest corresponding value δ,

$$\delta = \min_{j} \{d_j\} \tag{22.33}$$

3. Calculate actual slacks SL_{ikj} as

$$SL_{ijk} = d_j - \delta \tag{22.34}$$

The reader can easily verify that the above equations yield the slacks given in Figures 22.3b and c.

Determining the optimal cycle sequence of NB individual batches can be viewed as a traveling salesman problem (TSP) as shown in Figure 22.4 in which the nodes correspond to the individual batches 1,2,..5, and the two-way edges between every pair ℓ and m

Sec. 22.6 Cyclic Scheduling in Flowshop Plants

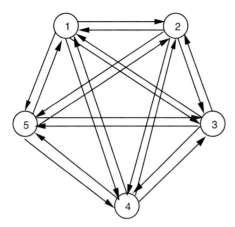

FIGURE 22.4 Traveling salesman representation for flowshop scheduling.

represent potential transitions ℓ to m or m to ℓ (Gupta, 1976; Pekny & Miller, 1991). The mathematical model is as follows. Let

$$y_{\ell m} = \begin{cases} 1 \text{ if batch } \ell \text{ followed by batch } m \\ 0 \text{ otherwise} \end{cases}$$

Note that each pair of batches ℓ, m, can correspond to the same product or to a different one. For the cycle time CT, we only need to analyze one stage, say stage 1. The cycle time is given by:

$$CT = \sum_{\ell=1}^{NB} \tau_{\ell 1} + \sum_{\ell=1}^{NB} \sum_{m=1}^{NB} SL_{\ell m 1} y_{\ell m} \tag{22.35}$$

where the second term takes into account the forced idle times. Also, for transitions of batch ℓ to itself, we set $SL_{\ell \ell 1} = \infty$ to make such choices infeasible in Eq. (22.35).

The selection of the optimal cyclic sequence can then be formulated as the following integer programming problem:

$$\min \ CT = \sum_{\ell=1}^{NB} \tau_{\ell 1} + \sum_{\ell=1}^{NB} \sum_{m=1}^{NB} SL_{\ell m 1} y_{\ell m}$$

$$\text{s.t.} \ \sum_{m=1}^{NB} y_{\ell m} = 1 \quad \ell = 1...NB$$

$$\sum_{\ell=1}^{NB} y_{\ell m} = 1 \quad m = 1,..NB \tag{22.36}$$

$$\sum_{\ell \in Q} \sum_{m \in \overline{Q}} y_{\ell m} \geq 1 \quad \forall Q \subseteq B, \ Q \neq \emptyset \quad y_{\ell m} = 0,1 \quad \ell, m = 1,...NB$$

The first two constraints above correspond to assignment constraints that ensure that every batch ℓ is followed by exactly one batch m, and every batch m is preceded by exactly one batch ℓ. These constraints, however, are not sufficient to ensure closed cycles. Therefore, the last set of constraints, known as subtour elimination constraints, must be considered. These simply state that for every subset Q of batches and its complement $\overline{Q}$ there must exist one link; B is the set of all batches, $B = \{1,2,..NB\}$.

Problem (22.36) is in principle very difficult to solve due to the fact that the number of subtour elimination constraints grows exponentially with the number of batches. Fortunately, problem (22.36) can in fact often be solved as an LP by removing the subtour elimination constraints and treating the variable $y_{\ell m}$ as continuous. When this is not possible, violated subtour elimination constraints are added sequentially to the LP.

From a design point of view a major difficulty is that we need to know the number of batches NB in advance, as well as their product identity. In design problems what we need to determine is in fact the number of batches n_i for every product i. To obtain a model that explicitly incorporates number of batches we will consider an aggregation of problem (22.36) in terms of NP products.

Let us define $NPRS_{ik}$ the number of changeovers from product i to product k. Also, let $B(i) = \{\ell \mid \text{batch } \ell \text{ corresponds to product } i\}$. Then we have the following relation with the 0-1 variables $y_{\ell m}$,

$$NPRS_{ik} = \sum_{\ell \in B(i)} \sum_{m \in B(k)} y_{\ell m} \tag{22.37}$$

By adding over the corresponding number of batches, we can aggregate the TSP problem in Eq. (22.36). So, for example, for the first assignment constraint we have

$$\sum_{\ell \in B(i)} \sum_{k=1}^{NP} \sum_{m \in B(k)} y_{\ell m} = \sum_{k=1}^{NP} NPRS_{ik} = n_i \quad i = 1..NP \tag{22.38}$$

where NP is the number of products. Proceeding in a similar manner with the second assignment constraint and the objective function, the aggregated model for minimization of cycle time by Birewar and Grossmann (1989a) is as follows:

$$\min \; CT = \sum_{i=1}^{NP} n_i \tau_{i1} + \sum_{i=1}^{NP} \sum_{k=1}^{NP} SL_{ik1} NPRS_{ik}$$

$$\text{s.t.} \; \sum_{k=1}^{NP} NPRS_{ik} = n_i \quad i = 1,..NP$$

$$\sum_{i=1}^{NP} NPRS_{ik} = n_k \quad k = 1,..NP \tag{22.39}$$

$$NPRS_{ii} \leq n_i - 1 \quad i = 1,..NP$$

$$NPRS_{i,k} = 0, 1, 2, 3, ...$$

Sec. 22.6 Cyclic Scheduling in Flowshop Plants

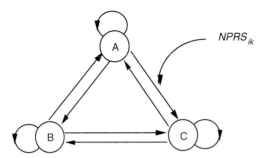

FIGURE 22.5 Aggregated TSP graph for flowshop scheduling.

In the above problem we have only added the simplest type of subtour elimination constraints to avoid subcycles involving only batches of product i. In most cases, the above problem can be solved as an LP yielding integer values for the variables $NPRS_{ik}$ and with no subcycles.

The other interesting feature of model (22.39) is its graph representation. As seen in Figure 22.5, nodes correspond to products and arcs to numbers of changeovers. Thus, the LP problem (22.39) will synthesize aggregated graphs from which detailed schedules can easily be derived. Instead of presenting a formal algorithm we will use a simple example.

As an example consider the case of a problem involving four products and 20 batches with $n_A = 7$, $n_B = 5$, $n_C = 3$, $n_D = 5$. Let us assume that the LP in Eq. (22.39) yields the graph in Figure 22.6. If we successively remove loops starting with arcs containing fewest changeovers, we can derive the sequence given in Figure 22.7. A simple interpretation of that sequence is that it represents a complete path that we can take on the aggregated graph in Figure 22.6.

It should also be noted that if subcycles are obtained using the LP model in Eq. (22.39), we can use subtour elimination constraints in a subsequent phase. So, for instance, if we synthesize the graph in Figure 22.8, we set $Q = \{A,B,C\}$ $\overline{Q} = \{D,E\}$, and add the constraint

$$\sum_{i \in Q} \sum_{k \in \overline{Q}} NPRS_{ik} \geq 1 \qquad (22.40)$$

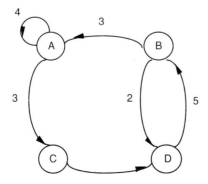

FIGURE 22.6 Example of four-product schedule with 20 batches.

734   Optimal Design and Scheduling of Multiproduct Batch Plants   Chap. 22

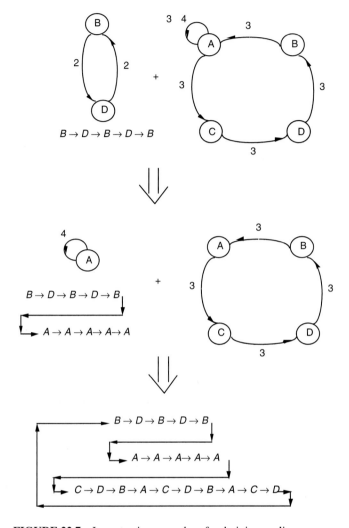

FIGURE 22.7   Loop tracing procedure for deriving cyclic sequence.

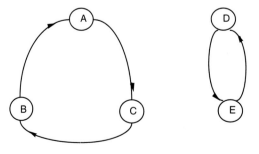

FIGURE 22.8   Two subcycles arising in five-product problem.

Sec. 22.7 NLP Design Model—Mixed Product Campaigns

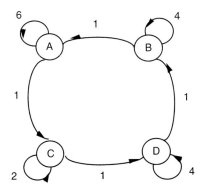

FIGURE 22.9 Aggregate graph for single-product campaigns.

Finally, single-product campaigns as in Figure 22.9 can be obtained by simply specifying the last inequality in Eq. (22.39) as an equality,

$$NPRS_{ii} = n_i - 1 \quad i = 1,..NP \tag{22.41}$$

22.7 NLP DESIGN MODEL—MIXED PRODUCT CAMPAIGNS

Having developed the aggregate LP model in Eq. (22.39) for cycle time minimization, the problem of determining continuous sizes can simply be formulated by treating the number of batches n_i as variables and by setting a constraint for the cycle time, $CT \leq H$, where H is the total horizon time. Following similar nomenclature and treatment as in sections 22.3, 22.4, and 22.5, the NLP model for the optimal design problem for mixed product campaigns and zero-wait is given by (Birewar and Grossmann, 1989b),

$$\min C = \sum_{j=1}^{M} \alpha_j V_j^{\beta_j}$$

$$\begin{aligned}
s.t. \quad & V_j \geq S_{ij} B_i & & i = 1,..,NP \quad j = 1,..M \\
& n_i B_i = Q_i & & i = 1,..,NP \\
& \sum_{k=1}^{NP} NPRS_{ik} = n_j & & i = 1,..,NP \\
& \sum_{i=1}^{NP} NPRS_{ik} = n_k & & k = 1,..,NP \\
& \sum_{i=1}^{NP} n_i \tau_{i1} + \sum_{i=1}^{NP}\sum_{k=1}^{NP} SL_{ik1} NPRS_{ik} \leq H
\end{aligned} \tag{22.42}$$

$$NPRS_{ii} \leq n_i - 1 \qquad i = 1..NP$$

$$V_j^L \leq V_j \leq V_j^u, \qquad n_i, B_i, NPRS_{ik} \geq 0$$

It can be shown that the above NLP has a unique solution. Its remarkable feature is that it is a model that accurately anticipates the effect of scheduling at the design stage. If only discrete sizes are available the above problem can easily be reformulated as an MILP as was done in section 22.4 (see Voudouris and Grossmann, 1992).

22.8 STATE-TASK NETWORK FOR THE SCHEDULING OF MULTIPRODUCT BATCH PLANTS

In the previous sections we have presented several different design models for flowshop batch plants. The reason the modeling of these problems was greatly facilitated is because we were able to anticipate the effect of scheduling with effective aggregate flowshop models for production cycle time. Deriving similar expressions for the more general job-shop or multipurpose plants is a much more difficult task. In this section we will not specifically address this problem, but instead we will introduce a very general MILP scheduling model that can be applied to a large number of batch processes that are specified by recipes. Also in contrast to the previous sections, we will be concerned with *short-term scheduling* in which demands of products are specified at various points in time in the form of deadlines.

The MILP model that we will describe is by Kondili et al. (1993), and it has the following three major capabilities:

1. Assignments of equipment to processing tasks need not be fixed.
2. Variable size batches can be handled with the possibility of mixing and splitting.
3. Different intermediate storage and transfer policies can be accommodated as well as limitations of resources.

The major assumption that will be made is that the time domain can be discretized in intervals of equal size. In practice, that will often mean having to perform some rounding to the original data. In addition, although this is not an inherent restriction, for the sake of simplicity in the presentation it will be assumed that changeover times can be neglected. The key aspect in the MILP model by Kondili et al. (1993), is the state-task network (STN) representation. This network has two types of nodes: (a) state nodes that correspond to feeds, intermediates, and final products; and (b) task nodes that represent processing steps. Figure 22.10 presents an example of a state-task network involving one raw material A for producing products F and C (E is a by-product). The specific steps are as follows:

1. Heat raw material A for 2 hours to produce intermediate B.
2. This intermediate is split so that one part follows reaction 1 for 3 hours (say with

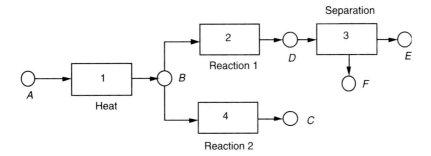

FIGURE 22.10 State-task network representation.

catalyst 1) to produce intermediate D, which is then separated in 1 hour in 80/20% fractions for producing products E and F.

3. The other part of intermediate B follows a different reaction 2 for 5 hours (say with catalyst 2) producing product C.

Note that the STN represents a recipe in terms of transfers and materials, and that different STNs may have to be considered for a plant processing different feeds. In addition, note that the STN has as many inputs (outputs) states as different input (output) materials, and that two or more streams entering same state have the same quality. A key point is that equipment is *not* represented in the STN because their assignment to tasks is treated as an unknown. As shown in Figure 22.11, we may have two reactors available as well as one batch distillation column. Clearly, since the reactors have a jacket, they can perform tasks 1, 2, and 4, while the column can only perform task 3. Finally, storage is represented as accumulation of material in the states.

Having introduced the STN representation, the MILP model will address the problem where given the STN representation of one or more feeds and the demands and their deadlines, we have to determine the timing of the operations, assignments of equipment to operations, and flow of material through the network. The objective is to maximize a given profit function. As for the discretization of the time domain, H time periods of equal size will be considered (see Figure 22.12).

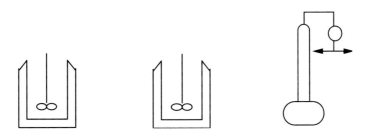

FIGURE 22.11 Available equipment for network in Figure 22.10.

FIGURE 22.12 Uniform time discretization in H intervals.

The following are the parameters for the MILP model:

Task i

$\underline{S}_i$ = Set of states inputs to task i
$\overline{S}_i$ = Set of states outputs of task i
$\underline{\rho}_{is}$ = Proportion input to task i from state s
$\overline{\rho}_{is}$ = Proportion output of task i for state s

(Note $\sum_s \underline{\rho}_{is} = 1$, $\sum_s \overline{\rho}_{is} = 1$)

p_i = Processing time for task i
K_i = Set of units j capable of processing task i

State s

$\underline{T}_s$ = Set of tasks receiving material from state s
$\overline{T}_s$ = Set of tasks producing material for state s
IP = Set of states s corresponding to products
IF = Set of states s corresponding to feeds
II = Set of states s corresponding to intermediates
d_{st} = Minimum demand for state $s \in IP$ at the beginning of period t
r_{st} = Maximum purchase for state $s \in IF$ at the beginning of period t
C_s = Maximum storage for state s

Equipment j

V_j = Maximum capacity
I_j = Set of tasks i for which equipment j can be used

As for the variables, we will require both 0-1 and continuous variables:

W_{ijt} = 1 if unit j starts processing task i at the beginning of period t
B_{ijt} = Amount of material starts task i in unit j at the beginning of period t
S_{st} = Amount of material stored in state s at the beginning of period t
U_{ut} = Demand of utility u over time interval t
R_{st}, D_{st} = Purchases and sales of state s at the beginning of period t

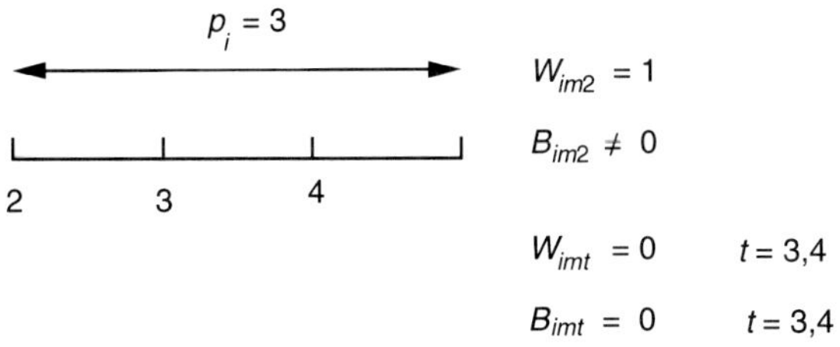

Sec. 22.8 State-Task Network for Scheduling of Multiproduct Batch Plant

$p_i = 3$

$W_{im2} = 1$
$B_{im2} \neq 0$

$W_{imt} = 0 \quad t = 3,4$
$B_{imt} = 0 \quad t = 3,4$

FIGURE 22.13 Definition of assignment and batch size variables for 3-hour task.

It is worth it to clarify according to the above definitions that the variables W_{ijt} and B_{ijt} are only non-zero at the start of the period, even if the unit and task continue to operate in subsequent periods. Figure 22.13 illustrates this point.

The constraints for the MILP model are as follows. First, we need to constrain the assignment of equipment j to tasks i over the various time periods t. As shown by Shah et al. (1993a) a "tight" MILP model can be obtained with the following assignment constraint which states that every equipment j can start at most one task i during times $\hat{t} = t$, $\hat{t} = t - 1..., \hat{t} = t - p_i + 1$, at every time t; that is,

$$\sum_{i \in I_j} \sum_{\hat{t}=t}^{t-p_i+1} W_{ij\hat{t}} \leq 1 \quad \forall j,t \tag{22.43}$$

Note if $W_{ijt} = 1$, this implies that unit j cannot be assigned to tasks other than i during the interval $[t - p_i + 1, t]$.

The capacity limits for equipment and storage tanks can be expressed as:

$$0 \leq B_{ijt} \leq V_j W_{ijt} \quad \forall i,t \quad j \in K_i \tag{22.44}$$

$$0 \leq S_{st} \leq C_s \quad \forall s,t$$

The mass balances for every state and time are as follows,

$$S_{st-1} + \sum_{i \in \overline{T}_s} \overline{\rho}_{is} \sum_{j \in K_i} B_{ijt-pi} + R_{st}$$

$$= S_{st} + \sum_{i \in T_s} \rho_{is} \sum_{j \in K_i} B_{ijt} + D_{st} \quad \forall s,t \tag{22.45}$$

That is, the initial, plus amount produced and purchased must equal the hold-up plus the amount consumed and sales. Also note that in the left-hand side we use B_{ijt-pi} and not B_{ijt}, because these variables are defined at the start of the operations. Also, for convenience we have written one single equation in Eq. (22.45). However, for products $s \in IP$, R_{st} should be removed, for feeds $s \in IF$, D_{st} should be removed, and for intermediates $s \in II$ both should be removed. Clearly, the following bounds also apply,

$$D_{st} \geq d_{st} \quad s \in IP$$

$$R_{st} \leq r_{st} \quad s \in IF \tag{22.46}$$

For the utility requirements, if we assume that the consumption of task i of utility u can be expressed by the equation

$$\alpha_{ui} W_{ijt} + \beta_{ui} B_{ijt} \qquad (22.47)$$

and the maximum amount of utility that is available is $U_u^{\max}$, the resource constraints for utilities, are given as follows,

$$U_{ut} = \sum_i \sum_{j \in K_i} \sum_{\theta=1}^{p_i - 1} \left(\alpha_{ui} W_{ij(t-\theta)} + \beta_{ui} B_{ij(t-\theta)} \right) \qquad \forall u, t \qquad (22.48)$$

$$0 \leq U_{ut} \leq U_u^{\max}$$

Finally, the profit function can be expressed as (sales − purchases + final inventory − utilities):

$$Z = \sum_s \sum_{t=1}^H C_{st}^D D_{st} - \sum_s \sum_{t=1}^H C_{st}^R R_{st} + \sum_s C_{sH+1} S_{sH+1} - \sum_u \sum_{t=1}^H C_{ut} U_{ut} \qquad (22.49)$$

where C_{st}^D, C_{st}^R, C_{sH+1}, and C_{ut} are appropriate cost coefficients.

The objective function in Eq. (22.49) subject to the constraints (22.43) to (22.48) correspond to an MILP problem that has a relatively modest LP relaxation gap. Therefore, provided the number of time intervals is not too large, this scheduling problem can be solved with reasonable computational expense.

The following features can be readily accomodated in the MILP scheduling model. The case of no intermediate storage is obtained by simply setting the capacity of states $C_s = 0$. Unlimited intermediate storage means placing no upper bound on C_s. Zero-wait policy can be imposed by adding constraints that specify that task $\hat{i}$ follows task i, that is

$$\sum_{j \in K_i} W_{ijt} = \sum_{j \in K_{\hat{i}}} W_{\hat{i}jt + p_i} \qquad \forall t \qquad (22.50)$$

Finally, multiple products in flowshop plants are represented by multiple STNs as explained before.

As an example of the application of the STN MILP model consider the recipe, available equipment, and storage capacity given in Table 22.2. The state-task network for that recipe is given in Figure 22.14.

Assuming that the time horizon is 9 hours and since all processing times are integer numbers, we will consider 9 time intervals each of 1 hour. The corresponding MILP model has 72 0-1 variables, 179 continuous variables, and 250 constraints. The optimal schedule is shown in Figure 22.15, where it can be shown how the equipment is being allocated to each task. Also, Figure 22.16 shows the storage profiles for each of the materials or states. Period 10 represents the final state.

Sec. 22.8 State-Task Network for Scheduling of Multiproduct Batch Plant

TABLE 22.2 Example STN Model

Recipe
- Task 1 (Heat): Heat A for 1 hour.
- Task 2 (Reac1): Mix 50% feed B and 50% feed C and react for 2 hours to form intermediate BC.
- Task 3 (Reac2): Mix 40% hot A and 60% intermediate BC and react for 2 hours to form intermediate AB (60%) and product 1 (40%).
- Task 4 (Reac3): Mix 20% feed C and 80% intermediate AB and react for 1 hour to form impure E.
- Task 5 (Separ): Distill impure E to separate pure product 2 (90%, after 1 hour) and pure intermediate AB (10% after 2 hours). Recycle intermediate AB.

Available Equipment
- Unit 1 (Heater): Capacity 100 Kg, suitable for task 1.
- Unit 2 (Reactor 1): Capacity 50 Kg, suitable for tasks 2, 3, 4.
- Unit 3 (Reactor 2): Capacity 80 Kg, suitable for tasks 2, 3, 4.
- Unit 4 (Still): Capacity 200 Kg, suitable for task 5.

Available Storage
- For feeds A, B, C (States 1, 2, 3): Unlimited
- For hot A (State 4): 1000 Kg
- For intermediate AB (State 5): 500 Kg
- For intermediate BC (State 6): 0 Kgr
- For impure E (State 7): 1000 Kg
- For products 1 and 2 (States 8, 9): Unlimited

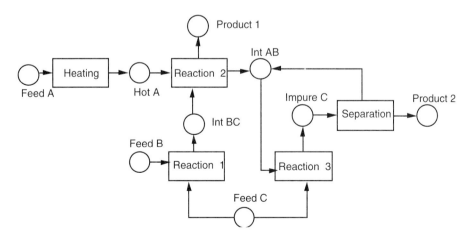

FIGURE 22.14 State-task network for numerical example.

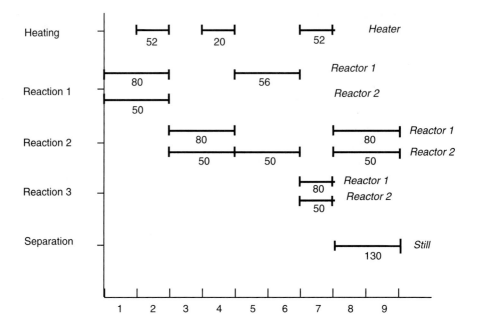

FIGURE 22.15 Optimal schedule for network in Figure 22.14.

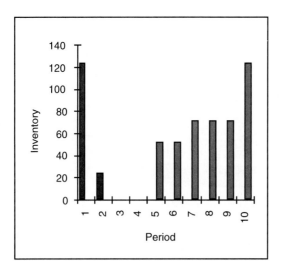

FIGURE 22.16 Storage for feed A (periods 1 and 2), product 2 (periods 5-10).

22.9 NOTES AND FURTHER READING

General reviews on optimization models for batch design and scheduling can be found in Reklaitis (1991, 1992), Pantelides (1994), and Rippin (1993). A review on mixed-integer optimization techniques for batch processing can be found in Grossmann et al. (1992), while a general classification of scheduling models has been outlined in Pinto and Grossmann (1995).

The MINLP model for flowshop plants in section 22.3 has been extended to the case of batch semi-continuous plants by Ravemark (1995) based on earlier work by Knopf et al. (1982). Effective TSP methods for flowshop models have been studied extensively by Pekny and co-workers (e.g., Gooding et al., 1994; Pekny and Miller, 1991). Scheduling models for continuous multiproduct plants have been reported by Sahinidis and Grossmann (1991) and Pinto and Grossmann (1994).

This chapter has not presented design models for multipurpose plants. A comprehensive MINLP model has been reported by Papageorgaki and Reklaitis (1990). Finally, a growing body of literature is evolving around the STN model and its variants. Examples of these papers include Shah et al. (1993a,b), Barbosa-Póvoa (1994), and Xueya and Sargent (1994).

REFERENCES

Barbosa-Póvoa, A. P. (1994). *Detailed design and retrofit of multipurpose batch plants*, Ph.D. Thesis, University of London, London (UK).

Birewar D. B., & Grossmann, I. E. (1989a). Efficient optimization algorithms for zero wait scheduling of multiproduct batch plants. *Ind. Eng. Chem. Res.*, **28,** 1333.

Birewar D. B., & Grossmann, I. E. (1989b). Incorporating scheduling in the optimal design of multiproduct plants. *Comp&Chem.Eng.*, **13**(1/2), 141.

Gooding, W. B., Pekny, J. F., & McCroskey, P. S. (1994). Enumerative approaches to parallel flowshop scheduling via problem transformation. *Computers Chem. Engng.*, **18**(10), 909.

Grossmann, I. E., & Sargent, R. W. H. (1978). Optimum design of multipurpose chemical plants. *Ind.Eng.Chem. Process Design and Dev.*, **18,** 343.

Grossmann, I. E., Quesada, I., Raman, R., & Voudouris, V. T. (1992). Mixed-integer optimization techniques for the design and scheduling of batch processes. Presented at the *NATO Advanced Study Institute—Batch Process Systems Engineering*, Antalya (Turkey).

Gupta, J. N. D. (1976). Optimal flowshop schedules with no intermediate storage space. *Naval Res. Logis. Q.*, **23,** 235.

Knopf, F. C., Okos, M. R., & Reklaitis, G. V. (1982). Optimal design of batch/semicontinuous processes. *Ind. Eng. Chem. Proc. Des. Dev.*, **21,** 79.

Kocis, G. R., & Grossmann, I. E. (1988). Global optimization of nonconvex MINLP problems in process synthesis. *Ind. Engng. Chem. Res.,* **27,** 1407.

Kondili, E., Pantelides, C. C., & Sargent, R. W. H. (1993). A general algorithm for short-term scheduling of batch operations-I. MILP Formulation. *Comp & Chem. Eng.*, **17**(2), 211.

Pantelides, C. C. (1994). Unified frameworks for optimal process planning and scheduling. In D. W. T. Rippin, J. C. Hale, & J. F. Davis (Eds.), *Foundations of Computer Aided Process Operations* (pp. 253–274). Austin, TX: CACHE.

Papageorgaki S., & Reklaitis, G. V. (1990). Optimal design of multipurpose batch plants-1. Problem formulation. *Ind.Eng.Chem.Res.*, **29**(10), 2054.

Pekny J. F, & Miller, D. L. (1991). Exact solution of the no-wait flowshop scheduling problem with a comparison to heuristic methods. *Comp & Chem. Eng.*, **15**(11), 741.

Pinto, J. M., & Grossmann, I. E. (1994). Optimal cyclic scheduling of multistage continuous multiproduct plants. *Computers Chem. Engng.*, **18**(9), 797.

Pinto, J. M., & Grossmann, I. E. (1995, submitted for publication). *Assignment and Sequencing Models for the Scheduling of Chemical Processes.*

Ravemark, D. (1995). *Design and Operation of Batch Processes.* PhD thesis, ETH, Zurich.

Reklaitis, G. V. (1991). Perspectives on scheduling and planning of process operations. Presented at the *Fourth International Symposium on Process Systems Engineering*, Montebello (Canada).

Reklaitis, G. V. (1992). Overview of scheduling and planning of batch process operations. *NATO Advanced Study Institute—Batch Process Systems Engineering*, Antalya (Turkey).

Rippin, D. W. T. (1993). Batch process systems engineering: A retrospective and prospective review. *Computers Chem. Engng.*, **17**(suppl. issue), S1–S13.

Sahinidis, N. V., & Grossmann, I. E. (1991). MINLP model for cyclic multiproduct scheduling on continuous parallel lines. *Computers Chem. Engng.*, **15**(2), 85.

Shah, N., Pantelides, C. C., & Sargent, R.W.H. (1993a). A general algorithm for short-term scheduling of batch operations. II. Computational issues. *Computers Chem. Engng.*, **17**(2), 229.

Shah, N., Pantelides, C. C., & Sargent, R. W. H. (1993b). Optimal periodic scheduling of multipurpose batch plants. *Ann. Oper. Res.*, **42**(1–4), 193.

Sparrow, R. E, Forder, G. J., & Rippin, D. W. T. (1975). The choice of equipment sizes for multiproduct batch plant. Heuristic vs. branch and bound. *Ind. Eng. Chem. Proc. Des. Dev.*, **14**(3), 197.

Voudouris, V. T, & Grossmann, I. E. (1992). Mixed integer linear programming reformulations for batch process design with discrete equipment sizes. *Ind. Eng. Chem. Res.*, **31**(5), 1314.

Xueya, Z., & Sargent, R. W. H. (1994). The optimal operation of mixed production facilities—A general formulation and some approaches to the solution. *Proceedings of the 5th Symposium on Process Systems Engineering*, Kyongju (Korea).

EXERCISES

1. Explain why the bounds in Eq. (22.12) for the cycle times and the batch sizes are valid.

2. Given is a multiproduct batch plant that consists of three processing stages: mixing, reaction, and centrifuge separation. Two products, A and B, are to be manufactured in such a plant using production campaigns of single products. The data for processing times, size factors for the units, demands, and cost data are given below. Assuming continuous sizes, and that the plant is operated with single-product campaigns determine the sizes of the units required at each processing stage, as well as the number of units that ought to be operating in parallel to minimize the investment cost.

Data

Demand A = 200,000 kg
Demand B = 150,000 kg
Horizon time = 6000 hrs

	Processing times (hrs)				Size factors (ℓ/kg)		
	Mixer	Reactor	Centrifuge		Mixer	Reactor	Centrifuge
A	8	20	4	A	2	3	4
B	10	12	3	B	4	6	3

Cost mixer = $250 $V^{0.6}$
Cost reactor = $500 $V^{0.6}$
Cost centrifuge = $340 $V^{0.6}$ (Volume V in liters)

Minimum size = 250 ℓ
Maximum size = 2500 ℓ

3. Resolve the MINLP model of problem 2 for the following cases:
   a. The demands of A and B are increased by 20%.
   b. For the above demands the time available for production is increased from 6000 hrs to 7500 hrs.

4. Assume that the mixer and reactor of problem 2 could be replaced by one single vessel so that the plant would consist of only two processing stages. If the cost of the new unit is 600V^{0.6}$, how would the optimal design of the plant be changed?
   (Hint: Assume that the processing time in the new vessel is the sum of mixing and reaction time for each products. Also, the size factor is the larger of the mixing and reaction steps for each product.)

5. How would you extend the MINLP model in Eq. (22.21) to account for fixed charges for the equipment cost?

6. Given is the NLP optimization model for the design of a multiproduct batch plant with one unit per stage and operating with single product campaigns:

$$\min \sum_{j=1}^{M} \alpha_j V_j^{\beta}$$

s.t. $\quad V_j \geq S_{ij}B_i \quad i = 1,..N, \quad j = 1,..M$

$$\sum_{i=1}^{M} \frac{Q_i}{B_i} T_{Li} \leq H$$

$V_j \geq 0 \quad j = 1,... M; \quad B_i \geq 0 \quad i = 1,...N$

where M is the number of stages and N the number of products. Show that if a feasible solution exists for this problem, the optimal design will be such that:

a. The horizon constraint (second inequality) will always be active

b. There will be at least $N + M$ active inequalities for the capacity constraints (first inequalities)

7. Assume that problem (22.36) is applied to a set of four batches $B = \{1,2,3,4\}$. Furthermore, assume that the problem is solved *without* subtour elimination constraints yielding two disjoint subcycles: $\{1,2\}$ and $\{3,4\}$. Show the explicit subtour elimination constraints in Eq. (22.36) that will ensure that these disjoint sets are connected.

8. Assume that the solution of the aggregated LP in Eq. (22.39) for minimizing cycle time in flowshop plants with zero-wait policy yields the following solution:

			$NPRS_{ik}$		
i/k	A	B	C	D	E
A		2	3		
B				3	
C					4
D		3			
E			4		

where $NPRS_{ik}$ is the number of changeovers from product i to product k. Determine if the above solution yields a valid cyclic schedule. If not, specify how would you modify the LP to accomplish this objective.

9. Given is a batch plant that manufactures four products A, B, C, D. It is desired to produce two batches of A, two batches of B, five batches of C, and four batches of D.

   a. Assuming a zero-wait policy, determine a cyclic sequence with minimum cycle time.

	Processing Times (hrs)		
	Stage 1	Stage 2	Stage 3
A	5	4	2
B	7	5	4
C	5	6	2
D	8	8	2

Exercises

Assume that cleanup times between different products can be neglected.

b. Repeat for the case in which 1 hour of cleanup is required between any change of products at any stage.

10. Repeat problem 2, assuming that the demands for $Q_A = 80,000$ kg, $Q_B = 50,000$ kg, and that only one unit per stage is allowed. Determine the sizes required for single-product campaigns, and for mixed-product campaigns with ZW and UIS policy. In all cases cleanup times can be neglected.

11. Repeat problem 10 for the case of single-product campaigns assuming that the equipment sizes are available as follows:

$$V = \{250, 750, 1000, 1500, 1750, 2500\} \text{ liters}$$

How does your solution compare with the one in which the sizes obtained in problem 10 are rounded to the next highest value?

12. Show that the NLP design model for flowshop plants can be reformulated as an MILP model if the equipment sizes are be available in discrete sizes v_{js}, $j = 1,2,...M$, $s = 1,2,...NDS$.

13. Develop an MILP model for the optimal design of multiproduct batch plants operating with single product campaigns and where parallel equipment may be involved in each stage. The equipment is assumed to be available in discrete sizes v_{js}, $j = 1,2,...M$, $s = 1,2,...NDS$.

SUMMARY OF OPTIMIZATION THEORY AND METHODS

This appendix will attempt to present in a very concise way basic concepts of optimization, optimality conditions, and an outline of the major methods that are used in Chapter 9 and in Part IV. A bibliography is given at the end of the appendix for readers who may wish to do further reading on this subject.

A.1 BASIC CONCEPTS

We will consider the following constrained optimization problem (Bazaraa and Shetty, 1979; Minoux, 1986):

$$\min f(x)$$
$$\text{s.t.} \quad h(x) = 0$$
$$g(x) \leq 0 \quad \text{(P)}$$
$$x \in R^n$$

where $f(x)$ is the objective function, $h(x) = 0$ is the set of m equations in n variables x, and $g(x) \leq 0$ is the set of r inequality constraints. In general, the number of variables n will be greater than the number of equations m, and the difference $(n - m)$ is commonly denoted as the number of degrees of freedom of the optimization problem.

Any optimization problem can be represented in the above form. For example, if we maximize a function, this is equivalent to minimizing the negative of that function. Also, if we have inequalities that are greater or equal to zero, we can reformulate them as in-

Sec. A.1   Basic Concepts

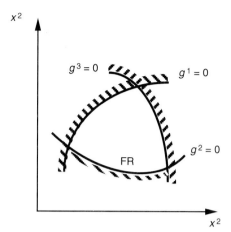

FIGURE A.1 Feasible region for three inequalities.

equalities that are less or equal than zero multiplying the two terms of the inequality by minus one, and reversing the sign of the inequality.

DEFINITION 1

The feasible region FR of problem (P) is given by

$$FR = \{ x \mid h(x) = 0, g(x) \leq 0, x \in R^n \}$$

Figure A.1 presents an example of a feasible region in two dimensions that involves three inequalities. Note that the boundary of the region is given by those points for which $g_i(x) = 0$, $i = 1,2,3$. Also, the infeasible side of a constraint is represented by dashed lines. In Figure A.2, if we add the equation $h(x) = 0$, the feasible region reduces to the straight line in boldface.

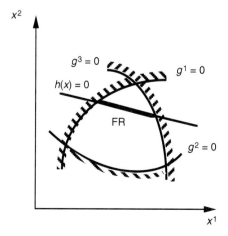

FIGURE A.2 Feasible region for three inequalities and one equation.

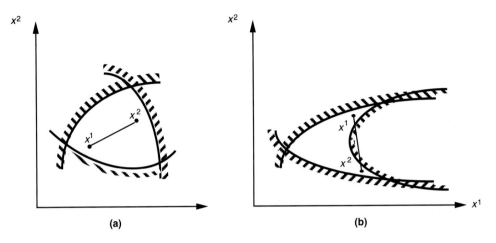

FIGURE A.3 (a) Convex feasible region; (b) nonconvex feasible region.

DEFINITION 2

FR is convex *iff* for any $x^1, x^2 \in FR$,

$$x = \alpha x^1 + (1 - \alpha) x^2 \in FR, \ \forall \ \alpha \in [0,1].$$

Figure A.3a presents an example of a convex feasible region; the region in Figure A.3b is nonconvex, since some of the points of the line that results from joining x^1 and x^2 lie outside the region *FR*.

The following is a useful sufficiency condition for the convexity of a feasible region.

PROPERTY 1

If $h(x) = 0$ consists of linear functions, and $g(x)$ of convex functions, then *FR* is a convex feasible region.

DEFINITION 3

$f(x)$ is a convex function *iff* for any $x^1, x^2 \in R$,

$$f(\alpha x^1 + [1 - \alpha] x^2) \leq \alpha f(x^1) + [1 - \alpha] f(x^2) \ \forall \ \alpha \in [0,1].$$

Figure A.4a presents an example of a convex function whose value is underestimated in the interval $[x^1, x^2]$ by the linear combination of the function values at the extremes of the interval. Figure A.4b presents an example of a function that is not convex. It should also be noted that if the above expression holds as a strict inequality for the points in the interval (x_1, x_2), then $f(x)$ is said to be strictly convex.

Sec. A.1   Basic Concepts

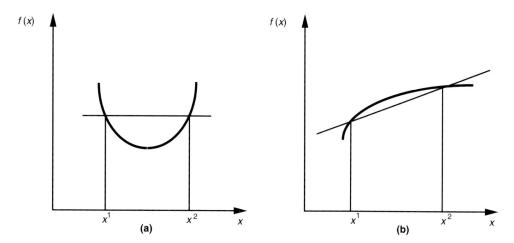

FIGURE A.4   (a) Convex function; (b) nonconvex function.

DEFINITION 4

$f(x)$ has a local minimum at $\hat{x} \in FR$, iff $\exists \delta > 0, f(x) \geq f(\hat{x})$ for $|x - \hat{x}| < \delta$, $x \in FR$.

If strict inequality holds the local minimum is a strong local minimum (see Figure A.5a); otherwise it is a weak local minimum (see Figure A.5b).

DEFINITION 5

$f(x)$ has a global minimum at $\hat{x} \in FR$, iff $f(x) \geq f(\hat{x})$ $\forall \, x \in FR$.

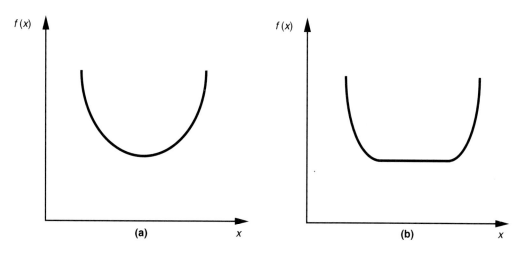

FIGURE A.5   (a) Function with strong local minimum; (b) function with weak local minimum.

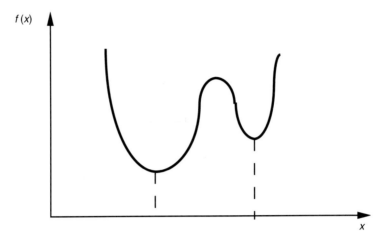

FIGURE A.6 Function with two local minima.

Clearly, every global minimum is a local minimum, but the converse is not true. Figure A.6 presents an example of a function with two strong local minima, one of them being the global minimum.

A.2 OPTIMALITY CONDITIONS

A.2.1 Unconstrained Minimization

Consider first the unconstrained optimization problem,

$$\min f(x)$$
$$x \in R^n$$

where $f(x)$ is assumed to be a continuous differentiable function.

First order conditions, which are necessary for a local minimum at $\hat{x}$, are given by a stationary point; that is, an $\hat{x}$ satisfying $\nabla f(\hat{x}) = 0$. This implies the solution of the following system of n equations in n unknowns,

$$\frac{\partial f}{\partial x_1} = 0$$
$$\frac{\partial f}{\partial x_2} = 0$$
$$\cdot$$
$$\cdot$$
$$\cdot$$
$$\frac{\partial f}{\partial x_n} = 0$$

Sec. A.2   Optimality Conditions

Second order conditions for a strong local minimum, which are sufficient conditions, require the Hessian matrix H of second partial derivatives to be positive definite. For two dimensions the matrix H is given by,

$$\begin{bmatrix} \dfrac{\partial^2 f}{\partial x_1^2} & \dfrac{\partial^2 f}{\partial x_1 \partial x_2} \\ \dfrac{\partial^2 f}{\partial x_2 \partial x_1} & \dfrac{\partial^2 f}{\partial x_2^2} \end{bmatrix}$$

Note that this matrix is symmetric.

The matrix H is said to be positive definite *iff* $\Delta x^T H \Delta x > 0$, $\forall\, \Delta x \neq 0$. The two following properties are useful for establishing in practice the positive definiteness of the Hessian matrix:

1. H is positive definite *iff* the eigenvalues $\rho_i > 0$, $i = 1, 2, \ldots n$.
2. If H is positive definite, then $f(x)$ is strictly convex.

That is, from property (1) we can establish the positive definiteness if the eigenvalues calculated from matrix H are all strictly positive. Property (2) simply states that functions whose Hessian matrix is positive definite are strictly convex functions. Therefore, analyzing the Hessian matrix of a function is one way to determine if a given function is convex.

The following is a useful sufficient condition for the uniqueness of a local minimum in an unconstrained optimization problem.

THEOREM 1

If $f(x)$ is strictly convex and differentiable, then if there exists a stationary point at $\hat{x}$, it will correspond to a unique local minimum.

A.2.2 Minimization with Equalities

Consider next the constrained optimization problem with only equalities:

$$\min f(x)$$

$$\text{s.t. } h(x) = 0$$

$$x \in R^n$$

In this case, the necesssary conditions for a constrained local minimum are given by the stationary point of the Lagrangian function

$$L = f(x) + \sum_{j=1}^{m} \lambda_j h_j(x)$$

where λ_j are the Lagrange multipliers. The stationary conditions are given by,

a. $\dfrac{\partial L}{\partial x} = \nabla f(x) + \sum_{j=1}^{m} \lambda_j \nabla h_j(x) = 0$

b. $\dfrac{\partial L}{\partial \lambda_j} = h_j(x) = 0 \quad j = 1, 2 \ldots m$

Note that (a) and (b) define a system of $n + m$ equations in $n + m$ unknowns (x, λ). Also, note that equation (a) implies that the gradients of the objective function and equalities must be linearly dependent, while equation (b) implies feasibility of the equalities. It must also be pointed out that for the above equations to be valid a "constraint qualification" (e.g., see Bazaraa and Shetty, 1979) must hold. In convex problems this qualification is always satisfied.

Second order sufficient conditions for a strong local minimum are satisfied when the Hessian of the Lagrangian is positive definite. That is, given an allowable direction p that lies in the null space, $\nabla h^T p = 0$, we have $p^T \nabla^2 L(x^*, \lambda^*) p > 0$, where $\nabla^2 L(x^*, \lambda^*) = \nabla^2 f(x^*) + \lambda^*_i \nabla^2 h_i(x^*)$.

A.2.3 Minimization with Equalities and Inequalities

Consider the constrained optimization problem with equalities and inequalities,

$$\min f(x)$$
$$\text{s.t. } h(x) = 0 \quad \text{(P)}$$
$$g(x) \le 0$$
$$x \in R^n$$

In this case the necessary conditions for a local minimum at $\hat{x}$ are given by the Karush-Kuhn-Tucker conditions:

a. Linear dependence of gradients

$$\nabla f(x) + \sum_{j=1}^{m} \lambda_j \nabla h_j(x) + \sum_{j=1}^{r} \mu_j \nabla g_j(x) = 0$$

b. Constraint feasibility

$$h_j(x) = 0 \quad j = 1, 2 \ldots m \qquad g_j(x) \le 0 \quad j = 1, 2 \ldots r$$

c. Complementarity conditions

$$\mu_j g_j(x) = 0, \quad \mu_j \ge 0 \quad j = 1, 2 \ldots r$$

where μ_j are the Kuhn-Tucker multipliers corresponding to the inequalities, and which are restricted to be non-negative. Note that the complementarity conditions in (c) imply a zero

Sec. A.2 Optimality Conditions

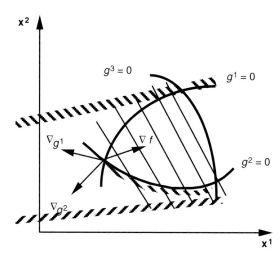

FIGURE A.7 Geometrical representation of a point satisfying the Karush-Kuhn-Tucker conditions.

value for the multipliers of the inactive inequalities (i.e., $g_j(x) < 0$), and in general a non-zero value for the active inequalities (i.e., $g_j(x) = 0$). Figure A.7 presents a geometrical representation of a point satisfying the Karush-Kuhn-Tucker conditions. Note that ∇f is given by a linear combination of the gradients of the active constraints ∇g_1, ∇g_2.

It can also be shown that the multipliers μ_j are given by

$$\mu_j = -\left(\frac{\partial f}{\partial g_j}\right)_{\delta g_i = 0, i \neq j}$$

In other words, they represent the decrease of the objective for an increase in the constraint function; or alternatively, the increase of the objective for a decrease in the constraint function. From the latter, it follows that active inequalities must exhibit a non-negative value of the multipliers.

The following is a useful sufficient condition on the uniqueness of a local optimum in constrained optimization problems.

THEOREM 2

If $f(x)$ is convex and the feasible region FR is convex, then if there exists a local minimum at $\hat{x}$,

i. It is a global minimum.
ii. The Karush-Kuhn-Tucker conditions are necessary and sufficient.

The difficulty with the equations in (a),(b),(c) for the optimality conditions of problem (P) is that they cannot be solved directly as is the case when only equalities are present. In general the solution to these equations is accomplished by an iterative active set strategy, which in a simplified form consists of the following steps:

Step 1: Assume no active inequalities. Set the index set of active inequalities $J_A = \emptyset$, and the multipliers $\mu_j = 0, j = 1,2,...r$.

Step 2: Solve the equations in (a) and (b) for x, the multipliers λ_j of the equalities, and the multipliers μ_j of the active inequalities (in 1st iteration there are none):

$$\nabla f(x) + \sum_{j=1}^{m} \lambda_j \nabla h_j(x) + \sum_{j \in J_A} \mu_j \nabla g_j(x) = 0$$

$$h_j(x) = 0 \quad j = 1,2...m \qquad g_j(x) = 0 \quad j \in J_A$$

Step 3. If $g(x) \leq 0$ and $\mu_j \geq 0, j = 1,2,...r$, STOP, solution found. Otherwise go to step 4.

Step 4: a. If one or more multipliers μ_j are negative, remove from J_A that active inequality with the largest negative multiplier.
  b. Add to J_A the violated inequalities $g_j(x) > 0$.
  Return to step 2.

The above is only a very general procedure and is suitable for hand calculations of small problems.

A.3 OPTIMIZATION METHODS

In this section we will present a brief overview of the different types of optimization methods covered in Parts II and IV. The emphasis will be on practical aspects, and only in the case of mixed-integer nonlinear programming we will present some more detail on the actual methods.

A.3.1 Linear Programming

When only linear functions are involved in problem (P), and the continuous variables x are restricted to non-negative values, this gives rise to the LP problem:

$$\min Z = c^T x$$
$$\text{s.t.} \quad A x \leqq a \qquad \qquad \text{(LP)}$$
$$x \geq 0$$

where the sign $\leqq$ denotes equalities and/or inequalities. Since linear functions are convex, from Property 1 and Theorem 2, the LP has a unique minimum. This may, however, be a weak minimum, for which alternate variable values may give rise to the same minimum objective function value.

The standard solution method is the simplex algorithm [Hillier and Lieberman, 1986] which exploits the fact that in an LP the optimum lies at a vertex of the feasible region (see Figure A.8). At this optimum, the Karush-Kuhn-Tucker conditions are satisfied.

Sec. A.3   Optimization Methods

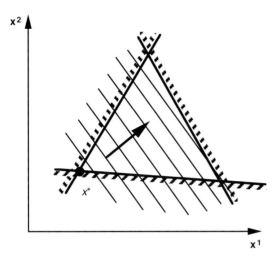

FIGURE A.8. Optimum lies at vertex x^* for LP problem.

Many refinements have been developed over the last three decades for the simplex method, and most of the current commercial computer codes (e.g., OSL, CPLEX, LINDO) are based on this method. Very large scale problems (thousands of variables and constraints) that are sparse (i.e., few variables in each constraint) can be solved quite efficiently. As a general guideline, the computational effort in the simplex algorithm is dependent mostly on the number of constraints (rows in LP terminology), not so much on the number of variables (columns). In problems with many rows and relatively few variables, it is advisable to solve the LP through its dual problem.

For variables x that can be positive and negative in an LP, these are replaced by $x = x^P - x^N$, where x^P and x^N are non-negative. If x^N is zero we get a positive value, and if x^P is zero we get a negative value. This manipulation should only be used when the variable x appears with a positive coefficient in the minimization of an objective function.

Recently, interior point methods for LP (Marsten et al., 1990) have been developed that are polynomially bounded in time. Although these methods are theoretically superior to the simplex algorithm, it is only for extremely large scale problems that substantial computational savings have been observed (e.g., problems with 100,000 constraints and variables).

As a final point, it is important to note that special classes of LP problems can be solved more efficiently than with standard LP codes. The best known case are network flow problems (see Minoux, 1986) where the matrix of coefficients involves only 0, 1, −1, elements. In this case the simplex method can be implemented with symbolic computations leading to order of magnitude reductions in computational time.

A.3.2 Mixed-Integer Linear Programming

This is an extension of the LP problem where a subset of the variables are restricted to integer values (most commonly to 0-1). The general form of the MILP problem is given by,

$$\min Z = a^T y + c^T x$$
$$\text{s.t.} \quad By + Ax \leqq b \qquad \text{(MILP)}$$
$$y \in \{0,1\}^t \quad x \geq 0$$

where y corresponds to a vector of t binary variables.

The MILP problem is very useful for modeling a number of discrete decisions with the binary variables y (see Chapter 15). Typical examples are the following:

a. **Multiple choice constraints**
   Select only one item:
   $$\sum_{j=1}^{t} y_i = 1$$
   Select at most one item:
   $$\sum_{j=1}^{t} y_j \leq 1$$
   Select at least one item:
   $$\sum_{j=1}^{t} y_j \geq 1$$

b. **Implication constraints.**
   If item k is selected, item j must be selected, but not vice versa: $y_k - y_j \leq 0$
   If a binary variable y is zero, an associated continuous variable x must also be zero:
   $$x - Uy \leq 0, \quad x \geq 0$$
   where U is an upper limit to x.

c. **Either-or constraints (disjunctive constraints)**
   Either constraint $g_1(x) \leq 0$ or constraint $g_2(x) \leq 0$ must hold:
   $$g_1(x) - U y \leq 0, \quad g_2(x) - U(1 - y) \leq 0$$
   where U is a large value.

A simple-minded approach to obtain the global optimum of the above MILP would be to solve the LPs that result from considering all the 0-1 combinations of the binary variables. However, the number of combinations is 2^t, which is too large for even modest number of variables (e.g., for 20 binaries there are 10^6 combinations).

A second approach is to relax the 0-1 constraints as continous variables that must lie between 0 and 1; that is, $0 \leq y_t \leq 1$. The problem is then solved as an LP. The difficulty here is that except for special cases (e.g., assignment problems), one or more binary vari-

Sec. A.3 Optimization Methods

ables will exhibit noninteger values at the optimum LP solution. The relaxed LP, however, is useful in providing a lower bound to the optimal mixed-integer solution.

In general, one cannot simply round the noninteger values of the binary variables in the relaxed LP solution to the nearest integer point. Firstly, because the rounding may be infeasible (see Figure A.9a), or secondly because it may be nonoptimal (see Figure A.9b). The standard method for solving MILP problems is the branch and bound method (Nemhauser and Wolsey, 1988), which was briefly outlined in Chapter 15 in the context of the synthesis of a separation sequence. For the MILP we start by solving first the relaxed LP problem. If integer values are obtained for the binary variables, we stop, as we have solved the problem. If, on the other hand, no integer values are obtained, the basic idea is then to examine through the use of bounds a subset of nodes in a binary tree to locate the global mixed-integer solution. In the tree the binary variables are successively restricted one by one to 0-1 values at each node where the corresponding LP is solved. This can be done quite efficiently by updating the successive LPs through few dual simplex iterations.

Nodes with noninteger solutions provide a lower bound, and nodes with feasible mixed-integer solutions provide an upper bound. The former nodes are fathomed whenever the lower bound is greater or equal than the current best upper bound. For the tree enumeration one has to consider branching rules to decide which binary variable is fixed next in the tree. These rules range from simply picking the first non-zero value to the use of penalties to estimate which binary produces the smallest degradation in the LP. Also, in a similar way as in the implicit enumeration described in Chapter 15, the tree can be enumerated through a depth-first method, a breadth-first method, or combination of the two.

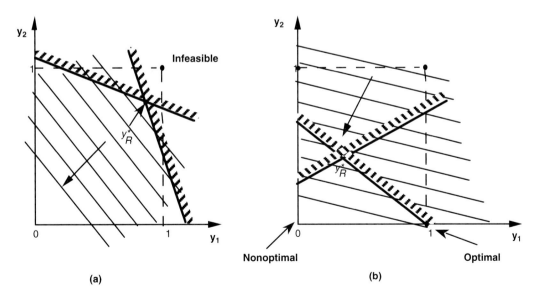

FIGURE A.9 (a) Infeasible rounding of relaxed integer solution; (b) nonoptimal rounding of relaxed integer solution.

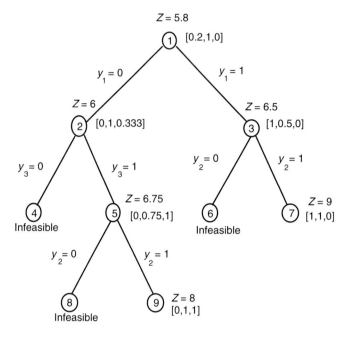

FIGURE A.10 Branch and bound tree for example problem (MIPEX).

The more advanced MILP packages allow the specialized user to specify the search option to be used. Figure A.10 presents an example of a tree search with branch and bound in the MILP problem:

$$\min \quad Z = x + y_1 + 3y_2 + 2y_3$$
$$\text{st.} \quad -x + 3y_1 + 2y_2 + y_3 \leq 0$$
$$-5y_1 - 8y_2 - 3y_3 \leq -9 \quad \text{(MIPEX)}$$
$$x \geq 0, \ y_1, y_2, y_3 = \{0,1\}$$

The branch and bound tree using a breadth-first enumeration is shown in Figure A.10. The numbers in the circles represents the order in which 9 nodes out of the 15 nodes in the tree are examined to find the optimum. Note that the relaxed solution (node 1) has a lower bound of $Z = 5.8$, and that the optimum is found in node 9 where $Z = 8$, $y_1 = 0$, $y_2 = y_3 = 1$, and $x = 3$.

Although the general performance of the branch and bound method can greatly vary from one problem to another, as a general guideline the computational expense tends to be proportional first to the number of 0-1 variables, secondly to the number of constraints, and thirdly to the number of continuous variables. Another criterion, which is often more relevant, is the gap between the objective function value of the relaxed LP and the optimal MILP solution. The smaller this gap the easier it is usually to solve the MILP problem since

A.3.3 Nonlinear Programming

In this case, the problem corresponds to:

$$\min f(x)$$
$$\text{s.t.} \quad h(x) = 0 \qquad \text{(NLP)}$$
$$g(x) \leq 0$$
$$x \in R^n$$

where in general $f(x)$, $h(x)$, $g(x)$, are nonlinear functions.

The more efficient NLP methods solve this problem by determining directly a point that satifies the Karush-Kuhn-Tucker conditions. As pointed out in Theorem 2, global minumum solutions can be guaranteed for the case when the objective and constraints are nonlinear convex functions, and the equalities are linear. Since the Karush-Kuhn-Tucker conditions involve gradients of the objective and constraints, these must be supplied by the user either in analytical form or through the use of numerical perturbations. However, the latter option is expensive for problems with large number of variables.

Currently the two major methods for NLP are the successive quadratic programming (SQP) algorithm (Han, 1976; Powell, 1978) and the reduced gradient method (Murtagh and Saunders, 1978, 1982). In the case of the (SQP) algorithm (see Chapter 9 for more details) the basic idea is to solve at each iteration a quadratic programming subproblem of the form:

$$\min \quad \nabla f(x^k)^T d + 1/2 \, d^T B^k d$$
$$\text{s.t.} \quad h(x^k) + \nabla h(x^k)^T d = 0 \qquad \text{(QP)}$$
$$g(x^k) + \nabla g(x^k)^T d \leq 0$$

where x^k is the current point, B^k is the estimation of the Hessian matrix of the Lagrangian, and d is the predicted search direction. The matrix B^k is usually estimated with the BFGS update formula, and the QP is solved with standard methods for quadratic programming (e.g., QPSOL routine). Since the point x^k will in general be infeasible, the next point x^{k+1} is set to $x^{k+1} = x^k + \alpha \, d$, where the step size α is determined so as to reduce a penalty function that tries to balance the improvement in the objective and the violation of the constraints.

An important point about the SQP algorithm is the fact that the QP with the exact Hessian matrix of the Lagrangian in B can be shown to be equivalent to applying Newton's method to the Karush-Kuhn-Tucker conditions. Thus, fast convergence can be achieved with this algorithm.

In the reduced gradient method, on the other hand, the basic idea is to solve a sequence of subproblems with linearized constraints, where the subproblems are solved by variable elimination. In the particular implementation of MINOS by Murtagh and Saunders, the NLP is reformulated through the introduction of slack variables to convert the inequalities into equalities; that is, the NLP reduces to

$$\min f(x) \quad \text{(NLP1)}$$
$$\text{s.t. } r(x) = 0$$

Linear approximations of the constraints are then considered with an augmented Lagrangian for the objective function:

$$\min \phi(x) = f(x) + (\lambda^k)^T [r(x) - r(x^k)] \quad \text{(NLP2)}$$
$$\text{s.t. } J(x^k) x = b$$

where λ^k is the vector of Lagrange multipliers, and $J(x^k)$ is the jacobian of $r(x)$ evaluated at the point x^k. Subproblem NLP2, which is a linearly constrained optimization problem, can be represented by

$$\min \phi(x)$$
$$\text{s.t. } A x = b$$

where A is a mxn matrix with m < n. The above problem can be solved with the reduced gradient method as follows. Firstly, the vector x is partitioned into the vector v of m dependent variables, and the vector u of $(n - m)$ independent variables. Likewise, the matrix A is partitioned into a (mxm) square matrix B, and a $mx(n - m)$ matrix C. The reduced gradient can then be computed from the equation

$$g_R = Z^T \nabla \phi(x^k)$$

where x^k is a feasible point satisfying the linear constraints, and Z is a transformation matrix given by

$$Z^T = [C^T B^{-T} \mid I]$$

With the reduced gradient the Newton step, Δu in the reduced space can be computed from

$$H_R \Delta u = -g_R$$

where H_R is the reduced Hessian matrix, which is estimated through a Quasi-Newton update formula (e.g., BFGS formula). The change in the dependent variables, Δv, is then obtained by solving the linear equations

$$B \Delta v = -C \Delta u$$

In summary, in the reduced gradient method the subproblem (NLP2) is solved as an inner optimization problem, while in the outer optimization the new point is set as $x^{k+1} = x^k + \alpha \Delta x$ where α is the step size that is used to reduce the augmented Lagrangian in (NLP2), and $\Delta x = [\Delta v \mid \Delta u]$

The importance of the reduced gradient method is that by efficient implementation for the solution of the above equations (see Murtagh and Saunders, 1982) and realizing that some of the tools for large-scale LP can be used, sparsity can be readily exploited. In this way large nonlinear optimization problems can be solved very effectively. In comparing the SQP algorithm and the reduced gradient method, the following general guidelines apply:

1. SQP requires fewer iterations than the reduced gradient method. However, there may be difficulties in applying it to large-scale problems since in general the matrix B^k, which is of dimension $n \times n$, will become dense due to the Quasi-Newton updates. The SQP method is best suited for "black-box" models (e.g., process simulators) that involve relatively few variables (e.g., up to 50) and where the gradients must be obtained by numerical perturbation. It should be noted, however, that the SQP algorithm can be effectively applied to large-scale problems that involve few decision variables by using decomposition techniques.
2. The reduced gradient method, as per the implementation in MINOS is best suited for problems involving a significant number of linear constraints, and where analytical derivatives can be supplied for the nonlinear functions. With this structure, MINOS can solve problems with several hundred variables and constraints. Compared to SQP, MINOS will require a larger number of function evaluations, but the computational time per iteration will be smaller. Furthermore, in the limiting case when all the functions are linear the method reduces to the simplex algorithm for linear programming.

A.3.4 Mixed-Integer Nonlinear Programming

MINLP problems are usually the hardest to solve unless a special structure can be exploited. The following particular formulation, which is linear in the 0-1 variables and linear/nonlinear in the continuous variables, will be considered:

$$\min Z = c^T y + f(x)$$
$$\text{s.t.} \quad h(x) = 0$$
$$g(x) \leq 0$$
$$A x = a \quad \text{(MINLP)}$$
$$B y + C x \leq d$$
$$E y \leq e$$
$$x \in X = \{x \mid x \in R^n, x^L \leq x \leq x^U\}$$
$$y \in \{0,1\}^t$$

As explained in Chapter 15, this special MINLP structure arises in process synthesis problems.

This mixed-integer nonlinear program can in principle also be solved with the branch and bound method presented in section A.3.2. The major difference here is that the examination of each node requires the solution of a nonlinear program rather than the solution of an LP. Provided the solution of each NLP subproblem is unique, similar properties as in the case of the MILP would hold with which the rigorous global solution of the MINLP can be guaranteed.

An important drawback of the branch and bound method for MINLP is that the solution of the NLP subproblems can be expensive since they cannot be readily updated as in the case of the MILP. Therefore, in order to reduce the computational expense involved in solving many NLP subproblems, we can resort to two other methods: Generalized Benders decomposition (Geoffrion, 1972) and Outer-Approximation (Duran and Grossmann, 1986). Below we first briefly describe the latter method with the equality relaxation variant by Kocis and Grossmann (1987).

The basic idea in the OA/ER algorithm is to solve an alternating sequence of NLP and MILP master problems. The NLP subproblems arise for a fixed choice of the binary variables, and involve the optimization of the continuous variables x with which an upper bound to the original MINLP is obtained (assuming minimization problem). The MILP master problem, on the other hand, provides a global linear approximation to the MINLP in which the objective function is underestimated and the nonlinear feasible region is overestimated. Furthermore, the linear approximations to the nonlinear equations are relaxed as inequalities. This MILP master problem accumulates the different linear approximations of previous iterations so as to produce an increasingly better approximation of the original MINLP problem. At each iteration the master problem predicts new values of the binary variables y and a lower bound to the objective function Z. The search is terminated when no lower bound can be found below the current best upper bound which then leads to an infeasible MILP.

The specific steps of this algorithm, assuming feasible solutions for the NLP subproblems, are as follows:

Step 1: Select an initial value of the binary variables y^1. Set the iteration counter $K = 1$. Initialize the lower bound $Z_L^0 = -\infty$, and the upper bound $Z_U = +\infty$.

Step 2: Solve the NLP subproblem for the fixed value y^k, to obtain the solution x^k and the multipliers λ^k for the equations $h(x) = 0$.

$$Z(y^k) = \min c^T y^k + f(x)$$

s.t.
$$h(x) = 0$$
$$g(x) \leq 0$$
$$A x = a$$
$$C x \leq d - B y^k$$
$$x \in X$$

Step 3: Update the bounds and prepare the information for the master problem:

Sec. A.3 Optimization Methods

a. Update the current upper bound; if $Z(y^K) < Z_U$, set $Z_U = Z(y^K)$, $y^* = y^K$, $x^* = x^K$.

b. Derive the integer cut, IC^K, to make infeasible the choice of the binary y^K from subsequent iterations:

$$IC^K = \{ \sum_{i \in B^K} y_i - \sum_{i \in N^K} y \leq |B^K| - 1 \}$$

where $B^k = \{ i \mid y_i^K = 1 \}$, $N^K = \{ i \mid y_i^K = 0 \}$

c. Define the diagonal direction matrix T^K for relaxing the equations into inequalities based on the sign of the multipliers λ^k. The diagonal elements are given by:

$$t_{jj}^K = \begin{cases} -1 \text{ if } \lambda_j^K < 0 \\ +1 \text{ if } \lambda_j^K > 0 \\ 0 \text{ if } \lambda_j^K = 0 \end{cases} \quad j = 1, 2 \ldots m$$

d. Obtain the following linear outer-approximations for the nonlinear terms $f(x)$, $h(x)$, $g(x)$ by performing first order linearizations at the point x^K:

$$(w^K)^T x - w_c^K = f(x^K) + \nabla f(x^K)^T (x - x^K)$$

$$R^K x - r^K = h(x^K) + \nabla h(x^K)^T (x - x^K)$$

$$S^K x - s^K = g(x^K) + \nabla g(x^K)^T (x - x^K)$$

Step 4: a. Solve the following MILP master problem:

$$Z_L^K = \min c^T y + \mu$$

s.t.
$$(w^k) x - \mu \leq w_c^k$$
$$T^k R^k x \leq T^k r^k \quad k = 1, 2, \ldots K$$
$$S^k x \leq s^k$$
$$y \in IC^k$$
$$By + Cx \leq d \quad \text{(MOA)}$$
$$Ax = a$$
$$Ey \leq e$$
$$Z_L^{K-1} \leq c^T y + \mu \leq Z_U$$
$$y \in \{0,1\}^t \quad x \in X \quad \mu \in R^1$$

b. If the MILP master problem has no feasible solution, stop. The optimal solution is x^*, y^*, Z_U.

c. If the MILP master problem has a feasible solution, the new binary value y^{K+1} is obtained. Set $K = K + 1$, return to step 2.

It should be noted that in step 2, there is the possibility that the NLP subproblem may not have a feasible solution for the selected value of the binary variable y^k. When this is the case, the value of x^K and λ^k can be obtained by solving the following NLP in which the infeasibility is minimized:

$$\min u$$
$$\text{s.t.} \quad h(x) = 0$$
$$g(x) \leq u$$
$$A x = a$$
$$C x - d - B y \leq u$$
$$x \in X \quad u \in R^1$$

Furthermore, the objective function value is set to $Z(y^k) = +\infty$

It should be noted that sufficient conditions to obtain the global optimum solution require convexity in the nonlinear terms $f(x)$, $g(x)$, and quasi-convexity in the relaxed nonlinear equations $T^k h(x)$. When these conditions are not met, there is the possibility that the master problem may cut off the global optimum solution as discussed below.

Also, as an interesting point it should be noted that for the limiting case when $f(x)$, $g(x)$, and $h(x)$ are linear, the MILP master problem provides an exact representation of the MINLP, and therefore the OA/ER algorithm would converge in no more than two iterations. For nonlinear problems, computational experience indicates that the master problems provide an increasingly good approximation with which convergence can be typically achieved in only 3 to 5 iterations.

In the Generalized-Benders decomposition the above steps are virtually identical except that the MILP master problem in step 4(a) (assuming feasible NLP subproblems) is given at any iteration K by:

$$Z_{GB}^K = \min \alpha$$
$$\text{s.t.} \quad \alpha \geq f(x^k) + c^T y + (\mu^k)^T [Cx^k + By - d] \quad k = 1, 2, \ldots K \quad \text{(MGB)}$$
$$\alpha \in R^1, y \in \{0,1\}^m$$

where α is the largest Lagrangian approximation obtained from the solution of the K NLP subproblems; x^k and μ^k correspond to the optimal solution and multiplier of the kth NLP subproblem; Z_{GB}^K corresponds to the predicted lower bound at iteration K.

Note that in both master problems the predicted lower bounds, Z_{GB}^K and Z_{OA}^K increase monotonically as iterations K proceed since the linear approximations are refined by accumulating the Lagrangian (in MGB) or linearizations (in MOA) of previous iterations. It should be noted also that in both cases rigorous lower bounds, and therefore convergence to the global optimum, can only be ensured when certain convexity conditions hold (see Geoffrion, 1972; Duran and Grossmann, 1986).

In comparing the two methods, it should be noted that the lower bounds predicted by the outer approximation method are always greater than or equal to the lower bounds

Sec. A.3 Optimization Methods

predicted by Generalized-Benders decomposition. This follows from the fact that the Lagrangian cut in GBD represents a surrogate constraint from the linearization in the OA algorithm (Quesada and Grossmann, 1992). Hence, the Outer-Approximation method will require the solution of fewer NLP subproblems and MILP master problems. On the other hand, the MILP master in Outer-Approximation is more expensive to solve so that Generalized Benders may require less time if the NLP subproblems are inexpensive to solve. As discussed in Sahinidis and Grossmann (1991), fast convergence with GBD can only be achieved if the NLP relaxation is tight.

As a simple example of an MINLP consider the problem:

$$\min Z = y_1 + 1.5 y_2 + 0.5 y_3 + x_1^2 + x_2^2$$
$$\text{s.t.} \quad (x_1 - 2)^2 - x_2 \leq 0$$
$$x_1 - 2 y_1 \geq 0$$
$$x_1 - x_2 - 4(1 - y_2) \leq 0$$
$$x_1 - (1 - y_1) \geq 0$$
$$x_2 - y_2 \geq 0 \tag{6}$$
$$x_1 + x_2 \geq 3 y_3$$
$$y_1 + y_2 + y_3 \geq 1$$
$$0 \leq x_1 \leq 4, \; 0 \leq x_2 \leq 4$$
$$y_1, y_2, y_3 = 0, 1$$

Note that the nonlinearities involved in problem (6) are convex. Figure A.11 shows the convergence of the OA and the GBD methods to the optimal solution using as a start-

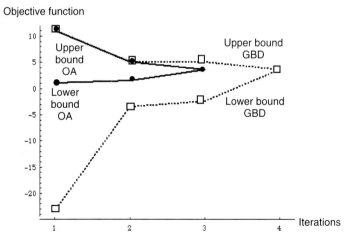

FIGURE A.11 Progress of iterations of OA and GBD for MINLP in (6).

ing point $y_1 = y_2 = y_3 = 1$. The optimal solution is $Z = 3.5$, with $y_1 = 0$, $y_2 = 1$, $y_3 = 0$, $x_1 = 1$, $x_2 = 1$. Note that the OA algorithm requires three major iterations, while GBD requires four, and that the lower bounds of OA are much stronger.

In the application of Generalized-Benders decomposition and Outer-Approximation, two major difficulties that can arise are the computational expense involved in the master problem if the number of 0-1 variables is large, and nonconvergence to the global optimum due to the nonconvexities involved in the nonlinear functions.

As for the question of nonconvexities, one approach is to modify the definition of the MILP master problem so as to avoid cutting off feasible mixed-integer solutions. Viswanathan and Grossmann (1990) proposed an augmented-penalty version of the MILP master problem for outer-approximation, which has the following form:

$$Z_L^K = \min \ c^T y + \mu + \sum_{k=1}^{K} (\rho^k)^T (p^k + q^k + r^k) \quad \text{(MOA)}$$

s.t.
$$\begin{aligned}
(w^k) x - \mu &\leq w_c^k + p^k \\
T^k R^k x &\leq T^k r^k + q^k \\
S^k x &\leq s^k + r^k \\
y &\in IC^k
\end{aligned} \right\} \ k = 1, 2, \ldots K$$

$$By + Cx \leq d$$
$$Ax = a$$
$$Ey \leq e$$
$$y \in \{0,1\}^t \quad x \in X \quad \mu \in R^1; \quad p^k, q^k, r^k \geq 0$$

in which the slacks p^k, q^k, r^k have been added to the function linearizations, and in the objective function with weights ρ^k that are sufficiently large but finite. Since in this case one cannot guarantee a rigorous lower bound, the search is terminated when there is no further improvement in the solution of the NLP subproblem. This version of the method together with the original version have been implemented in the computer code DICOPT++, which has shown to be successful in a number of applications. It should also be noted that if the MINLP is convex, the above master problem reduces to the original OA algorithm since the slacks will take a value of zero. For an updated review of MINLP methods see Grossmann and Kravanja (1995).

A.4 COMPUTER CODES AND REFERENCES

The following computer software can be used for solving different classes of problems:

1. For LP and MILP:
   - LINDO by Linus Schrage. Interactive program that is easy to use.

- ZOOM by Roy Marsten.
- OSL from IBM, CPLEX, and SCICONIC.
2. For NLP:
   - GINO by Leon Lasdon. Interactive program.
   - MINOS by Murtagh and Saunders.
   - CONOPT by Drud in Denmark.
3. For MINLP
   - DICOPT++/GAMS by Viswanathan and Grossmann.

The program GAMS by Brooke et al. (1988) provides a powerful computer interface that greatly facilitates the formulation and solution of LP, MILP, NLP, and MINLP problems. GAMS interfaces with OSL, CPLEX, ZOOM, MINOS, CONOPT, and DICOPT++.

CACHE distributes the case study "Chemical Engineering Optimization Problems with GAMS" (Morari and Grossmann, 1991), which contains about 20 optimization problems. A student version of GAMS that can solve LP, MILP, NLP, and MINLP problems is provided.

The following books deal with the basic concepts and methods for optimization covered in this Appendix, and they also include the references for computer software.

REFERENCES

Bazaraa, M. S., & Shetty, C. M. (1979). *Nonlinear Programming*. New York: Wiley.

Brooke, A., Kendrick, D., & Meeraus, A. (1988). *GAMS-A Users Guide*. Redwood City: Scientific Press.

Duran, M. A., & Grossmann, I. E. (1986). An outer-approximation algorithm for a class of mixed-integer nonlinear programs. *Mathematical Programming,* **36,** 307–339.

Geoffrion, A. M. (1972). Generalized Benders decomposition. *Journal of Optimization Theory and Applications,* **10**(4), 237–260.

Grossmann, I. E., & Kravanja, Z. (1995). Mixed-integer nonlinear programming techniques for process systems engineering. Supplement of *Computers and Chemical Engineering*, 19, S189–S204.

Han, S. P. (1976). Superlinearly convergent variable metric algorithms for general nonlinear programming problems. *Math Progr.,* 11, 263–282.

Hillier, F. S., & Lieberman, G. J. (1986). *Introduction to Operations Research*. San Francisco: Holden Day.

Kocis, G. R., & Grossmann, I. E. (1987). Relaxation strategy for the structural optimization of process flowsheets. *Industrial and Engineering Chemistry Research,* **26**(9), 1869–1880.

Liebman, J., Lasdon, L., Schrage, L., & Warren, A. (1986). *Modelling and Optimization with GINO.* Redwood City: Scientific Press.

Marsten, R., Saltzman, M., Lustig, J., & Shanno, D. (1990). Interior point methods for linear programming: Just call Newton, Lagrange and Fiacco and McCormick! *Interfaces,* **20**(4), 105–116.

Minoux, M. (1986). *Mathematical Programming: Theory and Algorithms.* New York: Wiley.

Morari M., & Grossmann, I.E. (Eds.). (1991). Chemical engineering optimization problems with GAMS. *CACHE Design Case Studies,* Vol. 6.

Murtagh, B. A., & Saunders, M. A. (1978). Large-scale linearly constrained optimization. *Mathematical Programming,* 14, 41–72.

Murtagh, B. A., & Saunders, M. A. (1982). A projected lagrangian algorithm and its implementation for sparse nonlinear constraints. *Mathematical Programming Study,* 16, 84–117.

Nemhauser, G. L., Rinnoy Kan, A. H. G., & Todd, M. J. (Eds). (1989). Optimization. In *Handbook in Operations Research and Management Science, Vol. 1,* North Holland.

Nemhauser, G. L., & Wolsey, L. A. (1988). *Integer and Combinatorial Optimization.* New York: Wiley.

Powell, M. J. D. (1978). A fast algorithm for nonlinearly constrained optimization calculations. In *Numerical Analysis,* Dundee, 1977. G. A. Watson (Ed.), *Lecture Notes in Mathematics 630,* Berlin: Springer-Verlag.

Quesada, I., & Grossmann, I. E. (1992). An LP/NLP based branch and bound method for MINLP optimization. *Computers and Chemical Engineering,* **16**.

Sahinidis, N. V., & Grossmann, I. E. (1991). Convergence properties of generalized benders decomposition. *Computers and Chemical Engineering,* **15,** 481.

Schrage, L. (1984). *Linear Integer and Quadratic Programming with LINDO.* Redwood City: Scientific Press.

Singal, J., Marsten, R. E., & Morin, T. (1987). Fixed-order branch and bound methods for mixed-integer programming: The ZOOM system. Working paper, Management Information Science Department, The University of Arizona, Tucson, Arizona.

Viswanathan, J., & Grossmann, I. E. (1990). A combined penalty function and outer-approximation method for MINLP optimization. *Computers and Chemical Engineering,* **14,** 769–782.

Williams, H. P. (1978). *Model Building in Mathematical Programming.* New York: Wiley-Interscience.

SMOOTH APPROXIMATIONS FOR MAX {0, f(x)}   B

The function $\phi(x) = \max\{0, f(x)\}$, which arises in model (18.24) of Chapter 18, is nondifferentiable at $f(x) = 0$ as shown in Figure B.1. We can, however, construct approximations to $\phi(x)$ that are continuous and differentiable everywhere.

Consider first the approximation proposed by Duran and Grossmann (1986). Let $\phi(x)$ be replaced by the exponential function $a \exp\{b\,f(x)\}$, for $f(x) \leq \varepsilon$, where a and b are parameters to be determined, and ε a small tolerance.

The parameters a and b we can select to insure continuity and differentiability at $f(x) = \varepsilon$. That is,

$$a \exp\{b\,\varepsilon\} = \varepsilon \tag{B.1}$$

$$a\,b \exp\{b\,\varepsilon\} \nabla f(\varepsilon) = \nabla f(\varepsilon) \tag{B.2}$$

From Eq. (B.2) it follows that

$$a\,b \exp\{b\,\varepsilon\} = 1 \tag{B.3}$$

Hence, combining with (B.1), $b = 1/\varepsilon$, and $a = \varepsilon/e$. Therefore, the function $\phi(x)$ can be approximated by:

$$\hat{\phi}(x) = \begin{cases} f(x) & \text{if } f(x) \geq \varepsilon \\ \varepsilon/e \exp\{f(x)/\varepsilon\} & \text{if } f(x) < \varepsilon \end{cases}$$

and is shown in Figure B.2. Too small a value at ε can cause ill-conditioning. Therefore, typical values should be between 0.0001 and 0.01.

Balakrishna and Biegler (1992) have proposed another smooth approximation that is similar in nature to the one described above, but is easier to implement, particularly in

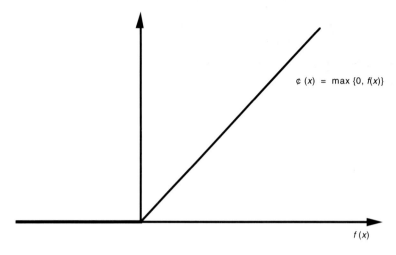

FIGURE B.1 Plot of max{0, f(x)} function.

equation-based systems. The function $\phi(x) = \max\{0, f(x)\}$ is simply replaced by the equation

$$\hat{\phi}(x) = 0.5[f(x)^2 + \varepsilon^2]^{1/2} + 0.5f(x) \tag{B.5}$$

It is easy to verify that for small values of ε the above equation yields an approximation similar to the one in Figure B.2. Equation (B.5) also exhibits ill-conditioning for small values of ε, and it introduces a small error at $f(x) \geq \varepsilon$.

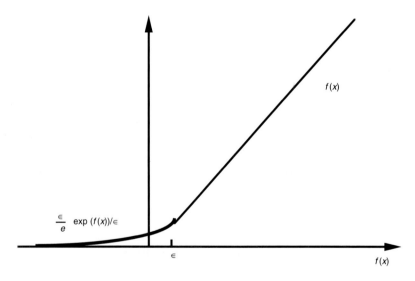

FIGURE B.2 Plot of smooth approximation scheme.

COMPUTER TOOLS FOR PRELIMINARY PROCESS DESIGN   C

This appendix presents a short list of computer software that can be used at the various stages of preliminary process design. A brief description for the software is given, as well as links to homepages or e-mail addresses where further information can be obtained on the computer tools. The appendix also includes at the end a list of design case studies, as well as a bibliography of articles that provides overviews on computer software.

C.1 COMPUTER SOFTWARE

C.1.1 Modeling Systems

Preliminary calculations for process design require tools that provide the capability for setting up quickly and easily simplified models of arbitary structure that can be effectively solved. Since these applications require relatively few data, fairly general purpose software tools can be used. These can be classified into spreadsheets for procedural calculations, and algebraic modeling systems that are suitable for equation models.

Spreadsheets

Excel
Microsoft: *http://www.microsoft.com/msexcel*
Lotus 1-2-3
Lotus: *http://www.lotus.com/123*

Equation Oriented

ASCEND Modeling system for formulating, *debugging* and solving and highly structured models expressed by algebraic equations and differential equations. Source code available for system. Allows parts of models to be switched on and off interactively for solving as when examining process alternatives.

ASCEND: http://www.cs.cmu.edu/afs/cs.cmu.edu/project/ascend/home/Home.html

GAMS. Modeling system that is best suited for formulating and solving optimization problems that are expressed by algebraic equations. Models include LP, MILP, NLP, and MINLP problems that are automatically linked to different optimization codes.

GAMS Dev: http://www.gams.com

gPROMS Equation based modeling system for steady state, dynamic, and distributed processes (Algebraic, ODEs, DAEs, and PDEs). In addition, it allows modeling of processes with both discrete and continuous characteristics, from purely continuous to purely batch.

Imperial College: http://www.ps.ic.ac.uk/gPROMS

SPEED-UP. Equation based modeling system for steady state and dynamic processes. Used for safety analysis, process control studies, prototype models involving ODEs, DAEs, and PDEs. Also includes NLP algorithms for optimization studies.

Aspen http://www.aspentech.com/products

C.1.2 Process Simulators

Due to space limitations, we provide information for process simulators from the top three process simulation vendors. There are several others and the interested reader is referred to the CEP Software Guide for more detailed information.

ASPEN-PLUS. This is a modular process simulation environment. Through the aid of Model Manager, it is easy to use through a graphical user interface. It is a comprehensive simulation package covering a full range of separation, reaction, transfer and flowsheeting tasks. *Aspen* http://www.aspentech.com/products

Corporate Headquarters:
Aspen Techology, Inc.
Ten Canal Park
Cambridge, MA 06141
Phone:+1-617/577-0100
Fax: +1-617/577-0303
email: info@aspentec.com

HYSIM and HYSYS. This is a modular process simulation environment entirely based on PCs. Integrated within the simulator is an easy to use graphical user inter-

C.1 Computer Software

face. It is a comprehensive simulation package covering a full range of separation, reaction, transfer and flowsheeting tasks, as well as dynamic analysis. *Hyprotech http://www.hyprotech.com*

> Corporate Headquarters
> Hyprotech Ltd.
> 300 Hyprotech Centre,
> 1110 Centre Street North
> Calgary, Alberta T2E 2R2
> CANADA
> Phone: (403) 520-6000
> Fax: (403) 520-6060

PRO/II, PROVISION and PROTISS. This product offers a comprehensive, easy-to-use and fully interactive simulation environment with a graphical user interface for building a full range of both simple and complex process models and flowsheets. The PROTISS package is also integrated into this environment for dynamic analysis. *Simulation Sciences: http://www.simsci.com*

> Corporate Headquarters
> Simulation Sciences, Inc.
> 601 S. Valencia Ave
> Brea, CA 92621
> Phone: 714-579-0412
> Fax: 714-579-7927

C.1.3 Data Banks

Two popular databanks for thermodynamic data are the DECHEMA Data Bank in Europe and the DIPPR data bank, developed in the US. Both contain comprehensive thermodynamic data for thousands of chemical components and cover phase equilibrium, enthalpy, volume, and transport properties. Both databanks are incorporated into several process simulation environments (see above) and can also be accessed through subscription.

DECHEMA Databank

This databank contains thermophysical data with more than 500 properties for pure compounds and mixtures and approximately 12,000 inorganic and organic substances. These include thermodynamic, multicomponent system, electric, transport, surface, and electrochemical properties; bibliographic information, indexing terms, property codes, substance information, abstracts, and CAS Registry Numbers are searchable.

> *http://www.cas.org/ONLINE/CATALOG/detherm.html*

DIPPR Databank

The DIPPR databank contains pure component and mixture physical property data for commercially important chemicals and substances. These data are compiled and

evaluated by a project of the Design Institute for Physical Property Data (DIPPR) of the American Institute of Chemical Engineers (AIChE). DIPPR also contains an interactive software package, TPROPS, that is started with the Messenger RUN command. TPROPS calculates temperature-dependent properties and plots the data of DIPPR substances, using regression equations.
   http://www.nist.gov/srd/dippr.htm

Also, a prototype online physical properties system is being developed at the University of Edinburgh. http://www.chemeng.ed.ac.uk/people/jack/physprops

C.1.4 Synthesis Tools

Most synthesis tools that are currently available are academic codes. The largest number are in the area of heat exchanger networks, followed by tools for flowsheet and distillation synthesis. Methodologies behind these programs include heuristics, hierarchical decomposition, pinch analysis, and mathematical programming.

Flowsheets

PIP. Interactive synthesis program implementing hierarchical decomposition technique for the conceptual design of petrochemical processes. The code identifies the decisions necessary based on heuristics to develop a flowsheet. The user can than go back and generate process alternatives.
   CACHE: http://www.che.utexas.edu/cache/product.html

PROSYN-HEU. Program that incorporates extensive heuristics and analysis capabilities for reaction and separation subsystems to sequentially integrate process flowsheets.
   Dortmund: schem@chemietechnik.uni-dortmund.de

PROSYN-MINLP. An equation based package for structural flowsheet optimization for a specified superstructures. The program includes a number of modules, simultaneous optimization models, and a package for physical properties.
   Carnegie Mellon: http://egon.cheme.cmu.edu/aturkay/list.html
   University of Maribor: kravanja@uni-mb.si

Separation

HYSYS Conceptual Design is devoted to the synthesis of nonideal separations problems. It is incorporated into the HYSYS framework, and includes the Mayflower package for azeotropic separation synthesis as well as libraries for phase equilibrium and enthalpic data from the Thermodynamic Research Center at Texas A&M University.
   Hyprotech http://www.hyprotech.com

SPLIT is AspenTech's package for the synthesis of nonideal distillation sequences. It deals with highly nonideal mixtures, including azeotropes and has numerous diagnostic, analysis, trouble-shooting and synthesis features, both for continuous and batch distillation operations.

Aspen http://www.aspentech.com/products

Heat Integration

ADVENT This is a process integration program that is based on pinch analysis. It includes targets, design and optimization capabilities for heat exchanger networks. It also includes modules for utility system. Also peforms exergy analysis using graphical diagrams.

Aspen http://www.aspentec.com/products/software/advent/advent.html

AUTOHEN. Program for automatic design of heat exchanger networks.

UMIST: http://www.cpi.umist.ac.uk/httpddoc/software.html

HERO. Targeting program based on pinch analysis for energy, area and number of units.

Institution of Chemical Engineers: http://icheme.chemeng.ed.ac.uk/soft.htm

HEXTRAN. Program primarily for simulating and rating of heat exchanger networks. Also includes limited synthesis capability.

SimSci: http://www.simsci.com

MAGNETS. Program that implements sequential synthesis strategy using the LP and MILP transhipment models, as well as NLP superstructure optimization.

Carnegie Mellon: http://egon.cheme.cmu.edu/aturkay/list.html

MATRIX. Selection of matches for retrofit of heat exchanger networks using a sequential technique with matrix method.

Chalmers: http://www.che.chalmers.se/inst/hpt/

PINCHLENI. Program based on pinch analysis. It performs exergy analyisis to aid evaluation of stream matches.

EPFL: http://leniwww.epfl.ch/pages/pinchy/

SPRINT. Program for simulation, optimization, control and flexibility of heat exchanger networks.

UMIST: http://www.cpi.umist.ac.uk/httpddoc/software.html

SUPERTARGET. Pinch analysis based program for targeting, design and optimization of heat exchanger networks. Can be used for grassroots and retrofit problems. Also includes exergy analysis.

Linnhoff March: vdhole@lm-uk.mhs.compuserve.com

SYNHEAT. Program for simultaneous MINLP synthesis of heat exchanger networks. Includes transhipment LP for utility optimization, and screening for reducing size of superstructure.

Carnegie Mellon: http://egon.cheme.cmu.edu/aturkay/list.html

THEN. The program is based on pinch analysis. It performs energy targeting and stream matching according to heuristic rules.

CACHE: http://www.che.utexas.edu/cache/product.html

C.1.5 Batch Processes

Simulation

BATCHES is a simulator for multiproduct, recipe-driven batch and semi-continuous processes. It has a modular representation and a graphical user interface. Process studies include process configurations and operating procedures as well as equipment sizing and evaluation of scheduling strategies. *Batch Process Technologies: girish@bptech.com*

Design

BATCHSPC, BATCHMPC Programs implementing MILP and MINLP models for determining sizes and number of parallel equipment of flowshop batch plants operating under single and mixed product campaigns.
   Carnegie Mellon : http://egon.cheme.cmu.edu/aturkay/list.html
SUPERIOR/Design Implements a decomposition approach in which detailed scheduling is included as part of the design model.
SUPERIOR/Schedule to solve the scheduling subproblems. *Advanced Process Combinatorics: info@combination.com*

Scheduling

gBSS. This program implements the resource task network, a variant of the state-task-network for short term scheduling. Discrete and continuous time models can be selected, as well as cyclic and aggregated scheduling models.
   Imperial College: e-mail: gBSS@ic.ac.uk
CYCLE. Aggregated LP traveling salesman model for determining the optimal sequence in flowshop plants with one unit per stage.
   Carnegie Mellon: http://egon.cheme.cmu.edu/aturkay/list.html
PARALLEL, MULTISTAGE. MINLP models for cyclic scheduling in continuous multiproduct plants with parallel lines, or plants with multiple stages separated by intermediate storage.
   Carnegie Mellon: http://egon.cheme.cmu.edu/aturkay/list.html
STBS. MILP models for short term scheduling of multistage plants consisting of parallel units at each stage. The objective is to minimize tardiness.
   Carnegie Mellon: http://egon.cheme.cmu.edu/aturkay/list.html
SUPERIOR/Schedule Implements an extension of a discrete time state-task network model with a customized solution method for solving the MILP problem.
   Advanced Process Combinatorics: info@combination.com

C.1.6 Information Management

Software systems now exist to aid teams of designers to manage information created while carrying out such activities as design projects. Within these systems engineers may store, organize and share information electronically. E-mail and bulletin boards are generically available on all computer systems. Many companies also set up and use internal World Wide Web facilities. Consulting companies supply their own document handling systems which allow companies to define document types, who should receive them and their updates, and who has to sign them. Other systems include:

> *BSCW (Basic Support for Cooperative Work):* A project of university researchers to develop tools to support cooperative work over the Web. Anyone with a browser can become a user of this system by registering. Users can readily share documents using this system.
> *GMD FIT:* http://www.bscw.gmd.de/

Exchange: A commercial product available from Microsoft. It supports both e-mail and groupware.
> *Microsoft:* http://www.windows95.com/connect/

Lotus Notes: A commercial product available from IBM. It supports both e-mail and groupware. Its document handling facilities support workflow. It aids electronic commerce with its security measures to protect information sent over the internet.
> *Lotus:* http://www.lotus.com/

n-dim: Created at Carnegie Mellon, *n*-dim supports information management by allowing users to capture, structure and share information kept in files, on the WWW, and in databases. It also supports tool integration.
> *Carnegie Mellon University:* http://www.ndim.edrc.cmu.edu/overview.html

C.2 DESIGN CASE STUDIES

CACHE Case Studies

Volume I: Separation System for Recovery of Ethylene and Light Products from a Naptha Pyrolysis Gas Steam

Volume II: Design of an Ammonia Synthesis Plant

Volume III: Design of an Ethanol Dehydration Plant

Volume IV: Alternative Fermentation Processes for Ethanol Production and Economic Analysis

Volume V: Retrofit of a Heat Exchanger Network and Design of a Multiproduct Batch Plant

Volume VI: Chemical Engineering Optimization Models with GAMS
> *CACHE:* http://www.che.utexas.edu/cache/product.html

Washington University Case Studies (partial list)

Ethylene Plant Design and Economics
Mixed Solvent Recovery and Purification

Analysis and Optimization of an Artificial Kidney System
A Distillate Desulfurizer
Bid Proposal for Star Oil Limited - Nevod Processing Plant
Evaluation of a Biphenyl Reactor
Dimethyl Formamide Recovery and Purification
Cellulose Triacetate Flake Plant to Support 20 MM lb/yr Fiber Plant
Contact: Prof. B. D. Smith, Chemical Engineering Department, Washington University, St. Louis, MO 63130

EURECHA Case Studies
Nonideal Separation Process Simulation
Methanol Synthesis Optimization
Reactor Modeling and Kinetic Parameter Estimation
Acrolein Process Design Studies
Safety Analysis
Control Studies
Contact: Dr. L. Murray Rose, The Old Vicarage, Beaminster, Dorset, ENGLAND DT8 3BU

REFERENCES

An interesting and comprehensive home page related to process design and analysis, with associated links to databases, software vendors, departments and research groups can be found on: *http://www.che.ufl.edu/WWW-CHE*

Biegler, L. T. (1989). Chemical process simulation, *Chemical Engineering Progress*, **85**, 10, p. 50.

Carnahan, B. (Ed.). (1997). *Past, Present and Future of Computing in Chemical Engineering Education,* CACHE Corp.

Chemical Engineering Progress Software Guide, published annually, American Institute of Chemical Engineers.

AUTHOR INDEX

Abbott, M. M., 210, 241
Achenie, L. E. K., 634, 658
Aggarwal, A., 590, 591
Agreda, V. H., 659
Aguirre, P., 418, 425
Ahmad, S., 36, 51
Andrecovich, M. J., 408, 425, 499, 521, 571, 575, 581, 582, 590, 591, 653, 659
Aris, R., 640, 659
Asbjornsen, O. A., 452
Ascher, U., 659
Astrom, K. J., 452
Au, Tung, 174

Baasel, W., 173
Bailey, J. K., 328, 332
Balakrishna, S., 611, 614, 622, 654, 659, 771
Balas, E., 520, 521
Barbosa-Povoa, A. P., 743
Barkeley, R. W., 278, 291
Bazaraa, M. S., 509, 521, 748, 754, 769
Beale, E. M. L., 308, 333
Berna, T., 333
Betts, J. T., 333
Biegler, L. T., 245, 290, 318, 320, 330, 333, 334, 442, 453, 596, 611, 613, 614, 615, 622, 654, 658, 659, 713, 715, 771, 780, 782
Birewar, D. B., 728, 732, 735, 743
Bischoff, K. B., 453
Black, J. H., 174
Blass, E., 418, 425, 488, 490
Bolio, B., 562
Boston, J. F., 222, 229, 241, 333
Bracken, J., 333, 336
Britt, H. I., 222, 241, 333
Brooke, A., 285, 291, 769

Carlberg, N., 420, 425
Carnahan, B., 627, 659, 780, 782
Cavalier, T. M., 517, 518, 521
Cerda, J., 535, 562
Chen, H-S, 333
Chen, J. J. J., 562
Chitra, S. P., 452, 630, 659
Christensen, J. H., 278, 291
Ciric, A. R., 547, 551, 553, 561, 562
Clocksin, W. F., 515, 522
Colberg, R. D., 561, 563
Colmenares, T. R., 686
Conti, G. A. P., 645, 659
Coon, A. B., 290, 291

Coulson, J. M., 241
Crowe, C., 269, 270, 291, 440, 452, 659
Cunningham, W. A., 14, 20

D'Couto, G. C., 333
Daichendt, M. M., 35, 39, 51, 520, 522, 562, 685, 686
Dennis, J. E., 264, 268, 290, 291, 333
Dhole, V. R., 425
Diaz, H. E., 110, 139
Diwekar, U. M., 686
Doherty, M. F., 477, 490
Domenech, S., 590, 592, 659
Douglas, J. M., 36, 38, 39, 41, 43, 51, 82, 104, 111, 139, 173, 430, 452, 645, 659, 666, 687
Droge, T., 453
Drud, A., 769
Duff, I., 290, 291, 596, 604, 605, 611, 612, 614, 659, 687

Edahl, R., 333
Edmister, W., 81, 104
El-Halwagi, M., 561, 562
Eliceche, A. M., 591
Erisman, A., 290, 291
Evans, L. B., 333

Fair, J. R., 139
Fein, G. A. F., 488, 490
Feinberg, M., 635, 641, 659
Fenske, R., 71, 104
Fjeld, M., 447, 452
Flatz, W., 199
Fletcher, R., 333
Floquet, P., 590, 592, 659
Floudas, C. A., 547, 551, 553, 561, 562, 581, 590, 591, 630, 641, 659
Flower, J. R., 592
Fogler, H. S., 452
Fonyo, Z., 418, 425
Forder, G. J., 719, 744
Foster, D., 687
Fredenslund, A., 214, 240
Frey, C. M., 686
Froment, G. F., 452

Gaminibandara, K., 590, 592
Garfinkel, R., 276, 278, 291

Geankoplis, C. J., 241
Geoffrion, A. M., 763, 766, 769
Gill, P. E., 333
Glasser, B., 453
Glasser, D., 432, 440, 452, 634, 645, 659
Gmehling, J., 240
Gooding, W. B., 743
Govind, R., 452, 630, 659
Grant, E., 174, 453
Green, D. W., 105, 139, 241
Grens, E. A., 278, 282, 291
Grossmann, I. E., 36, 39, 51, 509, 511, 514, 515, 520, 522, 562, 563, 529, 532, 535, 539, 542, 546, 547, 551, 553, 558, 561, 578, 587, 589, 591, 592, 596, 603, 604, 605, 611, 612, 613, 614, 615, 659, 665, 666, 670, 673, 676, 677, 681, 682, 685, 686, 687, 690, 698, 701, 702, 704, 706, 713, 714, 715, 719, 722, 726, 728, 732, 735, 736, 743, 744, 763, 766, 767, 768, 769, 770, 771
Gundersen, T., 290, 291, 342, 382, 561, 562
Gupta, J. N. D., 731, 743
Guthrie, K. M., 110, 133, 139

Halemane, K. P., 690, 698, 701, 713, 714
Han, S-P., 307, 333, 761, 769
Harada, T., 358, 382, 664, 687
Harriott, P., 241
Hartmann, K., 453, 659
Hawkins, R. B., 332
Heise, W. H., 659
Hendry, J. E., 498, 522
Henley, E. J., 241
Hertzberg, T., 290, 291
Hildebrandt, D., 440, 442, 453, 622, 635, 641, 659
Hillier, F. S., 508, 522, 756, 769
Hindmarsch, E., 342, 366, 382
Hirata, M., 241
Hohmann, E. C., 353, 382
Holmes, M. J., 241
Hooker, J. N., 518, 522
Horn, F. J. M., 432, 438, 453, 659
Howe-Grant, M., 14, 20
Hrymak, A. N., 332
Huffman, W. P., 333

Author Index

Hughes, R. R., 333, 498, 522, 712, 714
Hutchison, H. P., 290, 291
Ichikawa, A., 664, 687
Ireson, W. G., 174

Jackson, R., 659
Jelen, F. C., 174
Johns, W. R., 712, 714

Kabatek, U., 714
Kakhu, A. I., 592
Kalitventzeff, B., 687
Kan, R., 507, 522, 770
Kaplick, K., 453, 659
Karush, N., 333
Kelley, C. T., 290, 291
Kendrick, D., 285, 291, 769
King, C. J., 399, 401
Kisala, T. P., 333
Knopf, F. C., 743
Kocis, G. R., 665, 666, 673, 676, 677, 681, 687, 722, 744, 763, 769
Koehler, J., 418, 425
Kokossis, A. C., 630, 659
Kondili, E., 736, 743
Kramers, H., 453
Kravanja, Z., 553, 563, 596, 613, 614, 615, 677, 681, 682, 685, 687, 769
Kremser, A., 82, 105
Kroshwitz, J. L., 14, 20
Kuhn, H. W., 333
Kurtz, M., 174

Lakshmanan, A., 626, 659
Lang, Y-D., 320, 333, 596, 613, 615
Lange, N. A., 40
Lapidus, L., 276, 291
Lasdon, L., 333, 769
Lee, K. Y., 640, 659
Leesley, M. E., 271, 291
Levenspiel, O., 433, 435, 436, 453
Lieberman, G. J., 508, 522, 756, 769
Liebman, J., 333, 338, 769
Lien, K., 659, 453, 645
Lim, H. C., 659
Linnhoff, B., 36, 51, 342, 366, 382, 425, 562
Liu, Y. A., 488, 490
Locke, M. H., 333

Lockhart, F., 353, 382
Lucia, A., 227, 241, 333
Lustig, J., 757, 770
Luther, C., 659

Malik, R. K., 712, 714
Maloney, J. O., 105, 139, 241
Manousiouthakis, V., 561, 562
Marketos, G., 712, 714
Marsten, R., 757, 769, 770
Mason, A. W., 562
Mattheiij, R., 659
Mazzuchi, T. A., 714, 715
McCabe, W. L., 241
McCormick, G., 333, 336
McCroskey, P. S., 743
McKetta, J. J., 14, 20
Meeraus, A., 285, 291, 769
Mellish, C. S., 515, 522
Miller, D. L., 731, 743, 744
Minoux, M., 507, 522, 748, 757, 770
Morari, M., 561, 563, 690, 702, 713, 714, 715, 769
Morin, T., 770
Motard, R. L., 278, 282, 291
Murray, W., 333
Murtagh, B. A., 286, 291, 322, 334, 761, 762, 763, 769, 770

Naess, L., 342, 382, 561, 562
Nemhauser, G. L., 276, 278, 291, 507, 508, 522, 759, 770
Nishida, N., 453
Nishio, N., 270, 291
Nocedal, J., 333

Ohe, S., 241
Okos, M. R., 743
Omtveit, T., 447, 453, 645, 659
Onken, U., 240
Orbach, O., 269, 291, 334
Otto, R., 245, 289, 291, 647
Overton, M., 334

Pantelides, C., 664, 687, 736, 739, 743, 744
Papageorgaki, S., 743, 744
Papoulias, S. A., 529, 532, 535, 539, 542, 561, 562

Park, C. S., 174
Partin, L. R., 659
Paterson, W., 645, 659
Paules, G. E., 581, 590, 591
Pekny, J. F., 731, 743, 744
Perkins, J. D., 477, 490
Perry, R. H., 105, 139, 241, 401, 456, 490
Peters, M., 139, 173
Pho, T. K., 276, 291
Pibouleau, L., 590, 592, 659
Pikulik, A., 110, 139
Pinto, J. M., 743, 744
Piret, E. L., 453, 630, 659
Pistikopoulos, E. N., 714, 715
Poellmann, P., 488, 490
Poling, B. E., 51, 53, 105, 139, 241, 590, 592
Powell, M. J. D., 307, 334, 761, 770
Powers, G. J., 36, 51
Prausnitz, J. M., 51, 53, 105, 139, 241, 590, 592

Quesada, I., 561, 563, 591, 592, 743, 767, 770

Rachford, H. H., 219, 241
Raman, R., 514, 515, 520, 522, 578, 592, 743
Rasmussen, P., 240
Ravemark, D., 743, 744
Ravimohan, A., 659
Ray, W. H., 289, 291
Reeve, A., 180, 199
Reid, J., 291
Reid, R. C., 51, 53, 105, 139, 214, 241, 590, 592
Reklaitis, G. V., 195, 199, 743, 744
Rice, J. D., 219, 241
Richardson, J. F., 241
Rippin, D. W. T., 199, 712, 714, 719, 744
Rubin, E. S., 686
Rudd, D. F., 36, 51, 278, 291
Russell, R., 659

Saboo, A. K., 561, 563, 702, 715
Sahinidis, N. V., 744, 767, 770
Saltzman, M., 757, 770
Sargent, R. W. H., 272, 291, 334, 590, 591, 592, 713, 714, 719, 722, 736, 739, 743, 744
Saunders, M. A., 286, 291, 322, 334, 761, 762, 770

Schembecker, G., 453
Schittkowski, K., 334
Schmid, C., 330, 334
Schnabel, R., 264, 268, 290, 291, 333
Schrage, L., 333, 520, 522, 768, 769, 770
Schubert, S., 9, 20
Seader, J. D., 241, 400, 401
Seider, W. D., 488, 490, 686
Serafimov, L. A., 477, 490
Shah, N., 739, 743, 744
Shanno, D., 757, 770
Shetty, C. M., 509, 521, 748, 754, 769
Shiroko, K., 358, 382
Siirola, J. J., 36, 51
Simmrock, K., 453
Singal, J., 770
Smith, E., 664, 687
Smith, J. M., 51, 53, 210, 241
Soyster, A. L., 517, 518, 521
Sparrow, R. E., 719, 744
Stadtherr, M. A., 290, 291, 333
Stephanopoulos, G., 453
Straub, D., 690, 713, 714, 715
Swaney, R. E., 690, 700, 702, 706, 714, 715
Szekely, J., 289, 291

Tanskanen, J., 453
Taylor, R., 227, 232, 241
Terranova, B., 413, 425
Thompson, R. W., 399, 401
Timmerhaus, K., 173
Todd, M. J., 507, 522
Trambouze, P. J., 453, 630, 659
Treiber, S. S., 332
Trevino-Lozano, R. A., 333
Tsai, M. J., 659
Tucker, A. W., 333
Turkay, A., 562
Turkay, M., 520, 522, 686, 687

Umeda, T., 358, 382, 664, 687
Upadhye, R. S., 278, 282, 291

van de Vusse, J. G., 441, 453, 652, 659
Van Ness, H. C., 51, 53, 210, 241
van Winkle, M., 241
Varvarezos, D. K., 713, 715
Vasantharajan, S., 318, 334

Author Index

Viswanathan, J., 334, 558, 563, 587, 589, 592, 659, 768, 770
Voudouris, V. T., 726, 736, 743, 744

Waghmere, R. S., 659
Wahnschafft, O. M., 462, 488, 490
Wang, J. C., 241
Wang, Y. L., 241
Warren, A., 333, 769
Wegstein, J. H., 269, 291
Wehe, R. R., 592
Welty, J., 139, 235, 242
Westerberg, A. W., 14, 21, 139, 272, 278, 282, 291, 333, 400, 401, 408, 413, 420, 425, 453, 462, 488, 490, 499, 521, 535, 562, 571, 575, 581, 582, 590, 591, 592, 653, 659
Westerterp, K. R., 453
Westhaus, U., 453
Wicks, C. E., 139, 242
Widagdo, S., 488, 490

Wilcox, R. J., 546, 563
Wilkes, J., 659
Williams, H. P., 515, 520, 522, 770
Williams, T., 245, 289, 291, 647
Wilson, R. B., 307, 334
Wilson, R. E., 139, 242
Winter, P., 291
Wolsey, L. A., 508, 522, 759, 770
Wood, R. M., 546, 563
Wright, M. H., 333

Xu, J., 333, 453
Xueya, Z., 743, 744

Yee, T. F., 553, 561, 562, 563, 614, 615
Yeh, N. C., 195, 199

Zharov, W., 477, 490
Zitney, S. E., 290, 291
Zwietering, N., 636, 659

SUBJECT INDEX

Absorber, 88
Absorption factor
  definition, 80
  effective, 81
Abstraction, 34–35
Active set strategy, 704, 755, 756
Activity coefficient, 211, 389
Adiabatic flash
  ideal, 102
Adiabatic mixing, 98
Aggregated models, 604, 610, 611, 666, 732
Alcohols. *See* mixtures
Algorithm
  absorption, 82
  adiabatic flash
    ideal, 102
  Armijo line search, 259
  attainable region, 442
  flash
    ideal, 64
  generalized Benders decomposition, 509, 766, 767
  inside-out method, 224
  interior point, 757
  linear mass balance, 85
  Newton-Raphson, 256
  nonideal flash, 219–221
  outer-approximation, 509, 684, 764–768
  reactor network targeting, 625, 641
  reduced gradient, 762, 763
  rSQP, 326
  simplex, 757
  SQP, 311
Algorithmic synthesis methods, 497
Alternatives. *See* design alternatives
Ammonia synthesis, 319
Analysis, 6
Annualized payments, 151
Annuities, 148
Antoine equation, 62
Area estimation. *See* heat exchanger network synthesis
Armijo line search, 258
ASCEND, 774
ASPEN, 242, 245, 780
Assessing designs, 30–31
Attainable region (AR), 429, 432, 438–439, 440, 619
Autocatalytic reaction, 435, 446, 618
Average income on initial cost (AIIC), 145
Azeotropes
  detecting, 456–458

Subject Index

Azeotropic distillation, 20, 455–494
   acetone/chloroform/benzene, 455, 464, 465, 464–474
   ethyl alcohol/water, 455
   ethyl alcohol/water/toluene, 455
   general approach, 486–487
   n-pentane/acetone/methanol/water, 482–487
   water/n-butanol, 455, 456–464, 474

Base case design, 12
Base cost, 133, 134
Basic hens. See heat exchanger network synthesis
Basic problem. See heat exchanger network synthesis
Basic process design, 2
Batch, 38, 44, 181, 182
Batch processes,
   discrete sizes, 725
   NLP design model mixed product campaigns, 728, 735
   recipes, 181, 736, 737, 741
   equipment sizing, 190
   flowshop plant, 185
   jobshop plant, 185
   merging of tasks, 197–198
   MILP model flowshop plant, 726, 727
   MINLP model flowshop plants, 722
   multiproduct plant, 184
   single product plant, 180
   size factors, 190
   synthesis flowshop plants, 195
Batch scheduling
   aggregate LP model, 732
   changeover or clean-up times, 185
   cycle time, 183, 187, 189, 720, 721
   cyclic scheduling flowshop plants, 729
   effect intermediate storage, 187, 190
   effect parallel units, 187, 189
   Gantt chart, 182, 720, 722
   horizon constraints, 719, 721, 723, 728
   inventories, 193
   MILP model, 739, 740
   mixed product campaigns, 185, 186
   no intermediate storage (NIS), 186, 188
   single product campaigns, 185, 186, 719
   state-task-network, 736, 737
   transfer policies, 186
   unlimited intermediate storage (UIS), 187
   zero-wait (ZW) transfer, 186
Benzene. See styrene process
BFGS update, 310
Binary variables, 507, 514, 520, 541, 554, 572, 579, 588, 705
Bleed. See purge
Brainstorming, 10–11
Branch and bound, 33, 503, 507, 713, 759, 760
   breadth first, 504, 506
   depth first, 503, 505
   implicit enumeration, 503, 504, 759
Branching, 35
Breadeven time (BET), 167
Broyden, 255, 264, 308–309
Bubble point, 389, 416, 420
   ideal, 63
Buddle point calculation
   ideal, 67

Carbon dioxide. See styrene process
Cascaded heat. See heat exchanger network synthesis
Cascaded heat diagram. See distillation
Cauchy step, 261
CEP Software Guide, 782
Chemical abstracts, 27
Chemical Engineering Magazine, 26, 51
Chemical marketing reporter, 40
Chemical potential, 211
Coefficient of performance, 129
Cold shot cooling, 635
Cold stream definition. See heat exchanger network synthesis
Collocation, 489
Collocation points, 633
Column
   sizing, 118
   costing
      absorber, 124
      distillation, 122
      diameter, 120
      height, 121
Column design, 489. See also optimal design distillation columns

Column operation, 489
Column performance, 489
Column pressure, 73
Column stacking. *See* distillation
Combinatorial explosion, 32
Commissioning, 5
Composite curves. *See* heat exchanger network synthesis
Composition diagram, 492, 493
Composition space, 30
Compressors, 375
  centrifugal, 124
  nonideal, 234
  reciprocating, 128
  staged, 127
Computer software, 773–779
Concept generation, 6
Condenser
  partial, 74
  total, 74
Condenser duties. *See* distillation
Condensibles, 35
CONOPT, 645, 769
Conservation laws, 208
Constraint qualification, 303, 754
Constraints, 296, 508, 748
Construction, 5
Continuation method, 262
Continuous payments, 150
Continuous stirred tank reactor (CSTR), 431, 433, 619, 642
Continous variables, 508, 748
Contraction mapping theorem, 268
Control, 489
Controllability, 31
Conversion, 430
Convex combination, 438
Convex function, 724, 750
Convex hull, 443–444, 624, 652
Convex region, 750, 755
Convexity, 297
Cost comparison
  after tax, 159
  different lives, 153
  same lives, 152
Cost estimation, 111
Customer reaction, 3

CPLEX, 757, 761, 769
Critical parameter vale, 699, 701
Croton aldehyde. *See* ethyl alcohol process
Cycles. *See* heat exchanger network synthesis

Debottlenecking, 5, 12
Decision variables, 296
Decommissioning, 6
Decomposition strategies, 36–39
  bounding, 36–37
  Douglas, 38–39
  hierarchical, 38–39
  modeling-decomposition strategy. *See* flowsheet synthesis
Dependent variables, 296
Depreciation, 1986 tax code, 158
  declining balance, 156
  MACRS, 158
  straight line, 156
Design alternative generation, 6
Design alternatives, 12
Design calculation, 249
Design models, 209
Design teams, 8–10
Design under uncertainty, 712
  two-stage strategy, 712
Detailed engineering, 2, 21
Dew point, 389
  ideal, 63
Dew point calculation, 67
DICOPT, 558, 589, 769
Diethyl ether. *See* ethyl alcohol process
Differential sidestream reactor (DSR), 451
Direct fired heaters
  sizing, 116
Direct sequence, 400
Direct substitution, 268, 635, 637, 648, 652
Discounted cash flow, 152
Disjunctions, 406, 519, 520
  convex hull, 520
Distillation, 91
  azeotropic, 162, 168. *See also* azeotropic distillation
  cascaded heat diagram, 410
  column stacking. *See* distillation—cascaded heat diagram
  condenser duties, 410–411

Subject Index

heat flows, base case, 408–409, 412, 416, 421, 424
heat integration, 408–428
heuristics, 400–401
ideal. *See* ideal distillation
intercooling, 413–418
interheating, 407, 413–418
McCabe-Thiele diagram, 30
number of sequences, 397–399
operating lines, 413–415
pinch point, 402, 413–414, 417
pressure coupling, 422
qualitative four component example, 411–413
reachable products, 419, 489
reboiler duties, 410–411
reversible separation, 418
side enrichers, 420–425
side strippers, 420–425
simple sharp separators, 398–399
T vs. heat diagram. *See* distillation-cascaded heat diagram
thermal condition of feed, 419–420
Thompson and King formula, 399
Distillation boundaries, 489
Distillation calculations, 224–232
Distillation curves
acetone/chloroform/benzene, 466
definition, 466–468
sketching, 475–482. *See also* residue curves
Distillation methods
bubble point, 227, 477
Newton-Raphson, 228
sumrates, 227
Distillation model
split fraction, 70
Distillation optimization. *See* optimal design distillation columns
Distillation sequences. *See* optimal distillation sequences
Dominant eigenvalue (DEM), 269
Douglas hierarchical decomposition. *See* decomposition strategies

Eastman Chemical Company, 26
Economic evaluation, 30
Effect of pressure. *See* heat exchanger network synthesis

Efficiency
isentropic, 126
motor, 124
pump, 124
tray
overall, 121
turbine, 126
Eigenvalues, 488, 753
Eigenvectors, 488
EMAT, 556
Energy balance
ideal, 98–104
Energy integration. *See* heat exchanger networks
Enthalpy
liquid phase
ideal, 100
vapor phase
ideal, 98
Environment, 31
Equation of state (EOS) models, 214
Equation oriented simulation, 56
Equilibrium stage models, 209, 390
Equipment sizing, 111, 190
Ethanol process, 244, 252–254. *See also* ethyl alcohol process
Ethyl alcohol. *See* ethyl alcohol process; *see* mixtures
Ethyl alcohol process
aggregation levels, 28
design alternatives, 17–18
economic sensitivity analysis, 42
heat integration, 341
hierarchical decomposition, 34–35, 43
introduction, 13–18
liquid recovery, 49
maximum profit potential, 40
physical property data, 15
purge, 47–48
reactions, 14
recycle structure, 45
separation system synthesis, 45
synthesis strategies, 39–50
typical flowsheet, 16
vapor recovery, 46
Ethyl benzene. *See* styrene process

Ethylene. *See* ethyl alcohol process; *See* styrene process
Ethylene glycol. *See* mixtures
Evaluation, 6
  short cut, 19
Evaporator-condenser, 377
Evolutionary methods, 33
EXCEL. *See* spreadsheets
Excess properties, 212
Expected value
  investment, 171
Extent of conversion, 53, 778
Extractive distillation, 216, 485–486

Feasibility function, 698, 699
Feasible region, 749
Feed tray location, 489, 492
Fewest matches. *See* heat exchanger network synthesis
Fifty fifty split heuristic, 407
Finite difference approximation, 262
Finite elements, 622
First and second law of thermodynamics, 409
First order methods, 267–271
Five alcohols example. *See* mixtures
Fixed capital, 143, 415, 422
Fixed costs, 143
Fixed point problem, 251
Flash calculation
  ideal, 64–67
  nonideal, 217–224
Flash drums, 112, 254
Flash unit, 87
  ideal, 61
Flexibility, 20, 690
Flexibility analysis methods
  vertex solution, 701
  active set strategy, 31, 390, 704–712
Flexibility index, 696, 697, 700, 701, 707
Flexibility test, 697, 698, 701, 706, 710, 711, 795
Flooding velocity, 120–121
Flowsheet, 58, 86
Flowsheet optimization, 315
Flowsheet synthesis, 663. *See also* synthesis
  MINLP model, 673, 674
  superstructures, 317, 664–666, 682
  modeling/decomposition strategy, 672, 675–681
Flowsheeting, 19
Flowshop plant. *See* batch processes
FLOWTRAN, 319
Fugacity coefficient, 211
Furnaces
  sizing, 116
Future worth, 147

GAMS, 285, 769, 774
Gantt charts. *See* batch scheduling
Gas absorption, 79
Gaussian quadrature, 622
Generalized Benders decomposition, 287, 509, 645, 766, 767
Generalized disjunctive programming, 686
Generalized dominant eigenvalue (GDEM), 270, 627
Generating alternatives, 27
  heat exchanger networks, 32–33
Gibbs free energy, 32–34, 53, 210
Gibbs free energy minimization, 231
GNO, 459–462, 494, 769
Global minimum, 751, 755
Global optimizatioon, 511, 591
Goals, 10
Gradient, 255, 752, 754
Grand composite curve, 389. *See also* heat exchanger network synthesis
Grassroots design, 26
Guthrie's modular method, 133–138, 304

Hazop, 31
Heat and power integration, 341–386
Heat balance. *See* heat exchanger network synthesis
Heat duties
  condenser, 121
  reboiler, 121
Heat exchanger network synthesis, 19
  algorithmic approach, 528–561
  area estimation, 370–373
  basic problem, 341–372
  cascaded heat, 39, 356
  Chen's approximation, 556
  cold stream definition, 343

Subject Index

composite curves, 353–361
counter-example, 561
cycles, 350–352
effect of pressure, 343
fewest matches, 348–349
grand composite curve, 358–361
heat balance, 343
heat sink, 359
heat source, 359, 377–382
Hohmann/Lockhart composite curves, 29
hot stream definition, 343
inventing initial network, 349–350
minimum number of units, 541, 561
minimum temperature driving force, 346, 353–358, 368–373
minimum utility cost, 528
MINLP optimization model, 551, 554–557
NLP optimization model, 550, 551
optimal, 20
pinch design approach, 363–368
pinch point, 358
problem table, 346
right facing nose, 360, 361–363
sequential synthesis, 528, 559, 560
simultaneous synthesis, 551, 559, 560
stream splitting, 356, 366–368, 370
superstructures, 547, 548, 549, 551, 553
T vs heat diagram, 29, 381
temperature intervals, 346–348
transportation model, 535
transshipment model. *See* transshipment model
Heat exchanger networks (HENS), 648
Heat exchangers
sizing, 113–116, 235
Heat flows, base case, 650. *See also* distillation
Heat integrated distillation, 576. *See also* distillation
multieffect, 577
MILP model continuous temperatures, 578–581
MILP model discrete temperatures, 581–585
Heat integration, 596
simultaneous optimization, 596, 600, 612, 613, 648, 654, 655, 685

sequential optimization, 596, 599, 612, 613, 648, 654–655
See also distillation, 19
Heat pumps, 129, 373–382
investment costs, 379
right facing nose, 381
thermodynamic work, 378, 385
two stage, 376–377
using grand composite curve, 377–382
Heat recovery. *See* heat exchanger networks
Heat sink. *See* heat exchanger network synthesis
Heat source. *See* heat exchanger network synthesis
Heat Transfer and Fluid Flow Service, 27
Heat transfer coefficients, 114–115
Heat Transfer Research Institute, 27
Heavy key, 70
HENS. *See* heat exchanger network synthesis
Hessian, 255, 753
Heuristics, 401, 407, 431, 440, 507, 519, 528, 561
See distillation, 304
Hierarchical decomposition
Douglas, 44
ethyl alcohol process, 44, 407
Hohmann/Lockhart composite curves. *See* heat exchanger network synthesis
Hot stream definition. *See* heat exchanger network synthesis
HRAT, 528
HTFS. *See* Heat Transfer and Fluid Flow Service
HTRI. *See* Heat Transfer Research Institute
Hurdle, 168
Hydrogen. *See* styrene process
HYSIM, 242
HYSYS, 245

Ideal distillation, 387–407
design goals, 245, 389, 780
heuristics, 400–401, 780
marginal vapor flows, 393
minimum reboil, 390
minimum reflux, 390
minimum vapor flows, 390
product compositions, 391

Ideal distillation (cont)
    Underwood's method, 390–393, 419
    See also distillation, 489
Ill-posed, 10–13
Incidence matrix, 288, 403, 404
Infeasible path approach, 316
Infinite dilution activity coefficients, 459–462
    acetone/chloroform/benzene, 465
    n-pentane/acetone/methanol/water, 483
Infinite dilution K-values, 456–458
    acetone/chloroform/benzene, 464
    n/pentane/acetone/methanol/water, 482
    water/n-butanol, 457, 492
Inflation, 169, 481
Information gathering, 27
Initial points, 12
Input/output structure (Douglas), 44
Inside out methods, 222–224
Integer program, 276
Intercooling. See distillation
Interest rates
    continuous, 148
    effective, 148
    nominal, 148
Interheating. See distillation
Interior point methods, 757
Inventing initial network. See heat exchanger network synthesis
Investment alternatives analysis, 163
    loans required, 165
Investment risk, 170
Isobutane. See mixtures
Isopropyl alcohol. See ethyl alcohol process
Isothermal flash, 62

Jacobian, 228, 256
Jobshop plant. See batch processes

K-value
    ideal, 62
    nonideal, 212
Karush-Kuhn-Tucker (KKT) conditions, 218, 300–304, 705, 754–756, 761
Kirkpatrick award, 26

Langrange function, 307, 753
Lagrange multipliers, 676, 755
Levenberg-Marquardt method, 260

Life cycle, 2
Light key, 70
LINDO, 757, 761, 768
Linear fixed charge model, 380
Linear mass balance, 85
Linear programming (LP), 508, 509, 510, 513, 761–763
Linear programming relaxation, 758, 759
Liquid activity coefficient model, 212–214
Liquid liquid behavior, 494
    detecting, 459–462
Liquid liquid extraction, 483–484
Liquid recovery, 49
Local minimum, 488, 751, 755
Logic constraints, 514–521. See also propositional logic
Lotus 1-2-3. See spreadsheets
LP. See linear programming

MAGNETS, 551, 777
Maintenance, 4
Manufacturing capital, 143
Manufacturing costs, 144
Marginal vapor flows. See ideal distillation
Margules equation, 462
Margules model, 213
Mass balance, 57
Material and pressure factors
    compressor/turbine, 126
    direct fired heaters, 118
    furnaces, 117
    heat exchangers, 117
    pumps, 125
    refrigeration, 132
    tray stacks, 119
    vessel, 113
Materials of construction, 112
Mathematical programming, 296
Max function, 652
Maximum mixedness, 636
Maximum profit potential, 40
McCabe-Thiele diagram. See distillation
Membranes, 17
MERQ equations, 232
MESH equations, 226
Methane. See ethyl alcohol process, styrene process
Methanol. See mixtures

Subject Index

Methyl acetate process, 26
MILP. *See* mixed-integer linear programming
Minimum reboil. *See* distillation-ideal
Minimum reflux. *See* distillation-ideal
Minimum temperature driving force. *See* heat exchanger network synthesis
Minimum vapor flows. *See* distillation-ideal
MINLP. *See* mixed-integer nonlinear programming
MINOS, 286, 287, 314, 322, 330, 763, 769
MINPACK, 267, 619, 641, 652
Mixed-integer linear programming (MILP), 287, 314, 322, 330, 332, 508, 509, 513, 667, 763–768
Mixed-integer optimization, 498
Mixer, 59
Mixtures
   acetone/chloroform/benzene, 455
   ethyl alcohol/acetone, 86, 232, 401–402
   ethyl alcohol/water, 455
   ethyl alcohol/water/ethylene glycol, 470–474, 477, 480–481, 491, 492
   ethyl alcohol/water/toluene, 455
   ethylene/methane/propylene/isopropyl alcohol. *See* ethyl alcohol process
   five alcohols example, 395–396
   methanol/acetone/water, 491
   n-butane/n-pentane/n-hexane, 427
   n-pentane/acetone/methanol/water, 482–486
   n/pentane/n-hexane/isobutane/n-pentane, 425
   n/pentane/n-hexane/n-heptane, 387–395
   propane/propylene, 398
   styrene/ethyl benzene, 52
   water/n-butanol, 455
   water/toluene, 388
   water/toluene/pyridine, 490
Modular simulation mode, 56, 456–464, 474, 490
Module factors, 135
Multiperiod design problem, 713, 714
Multiple operating states, 244, 249–253, 489
Multistage compressors, 375
Myers Briggs, 8

N-butane. *See* mixtures
N-heptane. *See* mixtures
N-hexane. *See* mixtures
N-pentane. *See* mixtures
Net present value (NPV), 151
NETLIB, 267
Newton-Raphson
   descent property, 252, 258
NLP. *See* nonlinear programming
Nodes on ternary composition diagrams, 477
Non-convex optimization, 489
Noncondensibiles, 35, 255, 307
Nondifferentiability, 488, 652
Nondifferentiable function, 608, 611
Nonlinear programming (NLP), 296, 508, 509, 510, 513, 756, 757
   convexity, 297
   first order conditions, 300–303, 752–754
   global solution, 297
   local solution, 297
   second order conditions, 304, 754
Nonmanufacturing capital, 143
Nonrandom two liquid (NRTL) model, 213
Number of trays, 297, 489

Objective function, 295, 508, 748
Oil, Paint and Drug Reporter, 40
Operability, 690
Operating lines. *See* distillation
Operations research, 296
Optimal design distillation columns, 587
   MINLP model optimal feedtray, 588–590
   superstructure number of trays, 591
Optimal distillation sequences, 567
   MILP network model, 572, 573, 575
   sharp splits, 558, 567
Optimality conditions, 752–756
Optimization, 8, 295
Orthogonal collocation
   finite elements, 632, 657
OSL, 757, 761, 769
Outer-approximation algorithm, 509, 684, 764–768
Overall conversion, 430

P&ID. *See* piping and instrumentation diagrams
parallel reactions, 436
Partitioning, 271
Patents, 27

Payout time, 145, 167
Peng Robinson (PR), 215
Performance models, 209
Perpetuities, 150
Personality types, 8
PFD. *See* process flow diagrams
Phase behavior, 488
Phase equilibrium, 210
Phase separation
   ideal, 61
Physical properties, 208
Pinch candidates, 650
Pinch design approach. *See* heat exchanger network synthesis
Pinch point. *See* distillation, heat exchanger network synthesis
Pinch points, 650
Piping and Instrumentation Diagram, 26
Plate absorbers, 79
Plug flow reactor (PFR), 431, 433, 619
Powell dogleg method, 260–261
Power cycle, 374
Power law cost correlation, 132
Poynting correction factor, 211
Precedence ordering, 271
Preliminary design, 1–2, 25–26
Present value, 147, 166
Pressure
   setting levels, 94
Pressure coupling. *See* distillation
Pressure effects
   separation, 78
Pressure limits, 68
Pressure, effect of. *See* heat exchanger network synthesis
PRO/II, 242, 245, 780
Problem abstraction, 34. *See also* abstraction
Proceeds per dollar outlay (PDO)
   annual (APDO), 145
Process Flow Diagrams, 2
Process flowsheet, 245
Process representation, 27. *See also* representation
Product compositions. *See* distillation, ideal
Profit, 142
Project assessment, 166
Project manager, 4
Propane. *See* mixtures

Propositional logic, 514–516
   logic inference, 517
   conjunctive normal form (CNF), 515, 517
   DeMorgan's theorem, 515, 516
PROSYN-MINLP, 681, 682, 686, 776
Pseudocritical temperature, 68
Pumps, 233
Purge, 38, 47–48
Pyridine. *See* mixtures

Quadratic program (QP), 307
Qualitative four component example. *See* distillation
Quasi-newton, 255, 263

Raoult's law, 425
Rate of return, 151
Rating calculation, 249
Reachable products
   acetone/chloroform/benzene, 469
Reaction invariants, 447–448
Reaction path synthesis, 518
Reaction step, 26
Reaction vectors, 439–440
Reactive distillation, 26
Reactor, 86
   fixed conversion, 59–60
Reactor extensions, 623, 638
Reactor models
   equilibrium, 237
   kinetic, 238
   stoichiometric, 236
Reactor modules, 643
Reactor network synthesis
   targeting
     isothermal, 620–634
     nonisothermal, 635–640
   geometric concepts, 432
   graphical techniques, 432
   targeting, 429, 618
Reactor-energy synthesis, 651
Reactors
   sizing, 118
Readpert
   expert system, 450
Real-time optimization, 328–329
Reboil, 390

Subject Index

Reboiler
    partial, 75
    total, 75
Reboiler duties. See distillation
Recovery fraction, 81
Recycle reactor (RR), 433–434, 445, 642
Recycle structure, 39
Reduced gradient method, 762, 763
Reduced space SQP (rSQP), 323–327, 330
Reflux, 390. See also minimum reflux
Refrigerant, 129
Refrigeration, 128
Refrigeration cycles. See heat pumps
Relative volality, 62, 389
    mole fraction averaged, 402, 416
Representation, 27–30, 498
RESHEX, 562
Residence time, 434, 621
Residence time distribution, 620
Residue curves
    definition, 476
    sketching, 475–482. See also distillation curves
    topology
        equation for 3 component, 477
Retrofit design, 4
Retrograde condensation, 68
Return on investment (ROI), 145
Reversible separation. See distillation
Right facing nose. See heat exchanger network synthesis
Roadmap for book, 18–20
Routine design, 12

Saddle points on ternary composition diagrams, 477, 488
Safety, 4, 31, 390
Scenario of process design, 3–8
Schubert, S., 9
SCICONIC, 769
Searching among alternatives, 27, 32–34
Second law of thermodynamics. See first and second laws
Segragated flow, 620, 632
Selectivity, 430
Sensitivity analysis, 42
Separability factors (liquid/liquid)
    definition, 484

Separation, 20. See also distillation, ideal distillation, azeotropic distillation
Separation process synthesis
    n-pentane/acetone/methanol/water, 482–486
Sequential heat integration. See heat integration
Sequential modular, 244
Series reactions, 436
Set covering problem, 276
Side enrichers. See distillation
Side strippers. See distillation
Simple sharp separators, 404
Simplex method, 757
Simulation, 243, 244
    flowsheet, 56
Simulator, 210
Simultaneous heat integration. See heat integration
Simultaneous optimization and heat integration, 595
    linear model, 604
    nonlinear model, 604, 610, 611
    pinch location model, 605–610
    trade-off with raw material, 601, 612
Smooth approximation, 611, 771, 772
Soave Redlich Kwong (SRK), 215
Solvent feed, 489
Sparsity, 253, 254
Speedup, 245, 780
Split fraction model, 59
Splitter, 59, 602, 603, 668, 669
    single choice, 674, 675
Spreadsheets, 12, 54, 104, 392, 425
SRI international, 27
Start up, 2, 4, 5
Starting points, 12
Steepest descent method, 260
Stochastic flexibility, 714
Stream splitting. See heat exchanger network synthesis
Stripper model, 84
Structural flowsheet optimization, 663–666
Styrene. See mixtures
Styrene process, 52
Successive quadratic programming (SQP), 295, 306–307, 314, 761, 763
Sulfur dioxide oxidation, 640

Superstructure, 33, 500, 547, 553, 572, 586, 619, 641, 663–666, 671
   tree representation, 499, 501, 503, 504
   network representation, 499, 500, 501, 507
   decomposition, 677–681
SYNHEAT, 562, 777
Synthesis, 6–8
   basic steps, 26–30
   overview, 25–54
   strategies, 19
Synthesis utility plants, 669
   MILP model, 670
   superstructure, 671

T vs. Heat diagram, 29. *See also* heat exchanger network synthesis; *see* distillation
Targets, 33
Tasks, 29
Tear stream, 93
Tearing, 271, 274–284
Technical encyclopedias, 27
Temperature
   setting levels, 95
Temperature intervals. *See* heat exchanger network synthesis
Temperature limits, 68
Temperature-entropy diagram, 375
Ternary composition diagram. *See* composition diagram
Tests, 11
Thermal condition of feed. *See* distillation
Time value of money, 142, 147
Toluene. *See* mixtures, styrene process

Topology (distillation), 488
Total enumeration, 33
Transportation model, 535
Transshipment model, 530
   minimum utility lost, 532, 533
   constrained matches, 534, 536, 539, 540
   minimum number of units, 541, 542, 543, 544
   simultaneous optimization, 603
Traveling salesman problem, 731
Tree searching, 33, 34. *See also* searching alternatives
Turbines, 234, 375, 670–672

Uncertain parameters, 691, 697
Unconstrained optimization, 752
Underwood's method. *See* distillation, ideal
UNIFAC method, 214, 389, 457
UNIQUAC method, 213, 231
Unit models, 57, 208
Update factor, 133

Vapor and liquid recovery, 39
Variable costs, 143
Vessels, 112

Water. *See* mixtures
Well posed, 10
Wilson model, 213
Working capital, 143
World Wide Web, 27, 51
WWW. *See* World Wide Web

ZOOM, 761, 769